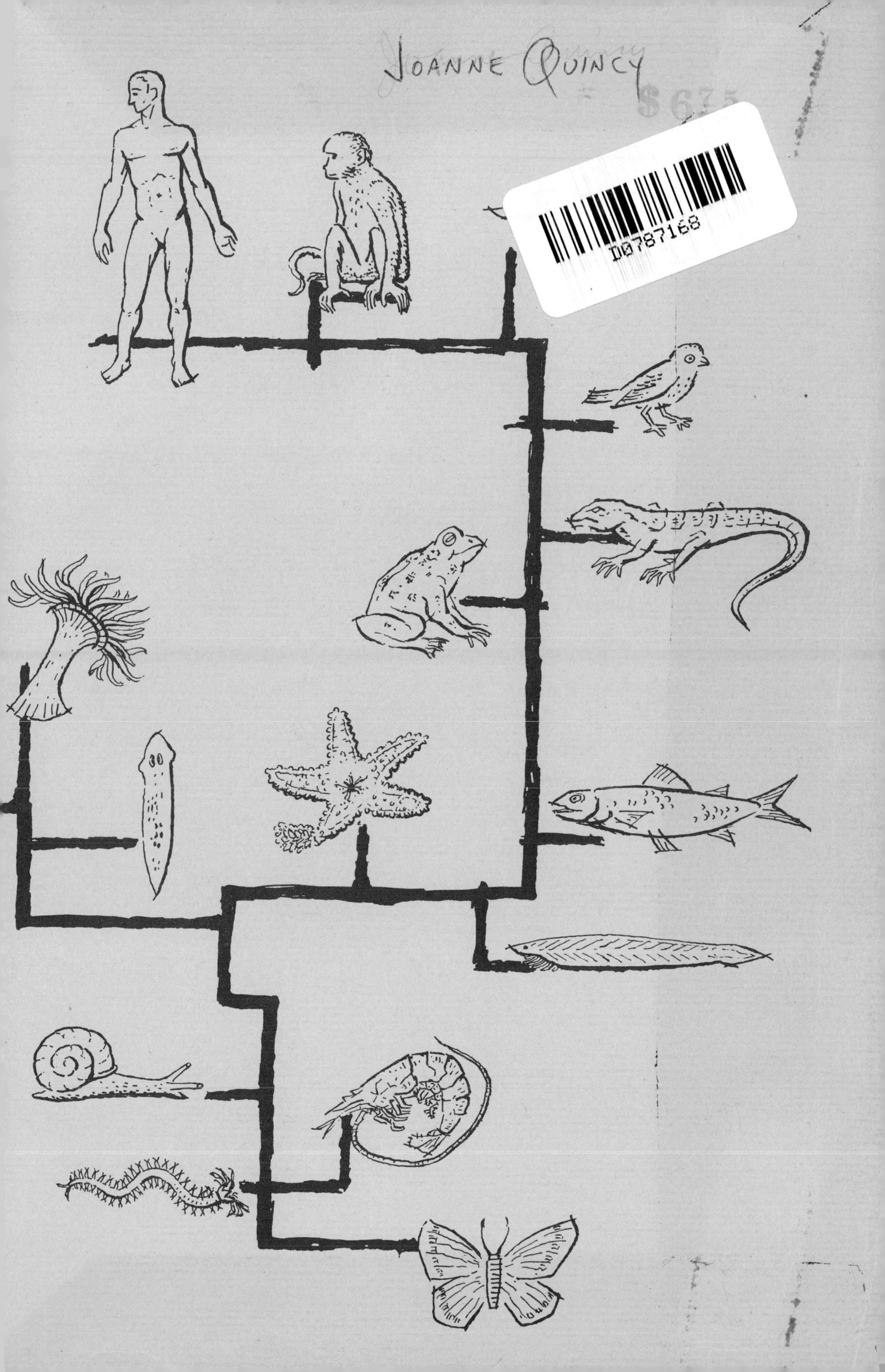

Claude A. Villee
Harvard University

BIOLOGY

Third Edition

W. B. Saunders Company
Philadelphia and London

REPRINTED AUGUST, 1957

LIBRARY OF CONGRESS CATALOG CARD NUMBER: 57–7051

Preface

THIS BOOK was written with the concept that biology is a definable body of facts and theories, concerned with all the myriad facets of all kind of living things, and that it is not simply a mixture, in some predetermined ratio, of botany and zoology, anatomy and physiology, heredity and evolution, or any other of the life sciences. To bring to the fore the biologic principles which are basic to the study of living things, this edition contains a new chapter (Chapter Two) in which some of the major generalizations of biology are briefly discussed. These, of course, cannot be fully appreciated at the first reading but they should be helpful in providing a frame of reference for the succeeding chapters. They could be reread with profit later in the course. This third edition contains, in addition to this new chapter, major revisions in the chapters on genetics, evolution, and human anatomy and physiology, and smaller changes in the others. A number of illustrations have been replaced and many new ones have been added. The new line drawings for this edition were made by R. Paul Larkin, Gail Limberg, and William Osburn.

In writing an introductory text it is difficult to steer a true course between the Scylla of superficiality and the Charybdis of overdetail. This text attempts to present the major facts and principles of modern biology without superficiality and yet without undue emphasis on detail. Most students find the facts of life so interesting that a superficial treatment of them is an affront to their intelligence. The book may contain somewhat more material than is covered in certain biology courses. This enables each instructor to emphasize the subjects he thinks most important, yet provides the interested student with the opportunity to read about topics which are omitted or considered only briefly in the lectures. The scope and content of this book were greatly influenced by the author's experience as an instructor in the biology course taught by Professor Richard M. Eakin at the University of California, Berkeley, and his subsequent experience teaching biology at the University of North Carolina, Chapel Hill. I am greatly indebted to the many instructors who have made suggestions for revisions based upon their experience in using previous editions of the text. My special thanks are due to Professor Ralph W. Lewis, of Michigan State University, for several helpful suggestions, and to Dr. Dwain Hagerman who read a number of chapters of the manuscript of this revision.

I want to express my thanks to the artists, R. Paul Larkin, Margaret Croup, Betsy Hulbert, Gail Limberg, and Muriel McLatchie, who transformed my rough sketches into finished line drawings, and William Osburn, who drew the end sheets. I am indebted to a number of scientists—Drs. Russell J. Barrnett, John Bonner, Lewis Dexter, William Duryee, R. F.

Escamilla, Don Fawcett, Sam Granick, Arthur Hertig, James Hillier, C. A. Knight, H. Lisser, John Luft, Daniel Mazia, E. Perry McCullagh, Dorothy Price, C. F. Robinow, and Betty Uzman who furnished me with original photographs to use as illustrations. The American Museum of Natural History, Australian News and Information Bureau, Buffalo Society of Natural Sciences, Chicago Natural History Museum, General Biological Supply House, General Electric Company, Life Magazine, Lilly Research Laboratories, New York Botanical Garden, Parke, Davis and Company, San Diego Zoo, E. R. Squibb and Sons, The Upjohn Company, the U. S. Department of Agriculture, the U. S. Forest Service, and the U. S. Soil Conservation Service have kindly permitted me to use certain of their copyrighted photographs. I want to thank Henry Holt and Company, the J. B. Lippincott Company, the Macmillan Company, the McGraw-Hill Book Company, the University of California Press, the University of Chicago Press, and John Wiley and Sons for permission to reproduce illustrations from their books. I am especially grateful to the W. B. Saunders Company, who have allowed me to reproduce illustrations from a number of their books and have cooperated in every way during the preparation of this manuscript. My thanks are due to Ann deNisco, Janet Loring, and Frederica Wellington for their assistance in reading proof and preparing the index. Finally, I want to express my deep appreciation to my wife, Dorothy, who, despite the demands of a busy career in medicine, found time to help me with all the stages of planning and writing the manuscript.

CLAUDE A. VILLEE

Contents

Chapter 1

INTRODUCTION: BIOLOGY AND THE SCIENTIFIC METHOD 1

1. Early History of Biology 1
2. The Biological Sciences 2
3. Sources of Scientific Information 2
4. The Scientific Method 3
5. Applications of Biology 6

Questions 7
Supplementary Reading 7

Chapter 2

SOME MAJOR GENERALIZATIONS OF THE BIOLOGICAL SCIENCES 8

6. Biogenesis 9
7. The Cell Theory 9
8. The Gene Theory 9
9. The Theory of Organic Evolution 10
10. Enzymes and Metabolism 11
11. Genic Control of Metabolism 12
12. Vitamins and Coenzymes 12
13. Hormones 12
14. Interrelations of Organisms and Environment 12

Questions 13
Supplementary Reading 13

Part One. Cell Structures and Functions

Chapter 3

PROTOPLASM 15

15. Characteristics of Living Things 16
    - Metabolism 16
    - Movement 17

Growth 19
Reproduction 19
Adaptation 19
16. Matter and Energy 20
17. The Structure of Matter 20
18. Types of Substances 20
19. Compounds 22
20. Organic Compounds in Protoplasm 24
21. Carbohydrates 25
22. Fats 26
23. Steroids 27
24. Proteins 27
25. Nucleic Acids 28
26. Physical Characteristics of Protoplasm 29
Questions 30
Supplementary Reading 31

Chapter 4

CELLS AND TISSUES 32
27. Cells 32
28. Methods of Studying Cells 38
29. Energy 38
30. The Motion of Molecules 39
31. The Speed of Diffusion 40
32. Exchanges of Material Between Environment and Cells 41
33. Tissues 43
34. Animal Tissues 43
Epithelial Tissues 43
Connective Tissue 44
Muscular Tissue 45
Blood Tissue 47
Nervous Tissue 47
Reproductive Tissue 47
35. Plant Tissues 48
Meristematic Tissue 48
Protective Tissue 48
Fundamental Tissues 49
Conductive Tissues 50
36. Organ Systems 51
37. Body Plan and Symmetry 51
Questions 52
Supplementary Reading 52

Chapter 5

CELLULAR METABOLISM 53
38. Chemical Reactions 53
39. Catalysis 54
40. Enzymes 54
41. Properties of Enzymes 55
Location of Enzymes in the Cell 56
Mode of Action of Enzymes 56

42. Factors Affecting Enzyme Activity 57
Temperature 57
Acidity 58
Concentration of Enzyme, Substrate and Cofactors 58
Enzyme Poisons 58
43. Respiration and Energy Relations 58
44. The Dynamic State of Protoplasm 61
45. Special Types of Metabolism 61
46. Bioluminescence 63
Questions 64
Supplementary Reading 64

Part Two. The World of Life: Plants

Chapter 6

BIOLOGIC INTERRELATIONSHIPS 65

47. The Classification of Living Things 65
48. Distinctions Between Plants and Animals 67
49. Modes of Nutrition 68
50. The Cyclic Use of Matter 69
The Carbon Cycle 69
The Nitrogen Cycle 70
The Water Cycle 72
The Phosphorus Cycle 72
The Energy Cycle 72
51. Ecosystems 73
52. Habitat and Ecologic Niche 73
53. Types of Interactions Between Species 75
Competition 76
Commensalism 76
Protocooperation 76
Mutualism 76
Amensalism 76
Questions 77
Supplementary Reading 78

Chapter 7

GENERAL PROPERTIES OF GREEN PLANT CELLS 79

54. Photosynthesis 79
55. Synthesis of Other Organic Substances 85
56. Plant Respiration 85
57. The Skeletal System of Plants 86
58. Turgor Pressure 86
59. Plasmolysis 88
60. Plant Digestion 88
61. Plant Circulation 88
62. Plant Sap 89
63. Plant Excretion 89
64. Plant Coordination 89

65. The Transmission of Impulses 92
66. Plant Hormones 93
67. Photoperiodism 97
68. Sleep Movements 99
Questions 99
Supplementary Reading 99

Chapter 8

THE STRUCTURES AND FUNCTIONS OF A SEED PLANT 100

69. The Root and Its Functions 100
70. The Environment of Roots: Soil 105
71. The Stem and Its Functions 106
72. The Leaf and Its Functions 111
73. Transpiration 113
74. The Movement of Water 114
75. The Storage of Food 115
76. The Economic Importance of Plants 115
Questions 115
Supplementary Reading 116

Chapter 9

TYPES OF PLANTS: BACTERIA 117

77. Koch's Postulates 118
78. Occurrence of Bacteria 118
79. Cell Structure 118
80. Reproduction of Bacteria 120
81. Adaptations to Unfavorable Environmental Conditions 121
82. Bacterial Metabolism 121
83. Methods for Studying Bacteria 122
84. Some Economic Uses of Bacteria 123
85. Parasitic Bacteria 124
86. Other Microorganisms 124
87. Filtrable Viruses 124
88. Bacteriophages 126
89. Rickettsias 127
90. Evolutionary Relationships 127
Questions 129
Supplementary Reading 129

Chapter 10

ALGAE AND FUNGI 131

91. The Blue-green Algae 133
92. The Euglenophyta 134
93. The Green Algae, Phylum Chlorophyta 134
94. The Golden-brown Algae, Phylum Chrysophyta 135
95. The Dinoflagellates, Phylum Pyrrophyta 136
96. The Brown Algae, Phylum Phaeophyta 136
97. The Red Algae, Phylum Rhodophyta 138
98. The Fungi 140
99. The Slime Molds, Phylum Myxomycophyta 140

100. The Eumycophyta, True Fungi 140
101. The Phycomycetes 142
102. The Ascomycetes 142
103. The Basidiomycetes 144
104. Lichens 145
105. Economic Importance of the Fungi 146
Questions 149
Supplementary Readings 149

Chapter 11

THE INVASION OF LAND BY PLANTS 150

106. The Bryophytes 151
107. The Phylum Tracheophyta 152
108. Subphylum Psilopsida 153
109. Subphylum Lycopsida 153
110. Subphylum Sphenopsida 155
111. Subphylum Pteropsida 156
112. Class Filicinae 156
113. Seed Plants 158
114. The Gymnosperms 158
115. The Angiosperms 161
Questions 161
Supplementary Readings 162

Chapter 12

THE EVOLUTION OF PLANT REPRODUCTION 163

116. Asexual Reproduction 163
117. The Evolution of Sex 164
118. The Life Cycle 168
119. The Life Cycle of a Moss 169
120. The Life Cycle of a Fern 170
121. The Life Cycle of a Gymnosperm 171
122. The Life Cycle of an Angiosperm 173
The Flower 173
Fruits 175
123. Germination of the Seed and Embryonic Development 176
124. Economic Importance of Seeds 178
125. Evolutionary Trends in the Plant Kingdom 178
Questions 179
Supplementary Reading 179

Part Three. The World of Life: Animals

Chapter 13

THE ANIMAL KINGDOM: LOWER INVERTEBRATES 181

126. The Basis for Animal Classification 183
127. The Protoplasmic Level of Organization 184
The Phylum Protozoa 184

128. The Cellular Level of Organization 187
The Porifera or Sponges 187
129. The Tissue Level of Organization 188
130. The Organ Level of Organization 192
131. The Organ System Level of Organization 194
The Nematoda 195
The Rotifera 196
The Gastrotricha 197
The Bryozoa 197
The Brachiopoda 198
Questions 198
Supplementary Reading 198

Chapter 14

THE HIGHER INVERTEBRATES 199

132. Problems of Terrestrial Life 199
133. The Annelida 200
134. The Arthropoda 202
135. Insect Metamorphosis 205
136. The Arthropod Body Plan 208
137. Colonial Insects 209
138. Insect Behavior 210
139. The Mollusca 211
140. The Echinodermata 213
Questions 215
Supplementary Reading 215

Chapter 15

THE PHYLUM CHORDATA 216

141. Acorn Worms or Hemichordates 216
142. Sea Squirts or Tunicates 216
143. Cephalochordates 218
144. The Vertebrates 218
145. The Origin of the Chordates 219
146. Jawless Fishes 219
147. Cartilaginous Fishes 220
148. Bony Fishes 220
149. The Amphibia 222
150. Reptiles 223
151. Birds 223
152. Mammals 224
Questions 225
Supplementary Reading 226

Part Four. The Organization of the Body

Chapter 16

BLOOD 227

153. Plasma 228
154. The Red Corpuscles 229

155. Hemoglobin and the Transport of Oxygen 229
156. Life History of Red Cells 230
157. Oxygen-Carrying Devices in Other Animals 231
158. White Corpuscles 232
159. Protective Functions of White Cells 233
160. The Life History of White Cells 234
161. Blood Platelets 234
162. The Clotting of Blood 234
163. Diseases of the Blood 236
Polycythemia 237
Leukemia 237
164. Medicolegal Tests for Blood 237
165. Blood Types and Transfusions 238
Questions 240
Supplementary Reading 240

Chapter 17

THE CIRCULATORY SYSTEM 241

166. The Blood Vessels 241
167. The Heart 243
The Heart Beat 245
Nodal Tissue 245
The Heart Cycle 246
The Heart Sounds 248
Electrical Changes in the Heart 248
Adaptation of the Heart Beat to Body Activity 248
168. Routes of the Blood Around the Body 250
Fetal Circulation and Changes at Birth 250
The Rate of Flow of Blood 252
Blood Pressure 255
The Role of Blood Pressure in the Exchange of Materials in the Capillaries 256
169. Heart and Vessel Disorders 258
170. The Lymph System 259
The Flow of Lymph 260
Functions of the Lymph System 260
171. Circulation in Other Organisms 261
Questions 264
Supplementary Reading 264

Chapter 18

THE RESPIRATORY SYSTEM 265

Direct Respiration 265
Indirect Respiration 266
172. Structure of the Human Respiratory System 266
173. The Mechanics of Breathing 268
174. The Quantity of Air Respired 270
175. Composition of Alveolar Air 270
176. Exchange of Gases in the Lungs 271
177. Transportation of Oxygen by the Blood 271
178. Transportation of Carbon Dioxide by the Blood 273
179. Asphyxia 274

Artificial Respiration 274
180. The Control of Breathing 274
The Effects of Training 275
181. The Evolution of the Human Lungs 276
182. Respiratory Devices in Other Animals 277
Questions 278
Supplementary Reading 278

Chapter 19

THE DIGESTIVE SYSTEM 279
183. The Mouth Cavity 279
The Tongue 279
The Teeth 281
The Salivary Glands 282
184. Microscopic Anatomy of the Digestive Tract 282
185. The Pharynx 283
Swallowing 283
186. The Esophagus 284
187. The Stomach 284
Vomiting 285
188. The Small Intestine 285
Intestinal Movements 286
189. The Liver 287
190. The Pancreas 288
191. The Absorption of Food 288
192. The Large Intestine and Rectum 289
193. Disorders of the Digestive Tract 291
194. The Chemistry of Digestion 291
Saliva 291
Gastric Juice 292
Pancreatic Juice 293
Intestinal Juice 293
195. Methods of Stimulating the Digestive Glands 294
196. Comparison of Digestive Systems of Other Animals 295
Questions 297
Supplementary Reading 297

Chapter 20

METABOLISM AND NUTRITION 298
197. The Basal Metabolic Rate 298
Energy Requirements 299
198. The Fuels 299
Carbohydrates 299
Fats 300
Proteins 300
199. The Metabolism of Carbohydrates, Fats and Proteins 301
Carbohydrate Metabolism 301
Fat Metabolism 302
Protein Metabolism 302
200. Other Components of the Diet 302
Minerals 302

Water 303
Condiments and Roughage 304
201. Vitamins 304
Vitamin A 305
Vitamin D 306
Vitamin C 306
Vitamin E (Alpha-tocopherol) 307
Vitamin K 307
The Vitamin B Complex 307
Thiamine (Vitamin B_1) 307
Riboflavin (Vitamin B_2 or G) 308
Niacin or Nicotinic Acid 309
Pyridoxine (Vitamin B_6) 310
Pantothenic Acid 311
Biotin 311
Folic Acid, Vitamin B_{12}, Choline, Inositol and Para-aminobenzoic Acid 311
202. Antimetabolites 311
203. Diet 312
Questions 313
Supplementary Reading 313

Chapter 21

THE EXCRETORY SYSTEM 317

204. The Kidney and Its Ducts 318
205. The Formation of Urine 319
206. The Regulatory Function of the Kidney 321
207. Substances Present in the Urine 322
208. Diseases of the Kidneys 322
209. Excretory Devices in Other Animals 323
Questions 324
Supplementary Reading 325

Chapter 22

THE INTEGUMENTARY AND SKELETAL SYSTEMS 326

210. The Skin 326
Parts of the Skin 327
Outgrowths of the Skin 328
211. The Skeleton 328
Types of Skeletons 328
Parts of the Skeleton 329
The Joints 330
Types of Locomotion 331
Questions 332
Supplementary Reading 333

Chapter 23

THE MUSCULAR SYSTEM 334

212. The Skeletal Muscles 334
213. Kinds of Contraction 337

The Single Twitch 338
Tetanus 338
Tonus 339
214. The Chemistry of Muscle Contraction 339
The Oxygen Debt 340
Fatigue 341
The Nature of Contraction 341
215. Cardiac and Smooth Muscle 342
216. The Muscles of Lower Animals 342
Questions 343
Supplementary Reading 343

Chapter 24

THE NERVOUS SYSTEM 344

217. The Neurons 344
218. The Nerve Impulse 345
219. Transmission at the Synapse 349
220. The Central Nervous System 350
The Spinal Cord 350
The Brain 351
221. Brain Waves 355
222. Sleep 355
223. Insanity and Neuroses 356
224. The Peripheral Nervous System 357
225. Reflexes and Reflex Arcs 358
226. Habit, Memory and Learning 360
Moods and Emotions 360
227. The Autonomic Nervous System 361
The Sympathetic System 361
The Parasympathetic System 363
228. The Nervous Systems of Lower Animals 363
Questions 364
Supplementary Reading 364

Chapter 25

THE SENSE ORGANS 365

229. The Stimulus-Receiving Process 366
230. The Perception of Sensations 366
231. The Localization of Stimuli 366
232. The Tactile Senses 367
Kinesthesis 367
Visceral Sensitivity 368
233. The Chemical Senses: Taste and Smell 368
234. Vision 369
The Human Eye 370
Defects in Vision 374
235. The Ear 375
236. Equilibrium 379
Questions 380
Supplementary Reading 380

Chapter 26

THE ENDOCRINE SYSTEM 381

237. The Thyroid Glands 383
238. The Parathyroid Glands 385
239. The Islet Cells of the Pancreas 386
240. The Adrenal Glands 388
    The Adrenal Medulla 388
    The Adrenal Cortex 389
241. The Pituitary 389
    The Posterior Lobe 390
    The Anterior Lobe 390
242. The Testes 394
243. The Ovaries 395
    The Estrus Cycle 396
    The Menstrual Cycle 396
244. The Placenta 398
245. Other Endocrine Glands 398
246. Interrelationships Between the Endocrine Glands 399
Questions 399
Supplementary Reading 399

Chapter 27

INFECTIOUS DISEASES AND THE BODY'S DEFENSES 401

247. How Microorganisms Cause Disease 402
248. Body Defenses Against Disease 402
    Acquired Immunity 403
    Natural Immunity 404
249. Allergy 405
250. Antibiotics 405
251. The Spread of Microorganisms 406
252. Some Common Infectious Diseases 407
    Poliomyelitis 407
    Typhus 407
    Syphilis 407
    Athlete's Foot 409
    Amebic Dysentery 409
    Tapeworm Infections 409
    Trichinosis 409
Questions 411
Supplementary Reading 411

Part Five. The Reproductive Process

Chapter 28

REPRODUCTION 413

253. Asexual Reproduction 413
254. Sexual Reproduction in Animals 415
    Hermaphroditism 415

Parthenogenesis ... 416
Types of Fertilization ... 416
255. Human Reproduction ... 417
The Male Reproductive Organs ... 417
The Female Reproductive Organs ... 419
Fertilization ... 420
Implantation ... 421
Nutrition of the Embryo ... 421
256. The Embryonic Membranes ... 422
257. The Placenta ... 423
258. Birth ... 424
259. Nutrition of the Infant ... 426
Questions ... 426
Supplementary Reading ... 427

Chapter 29

EMBRYONIC DEVELOPMENT ... 428

260. Types of Eggs ... 428
261. Cleavage and Gastrulation ... 429
Blastula Formation ... 430
Gastrulation ... 431
Cleavage and Gastrulation in Frog and Chick Eggs ... 431
Cleavage and Gastrulation in the Human Egg ... 432
262. Mesoderm Formation ... 433
The Notochord ... 435
263. Development of the Nervous System ... 436
264. Development of Body Form ... 437
265. Formation of the Heart ... 440
266. Development of the Digestive Tract ... 441
267. The Development of the Kidney ... 442
268. The Contributions of Each Germ Layer ... 443
269. Control of Development ... 444
270. Morphogenesis ... 444
271. Malformations ... 446
272. The Changes at Birth ... 447
Questions ... 448
Supplementary Reading ... 449

Part Six. The Mechanism of Heredity

Chapter 30

THE PHYSICAL BASIS OF HEREDITY ... 451

273. Heredity and Variation ... 451
274. Chromosomes and Genes ... 453
Chromosome Structure ... 453
Chromosome Number ... 453
275. Mitosis ... 454
Prophase ... 455
Metaphase ... 455
Anaphase ... 455

Telophase 455
Control of Mitosis 457
276. Meiosis 457
277. Spermatogenesis 460
278. Oogenesis 461
279. Genes and Alleles 462
280. A Monohybrid Cross 464
281. Phenotype and Genotype 465
282. Mendel's First Law 465
283. Test Crosses 465
284. Incomplete Dominance 466
285. Deducing Genotypes 467
Questions 468
Supplementary Reading 468

Chapter 31

GENETICS 469

286. Mendel's Second Law 469
287. Interactions of Genes 471
Complementary Genes 472
Supplementary Genes 473
288. Multiple Factors 473
289. Multiple Alleles 476
290. Lethal Genes 477
291. The Genetic Determination of Sex 477
292. Sex-Linked Characters 478
293. Sex-Influenced Characters 479
294. Linkage and Crossing Over 479
295. Penetrance and Expressivity 481
296. Inbreeding and Outbreeding 482
297. Population Genetics 483
298. Biochemical Genetics 484
The Chemical Nature of the Gene 484
Estimates of the Number and Size of Genes 485
Changes in Genes: Mutations 485
Gene Action 486
Questions 488
Supplementary Reading 489

Chapter 32

INHERITANCE IN MAN 490

299. Human Pedigrees 490
300. The Inheritance of Physical Traits 491
301. Heredity and Environment 493
302. Inheritance of Mental Disorders 494
303. Inheritance of General and Special Abilities 495
304. Eugenics 496
Questions 498
Supplementary Reading 498

Part Seven. Evolution

Chapter 33

PRINCIPLES AND THEORIES OF EVOLUTION 501
306. Organic Evolution 501
Jean Baptiste de Lamarck 503
Background for The Origin of Species 503
307. The Theory of Natural Selection 504
308. Modern Changes in the Theory of Natural Selection 504
Modifications and Mutations 504
Isolation 505
309. Genetic Drift 506
310. Preadaptation 506
311. Mutations: The Raw Material of Evolution 507
Mutation Theory of de Vries 507
Types of Mutations 509
Causes of Mutations 509
312. Straight-line Evolution 510
313. The Origin of Species by Hybridization 510
314. The Origin of Life 511
315. Principles of Evolution 514
Questions 514
Supplementary Reading 515

Chapter 34

THE FOSSIL EVIDENCE FOR EVOLUTION 516
316. Paleontology 516
317. The Geologic Time Table 517
318. The Geologic Eras 520
Archeozoic Era 520
Proterozoic Era 520
Paleozoic Era 521
The Mesozoic Era (The Age of Reptiles) 525
The Cenozoic Era (The Age of Mammals) 533
Questions 538
Supplementary Reading 538

Chapter 35

THE LIVING EVIDENCE FOR EVOLUTION 540
319. The Evidence from Taxonomy 540
320. The Evidence from Anatomy 541
Homologous Organs 541
Vestigial Organs 541
321. The Evidence from Comparative Physiology and Biochemistry 541
322. The Evidence from Embryology 543
323. The Genetic Evidence 545
324. Evidence from the Geographic Distribution of Organisms 545
The Biogeographic Realms 547

Questions 547
Supplementary Reading 548

Chapter 36

THE EVOLUTION OF MAN 549

325. Man and the Other Primates 549
326. Fossil Primates 552
327. The Man-apes 556
328. Fossil Ape-men 557
 The Java Man 557
 Peking Man 557
 Piltdown Man 558
329. Fossil Members of the Genus Homo 559
 Heidelberg Man 559
 Neanderthal Man 560
 Solo Man 561
 Rhodesian Man 561
 Modern Man 561
330. Cultural Evolution 565
331. The Present Races of Man 566
 The White Race 567
 The Negroid Race 568
 The Mongoloid Race 569
Questions 570
Supplementary Reading 570

Part Eight. Ecology

Chapter 37

PRINCIPLES OF ECOLOGY 571

332. Factors Regulating the Distribution of Plants and Animals 571
333. Food Chains 573
334. Populations and Their Characteristics 574
335. Population Cycles 577
336. Biotic Communities 578
337. Community Succession 579
338. Applications of Ecologic Principles 579
 Questions 582
 Supplementary Reading 582

Chapter 38

THE OUTCOME OF EVOLUTION: ADAPTATION 583

339. Adaptive Radiation 583
340. Convergent Evolution 583
341. Structural Adaptations 584
342. Physiologic Adaptations 585
343. Color Adaptations 586

344. Adaptations of Species to Species 587
345. Terrestrial Life Zones: Biomes 588
Tundra 590
Northern Coniferous Forest 590
Temperate Deciduous Forest 590
Broadleaved Evergreen Subtropical Forest 590
Grasslands 590
Deserts 590
Tropical Rain Forest 591
346. Marine Life Zones 592
347. Fresh-water Life Zones 594
348. The Dynamic Balance of Nature 594
Questions 594
Supplementary Reading 594

Appendix

A SURVEY OF THE PLANT AND ANIMAL KINGDOMS 595
I. The Plant Kingdom 595
II. The Animal Kingdom 597

BIBLIOGRAPHY 601

INDEX 605

Chapter 1

Introduction: Biology and the Scientific Method

IN ONE sense, biology is a very old science, for men began many centuries ago to study living things in attempts to solve the fascinating riddle of life. There was a considerable body of knowledge and theories about living things in the time of Aristotle (384–322 B.C.), and even in the older civilizations of Egypt, Mesopotamia and China much was known about practical uses of plants and animals. In fact, the cave men who lived 50,000 and more years ago must have been first rate biologists for they drew accurate and artistic pictures on the walls of their caves of the deer, cattle and mammoths that lived around them. The survival of early man depended on a knowledge of such fundamental biologic facts as which animals were dangerous and which plants could be safely eaten.

Yet in another sense biology is a young science. The major generalizations which are the foundations of any science have been made comparatively recently in biology and many of them are still being revised. The development of the electron microscope, for example, and the recent discovery of ways to prepare tissues for examination in this instrument, have revealed a whole new order of complexity in living matter.

1. EARLY HISTORY OF BIOLOGY

Biology as an organized body of knowledge can be said to have begun with the Greeks. They and the Romans described the many kinds of plants and animals known at the time. Galen (131–200 A.D.) described the anatomy of the human body and was the unchallenged authority for 1300 years. His descriptions, however, were based on dissections of apes and pigs and contained many errors. Galen was the first experimental physiologist and performed many experiments, mostly on pigs, to study the functions of nerves and blood vessels. Men such as Pliny (23–79 A.D.) prepared encyclopedias which were strange mixtures of facts and fiction about living things. In the succeeding centuries of the Middle Ages men wrote "herbals" and "bestiaries," cataloguing and describing plants and animals respectively. With the Renaissance interest in natural history revived and more accurate studies of the structure, functions and life habits of countless plants and animals were made. Vesalius (1514–1564), Harvey (1578–1657) and John Hunter (1728–1793) studied the structure and functions of animals in general and man in particular and laid the foundations of anatomy and physiology. With the invention of the microscope early in the seventeenth century, Malpighi (1628–1694), Swammerdam (1637–1680) and Leeuwenhoek (1632–1723) investigated the fine structure of a variety of plant and animal tissues. Leeuwenhoek was the first to describe bacteria, protozoa and sperm.

Biology expanded and altered greatly in the nineteenth century and has continued this trend at an accelerated pace in the twentieth. This is due in part to the broader scope and more detailed knowledge available today and in part to the new approaches made possible by the discoveries and techniques of physics and chemistry. In the past hundred years many biologists have been drawn to the level of inquiry represented by biophysics and biochemistry. This book is not primarily concerned with that level, but some knowledge of the ultramicroscopic world of atoms and molecules is necessary for a real understanding of even the simplest biologic processes.

2. THE BIOLOGICAL SCIENCES

The usual definition of biology as the "science of life" is only meaningful if we have some idea of what life and science mean. Life does not lend itself to a simple definition and its characteristics—growth, movement, metabolism, reproduction and adaptation—will be discussed in Chapter 3. Biology is concerned with the myriad forms that living things may have, with their structure, function, evolution, development and relations to their environment. It has grown to be much too broad a science to be investigated by one man or to be treated thoroughly in a single textbook, and most biologists are specialists in some one of the biological sciences. The **botanist** and **zoologist** study types of organisms and their relationships within the plant and animal kingdoms respectively. The sciences of **anatomy, physiology** and **embryology** deal with the structure, function and development of an organism; these can be further subdivided according to the kind of organism investigated: e.g., animal physiology, mammalian physiology, human physiology. The **parasitologist** studies those forms of life that live in and at the expense of other forms, the **cytologist** investigates the structure, composition and function of cells, and the **histologist** inquires into the properties of tissues. The science of **genetics** is concerned with the mode of transmission of the characteristics of one generation to another, and is closely related to the study of **evolution,** which attempts to discover how new species arise, as well as how the present forms evolved from previous ones. The study of the classification of plants and animals and their evolutionary relations is known as **taxonomy.** One of the newest biological sciences is **ecology,** the study of the relations of a group of organisms to its environment, including both the physical factors and other living organisms which provide food or shelter for it, or compete with or prey upon it.

There are also specialists who deal with one kind of living thing—ichthyologists, who study fish, mycologists, who study fungi, ornithologists, who study birds, and so on.

3. SOURCES OF SCIENTIFIC INFORMATION

Where, you may ask, do all the facts about biology described in this book come from? And how do we know they are true? The ultimate source of each fact, of course, is in some carefully controlled observation or experiment made by a biologist. In earlier times, some scientists kept their discoveries to themselves, but now there is a strong tradition that scientific discoveries are public property and should be freely published. It is not enough in a scientific publication for a man to say that he has discovered a certain fact; he must give all the relevant details by which the fact was discovered so that others can repeat the observation. It is this criterion of **repeatability** that makes us accept a certain observation or experiment as representing a true fact; observations that cannot be repeated by competent investigators are discarded.

When a biologist has made a discovery, he writes a report, called a "paper," in which he describes his methods in sufficient detail so that another can repeat them, gives the results of his observations, discusses the conclusions to be drawn from them, perhaps formulates a theory to explain them, and indicates the place of these new facts in the present body of scientific knowledge. The knowledge that his discovery will be subjected to the keen scrutiny of his colleagues is a strong stimulus for carefully repeating the observations or experiments before publishing them. He then submits his paper for publication in one of the professional journals in the particular field of his discovery (it is estimated that there are more than 7,000 of them published over the world in the various fields of biology!) and it is read by one or more of the board of editors of the journal, all of whom are experts in the field. If it is approved, it is published and thus becomes part of "the literature" of the subject.

At one time, when there were fewer journals, it might have been possible for one man to read them each month as they appeared, but this is obviously impossible now. Journals such as *Biological Abstracts* assist the hard-pressed biologist by publishing, classified by fields, very short reports or **abstracts** of each paper published—giving the facts found, and a reference to the journal. A considerable number of journals devoted solely to reviewing the newer developments in particular fields have sprung up in the past twenty-five years; some of these are *Physiological Reviews, The Botanical Review, Quarterly Review of Biology, Annual Review of Microbiology* and *Nutrition Reviews.* The new fact or theory thus becomes widely known through publication in a professional journal and by reference in abstract and review journals, and eventually may become a sentence or two in a textbook.

Other means for the dissemination of new knowledge are the annual meetings held by the professional societies of botanists, geneticists, physiologists and other specialists at which papers are read and discussed. There are, from time to time, national and international gatherings, called **symposia,** of specialists in a given field to discuss the newer findings and the present status of the knowledge in that field. The discussions of these symposia are usually published as books.

4. THE SCIENTIFIC METHOD

The facts of biology are gained by the application of the scientific method, yet it is difficult to reduce this method to a simple set of rules that apply to all the branches of science. One of the basic tenets of the scientific method is the **rejection of authority**—the refusal to accept a statement just because someone says it is so. The skeptical scientist wants confirmation of the statement by the independent observation of another.

The basis of the scientific method and the ultimate source of all the facts of science is careful, close observation and experiment, free of bias and done as quantitatively as possible. The observations or experiments may then be analyzed, or simplified into their constituent parts, so that some sort of order can be brought into the observed phenomena. Then the parts can be synthesized or reassembled and their interactions discovered. On the basis of these observations, the scientist constructs a **hypothesis,** a trial idea about the nature of the observation, or possibly the connections between a chain of events, or even cause and effect relationships between different events. It is in the construction of hypotheses that scientists differ most and that true genius shows itself. The ability to see through a mass of data and suggest a reason for their interrelations is all too rare.

It must be emphasized that science does not advance by the mere accumulation of facts, or by the mere postulation of hypotheses. The two go hand-in-hand in most scientific investigations: hypothesis, observation, revised hypothesis, further observation, and so on. When a scientist embarks upon an investigation he has the advantage of the relevant facts already known with which to build a "working hypothesis" to guide the design of his experiments. When a scientist makes an observation that does not agree with his hypothesis he may conclude either that his hypothesis or that his observation is wrong. He then repeats his observation, perhaps altering the design of his experiment to get at the relationship in a new way, or perhaps using a different technique. If he can satisfy himself that his observation is valid, he either discards his hypothesis or amends it to account for the new observation. In the final analysis, each new observation must either agree or disagree with the hypothesis to be useful.

Hypotheses are constantly being refined and elaborated. There are few scientists who consider any hypothesis, no matter how many times it may have been tested, as a statement of absolute and universal truth. The hypothesis is simply regarded as the best available approximation to the truth for some finite range of circumstances. The Law of the Conservation of Energy (p. 72), for example, was widely accepted until the work of Einstein showed that it had to be modified to allow for the possible interconversion of matter and energy. Although this might have seemed to be an inconsequential distinction at one time, for it has no importance at all in ordinary chemical processes, it is the theoretical basis of atomic power.

Once a hypothesis has been set up to explain a certain body of facts, the rules of formal logic can be used to deduce certain consequences. In a science such as physics, and to a lesser extent in biology, the hypotheses and deductions can be stated in mathematical terms and elaborate and far-reaching conclusions can be drawn. On the basis of these deductions the results of other observations and experiments can be predicted and the hypothesis can be tested by its ability to make valid predictions. If the hypothesis is a simple generalization, it may be enough simply to examine more examples and see if the generalization holds true. More complex hypotheses, that perhaps cannot be tested directly, can be tested by seeing whether certain logical deductions from the hypothesis hold true. A hypothesis must be subject to some sort of experimental test—it must make a prediction that can be verified in some way—or it is mere speculation.

A hypothesis that fits a large body of different types of observations becomes a **theory,** which is defined by Webster as "a scientifically acceptable general principle offered to explain phenomena; the analysis of a set of facts in their ideal relations to one another." A good theory relates, from one point of view, facts which previously appeared unrelated and which could not be explained on common ground. A good theory grows; it relates additional facts as they become known. Indeed, it predicts new facts and suggests new relationships between phenomena.

A good theory, by showing the relationship between classes of facts, simplifies and clarifies our understanding of natural phenomena. In the words of Einstein, "In the whole history of science from Greek philosophy to modern physics, there have been constant attempts to reduce the apparent complexity of natural phenomena to some simple, fundamental ideas and relations." Science is really the search for simplicity. William of Occam, a fourteenth century philosopher made the dictum, *"Essentia non sunt multiplicanda praeter necessitatem",* or "Entities should not be multiplied beyond necessity." This principle of parsimony (often called **Occam's razor** because it pares a theory to its bare essentials) means that no more forces or causes should be postulated than are necessary to account for the phenomena observed. In practice, this means that the simplest explanation which will account satisfactorily for all the known facts is to be preferred. A new theory in biology, by clearing away previous misconceptions and by pointing up new interrelations of phenomena, not only stimulates research in theoretical biology, it also provides the basis for a host of practical advances in medicine, agriculture, and similar fields.

A poor theory, in contrast, when its consequences are followed, will sooner or later lead to absurdities and clear, irreconcilable contradictions. It frequently happens that at some stage in our knowledge two, or even more, alternative theories provide equally good explanations for the data at hand. But as more observations or experiments are made, one or the other (or perhaps both!) are ruled out.

The scientific method, then, consists of making careful observations and arranging these observations so as to bring order into the observed phenomena. Then we try to find a hypothesis or a **conceptual scheme** which will explain not only the facts already observed but also new facts as they are discovered. Sciences differ widely in the extent to which they are predictable and there are some who claim that biology is not a science because it is not completely predictable. However, even physics, generally regarded as the most "scientific" of the sciences, is far from completely predictable. Although we can predict the occurrence of eclipses, we cannot make predictions in the field of quantum mechanics, nor can we predict an earthquake, or even tomorrow's weather.

In most scientific studies one of the ultimate goals is to explain the cause of some phenomenon, but the hard-and-fast proof that a cause and effect relationship exists between two events is extremely difficult to obtain. If the circumstances leading to a certain event always have a certain factor in common in a variety of cases, that factor may be the cause of the event. The difficulty lies in making sure that the factor under consideration is the *only* one common to all the cases. For example, it would be wrong to conclude from finding that Scotch and soda, bourbon and soda, and rye and soda all produce intoxication, that soda is the only factor in common and therefore the cause of the intoxication! This method of discovering the common factor in a variety of cases that may be the cause of the event (known as the **method of agreement**) can seldom be used as a valid proof because of this difficulty in being sure that it really is the only common factor. The finding that all people suffering from beriberi have diets which are low in thiamine is not proof that this deficiency causes the disease, for there may be many other factors in common.

Another method for unraveling cause and effect relations is the **method of difference:** If two sets of circumstances differ in only one factor, and the one containing the factor leads to an event and the other does not, the factor may be considered the cause of the event. For example, if two groups of rats are fed diets which are identical except that one contains all the vitamins and the second contains all but thiamine, and if the first group grows normally and the second group fails to grow and ultimately develops polyneuritis, this would be a strong suggestion, but not absolute proof, that polyneuritis or beriberi in rats is caused by a deficiency of thiamine. By using an inbred strain of rats that are as alike as possible in inherited traits, and by using litter mates (brothers and sisters) of this strain, one could make certain that there were no hereditary differences between the **controls** (the ones getting the complete diet) and the **experimentals** (the ones getting the thiamine-deficient diet). It could conceivably be that the diet without thiamine does not have as attractive a taste as the one with it, and the experimental group simply ate less food, failed to grow and developed the deficiency symptoms because they were partially starved. This source of error can be avoided by "pair-feeding," by pairing a control and an experimental animal, weighing the food eaten each day by each of the experimental animals and then giving only that much food to each control member of the pair.

A third way of detecting cause and effect relationships is the **method of concomitant variation:** If a variation in the amount of a given factor produces a parallel variation in the effect, the factor may be the cause. Thus if other groups of rats were given diets with varying amounts of thiamine and if the amount of protection against beriberi varied directly with the amount of thiamine in the diet, we could be reasonably sure that thiamine deficiency is the cause of beriberi.

It must be emphasized that it is seldom that we can be more than "reasonably sure" that X is the cause of Y. As more experiments and observations lead to the same result, the probability increases that X is the cause of Y. When experiments or observations can be made quantitative—when their results can be measured in some way—one can, by the methods of statistical analysis, determine the probability that X is the cause of Y, or the probability that Y follows X simply as a matter of chance. Scientists are usually satisfied that there is some sort of cause and effect relationship between X and Y if they can show that there is less than one chance in a hundred that the observed

X — Y relationship could be due to chance alone. A statistical analysis of a set of data can never give a flat yes or no to a question—it can only state that something is very probable or very improbable. It can also tell an investigator approximately how many more experiments he must do to reach a given probability that Y is caused by X.

Each experiment must contain a control group—one treated exactly like the experimental group in all respects but one, the factor whose effect is being tested. The use of controls in medical experiments raises the difficult question of the moral justification of withholding treatment from a patient who might be benefited by it. If there is sufficient evidence that one treatment is better than a second one, a physician would hardly be justified in further experimentation. However, the medical literature is full of treatments now known to be useless or even harmful, which were used for years but finally were abandoned as experience showed they were ineffective and that the evidence which had suggested their use originally was improperly controlled. There is a time in the development of any new treatment when the medical profession is not only morally justified but really morally required to do carefully controlled tests on human beings to be sure that the new treatment is better than the former one.

In such tests it is not sufficient simply to give a treatment to one group of patients and not to give it to another, for it is widely known that there is a strong psychologic effect in simply giving a treatment. For example, a group of students at a large western university served as subjects for a test of the hypothesis that daily doses of extra amounts of vitamin C might help prevent colds. This grew out out of the observation that people who drank lots of fruit juice seemed to have fewer colds. The group receiving the vitamin C showed a 65 per cent reduction in the number of colds contracted during the winter when they were receiving treatment compared to the previous winter when they were not receiving treatment. There were enough students in the group (208) to make this result statistically significant. In the absence of controls, one would have been led to conclude that vitamin C does help prevent colds. But a second group was given "placebos," pills identical in size, shape, color and taste to the vitamin C pills but without any vitamin C. The students were not told who was getting vitamin C and who was not, they only knew they were getting pills that might help prevent colds. The group getting placebos showed a 63 per cent reduction in the number of colds; thus, vitamin C had nothing to do with the result and the reported reductions in both groups were probably psychological effects.

In all experiments, the scientist must ever be on his guard against bias in himself, bias in the subject, bias in his instruments, and bias in the way the experiment is designed. The proper design of experiments is a science in itself, but one for which only general rules can be made.

A hypothesis that has been tested and found to fit the facts and capable of making valid predictions may then be called a **theory,** a **principle,** or a **law.** Although there is some connotation of greater reliance in a statement called a "law" than in one called a "theory," the two words are used interchangeably.

5. APPLICATIONS OF BIOLOGY

Some of the practical uses of a knowledge of biology will become apparent as the student reads on through this text—its applications in the fields of medicine and public health, in agriculture and conservation, its basic importance to the social studies, and its contributions to the formulation of a philosophy of life. There are esthetic values in a study of biology as well. A student cannot expect to learn all or even many of the names and characteristics of the vast variety of plants and animals, but a knowledge of the structure and functions of the major types will greatly increase the pleasure of a stroll in the woods or an excursion to the seashore. The average city-dweller gets only a small glimpse of the vast panorama of living things, for so many of them live in places where they are not easily seen—the sea, or parts of the earth that are not easily visited. Trips to botanical gardens, zoos, aquariums and museums will help give one an appreciation of the tremendous variety of living things.

It is impossible to describe the forms of life without reference to their habitats, the places in which they live. This brings us to one of the major unifying conceptual schemes of biology, that the living things of a given region are closely interrelated with each other and with the environment. The study of this is basic to sociology. The present forms of life are also related more or less closely by evolutionary descent. As we deal with each of the major life forms, the facts about them will be easier to understand and remember if we try to fit them into their place in the closely interwoven tapestry of life.

In our discussions of biologic principles we will focus our attention primarily on man, to gain an appreciation of man's place in the biologic world. It is only in man's somewhat biased opinion that he stands in the center of the universe, with other animals and plants existing only to serve him. In numbers, size, strength, endurance and adaptability he is inferior to many animals and in his adjustment to the environment—which, as we shall see, may be considered to be the most important biologic attribute of any living organism—he often fails. However, in a survey study of general biology, both practical considerations and interest demand that our discussions focus on man, for we are primarily concerned with such things as the human stomach ache, the human gestation period, and the endurance of the human body.

QUESTIONS

1. How would you define "science"?
2. Contrast a hypothesis and a law.
3. How would you go about testing the hypothesis that beriberi is caused by a deficiency of thiamine?
4. What would you consider to be proof that beriberi is caused by thiamine deficiency?
5. To which of the biologic sciences would you assign the following scientific papers:
   The Flora of Northern Michigan.
   The Fate of the Aortic Arches in the Development of the Chick.
   The Regulation of the Heart Rate.
   The Geographical Distribution of the Species of Wheat.
6. Describe in your own words the mode of operation of the scientific method.
7. Contrast the "method of agreement" and the "method of difference" as means of establishing cause and effect relationships.
8. What characteristics and attitudes do you think would be helpful for a career in science?
9. What is meant by a "controlled experiment"?

SUPPLEMENTARY READING

There are a number of fine books on the history of science: The development of the sciences in general is described in Sedgwick, Tyler and Bigelow's *A Short History of Science,* and a discussion of the role of science in society is given in J. B. Conant's *On Understanding Science.* The histories of the biologic sciences by Nordenskiöld and by Singer are well written and informative. The *History of Medicine* written by Douglas Guthrie describes the beginnings of anatomy, physiology and bacteriology.

The scientific method and its application to research problems are discussed in Conant's *Science and Common Sense* and Cohen's *Science, Servant of Man.* E. Bright Wilson's *An Introduction to Scientific Research* gives an excellent discussion in nontechnical terms of the methods of science and some of the problems involved in scientific investigation. W. B. Cannon's *The Way of an Investigator* gives some interesting examples of the scientfic method in medical research. *In the Name of Science,* by Martin Gardner, describes many pseudosciences and, in showing up their shortcomings, gives an appreciation for scientific evidence and standards.

Chapter 2

Some Major Generalizations of the Biological Sciences

BIOLOGY, like physics and chemistry, is a science composed of thousands of facts derived from a multitude of individual observations. Yet, to understand this science, the student need not memorize all of these facts, or even any considerable part of them. As in physics and chemistry, but perhaps not to the same extent, there are broad generalizations—hypotheses, theories, laws—which have been inferred from careful study and evaluation of these individual observations. Since these generalizations are the foundation of present-day biology, it seems worth while to discuss them briefly at this point before continuing with the more detailed presentation of plant and animal form and function. The subsequent, more detailed discussions should be more meaningful if viewed against the background of these theories. These generalizations, in turn, should become clearer and more firmly fixed in your mind as you examine the relevant observations and experimental results which will be presented in the succeeding chapters. We cannot discuss these generalizations fully at this time, for that would obviously require a large book, but the present summarization should be helpful to each student in getting a firm grasp of the broad picture of biology. It may, indeed, be helpful to reread this chapter after reading the subsequent ones.

As with most generalizations, there are exceptions to many of the statements in this chapter, some of them unimportant, others of considerable theoretical and practical importance. These exceptions will be dealt with in the later, more detailed discussions.

One of the basic tenets of modern biology is that the phenomena of life can be explained in terms of chemistry and physics; living systems are not distinguished from nonliving ones by some mysterious vital force. This idea has only recently gained acceptance in biology; only 35 years ago, for example, the German embryologist, Hans Driesch, postulated the existence of transcendent regulative principles, **entelechies,** which control the phenomena of life and development. His arguments were based on his experiments in which he found that isolated blastomeres of sea urchin embryos would develop into complete, perfect larvae—a part gave rise to the whole (p. 445). Further increases in our knowledge of embryology have shown that these phenomena can be explained in such physicochemical terms as gradient fields and organizers.

6. BIOGENESIS

There appear to be no exceptions to the generalization that "all life comes only fiom living things" (p. 19). The idea that large organisms, such as worms, frogs and rats, could arise by spontaneous generation was disproved in the seventeenth century. The experiments of Pasteur, Tyndall and others just a century ago finally provided convincing proof that microorganisms such as bacteria are also incapable of originating from nonliving material by spontaneous generation. Whether the submicroscopic, filtrable viruses should be considered as living or not is perhaps arguable, but it seems clear that they can originate only from preexisting viruses and not from nonviral material.

7. THE CELL THEORY

One of the broadest and most fundamental biological generalizations is the cell theory. The cell theory at present states that all living things, animals, plants and bacteria, are composed of cells and cell products, that new cells are formed by the division of preexisting cells, that there are fundamental similarities in the chemical constituents and metabolic activities of all cells, and that the activity of an organism as a whole is the sum of the activities and interactions of its independent cell units.

Like most broad theories this is not the product of any single person's research and thought. The German botanist, Schleiden, and zoologist, Schwann, are usually credited with this theory, for in 1838 they pointed out that plants and animals are aggregates of cells arranged according to definite laws. However, the French biologist, Dutrochet, clearly stated (1824) that "all organic tissues are actually globular cells of exceeding smallness, which appear to be united only by simple adhesive forces; thus all tissues, all animal organs are actually only a cellular tissue variously modified." Dutrochet recognized that growth is the result of increases in the volumes of individual cells and of the addition of new little cells. The presence of a nucleus within the cell, now recognized as an almost universal feature of cells, was first described by Robert Brown in 1831.

8. THE GENE THEORY

Once it had been established that a new organism originates from the union of an egg and a sperm, the question of how the parental characters are transmitted to the offspring through these tenuous bits of protoplasm arose. Among the many theories regarding this was the one stated by Charles Darwin, that each tissue or organ of the parent contributed some sort of models, or "pangenes," which were incorporated into the egg or sperm and thus transmitted to the offspring, where they guided development so as to produce in the offspring a duplicate of the organ from which they came. No evidence for such pangenes has ever been obtained. The theory of the "continuity of the germplasm" was formulated by August Weismann in 1889. His answer to the question of how a single germ cell, an egg or a sperm, can contain all the hereditary tendencies of the whole organism was that these germ cells are in turn derived from the parent germ cell and not from the body (somatic cells) of the individual. He postulated that from the very first division of the fertilized egg one line of cells, the germplasm, was distinct from the body cells or somatoplasm and that the germplasm was unaffected by the somatoplasm or by external influences. He realized, before chromosomes or genes were known, that heredity involves the transfer of particular molecular constitutions from one generation to the next. A corollary of Weismann's theory is that acquired characteristics are not inherited; only changes in the germplasm, not in the somatoplasm, can be transmitted to successive generations.

In certain invertebrate animals, the actual continuity of the germplasm from generation to generation is clear, for early in cleavage one cell can be distinguished as the precursor of the germ cells. In most animals the distinction between germplasm and somatoplasm is not clearly evident, and germ cells appear to arise from unspecialized somatic cells. As more has been learned about chromosomes and genes, it has become clear that the continuity from generation to generation lies in the chromosomes, present in all cells, and not in some peculiar property of the germ line itself.

A dramatic demonstration that the germ cells are indeed uninfluenced by the somatic cells was provided by W. E. Castle and J. C. Philips in 1909. They removed the ovaries from a white (albino) guinea pig and implanted in her an ovary from a black guinea pig. The animal was later mated with a white male guinea pig and produced offspring all of which were black.

The generalizations about the mechanism of inheritance are among the most exact and quantitative of biological theories, for they permit one to make predictions as to the probability that the offspring of two given parents will have a particular characteristic. These generalizations are called **Mendel's Laws,** for they were first enunciated by the Austrian abbot Gregor Mendel in 1868 from his careful breeding experiments with peas. Their importance was not recognized until 1900, when the principles were independently rediscovered by three different investigators, Correns, de Vries and von Tschermak. Mendel's First Law, the **Law of Segregation,** states that genes, the units of heredity, exist in individuals as pairs, and in the formation of gametes the two genes separate or segregate and pass into different gametes, so that each gamete has one and only one of each kind of gene. Mendel's Second Law, the **Law of Independent Segregation,** states that the segregation of each pair of genes in the process of gamete formation is independent of the segregation of the members of other pairs of genes, so that the members come to be assorted at random in the resulting gamete. Mendel's keen insight is truly remarkable, for he made these generalizations despite the fact that the details of chromosomes, meiosis and fertilization were unknown. Later, when chromosomes were discovered and genetic and cytologic evidence was available, the modern concept that the units of inheritance are arranged in linear order in the chromosomes was stated by Sutton in 1902 and Morgan in 1911.

The principle that a population of a given species of animals or plants is in genetic equilibrium, i.e., that it is genetically stable (in the absence of natural selection), and tends to have the same proportion of organisms with a given characteristic in successive generations, was arrived at independently by the mathematician, Hardy, and the physician, Weinberg, in 1908. They pointed out that the frequency of the possible combinations of a pair of genes in a population are described by the expansion of the binomial $(pA + qa)^2$. This **Hardy-Weinberg Law** (p. 483) has been of great importance in genetics, particularly human genetics, for it is the basis of statistical methods used in determining the mode of inheritance of a given trait in the absence of controlled test matings. It is also fundamental to the mathematical treatment of problems in evolution.

9. THE THEORY OF ORGANIC EVOLUTION

Another of the great generalizations of biology is the concept of organic evolution. The idea that all of the many kinds of plants and animals existing at the present time were not created *de novo* but have descended from previously existing, simpler organisms by gradual modifications which have accumulated in successive generations is one of the great unifying concepts of biology. Elements of this were implicit in the writings of certain Greek philosophers before the Christian era, from Thales to Aristotle. The theory of organic evolution was considered by a number of philosophers and naturalists from the fourteenth to the nineteenth centuries. However, not until Charles Darwin's publication in 1859 of *The Origin of Species* was the theory brought to general attention. In this book Darwin presented a wealth of detailed evidence and cogent argument to show that organic evolution had occurred. He also presented a theory, that of Natural Selection, to explain how evolution may occur.

According to Darwin's **theory of Nat-**

ural Selection, any group of animals or plants tends to undergo variation, more organisms of each kind are produced than can possibly obtain food and survive, there is a struggle for survival among the many individuals born, those individuals which possess characters that give them some advantage in the struggle for existence will be more likely to survive than those without them, and the survivors will pass these advantageous characters on to their offspring, so that successful variations will be transmitted to future generations. The core of Darwin's theory is the concept of the struggle for existence, the "survival of the fittest," and the inheritance of the advantageous characters by the offspring of the surviving individuals. This concept has had a central role in biological theory for the past century, and, with suitable amendments to bring it into line with subsequent discoveries in genetics and evolution, is held by most present-day biologists.

Studies of the development of many kinds of animals and plants from fertilized egg to adult have led to the generalization that organisms tend to repeat, in the course of their embryonic development, some of the corresponding stages of their evolutionary ancestors. This generalization, called the **Biogenetic Law,** or the **Law of Recapitulation,** was once interpreted as meaning that embryos have successively the appearance of the *adult* forms of their ancestors. Most biologists would now prefer the statement that embryos recapitulate some of the *embryonic* forms of their ancestors. The human being, at successive stages in development, resembles first a fish embryo, then an amphibian embryo, then a reptilian embryo, and so on.

10. ENZYMES AND METABOLISM

One of the characteristics of living things is their ability to metabolize, to carry on a great variety of chemical reactions. Our present-day generalizations about metabolism had their beginnings in 1780, when Lavoisier and LaPlace concluded that respiration is a form of combustion. They reached this conclusion from simple experiments comparing the utilization of oxygen and the production of carbon dioxide by animals and by candles kept in bell jars, even though their thinking was hindered by the current, erroneous "phlogiston" theory.

The concept that metabolism in all living organisms is mediated by specific organic catalysts, **enzymes,** synthesized by living cells, has gradually been crystallized since 1815, when Kirchhoff prepared an extract of wheat which would convert starch to sugar. A long argument between Liebig and Pasteur as to whether enzymes (or "ferments" as they were called then) themselves were living was resolved in Liebig's favor in 1897 when Eduard Büchner prepared a cell-free extract of yeast which would convert sugar to alcohol. Intensive research in enzymology has resulted in the isolation of many enzymes, the demonstration that they are all large protein molecules, and the generalization that each one controls a specific kind of chemical reaction because of the specific configuration of the surface of the enzyme molecule. The substance undergoing a chemical reaction (the **substrate**) unites with the enzyme to form a specific enzyme-substrate complex. In this way enzymes control the speed and specificity of essentially all the chemical reactions of living things.

The metabolic reactions of a wide variety of living things—animals, green plants, bacteria and molds—have been found to be remarkably similar in many respects. The continuation of life requires the expenditure of energy, and the ultimate source of all energy used by living things is sunlight. This energy, captured by green plants in the process of photosynthesis, is made available to the plants in further metabolism. Some of the energy may eventually be used by the animals that eat the plants, or by animals that eat the animals that ate the plants.

Metabolic processes are regulated so as to maintain the internal environment of the cell as constant as possible. External conditions tend to produce changes. Living organisms constantly adapt in ways which tend to resist the effects of these changes and keep the internal environment constant. This tendency toward constancy is called **homeostasis.** In the course of evolution, higher organisms have developed a greater degree of homeostatic control than the lower ones had.

11. GENIC CONTROL OF METABOLISM

One of the more recent important biological generalizations is the "one gene—one enzyme—one reaction" hypothesis stated by George Beadle and Edward Tatum in 1941. According to this widely accepted theory, each biochemical reaction in the development and maintenance of a particular organism is controlled by a particular enzyme, and the enzyme, in turn, is controlled by a single gene. A change **(mutation)** in the gene will result in an alteration or deficiency of the enzyme, a consequent failure of a particular metabolic step, and some particular change in the development of the organism.

12. VITAMINS AND COENZYMES

The discovery that substances other than salts, proteins, fats and carbohydrates are needed for adequate nutrition—substances called accessory food factors by F. G. Hopkins and **vitamins** by Casimir Funk in 1911—stimulated investigations into the role these substances play in metabolism and why they are needed in the diet of some organisms and not others. The generalization is now amply established that these substances are necessary for the metabolism of *all* organisms—bacteria, green plants and animals. Many organisms, however, are able to synthesize all of these substances that they require; the ones that cannot synthesize these materials must obtain them in their diet. The specific roles in metabolism of many of these vitamins are now known. In each instance they are known to become a part of a larger molecule which functions as a **coenzyme,** a partner of the enzyme and substrate, and is absolutely necessary for some particular reaction or reactions to occur. The deficiency diseases caused by the lack of the vitamins reflect the impaired metabolism caused by the deficient coenzyme.

13. HORMONES

The term **hormone** was originated by the British physiologist, E. H. Starling, in 1905 and was defined as "any substance normally produced in the cells of some part of the body and carried by the blood stream to distant parts, which it affects for the good of the body as a whole." The science of endocrinology can be said to have begun in 1849, when, on the basis of experiments in which testes were transplanted from one bird to another, Berthold postulated that these male sex glands secrete some blood-borne substance essential for the differentiation of the male secondary sex characteristics. This substance, testosterone, was finally isolated and synthesized in 1935. Our rapidly increasing knowledge of the many different hormones produced by vertebrate and invertebrate animals and by plants has led to the generalization that these are special chemical substances, produced by some restricted region of an organism which diffuse or are transported to another region where they are effective in very low concentrations in regulating and coordinating the activities of the cells. Hormones thus provide for chemical coordination which complements and supplements the coordination controlled by the activities of the nervous system.

14. INTERRELATIONS OF ORGANISMS AND ENVIRONMENT

The last major generalization we shall consider, and one of the major unifying concepts of biology today, comes from the field of ecology. From detailed studies of communities of plants and animals in a given area, the generalization has been made that all the living things in a given region are closely interrelated with each other and with the environment. This generalization includes the idea that particular kinds of plants and animals are not found at random over the earth but occur in interdependent **communities** of producer, consumer and decomposer organisms together with certain nonliving components. These communities can be recognized and characterized by certain dominant members of the group, usually plants, which provide both food and shelter for many other forms. Why certain plants and animals comprise a given community, how they interact, and how man can control them to his own advantage are major research problems in ecology.

The list of biological principles given here is not intended to be exhaustive, but rather to emphasize the fundamental unity of biological science and the many ways in which living things are interrelated and interdependent. These generalizations have been derived from careful observations and experiments made by many biologists over a long period of time. All of them have been tested repeatedly, and many have had to be revised as new information became available as the result of discoveries made with the aid of new techniques such as the electron microscope, radioactive isotopes for tracer studies, and the many other physical and chemical methods which are being used in biological research. Future studies may result in further revisions of some of these principles.

QUESTIONS

1. If the first interplanetary travelers report that they found living things on Mars, would you expect that the biological generalizations which hold true for living things on this planet would hold equally well for them?
2. If all living things originate from other living things, where do you suppose the very first living things came from?
3. Do you think it possible for the whole to be greater than the sum of its parts, i.e., for a whole organism to have properties which are more than the sum of the properties of its constituent cells?
4. What is meant by "the continuity of the germplasm"?
5. Why does it follow from Weismann's theory that acquired characters are not inherited?
6. Do you think that modern man evolved by natural selection? What characteristics will make man "fittest to survive" in the future?
7. Why may the vitamin requirements of man, mouse and mosquito be different?
8. In what ways do plants and animals interact with their environment?

SUPPLEMENTARY READING

For a physicist's look at some of the basic concepts of biology, read Schroedinger's *What Is Life?* The presentation of Darwin's views on evolution, Huxley's extension of them to sociology, education and ethics, and the impact of the theory of evolution on Victorian England in William Irvine's *Apes, Angels and Victorians* provides an interesting picture of the early days of one of biology's major generalizations. Some of the theories presented here are discussed further in Wightman's *The Growth of Scientific Ideas.* Some of the important ideas in biology, presented by extensive quotations from the original papers, are found in Gabriel and Vogel's *Great Experiments in Biology.*

Part One

Cell Structures and Functions

Chapter 3

Protoplasm

BIOLOGY is the science of living things; thus to define the field of biology we must first differentiate the living from the non-living. We use the word **organism** to refer to any living thing, plant or animal. It is relatively easy to see that a man, an oak, a rosebush and an earthworm are living, whereas rocks and stones are not, but it is more difficult to decide whether such things as viruses are alive.

The actual living material of all plants and animals is called **protoplasm.** This is not a single substance but varies considerably from organism to organism, among the various parts of a single animal or plant, and from one time to another within a single organ or part of an animal or plant. There are thus many kinds of protoplasm, but they share certain fundamental physical and chemical characteristics.

The protoplasm of the human body and of all plants and animals is nowhere present in a single large mass, but exists in discrete microscopic portions known as **cells.** These are the units of structure of the body, just as bricks may be the units of structure of a house. But they are more than mere building blocks; each is an independent, *functional* unit, and the processes of the body are the sum of the coordinated functions of its cells. These cellular units vary considerably in size, shape and function. Some of the smallest animals and plants have bodies made of a single cell; others, such as a man or an oak tree, are made of countless billions of cells fitted together.

We cannot see protoplasm on the surface of the human body because the living stuff is covered with a dead, protective layer of skin. In order to see what the fundamental stuff of life looks like, it is necessary to examine some simple animal or plant such as an ameba or slime mold, in which the protoplasm is naked and hence visible under the microscope. The protoplasm of such an organism is colorless, or perhaps faintly yellow, green or pink, and translucent. It has a thick, viscid, syrupy consistency and would feel slimy to the touch. When seen in the light microscope, protoplasm may appear to have granules or fibrils of denser material, droplets of fatty substances, or fluid-filled vacuoles, all suspended in the clear, continuous, semifluid "ground substance." Recent experiments indicate that protoplasm has an even more varied and complex structure beyond the limit of the microscope, one revealed by the electron microscope and by x-ray diffraction analyses.

15. CHARACTERISTICS OF LIVING THINGS

All living things have, to a greater or lesser extent, the properties of **specific size** and **shape, metabolism, movement, irritability, growth, reproduction** and **adaptation.** Although this list seems specific and definite, the line between the living and nonliving is rather tenuous, and whether we call things such as viruses living or nonliving becomes a matter of definition. Nonliving objects may show one or more of these properties, but not all of them. Crystals in a saturated solution may "grow," a bit of metallic sodium will move rapidly over the surface of water, and a drop of oil floating in glycerin and alcohol will send out pseudopods and move like an ameba.

Most biologists are agreed that all the varied phenomena of life are ultimately explainable in terms of the same physical and chemical principles which define nonliving systems. The idea that there is no fundamental difference between living and nonliving things is sometimes called the **mechanistic theory of life.** A corollary of this is that if we knew enough about the chemistry and physics of vital phenomena we might be able to synthesize living matter. An opposite view, widely held by biologists until the present century, stated that unique forces, not explainable in terms of physics and chemistry, are associated with and control life. The view that living and nonliving systems are basically different and obey different laws is called **vitalism.** Many of the phenomena of life that appeared to be so mysterious when first discovered have proved to be understandable without invoking a unique life force, and it is reasonable to suppose that future research will show that other aspects of life can also be explained by physical and chemical principles.

Specific Organization. Each kind of living organism is recognized by its characteristic shape and appearance; the adults of each kind of organism typically have a characteristic size. Nonliving things generally have much more variable shapes and sizes. Living things are not homogeneous, but are made of different parts, each with special functions; thus the bodies of living things are characterized by a specific, complex organization. The structural and functional unit of both plants and animals is the **cell,** the simplest bit of living matter that can exist independently. The cell itself has specific organization, for each type of cell has a characteristic size and shape, it has a **plasma membrane** which separates the living substance from the surroundings, and it contains a **nucleus,** a specialized part of the cell separated from the rest by a nuclear membrane. The nucleus, as we shall learn later, plays a major role in controlling and regulating the activities of the cell. The bodies of the higher animals and plants are organized in a series of increasingly complex levels: Cells are organized into **tissues,** tissues into **organs,** and organs into **organ systems.**

Metabolism. The sum of all the chemical activities of protoplasm which provide for its growth, maintenance and repair is called **metabolism.** The protoplasm of all cells is constantly changing by taking in new substances, altering them chemically in a variety of ways, building new protoplasm, and transforming the potential energy contained in large molecules of carbohydrates, fats and proteins into kinetic energy and heat as these substances are converted into other, simpler substances. This constant expenditure of energy is one of the unique and characteristic attributes of living things. Some kinds of protoplasm metabolize at a high rate; bacteria, for example, have very high metabolic rates. Other kinds, such as seeds and spores, have a barely detectable rate of metabolism. Even within a particular species or person metabolism may vary, depending on such factors as age, sex, general health, amount of endocrine secretion, and pregnancy.

Metabolic processes may be anabolic or catabolic. The term **anabolism** refers to those chemical processes in which simpler substances are combined to form more complex substances, resulting in the storage of energy and the production of new protoplasm and growth. **Catabolism** refers to the breaking down of these complex substances resulting in the release of energy and the wearing out and using up of protoplasm. Both types of processes occur continuously; indeed the two are intricately interdependent and difficult to distinguish. Complex compounds are broken down and their parts recombined in new ways to form different substances. The interconversions of carbohydrates, proteins and fats that occur continuously in human cells are examples of combined catabolism and anabolism. Since most anabolic processes require energy, some catabolic processes must occur to supply the energy to drive the reactions involved in building up the new molecules.

Both plants and animals have anabolic and catabolic phases of metabolism. Plants, however (with some exceptions), have the ability to manufacture their own organic compounds out of inorganic materials in the soil and air; animals must depend on plants for their food. Plant cells are simply better chemists than animal cells.

Movement. A third characteristic of living things is their ability to move. The movement of most animals is quite obvious—they wiggle, swim, run or fly. The movement of plants is much slower and less obvious, but is present nonetheless. A few animals—sponges, corals, oysters, certain parasites—do not move from place to place, but most of these have cilia or flagella to move their surroundings past their bodies and thus bring food and other necessities of life to themselves. Movement may be the result of muscular contraction, of the beating of the microscopic protoplasmic hairs called **cilia** or **flagella,** or of the slow oozing of a mass of protoplasm **(ameboid motion).** The streaming motion of the protoplasm in the cells of the leaves of plants is known as **cyclosis.**

Irritability. Living things are irritable; they respond to stimuli, physical or chemical changes in their immediate surroundings. Stimuli which are effective in evoking a response in most animals and plants are changes in the color, intensity or direction of light, temperature, pressure or sound, and changes in the chemical composition of the earth, water or air surrounding the organism. In man and other complex animals, certain cells of the body are highly specialized to respond to certain types of stimuli: the rods and cones in the retina of the eye respond to light, certain cells in the nose and in the taste buds of the tongue respond to chemical stimuli, and special cells in the skin respond to changes in temperature or pressure. In lower animals, and in plants, such specialized cells may be absent but the whole organism responds to stimuli. Single-celled animals and plants will respond by moving toward or away from heat or cold, certain chemical substances, light, or the touch of a microneedle.

The irritability of plant cells is not always so apparent as that of animal cells, but they are sensitive to changes in their environment. Protoplasmic streaming in plant cells may be speeded or stopped by changes in the amount of light. A few plants, such as the Venus flytrap of the Carolina swamps, have a remarkable sensitivity to touch and can catch insects. Their leaves are hinged along the midrib (Fig. 1), and the edges of the leaves are covered with hairs. The presence of an insect stimulates the leaf to fold, the edges

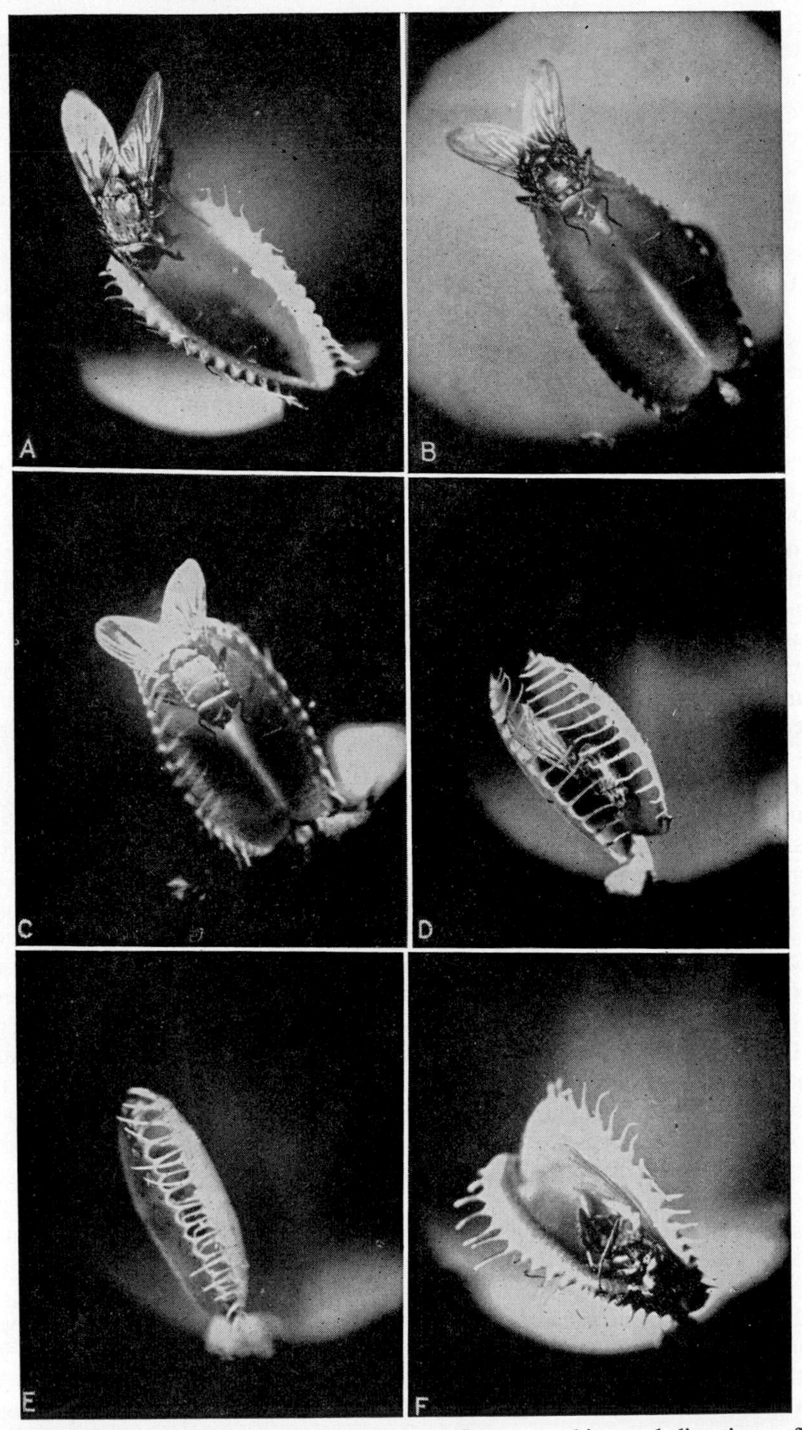

Figure 1. Photographs showing a leaf of the Venus flytrap catching and digesting a fly. (Copyrighted by the General Biological Supply House, Chicago.)

come together and the hairs interlock to prevent the escape of the prey. The leaf then secretes a material which kills and digests the insect. The development of fly-trapping is an adaptation which enables these plants to obtain part of the nitrogen they require for growth from the prey they "eat," since the soil in which they grow is deficient in nitrogen.

Growth. Another characteristic of living things, growth, is the result of anabolism. The increase in mass of protoplasm may be brought about by an increase in the *size* of the individual cells, by an increase in the *number* of cells, or both. An increase in cell size may occur by the simple uptake of water, but this swelling is generally not considered to be growth. The term growth is restricted to those processes which increase the amount of living substance of the body, measured by the amount of nitrogen or protein present. Growth may be uniform in the several parts of an organism, or it may be greater in some parts than in others so that the body proportions change as growth occurs. Some organisms—most trees, for example—will grow indefinitely. Most animals have a definite growth period which terminates in an adult of a characteristic size. One of the remarkable aspects of the growth process is that each organ continues to function while undergoing growth.

Reproduction. If there is any one characteristic that can be said to be the *sine qua non* of life, it is the ability to reproduce. As we shall see (p. 124) the simplest viruses do not metabolize, move or grow, yet because they can reproduce (and undergo mutations, p. 507), most biologists regard them as living. Although at one time worms were believed to arise from horse hairs in a water trough, maggots from decaying meat, and frogs from the mud of the Nile, we now know that each can come only from previously existing ones. One of the fundamental tenets of biology is that "all life comes only from living things."

The classic experiment disproving the **spontaneous generation of life** was performed by an Italian, Francesco Redi, about 1680. Redi proved that maggots do not come from decaying meat by this simple experiment: He placed a piece of meat in each of three jars, leaving one uncovered, covering the second with a piece of fine gauze, and covering the third with parchment. All three pieces of meat decayed but maggots appeared only on the meat in the uncovered jar. A few maggots appeared on the gauze of the second jar, but not on the meat, and no maggots were found on the meat covered by parchment. Redi thus demonstrated that the maggots did not come from the decaying meat, but hatched from eggs laid by blowflies attracted by the smell of the decaying meat. Further observations showed that the maggots develop into flies which in turn lay more eggs. Louis Pasteur, about two hundred years later, showed that bacteria do not arise by spontaneous generation but only from previously existing bacteria. The submicroscopic filtrable viruses do not arise from nonviral material by spontaneous generation but come only from previously existing viruses.

The problem of the original source of life will be discussed later (p. 511), but it is likely that billions of years ago, when chemical and physical conditions on the earth's surface were quite different from those at present, the first living things did actually arise from nonliving material.

The process of reproduction may be as simple as the splitting of one individual into two. In most animals and plants, however, it involves the production of specialized eggs and sperm which unite to form the fertilized egg or zygote, from which the new organism develops. Reproduction in certain parasitic worms involves several quite different forms, each of which gives rise to the next in succession until the cycle is completed and the adult reappears.

Adaptation. The ability of a plant or animal to adapt to its environment is the characteristic which enables it to survive the exigencies of a changing world. Each particular species can become adapted by seeking out an environment to which it is suited or by undergoing modifications to make it better fitted to its present surroundings. Adaptation may involve immediate changes which depend upon the irritability of protoplasm, or it may be the result of a long-term process of mutation and selection (p. 507). It is obvious that a single plant or animal cannot adapt to all the conceivable kinds of environment, hence there will be certain areas where it cannot survive. The list of factors that may limit the distribution of a species is almost endless: water, light, temperature, food, predators, competitors, parasites, and so on.

16. MATTER AND ENERGY

To get some idea of what living matter is and what it can do, we must consider not only the easily visible macroscopic aspects of life and those things visible under the microscope, but also the molecular patterns of protoplasm that lie far beyond the range of the microscope. To do this we must have at our command a certain basic minimum of physics and chemistry.

The universe is made up of two fundamental components—**matter** and **energy**—which under certain conditions are interconvertible. This is expressed by the famous Einstein equation: $E = mc^2$, where E = energy, m = mass and c = the velocity of light, which is a constant. This equation provides the theoretical basis for the conversion of mass to energy that occurs in an atom bomb or nuclear reactor. In the familiar, everyday world, however, matter and energy are separate, matter occupying space and having weight, and energy being the ability to produce a change or motion in matter—the ability to do work.

17. THE STRUCTURE OF MATTER

Regardless of the form—gaseous, liquid or solid—that matter may assume, it is always composed of small units called **atoms.** In nature there are ninety-two different kinds of atoms, ranging from hydrogen, the smallest, to uranium, the largest. In addition to these naturally occurring atoms, there are a half-dozen or more larger than uranium that have been man-made in a cyclotron or nuclear reactor. All of them, natural and synthetic, are much smaller than the tiniest particle visible under the microscope. In fact, no one has ever seen an atom; their structure and properties have been inferred from tests made with many types of elaborate apparatus.

Once the atom was believed to be the ultimate, smallest unit of matter. Now physicists know that atoms are divisible into even smaller particles organized around a central core, as our solar system of planets is organized around the sun. The exact number and kind of these particles and their arrangement in the atom are matters about which physicists are not in complete agreement. For our purposes we need consider only three types: **electrons,** which have a negative electric charge and an extremely small mass or weight; **protons,** which have a positive electric charge and are about 1800 times as heavy as electrons; and **neutrons,** which have no electrical charge, but have essentially the same mass as protons. The center of the atom, corresponding in position to the sun in our solar system, is the **nucleus,** containing protons, neutrons and perhaps electrons as well; it comprises almost the total mass of the atom. Just as most of the solar system is empty space, so is the atom, with electrons revolving around the nucleus in definite paths or **orbits.** The various types of atoms have different numbers of electrons circling in their orbits and different numbers of protons and neutrons in their nuclei. In all atoms, the number of protons in the nucleus equals the number of electrons circling around it, so that a state of electrical neutrality is maintained. There are only a few kinds of particles other than protons, neutrons and electrons; the different kinds of matter are produced by differences in the number and arrangement of these basic particles.

18. TYPES OF SUBSTANCES

An **element** is a substance composed of atoms all of which have the same number of protons in the atomic nucleus and therefore the same number of electrons circling in the orbits. A few elements can be found in nature as such—gold, silver, iron and copper, for example—but most elements have a strong tendency to unite with other elements to form compounds.

The unique property of protoplasm, its

aliveness, does not depend upon the presence of some rare or unique element. Four elements, carbon, oxygen, hydrogen and nitrogen, make up about 96 per cent of the material in the human body (Fig. 2). Another four, calcium, phosphorus, potassium and sulfur, constitute another 3 per cent of the body weight. Minute amounts of iodine, iron, sodium, chlorine, magnesium, copper, manganese, cobalt, zinc and perhaps a few other elements complete the list. All these elements, and especially the first four, are abundant in the atmosphere, the earth's crust, and the sea. Life depends upon the complexity of the interrelationships of these common, abundant elements.

Chemists have established a shorthand for their own convenience by assigning a symbol for each element. The symbol is usually the initial letter of the element's name: O, oxygen; H, hydrogen; C, carbon; N, nitrogen. When several elements have the same initial letter, a second letter is added to the symbol: C, carbon; Co, cobalt; Cl, chlorine; Cu, copper.

Physical research has shown that most of these elements are composed of two or more kinds of atoms, which differ in the number of neutrons in their nuclei. There are sixteen kinds of lead, three kinds of hydrogen, five kinds of carbon, and so on. These different types of atoms of an element are known as **isotopes** (*iso* = equal; *tope* = place) because they occupy the same place in the periodic table of the elements. All the isotopes of any given element have the same number of electrons in their orbits, and it is chiefly this electron number which is responsible for the chemical properties of an element.

With the development of the cyclotron and the nuclear reactor, it became possible to make artificially many new isotopes. The availability of these new isotopes in turn made possible a new type of biologic research, that of tracing particular elements and compounds through their many devious metabolic pathways, and of measuring the time required for any given component of the body to be replaced by new molecules of that substance. Al-

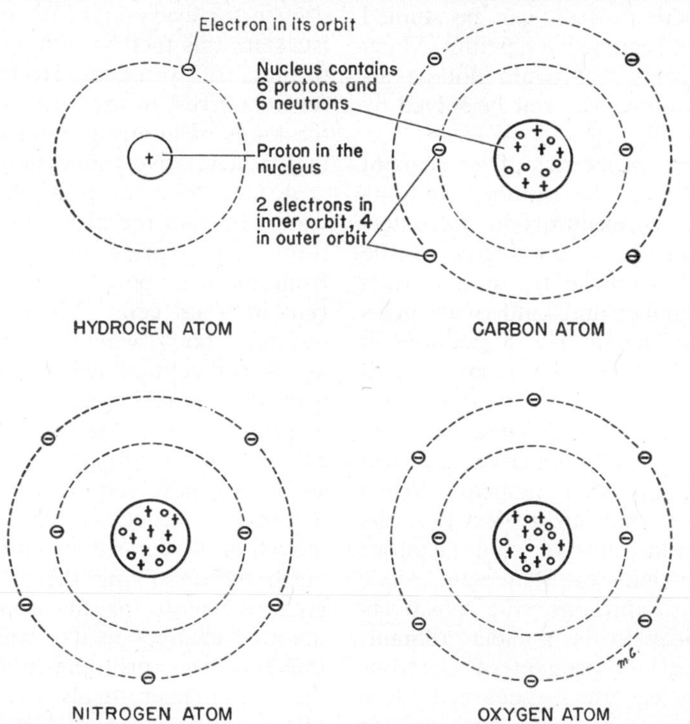

Figure 2. Diagrams of the atomic structure of the four chief elements of the body. The nitrogen atom nucleus contains seven protons (+) and seven neutrons (O); its inner orbit has two electrons, and its outer orbit, five. The oxygen atom nucleus contains eight protons and eight neutrons, and there are two and six electrons, respectively, in its inner and outer orbits.

though the isotopes of a given element have the same chemical properties, they can be differentiated physically. Some of them are radioactive and can be detected and measured by an instrument such as the **Geiger counter** by the kind and amount of radiation they emit. Others can be differentiated by the slight difference in the weight of the atoms caused by an extra neutron in the nucleus. Heavy hydrogen and heavy nitrogen can be detected and measured by a **mass spectrometer.** A tremendous insight has been gained into the intimate details of the metabolic activities of protoplasm by preparing some substance—sugar, for example—labeled with radioactive carbon (C^{11} or C^{14}) or heavy carbon (C^{13}) in place of ordinary carbon (C^{12}), and then feeding or injecting it into an animal or incubating cells in a solution containing it. In this way one can trace the pathway the substance follows in the body and determine the form in which it is finally excreted. The rate of formation of bone, and the effect of vitamin D and parathyroid hormone on this process, can be studied with the aid of radioactive calcium. Many biologic problems that could not be attacked in any other way can be solved by this method.

The chemical properties of an element depend chiefly on the number and arrangement of the electrons in the outermost orbit and to a lesser degree on the number of electrons in the inner orbits and on the number and kind of particles in the nucleus. The number of electrons in the outermost orbit varies from zero to eight in different atoms. If there are zero or eight electrons in the outer orbit the element is chemically inactive and will not combine with other elements. When there are fewer than eight electrons, the atom tends to lose or gain some in order to achieve an outer orbit of eight. Since the number of protons, the positively charged particles in the nucleus, remains the same, the loss or gain of electrons results in an electrically charged atom called an **ion.** Atoms with one, two or three electrons in the outer orbit tend to lose them to other atoms and thus become positively charged because of the excess protons in the nucleus. Atoms with five, six or seven electrons in the outer orbit tend to gain electrons from other atoms and thus become negatively charged because of the excess electrons revolving around the nucleus. Atoms with four electrons in the outer orbit (such as carbon) neither lose nor gain electrons, but share them with neighboring atoms. Both positively and negatively charged atoms are called ions. Because particles that have opposite electric charges attract each other, positive and negative ions tend to unite.

19. COMPOUNDS

Most elements are found in protoplasm as chemical compounds, substances composed of two or more different kinds of atoms or ions. The smallest particle of a compound having the composition and properties of a larger part of the substance is called a **molecule.** A **pure compound** is always made of two or more elements united in a definite ratio. Water, for example, always contains two atoms of hydrogen for every atom of oxygen. Chemists state this fact by writing the chemical formula of water as H_2O. A **chemical formula** gives, in the chemist's shorthand, the kinds of atoms present in a molecule and their relative proportion.

A large part of protoplasm is simply water. In man the percentage of water in protoplasm varies in different tissues, from about 20 per cent in bone to 85 per cent in brain cells. About two thirds of our total body weight is water; as much as 95 per cent of jellyfish protoplasm is water. Water serves a number of functions in protoplasm. Most of the other chemicals present are dissolved in it, and as we shall see, these require a water medium in order to react. Water aids in the removal of the waste products of metabolism by dissolving them. Water has a great capacity for absorbing heat with minimal changes in its own temperature; thus it protects protoplasm against sudden thermal changes. It also has the property of absorbing a great deal of heat as it changes from a liquid to a gas, thus enabling the body to dissipate excess heat by the evaporation of water. The characteristic high heat conductivity of water makes it possible for heat to be distributed evenly throughout the body tissues. Finally, water is indispensable as a lubricant, being present in body fluids wherever one organ rubs against another, and in the joints where one bone moves on another.

In contrast to a pure compound such as water, whose constituent parts are always present in a fixed ratio, a **mixture** is made of two or more kinds of atoms or molecules which may be combined in varying proportions. Water and alcohol may be mixed in all proportions and air is a mixture of varying amounts of oxygen and nitrogen plus small quantities of water vapor, carbon dioxide, argon, and other gases. It follows that a pure compound will exhibit certain fixed chemical and physical properties, whereas a mixture will have properties that vary with the relative abundance of its constituents.

Molecules may be composed of one, two, or many kinds of atoms. Oxygen and nitrogen, for example, exist in the atmosphere as O_2 and N_2, each molecule being made of two of the same kind of atom. A molecule of table salt, or sodium chloride, is made of one atom of sodium (Na, from its Latin name, *natrium*) and one of chlorine (Cl). A molecule of the simple sugar, glucose, is composed of three kinds of atoms: six carbon atoms, twelve hydrogen atoms and six oxygen atoms; its formula in chemical shorthand is $C_6H_{12}O_6$. Any larger portion of glucose —a pound, for instance—will also contain carbon, hydrogen and oxygen in these same proportions. If it were possible to divide a pound of glucose in half, then divide the half in half, then the quarters in half, and so on, eventually the glucose would be subdivided into its constituent molecules. Each of these molecules would have the same composition and properties as the original pound. However, if a glucose molecule were divided, the parts no longer would be sugar, but would be carbon, hydrogen and oxygen atoms, and these would have quite different properties.

There are two types of compounds found in protoplasm: inorganic and organic. The latter include all the compounds (other than carbonates) that contain the element carbon. Because its outer orbit contains four electrons, which can be shared in a number of different ways with adjacent atoms, carbon can form a wider variety of compounds than any other element. It was believed at one time that organic compounds were uniquely different and could be produced only by living matter. This hypothesis was disproved in 1828 when the German chemist Wöhler succeeded in synthesizing urea (one of the waste products found in human urine) from the inorganic compounds ammonium sulfate and potassium cyanate. Since then thousands of organic compounds have been prepared synthetically; some of these are complex molecules of great biological importance as vitamins, hormones, antibiotics and drugs.

The inorganic compounds found in living systems are acids, bases and salts. An **acid** is a compound which releases hydrogen ions (H^+) when dissolved in water. Acids turn blue litmus paper to red and have a sour taste. Hydrochloric (HCl) and sulfuric (H_2SO_4) are inorganic acids; lactic (from sour milk) and acetic (from vinegar) are two common organic acids. A **base** is a compound which releases hydroxyl ions (OH^-) when dissolved in water. Bases turn red litmus paper blue. Sodium hydroxide (NaOH) and ammonium hydroxide (NH_4OH) are common inorganic bases.

For convenience, the degree of acidity or alkalinity of a fluid, its hydrogen ion concentration, may be expressed in terms of ***p*H,** the negative logarithm of the hydrogen ion concentration. The protoplasm of most animal and plant cells is neither strongly acid nor alkaline, but contains an essentially neutral mixture of acidic and basic substances. Its hydrogen ion concentration is about 10^{-7} molar, and thus its *p*H is 7.0. Any considerable change in the *p*H of protoplasm is inconsistent with life. Since the scale is a logarithmic one, a solution with a *p*H of 6 has a hydrogen ion concentration 10 times as great as that of one with a *p*H of 7.

When an acid and a base are mixed,

the hydrogen ion of the acid unites with the hydroxyl ion of the base to form a molecule of water (H_2O). The remainder of the acid (anion) combines with the rest of the base (cation) to form a **salt.** Hydrochloric acid, for example, reacts with sodium hydroxide to form water and sodium chloride, common table salt:

$$H^+ Cl^- + Na^+ OH^- \rightarrow H_2O + Na^+ Cl^-$$

A salt may be defined as a compound in which the hydrogen atom of an acid is replaced by some metal.

When a salt, an acid or a base is dissolved in water it separates into its constituent ions. These charged particles can conduct an electric current, hence these substances are known as **electrolytes.** Sugars, alcohols, and the many other substances which do not separate into charged particles when dissolved, and therefore do not conduct an electric current, are called **nonelectrolytes.**

The protoplasm of any animal or plant contains a variety of mineral salts, of which sodium, potassium, calcium and magnesium are the chief **cations** (positively charged ions) and chloride, bicarbonate, phosphate and sulfate are the important **anions** (negatively charged ions). The body fluids of land vertebrates are quite similar to sea water in the kinds of salts present and in their relative concentrations. Many biologists believe that life, and hence protoplasm, originated in the sea, and it is reasonable to assume that this protoplasm incorporated into its own substance the salts of the surrounding sea, thus becoming adapted to this aspect of its environment. Later, many-celled marine animals evolved and developed body fluids surrounding their cells. These fluids were in equilibrium with the cells and with sea water and contained salts in similar proportions. When land forms evolved, their body fluids kept the same relative concentrations of salts as those of their marine ancestors. Professor Macallum of Toronto has suggested that the concentration of salts in the body fluids of present day mammals (about 0.9 per cent by weight) may reflect the salt concentration of the sea in the Cambrian period, when the ancestors of the land animals left the sea. In the more than half billion years since then, enough salt has been washed into the sea to bring its salt content up to 3.4 per cent by weight.

Although the salts are present in low concentration in the body fluids, they have a marked influence on cell function. The concentration of the various salts is extremely constant under normal conditions, and any great deviation from the normal causes marked effects, even death. A decrease in the concentration of calcium ions in the blood of mammals results in convulsions and death. Heart muscle can contract normally only in the presence of the proper balance of sodium, potassium and calcium ions. If a frog heart is removed from the body and placed in a pure sodium chloride solution, it soon stops beating, in the relaxed condition. If placed in a solution of potassium chloride or a mixture of sodium and calcium chloride, it will stop in the contracted state. It will continue to beat, however, if placed in a solution containing the proper balance of these three salts. This frog heart method is so sensitive that it can be used to measure the concentration of calcium ions in solutions.

In addition to these specific effects of particular salts on protoplasm, the mineral salts are important in maintaining the osmotic relationships between protoplasm and its environment.

20. ORGANIC COMPOUNDS IN PROTOPLASM

The major types of organic substances found in protoplasm are carbohydrates, proteins, fats, nucleic acids and steroids. Some of these are required for the structural integrity of the cell, others to supply energy for its functioning, and still others are of prime importance in regulating metabolism within the cell. The types of substances, and even their relative proportions, are remarkably similar for cells from the various parts of the body and for cells from different animals. A bit of human liver and the protoplasm of an ameba both contain about 80 per cent water, 12 per cent protein, 2 per cent nucleic acid, 5 per cent fat, 1 per cent carbohydrate and a fraction of 1 per cent of steroids and other substances. Certain specialized cells, of course, have unique patterns of chemical constituents; the brain, for instance, is rich in certain kinds of fats.

21. CARBOHYDRATES

Carbohydrates are compounds containing only carbon, hydrogen and oxygen. These are present in the ratio of 1 C : 2 H : 1 O. Sugars, starches and cellulose are examples of carbohydrates.

Some of the simplest carbohydrates found in protoplasm are the single sugars, with the formula $C_6H_{12}O_6$. **Glucose** (also called dextrose) and **fructose** are two single sugars with this formula. They differ slightly in the arrangement of their constituent atoms and these different arrangements give them slightly different chemical properties. Similar differences in the internal structure of the members of a group of similar compounds are frequently found in organic chemistry. Chemists show these internal structural differences by drawing **structural formulas** of such compounds (Fig. 3). In these, the atoms are represented by their symbols—C, H, O, etc.—and the chemical bonds, or forces that hold the atoms together, are indicated by connecting lines. Hydrogen has one bond to connect to other atoms; oxygen has two, and carbon, four. (See Fig. 2 for the structure of these atoms and the number of electrons they may gain or lose to form compounds.)

Glucose is the only single sugar found in any quantity in the body. The other carbohydrates we eat are converted by the liver into glucose. Glucose is an absolutely indispensable component of blood. Normally it is present in the blood and tissues of mammals in a concentration of about 0.1 per cent by weight. No particular harm results from a simple increase in the amount of glucose in the body, but a reduced concentration increases the irritability of certain brain cells, so that they respond to very slight stimuli. As a result of impulses from these cells to the muscles, muscular twitches, convulsions, unconsciousness and death may ensue. Brain cell metabolism requires glucose for fuel and a certain minimum concentration of glucose in the blood is necessary to supply this. There is an extremely complex mechanism, involving the nervous system, liver, pancreas, pituitary and adrenal glands, which maintains the proper concentration of glucose in the blood.

```
      H                     H
      |                     |
      C=O               H—C—O—H
      |                     |
  H—C—O—H                   C=O
      |                     |
 HO—C—H                 HO—C—H
      |                     |
  H—C—O—H               H—C—O—H
      |                     |
  H—C—O—H               H—C—O—H
      |                     |
  H—C—O—H               H—C—O—H
      |                     |
      H                     H
   Glucose               Fructose
```

Figure 3. Structural formulas of two simple sugars.

A second group of carbohydrates found in protoplasm is the double sugars, with the formula $C_{12}H_{22}O_{11}$. As their name implies, they are made of two single sugars joined together by the removal of a molecule of water. Both cane and beet table sugars are **sucrose,** a combination of one molecule of glucose with one of fructose. Several other double sugars are known, all with the formula $C_{12}H_{22}O_{11}$, but differing in the arrangement of their constituent atoms and hence in some of their chemical and physical properties. **Maltose** or malt sugar is composed of two molecules of glucose; **lactose,** or milk sugar, found in the milk of all mammals, is made of one molecule of glucose and one of galactose, a third kind of single sugar. These sugars differ markedly in the degree of their sweetness. Fructose is the sweetest of the common sugars. Lactose, the least sweet, is less than one tenth as sweet as fructose. Sucrose is intermediate

Table 1. RELATIVE SWEETNESS OF SOME OF THE COMMON SUGARS

SUGAR	RELATIVE SWEETNESS (SUCROSE 100)
Lactose	16.0
Galactose	32.1
Maltose	32.5
Glucose	74.3
Sucrose	100.0
Fructose	173.3
Saccharin	55,000

(Table 1). **Saccharin,** a synthetic sweetening agent, is much sweeter than any sugar and is used by people who want to sweeten food without using sugar.

The carbohydrates with the largest molecules are the compound sugars, starches and cellulose, composed of a large number of single sugars joined together. Since the exact number of sugar molecules joined to make a starch molecule is unknown, and indeed varies from one molecule to the next, the formula for starch may be written $(C_6H_{10}O_5)_x$, where x stands for the unknown, large number of single sugars which combine to make the starch molecule. Starches vary in the number and kind of sugar molecules present and are common constituents of both plant and animal protoplasm. Animal starch, which differs slightly from plant starch, is called **glycogen.** Carbohydrates are stored in plants as starches and in animals as glycogen; glucose could not be stored as such, for its small molecules would leak out of the cells. The larger, less soluble starch and glycogen molecules will not pass through the plasma membrane. In man and other higher animals, glycogen is stored in the liver and muscles. Liver glycogen is readily convertible into glucose and carried in this form in the blood to other parts of the body.

Most plants have a strong supporting outer wall of **cellulose,** an insoluble compound sugar resembling starch in that it is made of many glucose molecules. Chemical compounds of cellulose are extremely important commercially; explosives, rayon, celluloid, certain plastics, photographic film and varnishes are a few of these. The chief role of carbohydrates in protoplasm is as a readily available fuel to supply heat and energy for the many processes going on. Glucose is metabolized to carbon dioxide and water and energy is released:

$$C_6H_{12}O_6 + 6O_2 \rightarrow 6H_2O + 6CO_2 + \text{energy}$$

A few carbohydrates combine with other substances and enter into the structure of protoplasm and thus serve as building material as well as fuel.

22. FATS

True **fats** are also composed of carbon, hydrogen and oxygen, but have much less oxygen in proportion to the carbon and hydrogen than carbohydrates have. Fats have a greasy or oily consistency; some, such as beef tallow or bacon fat, are solid at ordinary temperatures, others such as olive oil or cod liver oil are liquid. Each molecule of fat is composed of one molecule of glycerol and three molecules of fatty acid; all fats contain glycerol but differ in the kinds of fatty acids present. A fat common in beef tallow, tristearin, $C_{57}H_{110}O_6$, has three molecules of stearic acid and one of glycerol.

Fats are important in protoplasm both as fuels and as structural constituents. Glycogen or starch is readily converted to glucose and metabolized to release energy quickly; the carbohydrates serve as short-term sources of energy. Fats yield more than twice as much energy per gram as do carbohydrates and thus are a more economical form for the storage of food reserves. Fats provide for the longer term storage of fuel. Carbohydrates can be transformed by the body into fats and stored in this form—a restatement of the generally known fact that starches and sugars are "fattening." The reverse may also occur to some slight extent: Fats or parts of the fat molecule may be converted into glucose and other carbohydrates. This has been shown by preparing fatty acids or glycerol labeled with radioactive or heavy carbon, feeding or injecting these into a rat or dog, and then isolating glucose from the blood or glycogen from the liver and demonstrating that these molecules now contain the labeled carbon atoms.

Fats are more important structural elements of the body than carbohydrates are. The plasma membrane around each cell and the nuclear membrane contain fatty substances as important constituents, and the myelin sheath around the nerve fibers (p. 344) has a high fat content. Fat is deposited in large amounts just under the skin, where it serves as an insulator against the loss of body heat. Whales, which live in cold water and have no insulating hair, have an especially thick layer of fat (blubber) just under the skin for this purpose. A certain amount of subcutaneous fat in man is necessary to keep the skin firm.

Besides the true fats, which contain glycerol and fatty acids, there are a number of related, fatlike substances with similar properties but which contain other things in addition to fatty acids. The phospholipids are important constituents of plant and animal cells in general and nerve cells in particular.

23. STEROIDS

Steroids are complex molecules containing carbon atoms arranged in four interlocking rings, three of which contain six carbon atoms each and the fourth of which contains five. Some steroids of biological importance are vitamin D, the male and female sex hormones, the adrenal cortical hormones, bile salts and cholesterol. Cholesterol is an important structural component of nervous tissue and other tissues, and the steroid hormones are of prime importance in regulating certain phases of metabolism.

24. PROTEINS

The **proteins** are compounds containing carbon, hydrogen, oxygen, nitrogen and usually sulfur and phosphorus. The characteristic element is nitrogen. Proteins are always present in protoplasm and are much more important as the basic building materials of protoplasm than are carbohydrates or fats.

Proteins are extremely complex substances, their molecules being the largest, most complicated and most varied of all the components of protoplasm. Chemists have been able to synthesize some carbohydrates and fats but no one has been able to synthesize a protein molecule. Thousands of atoms are present in each molecule of a protein. A typical protein of great importance to the human body is **hemoglobin,** the red pigment responsible for the color of blood. Some idea of the complexity of the hemoglobin molecule can be gained from its formula: $C_{3032}H_{4816}O_{872}N_{780}S_8Fe_4$. (Fe is the symbol for iron.) Large as this molecule is, it is only a small-to-medium sized protein. A large fraction of the proteins present in protoplasm are **enzymes,** substances which control the rates at which the many processes of the cell occur.

Protein molecules are made of simpler components, known as **amino acids.** At present, some thirty-five different amino acids have been found as the result of the chemical breakdown of proteins; the existence of about twenty-five of these has been confirmed by further investigations. Since each protein contains perhaps hundreds of amino acids combined in various proportions and in different orders, there is an almost infinite variety of protein molecules. Recently, analytic methods have been developed for determining the exact arrangement of the amino acids in a protein molecule. Insulin, the hormone secreted by the pancreas and used in the treatment of diabetes, was the first protein whose structure was elucidated. Not all proteins contain all the possible amino acids; some have only a few types. Each cell contains hundreds of different proteins and each kind of cell contains some proteins which are unique to it. There is evidence that every species of plant and animal has certain proteins in its protoplasm different from those of any other species. The degree of difference in the proteins of two species depends upon the evolutionary relationship of the forms involved. Organisms less closely related by evolution have proteins which differ more markedly. Investigations of the similarities and dissimilarities of the proteins of organisms have been useful in studies of evolution and have added strong confirmatory evidence to ideas of evolutionary relationships derived from other facts. This generalization is known as the theory of **species specificity:** The protoplasm of each species, due to its constituent proteins, is characteristic of the species, differing at least slightly from that of related species and more markedly from that of more distantly related species. Because of the interactions of unlike proteins, grafts of tissue taken from one species of animal usually will not grow when implanted into a host of a different species, but degenerate and are resorbed by the host.

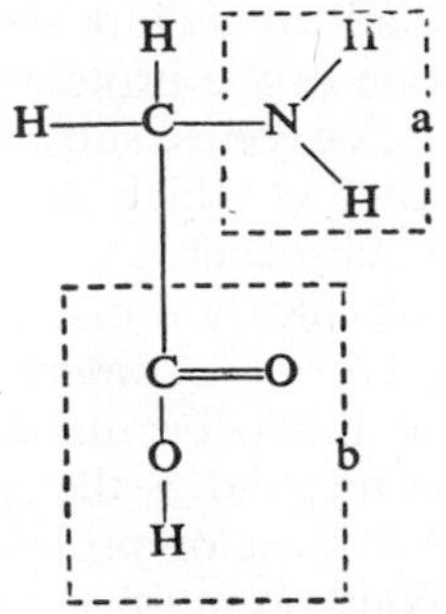

Figure 4. The structural formula of glycine, the simplest amino acid, showing, *a*, the amino group and, *b*, the acid (carboxyl) group.

The structural units of proteins, the amino acids, differ in the number and arrangement of their constituent atoms, but all contain an amino group (NH_2) and an acid group (COOH), whence their name (Fig. 4). The amino group enables the amino acid to act as a base and combine with acids; the acid group enables it to combine with bases. Because of this, amino acids and proteins serve as **buffers** and resist changes in acidity or alkalinity, thus protecting protoplasm. Amino acids are linked together to form proteins by a bond between the amino group of one and the acid group of another. Pure amino acids from proteins have a sweet taste. When proteins are eaten, they are broken up into amino acids before they can be absorbed into the blood stream. They are then carried to all parts of the body to be made into new protein or to be metabolized for the release of energy.

Although proteins are important in the body chiefly as the structural components of protoplasm, and as the functional constituents of enzymes and certain hormones, they may be used as fuel for the liberation of energy. Most diets include more protein than is necessary for the maintenance of protoplasm. The excess amino acids first lose their amino group (a process called **deamination).** The amino group reacts with other substances to form urea, and is excreted. The rest of the molecule may be changed, via a series of intermediate steps, into glucose and either used immediately as fuel or stored as glycogen. Again, this information about the conversion of protein to carbohydrates (and to fat, as well) is derived from experiments with substances labeled with isotopes of carbon, hydrogen and nitrogen. In prolonged fasting, after the glycogen and stored fats are exhausted, the proteins from protoplasm itself may be used as fuel. The evidence available at present indicates that the human body (and animal protoplasm in general) can manufacture some, but not all, of the amino acids if the proper raw materials are present. Those which cannot be made by the animal body must be obtained directly or indirectly from plants as food or perhaps from the bacteria (microbes) that live in our intestines. Plants can synthesize all the amino acids from simpler substances. The ones which animals cannot synthesize, but must obtain from plants, are known as **essential amino acids.** It must be understood that these amino acids are no more essential as components of proteins than are any other ones, they are simply essential *in the diet,* since they cannot be synthesized.

25. NUCLEIC ACIDS

The biologic importance of the nucleic acids has been fully appreciated only in recent years. These complex molecules, as large as or larger than most proteins, were discovered in 1870, when Miescher isolated them from the nuclei of pus cells. Nucleic acids contain carbon, oxygen, hydrogen, nitrogen and phosphorus; they gained their name from the fact that they are acidic and were first identified in nuclei. For a long time it was believed that there were but two kinds of nucleic acid —one containing the sugar ribose, and called **ribose nucleic acid** or **RNA,** and one containing the sugar desoxyribose, and called **desoxyribose nucleic acid** or **DNA.** It is now clear that there are many different kinds of RNA and of DNA which differ in their structural details. It is currently believed that DNA is responsible for a large part, perhaps all, of the specificity and chemical properties of the genes, the units of heredity located in the nucleus. Ribonucleic acid plays an important role in the complex process by which proteins are synthesized in the cell.

26. PHYSICAL CHARACTERISTICS OF PROTOPLASM

The properties of protoplasm depend not only on the kinds and quantities of substances present but on their physical state as well. A mixture of a substance with water may result in a true solution, a suspension or a colloidal solution, differentiated by the size of the dispersed particles. In one type, called a **true solution,** the ions or molecules of the dissolved substance (called the **solute)** are dispersed among those of the dissolving liquid (the **solvent).** These dispersed solute particles are extremely tiny with diameters less than 0.0001 micron* and are either atoms or small molecules. True solutions are transparent and have a higher boiling point and a lower freezing point than pure water. Most acids, alkalis, salts and some nonelectrolytes, such as sugar, form true solutions when added to water.

Other materials, such as clay, break up into relatively large particles and form a **suspension.** The dispersed particles in a suspension are made of aggregates of large numbers of molecules, which settle out if the suspension is allowed to stand. Thus the particles of clay in muddy water will sink to the bottom if the water is not stirred constantly. Suspensions are opaque, not transparent, and have the same boiling and freezing point as pure water.

If the dispersed particles are intermediate in size (between 0.1 and 0.0001 micron in diameter), too small to settle to the bottom but too large to form a true solution, the mixture is known as a **colloid,** or colloidal solution. Many common substances are colloids—mayonnaise, cream, butter, jello, glue, fog, and soap are just a few. There are several types of colloids: a solid in a liquid, such as glue; a solid in a gas, such as smoke; and a liquid in a gas, such as fog. A type of colloid in which a liquid is suspended in a liquid—such as cream, which consists of droplets of oil dispersed in water—is called an **emulsion.** All these are colloids because of the size of the dispersed particles. Colloids are transparent or translucent, have about the same boiling and freezing points as pure water, and are stable—that is, they do not separate into their constituent parts on standing. The particles of a colloidal solution typically have the same kind of electric charge and thus tend to repel each other; this keeps the particles dispersed.

Colloids are unique in their ability to change from a liquid condition, known as a **sol,** to a solid or semisolid condition known as a **gel.** For example, when the contents of a package of gelatin (a protein) are dissolved in hot water, the particles of gelatin are dispersed throughout the water and a liquid colloidal solution, a sol, results. As the gelatin cools, the gelatin particles become the continuous phase, the water particles become dispersed as tiny droplets throughout the gelatin, and a semisolid gel, the "jello" results. When reheated, the gel again becomes a sol. Thus the colloid is a sol when it consists of particles of gelatin surrounded by water and a gel when it consists of droplets of water surrounded by gelatin. The change from sol to gel is brought about by changing the temperature. In other colloidal systems the sol to gel change may be effected by changing the *p*H, the ions present, or by mechanical

* The system of measurement used in science is based on the meter, a certain fraction of the circumference of the earth, and equal to 39.37 inches. Other units are decimal parts of this: 1 centimeter = 1/100 meter and 1 millimeter = 1/1000 meter (about 1/25th of an inch). A micron is 1/1000 millimeter (1/25,000 inch) and a millimicron = 1/1000 micron. Units of volume and weight are derived from these units of length; the cubic centimeter is a unit for small volumes and a gram, the weight of 1 cubic centimeter of water under certain standard conditions, is a unit for small weights. A liter = 1000 cubic centimeters, slightly more than a quart and a kilogram = 1000 grams, about 2.2 pounds. The superiority of the metric system over the English system is apparent: in order to change 0.58 liter to cubic centimeters, one simply moves the decimal point three places to the right—580 cc. But to change 0.58 quart into ounces, one must multiply by 32, since there are 32 ounces in a quart.

means—whipping cream, for example, changes it from a sol to a gel. Not all colloidal systems are reversible; some are destroyed by extreme changes in acidity, alkalinity, or temperature. The particles coagulate, aggregate into larger particles (thus becoming a suspension) and finally settle out. Egg white, for example, is irreversibly changed from a sol to a gel when it is heated.

Many of the properties of colloids depend upon the enormous amount of surface area between the dissolved particles and the dissolving medium. The surface area obviously increases manyfold as a substance is divided into finer particles. A cube 1 cm. on each edge has a surface area of 6 square cm. When this is cut into eight cubes, each 0.5 cm. on an edge, each cube has a surface of 1.5 square cm. or a total area of 12 square cm. for the eight cubes. If the cube were divided into particles 100 microns on an edge, the total surface area would be 600 square cm.; and if the cube were divided into particles of colloidal size, say 0.01 micron on an edge, the total surface area would be 6,000,000 square cm. Since many chemical reactions can occur only at a surface, a colloidal system is a much better medium for frequent and rapid reactions than is any other type of mixture.

Most of the peculiar qualities of protoplasm depend upon the fact that it is a colloidal system, a mixture of protein particles in water. The protein molecules are too large to form a true solution with water, and too small to settle out, but are dispersed throughout the medium and form a colloidal suspension. Protoplasm is constantly and rapidly changing from the sol to the gel state and back; indeed this is one expression of the "aliveness" of protoplasm. If the temperature is raised beyond a certain point, or if certain chemicals are applied, protoplasm is irreversibly changed into a gel or sol and dies.

In our discussion of the composition of protoplasm we noted that it contains an amazing amount of water—muscle, for example, contains about 80 per cent water, yet we know that muscle is not liquid, but is a semisolid. The explanation, of course, is that muscle, like all protoplasm, is a colloidal system and thus can bind a lot of water in the gel state. As a muscle contracts it rapidly and reversibly changes from a sol to a gel. Shortly after death muscles are irreversibly changed into the gel state and contract, causing the phenomenon we call *rigor mortis*.

In summary, protoplasm consists of water, inorganic salts and organic substances—proteins, carbohydrates and fats, as well as many other things. Carbohydrates and fats have only a small role in the structure of protoplasm but are important as sources of fuel; carbohydrates are readily available fuel, fats are more permanently stored supplies of energy. Proteins are most important as structural and functional constituents of protoplasm, but may serve as fuel after deamination. The body can convert each of these substances into the others to some extent. Protoplasm is a colloidal system, with protein molecules and water forming the two phases, and many of the properties of protoplasm—muscle contraction, ameboid motion, and so on—depend on the rapid change from sol to gel state and back.

QUESTIONS

1. Discuss the characteristics of living things. Can you think of any which should be added to the list? Any which do not seem essential?
2. In what ways does the analogy between the cells of the body and the bricks of a house fail?
3. Are you inclined to believe the mechanistic or the vitalistic theory of life? Why?
4. Differentiate between an element and a compound.
5. What is the exact meaning of each of the following terms: atom, isotope, ion? Could a single particle of matter be all three simutaneously?
6. Explain why water is essential to protoplasm.
7. Why does the mixing of alcohol and water not constitute a chemical reaction?
8. What is the meaning of the symbol *p*H? Why is it important in biologic systems?
9. Differentiate between acids, bases and salts.

10. What are the functions of salts in protoplasm?
11. State the chief functions of each of the five chief types of organic compounds found in protoplasm.
12. Why do colloids afford a particularly good base for chemical reactions?
13. Describe three examples of colloidal solutions other than those described in the text. Can they be changed reversibly from sol to gel?

SUPPLEMENTARY READING

Some of the chemical aspects of protoplasm are discussed in Gerard's *Unresting Cells. The Cell and Protoplasm,* edited by F. R. Moulton, contains short papers, each by an authority on the subject, on a variety of subjects related to protoplasm. The subject of atoms, neutrons and isotopes is entertainingly discussed in A. K. Solomon's *Why Smash Atoms?* Further discussion of acids, bases, salts and other chemical compounds can be found in any introductory chemistry text.

Chapter 4

Cells and Tissues

SOME THREE hundred years ago, Robert Hooke used the newly invented microscope to look at the fine structure of plants. He observed that cork was not a homogeneous material, but consisted of tiny, boxlike cavities which he called **cells.** We now know that what he saw were the cellulose walls of dead cells and that the important part of the cell is not the wall but its contents. In 1839, the Bohemian physiologist, Purkinje, introduced the term **protoplasm** for the living contents of the cell. About the same time two Germans, Schleiden, a botanist, and Schwann, a zoologist, formulated the generalization which has since developed into the **cell theory.** The bodies of all plants and animals are composed of cells and these are the fundamental units of life. New cells can come into being only by the division of previously existing cells, a generalization first stated by Virchow in 1855. The corollary of this, that all the cells living today can trace their ancestry back to ancient times, was pointed out by August Weismann about 1880. The cell theory includes the concept that the cell is the fundamental unit of *function* as well as structure—the fundamental unit that shows all the characteristics of living things.

Each cell is a mass of protoplasm containing a nucleus and surrounded by a plasma membrane. Mammalian red blood cells lose their nucleus in the process of maturation, and the cells of skeletal muscles have several nuclei per cell, but these are rare exceptions to the general rule of one nucleus per cell. In the simplest plants and animals, all of the protoplasm is found within a single plasma membrane. These organisms may be considered to be unicellular, i.e., single-celled, or acellular, with bodies not divided into cells. They may have a high degree of specialization of form and function within this single cell, and the cell may be quite large, larger than the whole body of some multicellular organisms. Thus it is wrong to infer that a single-celled animal is necessarily smaller or less complex than a many-celled one.

27. CELLS

A single cell, if placed in the proper environment, will grow and eventually divide to form two cells. It is fairly easy to find an environment that will let single-celled plants and animals grow and multiply; for many of these, a drop of pond water will suffice. It is more difficult to prepare a medium that will support the growth and division of cells taken from the body of a man, chick or salamander. This was first accomplished by the American zoologist, Ross Harrison, who in 1907 was able to culture salamander cells in an artificial medium outside the body. Since then, many types of plant and animal cells have been cultured *in vitro** and many important discoveries about cell physiology have been made in this way.

The cells of different plants and animals, and of different organs within a single plant or animal, present a bewildering variety of sizes, shapes, colors and internal structures, but all have certain features in common. Each cell is surrounded by a plasma membrane, contains a nucleus, and has in its cytoplasm mitochondria, microsomes and Golgi bodies (Fig. 5).

The **plasma membrane** is made of protoplasm and is a living, functional part of the cell, extremely important in regulating its contents. All food entering the cell and all waste products or secretions leaving it must pass through this membrane. Nearly all plant cells (but not most animal cells) have a thick **cell wall** made of cellulose, outside of, and in addition to, the plasma membrane. This cell wall is nonliving, produced as a secretion of the living protoplasm. The cell wall is pierced in many places by tiny holes through which the protoplasm of one cell connects with that of the adjacent cells. Through these perforations materials can pass from one cell to another. These tough, firm cell walls provide support for the plant body.

Each cell contains a small, usually spherical or oval, body known as the **nucleus.** In some cells this has a relatively fixed position and is found near the center; in others it may move around freely and be found almost anywhere in the cell.

* *In vitro* (Latin, in glass) refers to an experiment performed outside the plant or animal body, typically in a glass vessel of some sort. In contrast to this, *in vivo* refers to an experiment using an intact, living animal or plant body. If we inject some radioactive glucose (i.e., labeled with radioactive carbon) into a rat's vein and then measure the amount of radioactive carbon in the rat's breath and urine, we are performing an *in vivo* experiment. But if we culture some muscle cells in a solution of radioactive glucose in a glass vessel, and do analyses to determine the fate of the radiocarbon, we are performing an *in vitro* experiment.

The nucleus is an important center of control of the cell, directing cellular activity and containing the hereditary factors (genes) responsible for the traits of the animal or plant. The protoplasm of the nucleus is called nucleoplasm.

The role of the nucleus can be studied by removing it and observing the consequences. By using microneedles, the nucleus of the single-celled animal *Amoeba* can be removed. After the operation, the cell continues to live and move, but it cannot grow and dies after a few days. The nucleus, we conclude, is necessary for the metabolic processes that provide for growth and cell reproduction. But, you may object, what if it were the operation itself, not the loss of the nucleus, that caused the ensuing death? This can only be decided by a **controlled experiment,** in which we subject two groups of animals to the same operative trauma, but have them differ in the presence or absence of a nucleus. We could do this by sticking a microneedle into some of the amebas, perhaps pushing it around inside to simulate the operation of removing a nucleus, but then withdraw the needle leaving the nucleus inside. This is called a "sham" operation. Amebas treated in this way will recover from the operation and grow and divide, demonstrating that it is the removal of the nucleus, not the operation, that brought about the death of the first group of animals.

A series of experiments demonstrating the importance of the nucleus in controlling the growth of the cell was performed by Hämmerling, using the single-celled plant *Acetabularia.* This marine alga, which may be as large as 5 cm., superficially resembles a mushroom, having "roots" and a stalk surmounted by a large, disc-shaped umbrella. The entire plant is a single cell and has but one nucleus, located near the base of the stalk. When Hämmerling cut across the stalk (Fig. 6) he found that the lower part could live, regenerate an umbrella, and recover completely from the operation. The umbrella part, which lacked a nucleus, lived for a considerable time but eventually died without being able to re-

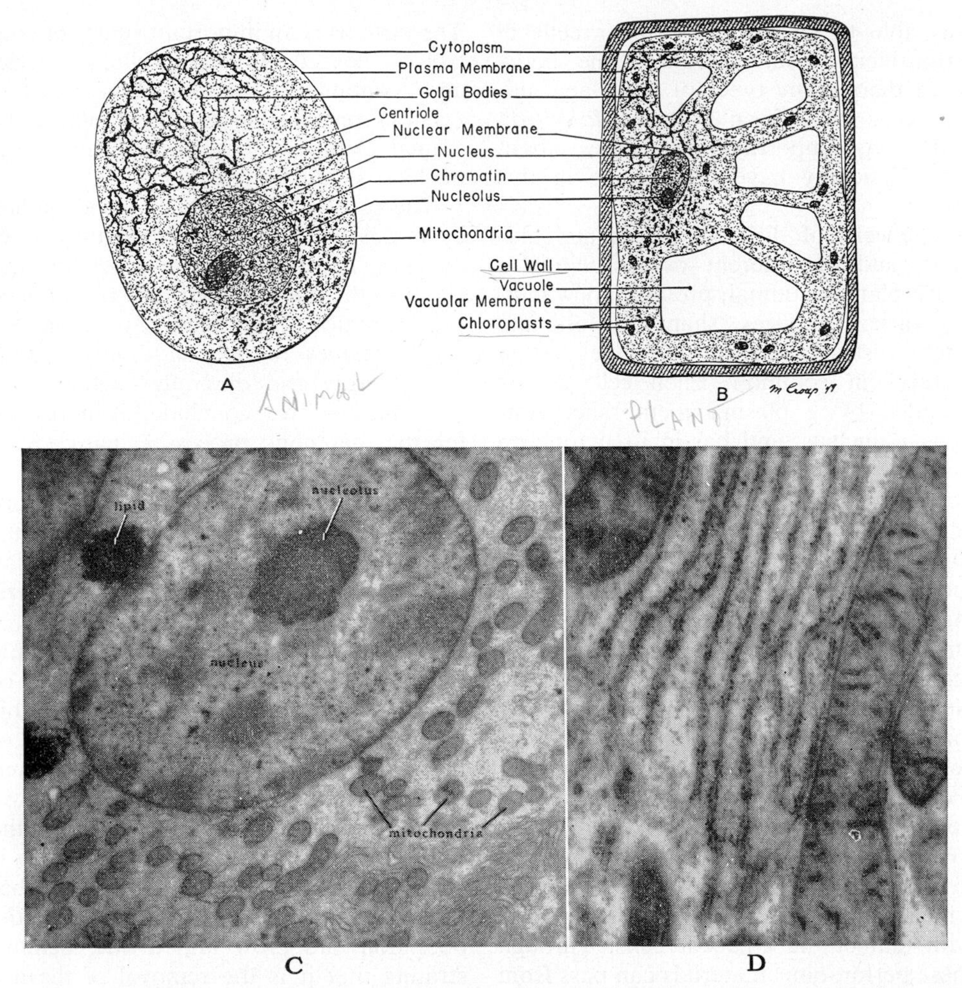

Figure 5. The structure of cells. *A,* Diagram of a typical animal cell. *B,* Diagram of a typical plant cell. *C,* Electron micrograph of the nucleus and surrounding cytoplasm of a frog liver cell. The spaghetti-like strands of the microsomes are visible in the lower right corner. Magnified 16,500 ×. *D,* High power electron micrograph of mitochondria and microsomes within a rat liver cell. Granules of ribonucleoprotein are seen on the strands of microsomes, and structures with double membranes are evident within the mitochondria in the upper left corner and on the right. Magnified 65,000 ×. (Electron micrographs courtesy Dr. Don Fawcett.)

generate the lower part. In *Acetabularia* as in the ameba, then, the nucleus is necessary for those metabolic processes underlying growth. (Regeneration is, of course, a form of growth.) In further experiments, Hämmerling first severed the stalk just above the nucleus (cut 1, Fig. 6), then made a second cut just below the umbrella (cut 2). The isolated section of stalk, when replaced in sea water, was able to regenerate a partial or complete umbrella. This might seem to show that a nucleus is not necessary for regeneration; however, when Hämmerling removed the second umbrella (cut 3) the stalk was unable to form a third one. From these and similar experiments, Hämmerling concluded that the nucleus produces some substance necessary for umbrella formation. This substance (we could call it an "umbrella hormone") passes by diffusion up the stalk and instigates umbrella growth. In the experiments just described, enough of this substance remained in the stalk after cuts 1 and 2 to produce one new umbrella. After that had been exhausted in the formation of one umbrella, no second regeneration was possible in the absence of a nucleus. This is one of the clearest examples of the production by the nucleus of a substance which regulates cell growth.

The membrane surrounding the nucleus and separating it from the adjacent protoplasm is called the **nuclear membrane.** Like the plasma membrane, it is made of protoplasm. It regulates the constant flow of materials in and out of the nucleus.

When a cell is killed by fixation in the proper chemicals and stained with appropriate dyes, several structures—strands of chromatin and one or more nucleoli—are visible within the nucleus. These are difficult to see in the living cell with an ordinary light microscope, but are readily evident by phase microscopy. Strands of chromatin, composed of proteins and nucleic acids, run irregularly through the nucleus and exhibit a netlike or granular

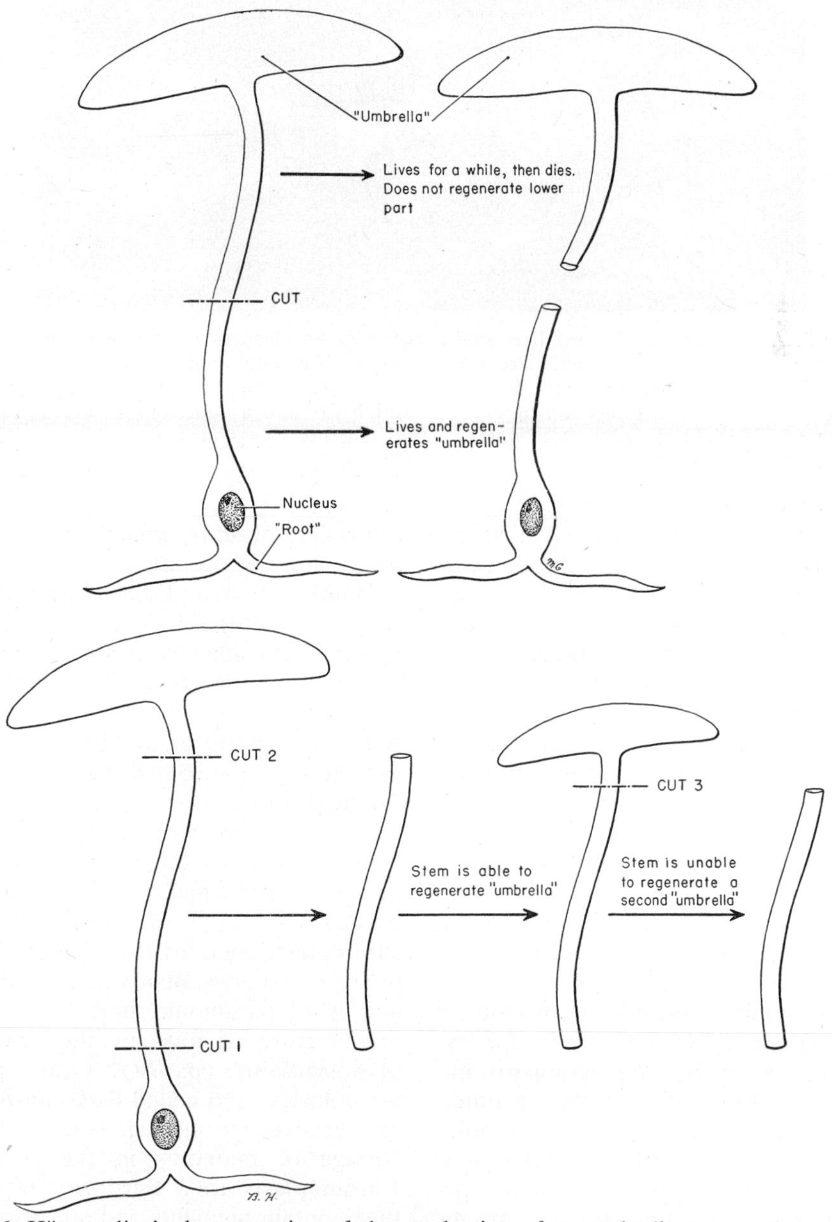

Figure 6. Hämmerling's demonstration of the production of an umbrella-regenerating substance by the nucleus of *Acetabularia.* See text for discussion.

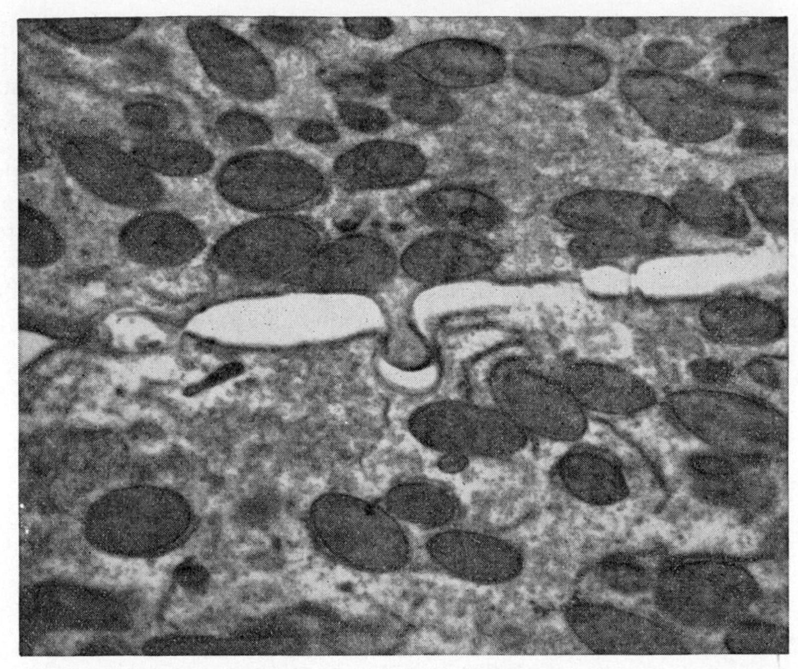

Figure 7. A photograph of a rat liver cell made with an electron microscope. The numerous dark, round, oval or elongate bodies are mitochondria. The light area across the picture is the boundary between two adjacent cells. In this boundary are visible, to the left, a bile canal and, in the center, a button, a projection of one cell which fits into a concavity in the adjacent cell and by means of which the cells of a tissue are held together. Magnification 16,500 ×. (Courtesy of Dr. Don Fawcett.)

appearance. At the time of cell division, these chromatin threads form the **chromosomes,** rod-shaped bodies which contain the hereditary units, the **genes.** The **nucleolus** is a spherical body found within the nucleus. It is extremely variable in some cells, appearing and disappearing, changing its form and structure. There may be more than one nucleolus in a nucleus, but the cells of any given animal or plant have a definite number of nucleoli. The nucleoli disappear when a cell is about to divide and reappear afterwards; their function is not well understood but they may play a role in the synthesis of proteins and ribonucleic acids.

The protoplasm outside the nucleus is called **cytoplasm.** In some cells, for example in an ameba, the cytoplasm has two more or less distinct parts: an outer, clear ectoplasm and an inner, granular endoplasm. In most cells this division is not visible. The cytoplasm contains specialized structures to perform specific functions. Some of these inclusions are centrioles, plastids, mitochondria, microsomes and Golgi bodies.

Within the cytoplasm of animal cells, lying near the nucleus, is a small, dark-staining inclusion, the **centriole.** The cells of higher plants lack these bodies, although the lower plants have them. The centriole is active at the time of cell division (Chap. 30), but it apparently is not essential for cell division, since plant cells manage to divide without it.

All plants, except fungi, have small protein bodies called plastids in the cytoplasm. One type of plastid, called a **chloroplast,** contains the pigment chlorophyll, which gives plants their green color and is of paramount importance in the manufacture of food by the process of photosynthesis (p. 79). Other plastids are colorless and called **leukoplasts.** They are believed to act as centers for the storage of materials in the cytoplasm. **Chromoplasts** are a third type of plastid; they contain pigments and are responsible for the colors of flowers and fruits.

Almost all cells have in their cytoplasm tiny bodies, known as **mitochondria,** which may appear as threads, granules or rods (Fig. 7). These bodies may change their size, appearance and position in the cell, and are usually concentrated in the part of the cell with the highest rate of metabolism. They are made of proteins, ribonucleic acids and phospholipids. Lying in or on these bodies is a complex group of enzymes (see Chap. 5) by means of which carbohydrates, fatty acids and amino acids are metabolized to carbon dioxide and water; energy is released in this process. Recently biochemists have been able to homogenize cells and then, by high speed centrifuges, to separate mitochondria from the other cytoplasmic particles. These purified mitochondria can then be incubated *in vitro* and their metabolic properties determined. Such isolated mitochondria are able to metabolize carbohydrates, fatty acids and amino acids to carbon dioxide and water, and so are still living. In plants, mitochondria are found in clusters at the point of formation of the plastids and are probably involved in providing energy for the synthesis of new plastids. Mitochondria can be recognized in living cells by the fact that they are selectively stained by a particular stain, Janus green.

Microsomes are too small to be seen with an ordinary microscope, but appear as long, spaghetti-like strands of dense material when viewed at very high magnification in the electron microscope. When separated from other cell components by high speed centrifugation and studied, the microsomes were found to be complexes of enzymes concerned with the synthesis of proteins, steroids and nucleic acids.

Golgi bodies, another type of living inclusion, are found in the cytoplasm of all cells except mature sperm and red blood cells, and are believed to be necessary for the synthesis of proteins. They, too, may appear as granules, threads or rods, or as canals, and are distinguished from mitochondria by the fact that they stain with neutral red instead of Janus green. Prof. H. W. Beams, of the University of Iowa, centrifuged cells in a high speed centrifuge and showed that mitochondria are heavier, and Golgi bodies lighter, than the rest of protoplasm.

In addition to these living elements, cytoplasm may contain vacuoles or cavities filled with watery liquid and separated by a vacuolar membrane from the rest of the cytoplasm. Vacuoles are common in plant cells and lower animals, but rare in those of higher animals. Most protozoa have food vacuoles, containing food in the process of being digested, and contractile vacuoles, which pump excess water out of the cell. Finally, cytoplasm may contain granules of stored starch or protein, or droplets of oil.

There are three chief structural differences between animal and plant cells: Animal cells, but not the cells of higher plants, have a centriole; plant cells, but not animal cells, have plastids in the cytoplasm; and plant cells have a stiff cell wall of cellulose which prevents their changing position or shape, while animal cells usually have only the thin plasma membrane and thus are able to move and alter in shape.

Most cells, both plant and animal, are too small to be seen with the naked eye. Their diameters range from about 1 micron to 100 microns, and a speck 100 microns in diameter is about at the lower limit of visibility. A few species of amebas are a millimeter or two in diameter; some single-celled plants such as *Acetabularia* may be a centimeter or more long. Some of the largest single cells are the egg cells of fishes and birds. The egg cell of a large bird may be several centimeters in diameter. Only the yolk of the egg is the true egg cell; the white of the egg is a noncellular material secreted by the hen's oviduct.

The limit to the size of a cell is set by the physical fact that as a sphere increases in size, the volume increases as the *cube* of the radius whereas the surface increases only as the *square* of the radius. Since cellular metabolism requires oxygen and nutrients that can enter only through the cell surface, it is clear that there is a limit to the size of the cell above which the surface area is too small to supply the metabolic activities of the enclosed protoplasm. The actual size of the limit will depend on the shape of the cell and the rate at which metabolism occurs. When the limit is reached, the cell must either stop growing or divide.

28. METHODS OF STUDYING CELLS

Many of the basic principles of biology have been discovered by observation and experiments with single cells. Living cells can be examined in a drop of fluid using a microscope and one can study the movement of amebas, or the beating of the hairlike projections of cilia that cover the body of paramecia. Since the discovery of methods of culturing cells removed from the body of a higher animal or plant, many new facts about cell function and structure have been discovered by observing and photographing such cells under the microscope. In culturing cells, a special complex nutritive medium, made of blood plasma, an extract of embryonic tissues, and added vitamins and other chemicals, is prepared and sterilized, then placed in a small cavity in a sterile glass slide. A bit of living tissue is placed in this nutritive medium and the cavity is sealed with another piece of glass. The tissue will live indefinitely if the medium is renewed and oxygen is supplied; cells from a chick's heart were kept alive in this way for more than twenty years at the Rockefeller Institute in New York. From such experiments it was found that cells in tissue culture do not grow old—at the end of twenty years the cells were just as vigorous and grew just as fast as the original ones. Cells isolated from a sarcoma (a type of cancer) grow with unusual vigor in tissue culture, and they grow more rapidly in blood plasma from a healthy person than in that from a person with sarcoma. This suggests that the sarcoma cells in the body stimulate some of the healthy cells to manufacture a substance which inhibits the malignant growth to some extent.

Cell morphology may be studied by using a bit of tissue that has been killed quickly by a special "fixative" that does not destroy cell structure, then sliced with a machine called a **microtome,** and stained with special dyes. The stained slices, mounted on a glass slide and covered with a glass cover slip, are then ready to be observed under the microscope. Since the nucleus, mitochondria, and other specialized parts of the cell are chemically different, they combine with different dyes and are stained characteristically. Tissues are prepared for electron microscopy by fixation with osmic acid, mounted in acrylic plastic to be cut into extremely thin sections, and placed on a fine grid to be inserted into the path of the beam of electrons.

How does a cell get its nutrients? Every cell, whether it is a single-celled plant or animal, or one of the billions of cells in an oak tree or man, must have a constant supply of nutrients, for it is constantly using up, "burning up" substances to obtain energy for its myriad metabolic activities. To understand this fundamentally important biologic problem, we must first have a firm grasp of the underlying physical concepts of energy, molecular motion and diffusion.

29. ENERGY

Energy has been defined as the ability to do work, to produce a change in matter; it may take the form of heat, light, electricity or motion. Physicists recognize two kinds: **potential energy,** the ability to do work owing to the position or state of a particle; and **kinetic energy,** the energy of motion. A rock at the top of a hill has potential energy; as it rolls down hill, the potential energy is converted to kinetic energy. The potential energy of water at the top of a dam may be transformed into kinetic energy if it is harnessed to a water wheel. Energy, derived ultimately from sunlight, is stored in the molecules of food as the chemical energy of the bonds connecting its constituent atoms. This chemical energy is a kind of potential energy. Then, inside the body, chemical reactions occur which change this potential energy into heat, motion, or some other kind of kinetic energy. All forms of energy are at least partially interconvertible, and living organisms are constantly transforming potential energy into kinetic energy or vice versa. Under experimentally controlled conditions, the amount of energy entering and leaving any given system may be measured and compared. It is always found that energy is neither created nor destroyed, but only transformed from one form to another. This is an expression of one of the fundamental laws of physics, the Law of the Conservation of Energy. Living things as well as nonliving systems obey this law.

30. THE MOTION OF MOLECULES

The molecules which make up all kinds of substances are constantly in motion. Despite the fact that wood, stone and steel seem very solid, their component molecules vibrate continuously within a very restricted space. The chief difference between the three states of matter—solid, liquid and gas—is simply the freedom of movement of the molecules present. The molecules of a solid are relatively closely packed and the forces of attraction between molecules will allow them to vibrate but not move around. In the liquid state, the molecules are relatively farther apart, the intermolecular forces are weaker, and the molecules move about with considerable freedom. Finally, in the gaseous state, the molecules are so far apart that the intermolecular forces are negligible and the movement of the molecules is restricted only by external barriers. The movement of the molecules in all three states of matter is caused by the inherent heat energy of the molecules. By increasing this molecular heat energy, one can change matter from one state to another—when ice is heated it changes to water, and when water is heated it becomes water vapor.

When a drop of water is examined under the microscope, the motion of the water molecules is not evident, but if a drop of India ink (which contains fine particles of carbon) is added, the carbon particles move about continually in aimless zig-zag paths. The carbon particles are constantly being bumped by water molecules and the recoil from this bump gives the carbon particles their motion. This motion of small particles is called **brownian movement,** after Robert Brown, an English botanist, who first observed it when he looked with a microscope at some pollen grains in a drop of water.

It is characteristic of molecules in liquids and gases to diffuse, or move, in all directions until they are spread evenly through the available space. **Diffusion** may be defined as the movement of molecules from a region of high concentration to one of lower concentration, brought about by their inherent heat energy. The rate of diffusion is a function of the size of the molecule and the temperature.

This physical process of diffusion is fundamental to many biologic phenomena, and examples of it are common in everyday life. If a room is tightly closed to prevent air currents, and a bit of perfume is released in one corner of the room, the molecules of perfume will spread out from the original corner where there is a high concentration of perfume, to all other parts of the room, regions of lower concentration. Or, if a lump of sugar is dropped in a beaker of water, the sugar dissolves and the individual sugar molecules diffuse from their original position in the beaker (Fig. 8 *A*) and come to be spread equally throughout the liquid (Fig. 8 *B*). The individual molecules move in a straight line until they bump into something—another molecule or the side of the container—then they rebound and move in another direction. That the sugar molecules have diffused from their original corner to all parts of the beaker could be proved by tasting a few drops of liquid taken from another region. The molecules continue to move after they have spread throughout the beaker; however, as fast as some molecules move from left to right, others move from right to left, so that an equilibrium is maintained.

Any number of substances will diffuse independently of each other. If a lump of salt is dropped into another corner of the same beaker (Fig. 8 *C*), its molecules will move out in all directions, unaffected by the sugar molecules, so that each drop of water in the beaker will come to have some salt and some sugar molecules (Fig. 8 *D*). Diffusion of course occurs more rapidly in gases than in liquids.

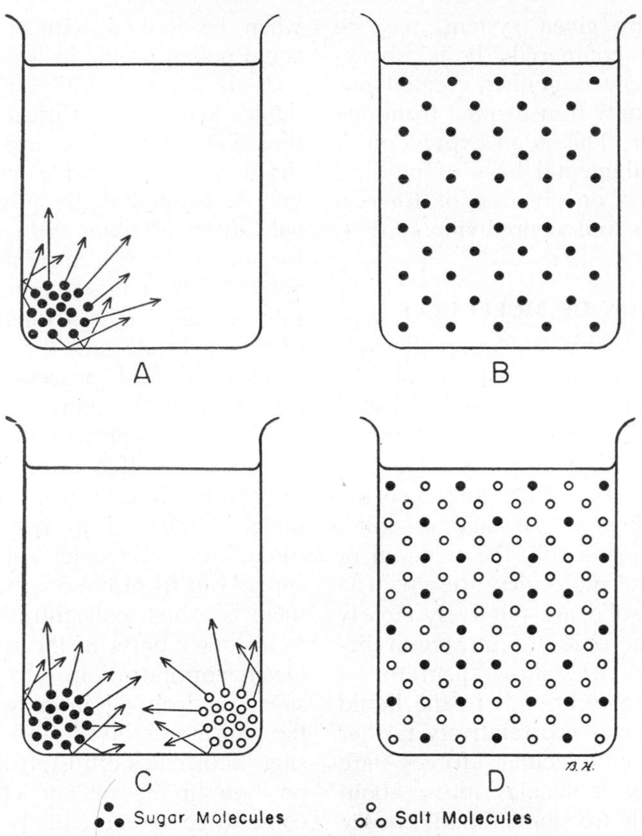

Figure 8. Diffusion. *A,* When a lump of sugar is dropped in a beaker of water, its molecules dissolve and begin to diffuse. *B,* As a result of diffusion, the sugar molecules are dispersed evenly throughout the water in the beaker. *C,* When lumps of both sugar and salt are placed in a beaker of water, each type of molecule diffuses independently of the other. *D,* As a result, both sugar and salt molecules are spread evenly throughout the beaker of water.

31. THE SPEED OF DIFFUSION

Although the speed of individual molecules is on the order of several hundred meters per second, each molecule can go only a fraction of a millimicron before bumping into another molecule and rebounding. Thus the progress of a molecule in a straight line is rather slow. This can be made visible by putting a bit of some dye at the bottom of a glass cylinder filled with water. As days and weeks go by, the colored substance will gradually move upward but it will take months before the dye is uniformly distributed throughout the cylinder. Thus although diffusion occurs rapidly over short distances, it takes a long time for a molecule to travel distances measured in inches.

This fact has important biologic implications, for it limits the number of molecules of oxygen and nutrients that will reach an organism by diffusion alone. Only a very small organism that needs relatively few molecules per second can survive if it sits in one place and lets molecules come to it by diffusion. A larger organism must either have some means of movement to a new region, or have some mechanism for stirring up its environment to bring molecules to it. Or it can live in some place where the environment is constantly moving past it—in a river or at the intertidal level on the shore. The larger land plants have solved the same problem by developing a tremendously branched root system, thus getting their needed raw materials from a large area of the surrounding environment.

32. EXCHANGES OF MATERIAL BETWEEN ENVIRONMENT AND CELL

Each cell, as we have noted, is surrounded by a plasma membrane, and all nutrients and waste products must pass through this membrane to get in or out of the cell. Cells are almost invariably surrounded by a watery medium. This might be the fresh or salt water in which an organism lives, the tissue sap of a higher plant, or the plasma or extracellular fluid of a higher animal. Usually, only dissolved substances can pass through the plasma membrane, but not all dissolved substances can penetrate this membrane equally well. The membrane behaves as though it had ultramicroscopic pores through which the substances pass and the size of these pores determines the maximum size of molecule that can pass. Factors other than molecular size, such as the electric charge, if any, carried by the diffusing particles, the number of water molecules, if any, bound to the surface of the diffusing particle, and their solubility in fats, may also be important in determining whether the substance will pass through the membrane.

Whether or not a membrane will permit a certain substance to pass through it depends on its structure, on the size of the pores present. A membrane is said to be **permeable** if it will permit any substance to pass through, **impermeable** if it will permit no substance to pass, and **semipermeable** or **differentially permeable** if it will allow some but not all substances to diffuse through. Permeability is a property of the membrane, not of the diffusing substance. All the membranes surrounding the cells, nuclei and vacuoles are semipermeable or differentially permeable.

The diffusion of a dissolved substance through a semipermeable membrane is called **dialysis.** If a pouch made of collodion, cellophane or parchment is filled with a sugar solution and placed in a beaker of water, the sugar molecules will dialyze through the membrane (if the pores are large enough) and eventually the concentration of sugar molecules in the water outside the pouch will equal that within the pouch. The molecules then continue to diffuse but there is no net change in concentration, for diffusion into and out of the pouch occurs at equal rates.

If the pouch is prepared with smaller pores, so that it is permeable to water molecules but not to the larger sugar molecules, a different phenomenon is observed. A sugar solution is placed in the pouch, the pouch is fitted with a cork through which is a glass tube, and placed in a beaker of water (Fig. 9 *A*). The sugar molecules are unable to penetrate the membrane and so remain inside the bag. The water molecules, however, diffuse through the membrane into the sugar solution. Since the liquid inside the membrane is 5 per cent sugar, it is only 95 per cent water. The liquid outside the membrane is 100 per cent water. Therefore the water molecules move from a region of higher concentration (100 per cent, outside the bag) to a region of lower concentration (95 per cent, inside the bag). This diffusion of water or solvent molecules through a membrane is called **osmosis.**

As osmosis occurs, the water rises in the glass tube (Fig. 9 *B*). If an amount of water equal to the amount originally inside the bag passes through the membrane, the sugar solution will be diluted to 2.5 per cent sugar and 97.5 per cent water, but the concentration of water on the outside will still be higher than that on the inside and osmosis will continue. Eventually the water in the glass tube will rise to a height such that the weight of the water in the tube will exert a pressure just equal to the tendency of the water to enter the bag. Then there will be no net change in the amount of water within the bag; osmosis will occur in both directions through the semipermeable membrane with equal speed. The pressure of the column of water is called the **osmotic pressure** of the sugar solution. The osmotic pressure is brought about by the tendency of the water molecules to pass through the semipermeable membrane and equalize the concentration of water molecules on the two sides. A more concentrated sugar solution would have a

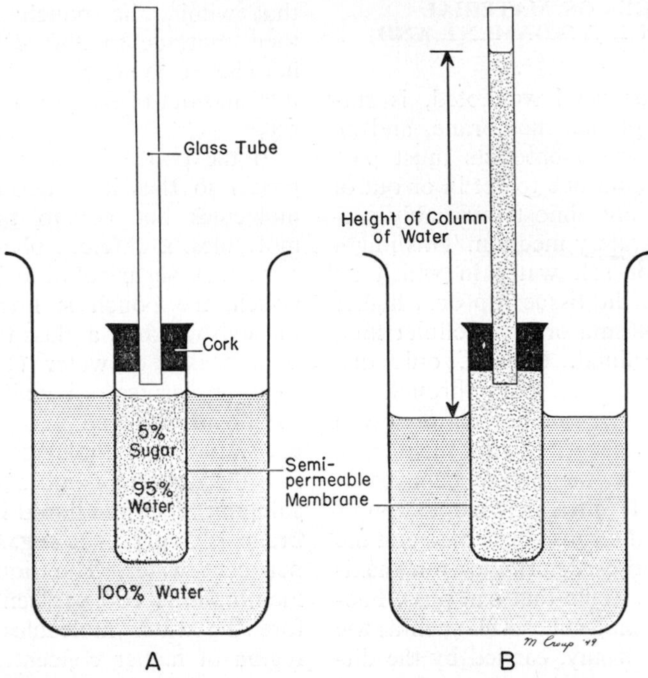

Figure 9. Diagram illustrating osmosis. *A,* When a 5 per cent sugar solution is placed in a sac made of a semipermeable membrane, such as collodion, and suspended in water, the water molecules diffuse into the sac, causing the column of water in the glass tube to rise. *B,* When equilibrium is reached, the pressure of the column of water in the tube just equals, and is a measure of, the osmotic pressure of the sugar solution.

greater osmotic pressure and thus would "draw" water to a higher level in the tube. A 10 per cent sugar solution would cause water to rise approximately twice as high in the tube as a 5 per cent solution.

Dialysis and osmosis are simply two special forms of diffusion. Diffusion is the general term for the movement of molecules from a region of high concentration to one of lower concentration, brought about by their inherent heat energy. Dialysis is the diffusion of dissolved molecules (solutes) through a semipermeable membrane and osmosis is the diffusion of solvent molecules through a semipermeable membrane. In living systems the solvent molecules are almost always water.

In the fluid of every living cell are dissolved salts, sugars and other substances which give the fluid a certain osmotic pressure. When a cell is placed in a fluid with the same osmotic pressure as its own, water does not enter or leave the cell (i. e., the cell neither swells nor shrinks) and the fluid is said to be **isotonic** or **isosmotic** with the cell. Normally the blood plasma and all the body fluids are isotonic; they contain the same amount of dissolved material as the body cells. If the surrounding fluid contains more dissolved substances than the protoplasm, water tends to pass out of the cell, and the cell shrinks. Such a fluid is said to be **hypertonic** to the cell. If a fluid has less dissolved material than the cell, it is said to be **hypotonic:** water tends to enter the cell and cause it to swell. A solution of 0.9 per cent sodium chloride (sometimes called "physiological saline") is isotonic to human cells. Red blood cells placed in a solution of 0.6 per cent sodium chloride swell and burst, whereas in a 1.3 per cent solution they shrink.

When a cell is placed in a solution that is not isotonic, it may adjust to the changed environment by undergoing a change in its water content, so that it finally reaches the same concentration of solutes as the environment. Many cells have the ability to pump water or certain solutes in or out through the plasma membrane and in this way can maintain an osmotic pressure that differs from that of the surrounding medium. Amebas, paramecia, and many other protozoa that live in pondwater, which is very hypotonic, have evolved **contractile vacuoles** (Fig. 156) which collect water from the protoplasm and pump it to the outside.

Many organisms that live in the sea have phenomenal powers to accumulate selectively certain substances from the sea water. Seaweeds can accumulate iodine so that the concentration inside the cell is 2,000,000 times that in the sea water. Certain primitive chordates, the tunicates, can accumulate vanadium so that it is also about 2,000,000 times as concentrated within the cell as in the sea water. The pumping of water or solutes in or out of the cell against a concentration gradient is physical work and requires the expenditure of energy. A cell is able to move molecules against a gradient only as long as it is alive and carrying on metabolic activities that yield energy. If the cell is treated with some metabolic poison, such as cyanide, it loses its ability to produce and maintain concentration differences on the two sides of its plasma membrane.

Plants that live in fresh water also have the problem of dealing with the water that enters the cell by osmosis from the surrounding hypotonic environment. Plant cells have no contractile vacuole to pump out the water but take advantage of the firm cellulose wall that surrounds the cell to prevent undue swelling. As water enters, the protoplasm is pressed against the cell wall and an internal pressure, called **turgor pressure,** is generated which prevents the entrance of any additional water. Turgor pressure is characteristic of plant cells in general (see p. 86) and is responsible in part for the support of the plant body. A flower wilts because the turgor pressure in its cells has decreased, which in turn was caused by a lack of water.

33. TISSUES

One of the major trends in the evolution of both plants and animals has been that leading to specialization and division of labor of the constituent cells. The cells that make up the body of a tree or a man are not all alike but are specialized to carry out certain functions. This specialization allows the cells to function more efficiently but also makes the parts of the body more interdependent: the injury or destruction of one part of the body may result in death of the whole. The advantages of specialization are obvious, however, and outweigh the disadvantages. In fact, it is difficult to imagine how an unspecialized large land plant or animal could exist at all.

A **tissue** may be defined as a group or layer of similarly specialized cells which together perform certain special functions. The study of the structure and arrangement of tissues is known as **histology.** Each tissue is composed of cells which have a characteristic size, shape and arrangement. Tissues may consist of more than living cells; connective and blood tissues, for example, contain some nonliving material between the cells.

34. ANIMAL TISSUES

Biologists differ somewhat in their ideas of how the various types of tissue should be classified and, consequently, of how many types of tissue there are. We shall classify animal tissues in six groups: epithelial, connective, muscular, blood, nervous and reproductive.

Epithelial Tissues. Epithelial tissues are composed of cells which form a continuous layer or sheet covering the body surface or lining cavities within the body. They may have one or more of the following functions: protection, absorption, secretion and sensation. The epithelia of the body protect the underlying cells from mechanical injury, from harmful chemicals and bacteria, and from drying. The epithelia lining the digestive tract absorb food and water into the body. Other epithelia give off a wide variety of substances as waste products or for use elsewhere in the body. Finally, since the body is entirely covered by epithelium, it is obvious that all sensory stimuli must penetrate an epithelium to be received. Examples of epithelial tissues are the outer

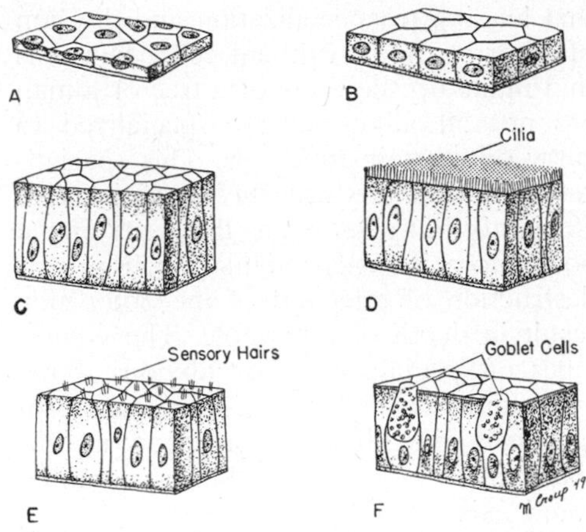

Figure 10. Types of epithelial tissue. *A,* Squamous epithelium; *B,* cuboidal epithelium; *C,* columnar epithelium; *D,* ciliated columnar epithelium; *E,* sensory epithelium (cells from the lining of the nose); *F,* glandular epithelium: two single-celled glands ("goblet" cells) in the lining of the intestine.

layer of the skin, the lining of the digestive tract, the lining of the windpipe and lungs and the lining of the kidney tubules. Epithelial tissues are divided into six subclasses according to their shape and function.

Squamous epithelium is composed of flattened cells the shape of pancakes or flagstones (Fig. 10 *A*). Squamous epithelium is found on the surface of the skin and the lining of the mouth, esophagus and vagina. In man and the higher animals, squamous epithelium usually occurs as several layers of these flat cells piled one on top of another, a condition called stratified squamous epithelium.

Cuboidal epithelium contains cells that are cube-shaped, resembling dice (Fig. 10 *B*). This tissue lines the kidney tubules.

The cells of **columnar epithelium** are elongated, resembling pillars or columns; the nucleus is usually located near the base of the cell (Fig. 10 *C*). The stomach and intestines are lined with columnar epithelium.

Ciliated epithelium is shown in Figure 10 *D*. Columnar cells may have on their free surface small protoplasmic projections called **cilia,** which beat rhythmically and move materials in one direction. Most of the respiratory system is lined with ciliated columnar epithelium whose cilia function to remove particles of dust and other foreign material.

Sensory epithelium has cells specialized to receive stimuli (Fig. 10 *E*). The cells lining the nose—the olfactory epithelium, responsible for the sense of smell—are an example.

Glandular epithelium cells (Fig. 10 *F*) are specialized to secrete substances such as milk, wax or perspiration. They are either columnar or cuboidal in shape.

Connective Tissue. Connective tissue, which includes bone, cartilage, tendons, ligaments and fibrous connective tissue, supports and holds together the other cells of the body. The cells of these tissues characteristically secrete a large amount of nonliving material, called **matrix,** and the nature and function of the particular connective tissue is determined largely by the nature of this intercellular matrix. The cells thus perform their functions indirectly by secreting a matrix which does the actual connecting and supporting.

In **fibrous connective tissue** the matrix is a thick, interlacing, matted network of fibers secreted by and surrounding the connective tissue cells (Fig. 11 *A*). Such tissue occurs throughout the body and holds skin to muscle, keeps glands in position and binds together many other structures. Tendons and ligaments are specialized types of fibrous connective tissue. **Tendons** are not elastic but are flexible, cable-like cords that connect muscles to each other or to bone. **Ligaments** are somewhat elastic and connect one bone to another. There is an especially thick mat of connective tissue fibers just below the skin. When this is treated chemically (tanned) it becomes leather. It would be possible to make leather from human tissue; indeed in some primitive tribes it was quite the thing to "tan the hide" of one's enemy and make a pair of gloves.

The supporting skeleton of vertebrates is composed of **cartilage** or **bone.** Cartilage is the supporting skeleton in the embryonic stages of all vertebrates, but is largely replaced in the adult by bone in all but the sharks and rays. In the human body, cartilage may be felt in the supporting structure of the ear flap (pinna) and in the tip of the nose. It is firm yet elastic. Cartilage cells secrete this hard, rubbery matrix around themselves and come to lie in groups of one, two or four in small cavities in the homogeneous, continuous matrix (Fig. 11 *B*). The cartilage cells embedded in the matrix remain alive; some of them secrete fibers which become embedded in the matrix and strengthen it.

Bone cells remain alive and secrete a bony matrix throughout a person's life. The matrix contains calcium salts and organic matter which are responsible for hardness and lack of brittleness, respectively, enabling bone to stiffen and support the body against the pull of gravity. The bones of elderly people have more calcium and less organic matter than those of young people, and for this reason are more brittle. Contrary to appearances, bone is not a solid structure. Most bones have a large cavity, the **marrow cavity,** in the center (Fig. 11 *C*), which may contain yellow marrow, mostly fat, or red marrow, the tissue in which red and certain white blood cells are made. Running through the matrix of the bone are channels (haversian canals) in which lie blood vessels and nerves to supply and control the bone cells. The bone matrix is secreted in concentric rings (lamellae) around the canals, and the cells become embedded in cavities in these rings (Fig. 11 *D*). Bone cells are connected to each other and to the haversian canals by protoplasmic extensions of the bone cells which lie in minute canals (canaliculi) in the matrix. The bone cells obtain oxygen and raw materials and eliminate wastes by way of these minute canals. Bones also contain cells which destroy bone, so that they gradually change their shape in response to continued stresses and strains.

Muscular Tissue. The movements of most animals result from the contraction of elongated, cylindrical or spindle-shaped cells, each of which contains many small, longitudinal, parallel, contractile fibers called **myofibrils.** Muscle cells perform mechanical work by contracting, by getting shorter and thicker; they cannot push. There are three distinct types of this tissue in the human body: skeletal, smooth and cardiac (Fig. 12). **Cardiac muscle** is found in the walls of the heart, **smooth muscle** in the walls of the digestive tract and certain other internal organs, and **skeletal muscle** makes up the large

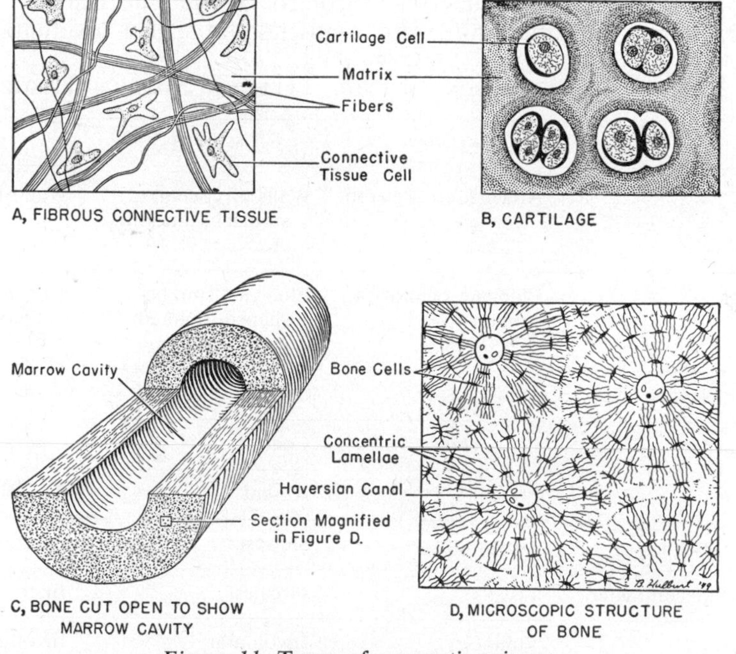

Figure 11. Types of connective tissues.

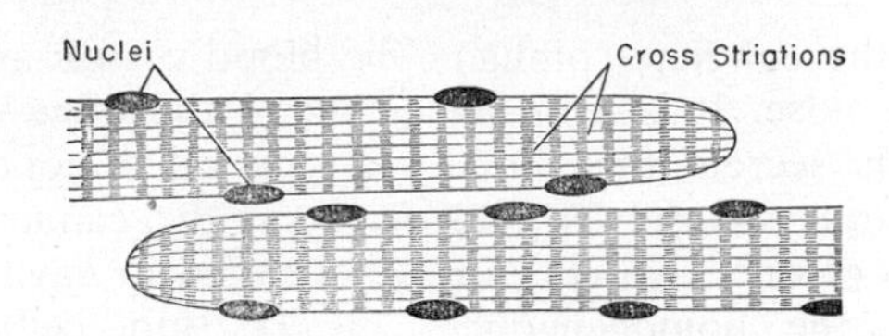

A, SKELETAL MUSCLE FIBERS

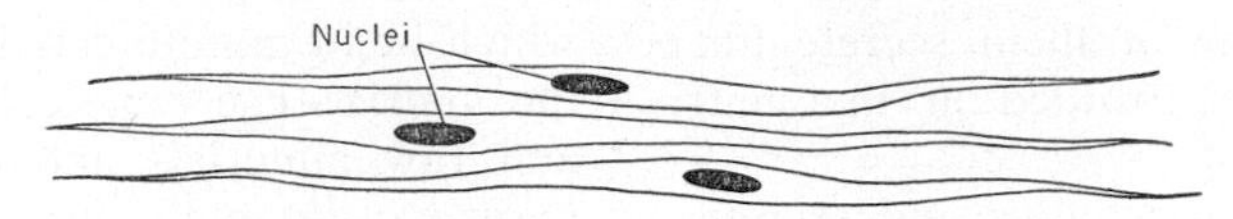

B, SMOOTH MUSCLE FIBERS

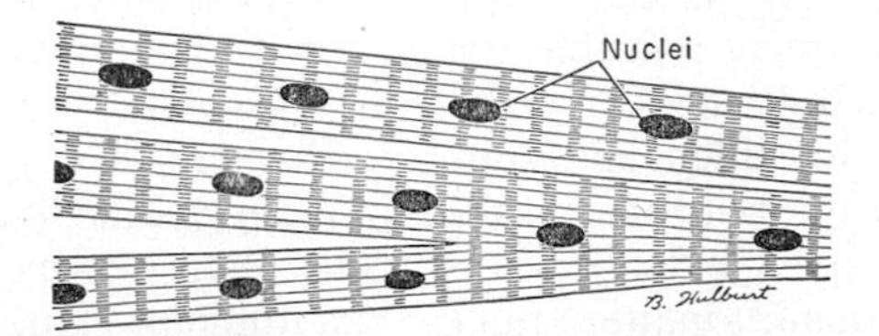

C, CARDIAC MUSCLE FIBERS

Figure 12. Types of muscle tissue.

muscle masses attached to the bones of the body. Skeletal and cardiac fibers are exceptions to the rule that cells have only one nucleus; each fiber has many nuclei. The nuclei of the skeletal fibers are also unusual in their position: they lie peripherally, just under the cell membrane; presumably this is an adaptation to increase the efficiency of contraction. Under the microscope, skeletal and cardiac fibers are seen to have alternate light and dark transverse stripes or striations. These stripes appear to be involved in contraction, for they change their relative sizes during contraction, the dark stripes decreasing and the light stripes increasing in width. Skeletal muscle is sometimes called voluntary muscle, because it is under the control of the will. Cardiac and smooth muscles are called involuntary, because

Table 2. COMPARISON OF THE TYPES OF MUSCLE TISSUE

	SKELETAL	SMOOTH	CARDIAC
Location	Attached to skeleton	Walls of viscera, stomach, intestines, etc.	Wall of heart
Shape of fiber	Elongate, cylindrical, blunt ends	Elongate, spindle-shaped, pointed ends	Elongate, cylindrical, fibers branch and fuse
Number of nuclei per fiber	Many	One	Many
Position of nuclei	Peripheral	Central	Central
Cross striations	Present	Absent	Present
Speed of contraction	Most rapid	Slowest	Intermediate
Ability to remain contracted	Least	Greatest	Intermediate
Type of control	Voluntary	Involuntary	Involuntary

they cannot be regulated by the will. Table 2 summarizes the features which distinguish the three types of muscle tissue.

Blood Tissue. Blood tissue includes the red and white blood cells and the liquid, noncellular part of the blood, the **plasma.** Many biologists classify blood with the connective tissues because they originate from similar cells.

Nervous Tissue. Nervous tissue is made of cells, called **neurons,** specialized for conducting impulses. Each neuron has an enlarged structure, the **cell body,** which contains the nucleus, and two or more nerve fibers extending from the cell body (Fig. 13). The nerve fibers are made of cytoplasm and are covered by a cell membrane. Some are extremely long—those stretching from the spinal cord down the arm or leg may be 3 feet or more in length. All are microscopic in width. The neurons are connected together in chains to pass impulses for long distances through the body. Two types of nerve fibers, **axons** and **dendrites,** are differentiated on the basis of the direction in which they normally conduct a nerve impulse: axons conduct nerve impulses away from the cell body; dendrites conduct them toward the cell body. The junction between the axon of one neuron and the dendrite of the next is called a **synapse.** The axon and dendrite do not actually touch at the synapse; there is a small gap between the two. An impulse can travel across the synapse only from an axon to a dendrite; the synapse serves as a valve to prevent the backflow of impulses. Neurons have many sizes and shapes but all have the same basic plan.

Reproductive Tissue. Reproductive tissue is composed of cells modified to produce offspring—namely, egg cells in females and sperm cells in males (Fig. 14). Egg cells are usually spherical or oval and are nonmotile. The cytoplasm of the eggs of most animals, but not of the

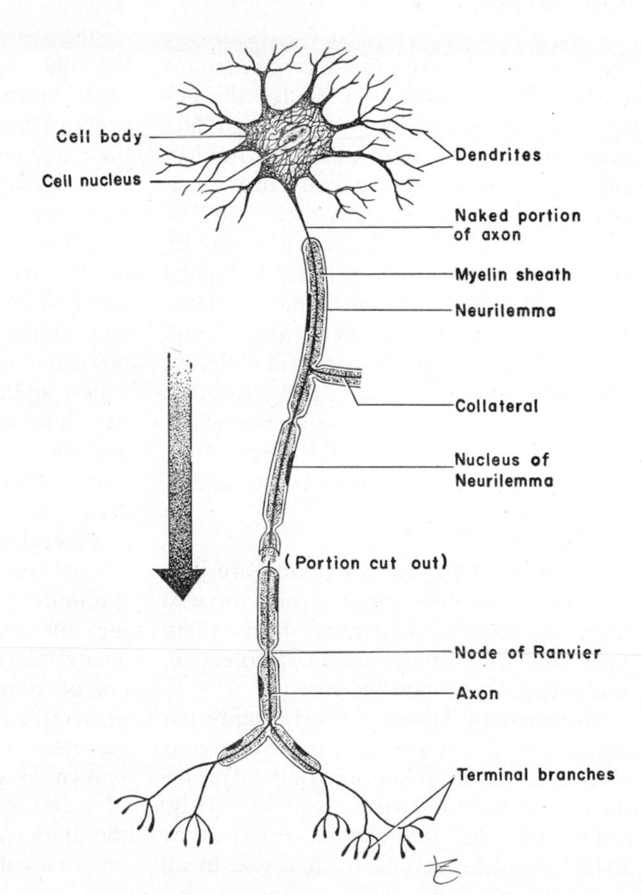

Figure 13. Diagram of a single neuron and its parts. The arrow indicates the direction of the normal nerve impulse.

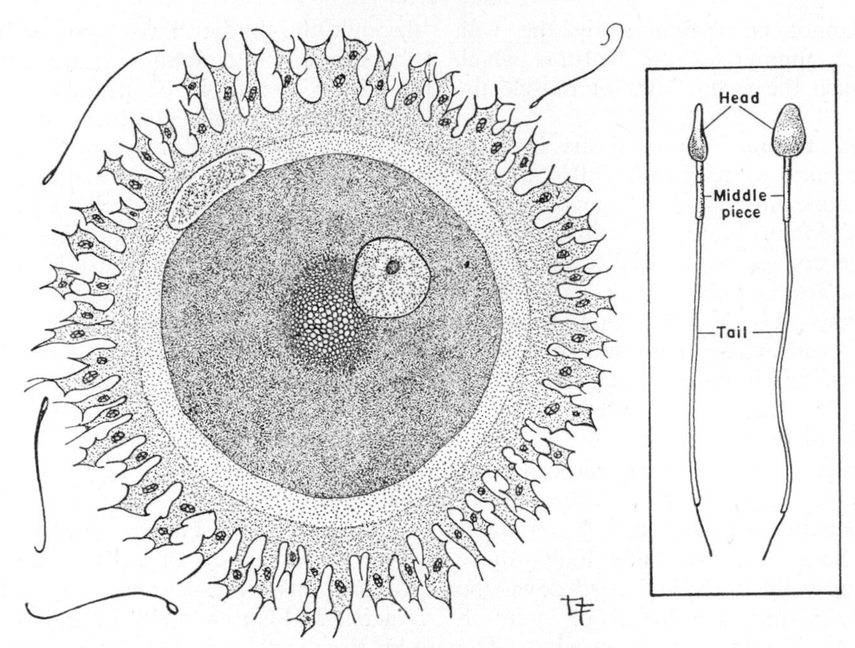

Figure 14. Left, Human egg and sperm magnified 400 ×. Note the large nucleus and nucleolus in the egg. A polar body is visible as a flattened oval body between the egg and the surrounding corona radiata cells. Compare with Figure 235. *Right,* Side and top views of a sperm, magnified about 2000 ×.

higher mammals, contains a large amount of **yolk** which serves as food for the developing organism from the time of fertilization until it is able to obtain food in some other way. Sperm cells are much smaller than eggs; they have lost most of their cytoplasm and developed a tail by which they propel themselves. A typical sperm consists of a **head,** which contains the nucleus, a **middle piece** and a **tail.** The shape of the sperm varies in different species of animals (Fig. 262). Because eggs and sperm develop from epithelial-like tissue in the ovaries and testes, some biologists classify them with those tissues.

35. PLANT TISSUES

The cells of the higher plants are also organized and differentiated into tissues. Plant biologists recognize four main types of tissue: meristematic, protective, fundamental and conductive.

Meristematic Tissue. Meristematic tissues are made of small, thin-walled cells with large nuclei. They are rich in protoplasm and have few or no vacuoles (Fig. 15). Their chief function is growth; they divide and differentiate to give rise to all the other types of tissue. An embryonic plant starts development composed entirely of meristem; as it develops, most of the meristem becomes differentiated into other tissues, but even in an adult tree there are regions of meristem which provide for continued growth. Meristematic tissues are found in the rapidly growing parts of the plant—the tips of the roots and stems, and in the cambium. The meristem in the tips of roots and stems, called **apical meristem,** is responsible for the increase in length of roots and stems, and the meristem in the **cambium,** called lateral meristem, makes possible the increase in diameter of stems and roots.

Protective Tissue. Protective tissues consist of cells with thick walls to protect the underlying thin-walled cells from drying out and from mechanical abrasions. The epidermis of leaves and the cork layers of stems and roots are examples of protective tissues. The epidermis of leaves secretes a waxy, waterproof material known as **cutin,** which prevents the loss of water from the leaf surface. Some of the epidermal cells of the roots have outgrowths called root hairs that increase the absorptive surface for the intake of water and dissolved minerals from the soil. Stems and roots are covered by layers of cork cells, produced by a separate cork cambium. Cork cells are closely packed and their cell walls contain another waterproof material, **suberin.** Since the suberin prevents the entrance of water into the cork cells themselves, they are short-lived and all mature cork cells are dead.

Fundamental Tissues. The **fundamental tissues** make up the great mass of the plant body, including the soft parts of the leaf, the pith and cortex of stems and roots, and the soft parts of flowers and fruits. Their chief functions are the production and storage of food. The simplest of the fundamental tissues is **parenchyma,** in which the cells have a thin wall and a large vacuole (Fig. 16 *A*). The

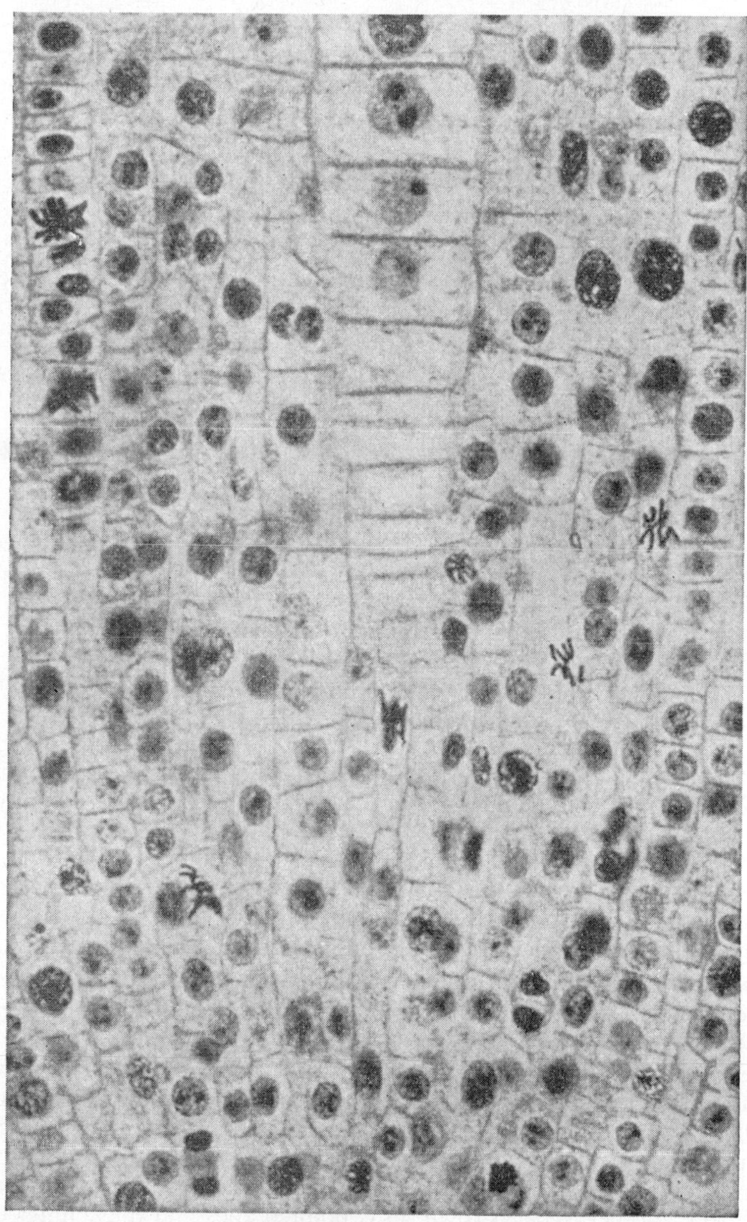

Figure 15. Photograph of a lengthwise section of an onion root tip, showing meristematic tissue. A number of the cells can be seen in various stages of cell division. (Holman and Robbins: Textbook of General Botany, John Wiley and Sons.)

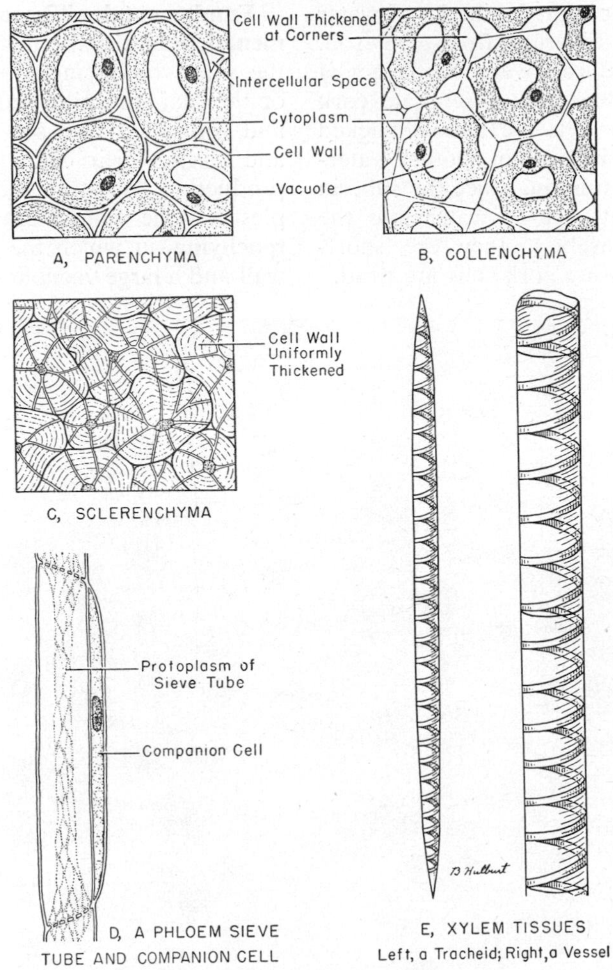

Figure 16. Some types of plant tissues: parenchyma, collenchyma, sclerenchyma, tracheids and sieve tubes.

protoplasm is present as a thin layer lining the cell walls. Chlorenchyma is a modified parenchyma containing green cell bodies called chloroplasts in which photosynthesis occurs. The chlorenchyma cells are loosely packed and make up most of the interior of leaves and some stems. They are characterized by thin cell walls, large vacuoles, and the presence of chloroplasts.

In some fundamental tissue the corners of the cells are thickened to provide the plant with support. Such tissue, called **collenchyma** (Fig. 16 *B*), occurs just beneath the epidermis of stems and leaf stalks. In still another type, known as **sclerenchyma** (Fig. 16 *C*), the entire cell wall becomes greatly thickened. These cells, which provide support and mechanical strength, are found in many stems and roots. They sometimes take the form of long thin fibers. Spindle-shaped sclerenchyma cells called **bast fibers** are found in the phloem of many stems. Rounded sclerenchyma cells called **stone cells** are found in the hard shells of nuts.

Conductive Tissues. There are two types of conductive tissue in plants: **xylem,** which conducts water and dissolved salts, and **phloem,** which conducts dissolved foods such as glucose. In all higher plants, the first xylem cells to develop are long ones, called **tracheids,** with pointed ends and thickenings on the walls in a circular, spiral or pitted pattern (Fig. 16 *E*). Later, other cells join end-to-end to form xylem vessels. As the vessels develop, the end walls dissolve and the side walls thicken, leaving a long cellulose tube for the conduction of water. These vessels may be as much as 10 feet long. In both tracheids and vessels the cytoplasm eventually dies, leaving the tubes, which continue to function. The thickening, which involves the deposition of **lignin,** the substance responsible for the hard, woody nature of plant stems and roots, enables xylem to act as a supportive as well as a conductive tissue. A similar end-to-end fusion of cells gives rise to the **sieve tubes** of the phloem (Fig. 16 *D*). The ends of the cells do not disappear, but remain as a perforated plate, the sieve plate. Unlike the tracheids and vessels of the xylem, the sieve tubes remain alive and have an abundance of cytoplasm. The streaming movements or cyclosis of the protoplasm of the sieve tubes is important in speeding up the transport of dissolved foods by the sieve tubes. Sieve tubes are found in woody stems, in the soft bark just outside the cambium layer.

36. ORGAN SYSTEMS

The bodies of single-celled animals and plants are not, of course, organized into tissues and organs; all the life functions are carried on by the one cell. In more complex organisms a division of labor has occurred and special systems have evolved to perform each of the principal life functions. In man, for example, the circulatory system is made of organs—heart, arteries and veins; the heart is made of several types of tissue—cardiac muscle, fibrous connective tissue, nerves, etc.; and each type of tissue is composed of millions of individual cells.

In man, eleven organ systems are recognized:

The **circulatory system,** which transports materials around the body.

The **respiratory system,** which provides a means for oxygen to enter, and carbon dioxide to leave, the blood.

The **digestive system,** which takes in food, breaks it up chemically into small molecules and then absorbs it into the blood.

The **excretory system,** which eliminates the waste products of metabolism.

The **integumentary system,** which covers and protects the entire body.

The **skeletal system,** which supports the body and provides for movement and locomotion.

The **muscular system,** which functions with the skeletal system in movement and locomotion.

The **nervous system,** which conducts impulses around the body and integrates the activities of the other systems.

The **sense organs,** which receive stimuli from the outer world and from various regions of the body.

The **endocrine system,** which is an additional coordinator of the body functions.

The **reproductive system,** which provides for the continuation of the species.

37. BODY PLAN AND SYMMETRY

In referring to parts of the body, biologists use the following terms: **anterior,** referring to the head end of the body; **posterior,** to the tail end of the body; **dorsal,** to the back side; **ventral,** to the belly side; **medial,** to the midline of the body; and **lateral,** to the side. These terms are also used to indicate relative position. For example, the neck is anterior to the chest; the ribs are posterior to the collar bone; the spinal cord is dorsal to the body cavity.

Animal and plant bodies may be organized according to one of three types of symmetry. A body is said to be symmetrical if it is possible to cut it into two equivalent halves. If a body is spherical, and is completely homogeneous, like a rubber ball, so that it is possible to cut it in any plane through the center and get two equal halves, it is said to have **spherical symmetry.** Only a few of the lowest plants and animals have this type of organization. In **radial symmetry** two sides are distinguishable, a top and a bottom, as in a starfish or a mushroom. A mushroom can be cut into two equal halves by any plane which includes the line or axis running through the center from top to bottom. Human beings and all the higher animals have **bilateral symmetry,** in which only one special cut will divide the body into two equivalent halves. In bilaterally symmetrical animals, anterior, posterior, dorsal and ventral sides can be distinguished. In man, for example, only a plane from head to foot exactly in the center will divide the body into equivalent halves, right and left. And, as we shall see when we study some of the details of human structure, even the right and left halves of the human body are not completely the same.

In a bilaterally symmetrical animal, three planes or sections are possible:

Transverse planes, which include a dorso-ventral axis and a left-right axis, but are at right angles to the anterior-posterior axis.

A **sagittal** plane, which includes the dorso-ventral axis and the anterior-posterior axis, but is at right angles to the left-right axis.

A **frontal** plane, which includes the anterior-posterior axis and the left-right axis, but is at right angles to the dorso-ventral axis.

In learning to differentiate between these body planes, it is helpful to make a rough model of some bilaterally symmetrical animal such as a fish, using modeling clay or some similar material. Practice making the various types of sections until you are completely familiar with these terms.

QUESTIONS

1. How would you define a cell?
2. What are the chief differences between plant and animal cells?
3. What are the advantages to an organism of having its body organized in cells?
4. Contrast the meaning of the term "cell" in the time of Robert Hooke, of Schleiden and Schwann, and at present.
5. What is the function of the nucleus? Discuss the evidence pertaining to the role of the nucleus in cell metabolism.
6. What is meant by *in vitro* and *in vivo* experiments? Give an example of each.
7. What factor limits the size of a cell?
8. Differentiate clearly between diffusion, dialysis and osmosis.
9. In what ways do gases, liquids and solids differ?
10. Of what importance to living things is the phenomenon of diffusion?
11. Which tissues of plants and animals are comparable? Which are peculiar to one or the other?
12. What kinds of tissue make up the following: the eyeball, the lung, the intestines, the sweat glands, the liver?
13. Define the terms anterior, ventral and medial.
14. What kind of symmetry do the following organisms have: earthworm, snail, pine tree, *Volvox,* clam?
15. How would you describe the position of the hump of a camel? Of a cat's eye in relation to its nose?

SUPPLEMENTARY READING

The development of the cell theory is presented in Hall's *A Source Book in Animal Biology* by means of long quotations from some of the original scientific papers. Further discussion of the properties of cells and protoplasm, together with a number of electron micrographs of tissues, will be found in *General Cytology* by De Robertis, Nowinski and Saez. Maximow and Bloom's *Textbook of Histology* is a detailed, technical discussion of the tissues of the human body and includes many fine illustrations of each type of tissue.

Chapter 5

Cellular Metabolism

ALL THE chemical activities of protoplasm, which provide for its irritability and movement, its growth, maintenance and repair, and its reproduction, are called **metabolism.** Modern biochemical research has shown that the metabolic activities of animal, plant and bacterial cells are remarkably similar, despite the differences in the appearances of these organisms. In all cells, glucose and related substances are constantly being broken down, by way of a long series of intermediate steps which make energy available to the cell for other processes. If this takes place in the presence of air (oxygen) the final products are carbon dioxide and water and the process is called **respiration.** If this takes place in the absence of air, the final product is lactic acid, alcohol or some other substance, and the process is called **fermentation.** One of the important metabolic differences between plants and animals is the ability of green plants to carry on photosynthesis, the ability to trap the energy of sunlight (or artificial light) and use it to synthesize compounds—to incorporate carbon dioxide molecules into organic compounds. In this way plants synthesize carbohydrates and, from them, proteins, fats and other substances. Bacteria and animal cells also have the ability to "fix" carbon dioxide, to incorporate carbon dioxide into a variety of organic compounds; but only green plants have the ability to utilize the radiant energy of light to run the process; animals and bacteria must get the energy for the process from other energy-yielding processes, such as the breakdown of glucose.

38. CHEMICAL REACTIONS

A chemical reaction is a change involving the molecular structure of one or more substances; matter is changed from one substance, with its characteristic properties, to another with new properties, and energy is released or absorbed. Hydrochloric acid reacts with the base, sodium hydroxide, to yield water and the salt, sodium chloride. In the process energy is released as heat. The chemical properties of HCl and NaOH are very different from those at H_2O and NaCl. In chemical shorthand a plus sign connects the symbols of the reacting substances, HCl and NaOH, and the products of the reaction, H_2O and NaCl. An arrow indicates the direction of the reaction:

$$HCl + NaOH \longrightarrow NaCl + H_2O$$

Note that there is the same number of each kind of atom on each side of the arrow, since atoms are neither destroyed nor created in a chemical reaction, but simply change partners. This is an expression of one of the basic tenets of physics, the Law of the Conservation of Matter. Most chemical reactions are reversible. The energy relations of the several chemicals involved, their relative concentrations, and their solubility are some of the factors which determine whether or not a reaction will occur, and whether it will go from right to left or left to right. Reversible reactions are indicated by a double arrow: $\rightleftharpoons$.

The rate at which chemical reactions occur is determined by a number of factors, one of which is temperature. Each increase of 10° C. approximately doubles the rate of most reactions. This is true of biologic processes as well as of reactions in a test tube, which indicates that the chemical reactions of living things are not fundamentally different from those of nonliving ones.

We can now write the formula for the net changes that occur when one of the cells of our body breaks glucose down in the presence of oxygen:

$$C_6H_{12}O_6 + 6O_2 \rightarrow 6H_2O + 6CO_2 + \text{energy}$$

Note again that there is the same number of each kind of atom on each side of the arrow; the reaction is said to be balanced.

In this reaction energy is given off as the glucose molecule is broken down. To reverse the reaction and synthesize glucose, as in photosynthesis, an equivalent amount of energy must be supplied to the plant. There are a number of different units of energy, but the one most used in biologic studies is the Calorie (kilocalorie or Calorie, written with a capital C), which is the amount of heat required to raise one kilogram of water one degree Centigrade (strictly, from 14.5° to 15.5° C.). Other forms of energy—light, electricity, energy of motion or position—can be converted to heat and measured by their effect in raising the temperature of water.

The amount of energy given off is proportional, of course, to the amount of glucose metabolized to carbon dioxide and water, each gram of glucose yielding about 4 Calories (actually 3.74).

39. CATALYSIS

Many of the substances that are rapidly metabolized by living cells are remarkably inert outside the body. A glucose solution will keep indefinitely in a bottle if it is kept free of bacteria and molds; it must be subjected to high temperature, strong acids or bases before it will break down. Protoplasm cannot use these extreme conditions to break down the glucose, for the protoplasm itself would be destroyed long before the glucose. The reactions are brought about by special agents called **enzymes,** which belong to the class of substances known as **catalysts.** A catalyst is a substance which regulates the speed at which a chemical reaction occurs without affecting the end point of the reaction and without being used up as a result of the reaction. Almost any substance may be a catalyst for some reaction. Water is an excellent one. For example, pure, dry hydrogen gas and chlorine gas may be mixed without result, but if a slight amount of water is present, the hydrogen and chlorine react with explosive violence to form hydrogen chloride. Finely divided metals—iron, nickel, platinum, palladium, and so on—are catalysts widely used in industrial processes such as the hydrogenation of cottonseed oil to make margarine or the "cracking" of petroleum to make gasoline. Since a catalyst is not used up in a reaction, it can be used over and over again, so that a very small amount of catalyst will speed up the reaction of vast quantities of reactants.

40. ENZYMES

Enzymes are catalysts, produced by living cells, that regulate the speed and specificity of the thousands of chemical reactions that occur in protoplasm. An enzyme does not have to be inside a cell to act as a catalyst; many of them have been extracted from cells with their activity unimpaired. They can then be purified, crystallized and their catalytic abilities can be studied. Enzyme-controlled reactions are basic to all the phenomena of life: respiration, growth, muscle contraction, nerve conduction, photosynthesis, nitrogen fixation, deamination, digestion, and so on. There is no need to postulate some mysterious vital force to account for these phenomena.

41. PROPERTIES OF ENZYMES

All the enzymes isolated and crystallized to date have been found to be proteins. They are usually colorless, but they may be yellow, green, blue, brown or red. Most enzymes are soluble in water or dilute salt solution, but some, for example, the enzymes present in the mitochondria, are bound together by lipoprotein (a phospholipid-protein complex) and are insoluble in water. Enzymes are usually named by adding the suffix *-ase* to the name of the substance acted upon, e.g., sucrose is acted upon by the enzyme sucrase.

The catalytic ability of some enzymes is truly phenomenal. For example, one molecule of the enzyme **catalase,** extracted from beef liver, will bring about the decomposition of 5,000,000 molecules of hydrogen peroxide (H_2O_2) per minute at 0° C. The substance acted upon by an enzyme is known as its **substrate;** thus, hydrogen peroxide is the substrate of the enzyme catalase. The number of molecules of substrate acted upon by a molecule of enzyme per minute is called the **turnover number** of the enzyme. The turnover number of catalase is thus 5,000,000. Most enzymes have high turnover numbers, which explains why they can be so effective although present in such minute amounts. Hydrogen peroxide is a poisonous substance and is produced as a by-product in a number of enzyme reactions. Catalase protects the cell by destroying the peroxide.

Enzymes differ in their specificity, in the number of different substrates they will attack. A few enzymes are absolutely specific; **urease,** which decomposes urea to ammonia and carbon dioxide, will attack no other substance, and a specific enzyme is necessary to split each of the three common double sugars, sucrose, maltose and lactose. Other enzymes are relatively specific and will work upon only a few, closely related substances. **Peroxidase** will decompose several different peroxides including hydrogen peroxide. Peroxidase is found in a wide variety of plant and animal tissues. Peroxidase activity can be demonstrated by mincing some raw potato and adding some hydrogen peroxide. A vigorous bubbling will ensue as peroxidase converts the peroxide to water and oxygen.

Finally, a few enzymes are specific only in requiring that the substrate have a certain kind of chemical bond. The **lipase** secreted by the pancreas will split the ester bonds connecting the glycerol and fatty acids of a wide variety of different fats.

Theoretically, enzyme-controlled reactions are reversible and the enzyme does not determine which way the reaction will go; it simply accelerates the rate at which the reaction reaches equilibrium. This equilibrium point depends upon complex thermodynamic principles, a discussion of which is beyond the scope of this book. Since many reactions give off energy when going in one direction, it is obvious that the proper amount of energy in a usable form must be supplied to drive the reaction in the other direction.

To run an energy-requiring reaction, some energy-yielding reaction must occur at about the same time. In most biologic systems, energy-yielding reactions result in the synthesis of "high energy" phosphate bonds, such as the terminal bonds of adenosine triphosphate (abbreviated as ATP). The energy of these high energy phosphate bonds can then be used by a cell to conduct a nerve impulse, contract muscle, synthesize proteins, and so on. Biochemists use the term "coupled reactions" for two reactions which must take place together so that one can furnish energy, or one of the reactants, needed by the other.

Enzymes usually work in teams, with the product of one enzyme-controlled reaction serving as the substrate for the next. We can picture the inside of a cell as a factory with many different assembly lines (and disassembly lines) operating simultaneously. Each of the assembly lines is composed of a number of enzymes, each of which carries out one step such as changing molecule A into molecule B and then passes it along to the next enzyme, which converts molecule B into molecule C, and so on. From germinating barley seeds one can extract enzymes that will convert starch to glucose. This consists of two enzymes: the first, called **amylase,** breaks starch down to maltose and the second, called **maltase,** breaks maltose down to glucose. Eleven different enzymes, working consecutively, are required to convert glucose to lactic acid. The same series of eleven enzymes is found in human cells, in green leaves and in bacteria.

Some enzymes, such as pepsin, consist solely of protein. Many others consist of two parts, one of which is protein (and called an **apoenzyme**) and the second (called a **coenzyme**) is made of a smaller organic molecule, usually containing phosphate. Coenzymes can usually be separated from their enzymes and, when analyzed, have proved to contain some vitamin as part of the molecule—thiamine, niacin, riboflavin, pyridoxine, and so on. This finding has led to the generalization that *all vitamins function as parts of coenzymes in body cells.* Neither the apoenzyme nor the coenzyme alone has any catalytic activity; only when the two are combined is activity present. Other enzymes require for activity, in addition to a coenzyme, the presence of some ion. Several of the enzymes involved in the breakdown of glucose require magnesium (Mg^{++}). **Ptyalin,** the enzyme of saliva, requires chloride ion (Cl^{-}) for activity. Most, if not all, of the elements needed by plants and animals in very small amounts—the so-called **trace elements,** manganese, copper, cobalt, zinc, iron, etc.—function as such enzyme activators.

Location of Enzymes in the Cell. Many of the enzymes are present simply dissolved in the cytoplasm of the cell. It is possible to make a water extract of ground liver that contains all the enzymes necessary to convert glucose to lactic acid. Other enzymes are tightly bound to certain cell bodies. The respiratory enzymes which catalyze the metabolism of lactic acid (and also substances derived from amino acids and fatty acids) to carbon dioxide and water are bound to the mitochondria; in fact, perhaps the mitochondria are in large part made of these enzymes. Certain other enzymes are integral parts of smaller cytoplasmic particles, the **microsomes.**

The location and functioning of enzymes within the cell can be studied histochemically. The tissue is fixed and sliced by methods which do not destroy enzyme activity. Then the proper chemical substrate for the enzyme is provided and, after a specified period of incubation, some substance is added which will form a colored compound with one of the products of the reaction mediated by the enzyme. The regions of the cell which have the greatest enzyme activity will have the largest amount of the colored substance (Fig. 17).

Mode of Action of Enzymes. An enzyme can speed up only those reactions that would occur to some extent, however slight, in the absence of the enzyme. Many years ago the German chemist Emil Fischer suggested that the specificity of the relationship of an enzyme to its substrate indicated that the two must fit together like a lock and key. The idea that the enzyme combines with its substrate to form an intermediate enzyme-substrate complex, which subsequently decomposes to release the enzyme and the reaction products, was formulated mathematically by Leonor Michaelis more than forty years ago. By brilliant inductive reasoning, he assumed that such a complex does form, and then calculated how the speed of the reaction should be affected by varying the concentrations of enzyme and substrate. Exactly these relationships are observed experimentally, which is strong evidence that Michaelis' assumption, that an enzyme-substrate complex forms as an intermediate, is correct. Direct evidence of the existence of an enzyme-substrate complex was obtained by David Keilin of Cambridge University and Britton Chance of the University of Pennsylvania. Chance isolated a brown-colored peroxidase from horseradishes. When he mixed this with hydrogen peroxide, he found a green-colored enzyme-substrate complex was formed. This was then changed to a second, pale red enzyme-substrate complex which then split to give the original brown enzyme plus the breakdown products. By following the changes of color, Dr. Chance was able to calculate the rate of formation of the enzyme-substrate complex and its rate of breakdown.

Although it is clear that the substrate is much more reactive when part of an enzyme-substrate complex than when it is free, it is not clear why this should be true. One current theory postulates that the enzyme unites with the substrate at two or more points, and the substrate is held in a position which strains its molecular bonds and makes them more likely to break.

42. FACTORS AFFECTING ENZYME ACTIVITY

Temperature. Enzymes are inactivated by moderate heat—temperatures of 50 to 60° C. rapidly inactivate most enzymes. The inactivation is irreversible, for activity is not regained upon cooling. This explains why most organisms are killed by a short exposure to high temperature: their enzymes are inactivated and they are unable to continue metabolism. A few remarkable exceptions to this rule exist: there are some species of primitive plants, blue-green algae, that can survive in hot springs, such as the ones in Yellowstone National Park, where the temperature is

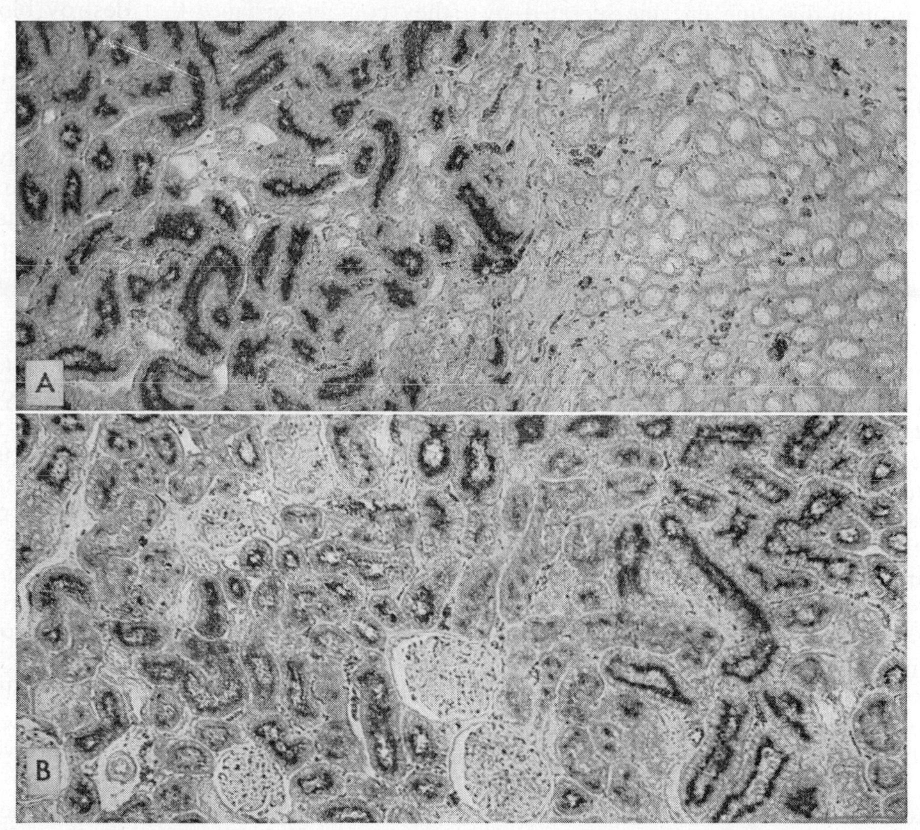

Figure 17. Histochemical demonstration of the location of an enzyme (alkaline phosphatase) within the cells of the rat's kidney. The tissue is carefully fixed and sectioned by methods which do not destroy the enzyme's activity. The tissue section is incubated at the proper *p*H with a naphthyl phosphate. Some hydrolysis of the naphthyl phosphate occurs wherever the phosphatase enzyme is located. The naphthol released by the action of the enzyme couples with a diazonium salt to form an intensely blue, insoluble azo dye which remains at the site of the enzymatic activity. The photomicrographs reveal the sites at which the azo dye is deposited—the sites of phosphatase activity. The "brush borders" of the cells of the proximal convoluted tubules in the kidney cortex (upper photograph) show thick deposits of azo dye, i.e., intense phosphatase activity. The tubules in the kidney medulla (the loop of Henle) have no activity (lower photograph). (Courtesy of Dr. R. J. Barrnett.)

almost 100° C. These algae are responsible for the brilliant colors in the terraces of the hot springs. Below the temperature where enzymes are inactivated (about 40° C.) the rates of most enzyme-controlled reactions, like other chemical reactions, are about doubled by each increase of 10° C.

Enzymes are not inactivated by freezing; their reactions go on very slowly or not at all at low temperatures but their catalytic activity reappears when the temperature is raised to normal.

Acidity. Enzymes are also sensitive to changes in *p*H, changes in acidity and alkalinity of the reaction medium. **Pepsin,** the protein-digesting enzyme secreted by the stomach lining, is remarkable in that it will work only in a very acid medium, and works optimally at *p*H 2. **Trypsin,** the protein-splitting enzyme secreted by the pancreas, is an example of an enzyme that works optimally in an alkaline medium, at about *p*H 8.5. The majority of intracellular enzymes have *p*H optima near neutrality and will not work at all in an acid or alkaline medium; stronger acids or bases will irreversibly inactivate them.

Concentration of Enzyme, Substrate and Cofactors. If the *p*H and temperature of an enzyme system are kept constant, and if an excess of substrate is present, the rate of the reaction is directly proportional to the amount of enzyme present. This is used to measure the amount of some particular enzyme present in a tissue extract. If the *p*H, temperature and enzyme concentration of a system are kept constant, the initial rate of reaction is proportional to the amount of substrate present, up to a limiting value. If the enzyme system requires a coenzyme or specific activator ion, the concentration of this substance may, under certain circumstances, determine the over-all rate of the reaction.

Enzyme Poisons. Certain enzymes are particularly susceptible to certain poisons —cyanide, iodoacetic acid, fluoride, lewisite, etc.—and even very low concentrations of these poisons inactivate the enzymes. Cytochrome oxidase, one of the respiratory enzymes, is especially sensitive to cyanide, and a person dies of cyanide poisoning because his cytochrome enzymes have been inactivated. One of the enzymatic steps in the breakdown of glucose is inhibited by fluoride and another one is inhibited by iodoacetic acid; biochemists have used these inhibitors as tools to investigate the properties and sequences of many different enzyme systems.

Enzymes themselves can act as poisons if they get into the wrong place. For example, as little as 1 milligram of crystalline trypsin will kill a rat if it is injected intravenously. Several types of snake, bee and scorpion venoms are harmful because they contain enzymes that destroy blood cells or other tissues.

43. RESPIRATION AND ENERGY RELATIONS

The term respiration was originally a synonym of breathing and simply meant inhaling and exhaling; the term "artificial respiration" reflects this usage. Later it came to mean the exchange of gases between a cell and its environment, the intake of oxygen and the release of carbon dioxide. More recently, as the details of cellular metabolism have become known, it has come to mean those enzymatic reactions in which oxygen is utilized and most of the energy is made available to the cell. The term "fermentation," originally defined by Pasteur as "life without air," is now used to refer to the chemical reactions which lead to the release of energy in the absence of oxygen.

The energy required by each cell is obtained by releasing the potential energy in some foodstuff molecule and converting it into a form that is usable by the cell for its various functions. The energy released is converted into "energy-rich" compounds, of which adenosine triphosphate, ATP, is of prime importance. The energy-rich compounds do not, in general, pass from one cell to another, but are formed and used within the same cell. Thus, the energy for muscle contraction is not released from food molecules in the stomach or liver and carried as "energy" to the muscle. Instead, food molecules such as glucose are carried by the blood to all the cells of the body. Within each cell, glucose is metabolized, energy is released, and energy-rich compounds are produced for use within that cell.

As biochemists have investigated the details of cellular metabolism in such diverse living things as green plants, rats, yeast, bacteria and sea urchins it has become apparent that the fundamental enzyme reactions in all cells are very similar. The steps by which glucose is broken down, called the **glycolytic cycle,** are the same not only in man and mouse, but in moss and mold as well. This similarity of enzyme systems may be due to the fact that all living things are related by evolutionary descent; the system of respiratory enzymes became established in the most primitive forms of life and has been transmitted to all the forms derived from it by evolutionary descent. Or, it may be that the types of chemical reactions that will support life are limited in number, and in the course of evolutionary history other methods have been tried and have failed.

Cells cannot use glucose as such; it must first be converted to glucose phosphate by an enzyme (glucokinase). Another enzyme converts fructose to fructose phosphate, and other enzymes convert starch or glycogen to glucose phosphate (Fig. 18). The glucose phosphate, whatever its source, and the fructose phosphate are converted to fructose diphosphate (a fructose molecule with two molecules of phosphate attached). Fructose diphosphate is then split by another enzyme into two molecules, each containing three carbons and a phosphate. A series of enzymes converts this 3-carbon phosphate molecule into pyruvic acid and in the course of these reactions, two high energy phosphate bonds are produced for each molecule of 3-carbon phosphate converted to pyruvic acid. This is only about 5 per cent of the total energy of a glucose molecule. Most of the energy is obtained in the oxidation of pyruvic acid to carbon dioxide and water. This is accomplished by a group of enzymes in the mitochondria which mediate a series of reactions called the **Krebs citric acid cycle,** studied by the English biochemist H. A. Krebs. The first step involves the conversion of pyruvic acid to an acetic acid-coenzyme A compound (2 carbons) and the combination of this acetyl coenzyme A with another substance, oxaloacetic acid (4 carbons), to form citric acid (6 carbons) (found in large amounts in citrus fruits). The enzymes in the mitochondria break down citric acid stepwise to oxaloacetic acid, which is then ready to combine with another molecule of acetyl coenzyme A. In this cycle (Fig. 18), carbon dioxide is given off by decarboxylase enzymes, hydrogen atoms are removed by dehydrogenases, the electrons of the hydrogen atoms are transferred by the electron-transmitting enzymes, the **cytochromes,** to oxygen which then unites with the hydrogen ions to form water. In the Krebs cycle and in the electron-transmitting enzymes about 36 more high energy phosphate bonds are made and the rest of the energy originally in the glucose molecule is made available as adenosine triphosphate—ATP—to run the many energy-requiring processes of metabolism. The Krebs cycle has been called the **intracellular energy wheel,** taking in acetyl coenzyme A, spewing off carbon dioxide and hydrogen, and trapping as ATP the energy released.

From this it can be seen that fermentation, the conversion of glucose to pyruvic acid in the absence of air, extracts only a small portion of the energy of the glucose molecule. When yeast cells carry on fermentation they convert the pyruvic acid to alcohol and carbon dioxide. When certain bacteria carry on fermentation, they convert the pyruvic acid to lactic acid. In the souring of milk by lactic acid bacteria, milk sugar, lactose, is converted via intermediates including glucose phosphate and 3-carbon phosphate to pyruvic acid, and finally pyruvic acid is converted to lactic acid.

The Krebs cycle is the final common pathway for the oxidation of fatty acids and amino acids as well as for carbohydrates and thus is the chief source of chemical energy for the cell. The fatty acids found in tissues are composed of long chains of carbon atoms—14-, 16-,

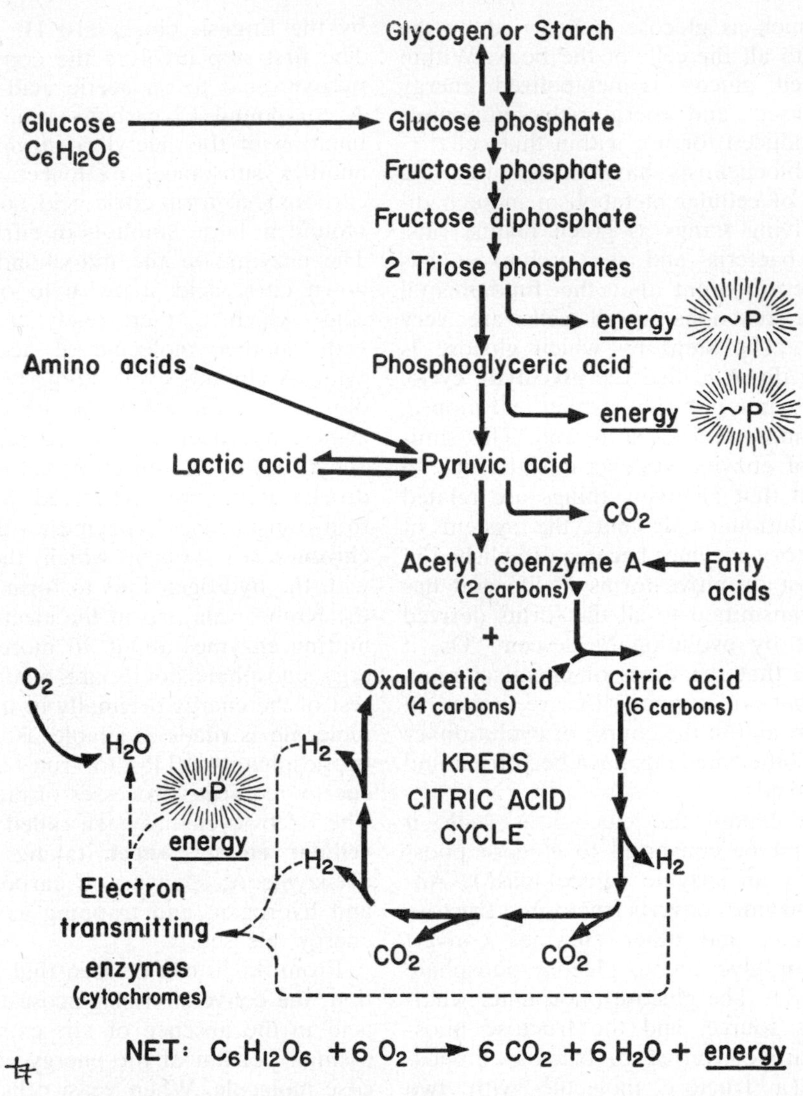

Figure 18. A diagram of some of the steps in the glycolytic cycle (glucose to pyruvic acid), the citric acid cycle, and the cytochrome system. The symbol ~P refers to energy-rich phosphate bonds such as those in adenosine triphosphate (ATP) which can yield their energy to drive cellular mechanisms. From this some appreciation can be gained of the tremendous oversimplification involved in writing the over-all formula for the oxidation of glucose given below.

and 18-carbon fatty acids are most common. These are broken down enzymatically to give 2-carbon compounds—the same acetyl coenzyme A formed from pyruvic acid. This is then metabolized in the Krebs cycle. Many amino acids can be converted enzymatically to pyruvic acid and others are converted to other substances in the Krebs cycle; thus, in a variety of ways, the breakdown products of proteins finally enter the Krebs cycle and are oxidized to yield carbon dioxide, water and energy.

The idea that we breathe in oxygen and breathe out carbon dioxide is so familiar that it is perhaps only natural to infer that the oxygen atoms in the molecules of carbon dioxide (CO_2) are the same ones that entered the body as gaseous oxygen. This is not true, however, as an examination of Figure 18 will make clear. The oxygen atoms that enter as oxygen unite with hydrogen to form water, and leave the body as water. The oxygen atoms that leave as carbon dioxide entered the body, by and large, in some substrate molecule such as glucose. The carbon and oxygen molecules are removed from the substrate molecule together, as carbon dioxide, by a process called **decarboxylation.**

Some interesting calculations of the over-all energy changes involved in metabolism in the human body have been made by E. G. Ball of Harvard University. Since the conversion of oxygen to water involves the participation of hydrogen atoms and electrons, the total flow of electrons in the human body can be calculated in terms of amperes. From the oxygen consumption of the body at rest, 264 cc. per minute, and the fact that each oxygen atom requires two hydrogen atoms and two electrons to form a molecule of water, Dr. Ball calculated that 2.86×10^{22} electrons are flowing from foodstuff via dehydrogenases and the cytochromes to oxygen each minute in all the body cells. Since an ampere equals 3.76×10^{20} electrons per minute, this current amounts to 76 amperes. This is quite a bit of current, for an ordinary 100 watt light bulb uses just a little less than 1 ampere. From the number of calories used per minute at rest, 1.27, Dr. Ball then calculated the number of watts of energy used per minute. This proved to be 88.7. Since, in electrical units, watts divided by amperes equals volts, 88.7 divided by 76 equals 1.17 volts. The body, then, utilizes energy at about the same rate as a 100 watt light bulb, but differs from it in having a much larger flow of electrons passing through a much smaller voltage change.

44. THE DYNAMIC STATE OF PROTOPLASM

The body of a plant or animal appears to be unchanging as days and weeks go by. It was inferred from this that the cells of the body, and the component molecules of the cells, are equally unchanging. Until about 20 years ago it was believed that the food molecules not used to increase the total mass of protoplasm are rapidly used to provide energy. It followed from this that two kinds of molecules could be distinguished: relatively static ones which constituted the cellular "machinery," and ones which were rapidly metabolized and thus corresponded to cellular "fuel."

However, experiments in which amino acids, fats, carbohydrates and water, each suitably labeled with some radioactive or heavy isotope, were fed to rats or other animals, have shown that protoplasm is in a much greater state of flux than was formerly believed to be true. Labeled amino acids are rapidly incorporated into body proteins, and labeled fatty acids are rapidly incorporated into fat deposits, even though there is no increase in the total amount of protein or fat. The proteins and fats of the body—even the substance of the bones—are constantly and rapidly being synthesized and broken down. In the adult the rates of synthesis and of degradation are essentially equal so that there is little or no change in the total mass of the body. Thus the distinction between "machinery" molecules and "fuel" molecules becomes less sharp, for some of the machinery molecules are constantly being broken down and used as fuel.

From the rate at which the labeled atoms are incorporated it has been calculated that one half of all the tissue proteins of man are broken down and rebuilt every eighty days. This is an average figure; some proteins are replaced much more rapidly, others more slowly. The proteins of the liver and of blood serum are replaced very rapidly, one half of them every ten days; muscle proteins are replaced more slowly, one half of them every 180 days. You are not the same person, chemically speaking, that you were yesterday!

45. SPECIAL TYPES OF METABOLISM

In addition to the general metabolic activities described above, certain animals and plants have special metabolic activities: green plants can photosynthesize; certain bacteria, molds and animals can produce light enzymatically; a few fish such as the electric eel can produce shocking amounts of electricity; certain plants produce a wide variety of substances—flower pigments, perfumes, many types of

Figure 19. A school of luminescent squid, *Watasenia scintillans.* (After Dahigren.)

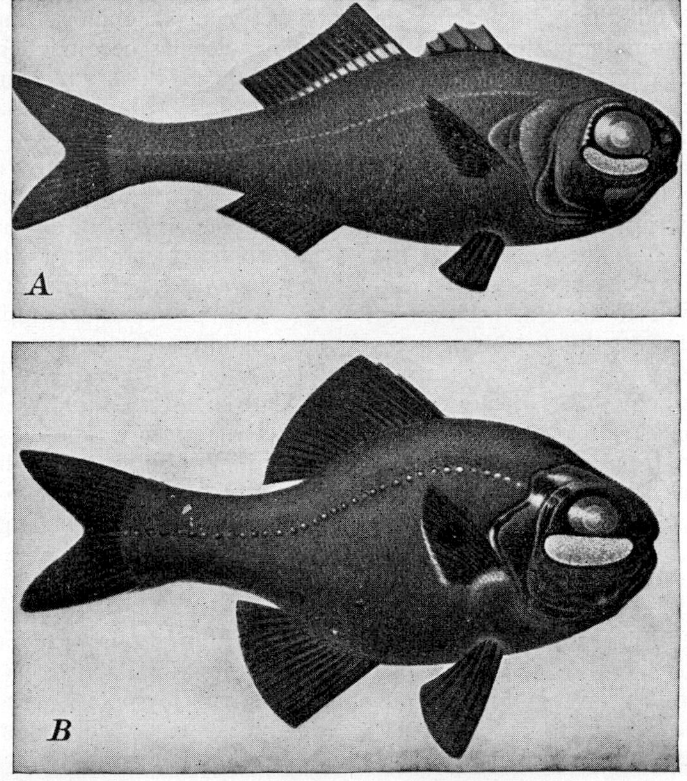

Figure 20. Two species of luminescent fish from the waters of the Malay Archipelago: *A, Anomalops katoptron,* and *B, Photoblepharon palpebratus.* The half-moon-shaped luminescent organs just ventral to the eyes are equipped with reflectors. (After Steche.)

drugs—and bacteria and molds, perhaps the best chemists of them all, can make everything from deadly poisons to antibiotics. The details of photosynthesis will be discussed in Chapter 7 and some aspects of bacterial metabolism in Chapter 9.

46. BIOLUMINESCENCE

Although the firefly and glowworm are the most conspicuous light-emitting organisms, a number of other animals and some bacteria and fungi, but no green plants, also have this ability (Fig. 19). Luminescent animals are found among the protozoa, sponges, coelenterates, ctenophores, nemerteans, annelids, crustaceans, centipedes, millipedes, beetles, echinoderms, molluscs, hemichordates, tunicates and fishes. There appears to be no single evolutionary line of luminescent forms; the ability to emit light has appeared independently a number of times. It is sometimes difficult to establish the fact that an organism is itself luminescent; in a number of instances, the light has been found to be emitted not by the organism but by bacteria. Several exotic East Indian fish have light organs under their eyes in which live luminous bacteria (Fig. 20). The light organ contains special long cylindrical cells, well supplied with blood vessels to supply the bacteria with adequate amounts of oxygen. The bacteria emit light continuously and the fish have a black membrane, like an eyelid, that can be drawn up over the light organ to turn off the light. No one knows how the bacteria come to collect in the fish's light organ, as they must in each newly hatched fish.

Some species, such as the shrimp, have accessory lenses, reflectors, and color filters with the light-emitting organ, so that the whole assembly is like a lantern.

The amount of light produced by some luminescent animals is amazing. Many fireflies produce as much light, in terms of lumens per square centimeter, as modern fluorescent lamps. Different animals emit lights of different colors—red, green, yellow, or blue. The "railroad worm" of Uruguay, the larva of a beetle, is remarkable in being able to produce two different colors: it has a row of green lights along each side of the body, and a pair of red lights at the head end. The light produced by luminescent organisms is entirely in the visible spectrum; no ultraviolet or infrared light is produced. Bioluminescence is sometimes called "cold light" since very little heat is given off.

The production of light is an enzyme-controlled reaction, the details of which differ in different species. Bacteria and fungi produce light continuously if oxygen is available. Most luminescent animals give out flashes of light only when their luminescent organs are stimulated. The names **luciferin** (the substrate) and **luciferase** (the enzyme) are given to the two components of the light-emitting system, but the luciferin and luciferase from one species of animal may be chemically quite different from those in another. The luciferin-luciferase reaction is a form of oxidation and can occur only in the presence of oxygen. It is possible to extract luciferin and luciferase from a firefly, mix the two in a test tube with added magnesium and adenosine triphosphate, and get light. The ATP supplies the energy for the reaction.

Two varieties of the fungus *Panus stipticus* are known, an American one which is luminescent and a European one which is not. When the two varieties are crossed, it is found that the ability to luminesce is inherited by a single dominant gene.

What good the ability to emit light may be to an organism is not clear. For deep sea animals, living in perpetual darkness, light organs would conceivably be useful, to enable members of a species to recognize each other or to serve as a lure for prey or a warning to would-be predators. It is known that the light emitted by fireflies serves as a signal to bring the sexes together for mating. The light emitted by bacteria and fungi probably serves no useful purpose to the organism, but is simply a by-product of oxidative metabolism, just as heat is a by-product of metabolism of other plants and animals.

QUESTIONS

1. Differentiate between fermentation and cell respiration.
2. In what ways are the metabolic processes of plant and animal cells similar? In what ways do they differ?
3. What factors regulate the rate of chemical reactions in a test tube? In a living cell?
4. Define an enzyme. In what ways are enzymes important in a bean plant? In the human body?
5. What additional substances may be necessary for an enzyme to function?
6. Does a yeast cell metabolize more efficiently in the presence or in the absence of oxygen? Explain.
7. What is the evidence that the chemical compounds of our cells are in a "dynamic state"? How would you set up an experiment to determine whether the proteins and fats of a bean leaf are also in a dynamic state?
8. Suppose you discovered a new bioluminescent worm. How could you prove that it was the worm itself and not some contaminating bacteria that was producing the light?
9. What functions may be served by bioluminescent organs?

SUPPLEMENTARY READING

A series of papers on the many different fields of biology in which enzymes play a role is found in *Enzymes: Units of Biological Structure and Function,* edited by O. A. Gaebler. Baldwin's *Dynamic Aspects of Biochemistry* gives a technical but extremely interesting discussion of the details of cellular metabolism. A description of the classic experiments demonstrating the rapid renewal of the chemical constituents of tissues is found in Rudolf Schoenheimer's *The Dynamic State of the Body Constituents.* A detailed discussion of our present knowledge of the phenomenon of bioluminescence is to be found in E. N. Harvey's *Bioluminescence* and, at a more elementary level, in his *Living Light.* L. J. Henderson, in his *Fitness of the Environment,* advanced the thesis that the environment had to have certain chemical and physical characteristics for life to develop.

Part Two

The World of Life: Plants

Chapter 6

Biologic Interrelationships

At first glance, the world of living things appears to be made up of a bewildering variety of plants and animals, all quite different and each going its separate way at its own pace. Closer inspection reveals, however, that all organisms, whether plant or animal, have the same basic needs for survival, the same problems of getting food for energy, getting space to live, producing a new generation, and so on. In solving these problems, plants and animals have evolved into a tremendous number of different forms, each adapted to live in some particular sort of environment. Each has become adapted not only to the physical environment—has acquired a tolerance to a certain range of moisture, wind, sun, temperature, gravity, and so on—but also to the biotic environment, all the plants and animals living in the same general region. The study of the interrelations between living things and their environment, both physical and biotic, is known as **ecology.** Living organisms are interrelated in two main ways, by evolutionary descent (see Chaps. 33–36) and ecologically. One organism may provide food or shelter for another, produce some substance harmful to the second, or the two may compete for food or shelter. To study ecology in detail requires knowledge of the structure and functions of a wide variety of plants and animals. In the latter part of this book, after we have considered some of the details of animal and plant physiology, heredity and evolution, we will come back to a discussion of ecology as one of the major unifying concepts of biology.

47. THE CLASSIFICATION OF LIVING THINGS

To deal with these myriad forms of life and describe their characters, biologists first had to name and classify them. Animals, for example, were classified by St. Augustine in the fourth century as useful, harmful or superfluous—to man. The herbalists of the Middle Ages classified plants as to whether they produced fruit, vegetables, fibers or wood. The classification of plants and animals by structural similarities was placed on a firm systematic basis by the Swedish biologist, Carl von Linné, or Linnaeus. He catalogued and described plants in *Species Plantarum* (1753) and animals in *Systema Naturae* (1758). With the acceptance of the theory of evolution, botanists and zoologists have tried to set up systems of classification based on natural relationships, putting into a single group those organisms which are closely related in their evolutionary origin. Since many of the structural similarities depend on evolutionary relations, the modern classification of organisms is similar in many respects to the one of Linnaeus based on logical structural similarities.

The unit of classification for both plants and animals is the **species.** It is difficult to give a definition of this term which will apply uniformly throughout the animal and plant kingdoms, but a species may be defined as a population of similar individuals, alike in their structural and functional characteristics, which in nature breed only with each other, and have a common ancestry.

Closely related species are grouped together in the next higher unit of classification, the **genus** (plural, *genera*). The scientific names of plants and animals consist of two words, the genus and the species, given in Latin. This system of naming organisms, called the **binomial** (two name) **system,** was first used consistently by Linnaeus. In accordance with it, the scientific name of the domestic cat, *Felis domestica,* applies to all the varieties of tame cats—Persian, Siamese, Manx, Abyssinian and plain tabby—since they all belong to the same species. Related species of the same genus are *Felis leo,* the lion, *Felis tigris,* the tiger, and *Felis pardus,* the leopard. The dog, which belongs to a different genus, is named *Canis familiaris.* Notice that in each of these names, the genus is given first, and is capitalized, the species name is given second and is not capitalized (some species names of plants are capitalized). The use of Latin rather than a modern language in naming species is a carryover from the days when Latin was the language of science.

Why, you may ask, bother to give Latin names to plants and animals? Why call a sugar maple *Acer* (maple) *saccharum* (sugar)? The primary reason is to be accurate and avoid confusion,* for in some parts of America this same tree is called either hard maple or rock maple. The tree most of us call white pine is *Pinus strobus,* but some people also refer to *Pinus flexilis* and *Pinus glabra* as white pines and still other people call *Pinus strobus* northern pine, soft pine or Weymouth pine. There are thousands of other instances of confusing common names, but these examples should make it clear that exact scientific names are really necessary and not simply scientific double-talk.

Just as several species may be grouped together to form a genus, a number of related genera constitute a **family,** and families may be grouped into **orders,** orders into **classes,** and classes into **phyla**

* The scientific names of organisms are not necessarily fixed forever, for new research sometimes shows that the relationships of certain genera and species differ from our previous ideas about them. This may necessitate changing the names of the organisms concerned, much to the distress of other biologists who are used to calling a particular animal by a particular scientific name. George Wald, in a discussion of Biochemical Evolution (*Trends in Physiology and Biochemistry,* Academic Press, New York, 1952) describes his difficulty in finding out which animals actually were referred to under the names *Cynocephalus mormon* and *Cynocephalus sphinx* in a paper published in 1904: "I have learned since that one is the mandrill, the other the guinea baboon. Since Nuttall wrote in 1904, these names have undergone the following vagaries. *Cynocephalus mormon* became *Papio mormon,* otherwise *Papio maimon,* which turned to *Papio sphinx.* This might well have been confused with *Cynocephalus,* now become *Papio sphinx,* had not the latter meanwhile been turned into *Papio papio.* This danger averted, *Papio sphinx* now became *Mandrillus sphinx,* while *Papio papio* became *Papio comatus.* All I can say to this is, thank heavens one is called the mandrill, the other the guinea baboon."

(singular, *phylum*). The phyla, then, are the large, main divisions of the plant and animal kingdoms, as the species are the fundamental small units. A complete classification of a white oak is phylum Tracheophyta, subphylum Pteropsida, class Angiospermae, subclass Dicotyledonae, order Sapindales, family Fagaceae, genus *Quercus,* species *alba.* Man is a member of the phylum Chordata, subphylum Vertebrata, class Mammalia, subclass Eutheria, order Primates, family Hominidae, genus *Homo,* species *sapiens.*

Many plants and animals fall into easily recognizable, natural groups, and their classification presents no difficulty, but other forms, which seem to lie on the borderline between two groups and have some characteristics in common with each, are difficult to assign to one or the other. The number and inclusiveness of the principal groups vary according to the basis of classification used and the judgment of the scientist making the classification. Some taxonomists like to group things together in already existing units; others prefer to establish separate categories for forms that do not fall naturally into one of the recognized classifications. Thus, different taxonomists consider that there are from ten to thirty-three animal phyla and from four to twelve plant phyla.

48. DISTINCTIONS BETWEEN PLANTS AND ANIMALS

Biologists ever since the time of Aristotle have divided the living world into two kingdoms, one of plants and one of animals. The word "plant" suggests trees, shrubs, flowers, grasses and vines—large and familiar objects of our everyday world. And "animal" suggests cats, dogs, lions, tigers, birds, frogs and fish. Further thought brings to mind such forms as ferns, mosses, mushrooms and pond scums, quite different but recognizable as plants, and insects, lobsters, clams, worms and snails that are definitely animals. But if you have ever had the pleasure of climbing over the rocky shore of the sea coast, looking at the organisms that cling to the rocks or live in a tide pool, you undoubtedly found some things that were difficult to recognize as animals or plants. And many of the one-celled organisms visible under the microscope cannot easily be assigned to the plant or animal kingdom. Some biologists, indeed, have suggested that a third kingdom be set up to include these intermediate forms. The single-celled *Euglena* moves around like an animal but contains chlorophyll and can carry on photosynthesis like a plant.

Fundamentally, plants and animals are alike in many ways; both are made of cells as structural and functional units and both (cf. Chap. 5) have many metabolic processes in common. But there are some obvious ways and some obscure ways in which they differ.

Plant cells, in general, secrete a hard outer cell wall of **cellulose** which encloses the living cell and supports the plant, while animal cells have no outer wall and hence can change their shape. However, there are some plants that have no cellulose walls, and one group of animals, the primitive chordates known as sea squirts or tunicates, do have cellulose walls around their cells.

Secondly, plant growth generally is indeterminate; that is, plants keep on growing indefinitely because some of the cells remain in an actively growing state throughout life. Many tropical plants grow throughout the year; those in the temperate regions grow primarily in the spring and summer. But although the cells of animals are replaced from time to time, the ultimate body size of most animals is established after a definite period of growth.

A third difference between the two types of living things is that most animals are able to move about, while most plants remain fixed in one place, sending roots into the soil to obtain water and salts and getting energy from the sun by exposing broad flat surfaces. A little thought will bring to mind exceptions to both of these distinctions.

The most important difference between the two is their mode of obtaining nourishment. Animals move about and obtain their food from organisms in the environment, but plants are stationary and manufacture their own food. Plants have the green pigment chlorophyll which enables them to carry on photosynthesis, to utilize light energy to split water and reduce carbon dioxide to carbohydrate. There are exceptions to this rule, too; fungi and bacteria are plants that lack chlorophyll. Thus, there are no hard and fast rules for distinguishing plants and animals.

49. MODES OF NUTRITION

Organisms that can synthesize their own food are said to be **autotrophic** (self-nourishing). Autotrophs can live if furnished only water, carbon dioxide, inorganic salts and a source of energy. There are two main types of autotrophs: (1) **photosynthetic,** green plants and purple bacteria, which get the energy for the synthesis of organic molecules from sunlight, and (2) **chemosynthetic,** a few bacteria which can get energy by oxidizing certain inorganic substances. Certain species of bacteria have enzyme systems that enable them to carry out particular oxidations as a source of energy. For example, nitrite bacteria (*Nitrosomonas*) oxidize ammonia to nitrites; nitrate bacteria (*Nitrobacter*) oxidize nitrites to nitrates; iron bacteria oxidize ferrous to ferric iron, and still other bacteria oxidize hydrogen sulfide to sulfates. The energy derived from these oxidations is utilized to synthesize all the organic materials necessary to maintain life and produce new protoplasm for growth. The nitrite and nitrate bacteria are also important in the cyclic use of nitrogen, for together they convert ammonia to a form readily used by green plants, nitrates.

Purple bacteria have pigments that can utilize the energy of sunlight to "fix" carbon dioxide as carbohydrate. However, in this reaction oxygen is not produced and the organisms use hydrogen sulfide, hydrogen, or certain organic substances as the source of hydrogen instead of the water used in green plant photosynthesis.

Organisms which cannot synthesize their own food from inorganic materials and therefore must live either at the expense of autotrophs or upon decaying matter, are called **heterotrophs,** and their mode of nutrition is called heterotrophic. All animals, fungi and most bacteria are heterotrophs.

There are several types of heterotrophic nutrition. When food is obtained as solid particles which must be eaten, digested and absorbed, as in most animals, the process is known as **holozoic** nutrition. Holozoic organisms must constantly find and catch other organisms; to do this animals have evolved a variety of sensory, nervous and muscular structures to find and catch food and several types of digestive systems to convert this food into molecules small enough to be absorbed.

Yeasts, molds and most bacteria cannot ingest solid food, but must absorb their required organic nutrients directly through the cell membrane. This type of heterotrophic nutrition is known as **saprophytic** nutrition. Saprophytes can grow only in places where there are decomposing bodies of animals or plants or masses of plant and animal by-products.

Yeasts are good examples of saprophytic plants. They need only inorganic salts, oxygen and some kind of sugar. From the last they can derive energy and can make all the other substances needed for life—proteins, fats, vitamins, and so on. When plenty of oxygen is available, the yeasts obtain energy by oxidizing the glucose completely to carbon dioxide and water via the Krebs citric acid cycle. But in the absence of enough oxygen to do this, they ferment the glucose and form alcohol and carbon dioxide. This, as we have seen, yields only about one twentieth as much energy as the complete oxidation of glucose, and therefore yeasts grow very slowly in the absence of oxygen.

The only practicable way in which ethyl alcohol can be obtained is through this action of yeast, and yeast is used in the manufacture of all alcoholic beverages. It is also used in industries which manufacture alcohol for use as a solvent, or as a raw material for the production of other substances such as synthetic rubber. Yeasts have a remarkable resistance to the toxic effects of alcohol, and continue to produce it until a concentration of about 12 per cent is reached, at which point the yeast organisms are inhibited. To produce beverages of higher alcoholic content, the wine or mash must be distilled. When yeast is mixed with bread dough, it ferments some of the sugar present; most of the alcohol is dissipated during the baking process, but the bubbles of carbon dioxide trapped in the dough expand and raise it, making the bread porous.

Another saprophyte, the baker's mold, *Neurospora,* requires the vitamin biotin in addition to salts and sugar; other saprophytes may require many different organic compounds.

A third type of heterotrophic nutrition, found among both plants and animals, is **parasitism.** A parasite lives in or on the living body of a plant or animal (called the **host)** and obtains its nourishment from it. Almost every living organism is the host for one or more parasites. A few plants, such as the mistletoe, are in part parasitic and in part autotrophic, for although they have chlorophyll and make some of their food, their roots grow into the stems of other plants, and they absorb some of their nutrients from their hosts.

Parasites may obtain their nutrients by ingesting and digesting solid particles or by absorbing organic substances through their cell walls from the body fluids or tissues of the host. Some parasites cause little or no harm to the host. Others produce definite diseases, harming the host by destroying cells or by producing toxic waste products. The pathogenic (disease-producing) parasites of man and other animals include viruses, bacteria, fungi, protozoa, and an assortment of worms. Most plants diseases are caused by parasitic fungi; a few are due to viruses, worms or insects.

Parasites are usually restricted to one or a few species of hosts; for example, most of the organisms that infect man will not infect other animals, or will infect only animals such as apes and monkeys which are closely related to man by evolution. A few human parasites do have wider host ranges and will infect more distantly related mammals and even birds. Saprophytes such as yeast or bread molds are easily grown in the laboratory, for they require only inorganic salts, glucose, and perhaps a vitamin or two, and will grow over a considerable range of temperature. But parasitic bacteria usually require a temperature near that of their normal host, and a complex medium containing sugars, amino acids and vitamins. Some, indeed, will grow only if provided with blood, liver or yeast extracts which contain one or more unknown growth factors. Finally, a few parasites, such as rickettsias and viruses, can only be grown in the presence of living cells. Until recently, the poliomyelitis virus could be grown only by infecting some experimental animal such as a monkey, but in 1952, John Enders of Harvard discovered methods by which the virus could be grown on tissue cultures of human cells. This work, for which Enders and his colleagues received a Nobel Prize in 1954, paved the way for the polio vaccine devised by Dr. Jonas Salk.

50. THE CYCLIC USE OF MATTER

The total amount of protoplasm that has ever lived in the billion or so years since life began on the earth is many times greater than the mass of the entire planet. According to the Law of Conservation of Matter, matter is neither created nor destroyed; obviously, matter must have been used over and over again in the course of time. The earth neither receives any great amount of matter from other parts of the universe nor does it lose significant amounts of matter to outer space. Each element, carbon, hydrogen, oxygen, nitrogen and the rest, is taken from the environment, made a part of living protoplasm and finally, perhaps by a quite circuitous route involving several other organisms, is returned to the environment to be used over again. An appreciation of the roles of green plants, animals and bacteria in this cyclic use of the elements can be gained from considering the details of the more important cycles.

The Carbon Cycle. Calculations show that there are about six tons of carbon (as carbon dioxide) in the atmosphere over each acre of the earth's surface. Yet each year an acre of plants, such as sugar cane, will extract as much as twenty tons of carbon from the atmosphere and incorporate it into the plant bodies. At an average rate of carbon dioxide utilization, the green plants would use up the entire

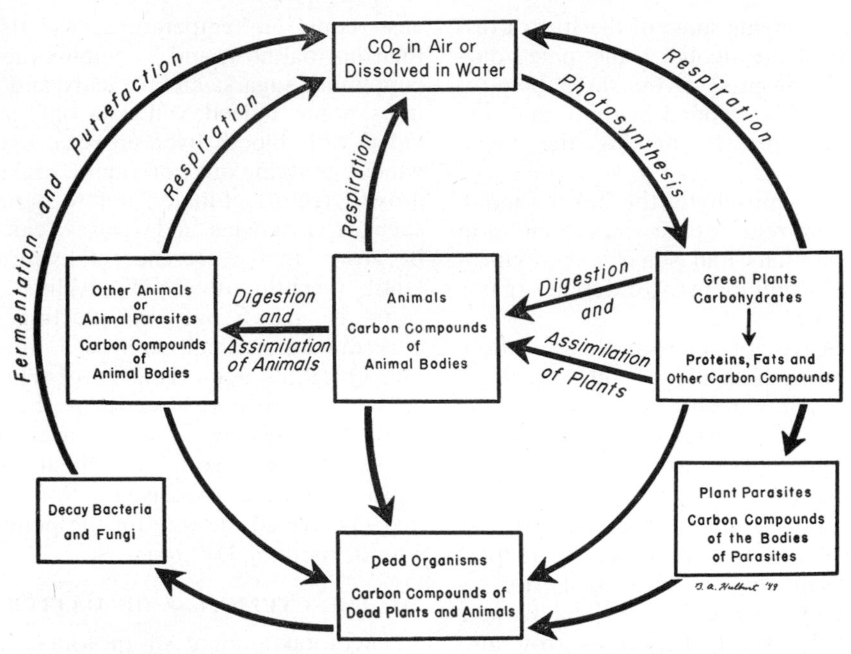

Figure 21. The carbon cycle in nature. See text for discussion.

atmospheric supply of carbon in about thirty-five years. Carbon dioxide fixation by bacteria and animals is another, but quantitatively minor, drain on the carbon dioxide supply. Carbon dioxide is returned to the air by respiration. Plants carry on respiration continuously, and green plant tissues are eaten by animals who, by respiration, return more carbon dioxide to the air. But respiration alone would be unable to return enough carbon dioxide to the air to balance that withdrawn by photosynthesis; vast amounts of carbon would accumulate in the dead bodies of plants and animals. The cycle is balanced by the decay bacteria and fungi which, by putrefaction and fermentation, break down the carbon compounds of dead plants and animals and convert the carbon to carbon dioxide (Fig. 21).

A group of investigators at the University of Notre Dame have been able to raise some bacteria-free animals in special incubators. This has been done to study problems such as the synthesis of vitamins by the bacteria normally present in the intestines. In time, these animals become old and die, but their bodies do not decompose—the carbon atoms have been withdrawn (temporarily) from the cycle.

When the bodies of plants are compressed under water they are not decayed by bacteria but undergo a series of chemical changes to form **peat,** then brown coal or lignite, and finally coal. The bodies of certain marine plants and animals may undergo somewhat similar changes to form oil. These processes remove some carbon from the cycle but eventually geologic changes or man's mining and drilling bring the coal and oil to the surface to be burned to carbon dioxide and restored to the cycle.

Most of the earth's carbon is present in the rocks as carbonates such as limestone and marble. The rocks are very gradually worn down and the carbonates in time are added to the carbon cycle. But other rocks are forming at the bottom of the sea from the sediments of dead animals and plants, so that the amount of carbon in the carbon cycle remains about the same.

The Nitrogen Cycle. The source of nitrogen for the synthesis of amino acids and proteins is the nitrates of the soil and water. These nitrates are absorbed by plants and synthesized into the amino acids and proteins of plant protoplasm (Fig. 22). The plants may be eaten by animals that in turn use the amino acids from the plant proteins in synthesizing their own protoplasm. When animals and plants die, the decay bacteria convert these nitrogen compounds into ammonia. Animals excrete several kinds of nitrogen-containing wastes—urea and uric acid as well as ammonia—and the decay bacteria convert these wastes to ammonia. Most of the ammonia is converted by nitrite bacteria to nitrites and this in turn is converted by nitrate bacteria to nitrates, thus completing the cycle. Denitrifying bacteria convert some of the ammonia to atmospheric nitrogen. Atmospheric nitrogen can be fixed, converted to organic nitrogen compounds such as amino acids, by some blue-green algae (*Nostoc*) and by the soil bacteria *Azotobacter* and *Clostridium*. Other bacteria of the genus *Rhizobium*, although unable to fix atmospheric nitrogen themselves, can do this when in combination with cells from the roots of legumes such as peas and beans. The bacteria invade the roots and stimulate the formation of root nodules, a sort of harmless tumor (Fig. 23). The combination of legume cell and bacteria is able to fix atmospheric nitrogen (something neither one can do alone) and for this reason legumes are often planted to restore soil fertility by increasing the content of fixed nitrogen. Nodule bacteria may fix as much as 5 pounds of nitrogen per acre per year, and free soil bacteria fix between 1 and 6 pounds per acre per year. Atmospheric nitrogen can also be fixed by electrical energy, lightning or man-made. Although 80 per cent of the atmosphere is nitrogen, no animals and only these few plants can utilize it in this form. When

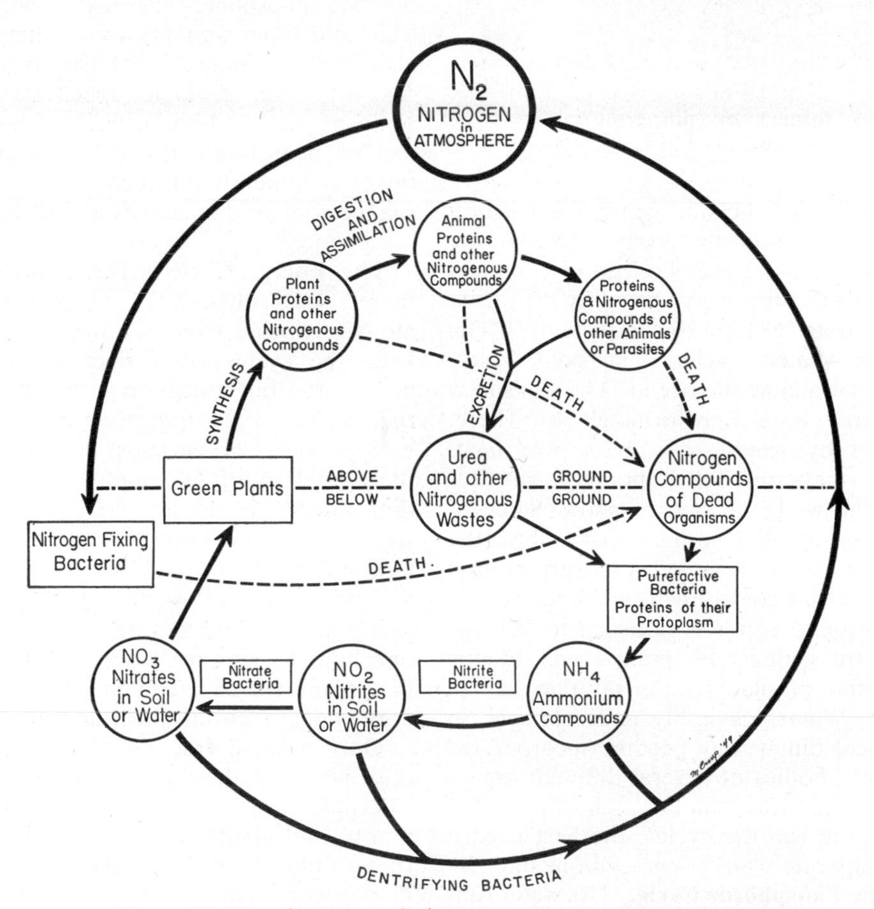

Figure 22. The nitrogen cycle in nature. See text for discussion.

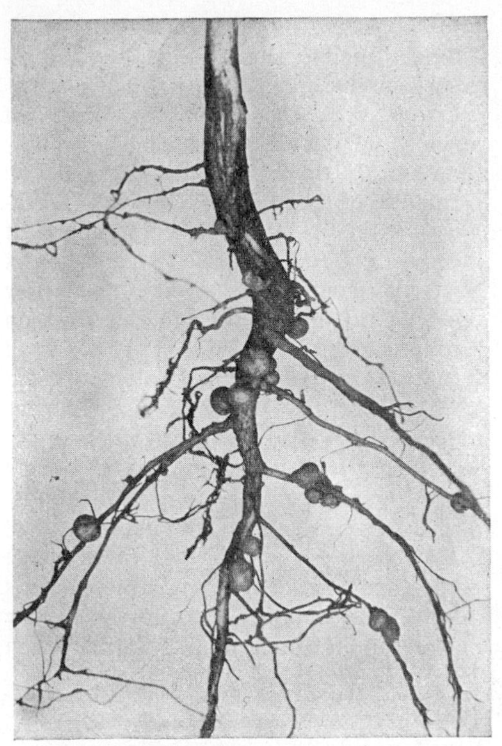

Figure 23. Roots of a soy bean plant with root nodules formed by nitrogen-fixing bacteria. (Weatherwax: Botany.)

the bodies of the nitrogen-fixing bacteria are acted upon by decay bacteria, the amino acids are metabolized to ammonia and this in turn is converted by the nitrite and nitrate bacteria to complete the cycle.

The Water Cycle. The great reservoir of water is the ocean. The sun's heat vaporizes water and forms clouds. These, moved by winds, may pass over land, where they are cooled enough to precipitate the water as rain or snow. Some of the precipitated water soaks into the ground, some runs off the surface into streams and goes directly back to the sea. The ground water is returned to the surface by springs, by pumps and by the activities of plants (transpiration; see p. 113). Water inevitably ends up back in the sea, but it may become incorporated into the bodies of several different organisms, one after another, en route. The energy to run the cycle—the heat needed to evaporate water—comes from sunlight.

The Phosphorus Cycle. As water runs over rocks it gradually wears away the surface and carries off a variety of minerals, some in solution and some in suspension. Some of these minerals, such as phosphates, sulfates, calcium, magnesium and others, are necessary for the growth of plants and animals. Phosphorus, an extremely important constituent of all protoplasm, is taken in by plants as inorganic phosphate and converted to a great variety of organic phosphate compounds (which are intermediates in carbohydrate and other metabolism). Animals get their phosphorus as inorganic phosphate in the water they drink or as inorganic plus organic phosphates in the food they eat. The phosphorus cycle is not completely balanced, for phosphates are being carried into the sediments at the bottom of the sea faster than they are being returned by the actions of marine birds and fish. Sea birds play an important role in returning phosphorus to the cycle by depositing phosphate-rich guano on land. Man and other animals, by catching fish, also recover some phosphorus from the sea. In time, geologic upheavals bring some of the sea bottom back to the surface as new mountains are raised and in this way minerals are recovered from the sea bottom and made available for use once more.

The Energy Cycle. The cycles of all these types of matter are closed: the atoms are used over and over again. To keep the cycles going does not require new matter but it does require energy, for *the energy cycle is not a closed one.* The Law of the Conservation of Energy, or the First Law of Thermodynamics, states that energy is neither created nor destroyed but only transformed from one kind to another (p. 38). However, the Second Law of Thermodynamics states that whenever energy is transformed from one kind to another, there is a decrease in the amount of useful energy; some energy is degraded into heat and dissipated.

Only a small fraction of the light energy reaching the earth is trapped; considerable areas of the earth have no plants, and plants can utilize in photosynthesis only about 3 per cent of the incident energy. This energy is converted into the potential energy of the chemical bonds of the organic substances made by the plant. When an animal eats a plant (or when bacteria decompose it) and these organic substances are oxidized, the energy liberated is just equal to the amount of energy used in synthesizing the substances (First Law of Thermodynamics), but some of the energy is heat and not useful energy (Second Law of Thermodynamics). If this animal in turn is eaten by another one, a further decrease in useful energy occurs as the second animal oxidizes the organic substances of the first to liberate energy to synthesize its own protoplasm.

Eventually, all the energy originally trapped by plants in photosynthesis is converted to heat and dissipated to outer space and all the carbon of the organic compounds ends up as carbon dioxide. The only important source of energy on earth is sunlight, energy derived from atomic disintegrations occurring at extremely high temperatures in the interior of the sun. When this energy is exhausted and the radiant energy of the sun can no longer support photosynthesis, the carbon cycle will stop, all plants and animals will die and organic carbon will be converted to carbon dioxide. There is no immediate cause for alarm, however; the sun will continue to shine for several billions of years!

51. ECOSYSTEMS

In the chapters that follow, we shall discuss the structures and activities of a variety of plants and animals. As we learn more about what each species is and does, it will be apparent that they are not independent of other living things but are interacting and interdependent parts of larger units. Ecologists use the term **ecosystem** to indicate a natural unit of living and nonliving parts which interact to produce a stable system in which the exchange of materials between living and nonliving parts follows a circular path. Ecosystems may be of different sizes: a lake, a tract of forest, or one of the cycles just discussed are examples of ecosystems. A balanced aquarium of tropical fish, green plants and snails is a very small ecosystem.

A classic example of an ecosystem is a small lake or pond (Fig. 24). The nonliving part of the lake includes the water, dissolved oxygen, carbon dioxide, inorganic salts such as phosphates and chlorides of sodium, potassium and calcium, and a multitude of organic compounds. The living part of the lake can be subdivided according to the functions of the organisms, i.e., what they contribute toward keeping the ecosystem operating as a stable, interacting whole. First of all, there are **"producer"** organisms—the green plants that can manufacture organic compounds from simple inorganic substances. In a lake there are two types of producers, the larger plants growing along the shore or floating in shallow water, and the microscopic floating plants, most of which are algae, that are distributed throughout the water, as deep as light will penetrate. These tiny plants, collectively known as **phytoplankton,** are usually not visible unless they are present in great abundance, when they give the water a greenish tinge. They are usually much more important as food producers for the lake than are the visible plants.

Secondly, there are **"consumer"** organisms, insects and insect larvae, crustacea, fish, and perhaps some fresh-water clams. Primary consumers are the plant eaters and secondary consumers are the carnivores that eat the primary consumers, and so on.

Finally, there are **"decomposer"** organisms, bacteria and fungi, which break down the organic compounds of dead protoplasm into inorganic substances that can be used by green plants.

Any ecosystem, no matter how large and complex, can be divided into these same major parts: producer, consumer, and decomposer organisms and nonliving components.

52. HABITAT AND ECOLOGIC NICHE

Two basic concepts useful in describing the ecologic relations of organisms are the habitat and the ecologic niche. The **habitat** of an organism is the place where it lives, a physical area, some specific part of the earth's surface, air, soil or water. It may be as large as the ocean or a

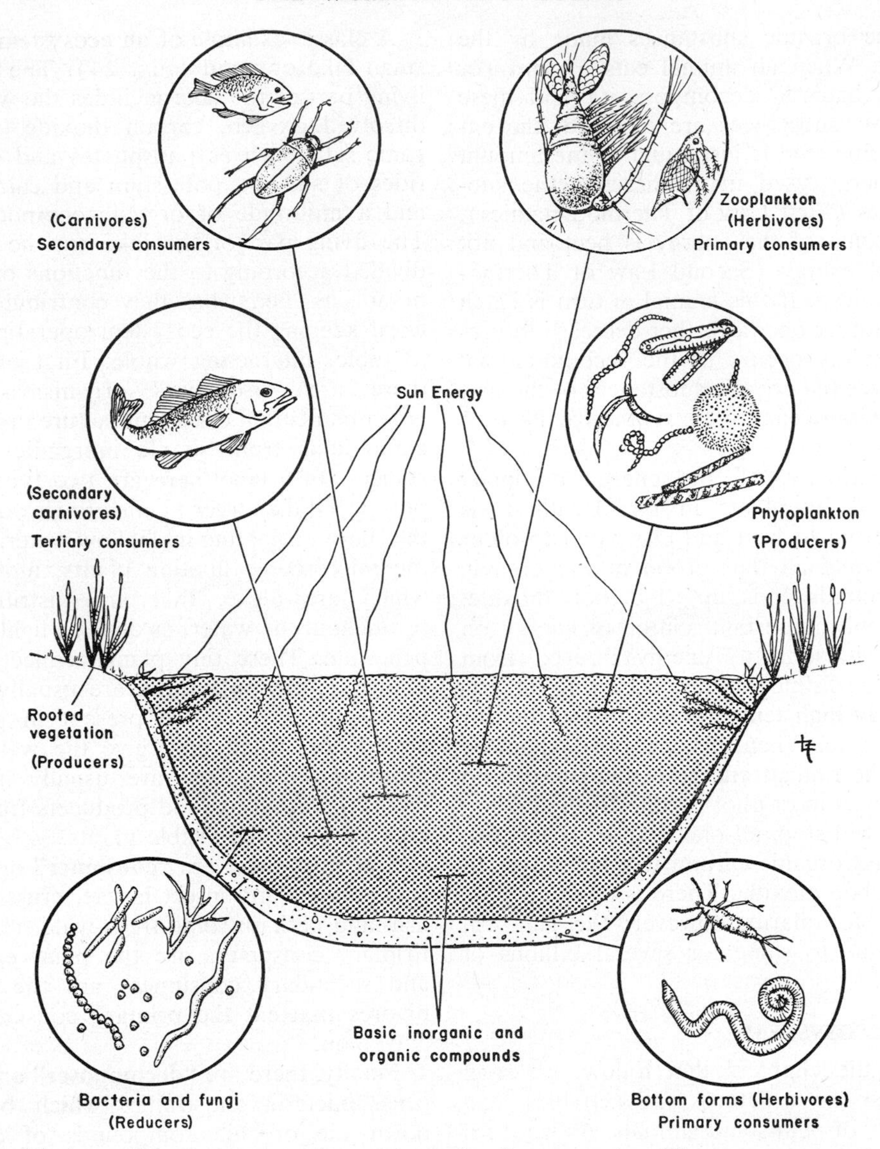

Figure 24. A small fresh water pond as an example of an ecosystem. The component parts, producer, consumer and decomposer or reducer organisms plus the nonliving parts are indicated.

prairie or as small as the underside of a rotten log or the intestine of a termite but it is always a tangible, physically demarcated region. More than one animal or plant may live in a particular habitat.

In contrast, the **ecologic niche** is the status of an organism within the community or ecosystem. It depends on the organism's structural adaptations, physiologic responses and behavior. E. P. Odum has made the analogy that the habitat is an organism's "address" and the ecologic niche its "profession," biologically speaking. The ecologic niche is an abstraction that includes all the physical, chemical, physiologic and biotic factors that an organism requires to live. To describe an organism's ecologic niche, one must know what it eats and what eats it, its range of movement and its effects on other organisms and on the nonliving parts of the surroundings.

The difference between these two concepts may be made clearer by an example. In the shallow waters at the edge of a lake one could find many different kinds of water bugs, all of which have the same habitat. But some of these, such as the "backswimmer" *Notonecta,* are predators, catching and eating other animals of about its size, while others, such as *Corixa,* feed on dead and decaying organisms. Each has quite a different role in the biologic economy of the lake and each thus occupies an entirely different ecologic niche.

A single species may occupy somewhat different niches in different regions, depending on such things as the available food supply and the number of competitors. Some organisms, such as animals with distinct stages in their life history, occupy in succession different niches. A tadpole is a primary consumer, feeding on plants, but an adult frog is a secondary consumer, feeding on insects and other animals.

53. TYPES OF INTERACTIONS BETWEEN SPECIES

Populations of different species may interact in several different ways. If neither population is affected by the presence of the other so that there is no interaction, the situation is termed **neutralism.** If each population is adversely affected by the other in its search for food, space, or some other need, the interaction is one of **competition.** If each population is benefited by the presence of the other but can survive in its absence, the relationship is termed **protocooperation.** But if each population is benefited by the other and cannot survive in nature without it, the relationship is known as **mutualism.** The term **commensalism** is given to a relationship in which one species is benefited and the second is not affected by existing together, and **amensalism** to the relationship where one species is inhibited and the second unaffected by the presence of the first. Finally, where one species adversely affects the second but cannot live without it the relationship is termed either **parasitism** or **predation.** The term parasitism is used if the one species lives in or on the body of the second and predation if the first species catches, kills and feeds upon the second.

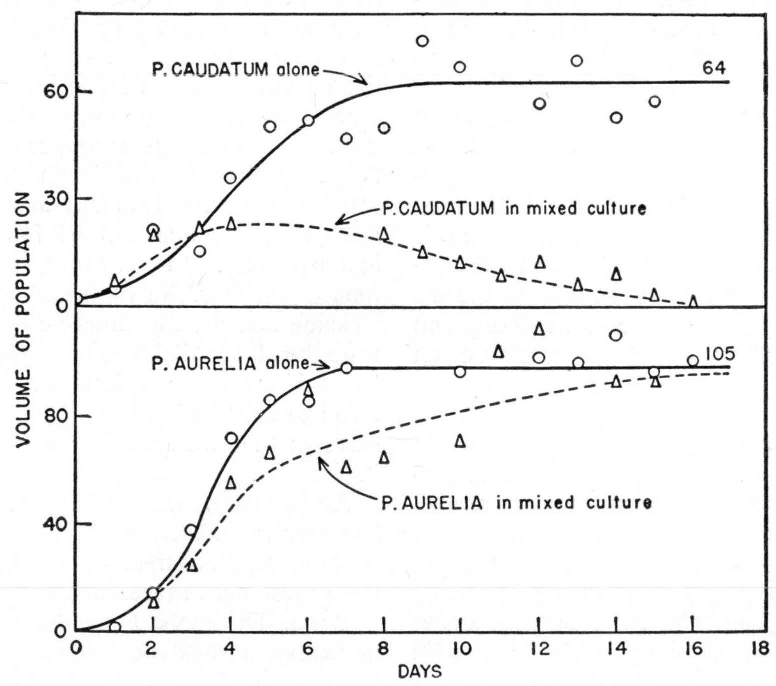

Figure 25. An experiment to demonstrate the competition between two closely related species of paramecia which have identical niches. When grown separately in controlled cultures with a fixed supply of food (bacteria), both *Paramecium caudatum* and *Paramecium aurelia* show normal S-shaped growth curves (solid lines). When grown together, *Paramecium caudatum* is eliminated (dotted lines). (After Gause, from Allee *et al.,* Principles of Animal Ecology.)

Competition. One of the classic examples of competition is provided by the experiments of Gause with populations of the protozoan *Paramecium* (p. 185). When either of two closely related species, *Paramecium caudatum* or *Paramecium aurelia,* was cultured separately on a fixed amount of food (bacteria) it multiplied and finally reached a constant level (Fig. 25). But when both species were placed in the same culture with a limited amount of food, only *Paramecium aurelia* was left at the end of sixteen days. The *P. aurelia* had not attacked the other species nor secreted any harmful substance, it simply had a slightly greater growth rate and had thus been more successful in competing for the limited food supply.

Commensalism. Commensalism, the relationship in which two species habitually live together, one of the species (called the commensal) deriving benefit from the association and the other being unharmed, is especially common in the ocean. Practically every worm burrow and shellfish contains some uninvited guests that take advantage of the shelter provided by the host organism but do it neither good nor harm. Some flatworms live attached to the gills of the horseshoe crab and get their food from the scraps of the crab's meals. They receive shelter and transportation from the host but apparently do not injure it. One of the more startling examples of commensalism is that of a small fish that lives in the posterior end of the digestive tract of the sea cucumber (an echinoderm), entering and leaving it at will. They are quickly eaten by other fish if removed from their sheltering host.

Protocooperation. If both species gain from an association, but are able to survive without it, the association is termed **protocooperation.** A number of kinds of crabs put coelenterates of one sort or another on top of their shells, presumably as camouflage. The coelenterates benefit from the association by obtaining particles of food when the crab captures and eats an animal. Neither crab nor coelenterate is absolutely dependent on the other.

Mutualism. When both species gain from an association and are unable to survive separately, the association is called **mutualism.** The classic example of this is the relationship of termites and their intestinal flagellates. Termites are famous for their ability to eat wood yet they have no enzymes to digest it. In their intestines, however, live certain flagellate protozoa that do have the enzymes to digest the cellulose of wood to sugars. Although the flagellates use some of their sugar for their own metabolism, there is enough left over for the termite. Termites cannot survive without their intestinal inhabitants; newly hatched termites instinctively lick the anus of another termite to obtain a supply of flagellates. Since a termite loses all its flagellates at each molt, termites must live in colonies so that a newly-molted individual will be able to get a new supply of flagellates from its neighbor. The flagellates also benefit by this arrangement: they are supplied with plenty of food and the proper environment; in fact, they can survive only in the intestines of termites.

Other examples commonly called mutualism are the relations of nitrogen-fixing bacteria and legumes (p. 71), and algae and fungi in lichens (p. 145). Since both the bacteria and the legume can survive separately, their association should perhaps be termed protocooperation. The fungus partner of the lichen association usually cannot survive alone, but the algae are all species that can be found living independently. In a sense, then, the fungus is a parasite of the alga; indeed, in some lichens, the fungi actually penetrate the algal cells. The alga also derives benefit from the association, for it is enabled to exist in places such as bare rock surfaces that would otherwise be denied it.

Amensalism. Examples of **amensalism** are provided by many different mold-bacteria relationships. Many molds produce substances that inhibit the growth of bacteria. The mold *Penicillium* produces **penicillin,** a substance which will inhibit the growth of a variety of bacteria. The mold presumably benefits by having a greater food supply when the competing bacteria have been removed. Man, of course, takes good advantage of this and cultures *Penicillium* and other antibiotic-producing molds in huge quantities to obtain bacteria-inhibiting substances to combat bacterial infections. The use of these bacteria-inhibiting agents has had the unexpected effect of increasing the incidence of fungus-induced diseases; these are apparently kept in check normally by the presence of bacteria. When the bacteria are killed off by antibiotics, pathogenic fungi have a golden opportunity.

We would be quite wrong if we assumed that the host-parasite or predator-prey relationship was invariably harmful to the host or prey as a species. This is usually true when such relationships are first set up, but in time, the forces of natural selection tend to decrease the detrimental effects. If the detrimental effects continued, the parasite would eventually kill off all the hosts and, unless it found a new species to parasitize, would die itself.

A striking example of the result of upsetting a long-standing predator-prey relationship occurred on the Kaibab plateau, on the north side of the Grand Canyon of the Colorado River. In this area in 1907 there were some 4,000 deer and a considerable population of their predators, mountain lions and wolves. When a concerted effort was made to "protect" the deer by killing off the predators, the deer population increased tremendously. By 1925 there were some 100,000 deer on the plateau, far too many for the supply of vegetation. The deer ate everything in reach, grass, tree seedlings and shrubs, and there was marked damage to the vegetation. There was no longer enough vegetation to support the deer population over the winter, and in the next two winters vast numbers of deer starved to death. Finally the deer population fell to about 10,000. The original predator-prey interaction had been maintaining a fairly stable equilibrium, with the number of deer being kept at a level within the available food supply.

Studies of hundreds of different examples of parasite-host and predator-prey interrelations show that in general, where the associations are of long standing, the long-term effect on the host or prey is not very detrimental and may even be beneficial. Conversely, newly acquired predators or parasites are usually quite damaging. The plant parasites and insect pests that are most troublesome to man and his crops are usually those which have recently been carried into some new area and thus have a new group of organisms to attack.

In the following survey of plants and animals the student should keep in mind not only how they are classified, from phyla to species, but where they live and, even more important, what they do—in what ways they contribute to the ecosystems of which they are members. In this way one can gradually build up a picture of the world of animals and plants, all related closely or distantly by evolutionary descent, and bound together in a variety of interspecific interactions. Without these unifying concepts of evolutionary relationship and ecologic interrelations, the world of living things is simply a bewildering hodgepodge.

QUESTIONS

1. How would you define a species? A class? A phylum?
2. Distinguish between natural and artificial systems of classification.
3. What are the prime differences between plants and animals? Name some organisms which are difficult to assign to either kingdom.
4. What is meant by an autotrophic organism? Which organisms are autotrophic?
5. Describe the several types of heterotrophic nutrition and give an example of each.
6. How do saprophytes differ from parasites?
7. Are parasites necessarily pathogenic?
8. Why are parasites usually restricted to one or a few host species?
9. In what ways are parasites adapted for their parasitic life?
10. Describe the role of bacteria in (*a*) the carbon cycle and (*b*) the nitrogen cycle.
11. How does the phosphorus cycle differ from the carbon and nitrogen cycles?
12. Explain why all life on the earth is dependent on a continued supply of sunlight.
13. Define the terms ecosystem, habitat and ecologic niche.
14. Discuss an aquarium as an example of an ecosystem.
15. Discuss the several types of interactions between species and give an example of each.

16. Differentiate between commensalism, protocooperation and mutualism.

SUPPLEMENTARY READING

E. P. Odum's *Fundamentals of Ecology* discusses the cycles of substances in nature and the types of relations between different species of organisms. For more detailed discussions of plant ecology see Weaver and Clement's *Plant Ecology,* and of animal ecology see Allee, Emerson, Park, Park and Schmidt's *Principles of Animal Ecology.*

Chapter 7

General Properties of Green Plant Cells

THE PRIMARY producers of the living world are the green plants. Because of the unique properties of chlorophyll, the pigment that gives them their green color, they are able to utilize the radiant energy of sunlight to synthesize energy-rich compounds such as glucose from water and carbon dioxide, substances which have no physiologic value as fuels. This process, called **photosynthesis,** is the only significant way in which energy from the sun is captured and made available for life on this planet. From the discussion of the energy cycle (p. 72) it is evident that without green plants life, except for a few chemosynthetic bacteria, would disappear from the earth. The details of the photosynthetic process are much the same, if not identical, in all green plants from the smallest single-celled alga to the largest redwood tree. The nature of many of the chemical reactions involved in the process have been clarified by the work of Calvin at Berkeley and Gaffron at Chicago. Most of their research has been done with the unicellular green algae *Chlorella* or *Scenedesmus,* but comparable experiments with the leaves of bean, barley, tobacco and cantaloupe plants have shown that the same series of reactions takes place in higher plants.

54. PHOTOSYNTHESIS

A tremendous amount of carbon (estimates place the actual amount at about 300 billion tons) is made into organic material each year by green plants. Land plants synthesize about one tenth of the total and the marine plants, mostly microscopic algae, manufacture the rest.

The first studies of photosynthesis date back to 1630, when van Helmont, a Flemish botanist, showed that plants make their own organic materials and do not get them from the soil. He weighed a pot of soil and the willow tree planted in it, and showed that the tree gained 164 pounds in five years but the soil weighed only two ounces less. Van Helmont concluded that the rest of the substance came from the water he had added; we now know that carbon dioxide removed from the air contributed largely to the plant material synthesized. Joseph Priestley showed in 1772 that a sprig of mint would "restore" air that had been "injured" by the burning of a candle. Seven years later Jan Ingenhousz showed that vegetation could restore bad air only if the sun was shining and that the ability of the plant to restore air was proportional to the clearness of the day and to the exposure of the plant to the sun. In the dark, plants give off air "hurtful to animals."

The next major step in understanding photosynthesis came in 1804, when de Saussure weighed both air and plant before and after photosynthesis and showed that the increase in the dry weight of the plant was greater than the weight of carbon dioxide removed from the air. He concluded that the other substance contributing to the gain in weight was water. Thus, 150 years ago, the broad outlines of the photosynthetic process were understood:

Carbon dioxide + water + light energy
→ oxygen + organic material

Ingenhousz concluded from his experiments that photosynthesis occurred only in the leaves and green stalks of plants; flowers, fruits and roots "rendered air noxious to animals." Since then it has been found that only the green parts of leaves can photosynthesize. For example, in the leaves of variegated ivy, which have a pattern of green and white areas, photosynthesis occurs only in the green and not in the white parts of a single leaf. Roots, which ordinarily are not exposed to light and do not contain chlorophyll, nevertheless can develop it if exposed to light and then can carry on photosynthesis.

Microscopic examination of a bit of leaf shows that chlorophyll does not occur everywhere within the protoplasm but is confined to small bodies called **chloroplasts.** These bodies have the ability to grow and divide to form two daughter chloroplasts. The electron microscope has shown that within the chloroplasts are still smaller bodies, known as **grana,** which contain the chlorophyll (Fig. 26).

The chlorophyll molecule is made of many atoms of carbon and nitrogen joined together in a complicated ring. In the center of the ring, bound to the nitrogen atoms, is an atom of magnesium. The chlorophyll molecule is strikingly like the red pigment hemoglobin found in red blood cells but the latter has an atom of iron instead of magnesium in the center of the ring. Both chlorophyll and hemoglobin are joined to protein molecules, and are physiologically active only in this form. Chlorophyll differs from hemoglobin in having a substance called **phytol** (a long chain alcohol) joined to the ring. In the higher plants there are two forms of chlorophyll, called a and b; chlorophyll b has one oxygen more and two hydrogens less than chlorophyll a. Some algae have chlorophyll a and b, others have slightly different forms, c, d and e.

Plants contain, in addition to chlorophyll, many pigments which give them their great variety of colors. Some of these may also absorb light energy and transfer it to chlorophyll to be used in photosynthesis. Most plants have a deep orange pigment, **carotene,** which can be converted in the animal body to vitamin A, and a yellow pigment, **xanthophyll.** The red algae have a red pigment, **phycoerythrin,** which is more effective in photosynthesis in these red algae than are the chlorophylls present.

A balanced equation for the synthesis of glucose in photosynthesis is as follows:

$$6\ H_2O + 6\ CO_2 + \text{energy} \rightarrow C_6H_{12}O_6 + 6\ O_2$$

To make 180 gm. of glucose (1 gram-molecular weight) requires 673 Calories of light energy. In this reaction, light energy, which is otherwise useless to living things, is transformed into potential chemical energy in the glucose molecule. The process also replenishes the supply of oxygen in the atmosphere. Chlorophyll functions in the reaction by absorbing the radiant energy of sunlight and making it available to the reacting chemicals. Ordinarily, only about 3 per cent of the absorbed light energy is converted to chemical energy in the reaction, but with weak light and ample amounts of carbon dioxide the amount of energy converted may reach 30 per cent. There is no decrease in the concentration of chlorophyll in the leaves after intense photosynthesis; the pigment is a true catalyst and is not used up in the process.

Land plants absorb the water necessary for photosynthesis through their roots;

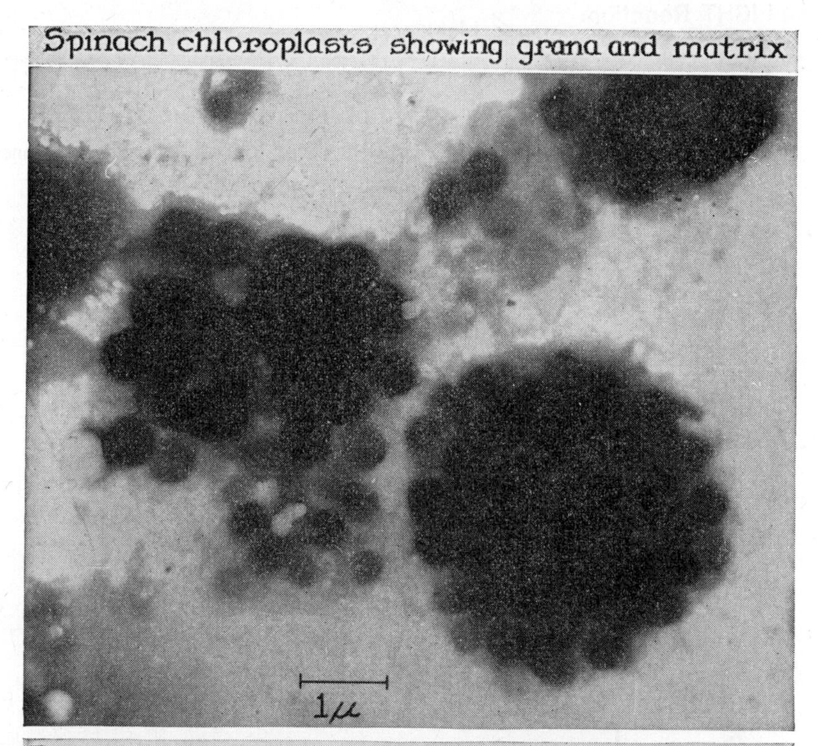

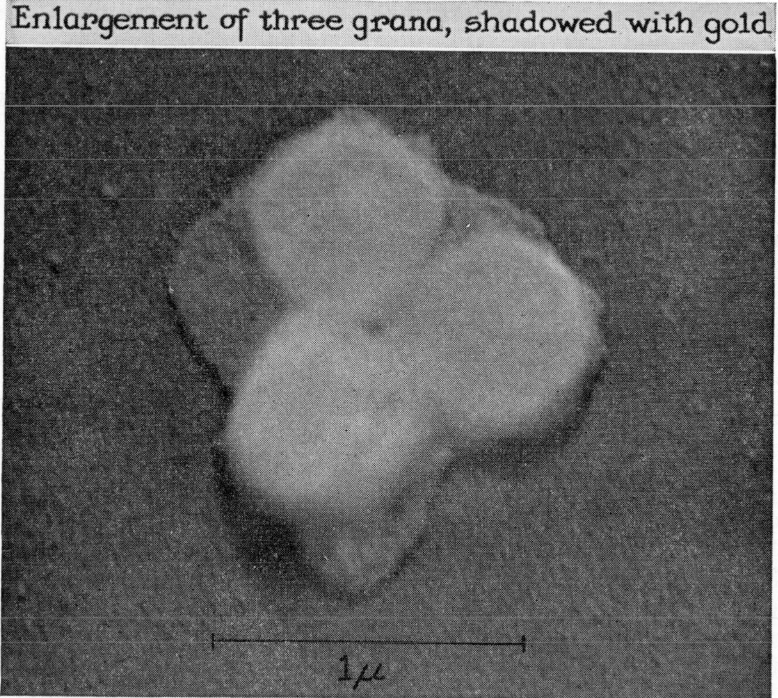

Figure 26. Electron micrographs of individual chloroplasts from spinach leaves, showing the tiny, spherical bodies, called grana, within each chloroplast. (Courtesy of Dr. S. Granick, Rockefeller Institute for Medical Research.)

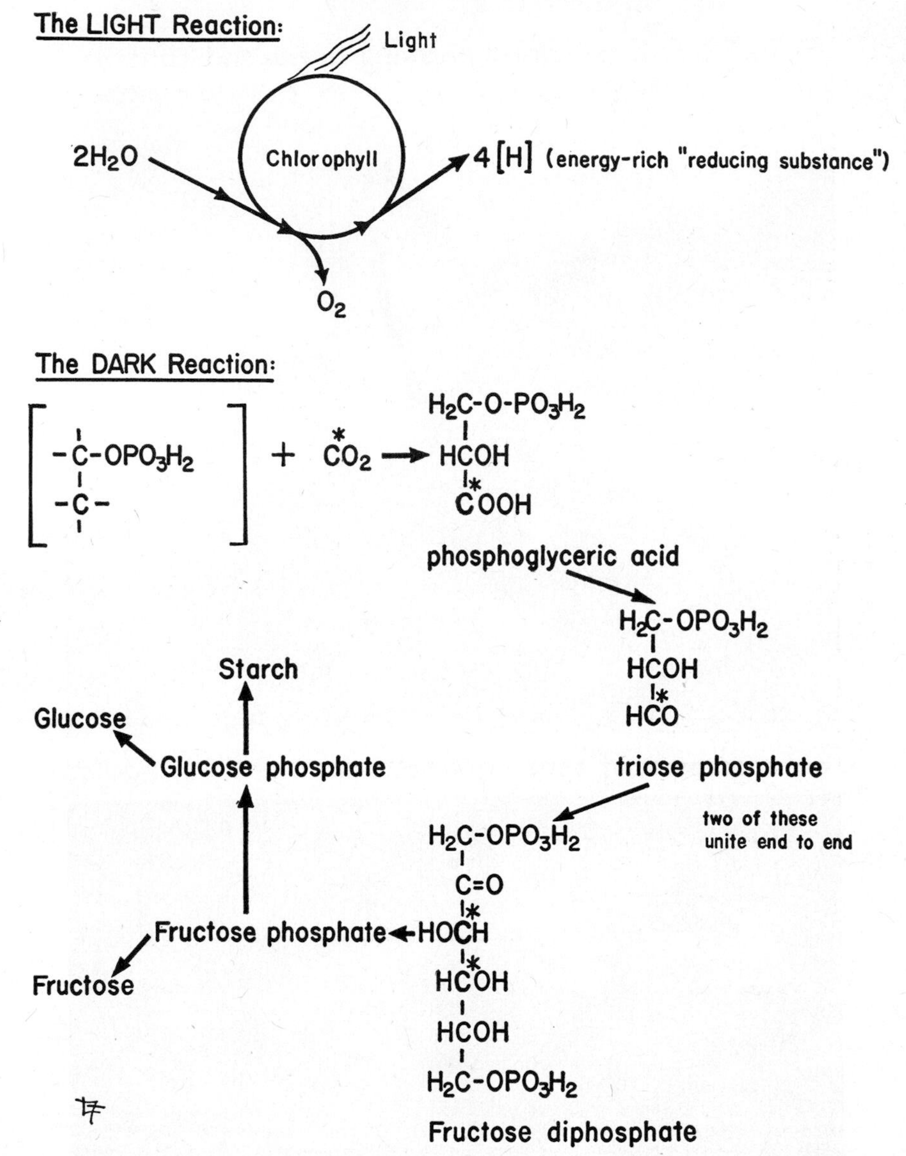

Figure 27. A diagram of some of the chemical reactions involved in photosynthesis. In the "light reaction" light energy is used to split water to yield oxygen and an energy-rich reducing substance, [H]. The energy from this substance is utilized to drive the carbon dioxide fixation reactions of the "dark reaction." Note the similarity between the steps from phosphoglyceric acid to glucose and the steps in glycolysis (Fig. 18).

aquatic plants receive it by diffusion from the surrounding medium. The necessary amount of carbon dioxide diffuses into the plant by way of small holes (stomata) in the surface of the leaves. Since carbon dioxide is steadily used up in photosynthesis, its concentration within the cell is always slightly lower than that in the atmosphere and carbon dioxide constantly diffuses into the leaf cells. As oxygen is liberated in the process, the intracellular oxygen concentration increases and oxygen tends to diffuse out of the cell. The glucose formed also tends to diffuse away from the site of formation to other regions of lower concentration.

Plants need vast quantities of air to carry on photosynthesis, for air contains only 0.03 per cent carbon dioxide. Thus, 10,000 cubic feet of air are needed to yield 3 cubic feet of carbon dioxide. Plants generally grow better and faster in air with a higher carbon dioxide content, and some greenhouses are artificially maintained with an atmosphere containing from 1 to 5 per cent of it. There are a number of reasons for believing that in past geologic ages the air contained more carbon dioxide than it does at present, and that plants grew more rapidly. The phenomenal growth of the giant tree ferns of some 250,000,000 years ago, whose remains were converted into coal, probably resulted from a high concentration of carbon dioxide in the atmosphere.

The actual chemical process involved in photosynthesis is complex and involves a number of different steps (Fig. 27). It has been known since 1905 that the photosynthetic process is composed of two parts, one of which can occur in the dark, and is therefore called the "dark reaction," and one of which can occur only in the light.

The light reaction occurs first and consists of the utilization of light energy to split water to yield oxygen (Fig. 28), and an unstable, intermediate, hydrogen-containing substance:

$$2H_2O + \text{light} \rightarrow 4[H] + O_2$$

The hydrogen is not released as a gas but is taken up by other substances, called hydrogen acceptors, which pass it on in a bucket brigade fashion so that eventually it is combined with the carbon dioxide.

The light reaction thus splits off oxygen from water and passes the hydrogen on to certain unstable intermediate compounds. The dark reaction follows and, by a series of steps, combines these intermediates with carbon dioxide to form an organic molecule. The dark reaction occurs much more slowly than the light one and limits the rate at which the whole reaction can occur.

By using isotopic tracers, investigators have shown that the oxygen released in photosynthesis is derived entirely from the decomposition of water; none of it comes from carbon dioxide. Ruben, of the University of California, used either water or carbon dioxide labeled with heavy oxygen (O^{18}) to detect the source of the oxygen

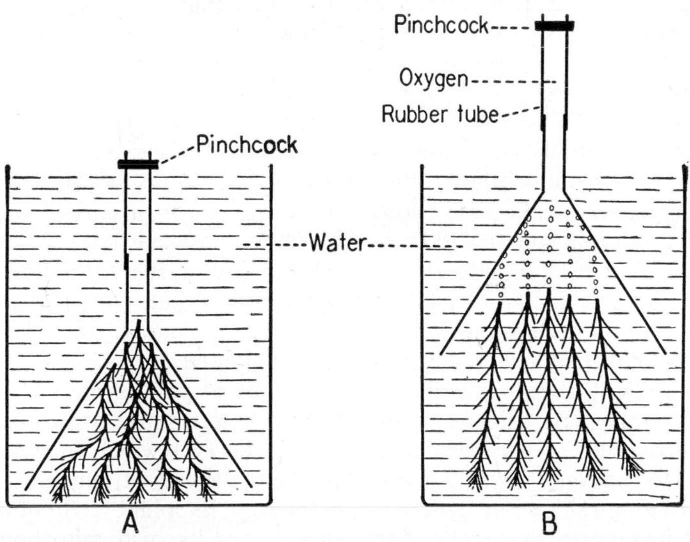

Figure 28. Experiment to show that oxygen is released in photosynthesis. *Elodea* or some other aquatic plant is placed in a beaker of water and covered by a glass funnel (*A*). The funnel is then raised as in (*B*) and the plant is exposed to light. The bubbles of gas given off collect in the stem of the funnel. To test the gas for oxygen a glowing splinter of wood is placed over the end of the stem and the pinchcock is slowly opened. If oxygen is present the splinter will burst into flame. (Weatherwax: Botany.)

given off by photosynthesizing algae. In each case the oxygen given off had the same content of heavy oxygen as that of the water; thus, all the oxygen given off comes from H_2O. By analogous experiments, it has been found that the carbon of the glucose comes, of course, from the carbon dioxide, the oxygen of the glucose comes from carbon dioxide and the hydrogen of glucose comes from water.

In their attempts to trace the pathway of carbon in photosynthesis, investigators such as Calvin and Gaffron have incubated suspensions of algae in the presence of carbon dioxide labeled with radioactive carbon (C^{14}). In their first experiments, the algae were incubated for periods of one to five or ten minutes, then the algae were quickly killed with boiling alcohol and by a variety of techniques the individual chemicals of the cell were isolated. With incubations of this length—even with one minute experiments—many of the substances isolated were found to contain radiocarbon: sugars, amino acids and intermediate substances. Finally, by decreasing the incubation time to five seconds or less, they were able to show that 80 per cent or more of the radiocarbon was in **phosphoglyceric acid,** a substance with 3 carbon atoms. They now believe that carbon dioxide is added on to some 2-carbon substance to give the phosphoglyceric acid. With a five second incubation, the only substances that contain detectable amounts of radiocarbon, other than phosphoglyceric acid, are triose phosphate, fructose diphosphate and fructose phosphate, all substances which are intermediate in the breakdown of glucose to pyruvic acid (p. 59). Phosphoglyceric acid itself is an intermediate in the glycolytic cycle. Thus it seems reasonable that, once phosphoglyceric acid has been made in photosynthesis, the reactions leading to the formation of glucose and other sugars are simply the glycolytic enzymes running in reverse (Fig. 27). The hydrogen-containing substance formed by the light reaction supplies the driving force for the cycle by reducing organic acids to aldehydes and alcohols. Research is now concentrating on the nature of the 2-carbon substance that unites with carbon dioxide to form phosphoglyceric acid, and the metabolic pathway by which this 2-carbon substance is made.

Until recently, photosynthesis was thought to be the only chemical process in nature by which carbon dioxide is converted to organic substances. But in 1935 it was discovered that bacteria, too, under certain conditions, can convert carbon dioxide into organic materials such as glucose. Shortly after that the same chemical process was demonstrated in animal tissues such as the liver and kidney. In fact, more than half a dozen different carbon dioxide-fixing reactions are now known to occur in animal tissues. Animal cells, however, must obtain energy for these carbon dioxide-fixing reactions from some energy-yielding reaction such as the oxidation of glucose; only plant cells can trap light energy to drive the reaction. The unique characteristic of photosynthesis, then, is the light-induced splitting of water to give gaseous oxygen and a hydrogen-containing reducing substance.

A striking example of the fundamental biochemical unity of living matter was provided by the experiments of Ochoa and Vishniac at New York University. They prepared grana from spinach chloroplasts and mitochondria from ground up rat liver cells. When these two were mixed and illuminated the mixture could function and produce high energy phosphate bonds, the energy of which was available for the fixation of carbon dioxide. These investigators used radioactive phosphorus, P^{32}, to follow the formation of the high energy phosphate bonds of ATP (adenosine triphosphate). In this way a clue was obtained to one of the important problems of photosynthesis, how light energy is trapped and converted to chemical energy (ATP) which can then be utilized in the synthesis of sugar and other substances. The plant cell, of course, normally uses its own mitochondria. Ochoa and Vishniac were able to prepare mitochondria from mung beans that worked in this system when mixed with spinach grana. The fact that enzymes from rat liver mitochondria can be substituted for those from plant cell mitochondria is good evidence that the two are very similar, if not identical. It also provides a glimpse of just how similar the intimate details of plant and animal cellular metabolism may be.

55. SYNTHESIS OF OTHER ORGANIC SUBSTANCES

The first result of photosynthesis is the formation of phosphoglyceric acid. From this the plant cell can synthesize, via a long series of intermediate organic acids, glucose and fructose and then combine the two to make sucrose (our table sugar) (Fig. 27). Or it may combine many molecules of glucose to make starch for storage or cellulose. Starch, unlike glucose or sucrose, is insoluble in water and will not diffuse away from the storage site. Some plants store the carbohydrate made by photosynthesis mostly as starch, others store it mostly as sugar.

Other enzymes of the plant cell convert the phosphoglyceric acid, via another series of intermediates, into glycerol and fatty acids which can then be combined to form fats. Yet other enzymes convert the carbon chain of phosphoglyceric acid into the different amino acids and these can then be combined to form proteins. Other sets of enzymes put together, step by step, the compounds we call vitamins (p. 304). Energy is required to drive all of these synthetic reactions and the necessary energy can be derived from the oxidation, in the Krebs citric acid cycle (p. 59), of some of the phosphoglyceric acid.

Cells from certain species of plants have other enzymes which can synthesize a wide variety of chemicals, many of which are of great economic importance to man: rubber, drugs such as quinine and morphine, flavorings and perfumes. Plant and animal cells have many enzyme systems in common, and when the details of metabolism are studied, it is found that yeasts, green plants and animals all use the same enzymatic pathway to break down or build up a glucose molecule. Plant cells, however, are better synthetic chemists than animal cells for they can (1) carry on photosynthesis, (2) synthesize all their needed amino acids and vitamins, and (3) synthesize many other substances. Animal cells are unable to synthesize some eight or ten of the twenty-odd amino acids. Thus animals must get their vitamins, certain of their amino acids, and their carbohydrates for energy either directly or indirectly from plants.

56. PLANT RESPIRATION

The process of respiration occurs in every living cell. It is basically the same in all organisms and involves the taking in of oxygen (Fig. 29), the giving off of carbon dioxide, and the releasing of energy from glucose and other foods in a form in which it can be used in cellular activity. The overall chemical reaction of respiration is just the reverse of that of

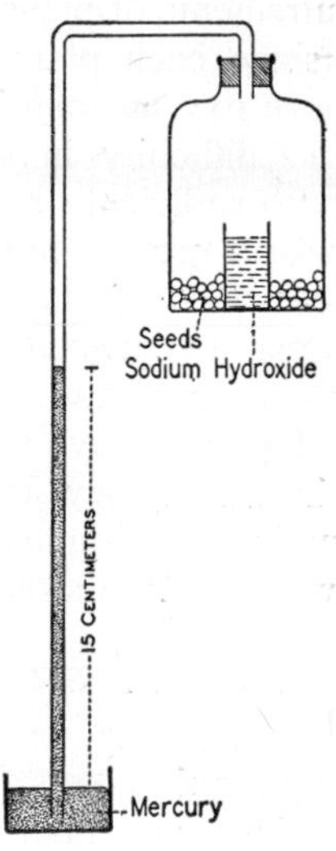

Figure 29. Experiment showing that plant material respires, uses oxygen and produces carbon dioxide. Germinating seeds are placed in a stoppered jar containing a beaker of sodium hydroxide and connected to a mercury manometer. The utilization of oxygen is demonstrated by the rise of mercury in the manometer tube. The carbon dioxide produced is absorbed by the sodium hydroxide and can be measured by titration. If the plant used all the air, both oxygen and nitrogen, the mercury would rise 76 cm. (atmospheric pressure). The mercury rises only prices one-fifth of the air, we infer that only one-fifth of the air is used. Since oxygen comprises one-fifth of the air, we infer that only oxygen is used. (Weatherwax: Botany.)

photosynthesis. The two processes are compared in the following table:

	PHOTOSYNTHESIS	RESPIRATION
Occurs in:	Only those cells of green plants which contain chlorophyll	All living cells
Raw materials:	Water and carbon dioxide	Oxygen and organic substances such as glucose
Time of occurrence:	Only when light shines on the cell	Continuously, night and day
Energy:	Stored by the process	Released by the process
Matter:	Results in an increase in the weight of the plant	Results in a decrease in the weight of the plant or animal
Products:	Oxygen and organic materials	Carbon dioxide and water

When a plant is illuminated, photosynthesis occurs at a rate ten to thirty times as great as that of respiration and the latter process is masked completely. Tracer experiments with heavy oxygen show that respiration continues at about the same pace whether or not photosynthesis is occurring. Oxygen is carried to the cells, and carbon dioxide is carried from them by simple diffusion. Since the oxygen requirements of plant cells are not large, and since each plant as a whole produces more oxygen than it uses, there is usually no difficulty in supplying all plant cells with sufficient oxygen. No special respiratory organs are present or needed: the bodies of the larger plants have air spaces between the loosely packed cells to facilitate gas diffusion. The roots of plants may be asphyxiated if the soil in which they are growing is packed together too tightly or if it is filled with water as in a swampy land.

57. THE SKELETAL SYSTEM OF PLANTS

One of the requirements for photosynthesis is sunlight, and plant bodies have evolved in a number of ways to insure their chlorophyll-containing parts—the leaves—a place in the sun. Plant cells, as we have seen, characteristically have a thick cell wall made of cellulose outside of, and in addition to, the plasma membrane. Plants have no separate skeletal system for support as many animals do. At the simplest level, the algae, which are almost entirely aquatic, have little need for specialized skeletal structures for their bodies are supported by the water. The land plants do need some structure strong enough to hold their leaves in position to receive sunlight. This has been achieved in two major ways: the cellulose wall can be very thick, as in the woody stems of trees and shrubs, and serve directly for the support of the plant body, or it can be rather thin and provide support indirectly by way of turgor pressure. Trees and shrubs have woody cells, tracheids and vessels, in the xylem for support. These cells secrete a very thick wall of cellulose impregnated with a complex chemical called **lignin,** the exact nature of which is unknown. Phloem or inner bark also contains thick fibers to help support the trunk.

The difference between the two methods of support can be seen when the twig of a tree and a crisp lettuce leaf are placed in salt water. The twig is unaffected and remains stiff, but the lettuce leaf wilts quickly and becomes limp and flaccid. The cells of the lettuce leaf still have their cellulose walls, but these walls are unable, by themselves, to provide sufficient support, and turgor pressure has been decreased by the salt treatment.

58. TURGOR PRESSURE

To understand what turgor pressure is, how it provides support, and why it is decreased when a plant is immersed in salt water requires knowledge of basic plant anatomy and of the physical process of osmosis (p. 41). The plant cell, inside its cellulose wall, has one or more large vacuoles filled with **cell sap.** This sap is a solution of a variety of salts, sugars and other organic substances in water. The plasma and vacuolar membranes, which separate the cell sap from the fluid outside the cell, are differentially permeable membranes, through which water passes much more freely than salts do. When the concentration of salts is higher in the cell sap than in the fluid outside, as it generally is, water tends to pass in. The water is moving by diffusion from a region of higher concentration to a region of lower concentration. The added water distends the vacuole, pressing the cytoplasm against the cellulose outer wall (Fig. 30). The cellulose wall is slightly elastic, and thus is stretched by the internal pressure. After a certain amount of water has entered, an equilibrium is reached in which the pressure exerted by the stretched cell wall equals the pressure exerted by the cell sap. After this, the number of water molecules entering the vacuole is equaled by the number leaving, and the total volume of the cell sap is constant.

Turgor pressure, then, is defined as the pressure exerted by the contents of the cell against the cell wall. This should not be confused with the osmotic pressure of the cell sap, which is the pressure that could be developed if the cell sap were separated from pure water by a membrane completely impermeable to all the solutes present in the cell sap. Turgor pressure is almost always less than the osmotic pressure of the cell sap because (1) the fluid outside the cells is usually not pure water, but a dilute salt solution, and (2) the cell membranes are permeable to the salts and organic materials of the cell sap. These can in time diffuse through the membranes and reduce the turgor pressure. Why then, you may ask, as time passes, don't the salts diffuse out of the cell sap to the fluid outside the cell wall, thereby reducing turgor pressure to zero? If the cell were a dead, strictly physicochemical system, this is exactly what would happen; but the living cell, by expending energy, can pump salts into the cytoplasm (the details of this process are as yet unknown). And by photosyn-

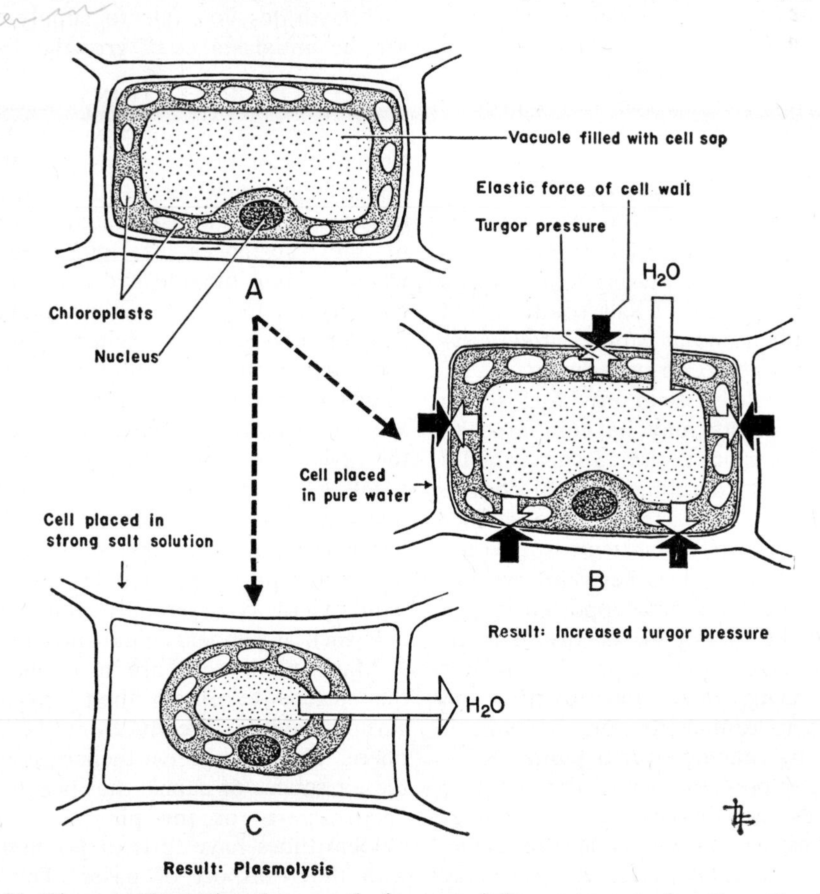

Figure 30. Diagram illustrating the osmotic forces and the movement of water molecules which result in increased turgor pressure (*B*) or in plasmolysis (*C*).

thesis it can produce new organic molecules to increase the concentration of solutes in the cell sap and thus increase turgor pressure.

In all nonwoody plants turgor pressure is important in maintaining the form of the plant body. In young cells, turgor pressure provides the force to stretch cell walls and makes possible cell growth.

59. PLASMOLYSIS

If the fluid outside the cell has a higher salt concentration than the cell sap, as when a lettuce leaf is placed in strong salt solution, water from the cell sap diffuses out of the cell, going from a region of higher water concentration to a region of lower water concentration. Finally the cell contents no longer exert a pressure against the cell wall, i.e., turgor pressure is zero, and the lettuce is wilted.

When the volume of the cell sap decreases due to the loss of water, the protoplasm of the cell is no longer pressed against the cellulose cell wall. Instead, it shrinks away from the cell wall, a process called **plasmolysis** (Fig. 30). If plant cells are exposed to a high salt concentration for too long, they die. If exposed for only a short time and then returned to pure water they can regain their turgidity. To demonstrate this, try putting half a dozen carrots in a beaker of salt solution and removing them one at a time, after varying lengths of time, to pure water.

The role of cellulose in providing for the support of plants may be compared to that of rubber in a tire. Some vehicles have tires of solid rubber, and of course, the vehicle is supported by the rubber itself, just as woody plants are supported by the cellulose itself. Other vehicles have pneumatic tires and are supported by the air within the slightly elastic tire. The air within the inner tube, by pressing against the tire casing, makes the tire firm and enables it to resume its normal shape if distorted by running over a bump. Similarly, turgor pressure against the cellulose wall makes the plant cell firm and enables it to resume its normal shape after being bent by some external force. (To carry the analogy further, we could "plasmolyze" a tire by removing the air from the inner tube!)

60. PLANT DIGESTION

Plants need no digestive system, since their food is either made within the cells or absorbed directly through the cell membranes. The food is either used at once or transported to another part, such as the stem or root, where it is stored for later use. The few insect-eating plants, although without an organized digestive system, do secrete digestive enzymes similar to those secreted by animals.

Plants accumulate reserves of organic materials for use during those times when photosynthesis is impossible—at night, or over the winter. Since an embryo plant cannot make its own food until the seed has sprouted and the embryo has developed a functional root, leaf, and stem system, seeds contain a rich reserve of carbohydrates and fats to supply energy for the initial stages of growth.

61. PLANT CIRCULATION

The simpler plants which consist of single cells or small groups of cells have no circulatory system and no need for one, for simple diffusion suffices to bring in the substances the plant requires: water, carbon dioxide and salts. The circulatory systems of higher plants are simpler than those of higher animals and constructed on an entirely different plan. Plants have no heart and no blood vessels. Transportation is accomplished by the xylem and phloem systems; a few plants have an additional latex system to assist in circulation. **Latex** is a milky material, rich in food substances (carbohydrates and proteins). The latex of certain plants yields commercially valuable products such as rubber, chicle and opium.

Many plant cells are near enough to the surface to obtain their oxygen and carbon dioxide directly from the atmosphere. The rest receive their oxygen from the air spaces or canals that penetrate the deeper parts of the plant; xylem and phloem tubes have little or nothing to do with the transport of gases. The xylem tubes are chiefly concerned with transporting water and minerals from the roots up the stem to the leaves, while phloem tubes transport food which has been manufactured in the leaves down the stems for storage and use in the stems and roots. Phloem transports nutrients up as well as down; in the spring, for example, substances pass from their places of storage to the buds to supply energy for growth. Foods and minerals travel primarily by diffusion; water rises in the xylem tubes because of the combined forces of transpiration and root pressure.

The connections of the phloem and xylem vessels are fundamentally different from those in arteries and veins. In the latter a large tube branches successively into smaller and smaller tubes. All xylem and phloem tubes are small and occur in bunches called vascular bundles. In the lower part of the stem there are many vessels per bundle; in the upper part there are fewer per bundle, the others having entered the branches of the stem.

62. PLANT SAP

The material in the xylem, phloem and latex tubes of the higher plants is called **plant sap;** it is somewhat analogous to the blood plasma of man and higher animals. Plant sap is a complex solution of many substances, both organic and inorganic, which are transported from one part of the plant to another. The substances present and their concentrations vary greatly in different plants and in various parts of the same plant. Water is absorbed by the epidermal cells of the roots and moved to all parts of the plant. Sugars are manufactured in the leaves and carried to other cells to be used for the release of energy or for the production of more complex carbohydrates, fats or proteins.

Plant saps are much more variable in their composition than is blood plasma. As much as 98 per cent of sap may be water. Salts of sodium, potassium, calcium and magnesium are most abundant, as in plasma, and traces of others are also present. Saps contain widely varying amounts of glucose and other sugars, depending on the species, time of year, part of the plant, and other factors. They also contain proteins, enzymes, plant hormones, dissolved oxygen and carbon dioxide, organic acids, and waste products of metabolism. Citric and malic acids are common organic acids, both of which are constituents of the Krebs citric acid cycle, fundamental to the metabolism of all living cells. Citric acid was first isolated from citrus fruits and malic acid from apples. Plant saps differ from plasma in that they are usually acid rather than alkaline; most of them have a *p*H between 4.6 and 7. The most acid plant sap known is that of the begonia, with a *p*H of 0.9.

63. PLANT EXCRETION

A striking difference between plants and animals is that plants excrete little or no nitrogenous wastes. Those excreted by animals—urea, uric acid and ammonia—come from the breakdown of proteins and other substances. Similar nitrogenous compounds are released during the metabolism of plant protoplasm, but instead of being excreted as wastes, they are reutilized in the synthesis of new protoplasm. Since plants neither ingest proteins nor carry on muscular activity (the two largest sources of metabolic wastes in animals), the total amount of nitrogenous waste is small and can be eliminated by diffusion as ammonia through the pores of the leaves or by diffusion as nitrogen-containing salts from the roots into the soil. Since there is little for plants to excrete, no special excretory system is needed. In some plants, a few waste products accumulate and remain as crystals within the cells; for example, spinach leaves contain about 1 per cent oxalic acid. Wastes which are deposited in the leaves are eliminated when the leaves are shed.

64. PLANT COORDINATION

The activities of the various parts of a plant are much more autonomous than are those of the parts of an animal. The coordination between parts that does exist is achieved largely by direct chemical and physical means, since plants have developed no specialized sense organs and no nervous system.

Actively growing plants can respond to a stimulus coming from a given direction

Figure 31. Phototropism in young radish plants. *A,* In the dark or in uniform light, the plants grow straight upward. *B,* When they are exposed to light coming from a single direction, they quickly bend toward it—that is, they are positively phototropic. The photograph on the right was taken half an hour after the one on the left. (Weatherwax: Botany.)

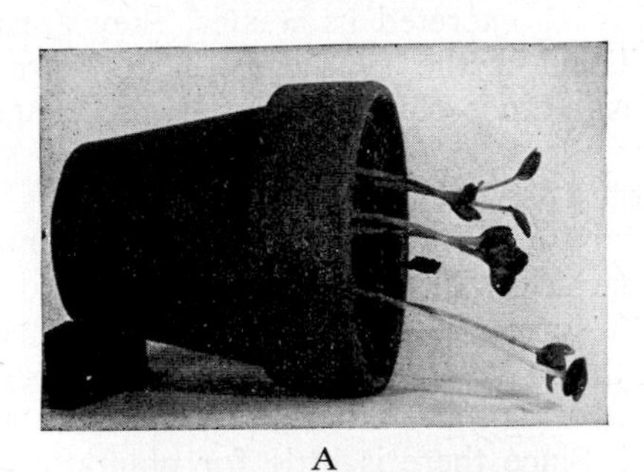

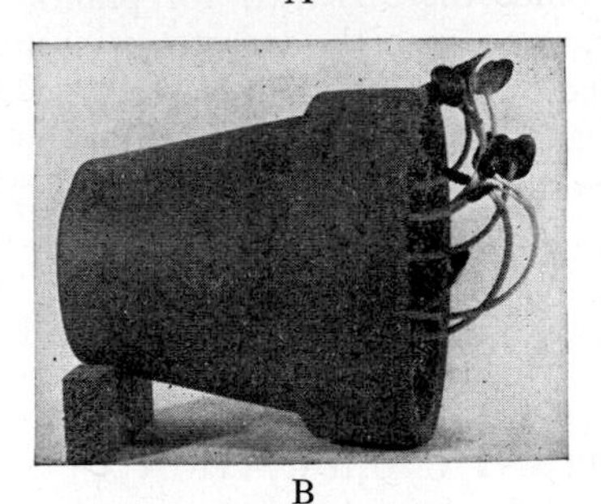

Figure 32. Geotropism in young radish plants. *A,* Pot of straight radish seedlings placed on its side and kept in the dark to eliminate phototropism. *B,* Within thirty minutes the seedlings had bent; since they bend away from the direction of the force of gravity, they are negatively geotropic. (Weatherwax: Botany.)

by growing more rapidly on one side, and hence bending toward or away from the stimulus. Such a growth response is called a **tropism;** it can occur only in those parts of the plant that are growing and elongating, but the stimulus for it may be received in some distant part of the plant. If an organism is motile, it may respond to a stimulus by moving toward or away from it. An orientation movement in response to a stimulus is called a **taxis.** Tactic responses are limited to cells capable of movement: animals, some lower plants, and the sex cells of mosses or ferns. Tropisms and taxes are named for the kind of stimulus eliciting them: **phototropism** (a growth response to light, Fig. 31), **geotropism** (a growth response to gravity, Fig. 32), **chemotropism** (a growth response to some chemical) and **thigmotropism** (a growth response to contact, Fig. 33). Ivy plants are famous for their thigmotropism—they grow so as to maintain contact with a wall, tree or some other supporting object.

Tropisms and taxes may be either positive or negative—the response may be either toward or away from the stimulus. For example, no matter how a seed may be oriented in the ground, the primitive root and shoot of the developing embryo grow so that the root grows downward and the shoot (stem) grows upward (Fig.

34). Thus, the root is positively geotropic and the shoot is negatively geotropic. A growing plant can be fooled by fastening it to a disc and spinning the disc, thus imposing a field of centrifugal force on the growing plant. The plant will respond to this force that has replaced gravity and the shoot will grow toward the center of the disc (which corresponds to up) and the roots will grow toward the periphery.

A clue has recently been obtained to the mechanism by which plant cells receive the light stimuli that bring about phototropisms. When plants are grown in the dark, they will grow for a while, using the energy stored in the seed, and will form stems and leaves that contain no chlorophyll at all. Such plants, called **etiolated,** will still respond to a beam of light by growing toward it, so we can conclude that the phototropic response is not brought about by chlorophyll. By shining light of different colors (wavelengths) on etiolated plants and noting the resulting bending, it was found that blue or violet light was much more effective in eliciting a phototropic response than a green, yellow or red light. (In contrast, yellow-green light is most effective in photosynthesis.) This indicated that the pigment responsible was a yellow or orange one. Detailed experiments showed that carotene and xanthophyll are the pigments involved. A discussion of these experiments, in which the action spectrum of light was compared to the absorption spectra of xanthophyll and other pigments, is beyond the scope of this book.

Since tropistic responses require the

Figure 33. Response of a squash tendril to touch. The straight tendril (*A*) is stroked with a stick (*B*). Five minutes later (*C*) it is beginning to bend, and within 20 minutes it has completed one coil (*D–F*) (Weatherwax: Botany).

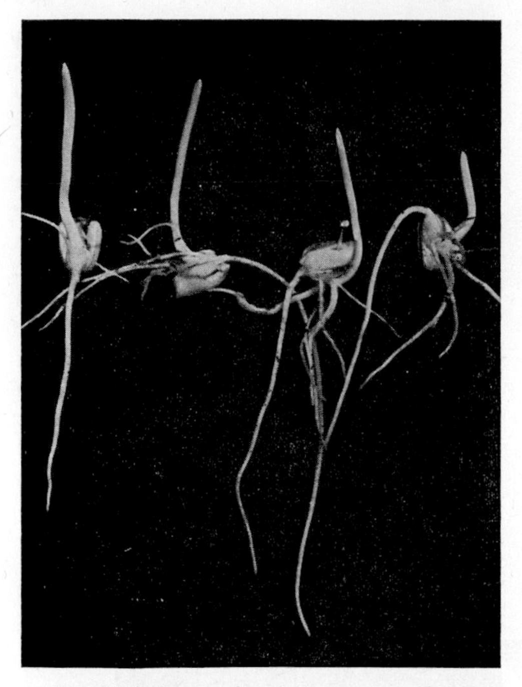

Figure 34. Geotropism in the roots and shoots of corn seedlings. The grains were planted in various positions, but the roots tend to grow downward and the shoots upward. (Weatherwax: Botany.)

actual growth of cells, one side of the plant growing faster than the other, they are necessarily slow and require from an hour to perhaps a week for completion. This differential acceleration of growth, as well as normal uniform growth, depends upon the stimulation of cells by plant growth hormones or **auxins** (p. 93). As yet we have no good explanation as to how light, gravity or any of the other effective stimuli bring about the secretion of auxins nor how the auxins are distributed differentially to the two sides of the plant. It is obviously of advantage to a plant to grow toward the light and to have its roots grow toward moisture, but no one supposes that the plant *purposefully* grows toward the light because it needs light for photosynthesis.

65. THE TRANSMISSION OF IMPULSES

All living cells, whether animal or plant, exhibit irritability to some extent and can transmit an excitation, even though only slowly. The excitation produced by the penetration of an egg by a

Figure 35. A, The sensitive plant, *Mimosa pudica,* before being disturbed. *B,* The plant five seconds after being touched; note how the leaves have folded and drooped. (Courtesy of the General Biological Supply House, Chicago.)

sperm travels across the surface of the egg at a rate of about 1 cm. per hour. The unspecialized cells of sponges transmit excitations at about 1 cm. per minute. The protoplasm of plant cells also can transmit excitations, although usually the rate of propagation is so slow that the results are not easily observed. In a few plants, however, responses to stimuli do occur rapidly enough to be readily seen. One of these is the response of the Venus fly trap (Fig. 1) to the presence of an insect on the leaf, and another is the response of the "sensitive plant" *Mimosa pudica* to touch (Fig. 35). Normally the leaves of the latter plant are horizontal, but if one of them is lightly touched, all the leaflets fold within two or three seconds. Touching one leaf sharply causes not only the stimulated leaf, but also the neighboring leaves, to fold and droop. After a few minutes the leaves return to their original position. The wilting is due to a decrease in the turgor pressure of the cells at the base of the leaf (Fig. 36), but the excitation is transmitted along the sieve tubes of the leaves and stems. The rate of transmission in *Mimosa* is about 5 cm. per second, compared to about 120 meters per second in the nerves of the higher vertebrates, but the nature of the impulse is fundamentally the same. The excitation is accompanied by electrical phenomena, increased permeability of the excited cells and a temporary change in their metabolism. The response can be altered by various drugs.

66. PLANT HORMONES

Plant hormones, like animal hormones, are organic compounds which can produce striking effects on cell metabolism and growth even though present in extremely small amounts. The plant hormones are produced primarily in actively growing tissue, especially by meristem tissue in the growing points at the tip of stems and roots. Like animal hormones, the plant hormones usually exert their effects on parts somewhat removed from the site of production. The plant hormones have many different types of effects on metabolism and cell division, but one outstanding effect is on the elongation of cells; they regulate the lengthwise growth of individual cells in the growing part of the plant. Other effects are the induction of new roots, the initiation of flower development, the stimulation of cell division in the cambium, the inhibition of lateral buds and the inhibition of formation of abscission regions (p. 113) and hence the prevention of the fall of leaves or fruit.

The plant growth-promoting substances are called **auxins.** The best known auxin, and the one believed to be responsible for most if not all of the growth phenomena of plants is **indoleacetic acid,** also known as **heteroauxin.** Two other substances with growth-promoting properties have been isolated from plants: auxin a (auxentriolic acid) and auxin b (auxenolonic acid). In addition to these naturally occurring hormones, a great variety of chemicals have been synthesized which

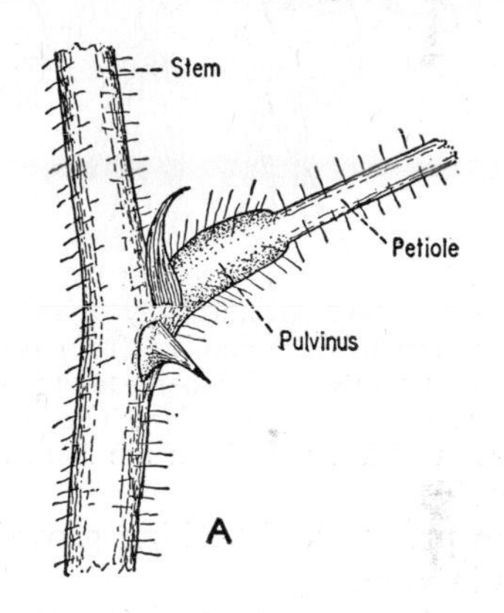

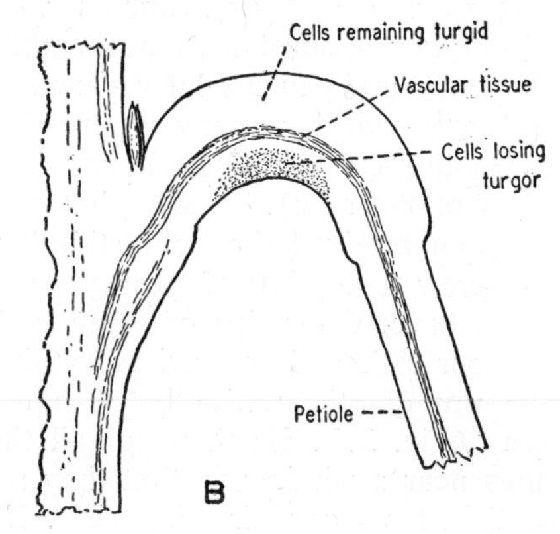

Figure 36. The mechanism of response in *Mimosa. A,* The base of the petiole showing the pulvinus. *B,* Section through the pulvinus showing cells which lose turgor to produce folding of the leaves. (Weatherwax: Botany.)

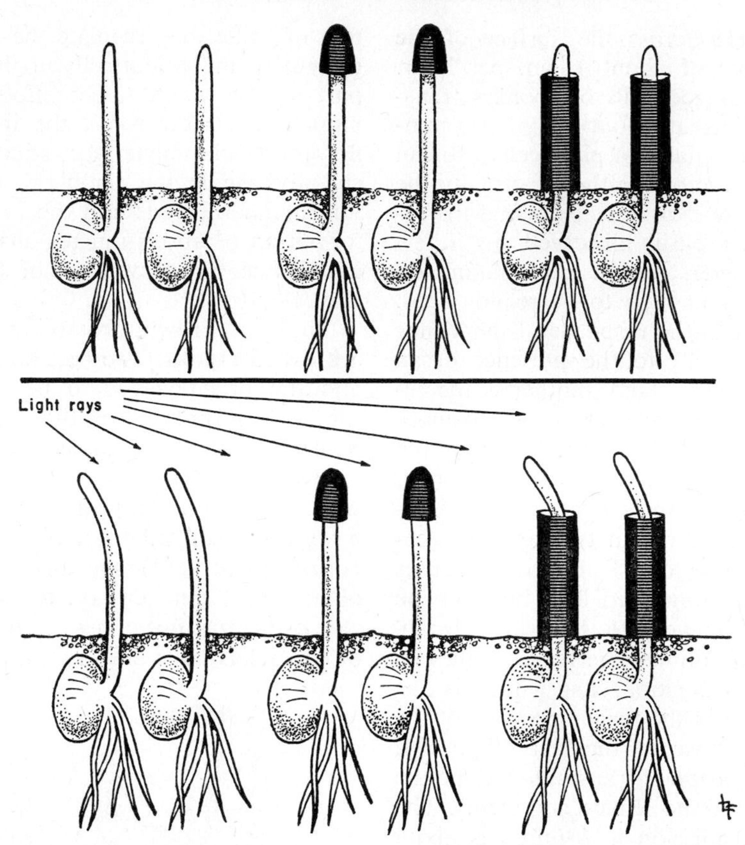

Figure 37. Darwin's experiment with canary grass seedlings. *Upper row:* Some plants were uncovered, some were covered only at the tip and others were covered everywhere but at the tip. After exposure to light coming from one direction (*lower row*) the uncovered plants and the plants with uncovered tips bent toward the light; the plants with covered tips (center) grew straight up. Darwin's conclusion: The tip of the seedling is sensitive to light and gives off some "influence" which moves down the stem and causes the bending.

have the property of eliciting growth responses in plants.

Some of the first experiments with growth-promoting substances were made by Charles Darwin in his later years. It had been known for many years that plants would grow toward the light (were positively phototropic). To see what part of the plant received the light stimulus, Darwin grew a number of canary grass seedlings, covered the tips of some with black paper caps and covered everything but the tips of others with black paper cylinders (Fig. 37). He then put all the seedlings near a window so that they received light from one direction only. By the next day the seedlings with no cover at all and the seedlings with everything but the tip covered were both bent very strongly toward the light, but the seedlings which had been capped had grown straight up. From these experiments Darwin concluded that the light was received by the tip of the plant and that some "influence" moved down the stem from the tip to cause the plant to bend.

The investigations of Boysen-Jensen of Denmark, Frits Went of Holland and others from 1910 until about 1930 clarified the mechanism underlying tropistic responses and showed that this "influence" was a plant growth hormone. These classic experiments were performed using the coleoptile of the oat seedling. The coleoptile is a hollow, practically cylindrical organ that envelops the unexpanded leaves like a sheath. After a certain stage of growth, its further increase in length is due almost entirely to cell elongation. In 1910 it was discovered that if the coleoptile tip is cut off at this stage, the decapitated coleoptile immediately stops elongating. If the tip is replaced, the coleoptile again begins to grow (Fig. 38). Other experiments showed that a thin sheet of agar gel could be placed between the tip and the rest of the coleoptile without interfering with the growth of the coleoptile. Indeed, the decapitated tip could be placed for a time on a block of agar, then removed, and the bit of agar, when placed on the coleoptile, would stimulate it to grow (Fig. 38*D*). These experiments showed that the growth of the coleoptile is controlled by some substance produced in the tip that normally passes downward and stimulates the coleoptile cells to elongate. By leaving a tip on an agar block for varying lengths of time it was shown that the amount of growth substance that diffused into the agar was proportional to the length of time they were in contact. The amount of growth substance was measured by its effect on the elongation of the coleoptile.

Further experiments by Frits Went showed that if the agar block is placed on one side of the decapitated coleoptile, growth is asymmetrical; the coleoptile bends *away* from the side on which the block was placed, i.e., growth is more rapid in the cells directly under the agar block (Fig. 38*E*). It was found that this test was extremely sensitive and Went could measure the growth-promoting substance in terms of "curvature units," the number of degrees of bending produced in the coleoptile. This is an example of a

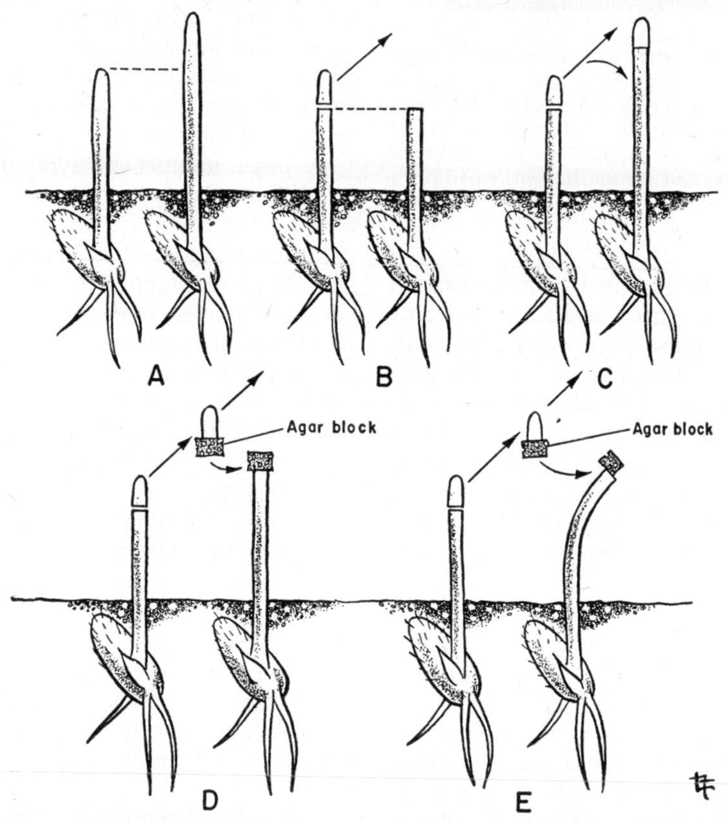

Figure 38. A series of experiments which demonstrate the existence and mode of action of plant growth hormones in oat coleoptiles. In each pair of drawings, the figure on the left indicates the experiment performed and the figure on the right the growth after a period of time. *A,* Control: No operation performed, normal growth. *B,* If the tip of the coleoptile is cut off and removed, no growth occurs. *C,* If the tip is cut off and then replaced, normal growth ensues. *D,* If the tip is cut off, placed on a block of agar for a time and then the agar block, but not the coleoptile tip, is placed on the seedling, growth occurs. *E,* If the tip is placed on an agar block for a time and then the agar block is placed asymmetrically on the seedling, curved growth results.

A

B

Figure 39. The effect of indoleacetic acid in stimulating the formation of roots in cuttings from lemon trees. The cuttings in *A* had been soaked for eight hours in a dilute solution (500 parts per million) of indoleacetic acid eighteen days previously, and the cuttings in *B* had been soaked in plain water for a similar length of time.

bioassay, a test in which some chemical substance is measured or assayed in terms of its effect on some biologic system. When it was found that the growth-promoting substance was indoleacetic acid, the test was standardized in terms of this substance: the coleoptile test can detect as little as one ten-millionth of a gram of indoleacetic acid. Other natural and synthetic growth-promoting substances can be compared by this bioassay. Another test utilizing the oat coleoptile is the "comb" test in which coleoptile sheaths are decapitated and cut to standard lengths. These cylindrical lengths are placed on the teeth of a fine wire comb and immersed in the solution to be tested. By comparing the growth of the coleoptile sheaths in the unknown solution with that of the sheaths placed in a standard solution of indoleacetic acid, the amount of auxin activity in the unknown solution can be determined.

The growth response of the coleoptile sheath is a metabolic process, for it will occur only in the presence of oxygen and growth is improved if some glucose is supplied as a source of energy.

Plant hormones also determine the growth correlations of different parts of plants. The terminal bud of a stem normally inhibits the development of lateral buds. If the terminal bud is cut off, the lateral buds, freed of the inhibition induced by the auxin given off by the terminal bud, begin to develop. That this inhibition is indeed caused by auxin can be shown by an experiment in which the terminal bud is cut off and replaced by a suitable amount of indoleacetic acid held in place by a bit of wool fat. The lateral buds of a stem treated in this way remain inhibited. The proper control experiment is to cut off a terminal bud and replace it with a bit of wool fat. The fact that the lateral buds are now free of inhibition shows that the inhibition is really due to the indoleacetic acid, and not to the wool fat or to some other substance that might be present in the wool fat.

A large number of synthetic growth-regulating substances have been made and tested. One of these, **"2, 4-D"** (2, 4-dichlorophenoxyacetic acid), is a potent stimulator of plant metabolism. Dicotyledonous plants (most of the common weeds are dicotyledons) are more sensitive to this than are monocotyledonous plants, such as grasses. Thus, spraying a lawn with the proper concentration of 2, 4-D, enough to stimulate the weeds but not the grass, will cause the former to metabolize at a tremendous rate, consume their own protoplasm, and die. It is widely used as a weed killer on lawns, and even more important economically, on field crops. Since many of the important crop plants are monocotyledons—corn, oats, rye, barley and wheat—millions of acres of crops are now treated each year with 2, 4-D to eliminate weeds and increase yields.

Natural and synthetic auxins have been found to have a variety of practical uses and are of tremendous economic importance. For example, indoleacetic acid and indolebutyric acid are used to stimulate the growth of roots from cuttings. Rose bushes, holly and a number of ornamental shrubs that are ordinarily difficult to induce to root will root nicely in a week or two after an eight hour soaking in an indolebutyric acid solution (Fig. 39). This substance is also used to produce **parthenocarpic** fruits, ones formed without pollination and therefore without seeds. Seedless tomatoes, cucumbers, squash and strawberries have been produced in this way. Other hormones are used to improve fruit-set, to hasten the process of ripening, to prevent the dropping of apples and pears before harvest and to inhibit the sprouting of stored potatoes.

67. PHOTOPERIODISM

It had, of course, been known for a long time that different kinds of plants flower at different seasons of the year and that the amount of daylight per day varies with the season: in the northern hemisphere, June 21 has the most and December 21 the fewest hours of daylight. But the role of the ratio of daylight and darkness in determining the time of flowering of plants was unknown until 1920, when Garner and Allard, of the U.S. Department of Agriculture, demonstrated that the time of flowering of tobacco plants could be altered by changing the **photoperiod,** the number of hours per day that the plant was exposed to light. Further observations showed that some species of plants (e.g., asters, cosmos, chrysanthemums, dahlias, poinsettias and potatoes) will produce flowers only when the photoperiod is less than about fourteen hours per day. Such plants are called **"short-day" plants** and normally flower in the early spring or late summer or fall. They can be made to flower earlier than usual by decreasing their daily exposure to light—by covering them—or they can be kept from flowering by giving them artificial illumination (Fig. 40). Other species of plants (e.g., beets, clover, coreopsis, corn, delphinium and gladiolus) will produce flowers only when the photoperiod is more than fourteen hours per day. These **"long-day" plants** normally

Figure 40. Two cosmos plants grown in Washington, D. C., from seeds planted May 26. Normally, the plant begins to flower in late August or September, but by artificially shortening the exposure to light to ten hours a day, the plant on the left was induced to flower July 8. The plants which were exposed to the normal fourteen or fifteen hours of daylight per day are shown to the right. This response to the relative length of day and night is known as photoperiodism. (Weatherwax: Botany; photograph courtesy of Dr. H. A. Allard and the United States Department of Agriculture.)

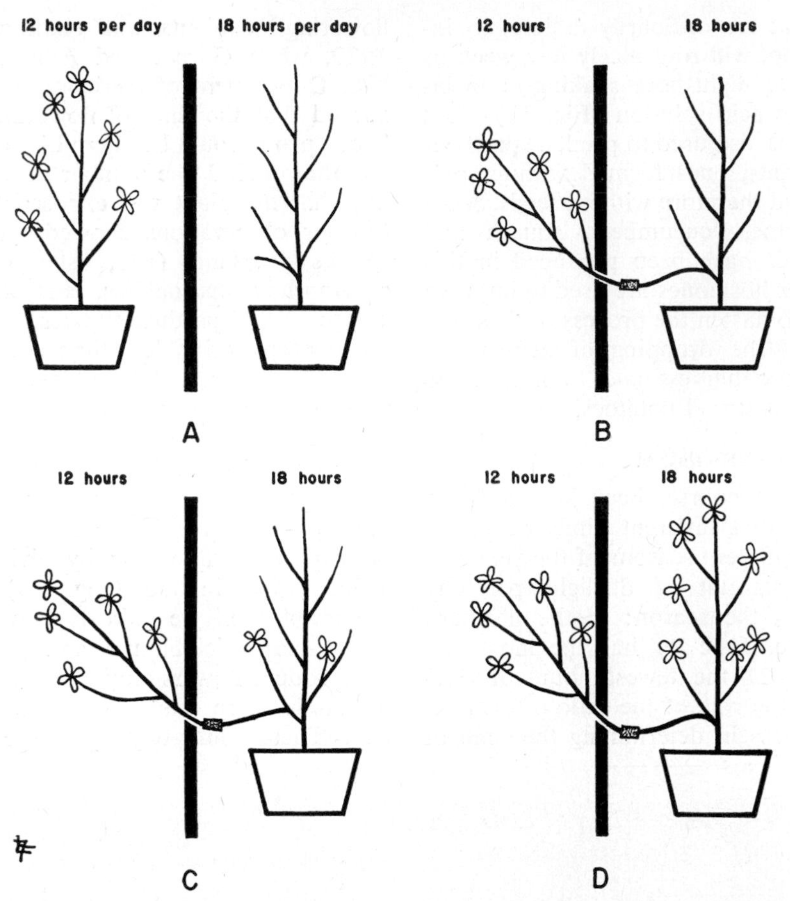

Figure 41. An experiment to demonstrate the existence of a flower-inducing hormone. *A,* Two cocklebur plants are grown in pots separated by a light-tight partition, exposed to twelve and eighteen hours, respectively, of light per day. The twelve-hour plant has flowered; the eighteen-hour plant has not. *B,* The twelve-hour plant is cut off, inserted through a light-tight hole in the partition and grafted to the eighteen-hour plant. The two parts continue to receive twelve and eighteen hours of light respectively. The eighteen-hour plant gradually develops flowers, first on the twigs nearest the graft (*C*) and eventually on all twigs (*D*). If no graft had been made, the eighteen-hour plant would not have developed flowers.

flower in the late spring and early summer. If they are covered part of each day, so that their daily exposure to light is less than thirteen to fourteen hours, flowering is greatly delayed or prevented entirely. Carnations, cotton, dandelions, sunflowers and tomatoes are examples of **indeterminate plants** which flower at a given time and are relatively unaffected by the amount of daylight per day. The time of flowering is not controlled solely by the photoperiod, for temperature, moisture, soil nutrients and the amount of crowding may also play a role.

The mechanism by which the length of daylight affects time of flowering is not known in detail, but the results of some experiments suggest that a flower-producing hormone is involved. A typical experiment, using cocklebur, a short-day plant, is as follows (Fig. 41): One plant is grown exposed to twelve hours of light per day until it is producing flowers. It is then grafted to another plant that had been grown exposed to eighteen hours of light per day (and thus had been inhibited from producing flowers). The two parts, though grafted, are separated by a light-tight partition and the first part continues to receive twelve hours and the second eighteen hours of daylight. The short-day part of the plant continues to produce flowers and, in time, the long-day part of the plant also produces flowers, usually beginning at the point nearest the graft. This is taken as evidence for a diffusible, flower-inducing hormone. Nothing is known of the chemical composition of this hormone nor of how it might act to induce flowering.

A knowledge of the phenomenon of photoperiodism is an intensely practical thing for commercial plant growers, for by altering the amount of light per day in their greenhouses they can speed up or retard the flowering of their plants so that they come into full bloom at just the right time for the Christmas or Easter season.

In general, short-day plants are originally natives of tropical or subtropical regions, where there is no more than thirteen or fourteen hours of daylight per day. Most long-day plants are originally natives of the higher latitudes. Indeterminate plants may be found almost anywhere.

A number of plant functions other than flowering may also be affected by the daily photoperiod: the formation of tubers by Irish potatoes is accelerated when the daily exposure to light is shortened. Since the growth of the tuber (the part we know as the potato) involves the deposition of starch, the photoperiod must in some way stimulate the transfer of carbohydrates from the leaves to the tuber.

68. SLEEP MOVEMENTS

Many plants change the position of their leaves or flower parts in the late afternoon or evening and these parts return to their original position in the morning. The leaves of peas, beans, clover and a number of other plants fold together in darkness and return to an expanded position in daylight. These changes in position have been termed **sleep movements,** although they are in no way related to the sleep of animals. Several kinds of flowers close at night and open in the morning.

QUESTIONS

1. What, exactly, is the unique characteristic of photosynthesis?
2. How would you prove that oxygen is given off by green plants during photosynthesis?
3. What pigments may be present in plant cells? What are the functions of these pigments?
4. What is meant by the "light reaction"? What are the products of this reaction?
5. Compare the processes of photosynthesis and cellular respiration.
6. Why would it be incorrect to say that photosynthesis is carbon assimilation?
7. What organisms are capable of "fixing" carbon dioxide?
8. Compare the skeletal systems of plants and animals.
9. Describe the physiologic mechanism underlying turgor pressure.
10. What are the functions of the xylem, phloem and latex systems?
11. Describe the physiologic basis of tropisms. Where in a plant can tropistic responses occur?
12. What effects may be produced by plant hormones?
13. Describe an experiment designed to prove that some new synthetic chemical has auxin-like properties.
14. What are some practical applications of plant hormones?
15. What is meant by photoperiodism? How would you determine experimentally whether a particular plant is a "short-day," a "long-day," or an "indeterminate" plant?

SUPPLEMENTARY READING

For further information about the functions of plant cells consult a textbook of general botany such as that by Fuller and Tippo, by Hylander and Stanley, or by Smith, Gilbert, Bryan, Evans and Stauffer. More technical discussions of plant physiology will be found in Bonner and Galston or Curtis and Clark. C. M. Wilson's *Trees and Testtubes* describes the use of latex in the preparation of rubber. The effects of plants on human existence is considered in Anderson's *Plants, Life, and Man.* Plant growth hormones and some of their practical and scientific uses are discussed in *Growth Hormones and Tissue Growth in Plants* by P. R. White, in *The Control of Plant Growth by the Use of Special Chemicals* by L. G. Nickell and in A. C. Leopold's *Auxins and Plant Growth.*

Chapter 8

The Structures and Functions of a Seed Plant

IN THE previous chapter the basic functions common to most, if not all, green plant cells were discussed. In the more primitive plants these functions may all occur in a single cell, but in the higher plants cellular specialization has occurred and we can differentiate the several parts —root, stem and leaf—of a plant (Fig. 42). The evolution of conducting tissues, xylem and phloem, and the specialization of regions of the body enabled plants to survive on land and to grow to large size. Since these higher seed plants are the most widespread and familiar as well as the most useful plants for man, let us examine some of the details of seed plant structure and see how certain functions are localized in particular parts of the plant.

69. THE ROOT AND ITS FUNCTIONS

The most obvious function of the root is to anchor the plant and hold it in an upright position; to do this, it branches and rebranches extensively through the soil. Although the depth of the roots seldom equals the height of the stem, the roots frequently extend farther laterally than the stem's branches and the total surface of the root usually exceeds that of the stem. The second and biologically more important function of the root is the absorption of water and minerals from the soil, and the conduction of these substances to the stem. In some plants—for example, carrots, sweet potatoes, beets— the roots have still another function as storage places for large quantities of food.

The tip of each root is covered by a protective **root cap,** a thimble-shaped covering of cells which fits over the rapidly growing embryonic (meristematic) region (Fig. 43). The outer part of the root cap is rough and uneven because its cells are constantly being worn away as the root pushes through the soil. The growing point consists of actively dividing meristematic cells, from which all the other tissues of the root are formed. The growing point also gives rise to new root cap cells to replace the ones worn away. Immediately behind the growing point is the **zone of elongation;** here the cells remain undifferentiated but grow rapidly in length by taking in large amounts of water. The growing point is about 1 mm. in length and the zone of elongation is 3 to 5 mm. long; these two are the only parts of the root that account for the continued elongation of the root.

Above the zone of elongation is the **root hair zone** (Fig. 43), characterized externally by a downy covering of whitish root hairs. It is in this zone that the cells differentiate into the permanent tissues of the root. Each root hair is a slender, elongated, lateral projection from a single epidermal cell, through which most of the water and minerals are absorbed. Any epidermal cell may absorb substances, but the surface exposed by a root hair is much larger and hence more effective than that of a single epidermal cell. Root hairs are very delicate and short-lived; new hairs are constantly formed just behind the zone of elongation and old hairs farther back wither and die as the root elongates (Fig. 44). Only a short segment of the root, perhaps 1 to 6 cm. long has root hairs.

Some measurements made by Dittmer give an idea of the astonishing amount of root surface available for the absorption of water and minerals. A single rye plant, grown in a box 12 inches square and 22

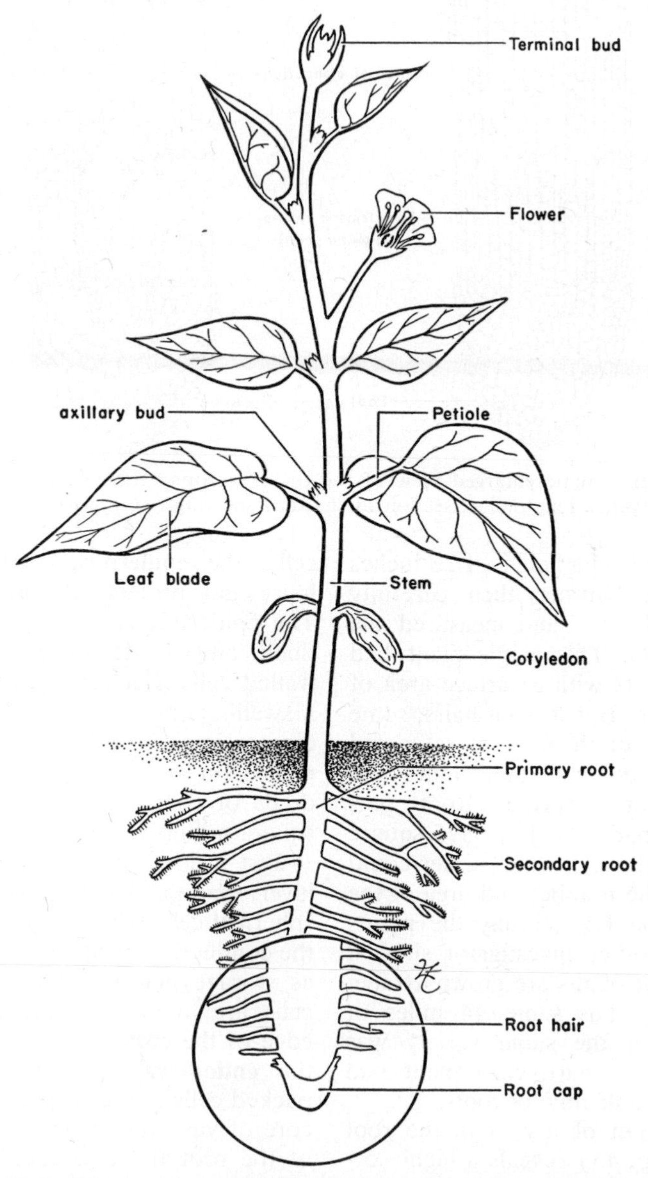

Figure 42. A diagram of a young bean plant showing the root, stem and leaf.

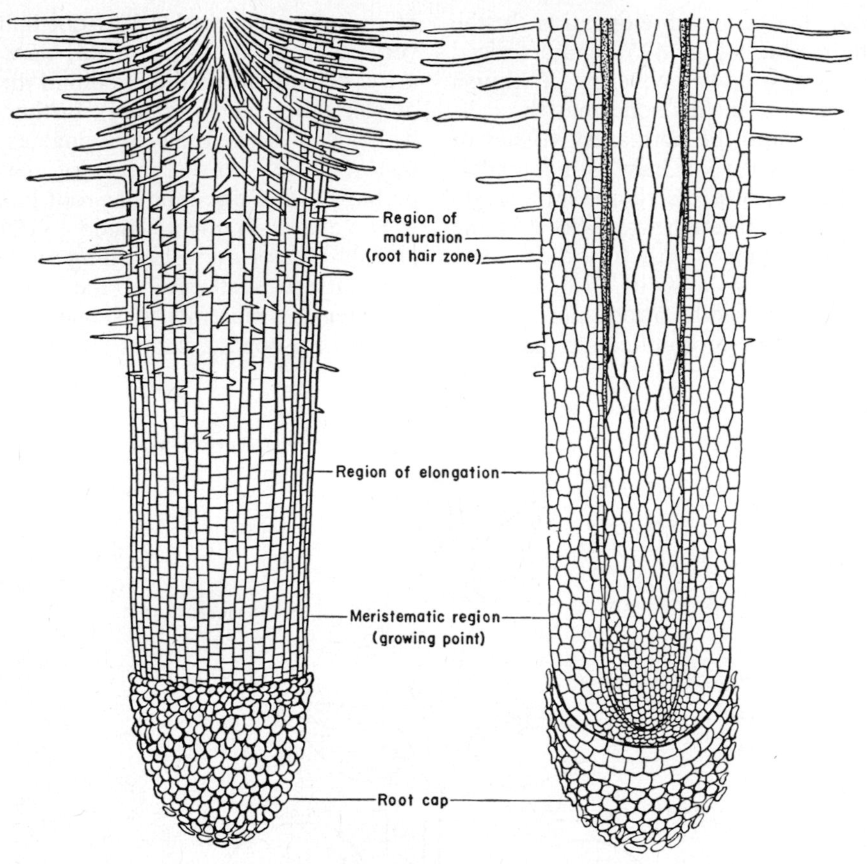

Figure 43. A diagrammatic enlarged view of the tip of a young root. *Left,* The surface of the root; *right,* a longitudinal section of the root showing the internal structure.

inches deep, reached a height of 20 inches in four weeks. Dittmer then carefully washed away the soil and measured the number of roots. This single plant had 387 *miles* of roots with a surface area of 2554 square feet. But its root hairs, some 14,500,000,000 of them, were estimated to have a total length of 6600 miles and 4321 square feet of surface. Roots plus root hairs thus had a total of 6875 square feet of surface through which water could be absorbed. The number and area of the roots depend on how closely the plants are grown. Another investigator showed that when wheat plants are grown 10 feet apart, each one has some 44 miles of roots, but when the same variety was planted 6 inches apart, each plant had only about one-half mile of roots.

A cross section of a root in the root hair region (Fig. 45) reveals a highly organized structure. The outermost layer of cells, the **epidermis,** produces the root hairs and protects the underlying cells. The epidermis is only one layer of cells thick and made of rectangular, thin-walled cells. Each root hair originates as a swelling on an epidermal cell which increases in size until it becomes a hairlike projection as much as 8 mm. long. Hundreds of root hairs are present on each square millimeter of surface.

Just inside the epidermis is the **cortex** layer, made of large, thin-walled, roughly spherical cells which serve as avenues for the conduction of minerals and water, and as storage places for food. Between the cells are many air spaces. At the inner edge of the cortex, a single layer of cells, the **endodermis,** separates the loosely-packed cells of the cortex from the central core of vascular tissue. In the older parts of the root the endodermis is thick and contains a waxy material which prevents

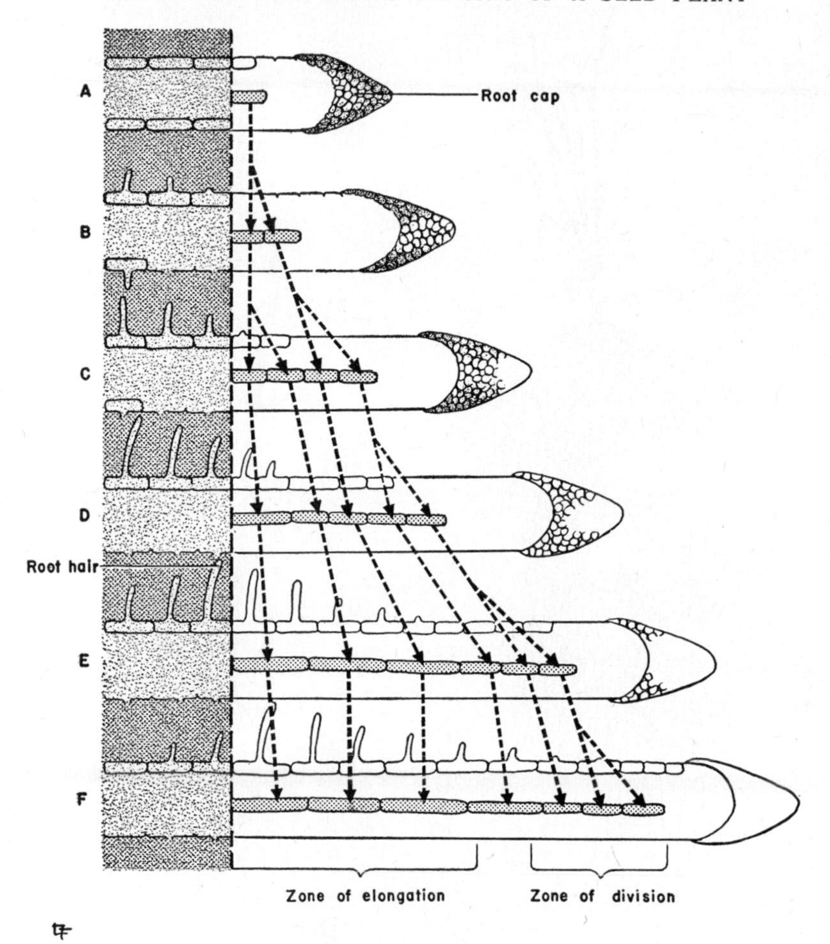

Figure 44. Diagram of successive stages in the growth of a root tip. The vertical dotted line is a landmark to indicate the same part of the root as growth occurs. Of the many cells which make up the interior of the root (see Fig. 43) only one is indicated in *A*. This divides longitudinally to form two (*B*) and four (*C*) cells. The cells nearest the tip continue to divide (*D, E, F*) and the ones farther back elongate. Root hairs grow out of the epidermal cells in the zone of elongation (*B, C, D*) and then wither and die as the growing tip moves on (*E, F*).

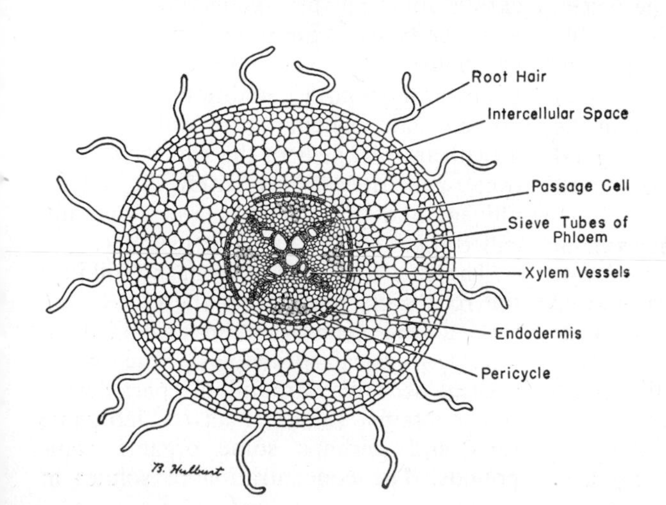

Figure 45. Cross section of a root near its tip where root hairs are present.

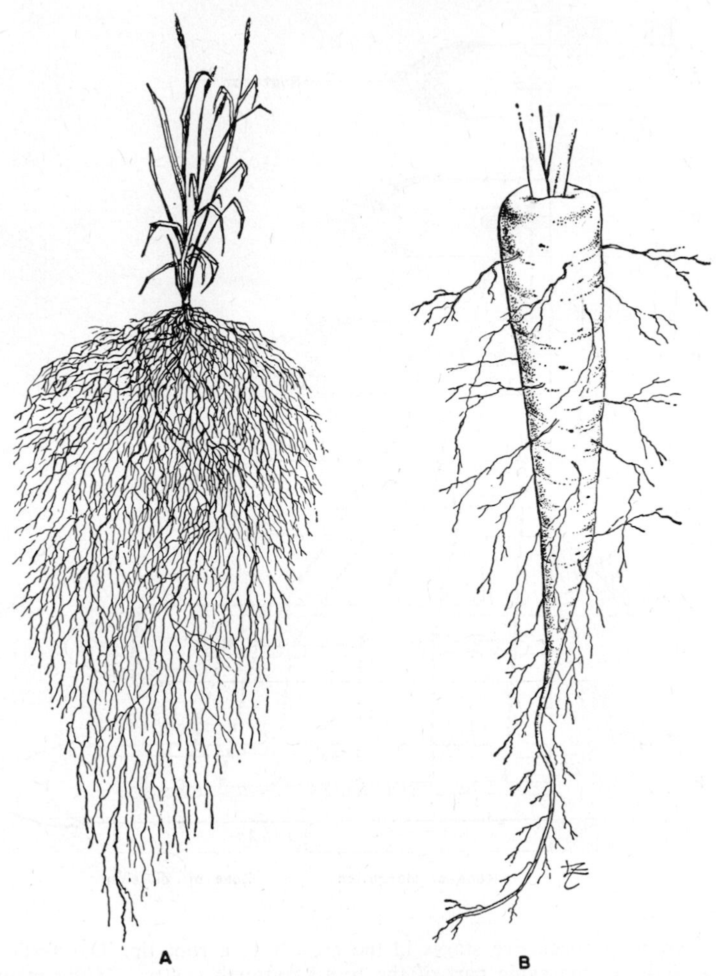

Figure 46. Types of root systems in plants. *A*, The diffuse root system of a grass; *B*, the tap root system of a carrot.

the diffusion of water into the cortex. Adjacent to the endodermis is the **pericycle,** a single layer of parenchyma cells capable of being transformed into meristem to give rise to branch roots. The fact that secondary roots originate deep within the original root in this way insures a good connection between the vascular tissues of branch and main root.

Within the pericycle lies the **xylem,** usually arranged in the form of a star, or like the spokes of a wheel, and composed of thick-walled, rounded, elongated conduction cells, tracheids and xylem vessels. Between the adjacent points of the xylem star are small groups of **phloem** cells, smaller and thinner-walled than the xylem. Some of the phloem cells are specialized sieve tubes. The roots of trees, shrubs and other perennials typically have a single-celled **cambium** layer between the phloem and xylem which gives rise, by cell division, to additional layers of phloem and xylem to provide for the growth in thickness of the root. The pericycle and the tissues within it, xylem, phloem and parenchyma, are commonly called the **stele,** or vascular cylinder.

The transfer of water from the soil into the root hairs and across the cortex and stele to the xylem can be explained on purely physical principles. The water present as a film around the particles of soil contains some dissolved inorganic salts and perhaps some organic compounds. The concentration of solutes in this water is low, however, and the solution is hypotonic to the protoplasm of the root hairs. Water therefore passes by diffusion from a region of higher concentration (of water) to a region of lower concentration inside the root hairs. Since the water is passing through a differentially permeable membrane, the plasma membrane of the root hair cell, the process is one of osmosis. As the root hair cells take in water, they become, in turn, hypotonic to the underlying cells of the cortex, and water passes inward, via the various tissues of the root, to the ducts of the xylem. The liquid in the xylem, the plant sap, contains sugars and salts and is hypertonic to the surrounding tissues. Water therefore tends to pass into the xylem vessels and increases the pressure of the liquid within the ducts, just as water entering a sugar-filled cellophane membrane does (p. 41). This **root pressure** is one of the factors moving sap upward through the root and stem. Since water is constantly escaping from the leaves (p. 113), and new sugars are formed in the leaves and sent via the phloem to the roots, the hypertonicity of the root sap is maintained and further absorption of water is possible.

The entrance of inorganic salts is accomplished in part by simple diffusion, for any ion present in greater concentration in the soil water than in the root hair protoplasm will tend to diffuse into the root hair. But certain roots, at least, are able to take in inorganic salts by processes, the details of which are not understood, which require the expenditure of energy. An increase in the rate of cellular metabolism can be detected in such roots when they are absorbing inorganic ions against a concentration gradient. The concentration of inorganic salts in plant sap is rather low and so large volumes of it must flow to the stems and leaves to provide the necessary nutrients.

Botanists distinguish two common types of roots: **diffuse roots,** made of many slender branches of nearly equal size, and **tap roots,** large single roots, usually growing directly downward, with many, much smaller, secondary rootlets (Fig. 46). In many plants, additional roots may grow from the stem or leaf. Roots which develop from any structure other than the primary root or one of its branches are called **adventitious roots.**

70. THE ENVIRONMENT OF ROOTS: SOIL

The soil provides a solid, yet penetrable foundation in which plants can anchor themselves and also serves as a reservoir for the water and minerals needed by plants. Plant growth requires (in addition to carbon, hydrogen, oxygen and nitrogen) the minerals calcium, iron, magnesium, potassium, phosphorus and sulfur plus the trace elements boron, copper, cobalt, manganese and zinc. A deficiency of any one of these will limit plant growth even though all the others are present in optimal amounts. The mineral portion of soils is made up of particles varying in size from large rocks to fine clay, broken off of the rock in the earth's crust by the physical and chemical processes of weathering or by the action of plants and animals. Soils made solely of minerals are unable to support plant growth. A productive soil must contain organic material, known as **humus,** as well. Humus is derived from the decaying remains of plant and animal bodies; it gives soil a brown or black color. Humus, in addition to supplying plant nutrients, increases the porosity of soil so that proper drainage and aeration can occur, and increases the ability of the soil to absorb and hold water. In addition to humus, soil contains innumerable bacteria and fungi which bring about decay and are of prime importance in keeping the soil in good condition. The soil also contains a host of animals, ranging from microscopic forms to moles, gophers and field mice. One of the most important soil animals is the earthworm which, by its burrowing, keeps turning soil over and mixing in organic material. It has been estimated that an acre of good soil will contain about 50,000 worms and that these individuals, in the course of a growing season, will take in some 18 tons of earth and deposit it on the surface. The soil, then, is another major ecosystem containing a large number of different kinds of animals, bacteria and plants that comprise an interrelated biologic complex. The productiveness of a soil depends on such things as its chemical composition and porosity, its content of air and water, and its temperature.

Soils are classified according to the particle size of their mineral constituents, from coarse gravel (over 2 mm. in diameter) through several classes of gravel, sand and silt to clay (particle size less than 0.002 mm. in diameter). A soil is called **loam** if it has about one-half sand, one-fourth silt and one-fourth clay. Soil texture determines the amount of air and water a soil can contain and hence is an important property of a soil. Since soil water is mostly present as a capillary film on the surface of the mineral particles, a clay soil, with a large number of small particles will have a high water content. A good porous soil can provide sufficient air for root growth but a tightly packed or wet soil will result in the death of the plants present.

One of the major conservation problems in this country is to decrease the amount of valuable topsoil carried away each year by wind and water. Reforestation of mountain slopes, building check dams in gullies to decrease the speed of the run-off water, contour cultivation, terraces and the planting of windbreaks are some of the methods that are currently being used successfully to protect the topsoil against erosion.

About one hundred years ago, when botanists began experimenting to determine the kinds and amounts of nutrients necessary for plant growth, they attempted to grow plants in water without soil. After trying many different methods, it was found that plants could be grown in liquid media if they were supported either by screening or sand. This soilless culture of plants (**hydroponics**) makes possible carefully controlled experiments on the requirements of plants for trace elements, but it is not an economically practicable way to grow plants in quantities. The costs of circulating and aerating the water prohibit any large-scale commercial development of soilless agriculture, but it was used by the Navy in World War II to produce food on certain Pacific islands which could not have supported agriculture otherwise.

71. THE STEM AND ITS FUNCTIONS

The stem, which in a tree includes the trunk, branches and twigs, is the connecting link between the roots, where water and minerals enter the plant, and the leaves, which manufacture food. The vascular tissues of the stem are continuous with those of root and leaf and provide a pathway for the exchange of materials. The stem and its branches support the leaves so that each leaf is exposed to as much sunlight as possible. Stems also support flowers and fruits in the proper position for reproduction to occur. The stem is the source of all the leaves and flowers produced by a plant, for its growing points produce the primordia of leaves and flowers. No seed plant consists solely of roots and leaves, even though all the processes necessary for plant maintenance can occur in these organs. Short-stemmed plants such as dandelions are at a disadvantage, for they can grow only where not overshadowed by taller ones.

Roots and stems are sometimes confused, for many kinds of stems grow underground and some roots grow in the air. Ferns and grasses are examples of plants that have underground stems, called **rhizomes.** These grow just beneath the surface of the ground and give rise to above-ground leaves. Thickened underground stems, adapted for food storage, called **tubers,** are found in plants such as the potato. An onion bulb is an underground stem surrounded by overlapping, tightly packed scale leaves. Roots and stems are structurally quite different: stems, but not roots, have nodes which give rise to leaves. The tip of a root is always covered by a root cap whereas the tip of a stem is naked unless it terminates in a bud. There are, in addition, certain differences in the arrangement of xylem and phloem in the two structures.

Plant stems are either herbaceous or woody. The soft, green, rather thin **herbaceous** stems are typical of plants called **annuals.** Such plants start from seed, develop, flower, and produce seeds within a single growing season, dying before the following winter. Another type of herbaceous plant is the **biennial,** which has a two-season growing cycle. During the first season, while the plant is growing, food is stored in the root. Then the top of the plant dies and is replaced in the second growing season by a second top which produces seeds. Carrots and beets are examples of biennials. Plants whose stems are soft and perishable, and which are supported chiefly by turgor pressure, are called **herbs.**

Quite different from the herbaceous annuals and biennials are the **woody perennials,** which live longer than two years and have a thick tough stem, or trunk, covered with a layer of cork. A **tree** is a woody-stemmed perennial that grows some distance above ground before branching and so has a main stem or trunk. A **shrub** is a woody perennial with several stems of roughly equal size above the ground line. Some common shrubs are lilacs, oleanders and sagebrush.

It is generally believed that woody-stemmed plants are more primitive than the herbaceous ones, for the available evidence indicates that the first true seed plants were woody-stemmed perennials. In past geologic ages such plants grew as far north as Greenland, but with the change in climate toward the end of the Mesozoic era, some of these plants were killed by the advancing cold and the species were forced to retreat toward the equator. Still other woody plants adapted to the cold by evolving a life cycle in which growth and flowering were completed in the warm summer of a single

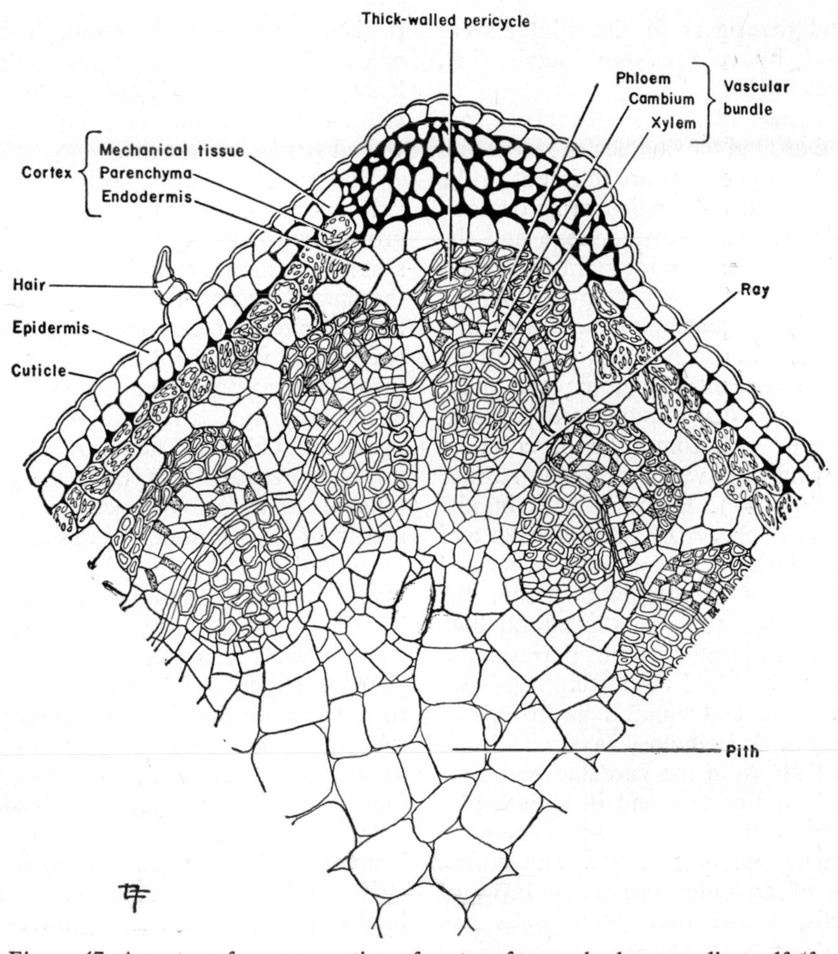

Figure 47. A sector of a cross section of a stem from a herbaceous dicot, alfalfa.

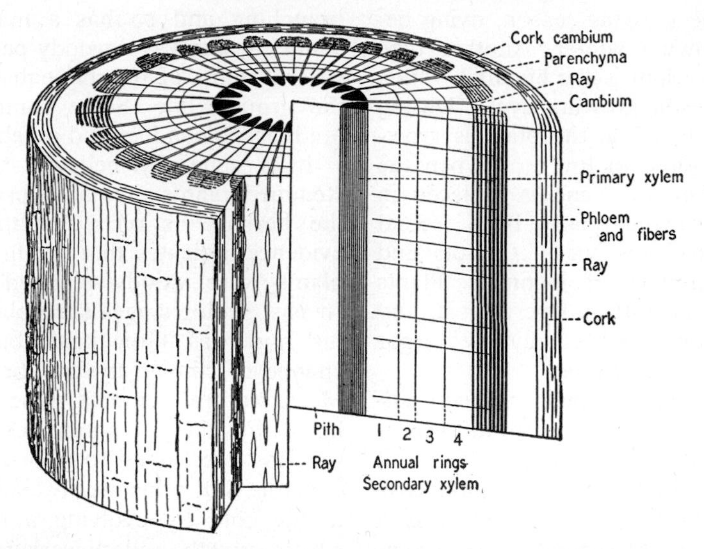

Figure 48. Diagram of a four year old woody stem, showing transverse, radial and tangential sections. (Weatherwax: Botany.)

year and the rigors of the winter were withstood by cold-resistant seeds; i.e., they became herbs.

The tissues of herbaceous stems are arranged around the bundles of xylem and phloem but the details are different in the two main groups of Angiosperms or flowering plants, the Dicotyledoneae and the Monocotyledoneae. In the stem of a dicot, such as the sunflower or clover, there is a circular arrangement of the xylem and phloem bundles which subdivides the stem into three concentric regions: the outer **cortex,** the **vascular bundles,** and the central core or **pith** composed of colorless parenchyma cells which serve as storage places (Fig. 47). Each vascular bundle has an outer cluster of phloem cells and an inner cluster of xylem cells, separated by a layer of meristematic tissue, the **cambium** (Fig. 47). On the lateral border of the phloem is the **pericycle,** a layer of thick-walled supporting cells. Between the vascular bundles lie groups of cells known as **medullary rays,** which extend radially from the vascular region to both pith and cortex, and distribute materials from the xylem and phloem to these inner and outer parts. The cortex consists of an outer protective layer of epidermis, whose outer cell walls are thickened and contain cutin. Inside the epidermis is a layer of thick-walled collenchyma cells, which are supporting tissues, and inside this is a layer of thin-walled parenchyma cells.

The stems of woody plants resemble herbaceous ones during their first year of growth but by the end of the first growing season, additional cambium has formed in the medullary rays so that there is a continuous circle of cambium extending between the vascular bundles as well as through them. In each successive year the cambium forms an additional layer of xylem and phloem. The phloem formed in this way eventually replaces the primary phloem and forms a continuous thin sheath of food-conducting tissue just outside the cambium. The yearly deposits of xylem form the **annual rings** (Fig. 48). These can be distinguished because the xylem vessels formed in the spring of the year are larger, and hence appear lighter, than those formed in the summer. Only the youngest, outermost layer of xylem, known as the **sapwood,** carries sap to the leaves; the inner layers of dead, hard xylem cells and fibers, known as the **heartwood,** increase the strength of the stem, and accommodate the increasing load of foliage as the tree grows.

The width of the annual rings varies according to the climatic conditions prevailing when the ring was formed, so that it is possible to infer what the climate was at a particular time, several hundred or even thousands of years ago, by examining the rings of old trees. An interesting application of this technique is the dating of the time of construction of certain Indian pueblos in the Southwest by an analysis of the annual rings in the logs used. By comparing the pattern of thick and thin annual rings in these logs with those of trees whose year of felling is known it has been possible to determine the year of construction of these pueblos and to deduce the weather conditions for the past hundreds of years.

The cambium is also important in the healing of wounds. When the stem's outer layer is removed through injury, the cambium grows over the exposed area and differentiates into new xylem, phloem and cambium, each of which is continuous with the same type of tissue in the uninjured part of the plant. Certain cells in the outer cortex of most woody plants become meristematic and form a second, or **cork cambium.** In some plants the divisions of the cork cambium occur unequally and irregularly, producing a bark that is rough and full of ridges. These outer cork cells become impregnated with a waterproof, waxy material and eventually die and fall off, partly under the stress of wind and rain, partly because of the outward pressure of the growing tissues within.

The stem of a monocot, such as corn, has an outer epidermis made of thick-walled cells and pierced by openings (stomata) similar to the ones in leaves. The epidermis and the cells of the cortex just beneath the epidermis become thick-walled and lignified and serve as supporting tissues. The vascular bundles are scattered throughout the stem instead of being arranged in a ring as in the dicots (Fig. 49). The bundles are smaller and more numerous in the outer part of the stem.

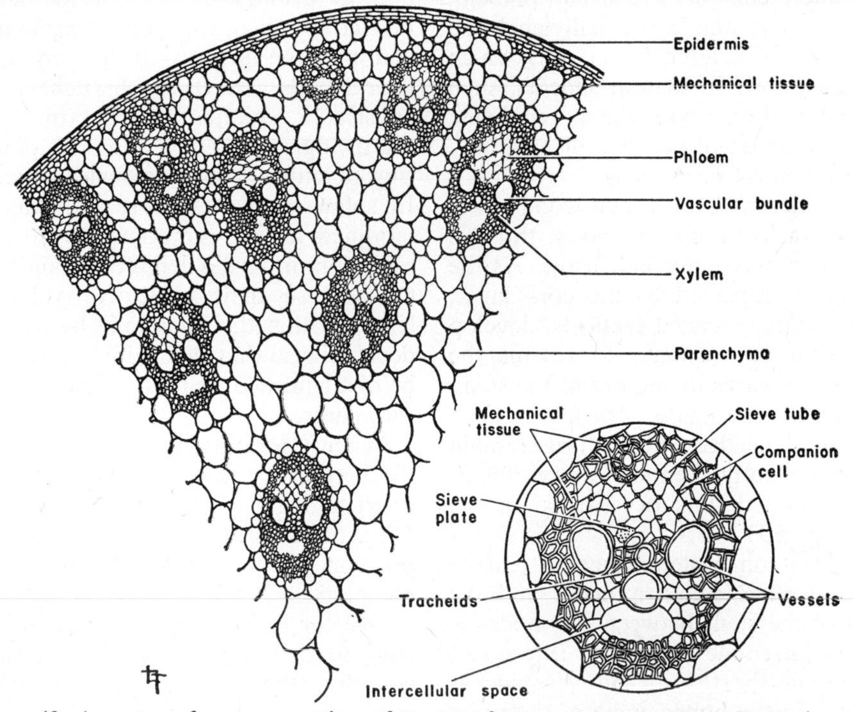

Figure 49. A sector of a cross section of a stem from a monocot, corn. *Inset:* An enlarged view of a vascular bundle containing phloem (sieve tubes and companion cells) and xylem (tracheids and vessels).

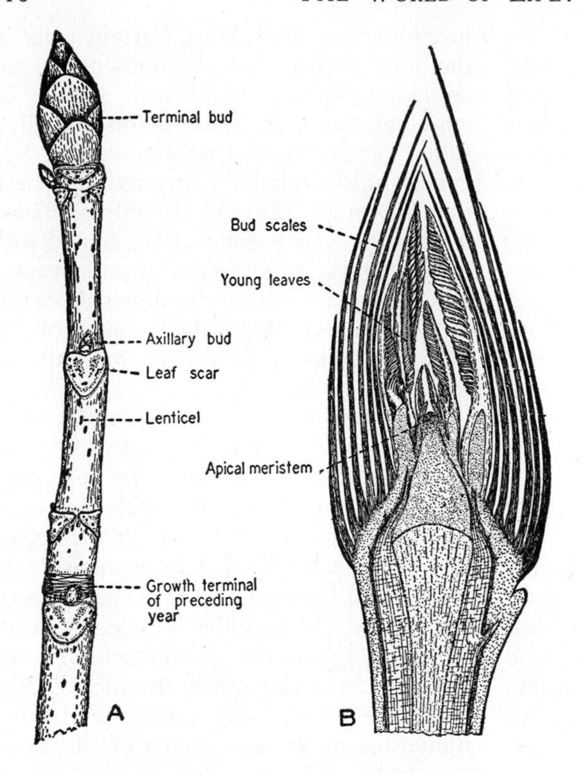

Figure 50. A, Buckeye twig, showing buds and scars; *B,* longitudinal section through a terminal bud from a hickory tree. (Weatherwax: Botany.)

Each bundle contains xylem and phloem, but has no cambium; it is usually enclosed in a sheath of sclerenchyma cells which provide support. In some monocots, such as wheat and bamboo, the parenchyma cells of the center of the stem disintegrate, leaving a central pith cavity.

The epidermis of a young woody twig has stomata like those of leaves, through which gases may enter and leave. As the epidermis is replaced by the cork layer, special openings, called **lenticels,** develop in place of the stomata to permit the movement of gases in and out of the stem. But unlike the stomata, which can open and close as needed, the lenticels remain open permanently. Lenticels are visible as slightly elevated dots or streaks on the bark (Fig. 50).

The point on a stem where a leaf or bud develops is called a **node,** and the section of the stem between two nodes is called an **internode** (Fig. 50). Internodes may be quite short or up to several inches in length. In the upper angle of the point of junction of a leaf with the stem, a bud usually appears, called a lateral or axillary bud to distinguish it from the terminal bud at the tip of the stem. Terminal buds continue the growth of the main stem; lateral buds give rise to branches. A bud consists of a number of embryonic leaves, a growing point, and (in woody plants) a ring of outer protective scales (Fig. 50). In some species these scales, which are modified leaves, are coated with a waxy secretion or have a dense covering of hairs to increase their protective value. The leaves within the buds may be fairly well developed so that their ultimate shape can be distinguished or they may be shapeless rudiments.

When a terminal bud begins to grow in the spring, its covering scales are forced apart and fall off, leaving a ring of scars (Fig. 50). These scars, which mark the position of the end of the stem at the completion of the season, may remain visible for several years, so that it is possible to determine the age of a twig by counting the number of terminal bud scars.

The over-all shape of a tree or shrub is determined by the position, arrangement and relative strength or activity of the terminal and axillary buds. In a tree with a strong terminal bud, such as a pine or poplar, the twig produced by the terminal bud is much more vigorous than those produced by lateral buds and a single, strong, straight main trunk results. Plants with vigorous lateral buds have strong horizontal branches and a spreading shape. Other factors influencing the shape of a tree are the direction and strength of the prevailing wind, and the presence of other trees nearby.

In addition to terminal and axillary buds, lenticels and bud scars, the surface of a twig may show certain other structures. These include **leaf scars** (Fig. 50) left when the stalk of a leaf breaks away from the twig, and **fruit scars,** produced by the breaking off of fruit.

It is possible to show experimentally that water and salts are carried up the stem in the xylem and that organic materials are carried in the phloem. If a cut is made all around a stem, deep enough to penetrate the phloem and cambium, but not the xylem, the leaves remain in good condition and do not wilt for some time, indicating that they are receiving sufficient water. They must be getting the water via the xylem since the phloem sieve tubes have been cut. The parts of the plant above the injury also continue to grow in diameter, showing that they are receiving food products from above, via the phloem, but the parts below the injury cease growing and eventually die when the stored organic materials of the lower part of the plant have been exhausted. By special techniques it is possible to cut the inner xylem and leave the outer phloem relatively intact. When this is done, the leaves wilt and die almost immediately, showing again that the water reaches them by the xylem and not via the phloem. Some plants have translucent stems and when one of these is cut and placed in a solution containing some colored dye, the course of the dye through the stem, petiole and leaf can be followed. When the stem is cut transversely and examined under the microscope, the dye can be shown to be present in the xylem.

72. THE LEAF AND ITS FUNCTIONS

The characteristics of leaves are correlated with the fact that each leaf is a specialized nutritive organ whose function is to carry on photosynthesis. Because of this, a leaf is broad and flat to present a maximum of surface to sunlight and to have a maximum surface area for the exchange of gases, oxygen, carbon dioxide and water vapor.

Leaves originate as a succession of lateral outgrowths, called leaf primordia, from the apical meristem at the tip of the stem. Each outgrowth undergoes cell division, growth and differentiation and finally a miniature, fully formed leaf is produced within the bud. In the spring, the leaves grow rapidly, forcing apart the bud scales and, largely by the absorption of water, unfold, enlarge, and reach their full size. Leaves have no meristematic tissue and thus do not live long—a few weeks in some desert plants, a few months for most trees, and up to three or four years for the needle-shaped evergreen leaves.

The leaf of a typical dicot consists of a stalk, called the **petiole,** by which it is attached to the stem, and a broad **blade,** which may be one simple structure or a compound one, with two or more parts. The petiole may be short and in some species is completely lacking. Like a stem in cross section, it is composed of vascular bundles, attached at one end to those of the stem and at the other end to the midrib of the blade. Within the blade the vascular bundles fork repeatedly and form the **veins.**

A microscopic section through a leaf (Fig. 51) shows it to be composed of several types of cells. The outer cells, both top and bottom, make up a colorless, protective layer called the **epidermis** which secretes a waxy material called **cutin.** The epidermal cells—thin, tough, firm-walled and translucent—are well adapted to give protection to the underlying cells and decrease water loss yet admit light. Scattered over the epidermal surface are many small pores, called **stomata,** each surrounded by two **guard cells.** These cells, by changing their shape, can change the size of the aperture and so control the es-

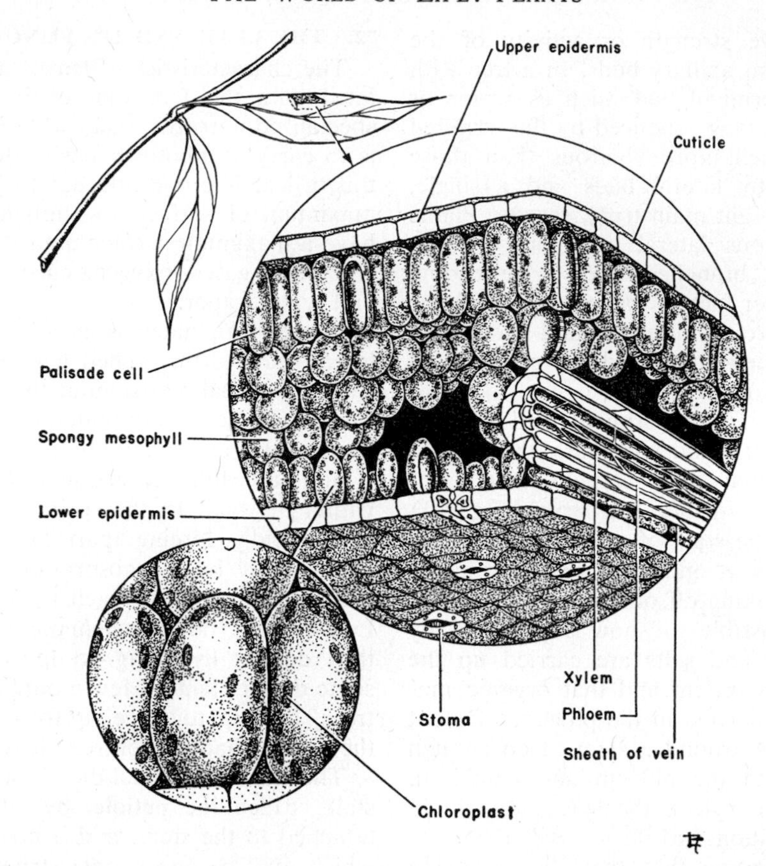

Figure 51. A diagram of the microscopic structure of a leaf. Part of a small vein is visible to the right.

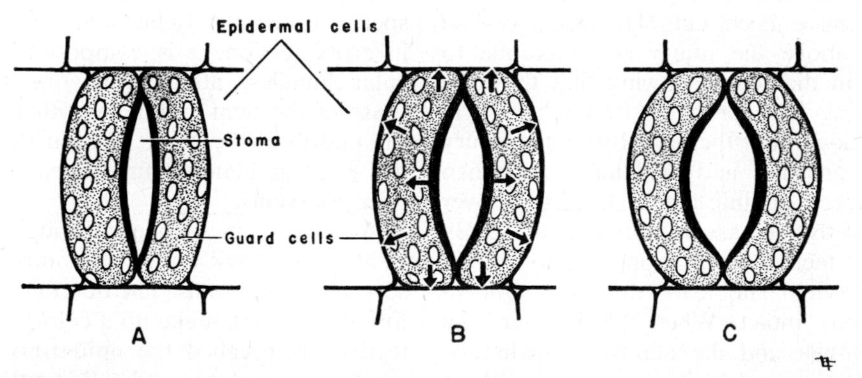

Figure 52. Diagrams illustrating the regulation of the size of the stoma by the guard cells. *A,* Partially closed condition. *B,* When osmotically active substances such as glucose are produced, water enters the guard cells, turgor pressure increases, and the guard cells buckle so as to increase the size of the stoma. *C,* Stoma open.

cape of water and the exchange of gases. In contrast to epidermal cells, guard cells contain chloroplasts. There are 50 to 500 stomata per square millimeter of leaf, many more on the lower than on the upper surface.

A pair of bean-shaped guard cells surrounds each stoma (Fig. 52 *A*). The guard cells have thicker walls on the side toward the stoma than on the other sides. In the light, photosynthesis occurs in these guard cells, glucose and other osmotically active substances are produced, turgor pressure increases, and the guard cells buckle so as to open the stoma (Fig. 52 *B*). The leaf is then able to carry on photosynthesis, for the open stomata permit the entrance of carbon dioxide. In the absence of light, photosynthesis ceases in the guard cells as in all cells, turgor pressure decreases and the stomata close. If, on a hot, dry day, the amount of water supplied by the roots is too low, the guard cells will be unable to maintain the turgid state and hence will close, effectively conserving the decreased supply of water.

Most of the space between the upper and lower layers of the leaf epidermis is filled with thin-walled cells, called **mesophyll,** which are full of chloroplasts. The mesophyll layer near the upper epidermis is usually made of cylindrical cells, called **palisade cells,** closely packed together and so arranged that their long axes are perpendicular to the epidermal surface. The rest of the mesophyll cells are very loosely packed together, with large air spaces between them.

The veins of a leaf branch and rebranch repeatedly to form an extremely fine network, so that no mesophyll cell is far from a vein. Each vein contains both xylem and phloem tissues. The xylem is on the upper side of the vein, the phloem on the lower. In the smallest veins there are only a few xylem vessels and tracheids and a few phloem sieve tubes.

Leaves may serve other functions in addition to that of manufacturing food. The leaves of a number of desert plants are thick and fleshy, and serve as storage places for water. The leaves of pond lilies and other aquatic plants have large air spaces to provide buoyancy. A few leaves, such as those of cabbages, store considerable amounts of food. The insect-trapping abilities of the leaves of plants such as the pitcher plant and the Venus fly trap have been discussed.

The fall of leaves in the autumn of the year is brought about by changes at the point where the petiole is attached to the stem. A special layer (called the **abscission layer**) of thin-walled cells, loosely joined together, extends across the base of the petiole, weakening the base of the leaf. The part next to the stem becomes corklike and forms a protective layer which will remain when the leaf falls off. When the abscission layer has formed the petiole is held on only by the epidermis and the easily broken vascular bundles, so that a high wind will bring about the fall of the leaf. The change in color of the leaves is effected partly by the decomposition of the green chlorophyll, which exposes the yellow xanthophyll and orange carotene, previously hidden by the green pigment, and partly by the formation of red and purple pigments—anthocyanins—in the cell sap.

73. TRANSPIRATION

The leaves of a plant are normally exposed to the air, and they will lose moisture by evaporation unless the air is saturated with water vapor. The sun's heat vaporizes the water from the surfaces of the mesophyll cells, and the resulting water vapor passes through the stomata and escapes. This loss of water, called **transpiration,** may occur in all parts of the plant exposed to the air, but most of it occurs in the leaves. If the plant has an adequate supply of water, the stomata remain open and an amazing amount of water is transpired. But if the plant is not getting sufficient water from its roots, the guard cells around the stomata will become less turgid and the stomata will close, thereby conserving water.

In sunlight an average plant will transpire about 50 cc. of water per square meter of leaf surface per hour. An average corn plant uses more than 50 gallons of water in the course of a growing season and a medium-sized tree will transpire that much in a single day. The amount transpired varies widely in different plants; for example, it is estimated that an acre of corn will transpire 350,000 gallons of water in a growing season whereas an acre of cactus in the Arizona desert will transpire no more than 275 gallons in a whole year. The amount of water vaporized from the leaves of trees in a forest is enough to influence significantly the rainfall, humidity and temperature of the region.

Transpiration contributes to the economy of the plant by assisting the upward movement of water through the stem, by concentrating in the leaves the dilute solutions of minerals absorbed by the roots and needed for the synthesis of new protoplasm, and by cooling the leaves, in a manner analogous to the evaporation of sweat in animals. Although the leaf absorbs some 75 per cent of the sunlight reaching it, only about 3 per cent is utilized in photosynthesis. The rest is transformed into heat which must be removed or it would kill the tissues of the leaf. Some of this heat is removed by the vaporization of water, for 540 Calories are required to convert a liter of water to water vapor.

When water is lost by evaporation from the surface of a mesophyll cell, the concentration of solutes in the cell water increases, the cell becomes slightly hypertonic. Water thus tends to pass into it from neighboring cells that contain more water. These cells in turn receive water from the tracheids and vessels of the leaf veins. During transpiration, then, water passes by the purely physical process of osmosis from the xylem vessels of the veins, through the intervening cells, to the mesophyll cells next to the air spaces of the leaf, where it is vaporized. In fact, a continuous stream of water passes from the soil into the vascular system of the roots, up through the stem and petiole to the veins of the leaf blade.

74. THE MOVEMENT OF WATER

The ascent of sap is brought about by transpiration and root pressure acting together. Root pressure is the positive pressure of the sap in the ducts at the junction of root and stem, generated by the hypertonicity of the sap in the roots to the water in the surrounding soil. Although it was formerly believed that root pressures were small, better techniques for measuring root pressures have shown that even a small plant such as the tomato can generate a root pressure of 12 atmospheres, enough to move water to a height of 384 feet. Root pressure is measured by cutting off the stem of a plant and fitting a mercury manometer to the stump. Only when the problem of fitting a leak-proof connection that would not injure the stump had been solved was it possible to get accurate measurements of root pressure. In the spring, before leaves have been formed, root pressure is the sole cause of the rise of sap.

Once leaves have developed, the continued ascent of water is brought about largely by the process of transpiration. The constant evaporation of water from the cells of the leaf and the production of osmotically active substances by photosynthesis combine to keep the leaf cells hypertonic to the sap in the veins. They constantly draw water from the upper ends of the xylem vessels and this tends to lift the column of sap upward in each duct. This lifting power can be demonstrated by attaching a cut branch to a glass tube by a watertight connection and putting the other end of the tube in a beaker of water. If an air bubble is introduced into the tube the rate of the water movement can be measured by the movement of the bubble. The water columns in the xylem vessels, being under tension from above, are slightly stretched, but water molecules have a strong tendency to cling together, and the slender column of water in the xylem vessel has a high tensile strength. Thus transpiration provides the pull at the top of the column, and the tendency of the water molecules to stick together, carrying this force through the length of the stem and roots, results in the elevation of the whole column of sap. This explanation of the movement of water in plants, called the **cohesion theory,** because it invokes the tendency of water molecules to cohere or cling together, was devised by Dixon and Joly in 1895.

Plants have become adapted to grow in environments with a wide range of water content, and botanists classify them according to their water preferences as hydrophytes, mesophytes and xerophytes. **Hydrophytes** grow in a very wet environment, either completely aquatic or rooted in water or mud but with stems and leaves above the water. Water lilies, pondweeds and cattails are common hydrophytes. **Mesophytes** are the common land plants that live in a climate with an average amount of moisture—beech, maple, oak, dogwood, and birch are typical mesophytic trees. **Xerophytes** are plants such as yuccas and cactuses that are adapted to live where soil water is scarce. By reducing the number of stomata, developing thick stems and leaves to store water, heavily cutinized surfaces and so on, these plants manage to survive with a limited amount of water. A moment's reflection will reveal that plants living in a salt water marsh on the seacoast are also xerophytes, for although water is present, it is largely unavailable to the plant. The concentration of salts in the water exceeds that in the plant tissues and the plant is unable to get water by simple diffusion.

75. THE STORAGE OF FOOD

In sunlight a green plant may produce more than twenty times as much food as it is using at the moment. At other times, during the night and over the winter season, it consumes more food than it makes. Each plant must therefore accumulate food reserves to tide it over periods when photosynthesis cannot occur. Plants deposit food stores in leaves, stems or roots. Leaves serve as temporary depots for food; they are not suitable for long-term storage for they are too easily and too rapidly lost. The stems of woody perennials serve as storage places for large amounts of food; other plants utilize underground fleshy stems for the purpose. Perhaps the most common storage organs are roots, for, being underground, they are somewhat protected from climatic changes and from the prying eyes of animals.

A plant which sheds its leaves each fall must put away a reserve of food in the stem or roots to carry it through the winter and to provide energy for the growth of new leaves the following spring. The stored food must be in the form of some insoluble substance to prevent its diffusing away, and the usual form of storage is starch. Plants also deposit rich stores of food in their seeds to provide energy for the development of the embryo until the new plant has developed a functional root, stem and leaf. Such seeds, rich in proteins, fats and starches, are an important source of food energy for man and other animals.

76. THE ECONOMIC IMPORTANCE OF PLANTS

We are completely dependent upon plants, both directly for food and indirectly for the food they supply to the animals we use as food. Man obtains foods, seasonings, beverages and a wide variety of drugs from all parts of plants: roots (radishes, sarsaparilla, sweet potatoes, carrots, tapioca, etc.); stems (garlic, sugar cane, white potatoes); leaves (lettuce, spinach, cabbage); flowers (artichokes); seeds (corn, nuts, cocoa, coffee, nutmeg, mustard); and fruits (berries, squash, apples, oranges, eggplant).

The list of nonfood products derived from plants and used in everyday life—including such things as lumber, paint, rubber, soap, cork, cotton and resins, derived from a wide variety of plant organs and secretions—is almost endless.

QUESTIONS

1. What are the functions of roots? Of adventitious roots? What are the parts of a typical root?
2. Describe the processes by which a root absorbs water and salts from the surrounding soil.
3. What are the constituents of a good, rich soil? What is the role of each in plant nutrition?
4. What measures can be taken to prevent the loss of topsoil?
5. What are the functions of stems? How are stems and roots differentiated?
6. Differentiate between herbaceous and woody plants, and annual and perennial plants. Give an example of each.
7. What are the features which distinguish dicots from monocots?
8. What are the functions of (*a*) the cambium, (*b*) stomata, (*c*) heartwood, (*d*) lenticels, (*e*) abscission layer, (*f*) cutin?
9. Describe experiments to show that water

is carried both up and down the stem by the xylem.
10. Discuss the mechanism by which the guard cells regulate the size of a stoma.
11. Discuss the functions of leaves. What is the role of transpiration in the plant?
12. What factors affect the rate of transpiration?
13. In what ways are leaves adapted for food manufacture?
14. What adaptations to the environment are evident in xerophytes?

SUPPLEMENTARY READING

The structure and functions of seed plants is discussed in greater detail in Fuller and Tippo, *College Botany,* Sinnott and Wilson's *Botany: Principles and Problems,* Smith, Gilbert, Bryan, Evans and Stauffer, *A Textbook of General Botany* and Weatherwax, *Botany.*

Chapter 9

Types of Plants: Bacteria

FOR MOST people the words disease and bacteria are associated to such an extent that they cannot hear one without thinking of the other. Yet many diseases are not caused by bacteria and most bacteria do not cause disease. The pathogenic (disease-causing) types are only a small fraction of the thousand or so different species known. The importance of bacteria in the carbon, nitrogen and other cycles has been discussed previously (p. 69).

Bacteria were probably first seen by Antonj van Leeuwenhoek (1632–1723) who was a draper in Delft, Holland. He became interested in the structure of the threads used in making linen and ground some crude microscope lenses to examine them. He found a number of ways to improve the grinding of lenses and with these microscopes, which magnified about 150 diameters, Leeuwenhoek examined almost everything at hand—pond water, sea water, vinegar, pepper solutions (he wanted to find out what made it hot), feces, saliva, semen and many other things. He described the objects he saw in letters written to the Royal Society of London. It was in a letter written to this society in 1683 that he described what were unquestionably bacteria—the size, shape and the characteristic motion of the organisms he described leaves no doubt that they were bacteria.

The extensive research of Louis Pasteur in the 1870's and 80's, which revealed the importance of bacteria as agents of disease and decay, stimulated other scientists such as Robert Koch, Ferdinand Cohn and Joseph Lister and the science of bacteriology blossomed rapidly in the latter part of the nineteenth century. Pasteur made his discovery while studying the "diseases" of souring wine and beer. These processes are caused, he found, by microorganisms which enter the wine or beer from the air and bring about undesirable fermentations which yield products other than alcohol. He found that by gently heating (a process now known as **pasteurization**) the grape juice or beer mash to kill the undesirable organisms and then seeding the cooled juice with yeast, he could prevent these diseases. Another of Pasteur's contributions to bacteriology was his unequivocal demonstration that bacteria cannot arise by spontaneous generation. After his study of the diseases of wine, Pasteur was asked by the French government to investigate another disease of economic importance, a disease of silkworms. When Pasteur found that this, too, was caused by microorganisms he reasoned that many animal and plant diseases might be caused by the invasion of "germs." During his investigations of anthrax, a disease of sheep and cattle, and chicken cholera, he devised a method of treatment, that of **inoculation,** which greatly reduced the death rate from these diseases.

Lord Lister, an English surgeon, was one of the first to understand the significance of Pasteur's discoveries and to apply the germ theory to the procedures of surgical operations. He initiated antiseptic techniques by dipping all his operating instruments into carbolic acid and by spraying the scene of the operation with that germicide. In this way he effected a marked decline in the number of fatalities following operations.

After Pasteur, important contributions were made to the new science of bacteriology by the German physician Robert Koch, who went to great pains to prove absolutely that anthrax is caused by a particular organism. To do this he developed methods for isolating pure cultures of bacteria and for growing these cultures outside the body of the normal host. By extracting blood from an animal suffering from anthrax, and growing the bacteria from it in artificial cultures—transferring them from culture to culture many times—he eliminated all possibility that the disease could be due to anything but the bacteria present. Bacteria from the last culture, when injected into a new animal, produced anthrax; it could no longer be doubted that these organisms were the true cause of the disease.

77. KOCH'S POSTULATES

The evidence necessary to prove that a particular microorganism is the cause of a disease is often condensed into a series of statements or postulates, commonly called Koch's postulates, because of his pioneer work in the field.

1. The microorganisms must be observed in the blood or tissues of the infected animal or plant, and must have a reasonable relationship to the disease symptoms and lesions.
2. The organisms must be isolated from the diseased host and grown outside the body in a pure culture.
3. A portion of this culture must be injected into a second, previously uninfected animal or plant, and symptoms and lesions similar to those in the first host must appear.
4. The microorganism must be observed in, and recovered from the experimentally diseased animal in pure culture.

78. OCCURRENCE OF BACTERIA

There are not many places in the world devoid of bacteria. They have been found as much as 16 feet deep in soil; they are most numerous in the top 6 inches of soil, where it is estimated that there are about 100,000 per cubic centimeter. They are found in fresh and salt water and even in the ice of glaciers. They are abundant in air, in liquids such as milk, and in and on the bodies of animals and plants, both living and dead.

79. CELL STRUCTURE

Bacterial cells are very small, from less than 1 to 10 microns in length and from 0.2 to 1 micron in width. The majority of bacterial species exist as single-celled forms, but some occur as filaments of loosely joined cells. Because of their small size and general similarity of structure, the classification of bacteria usually depends on physiologic or biochemical characters rather than morphologic ones. There are rodlike forms called **bacilli,** spherical forms called **cocci,** and spiral forms (Fig. 53). The bacilli may occur as single rods (Fig. 54) or, as in the bacillus causing anthrax, as long chains of rods joined together. Diphtheria, typhoid fever, tuberculosis and leprosy are all caused by bacilli. The spherical forms occur singly in some species; in groups of two (e.g., the gonococcus, the agent causing gonorrhea); in long chains (spherical bacteria which exist in long chains are called **streptococci);** or in irregular clumps, resembling bunches of grapes (spherical bacteria which occur in such clumps are called **staphylococci,** Fig. 55). There are two types of spiral forms: the **spirilla,** which are less coiled and sometimes resemble a comma (the one causing cholera looks like this); and the **spirochetes,** which are highly coiled and resemble a

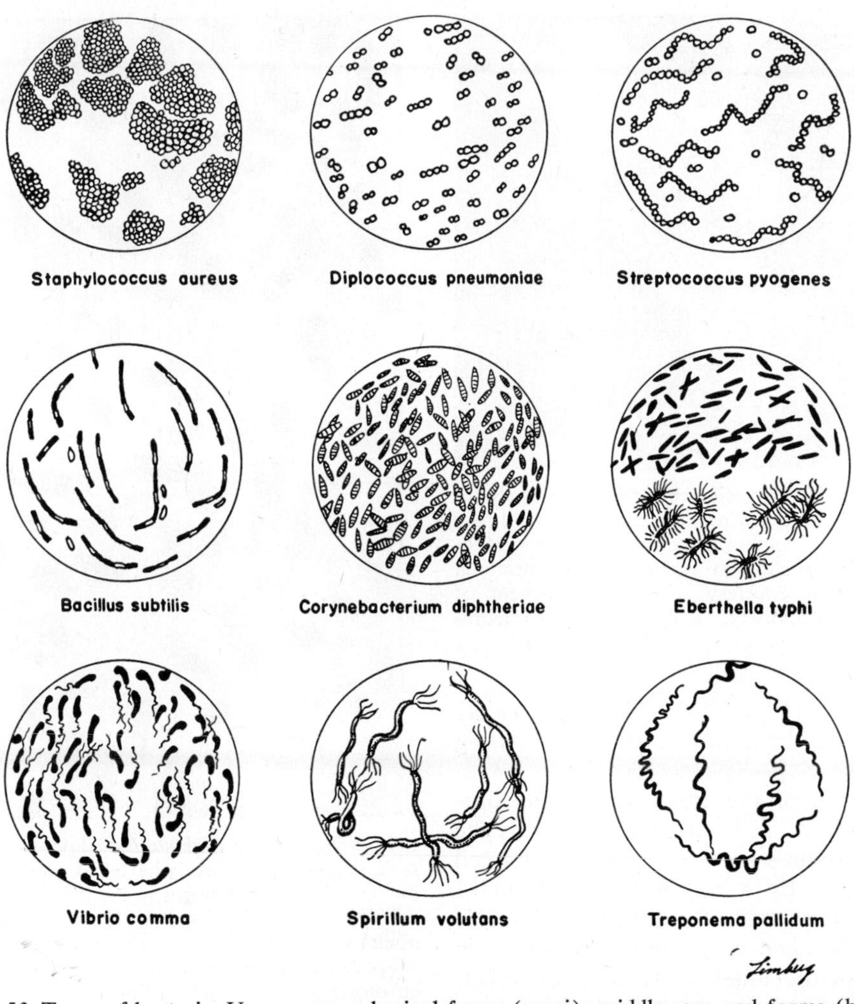

Figure 53. Types of bacteria. Upper row, spherical forms (cocci); middle row, rod forms (bacilli); lower row, spiral forms.

Figure 54. An electron microscope photograph of a bacillus, a rod-shaped bacterium. The bacillus was shadowed with a thin film of gold before being photographed. The thin whip-like flagella are clearly visible. (Photograph courtesy of Dr. C. F. Robinow and Dr. James Hillier of R.C.A.)

Figure 55. An electron microscope photograph of a group of staphylococci, spherical bacteria which occur in bunches like grapes. Magnification: × 25,000 reduced ⅕ in printing. (Photograph courtesy of the Department of Physical Chemistry, Lilly Research Laboratories.)

corkscrew (Fig. 225). The most widely known of the latter is the one causing syphilis.

Because of the small size of the bacterial cell, it has been difficult to study the details of its structure. The bacterial cell is covered by a cell wall which, unlike the cellulose wall of other plants, is made of a substance very similar to the chitin found in the tough body walls of insects and other arthropods. Most bacteria have a slimy capsule outside this cell wall which serves as an additional protective layer. The cytoplasm of bacterial cells is dense and contains granules of glycogen, proteins and fats. It is difficult to demonstrate the presence of nuclei in bacteria and a controversy raged for a long time as to whether these organisms really had an organized nucleus. But with improved methods of preparing and staining bacterial cells it has been possible to show that nuclei are present (Fig. 56). Cocci usually have one nucleus per cell, rodlike bacilli usually have two or more nuclei per cell.

Many bacteria are able to swim about by the beating of whiplike cellular outgrowths called **flagella.** Most rod and spiral-shaped bacteria have flagella; most spherical ones do not. Some bacteria can

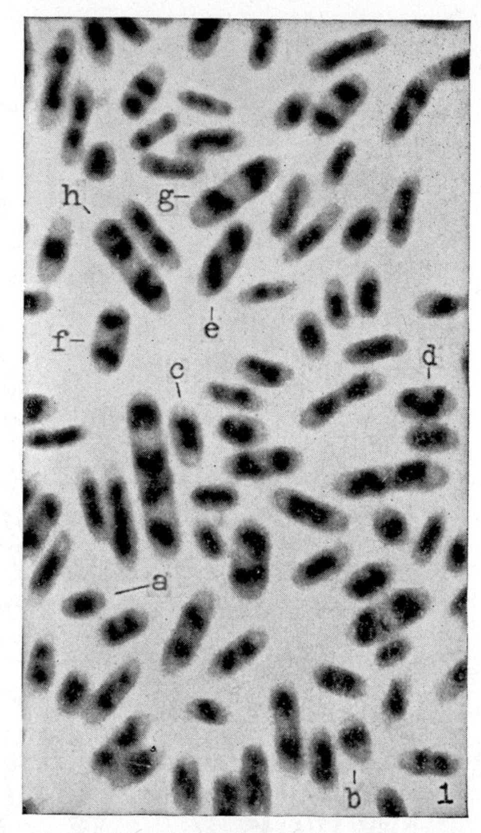

Figure 56. The dark-staining bodies of *Bacterium coli,* which are believed to represent the nucleus. Successive stages in the development of a single, oval-shaped cell with a broad, central nucleus or chromosome (*a*) into a long, rod-shaped, two-celled bacterium with two pairs of chromosomes (*h*). In *e,* a fine strand of nuclear material can be seen connecting the two recently divided chromosomes. (Robinow, C. F., in Cambridge J. Hygiene, Vol. 43.)

travel as much as 2,000 times their own length in an hour in this way. Bacteria can move only in liquids, but by becoming attached to a dust particle or water droplet they may travel considerable distances through the air.

80. REPRODUCTION OF BACTERIA

Bacteria generally reproduce asexually, by simple fission—the cell simply divides into two cells. It was thought at one time that this process was a simple splitting, a sort of pinching in of the sides of one cell to form two daughter cells. With the newer techniques for examining bacterial cells it has been found that the nucleus contains structures very much like the chromosomes of higher plants and animals and that cell division involves a true mitotic process similar to that of higher forms. In some species, the parent cell elongates before cell division but in others cell growth follows division. Cell division occurs in bacteria with remarkable speed, some bacteria dividing once every twenty minutes. At this rate, if there were plenty of food and nothing to interfere, one bacterium could give rise to about 500,000 bacteria within six hours. After twenty-four hours the resulting mass of bacteria would weigh 4,000,000 pounds and in less than a week this single bacterium could give rise to a mass of bacteria the size of the earth! This explains why the entrance of relatively few pathogenic bacteria into a human being can quickly result in disease symptoms. Fortunately for all other forms of life, bacteria cannot reproduce at this rate for a very long time, for they soon are checked by a lack of food, or by the accumulation of waste products. Certain cytologic observations and some genetic studies indicate that something like sexual reproduction, involving the fusion of two different cells and an exchange of hereditary factors, may occur infrequently in bacteria.

81. ADAPTATIONS TO UNFAVORABLE ENVIRONMENTAL CONDITIONS

Under unfavorable conditions such as the drying out of the environment, many bacteria become dormant. The cell loses water, shrinks somewhat, and remains quiescent until water is again available.

Some species form spores to survive in extremely dry, hot or cold environments. Spore formation is not a form of reproduction since only one spore is formed per cell—the total number of individuals does not increase. During spore formation, the cell shrinks, rounds up within the former cell membrane, and secretes a new, thicker wall inside the old one. When conditions are again suitable for growth, the spore absorbs water, breaks out of the inner shell, and becomes a typical bacterial cell. Anthrax bacilli have been found to be able to hatch out after thirty years in the spore form; fortunately, most disease-causing bacteria do not form spores.

82. BACTERIAL METABOLISM

Like other organisms, bacteria have a host of enzymes that mediate and regulate their metabolic processes. A few bacteria, as we have seen, are autotrophic, can synthesize their needed organic compounds from simple inorganic substances present in the environment. Most bacteria are either saprophytes, getting their food from the dead bodies of plants or animals or from organic substances produced by plants or animals, or parasites, living in or on the living body of a plant or animal.

The majority of bacteria, like animals, utilize atmospheric oxygen in respiration; such forms are called **aerobes.** Other bacteria can grow and multiply in the absence of free oxygen, getting energy by the anaerobic metabolism of carbohydrates or amino acids and accumulating a variety of partially oxidized intermediates—alcohol, glycerol or lactic acid. Some bacteria, called **obligate anaerobes,** will grow only in the absence of oxygen; they are quickly killed by molecular oxygen. Others, called **facultative anaerobes,** will metabolize equally well in the presence or absence of oxygen.

The various species of bacteria can utilize almost any organic compound as a source of energy, not only foods such as sugars, amino acids and fats, but waste products such as urea and uric acid found in urine and substances present in feces. Although many kinds of bacteria are killed by penicillin, there is one variety that has become adapted to using this substance for food!

The enzymatic anaerobic breakdown of carbohydrates is known as **fermentation** and the enzymatic anaerobic breakdown of proteins and amino acids is known as **putrefaction.** The foul smells associated with the decay of food and plant or animal bodies are due to the nitrogen- and sulfur-containing compounds which are formed in putrefaction. The substances produced by the activities of one kind of bacteria may be used as a source of energy by another kind of bacteria. The importance of bacteria in the carbon, nitrogen and other cycles has been discussed (p. 69). Without bacteria all the available carbon and nitrogen atoms would eventually be tied up in the dead bodies of plants and animals and life would cease for the lack of raw materials for the synthesis of new protoplasm.

83. METHODS FOR STUDYING BACTERIA

To identify bacteria and study their metabolic processes one must first set up a **pure culture,** one containing only one kind of bacteria, and take pains to insure that it does not become contaminated with other kinds of bacteria. Since bacteria and their spores are present almost everywhere, this is not easy. A nutritive broth is prepared containing salts and one or more organic compounds as a source of energy and then sterilized to kill off any bacteria that may have been present in the water or in the substances added. Sterilization is usually accomplished by putting the tubes containing the medium in an **autoclave,** a cylindrical container that can be sealed, and heating with steam. Culture media, surgical instruments, and so on can be sterilized by autoclaving for twenty minutes at 15 pounds pressure, a temperature of 121° C. Some culture media contain complex organic substances that are broken down by this temperature; such media can be sterilized by filtration through a very fine filter. The tops of tubes of culture media are stoppered with cotton plugs that permit air, but not bacteria, to enter.

With a supply of tubes containing sterile nutritive broth, a pure culture can be obtained by **serial dilution,** a method introduced by Joseph Lister. A bit of soil, feces or sputum contains millions of bacteria of anywhere from one to a score or more different species. If this is dissolved in 100 ml. of nutrient medium and mixed, and then 1 ml. of the mixture is removed and placed in another 100 ml. of medium, the second tube contains only 1/100th as many bacteria as the first. If this is repeated a sufficient number of times, taking 1 ml. of mixture from the second tube and putting it in 100 ml. of medium in a third tube, mixing, taking 1 ml. to put in a fourth tube, and so on, eventually a tube is obtained that has only 1 bacterium. These dilutions are carried out rapidly to prevent bacterial growth between transfers. The final tubes are incubated and then the bacteria in each of these tubes will have arisen by the reproduction of a single cell—it will be a pure culture. Once a pure culture has been established it is simply transferred from tube to tube as often as needed to provide the bacteria with fresh nutrients.

Another, somewhat simpler method for preparing pure cultures, invented by Robert Koch, takes advantage of the fact that bacteria cannot spread very far on a solid medium. A nutrient medium is made containing **agar,** a substance extracted from seaweeds which is somewhat like gelatin —it becomes liquid at about 90° C. and solidifies when cooled to about 40° C. The medium is sterilized by autoclaving and then cooled until just about to solidify. The drop of feces, sputum or blood to be analyzed is then mixed with the agar medium and poured in a very thin layer on a flat, covered glass dish. When the medium cools, it solidifies and the bacteria present are held in position. They are free to multiply, but the daughter cells that result are grouped in a small area and form a colony, one large enough to be seen with the naked eye (Fig. 57). With a sterile loop of wire, some of the members of this colony are transferred to a fresh dish of agar and incubated.

Once a pure culture of bacteria has been established, the species can be identified by appropriate tests. The morphologic differences between different species are small and there are many species that look exactly alike under the microscope. Microscopic observation can provide certain clues—whether the bacteria are cocci, bacilli or spiral forms, whether or not they are motile, and whether, in division, the daughter cells separate or remain together in chains or bunches. Other clues can be obtained by growing the bacteria on an agar medium and observing the characteristics of the colony visible with the naked eye—its shape, color, mode of spreading, and so on. Another method of differentiating species is by the use of stains such as the Gram stain devised by the Danish physician, Christian Gram. Bacteria are divided into two large groups, gram-positive and gram-negative, on the basis of whether they do or do not take up a violet stain under specified conditions.

Probably the most important methods of identifying bacteria are biochemical ones, by the kind of substances they need to grow, or the kind of substances they produce. The colon bacillus, *Escherichia coli,* is a normal inhabitant of the human colon; the typhoid bacillus and one of the food-poisoning bacilli, *Shigella,* are not. The normal and the pathogenic bacilli cannot be distinguished on any morphologic basis, but the colon bacillus can ferment lactose while the pathogenic forms cannot.

The fact that some bacteria will grow on a certain medium and others will not is also used to aid in isolating pure cultures. A bit of soil can be suspended in sterile water and samples of this are then placed on five or more different kinds of media. In each, a different spectrum of bacteria will appear which can then be further subcultured. Or, if an investigator wants to find a species of bacteria that can split acetylcholine (p. 349), he can prepare an incubation medium that contains only acetylcholine and salts. He then inoculates samples of this medium with cultures of various species of bacteria. If he finds one species that can grow, he knows it must be able to split acetylcholine to obtain a source of energy.

84. SOME ECONOMIC USES OF BACTERIA

The different species of bacteria produce a wide variety of substances as a result of their metabolic processes, some of which are very useful to man. Many industries are partly or wholly dependent on bacterial action. Large quantities of important chemicals such as ethyl alcohol, acetic acid, butyl alcohol and acetone are produced by specific bacteria. Bacterial action is necessary for curing tobacco, preparing hides for tanning, and for separating the fibers from flax and hemp to make linen and rope. Some other sub-

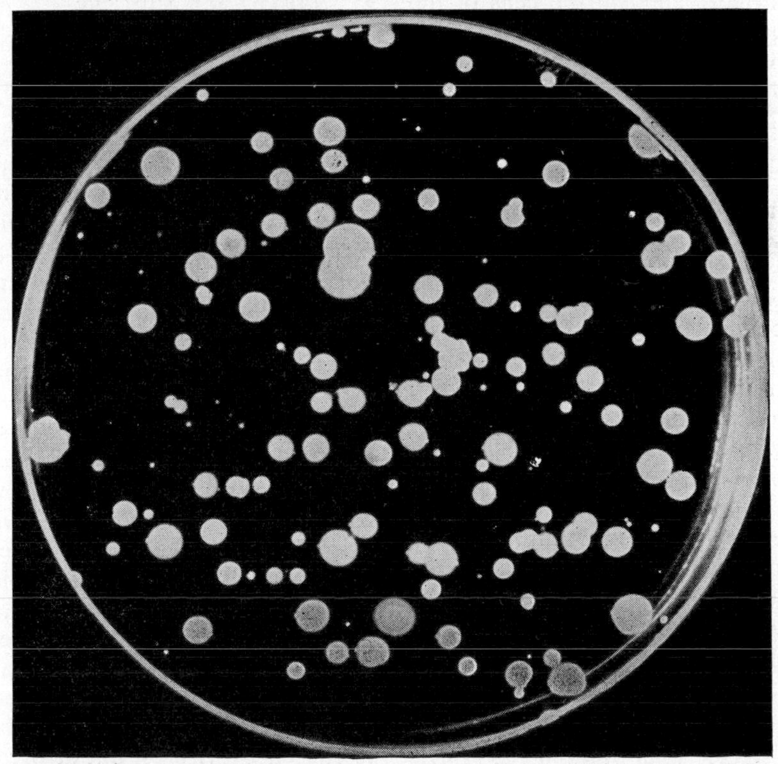

Figure 57. Photograph of colonies of bacteria growing on nutrient agar in a Petri dish. The photograph was taken through the glass cover of the dish. (Photograph courtesy of General Electric Co., Lamp Division, Cleveland, Ohio.)

stances whose production involves the use of bacteria are butter, cheese, sauerkraut, rubber, cotton, silk, coffee and cocoa.

Man also exploits bacteria in disposing of sewage. Sewage is allowed to pass slowly over beds of gravel and sand. The solid matter in the sewage settles out and is acted upon by a variety of bacteria and converted into material that is dried and used as fertilizer. The water gradually trickles through the sand and gravel and any disease-producing bacteria present are killed and digested by the decay bacteria.

85. PARASITIC BACTERIA

The importance of bacteria as the causative agents of diseases in animals will be discussed in Chapter 27. Bacteria are the cause of a number of diseases of plants; some of these diseases affect plants of great economic importance to man. A common plant disease caused by bacteria is **fire blight,** which affects apple, pear and related trees. The bacteria enter the tree through a wound or through a flower, multiply rapidly and kill the cells of the tree so quickly that they appear to be burned. The tree may produce a sticky liquid around the infection which contains bacteria. New trees become infected when drops of this liquid are transported by insects.

A number of kinds of vegetables can be attacked by the bacteria causing **soft rot.** This disease can affect both stored vegetables and ones in the field. Cabbage plants are subject to **black rot,** a bacterial disease. The bacteria enter the cabbage leaves through stomata, get into the xylem vessels, and multiply greatly, clogging the xylem vessels and causing the leaf to wilt for lack of water. A number of plants are hosts of the bacteria causing **crown galls.** The bacteria enter the plant through some wound and their presence stimulates the overgrowth of the host cells so that a tumor-like enlargement, a gall, results.

86. OTHER MICROORGANISMS

Much smaller than bacteria, indeed scarcely larger than very large single molecules, are other forms, called **viruses, bacteriophages** and **rickettsias.** With the exception of the last, these are too small to be seen with ordinary microscopes, and can be photographed only with an electron microscope. None of these can be classified as either plant or animal; their status in the world of living things is not clear. Some investigators believe that they are not living organisms at all, but large protein molecules which simulate some of the characteristics of life such as growth, reproduction and specific organization. The difficulty in deciding whether these forms shall be considered as living or nonliving is largely due to the difficulty in defining life. These forms exhibit some, but not all of the usual characteristics of living things. When we realize that we cannot really answer the question of whether they are living or nonliving, but only the question of whether they shall be *called* living or nonliving, the problem seems much less important. In the following discussion, we shall use the word "organism" for convenience and with certain reservations.

87. FILTRABLE VIRUSES

These ultramicroscopic forms, which take their name from the fact that they are tiny enough to pass through very fine-pored, porcelain filters, were discovered by the Russian botanist Iwanowski in 1892. Iwanowski found that a disease of tobacco plants (called mosaic disease because the infected leaves had a spotted appearance) could be transmitted to healthy plants by the sap of diseased ones, even after it had passed through filters fine enough to remove all bacteria. Today, many diseases of both plants and animals are known to be caused by viruses; a few economically important ones are hog cholera, swine influenza, chicken cancer and the foot-and-mouth disease which has plagued the cattle herds of Mexico and Argentina. The list of virus diseases of man includes smallpox, rabies, infantile paralysis, measles, yellow fever, warts, fever blisters and the common cold. A vast amount of research has been done to explore the possibility that human cancer is caused by viruses, but at present there is no evidence that it is. One type of breast cancer in mice has been found to be caused by a virus-like agent, but human cancers are not infectious, as one would expect a virus disease to be.

Viruses can easily be grown in an artificial medium containing living cells and apparently they can reproduce *only* in such an environment. Many attempts have been made to grow viruses in cell-free culture media containing all known vitamins and amino acids, but to date they have all been unsuccessful. This has given rise to the opinion that viruses do not really reproduce themselves, but are reproduced by other living cells. Viruses are commonly cultured for experimental purposes by injecting them into fertilized hens' eggs (Fig. 58). Some viruses develop attached to the embryonic membranes; others multiply while floating in the extraembryonic fluids. Many viruses which will not grow if injected into an adult chick will grow in chick embryos.

The tobacco mosaic virus was isolated and crystallized by W. M. Stanley in 1935 and since then a number of other viruses have been obtained as crystals. When these crystals are put back into tobacco plants, they multiply and produce the symptoms of tobacco mosaic disease. The fact that some viruses have been crystallized has provided another argument for considering them as nonliving things.

In 1956 Stanley announced that he had been able to separate a virus into its component parts, protein and nucleic acid, and then to recombine these to get an active virus. Moreover, he found he could combine the nucleic acid part of one kind of virus with the protein part of a second kind and get an active "hybrid" virus. This had the biological properties of the strain from which the nucleic acid came, which indicated that the essential nature of the virus is determined by its nucleic acid, not its protein, component.

Estimates of the size of viruses have been made in several different ways: from the size of the filter pores which permit them to pass, from the speed with which they settle when centrifuged, and from measurements of electron microscope photographs of them. These measurements show that viruses vary widely in size; one of the largest—the psittacosis virus, the cause of a disease transmitted by parrots and other birds—is about 275 millimicrons in diameter, and one of the smallest, the one causing foot-and-mouth disease of cattle, is 10 millimicrons in diameter. The electron microscope reveals that some viruses are spherical and others are rod-shaped (Fig. 59). Although individual virus particles cannot be seen, virus-infected cells frequently contain "inclusion bodies" which are visible with an ordinary microscope. These are believed to be huge colonies of viruses.

Each kind of virus usually attacks some specific part of the body; apparently the virus particles can reproduce only in certain kinds of cells and not in all the cells of the body. The virus particles of smallpox, measles and warts attack the skin, those of infantile paralysis and rabies attack the brain and spinal cord, and those of yellow fever attack the liver. Fortunately, many of the infections caused by viruses create lasting immunity against reinfection, and inoculations for smallpox,

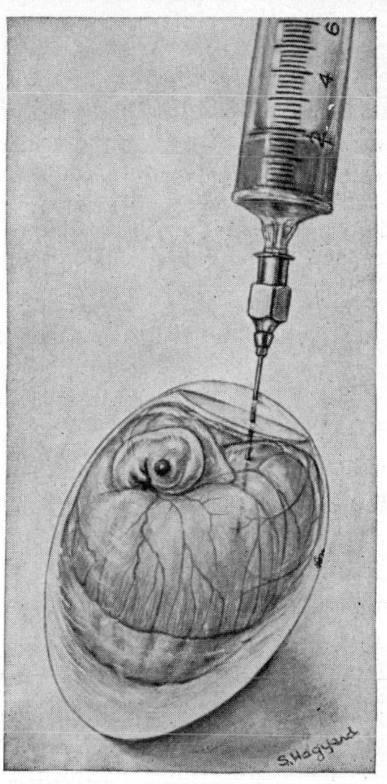

Figure 58. The culture of viruses on the embryonic membranes of the chick. A suspension of virus particles is being injected through a small hole in one end of the egg. (From Therapeutic Notes, October, 1942, Parke, Davis and Company.)

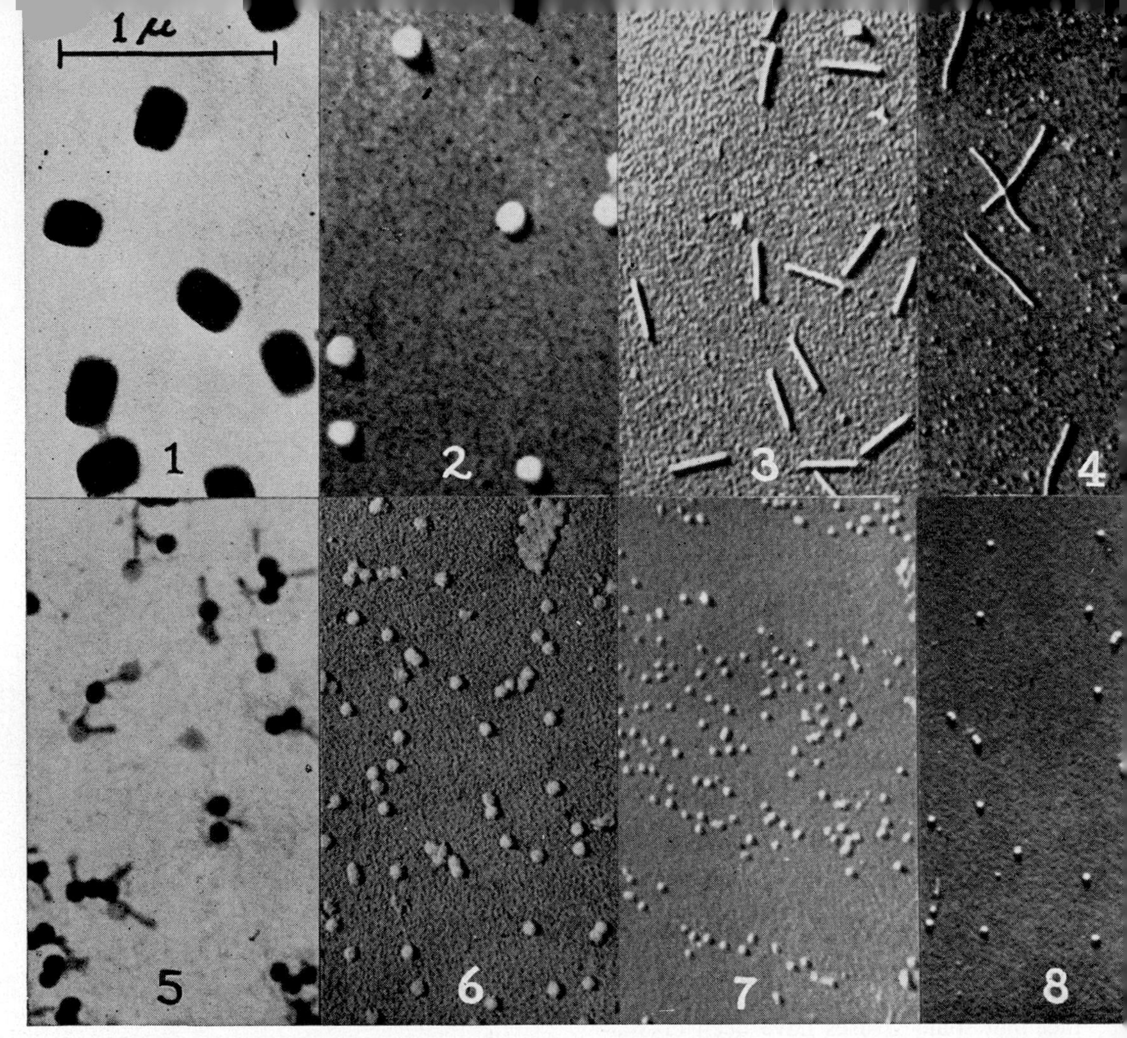

Figure 59. Photographs made with an electron microscope of a variety of viruses, *1,* Vaccinia virus (used in vaccinating for smallpox). *2,* Influenza virus. *3,* Tobacco mosaic virus. *4,* Potato mosaic virus. *5,* Bacteriophages. *6,* Shope papilloma virus. *7,* Southern bean mosaic virus. *8,* Tomato bushy stunt virus. Numbers 2, 3, 4, 6, 7 and 8 were shadowed with gold before being photographed in the electron microscope. (Courtesy of Dr. C. A. Knight.)

rabies and yellow fever are highly successful.

88. BACTERIOPHAGES

Viruses that parasitize bacteria, **bacteriophages,** were discovered in 1917 by the French scientist d'Herelle, who noticed that some invisible agent was destroying his cultures of dysentery bacilli. Like other viruses, bacteriophages are filtrable and will grow only in the presence of living cells—in cultures of bacteria, which they cause to swell and dissolve. These viruses are found in nature wherever bacteria occur and are especially abundant in the intestines of man and other animals.

There are many varieties of bacteriophages; usually each kind of bacteriophage will attack only one kind of bacteria. Electron micrographs (Fig. 60) show that some are about 5 millimicrons in diameter (they vary considerably in size) and that they may be spherical, comma-shaped, or they may have a tail and resemble a ping-pong paddle. The fact that bacteriophages can destroy bacteria suggests, of course, that they might be used to combat bacterial diseases. Many bacteriophage preparations have been given, without much effect, to patients suffering from such diseases as dysentery and staphylococcus infections. Such experiences, as well as laboratory evidence that bacteriophages are ineffective in the presence of blood, pus or fecal material, have led to the abandonment of their use for therapeutic purposes.

To obtain a culture of bacteriophage, an emulsion of feces, soil or sewage is prepared and passed through a fine filter. If bacteriophages are present, a drop of the filtrate added to a turbid bacterial culture will cause the death and dissolution of the bacteria, and the culture becomes clear (Fig. 61). When some of this clear culture is filtered in turn and a drop of the new filtrate placed on a second bacterial culture, the latter will also become clear. In this way, serial transfers of the bacteriophage can be made indefinitely.

89. RICKETTSIAS

These disease organisms were named after their discoverer, Howard Ricketts, who died in Mexico in 1910 of typhus fever while studying the organisms that cause it. They resemble viruses in that with a single exception (a nonpathogenic parasite of the sheep tick) they will multiply only within living cells. Their cellular structure is similar in most respects to that of bacteria (Fig. 62). Some are spherical, others rod-shaped, and they vary in length from 300 to 2000 millimicrons. They are larger than viruses and hence are nonfiltrable and just barely visible under the microscope.

Rickettsias are generally harmless parasites of insects and almost fifty different kinds of rickettsias have been found in the intestinal tracts and salivary glands of insects such as lice, bedbugs and ticks. Some of these are transmitted to man by insect bites; once inside the human body they multiply and produce the symptoms of disease. Only four kinds of rickettsias produce human diseases; the principal diseases in the United States caused by these organisms are typhus fever (not to be confused with typhoid fever, caused by a bacillus), transmitted by the bite of a louse, and Rocky Mountain spotted fever, transmitted by the bite of a tick. Rickettsias are now grown on chick embryos developing inside the egg. In this way large numbers of organisms can be grown to prepare vaccines. Formerly it required the rickettsias grown in 150 lice to prepare enough vaccine to immunize one man!

90. EVOLUTIONARY RELATIONSHIPS

The evolutionary relationships of the bacteria are not at all clear and because of this, their classification in the plant kingdom has undergone a number of changes. One recent scheme, which we will use, classifies them as the phylum Schizomycophyta in the subkingdom Thallophyta, which includes all plants not forming embryos during development, ones commonly called algae and fungi.

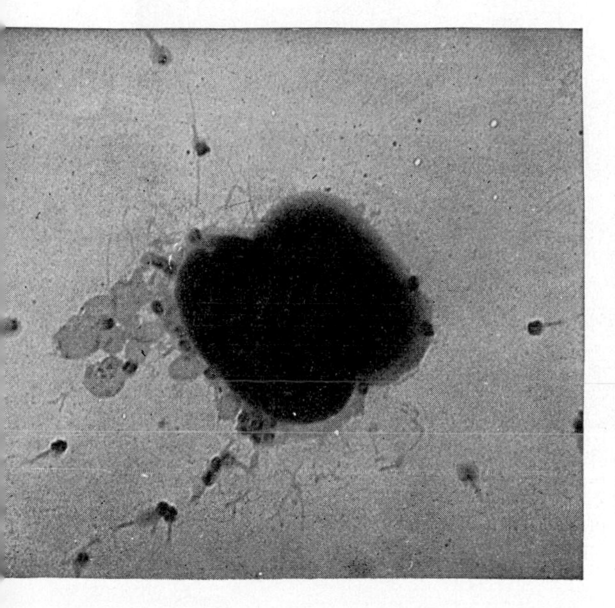

Figure 60. An early stage in the destruction of two colon bacilli (the large, dark bodies) by bacteriophages (the small particles shaped like ping-pong paddles). To the left of the bacilli are disc-shaped particles, remnants of bacterial cells destroyed previously. Some of the bacteriophage particles have already penetrated the outer layer of the bacterial cell. (Reduced in printing from an initial magnification of $\times$ 30,000; from Frobisher: Fundamentals of Bacteriology.)

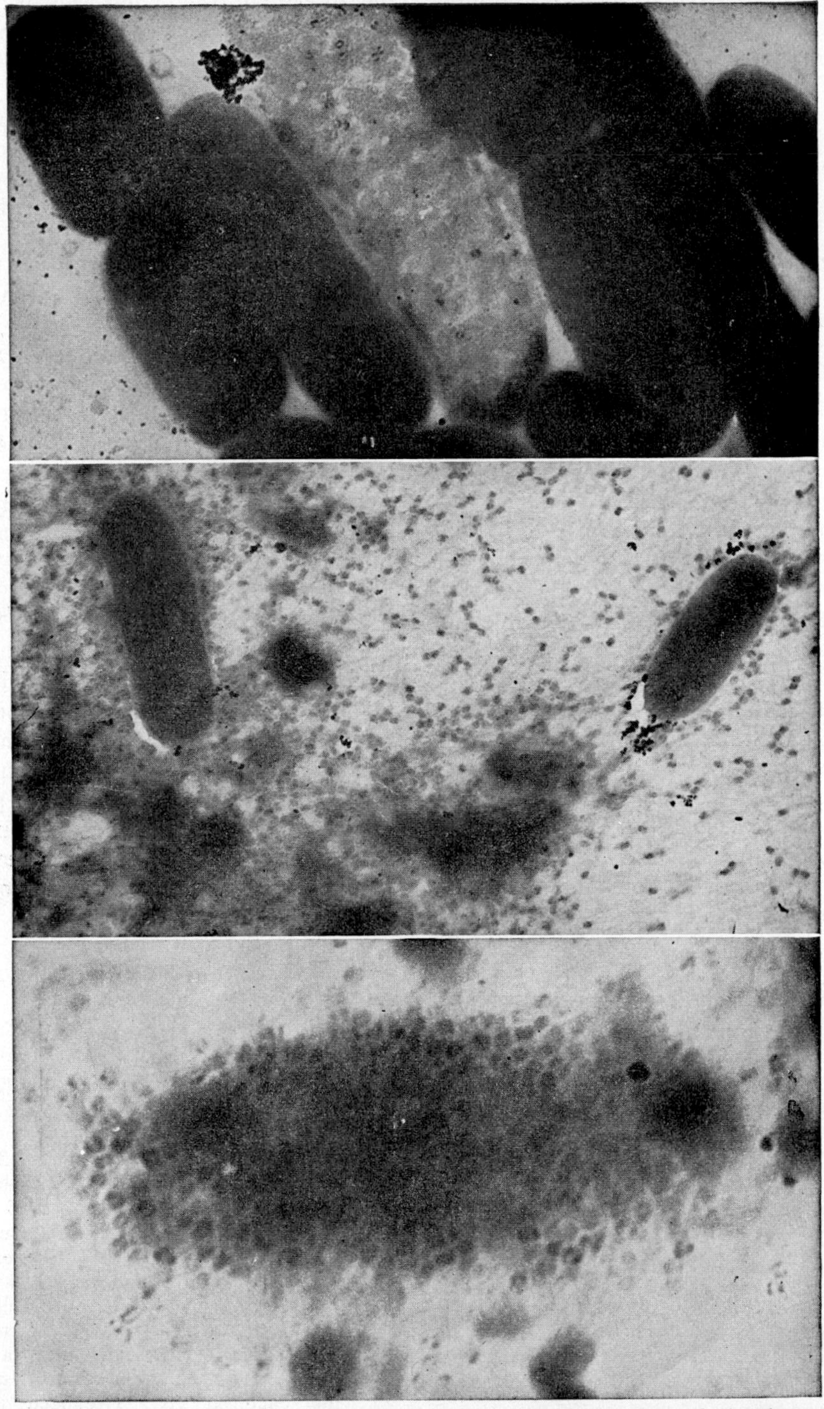

Figure 61. Electron micrographs of the destruction of *Bacterium coli* by bacteriophages. *Top,* Dark, sausage-shaped structures are normal bacteria; lighter one in middle has been attacked and destroyed by 'phage. 'Phage particles are evident within the cell. *Center,* Later stage with more 'phage particles visible and more bacteria destroyed. *Bottom,* Dense mass of 'phage particles occupying the space of the bacillus they have destroyed. (Burrows: Textbook of Microbiology.)

One widely held view is that bacteria have descended from the blue-green algae; after becoming adapted to a saprophytic or parasitic existence, they lost their chlorophyll. This view is based on the general similarity of the cell structure of these two forms. Other investigators believe that the fact that many bacteria have flagella indicates that these organisms descended from some simple flagellated form, perhaps one that also gave rise to the green algae. Still others believe that the present-day heterotrophic bacteria evolved from autotrophic ones like the present-day iron and sulfur bacteria. Autotrophic bacteria may be the most primitive of all plants and may have appeared before any of the chlorophyll-containing ones. It is also possible that different groups of bacteria have descended independently from different ancestors and that any two or even all three of these explanations are true.

Our knowledge of the evolutionary relationships of bacteria with higher organisms is just about as unsatisfactory. Bacteria may be a terminal group in evolution that has given rise to no other forms, or, if they did not arise from the blue-green algae, perhaps the blue-green algae have descended from them.

QUESTIONS

1. What contributions to bacteriology have been made by Leeuwenhoek, Pasteur, Koch and Lister?
2. In what ways do bacteria resemble animals? In what ways do they resemble green plants?
3. Describle the reproduction of bacteria.
4. What is the role of bacterial spores?
5. What is the difference between fermentation and putrefaction?
6. What is the role of bacteria in the nitrogen cycle?
7. Describe a method for obtaining a pure culture of a species of bacteria.
8. How are particular species of bacteria identified?
9. In what ways are bacteria economically important to man?
10. Distinguish between rickettsias and viruses.
11. Discuss the several theories of the evolutionary origin of the bacteria. Which seems most reasonable to you?

SUPPLEMENTARY READING

A well-illustrated exposition of the role of bacteria in our daily life is found in Madeleine Grant's *Microbiology and Human Progress.* The importance of the non-pathogenic bacteria is discussed in Rahn's *Microbes of Merit* and Henrici and Ordal's *Biology of the Bacteria. Rats, Lice and History* by Hans Zinsser and Smith's *Plague on Us* give fascinating accounts of the great plagues and their influence on historic events.

Antonj von Leeuwenhoek and His "Little Ani-

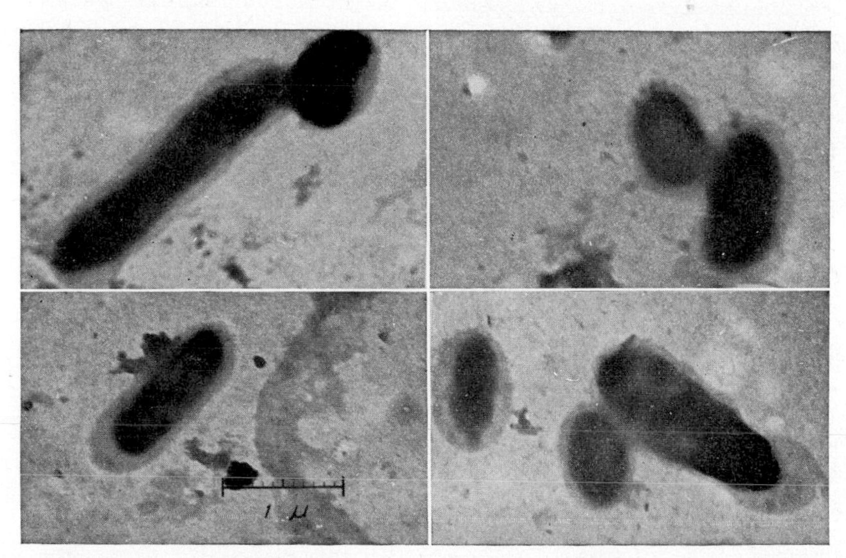

Figure 62. Electron micrographs of the rickettsia causing typhus fever. Note the variations in the size and shape of the cells and the less dense capsule which surrounds each cell. (Lilly Research Laboratories.)

mals," by Clifford Dobell, is a biography of Leeuwenhoek and includes his descriptions of the discovery of bacteria. A general history of the development of theories and knowledge about bacteria is given in William Bulloch's *The History of Bacteriology*. Both Vallery-Radot's and Rene Dubos' biographies of Pasteur are interesting reading. The use of bacteria in biologic warfare is described in *Peace or Pestilence,* by Theodor Rosebury.

Chapter 10

Algae and Fungi

THE MORE primitive plants, which neither form embryos during development nor have vascular tissues, are known as **thallophytes** and are classified in the subkingdom Thallophyta. There are more than 100,000 species in this group, widely distributed in fresh and salt water, on land, and as parasites on other plants or animals. The members of this group range in size from microscopic, single-celled plants to giant seaweeds or kelp that may be several hundred feet long. The body of these plants, called a **thallus,** may show some differentiation of parts, but has no true roots, stem or leaves.

The classification of the thallophytes is difficult and many different systems of classification have been proposed. Two kinds of plants are included in the thallophytes: those which have chlorophyll and can live independently, the **algae,** and those which lack chlorophyll and must live as saprophytes or parasites, the **fungi.** The separation of algae and fungi is to some extent an artificial rather than a natural one, for it separates some plants that are very much alike except in color. Some plants, like *Euglena,* that lose chlorophyll and live as saprophytes if put in the dark but regain it if returned to light, strain the classification even further.

Alga is the Latin word meaning "seaweed," but although most seaweeds are algae, there are many algae that are not seaweeds. Algae are primarily inhabitants of water, fresh or salt, but a few live on rock surfaces or on the bark of trees. The ones living in such comparatively dry places usually remain dormant when water is absent. Algae, by virtue of their tremendous numbers, are important food producers; almost all of the photosynthesis in the seas and most of that in fresh water is carried on by algae. Human beings do not commonly use algae directly as food, but a considerable fraction of human food is fish, which eat either algae or other organisms that depend on algae for food.

Although botanists are not in complete agreement as to how the algae and fungi should be subdivided, one widely accepted classification sets up seven phyla or major groups of algae and three of fungi. One of the important criteria in classifying these forms is their mode of reproduction. With the exception of the blue-green algae, and perhaps some of the bacteria, the members of all these phyla reproduce sexually at some time in their life cycle. Sexual reproduction fundamentally involves the fusion of two special sex cells, or **gametes,** to form a **zygote.** In some forms the two gametes which fuse are identical in size and structure, but usually one is larger and nonmotile and called an **egg** or ovum and the other is smaller and

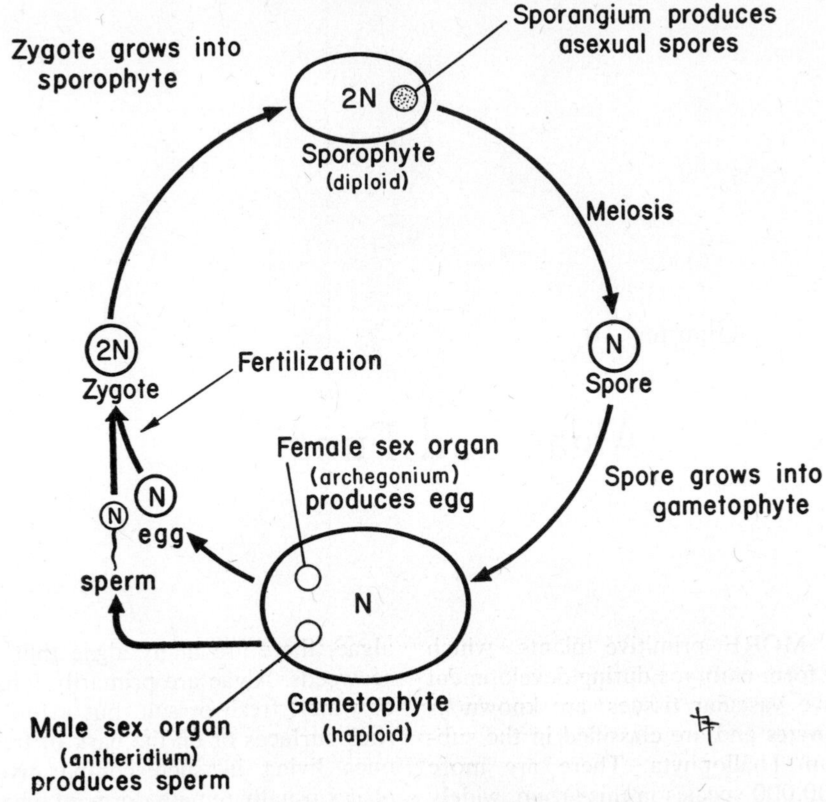

Figure 63. Diagram of the alternation of sporophyte (diploid) and gametophyte (haploid) generations in plants.

motile and called a **sperm.** When two gametes unite, their nuclei fuse together to form a single nucleus but the individual chromosomes within the nuclei remain distinct. The zygote therefore has twice as many chromosomes as either gamete; each gamete furnishes one set of chromosomes and the zygote has two sets. The number of chromosomes per set varies in different species from one or two to several hundred but is constant for any given species. Cells such as gametes, whose nuclei contain one set of chromosomes, are called **haploid;** those such as zygotes, whose nuclei contain two sets of chromosomes, are called **diploid.** Obviously, if a zygote has twice as many chromosomes as a gamete, there must be some mechanism for halving the chromosome number somewhere in the process of forming a gamete, else the chromosome number would double with each generation. A special kind of nuclear division, known as **meiosis,** results in daughter cells with the haploid number of chromosomes. The details of this process are described in a later section (p. 457).

The union of gametes and meiosis are landmarks which serve to divide the life cycle of a plant into two phases: one, from gamete union to meiosis, characterized by diploid cells, and the other, from meiosis to gamete union, characterized by haploid cells (Fig. 63). In many algae and some fungi the zygote nucleus divides by meiosis; thus only the zygote is a diploid cell and the haploid phase may be one-celled or many-celled. All the higher plants and some algae and fungi have a life cycle in which the zygote divides by **mitosis,** the usual cell division which results in daughter cells that have the same number of chromosomes as the parent cell. The diploid phase in these plants includes the zygote and all the cells derived from it by mitosis. Eventually some of these cells undergo meiotic division to produce haploid spores which in turn divide mitotically to produce a many-celled haploid generation. The haploid generation completes the life cycle by producing gametes which unite to form the zygote. The diploid, spore-producing generation is called the **sporophyte** and the haploid, gamete-producing generation is called the **gametophyte.**

Plants generally have a life cycle characterized by an alternation of generations, a haploid, gamete-producing generation alternating with a diploid, spore-producing generation. In the process of evolution from alga to rose bush the basic pattern of the cycle has remained the same, but there have been tremendous changes in the vegetative and reproductive organs and changes in the relative size and nutritional relations of the two generations.

91. THE BLUE-GREEN ALGAE

The 2,500 or so species of blue-green algae are probably the most primitive chlorophyll-containing plants in existence. The oldest fossil plants found so far appear to have been blue-green algae. The chlorophyll of the blue-greens is not present in discrete chloroplasts but is scattered through the cytoplasm as small granules. The blue-greens have a blue pigment, **phycocyanin,** in addition to chlorophyll, carotene and xanthophyll. These primitive algae resemble bacteria in a number of respects: It is difficult to distinguish a discrete nucleus in either and both apparently reproduce only asexually. Because of these and other resemblances, some botanists classify the blue-greens and bacteria together as members of the same phylum, called the Schizophyta. But because the blue-greens have chlorophyll and occur mostly as many-celled filaments or colonies whereas the bacteria lack chlorophyll and are predominantly unicellular, other botanists classify the blue-greens as a class, the Myxophyceae, in the phylum that contains all the algae. In recent years, many botanists have come to believe that the evolutionary differences between the blue-green algae and other forms are so great that the former should be considered as a separate phylum, the Cyanophyta.

Blue-green algae occur in fresh water pools and ponds; sometimes they occur in sufficient numbers to color the water and to give it an unpleasant taste and smell. Other species live in hot springs (p. 57) or in the ocean. From time to

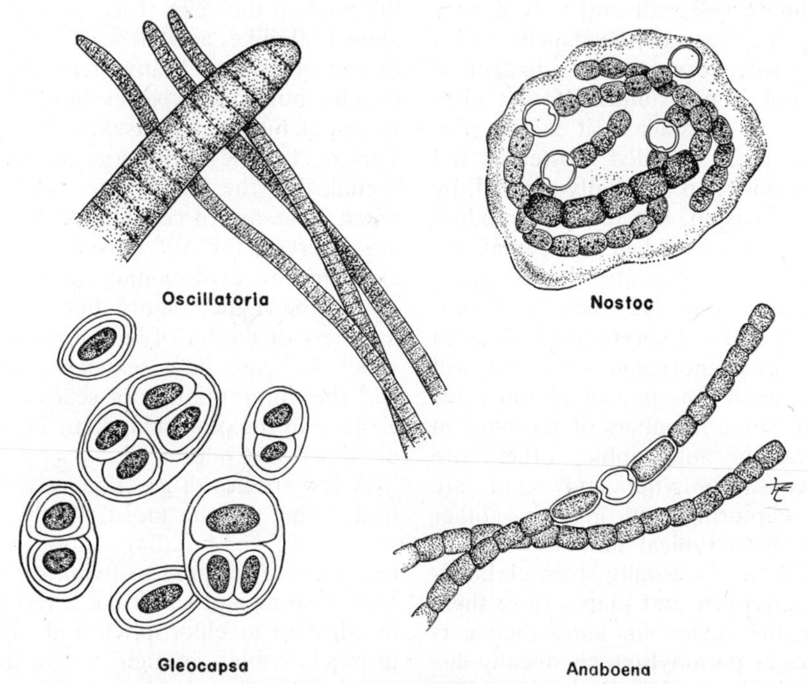

Figure 64. Some common species of blue-green algae.

time, members of the genus *Trichodesmium* which contain a red pigment occur in the Red Sea in such great numbers that they color the water. Some blue-greens live on the surface of soil and rocks in damp, shady places.

A few species of blue-greens are unicellular but most occur as long, many-celled filaments (Fig. 64). None of the blue-greens has flagella, but some of the filamentous ones are capable of a slow, back and forth oscillatory movement.

92. THE EUGLENOPHYTA

The unicellular alga *Euglena* and its relatives are a difficult group to classify, for they have a mixture of animal and plant characteristics. Botanists call them algae and set up a separate phylum, the Euglenophyta, for them; zoologists classify them as the Order Euglenoidina, Class Flagellata, Phylum Protozoa (p. 186). The euglenoids are more advanced than the blue-green algae, for they have a definite, easily stained nucleus and the chlorophyll is not scattered in granules, but is localized in chloroplasts as in higher plants. All the euglenoids have one or two flagella by means of which they swim actively. These organisms do not have an outer cellulose cell wall and they have a gullet near the base of the flagella and a red-pigmented eye spot. The pigment of the eye spot is **astaxanthin,** found elsewhere only in crustacea—it is the substance that gives a boiled lobster its red color. Reproduction is usually asexual, by simple cell division, but sexual reproduction has been observed in one genus. Although *Euglena* has plenty of chlorophyll it apparently cannot live solely by photosynthesis. It will not survive in a medium containing only inorganic salts, but will flourish if small amounts of amino acids are added. Some members of the phylum are completely autotrophic, others are completely saprophytic, and some are holozoic, capturing and ingesting other organisms in a typical animal mode of nutrition. Animals usually store carbohydrates as glycogen and plants store them as starch; the euglenoids store their carbohydrates as **paramylum,** chemically distinct from both starch and glycogen. The euglenoids, with their curious mixture of plant and animal characters, give us an idea of what early living things might have been like before plants and animals had evolved separately. Whether they should be called animals, plants, or perhaps members of a third kingdom, is a matter of definition.

93. THE GREEN ALGAE, PHYLUM CHLOROPHYTA

There are about 5,000 species of green algae; these live in a wide variety of habitats ranging from salt to fresh water. Most botanists believe that the higher plants probably evolved from some form very similar to the present-day green algae, for these algae have a number of characters in common with the higher plants: A definite nucleus is present, the chlorophyll is present in distinct chloroplasts, the pigments present are the same as those in the higher plants (chlorophylls a and b, carotene and xanthophyll), food is stored as starch and a cellulose cell wall is present.

The simplest green algae are unicellular; more advanced members of the phylum have many-celled bodies in the shape of filaments or flat, leaflike structures. Even in the more advanced members of the phylum the cells of the plant body are almost all alike, and there is little differentiation of tissues. Many green algae have flagella but some species are nonmotile. Reproduction may be asexual, by cell division or by the formation of spores, or sexual, by the union of gametes. The green algae are of considerable evolutionary interest, for not only are they very close to the evolutionary line that gave rise to the higher plants, but they exhibit a variety of modes of sexual reproduction which indicate how sexual reproduction and the differentiation of sexes may have evolved in the plant kingdom. This will be discussed in Chapter 12.

A few species of green algae are terrestrial, living on the moist, shady sides of trees, rocks and buildings. Another species has become adapted to living on the surface of snow and ice; it has a red pigment in addition to chlorophyll and may grow in patches thick enough to give the snow a red tinge. The fresh water many-celled green algae are the pond scums and similar forms which may grow very thickly in ponds and streams. A number of marine many-celled forms occur near the low-tide mark and in the upper 20 feet of water. An example of this is "sea lettuce," with a plant body a foot or so long but only two cells thick. It resembles a crinkled sheet of green paper. Some of the tropical marine forms have developed thickened plant bodies, the size of a moss or small fern plant, with parts that superficially resemble the roots, stems and leaves of higher plants. There are hundreds of species of nonmotile, single-celled fresh water green algae known as **desmids,** most of which have symmetrical, curved, spiny or lacy bodies with a constriction in the middle of the cell. When seen under the microscope they look like snowflakes (Fig. 65).

94. THE GOLDEN-BROWN ALGAE, PHYLUM CHRYSOPHYTA

The phylum Chrysophyta is composed of several subgroups, one of which is the "diatoms" — microscopic, usually single-celled forms found in fresh and salt water and constituting an important food source for animals. Diatoms have a number of unique characteristics: Their cell walls contain silica and are constructed in two overlapping halves, which fit together like the two parts of a pillbox. This wall is ornamented with fine ridges, lines and pores that are characteristic for each species. The markings are either radially symmetrical or bilaterally symmetrical on either side of the long axis of the cell (Fig. 66). Many of these markings are at the limit of resolution of the best light microscopes and are used as test objects to determine the quality of the lens. Dia-

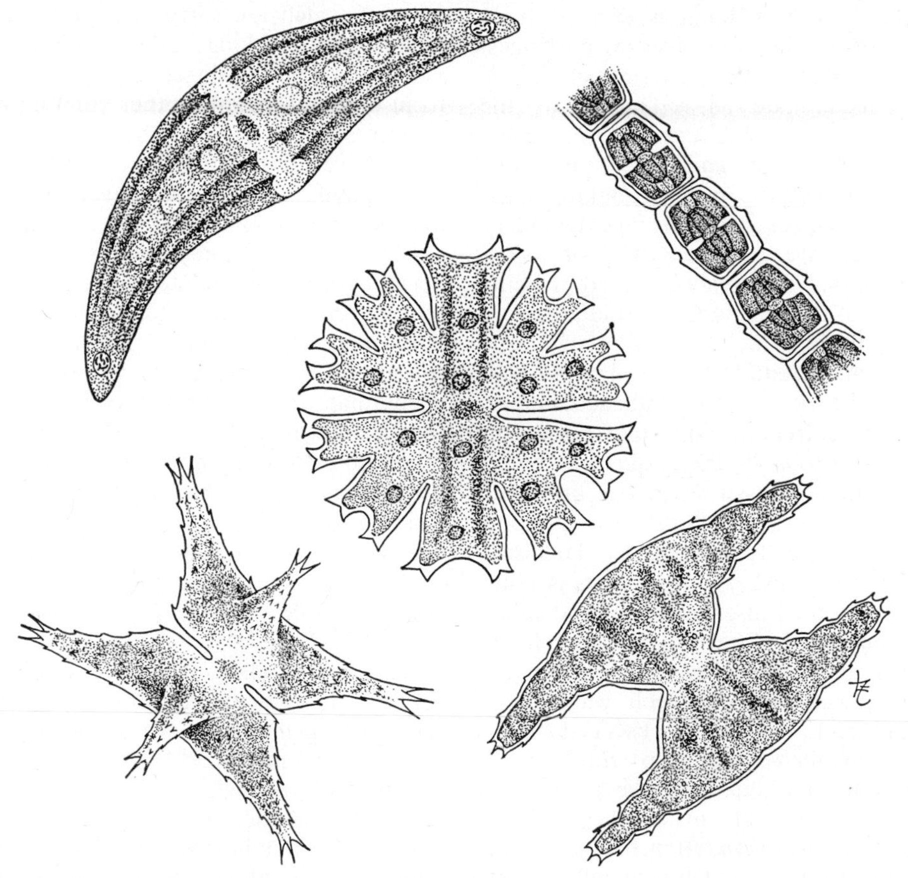

Figure 65. Several different species of desmids, unicellular green algae, highly magnified, showing the symmetry of the cells.

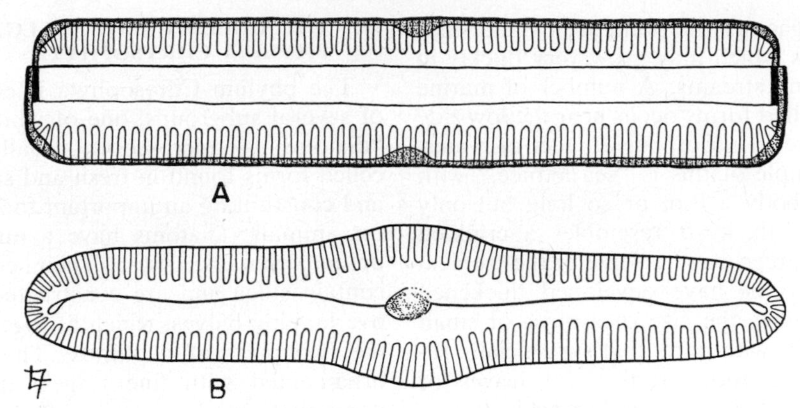

Figure 66. A side view (*A*) and a top view (*B*) of a typical diatom, highly magnified. Note the characteristic fine lines on the shells and the way the upper and lower shells fit together.

toms are capable of a slow, gliding movement, apparently produced by the streaming of protoplasm through the grooves on the surface of the cell wall. Another diatom characteristic is the storage of food as oil rather than as starch; it is widely believed that petroleum is derived from the oil of diatoms that lived in past ages.

The remains of the silica-containing cell walls accumulate as sediments in the oceans. Later geologic uplifts may bring these to the surface and the diatomaceous earth is mined and used in making insulating bricks, as a filtering agent, and as a fine abrasive (several kinds of toothpaste contain diatomaceous earth). Some deposits of diatoms in California are more than 1,000 feet thick.

Diatoms resemble brown algae in possessing the brown pigment **fucoxanthin.** They are extremely important photosynthesizers; probably three quarters of all organic material synthesized is produced by diatoms and dinoflagellates. Diatoms reproduce sexually or asexually. The asexual reproduction by cell division is complicated by the presence of the hard silica-containing wall. After the diatom cell has divided and formed two cells within the old cell wall, two new cell walls form, back to back, between the two cells. Thus each daughter cell ends up with two cell walls, one inherited from the parent and a new one that fits inside the old one. Each successive generation thus gets a little smaller because each new cell wall fits inside the old one. Finally, a special cell is formed which discards the old cell walls, enlarges and then forms a pair of new, large walls.

95. THE DINOFLAGELLATES, PHYLUM PYRROPHYTA

Dinoflagellates are single-celled algae, most of which are surrounded by a shell made of interlocking plates. They are all motile, having two flagella, one projecting from one end and the other running in a transverse groove (Fig. 67). Like diatoms, they have fucoxanthin in addition to chlorophyll. Most dinoflagellates are marine and are important photosynthesizers in the ocean. Occasionally vast numbers of them accumulate in some part of the sea, coloring the water red or brown. Some species of dinoflagellates are poisonous to vertebrates and when these accumulate, large numbers of fish in that region are killed. Other dinoflagellates are taken up as food by mussels. The mussels are not harmed by the dinoflagellates, but if a man eats some of these infected mussels he may become seriously ill.

96. THE BROWN ALGAE, PHYLUM PHAEOPHYTA

The brown algae include about 1000 species of multicellular forms ranging in size up to giant kelps, whose bodies may be several hundred feet long. They are the prominent brownish-green seaweeds that usually cover the rocks in the tidal zone and extend out into water 50 or so feet deep. These plants have considerable amounts of the golden-brown pigment **fucoxanthin,** which tends to mask the

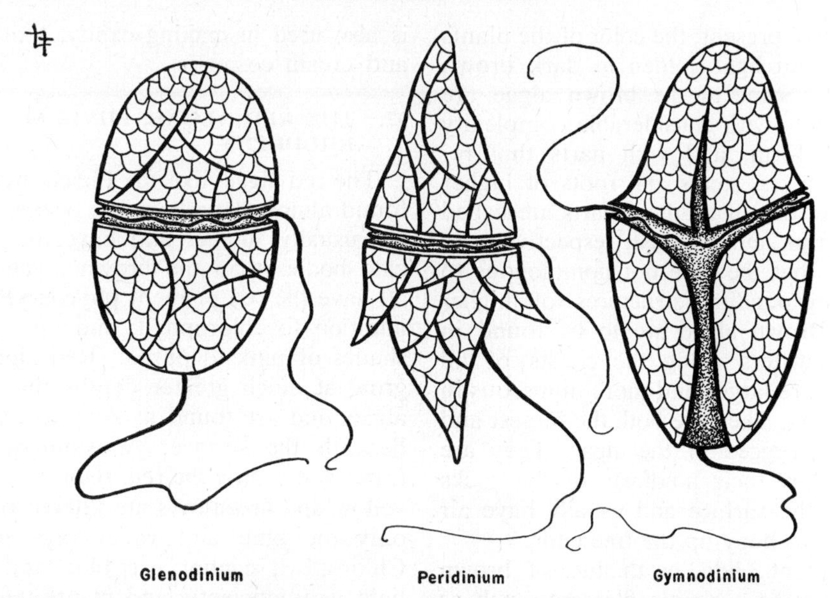

Figure 67. Three species of dinoflagellates. Note the plates which encase the single-celled body and the characteristic two flagella, one of which is located in a transverse groove.

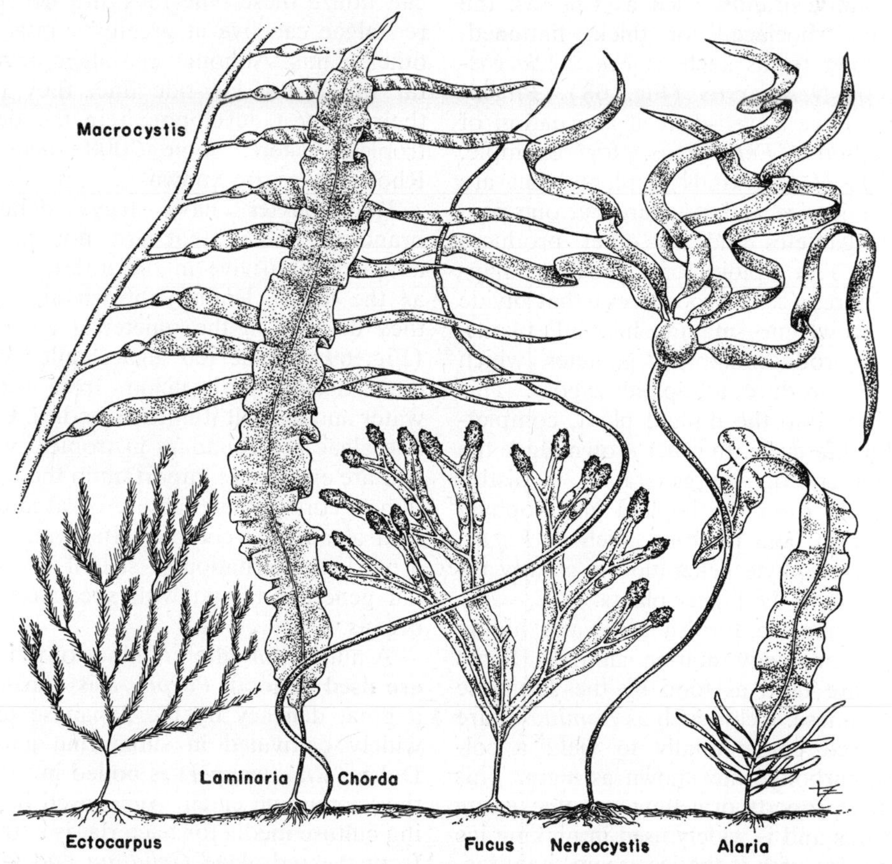

Figure 68. Some of the kinds of brown algae or kelps, all of which are multicellular marine plants.

chlorophyll present; the color of the plants ranges from light golden to dark brown or black. Some of the brown algae are large plants with considerable complexity of body form, and with parts that resemble leaves, stems and roots of higher plants. In the algae, these parts are called **blade, stipe** and **holdfast,** respectively, to indicate that they are not homologous to the corresponding structures of higher plants. Brown algae are to be found in shallow waters along the coasts of all seas but are larger and more numerous in cool waters. They are both the largest and the most rugged of the algae. They are attached by their holdfasts to the rocks beneath the surface and usually have air bladders to buoy up the free ends.

The plant body, or thallus, of brown algae may be a simple filament, such as the soft brown tufts of *Ectocarpus,* commonly found on pilings, or tough, ropelike, slimy strands, such as *Chorda,* the "Devil's shoelace," or thick, flattened, branching forms such as *Fucus, Sargassum* or *Nereocystis* (Fig. 68). Phaeophytes have a well-defined alternation of generations. *Ectocarpus,* for example, consists of two kinds of plants that are similar in size and structure but one produces gametes and the other produces spores. The diploid form produces haploid spores (called zoospores) that divide and grow into mature haploid plants. These produce haploid gametes which fuse to produce a diploid zygote. This develops into the diploid plant, completing the life cycle. In other brown algae the diploid sporophyte generation is distinguishable from the haploid gametophyte generation, and in some, such as *Fucus,* the gametophyte generation is greatly reduced, as in the higher plants.

Brown algae furnish food and hiding places for many marine animals. Some kelps are used as food by the Japanese and Chinese. Kelps such as *Laminaria* are processed commercially to yield a colloidal carbohydrate known as **algin.** This has the property of gelling and thickening mixtures and is widely used in making ice cream, for with it the ice cream manufacturer can use much less real cream and still have a smooth creamy product. Algin is also used in making candy, toothpaste and cream cosmetics.

97. THE RED ALGAE, PHYLUM RHODOPHYTA

The red algae, like the Phaeophyta, are found almost entirely in the oceans. They are usually smaller and have more delicate bodies than the brown algae. They all have the red pigment **phycoerythrin** in addition to chlorophyll, and are various shades of pink to purple. Red algae can grow at much greater depths than other algae, and are found as deep as 300 feet beneath the surface. As sunlight penetrates water, first the red, then the orange, yellow and green rays are filtered out and only the blue and violet rays remain. Chlorophyll cannot use blue and violet light very efficiently and plants that have only chlorophyll cannot carry on photosynthesis at these depths. Phycoerythrin can utilize these blue rays and hence the red algae can live at greater depths than other plants. Although red algae occur as far up as the low-tide line, they reach their greatest development in the deeper tropical waters. Some 3000 species of Rhodophytes are known.

Rhodophytes have lacy, delicately branched bodies that are not as well adapted to survive in the intertidal zone as the tough, leathery brown algae, but they do well in the quieter deep waters (Fig. 69). Some red algae, called corallines, accumulate calcium from the sea water and deposit it in their bodies. Coralline algae are abundant in tropical waters and are even more important in the formation of coral atolls than are coral animals. Red algae have complex life cycles, with a marked alternation of sexual and asexual generations and with specialized sex organs.

A number of different kinds of red algae are used as food. *Porphyra* is considered a great delicacy by the Japanese and is widely cultivated in submarine gardens. Dulse (*Rhodymenia*) is boiled in milk by the Scotch and eaten. **Agar,** used in making culture media for bacteria, is extracted from the red algae *Gelidium* and *Gracilaria.* Agar is extensively used in baking and canning. **Carrageenin,** extracted from

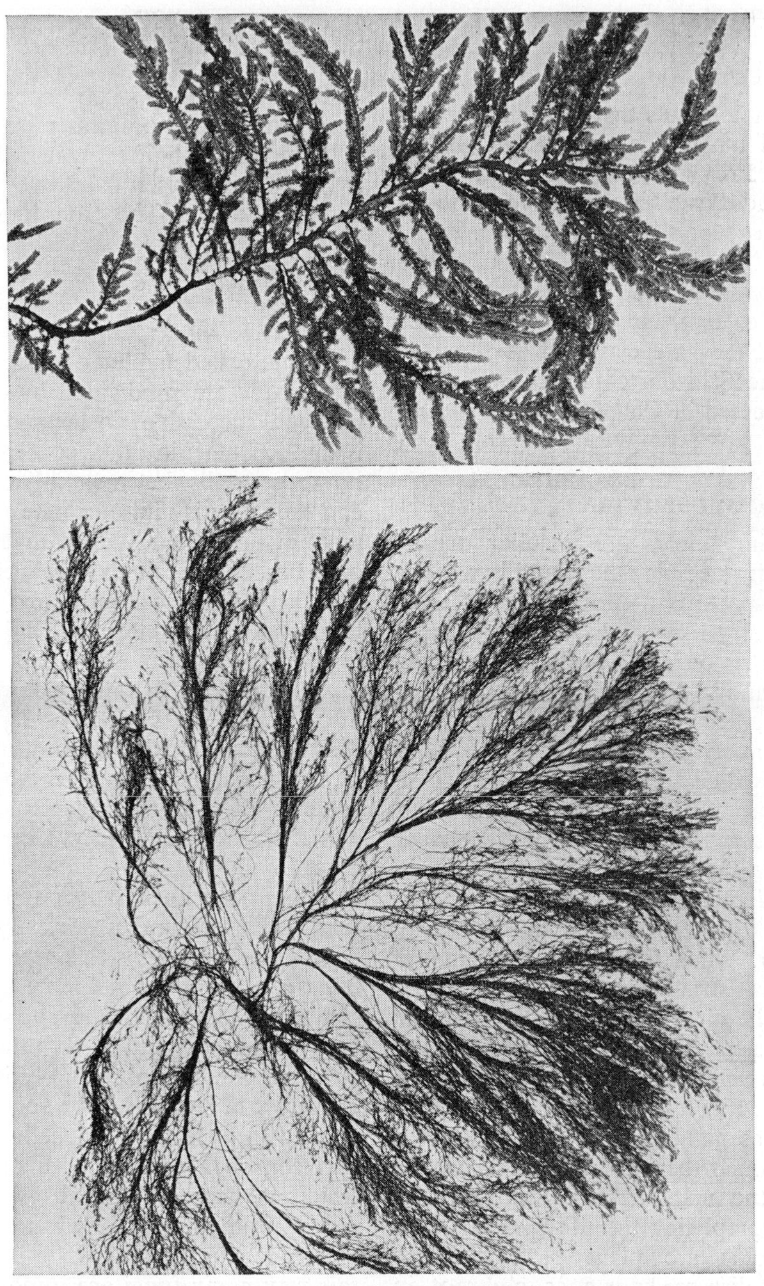

Figure 69. Two species of red algae. Note the lacy, delicately branched bodies. (Courtesy of the New York Botanical Garden.)

Irish moss, is used in the preparation of chocolate milk to keep the chocolate from settling out.

98. THE FUNGI

The simple plants that lack chlorophyll are called fungi. The three main groups of nongreen pants, the bacteria, slime molds, and "true" fungi, have little in common except the heterotrophic nutrition necessitated by the absence of chlorophyll. Because of this, and because of the fact that each group has evolved independently, they are classified as separate phyla. The Schizomycophyta or bacteria were discussed in Chapter 9.

99. THE SLIME MOLDS, PHYLUM MYXOMYCOPHYTA

The slime molds are peculiar organisms which resemble other fungi in a number of respects but during part of their lives are very like amebas. They exist as slimy masses on decaying leaves or lumber and move by sending out pseudopodia, as amebas do. Although each mass contains many nuclei, it is not divided into individual cells. Ordinarily these plants reproduce asexually, by fission, but one of the most curious of biologic phenomena is the way in which, at certain times, individual slime molds congregate to form a fruiting (spore-producing) body (Fig. 70). When the spores are released from this structure they are carried through the air. If a spore falls on some moist surface it will absorb water, split out of its wall, and divide to form ameboid cells. Finally some of these ameboid cells act as gametes and fuse to form a zygote, which then divides and grows to become the many-nucleate slimy mass, thereby completing the life cycle. A few slime molds are plant parasites; one causes a disease known as clubroot in cabbages.

The slime molds have many characters in common with animals such as amebas and flagellates and it is a matter of opinion whether they should be called animals or plants. They may have descended from flagellates; however, the formation of a fruiting body is characteristic of other fungi and for this reason they are usually classed as plants.

100. THE EUMYCOPHYTA, TRUE FUNGI

There are about 70,000 species of true fungi; these have a number of characters in common with the algae and are believed to have arisen from one or more of the algal phyla. The true fungi include such plants as yeasts, molds, mildews, rusts, smuts and mushrooms. A few of the true fungi are unicellular, but most have many-celled bodies made of branching filaments called **hyphae.** In some species the hyphae are subdivided by cross walls between successive nuclei and the plant is multicellular; in other species there are no cross walls between adjacent nuclei and the plant is multinucleate. The whole mass of branching hyphae that constitute one fungus is called a **mycelium.** The presence of a mycelium is one of the distinguishing characters of the Eumycophyta. In the common bread mold, this mycelium is visible as the cobwebby mass of fibers on the surface and penetrating the interior of the bread. In a fungus such as the mushroom, the mycelium is below ground; the mushroom cap that we eat is a fruiting body that grows out from the mycelium.

Fungi are either saprophytic or parasitic and are found universally wherever organic material is available; they grow best in dark, moist habitats. Some fungi can grow under what are apparently very unfavorable conditions — they have a strong resistance to plasmolysis, for they can grow in strong salt or sugar solutions (e.g., on jelly). As the mycelium branches and comes in contact with organic material, it secretes enzymes which break down proteins, carbohydrates and fats, and then absorbs the split products. In this way many fungi are important members of the carbon, nitrogen, and other cycles. Fungi cause serious and important diseases of man, his domestic animals and his crop plants, and are responsible in large measure for the deterioration of wood, leather, cloth, and similar materials. Reproduction occurs in a variety of ways: asexually by fission, budding or by

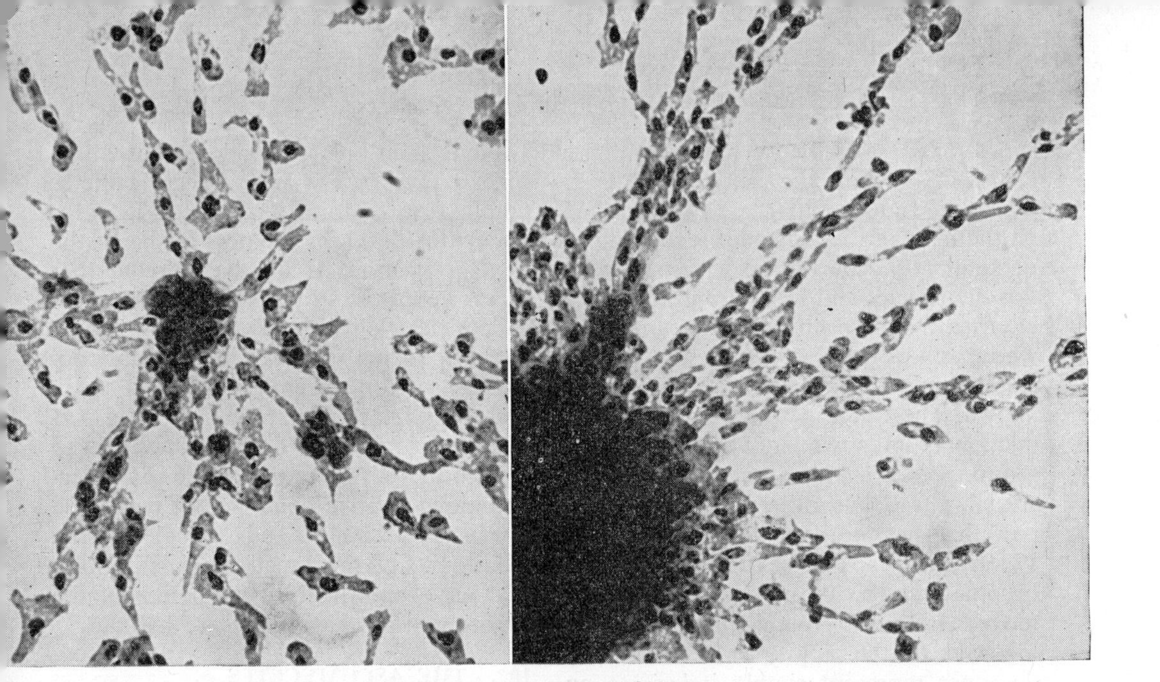

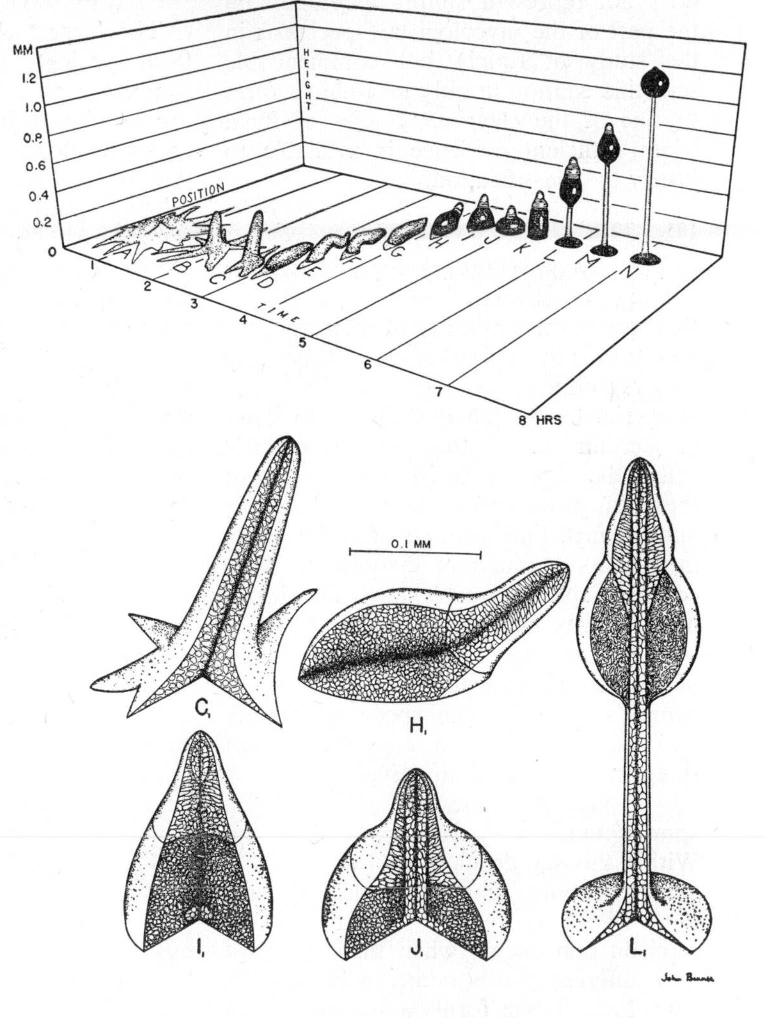

Figure 70. Reproduction in slime molds. *Upper left,* Photomicrograph showing individual organisms beginning to congregate as a fruiting body. *Upper right,* Photomicrograph showing a later stage in the process of congregation. *Below,* Diagram of the various stages in the process. *A–C,* The fruiting body is formed by the aggregation of hundreds of individual ameba-like slime molds; *D–H,* it crawls along the substrate surface for some time; *I–N,* it grows a stalk which lifts the spore-producing part off the surface. Finally, the spores are released. (Courtesy of Dr. John Bonner.)

spores, and sexually by means that are characteristic for the subgroups. Four classes of Eumycophyta are distinguished on the basis of the means of sexual reproduction: Phycomycetes, Ascomycetes, Basidiomycetes and Fungi Imperfecti. The latter is a heterogeneous group of fungi whose status is not completely understood, and whose life histories are unknown. The fungi in which sexual reproduction is unknown are assigned to this group. Some fungi may actually have no sexual phase but others probably have one that has not yet been discovered. When the means of sexual reproduction become known, the fungus can be removed from the Fungi Imperfecti and assigned to the appropriate class. This does not represent simply indecision on the part of the mycologists (specialists in the study of fungi) but commendable scientific caution in only assigning a form to one of the clear-cut classes of fungi when sufficient evidence is available to justify the classification.

101. THE PHYCOMYCETES

The phycomycetes are the smallest class of fungi, consisting of about 500 species. Because of the similarity of their filamentous bodies to those of algae, and because they reproduce asexually by motile spores and sexually by gametes similar to those of certain algae, they are sometimes called the alga-like fungi. The hyphae of these fungi have few or no cross walls and the mycelium consists of many nuclei in a common mass of cytoplasm. Some common phycomycetes are the bread mold, downy mildews and white rusts.

Bread becomes "moldy" when a spore of the black bread mold, *Rhizopus nigricans,* falls on it. The spore germinates and grows to form a tangled mass of threads, the mycelium. Eventually, certain hyphae grow upward and develop a **sporangium,** or spore sac, at the tip. Within this sac develop clusters of black, spherical spores which are released when the delicate spore sac ruptures. Sexual reproduction occurs when the hyphae of two different plants come to lie side by side. Each hypha forms a swelling which grows toward the other; the tip of the swelling then enlarges and pinches off to form a gamete. The two adjacent gametes finally fuse to form a zygote (Fig. 71). From this develops the new hypha of the next generation. It has been found that there are two strains of bread molds, called the "plus" and "minus" strains, and that sexual reproduction can only occur between one member of a plus strain and one of a minus strain. This is a sort of physiologic sex differentiation, even though there is no morphologic sex differentiation, and we could scarcely call the two strains "male" and "female." Only the zygote of the bread mold is diploid. Meiosis occurs in the germination of the zygote and all the hyphae are haploid.

102. THE ASCOMYCETES

The largest class of fungi (about 35,000 species), the Ascomycetes or sac fungi, are so called because their spores are produced in sacs called **asci.** Each ascus produces two to eight ascospores. Among the Ascomycetes are the yeasts, powdery mildews, the molds which appear on cheese, jelly and fruit, and the edible truffle. The ascomycete molds which appear on food may give it an unpleasant taste but they are not poisonous. The unique flavor of cheeses such as Roquefort and Camembert is produced by the action of ascomycetes. The ascomycete *Penicillium* produces the famous antibiotic **penicillin.**

The bodies of ascomycetes may be unicellular, as in yeasts; many-celled filamentous mycelia, as in powdery mildews; or thickened and fleshy, as in the truffle. Reproduction is accomplished asexually by spores or by budding (yeasts), and sexually by gametes which unite to form a "fruit" containing the ascus. Although the structures in which asci are produced are often large and fleshy and may superficially resemble true fruits, they have, of course, no relation to them.

The economically important ability of yeasts to produce ethyl alcohol from glucose in the absence of oxygen was discussed previously (p. 68). The yeasts used in wine-making are usually the wild yeasts normally present on the skins of the grapes; some of the differences in the taste of different kinds of wines are due to the kind of yeasts present in the wine-growing region. The yeasts used in baking and in brewing beer are cultivated yeasts, carefully kept as pure strains to prevent contamination.

The ascomycete *Neurospora crassa,* a saprophyte that occurs on pies and cakes, appearing as a cottony white fluff at first and turning pink as it develops asexual pink spores, has become an important tool in genetics and biochemistry (see p. 486). As in the black bread mold, *Rhizopus,* there are two mating types, indistinguishable in body form, and sexual reproduction will occur only between two hyphae of opposite types. The diploid cell that results from sexual reproduction divides by meiosis and then by mitosis to produce eight haploid **ascospores** within the ascus. Under favorable conditions, each ascospore will germinate to produce a new mycelium. It is possible to dissect out the individual ascospores under a microscope and establish pure strains for use in genetic and biochemical research.

Some characters of the ascomycetes have a resemblance to those of some phycomycetes, and some mycologists believe the former may have evolved from the latter. But in a number of other respects, the ascomycetes resemble the red algae, and many mycologists are inclined to believe that ascomycetes evolved from red

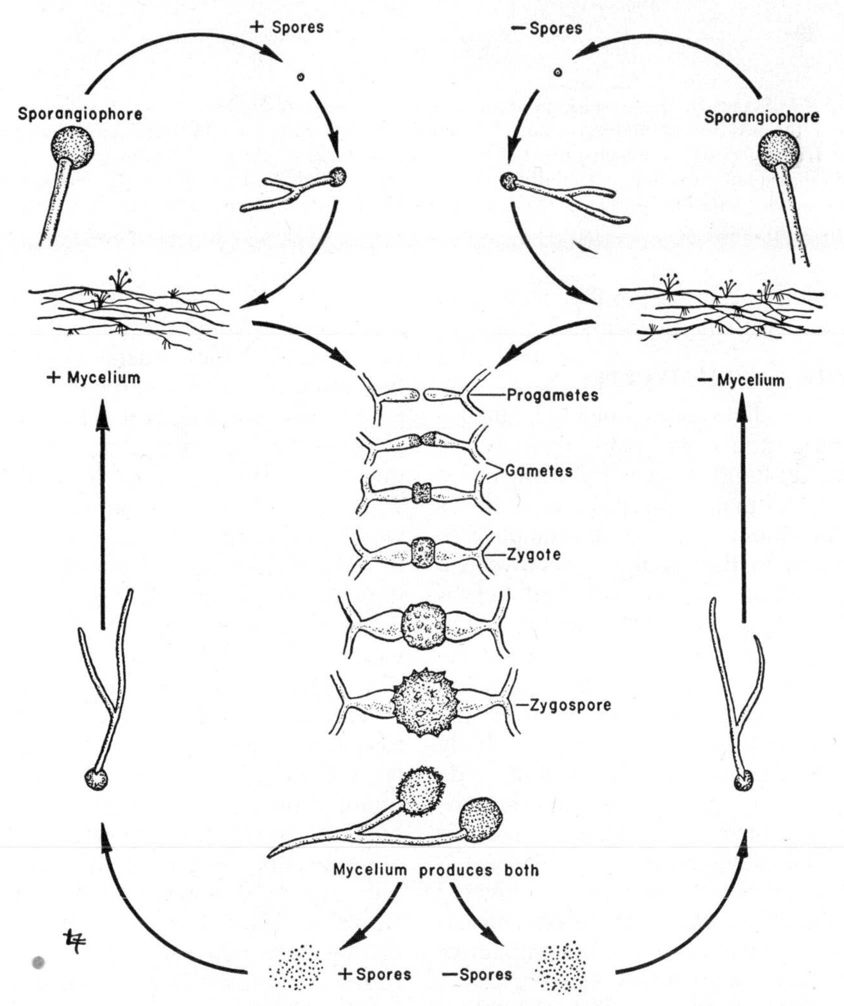

Figure 71. The life cycle of the black bread mold, *Rhizopus nigricans.* The upper circles indicate the asexual production of mycelia from spores. In the center is a series of stages in sexual reproduction. See text for discussion.

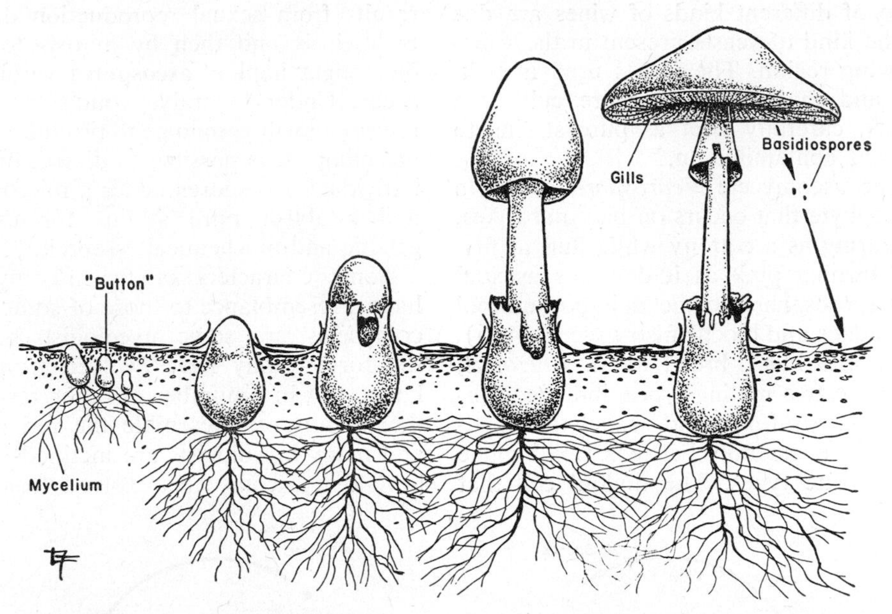

Figure 72. Stages in the development of a mushroom from the mycelium, the mass of white, branching threads found underground. A compact mass, called a "button," appears and grows into the fruiting body or mushroom. On the under surface of the fruiting body are "gills," thin perpendicular plates extending radially from the stem. Basidia develop on the surface of these gills and produce basidiospores which are shed and, if they reach a suitable environment, give rise to new mycelia.

algae which became saprophytic and lost their photosynthetic pigments.

103. THE BASIDIOMYCETES

The basidiomycetes include mushrooms, toadstools, puff balls, rusts, smuts and bracket fungi. They derive their name from the fact that they reproduce sexually by a **basidium,** a structure comparable in function to the ascus of ascomycetes. Each basidium is an enlarged, club-shaped, hyphal cell, at the tip of which develop four **basidiospores.** These are released and develop into new mycelia when they come in contact with the proper environment. The vegetative body of the plant consists of a mycelium made of many-celled hyphae. There are no motile cells formed at any stage of the life cycle of basidiomycetes.

The vegetative body of the cultivated mushroom, *Psalliota campestris,* consists of a mass of white, branching, threadlike hyphae that occur mostly below ground. After a time, compact masses of hyphae, called "buttons," appear at intervals on the mycelium (Fig. 72). The button grows into the structure we ordinarily call a mushroom, consisting of a stalk and an umbrella. On the underside of the umbrella are many thin perpendicular plates called **gills,** which extend radially from the stalk to the edge of the cap. The basidia develop on the surface of these gills (Fig. 73). Each basidium contains two nuclei which fuse to form a diploid nucleus. This in turn divides by meiosis to form four haploid **basidiospores.** Each plant produces millions of basidiospores, each of which can, if it falls in the proper environment, give rise to a new mycelium.

The word "mushroom" does not refer to any particular species of basidiomycetes but simply to the fruiting body of a number of forms. There are some two hundred edible kinds of mushrooms and about twenty-five poisonous ones (poisonous ones are sometimes called "toadstools"). There is no simple test which distinguishes edible and poisonous mushrooms; the only safe way is to identify the mushroom as a particular species, one known to be edible. This is a job for an expert; the simplest way to avoid mushroom poisoning is to eat only those grown commercially.

The evolutionary origin of the basidiomycetes is shrouded in mystery. They show no relations with any of the algae and it is generally presumed that they are derived from other fungi, possibly from the ascomycetes.

104. LICHENS

The lichens are curious plants which, although they look like individual plants, are really an intimate combination of an alga and a fungus (Fig. 74), and are a classic example of **mutualism** (p. 75). The algal component is either a green or a blue-green alga, and the fungus is usually an ascomycete; in some lichens from tropical regions the fungus partner is a basidiomycete. Lichens are resistant to extremes of temperature and moisture, and grow everywhere that life can be supported at all. They exist farther north than any other plants of the Arctic region, and are equally at home in the steaming equatorial jungle. The alga, by photosynthesis, produces food for both, while the fungus protects the alga and provides it with moisture and mineral salts. Lichens have an important role in the formation of soil, for they gradually dissolve and disintegrate the rocks to which they cling. The fungi of some lichens produce colored pigments. One of these, **orchil,** was used to dye woolens, and another, **litmus,** is widely used in chemistry laboratories as an acid-base indicator. The "reindeer moss" of the Arctic is a lichen. There are some 10,000 species of lichens, and they present a nasty problem in classification—should they be classified according to their algal component, their fungal component, or in some third way? The usual procedure, even though a lichen consists of an alga with one name and a fungus with another name, is to give it a third name and place it in a separate class of the phylum Eumycophyta.

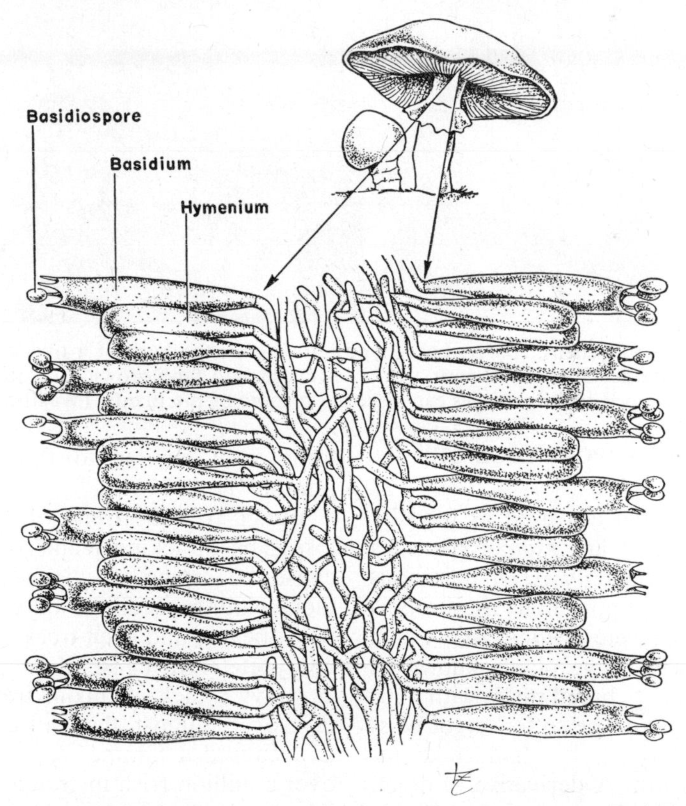

Figure 73. Section of a gill from the under side of a mushroom cap, magnified 500 times, to show the basidia and their basidiospores.

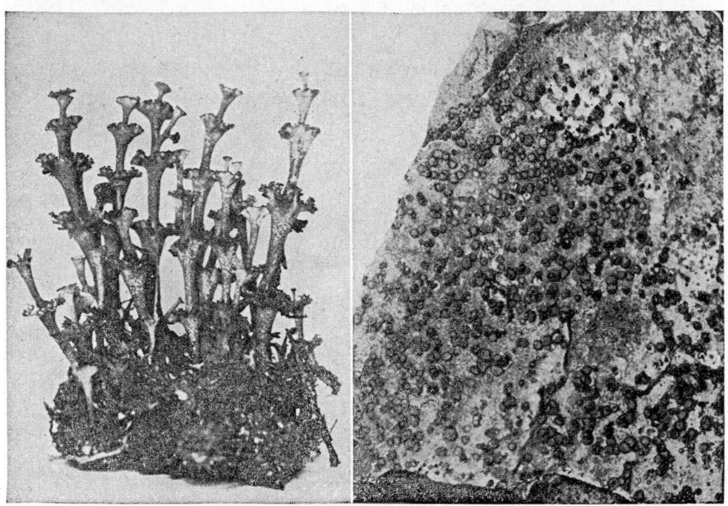

Figure 74. Types of lichens. *Above,* Leafy type growing on the bark of a tree. *Lower left,* The lichen known as "reindeer moss." *Lower right,* An encrusting type growing on the surface of a rock. Flat or cup-shaped fruiting bodies can be seen on some of the plants. (Weatherwax: Botany.)

105. ECONOMIC IMPORTANCE OF THE FUNGI

Only a few fungi are used as food by man and only few are human parasites (see p. 409). The only fungi poisonous to man are the few poisonous mushrooms and the ascomycete *Claviceps,* which causes a disease of rye plants known as **ergot.** If a man eats bread made with flour from diseased plants, he suffers ergot poisoning, characterized by hallucinations, insanity and death. A derivative of ergot, lysergic acid, has been used recently in experimental psychiatry to produce symptoms very similar to those of schizophrenia (p. 357).

Fungi cause tremendous economic losses by attacking plants. Phycomycetes cause "damping-off disease," which attacks young seedlings of corn, tobacco, peas, beans, and even trees. Another phycomycete is the cause of the potato blight. A heavy attack of this in Ireland in 1845 destroyed almost the entire potato crop and caused a famine. As a result of this over a million Irish migrated to the United States. A downy mildew which attacks grapes was introduced into France from

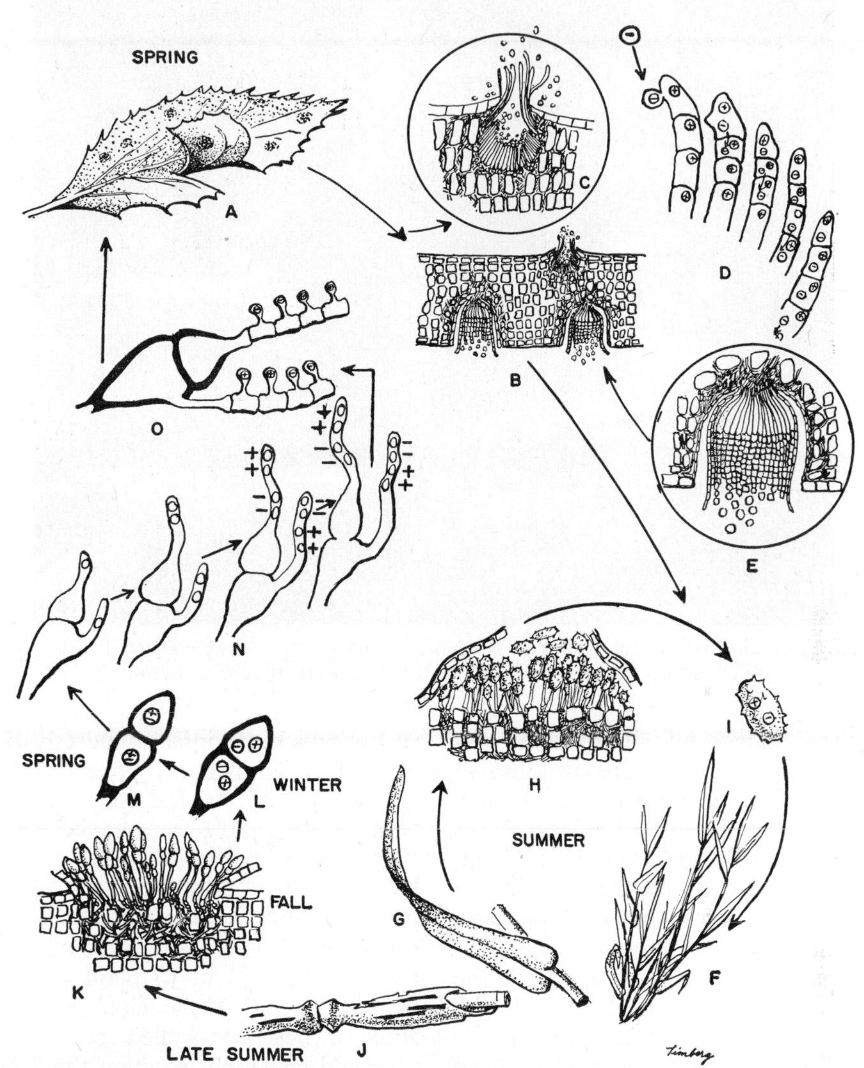

Figure 75. Life cycle of the wheat rust, *Puccinia graminis. A,* In the spring basidiospores from infected wheat plants of the previous year infect leaves of the barberry plant, forming pycnia containing clusters of spermagonia on the upper surface and cluster cups of aecia on the lower surface of the leaf. *B,* Section of barberry leaf showing pycnium on upper surface and two cluster cups on lower surface. *C,* Spermagonium. *D,* A spermatium of the opposite strain (—) is carried by an insect to a hypha in the spermagonium. Its nucleus enters the cell of the (+) strain and by nuclear division and migration forms binucleate N + N (not 2N, for the nuclei do not fuse) cells. *E,* An aecium on the under surface of a barberry leaf producing N + N aeciospores. *F,* In early summer aeciospores infect the leaves of young wheat plants. *G,* Leaf of mature wheat plant with clusters of red, single-celled uredospores ("red rust" stage). *H,* Section of wheat leaf showing N + N uredospores. *I,* Uredospores are released, infect other wheat plants and form more uredospores. *J,* In late summer uredospores develop into dark brown, two-celled teliospores on the stems and leaf sheaths of wheat plants ("black rust" stage). *K,* Section of leaf sheath showing N + N teliospores. *L,* The thick-walled N + N teliospores remain dormant over the winter. *M,* In the spring the N + N nuclei within each cell of the teliospore fuse to form a 2N nucleus. *N,* The teliospore, still attached to the wheat plant, germinates and undergoes meiosis. *O,* Four haploid (N) basidiospores are produced from each teliospore cell; these then infect a barberry leaf to complete the cycle.

Figure 76. A bracket fungus, *Polyporus.* The spores are borne on basidia located in pores on the underside of the fleshy, shelflike bracket. (Weatherwax: Botany.)

the United States and almost destroyed the French vineyards before an effective fungicide, called Bordeaux mixture, was discovered. Some important plant diseases caused by ascomycetes are chestnut blight, Dutch elm disease, apple scab, and brown rot, which attacks cherries, peaches, plums and apricots. Basidiomycetes include smuts and rusts which attack the various cereals, corn, wheat, oats, and so on—in general, each species of smut is restricted to a single host species. Some of these parasites, such as the stem rust of wheat and the white pine blister rust, have complicated life cycles which are passed in two or more different plants, and which involve the production of several kinds of spores. The white pine blister rust must infect a gooseberry or a red currant plant before it can infect another pine. The wheat rust must infect a barberry plant at one stage in its life cycle (Fig. 75). Since this has been known, the eradication of barberry plants in wheat-growing regions has effectively reduced infection with wheat rust, but the eradication must be complete, for a single barberry bush can support enough wheat rust organisms to infect hundreds of acres of wheat. The spores produced by the wheat rust in the fall, ones with thick walls which enable them to survive a cold winter, will grow only if they fall on a barberry. The usual, thin-walled spores produced most of the summer can infect other wheat plants directly, and infection spreads from plant to plant in a wheat field in this way. If the winter is very mild, some of these thin-walled spores may survive and cause an infection the following year even in the absence of barberry plants. Thus, even the complete eradication of barberry plants does not provide a final solution to the wheat rust problem.

Bracket fungi (Fig. 76) cause enormous losses by bringing about the decay of wood, both in living trees and in stored lumber. The amount of timber destroyed each year by these basidiomycetes approaches in value that destroyed by forest fires.

Some of the Fungi Imperfecti cause important diseases of man. *Monilia* causes a throat and mouth disease called "thrush," and also infects the mucous membranes of the lungs and genital organs. The Trichophytoneae infect the skin of man and other animals, causing ringworm, athlete's foot and barber's itch. Other members of the Fungi Imperfecti are parasites of higher plants and cause important diseases of fruit trees and crop plants.

QUESTIONS

1. What is meant by the term "thallophyte"?
2. Define the terms haploid, diploid, gamete, zygote, sporophyte and gametophyte.
3. Why are the blue-green algae regarded as the most primitive green plants?
4. Why do botanists believe that the higher plants evolved from the green algae?
5. Of what importance to man are desmids, diatoms and dinoflagellates?
6. How are red algae adapted to survive in deep water?
7. In what ways are slime molds like true fungi? In what ways do they resemble animals?
8. What is a "fruiting body"? How does it differ from a true fruit?
9. In what ways are fungi important economically to man?
10. What measures can be taken to keep bread from becoming "moldy"?
11. How do lichens differ from other plants? How are they classified?
12. What measures should be taken to eliminate white pine blister rust?
13. What is the difference between an ascus and a basidium? between a hypha and a mycelium?
14. What organisms are responsible for decay?

SUPPLEMENTARY READINGS

Rachel Carson's beautifully written *The Sea Around Us* discusses the importance of the algae as the primary producers of the sea. Tiffany's *Algae: The Grass of Many Waters* is a nontechnical description of these important plants. Popular accounts of the fungi are given in Rolfe's *Romance of the Fungus World* and in *Molds and Man.* A more technical presentation of the taxonomy and structure of fungi is found in *Introductory Mycology* by C. Alexopoulos.

Chapter 11

The Invasion of Land by Plants

IT IS generally believed that land plants and animals evolved from aquatic ancestors. The most primitive plants and animals living today—the algae and lower invertebrates—are aquatic and we assume that the ancestral primitive forms were also aquatic. In tracing the evolutionary history of certain plants and animals we may find that, having once become adapted to terrestrial life, they may return to an aquatic habitat — even reemerge later on and once again become terrestrial. But when such evolutionary lines are traced back as far as possible, the primitive ancestral forms are all aquatic and it is generally believed that life began in a watery environment.

The colonization of the land was a tremendous undertaking for both plants and animals. We believe that land animals were possible only after plants had successfully established themselves on shore and were available as food. Aquatic plants have few "problems" to solve for survival and can survive without many specialized structures. The surrounding water keeps them supplied with nutrients, prevents the cells from drying out, buoys up and supports the plant body so that special supporting structures are unnecessary, and serves as a convenient medium for the meeting of gametes in sexual reproduction and as a means of dispersal for asexual spores. In leaving the friendly water and taking up life on the barren land, the plants had to become adapted by developing new structures to take over the many functions previously served by the surrounding water. The conquest of the land must have been a long and difficult process, fraught with many failures. For the new forms that tried to live in the soil there were adequate salts, water and carbon dioxide, but no light for photosynthesis. For the ones that tried to live above ground there were light and carbon dioxide but not enough water and no salts. The plants which finally triumphed and became truly terrestrial were able to survive because they evolved specialized parts: (1) **leaves,** which extend into the air to absorb light and carry on photosynthesis, (2) **roots,** which extend into the soil to provide anchorage and absorb water and salts, (3) **stems,** which support the leaves in the sunlight and connect them with the roots to provide a two-way connection for the transfer of nutrients, and (4) some means of reproduction, such as flowers and seeds, by which male and female gametes can unite in the absence of a watery medium and by which the zygote can begin development protected from desiccation.

Just as the present-day amphibians of the vertebrate phylum — salamanders, newts and frogs—give us an idea of what the first land vertebrates may have looked like, the **bryophytes**—mosses, liverworts and hornworts — suggest the stages through which aquatic algae evolved to become fully terrestrial.

The algae have evolved body structures adapted to expose a maximum of surface for absorbing nutrients from the surrounding water. For survival on land, plants need a more compact body to decrease the water lost through the surface. Probably the first land plants lay flat, and exposed only one surface to the air. Plants were only successful on land after they developed a specialized epidermal tissue, with thickened cell walls impregnated with a waxy, waterproof material. The bryophytes have an epidermis which may be slightly thickened and waxy, and which is provided with pores to permit the diffusion of gases.

The bryophytes did not really solve the problem of reproduction in the absence of a watery medium; they got around the difficulty by evolving reproductive structures that would have a watery medium for the union of gametes (p. 169). All land plants, including the Bryophyta, have evolved a life cycle in which the zygote is retained within the female sex organ, where it obtains food and water from the surrounding parental tissues and is protected from drying while it develops into a multicellular **embryo.** For this reason the bryophytes and tracheophytes or higher plants are classified together in the subkingdom Embryophyta.

106. THE BRYOPHYTES

The phylum Bryophyta is made up of about 23,000 species of **mosses, liverworts** and **hornworts.** The name "moss" is erroneously applied to a number of plants which are not bryophytes: the moss on the bark of the north side of a tree is really an alga; "reindeer moss" is a lichen and "Spanish moss" hanging from trees in the Southern states is really a seed plant, a relative of the pineapple. Mosses in general form an insignificant part of the vegetation even though they are widely distributed. Some species can live only in damp places; others can survive in a dormant state in dry rocky places where enough moisture for growth is present only during a short part of the year. Liverworts, which are not as well protected against desiccation as mosses, are even more restricted and are found in the deep shade of forests or on the shady side of a cliff. The life cycle of bryophytes is characterized by a strongly marked alternation of sporophyte and gametophyte generations (Fig. 91).

The mosses are all rather similar in structure, consisting of a filamentous green body, or **protonema,** on or in the soil, from which grows an erect stem to which are attached a spiral whorl of one-cell thick leaves. From the base of the stem extend many colorless rootlike projections called **rhizoids.** Mosses are never more than 6 to 8 inches high owing to the inefficiency of the rhizoids as water absorbers. The height of the stem is also restricted by the absence of true vascular tissues and well developed supporting tissues.

The conspicuous, familiar moss plant is the sexual or **gametophyte** generation; the **sporophyte** or asexual generation develops as a partial parasite on the gametophyte—it can carry on photosynthesis but is dependent on the gametophyte for its supply of water and minerals. Many moss "plants" can be produced asexually by a single protonema.

Mosses, like lichens, can grow on bare places where few other plants could survive. Once mosses have become established and soil begins to accumulate, other plants can follow. An economically important plant is *Sphagnum,* which grows in boggy places. The remains of this plant accumulate under water and form **peat,** used as a fuel in many countries. Dried *Sphagnum* can absorb and retain vast quantities of water and is used as a packing material for live plants.

Liverworts, a second class of bryophytes, are simpler and more primitive than the mosses. Their bodies consist of flat, sometimes branched, ribbonlike structures that lie on the ground, in contrast to the upright habit of the mosses. This body, usually no more than an inch or two in diameter, is attached to the soil

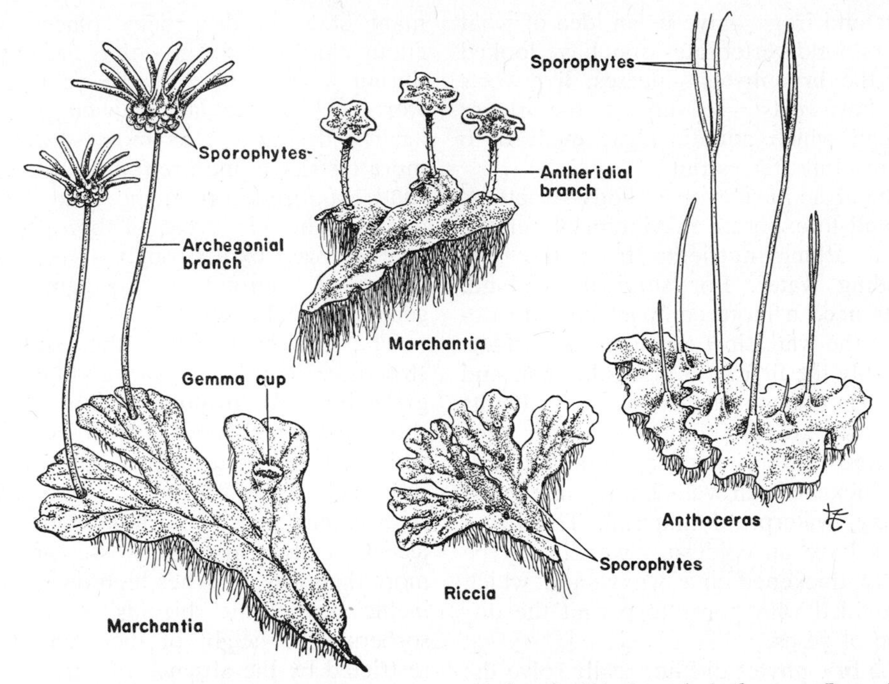

Figure 77. Sketches of some common liverworts, *Marchantia, Riccia* and *Anthoceras.* In each, the flat part of the body is the gametophyte, upon which the sporophyte develops.

by numerous rhizoids and lacks a stem. The epidermis of the upper surface is covered by a waxy cutin to prevent the loss of water, and is punctured by many pores for the exchange of gases. The recognizable plant is the gametophyte generation; as in mosses, the sporophyte grows as a parasite on the gametophyte (Fig. 77). The upper surface of the liverwort gametophyte may bear **gemma cups** (Fig. 77), within which are produced small, flattened, ovoid gemmae. These separate from the parent plant and grow into new gametophytes, a process of asexual reproduction.

Mosses and liverworts evolved from alga-like ancestors; they have many characters in common with the green algae and are generally believed to be derived from them. At one time it was believed that the higher, vascular plants evolved from the bryophytes, from a liverwort or a hornwort. But although there is fossil evidence of true vascular plants in the Silurian period (p. 522) some 360,000,000 years ago, the first evidence of the bryophytes dates from the Pennsylvanian period, which began about 100,000,000 years later. For this and other reasons, botanists are now inclined to believe that vascular plants evolved independently from the green algae and that the mosses represent the end of a separate branch of the evolutionary tree.

107. THE PHYLUM TRACHEOPHYTA

The phylum Tracheophyta is an ancient and diverse group and includes the dominant land plants of today. It is subdivided into four subphyla: the **Psilopsida,** the most primitive vascular plants, only three species of which are alive today; the **Lycopsida** or club mosses and quillworts; the **Sphenopsida** or horsetails; and the **Pteropsida,** an enormous group which includes all the familiar ferns, conifers and flowering plants. All the tracheophytes, like the mosses, have a life cycle with an alternation of gametophyte and sporophyte plants. However, the sporophyte of the higher plants is a free-living, independent plant and the gametophyte is either a small, independent plant or is contained within the sporophyte. The tracheophytes, as their name implies, are characterized by the presence of vascular tissues, xylem and phloem, in the sporophytes.

108. SUBPHYLUM PSILOPSIDA

The most primitive vascular plants known are the Psilophytales (Fig. 78), which lived in the Devonian period and probably even earlier, in the Silurian. These plants grew to a height of about 2 feet and had a creeping horizontal stem from which grew branching, erect, green stems. The smaller branches were coiled at the tip and probably unrolled as they grew, as present-day ferns do. They had no roots and were either leafless or had small scalelike leaves. Fossil remains of these plants have been found in Scotland in such a good state of preservation that details of the internal structures were visible. Only three species of Psilopsida are known to exist today; these living fossils are small, simple, subtropical plants, one of which, *Psilotum,* grows in southeastern United States. The sporophytes are rootless but have an underground, branching stem from which grow green, photosynthetic upright branches bearing tiny scalelike leaves. The gametophytes are small, underground, nongreen bodies bearing sex organs. The Psilopsida are widely believed to be the ancestors of the other vascular plants. The several fossil species known each suggest the beginning of the sort of specialization found more fully developed in one of the other subphyla.

109. SUBPHYLUM LYCOPSIDA

This subphylum, composed of club mosses, quillworts and their relatives, was widespread in the later Devonian and Carboniferous periods and many of them were tall, treelike forms, but today only four genera remain, all of which are small, usually less than a foot high. These inconspicuous plants consist of a creeping stem which gives off upright stems with thin, flat, spirally arranged leaves (Fig. 79). At the tip of the stem are specialized leaves arranged somewhat in the shape of a pine cone which bear spore-producing structures. One genus of living lycopsids, *Lycopodium,* has a life cycle rather like that of the Psilopsida. The sporophyte produces spores, all of which are alike, which germinate to form gametophytes. On these develop sex organs in which are produced eggs and sperm. After fertilization, the developing embryo is, for a time,

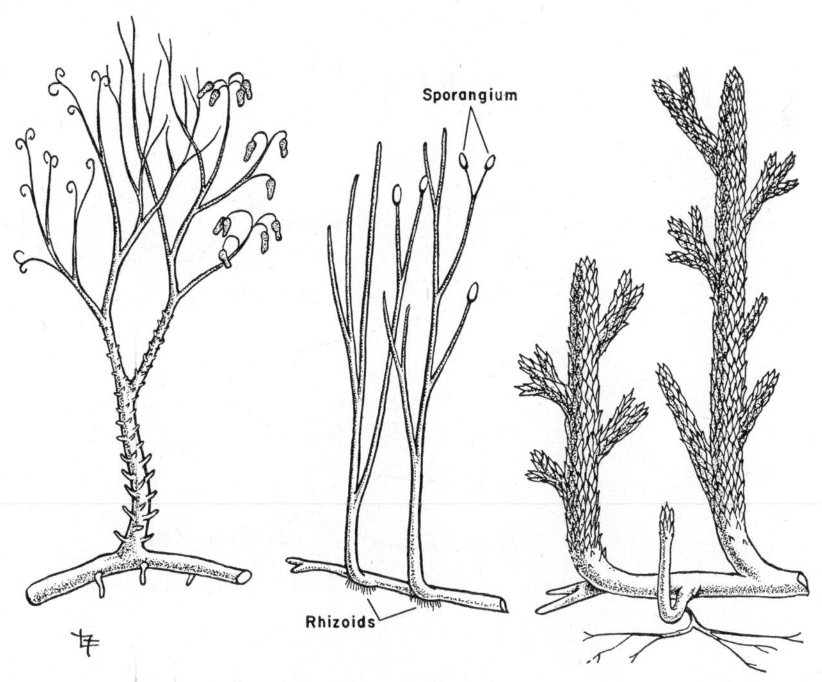

Figure 78. Three species of Psilophytales, the most primitive vascular plants, which lived during the Devonian period.

Figure 79. A, Photograph, half natural size, of a club moss, *Lycopodium,* sometimes used as a Christmas ornament. *B,* Enlarged view of a single stem, showing the arrangement of the leaves. (Weatherwax: Botany.)

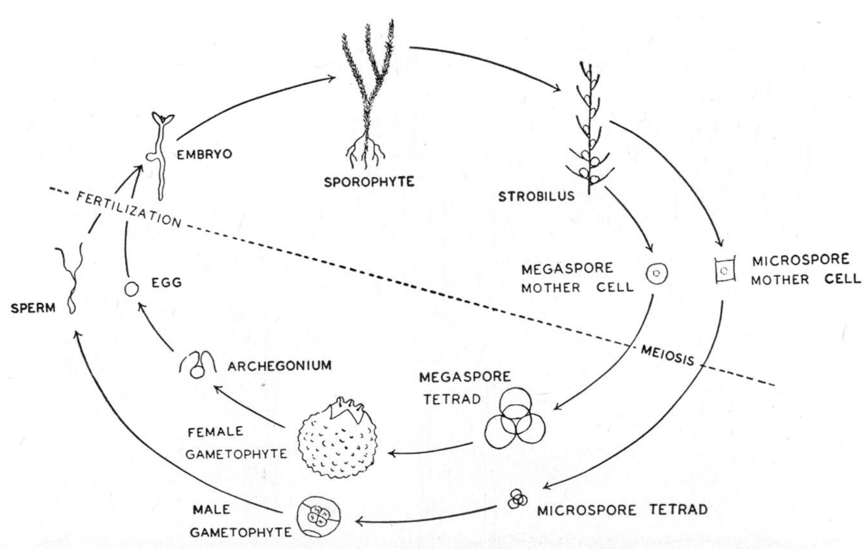

Figure 80. Diagram of the life cycle of *Selaginella.* (Weatherwax: Botany.)

dependent upon the gametophyte for nutrition. A second genus, *Selaginella*, commonly called spike mosses or little club mosses, shows an important evolutionary advance in having two types of spores, **megaspores,** which germinate to produce female gametophytes, and **microspores,** which germinate to produce male gametophytes (Fig. 80). The gametophyte generation is greatly reduced in size and is dependent on the sporophyte for nourishment. When the haploid microspores are released from the sporophyte they may drop near a megaspore. When wet by dew or rain, the microspore wall splits and the sperm within are free to swim to the megaspore and fertilize the haploid egg. As in the seed plants, the embryo is produced in the female gametophyte while it is still within the sporophyte. Although the reproductive cycle of these plants foreshadows that of the seed plants, they are not the ancestors of them but are a terminal group. The "resurrection plant" of the Southwest is a species of *Selaginella* that rolls up into a compact ball of apparently dead leaves during the dry season, then unrolls and carries on its normal activities when moisture is present. The **quillworts,** another group of Lycopsida, are superficially quite different, with slender, quill-like leaves resembling those of a bunch of garlic (Fig. 81). The leaves are attached by their broad bases to a short stem, out of which grow the roots. The leaves have spore-bearing organs at the basal end.

110. SUBPHYLUM SPHENOPSIDA

Like the two previous subphyla of tracheophytes, the subphylum Sphenopsida includes many more fossil forms than living ones. It arose during Devonian times and developed into a variety of species, some small, and some that were gigantic, treelike plants as much as 40 feet tall. The latter flourished especially during the Carboniferous period and their dead bodies, together with those of certain other plants, are the source of our present-day coal deposits (Fig. 82).

The present-day sphenopsids, called "horsetails," are wide-spread plants, usually less than 15 inches tall, which are found from the tropics to the Arctic on all continents except Australia, and in both boggy and dry places. The name "horsetail" is appropriate because of the multiple-branched, bushy structure of many species. Another popular name, "scouring rushes," is based on the fact that deposits of silica in the epidermis give the plants a harsh, abrasive quality. They were used to clean pots and pans for centuries before the invention of steel wool. The sporophyte plant is made up of a horizontal, branching, underground stem from which branch slender, branching roots and jointed aerial stems. The stem has many vascular bundles arranged in a circle around a hollow center. The stems have conspicuous nodes that divide them into jointed sections; at each node there

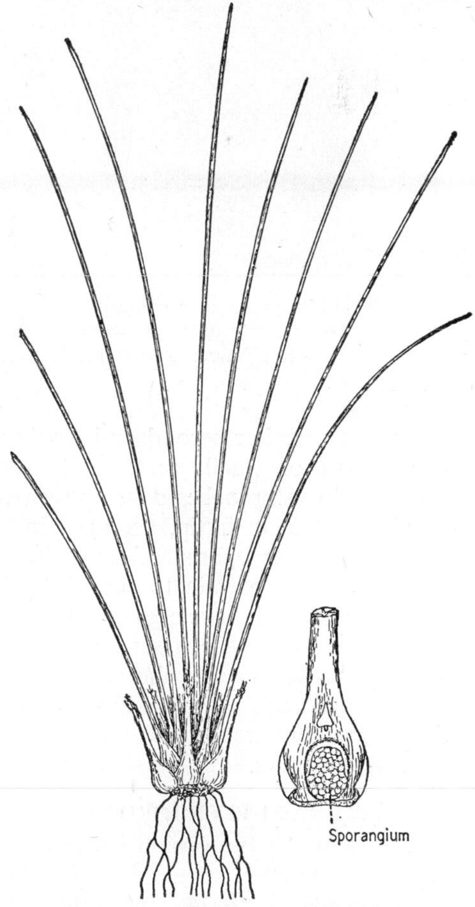

Figure 81. Sketch, natural size, of a quillwort, *Isoetes*. To the right is an enlarged view of the spore-bearing sporangium at the base of a leaf. (Weatherwax: Botany.)

Figure 82. Photograph of a calamite, a fossil sphenopsid. The whorls of long, linear leaves are clearly evident. (Fuller: The Plant World, Henry Holt & Company.)

is a whorl of smaller secondary branches and a whorl of small, scalelike leaves (Fig. 83). Some branches develop spore-bearing leaves at their tip, arranged in a sort of cone. The spores released from these structures germinate into green gametophytes which have egg- and sperm-producing organs. The zygote formed after fertilization develops into the sporophyte plant; at first this is parasitic on the gametophyte, but it quickly develops its own stem and roots.

111. SUBPHYLUM PTEROPSIDA

The subphylum Pteropsida is the largest group in the plant kingdom, composed of three classes: the **ferns,** class Filicinae; the **conifers,** class Gymnospermae; and the **flowering plants,** class Angiospermae. The sporophytes of these plants have well-differentiated roots, stems and leaves and are the important, complex generation. The gametophytes are small and either independent, small, short-lived plants, as in the ferns, or minute structures, composed of only a few cells, found within the spores of the conifers and flowering plants. The leaves of the Pteropsida are believed to have arisen from stems, by the flattening and transformation of the distal parts. Some of the primitive psilopsids have leaves formed in this way and may be the ancestors of the pteropsids. The specialized roots, stems and leaves of the pteropsids have enabled them to become truly adapted to terrestrial life, and the well-developed vascular and supporting tissues have enabled them to grow to a large size.

112. CLASS FILICINAE

There are about 9,000 species of ferns living today, widely distributed. Those in the temperate regions thrive best in cool, damp and shady places. Ferns are abundant in tropical rain forests; some of these are tall and superficially resemble palm trees, having an erect, woody, unbranched stem with a cluster of compound leaves or fronds at the top. The common, temperate-zone ferns have horizontal stems growing at, or just beneath, the surface of the soil. From these grow hairlike roots and erect fronds or compound leaves. The leaves of ferns characteristically are coiled in the bud and unroll and expand to form the mature leaf (Fig. 84).

The root of a fern has root cap, meristematic, elongation and mature zones like the root of a seed plant. Its stem has a protective epidermis, supporting and vascular tissues, and the leaves have veins, chlorenchyma, a protective epidermis, and stomata. In all these respects the ferns are similar to the seed plants and for this reason are grouped with them in the subphylum Pteropsida. They differ from the seed plants in that the spores are all alike and are produced on the under surfaces of the leaves. These haploid spores are released and will germinate into green gametophytes on moist soil. The gametophytes are heart-shaped structures about the size of a dime, which bear sex organs in which eggs and sperm are produced.

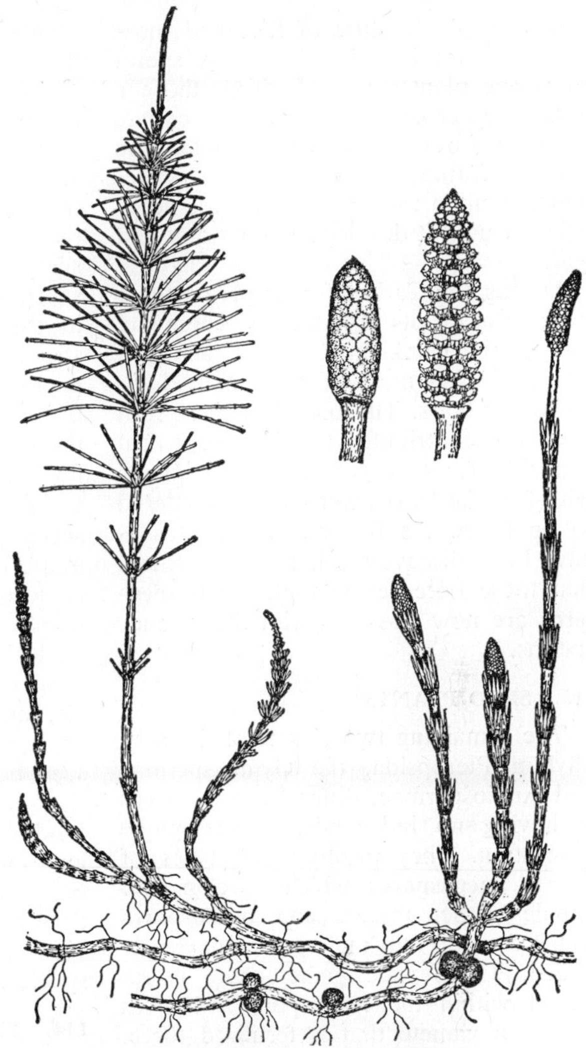

Figure 83. Sketch, half natural size, of a horsetail, *Equisetum. Inset,* natural size, the cones of this plant. (Weatherwax: Botany.)

Figure 84. Photograph of young fern plant showing young leaves uncoiling as they develop. (Weatherwax: Botany.)

The general structure of fern and moss gametophytes is quite similar. A sperm from one plant usually fertilizes the egg from another one and a new sporophyte develops from the resulting zygote. The embryo formed from the zygote is dependent upon the gametophyte for nourishment until it develops its own root and leaf.

During the Carboniferous period there were great forests of fern trees. These had tall, slender trunks made of the stem plus an enveloping mass of roots matted together by hairs. The bodies of these fern trees also contributed to our present coal deposits. Another group of fossil plants with fernlike leaves were long considered to be ferns. As better-preserved fossils have been discovered, it has been found that these bore seeds; these fossil **seed-ferns** are now classified with the gymnosperms.

113. SEED PLANTS

The remaining two classes of the subphylum Pteropsida, the Gymnospermae and Angiospermae, differ from the ferns in having no independent gametophyte generation. They produce two types of spores, **megaspores** which develop into female gametophytes, and **microspores** which develop into male gametophytes or pollen. The female gametophyte is retained within the megaspore and gives rise to a gamete that is fertilized there. The zygote develops into an embryo with the rudiments of leaves, stems and roots while still within the seed coat derived from the previous sporophyte. The **seed** thus consists of structures belonging to three distinct generations—the **embryo,** the new sporophyte; the **endosperm,** nutritive tissue derived from the female gametophyte; and the **seed coat,** derived from the old sporophyte. Under one system of classification, the gymnosperms and angiosperms comprise the phylum Spermatophyta, the seed plants, and the ferns, club mosses, horsetails, and psilopsids are grouped together as the phylum Pteridophyta. This system is gradually being abandoned as less natural than the present one.

There are over 200,000 species of seed plants, adapted to survive in a variety of terrestrial environments, and varying in size from the minute duckweed a few millimeters in diameter to giant redwood trees. They also are the plants of greatest value to man as sources of food, shelter, drugs and industrial products. There are two classes of seed plants, the Gymnospermae ("naked seeds") and the Angiospermae ("enclosed seeds"), which differ in the relationship of the seeds to the structures producing them. The seeds of angiosperms are formed inside a **fruit,** and the seed covering is developed from the wall of the ovule (see p. 175) of the flower. The angiosperms which apparently violate this rule and have exposed seeds — wheat, corn, sunflowers and maples, for example—really do have enclosed seeds, for the structure commonly called the "seed" actually is a fruit, enclosing the true seed. The seeds of gymnosperms are borne in various ways, usually on cones, but they are never really enclosed as are angiosperm seeds. The seed habit has contributed immensely to the success of the seed plants; the stored food nourishes the embryo until it can lead an independent life; the tough outer coat protects the embryo from heat, cold, drying, and from parasites; and they provide a means for the dispersal of the species.

114. THE GYMNOSPERMS

The gymnosperms are woody plants which do not differ greatly in their structural pattern from the woody angiosperms. They have no flowers, but bear their seeds on the inner sides of scalelike leaves that are usually arranged spirally to form a **cone.** There are only about 600 species living today, but there were many more in former geologic times. There are four orders of living gymnosperms, the **conifers,** the **cycads,** the **ginkgos** and the **gnetales.** Three other orders, the Cycadofilicales or **seed ferns,** the Bennettitales and Cordaitales are known only from fossil remains. In addition to the differences in reproductive methods, gymnosperms differ from angiosperms in that the xylem of the stem is composed only of tracheids and wood parenchyma; no wood fibers or vessels are present and as a result the gymnosperms are softwoods rather than hardwoods.

Of the living gymnosperms the conifers —pine, cedar, spruce, fir and redwood trees—are the most successful biologically. The needle-like leaves of these evergreens are well adapted to withstand hot summers, cold winters and the mechanical abrasion of storms. Under the thick, heavily cutinized epidermis layer is a layer of thick-walled sclerenchyma. The stomata are set in deep pits that penetrate the sclerenchyma. The conifers bear two kinds of cones, one producing pollen and one producing eggs; the male cones release the pollen, which is carried by the wind to the female cones, where the eggs are fertilized. In pines, as much as a year may elapse between pollination and fertilization, and several more years may elapse between fertilization and the shedding of the seeds.

The conifers are important economically as the source of more than 75 per cent of the wood used in construction and in the manufacture of paper and plastics. Some of them produce resins used in the production of turpentine, tar and oils. A few conifer seeds—pine "nuts," for example—are used as food, and the berries of juniper produce aromatic oils used for flavoring alcoholic beverages such as gin.

The **cycads** live mainly in tropical and semitropical regions and have either short, tuberous, underground stems or erect, cylindrical stems above the ground. Their large, compound, divided leaves (Fig. 85) make them look much like ferns or miniature palm trees, for which they are frequently mistaken. The only cycad na-

Figure 85. Photographs of cycads. *Above, Cycas; below left,* a male plant of *Zamia;* and *below right,* a mature, fruiting plant of *Zamia.* This plant, found from Florida to Mexico, is the only cycad native to the United States. (Weatherwax: Botany.)

Figure 86. Ginkgo twigs. *Left,* A cluster of mature seeds; *center,* a twig with leaves and young ovules; *right,* a twig with leaves and pollen cones. (Fuller, H. J.: The Plant World, Henry Holt and Co., Inc.)

tive to the United States is the sago palm, *Zamia,* found in Florida. In contrast to the pine, which produces two kinds of cones on the same tree, the cycad population consists of two kinds of trees, one of which produces only cones yielding pollen and the other produces only cones yielding eggs. These are not "male" and "female" trees, for they are sporophytes, asexually producing male and female gametophytes respectively.

The **ginkgo,** or maidenhair tree, is the only living representative of a once numerous and widespread order. It survived in China and Japan where it was cultivated as an ornamental tree, because of its distinctive, fan-shaped leaves, which it sheds in the fall (Fig. 86). Recently, it has been introduced into other countries. Like the cycads, a given ginkgo tree will produce only staminate (pollen-producing) cones or ovulate (egg-producing) cones. The ginkgo and cycads are the only seed plants that produce swimming sperm rather than pollen tube nuclei. When the egg is fertilized and matures, the inner seed covering becomes hard, while the outer covering becomes soft and pulpy, and has a rancid, foul odor. When a tree produces many seeds at once, the odor is extremely offensive. For this reason, when ginkgos are planted in parks care must be taken to ensure that all the trees in one region are of the same "sex," so that seeds will not be produced.

The Gnetales is a small order of gymnosperms that includes some very peculiar plants. The only one found in the United States is *Ephedra,* a low, much branched shrub with naked twigs and rudimentary scale leaves that superficially looks like one of the horsetails. *Welwitschia,* a plant that grows in the desert in southwest Africa, consists of a short thick trunk, partly imbedded in the soil, from which grow two long, flat, leathery leaves. Although the plant may live for centuries, it never produces more than these two leaves, which, subjected as they are to the sand and windstorms of the desert, eventually become very tattered.

In a number of respects the gymnosperms are more advanced, better adapted for land life, than the ferns. They have become independent of water for reproduction by evolving wind-borne pollen to transfer the male gametophyte to the female gametophyte, by developing a pollen tube to replace motile sperm as a means of effecting fertilization, and by evolving the seed habit. The development of a cambium and, from it, secondary wood, has made possible the large size of many of the seed plants.

115. THE ANGIOSPERMS

The **angiosperms,** or true flowering plants, are the largest class in the plant kingdom and include almost 200,000 species of trees, shrubs, vines and herbs adapted to almost every kind of habitat. Some live completely under water, others in extremely arid regions. The vast majority are autotrophic, but some, such as the orchid, Indian pipe and mistletoe, have little or no chlorophyll and so are partly or wholly parasitic. A few angiosperms have evolved devices for catching insects and other small animals, and hence are holozoic and carnivorous. Many angiosperms can complete an entire life cycle, from the germination of the seed to the production of new seeds, within a month, but others require twenty or thirty years to reach sexual maturity. Some live for a single growing season; others live for centuries. The stems, leaves and roots present a bewildering variety of forms, but all angiosperms develop flowers which have a fundamental similarity of pattern.

Angiosperms differ from gymnosperms in the abundance and prominence of the xylem vessels, in the formation of flowers and fruits, in the formation of a pistil through which the pollen tube grows to reach the ovule and egg (see p. 174) (in gymnosperms the pollen lands on the surface of the ovule and the pollen tube grows in directly), and in the further reduction of the gametophyte generation to a few cells completely parasitic on the sporophyte.

Fossil remains of angiosperms have been found in rocks from the Cretaceous period, and, while botanists agree that they probably arose from some primitive gymnosperm, there are no intermediate plants in older rocks to indicate which group of gymnosperms might have been the ancestor of the angiosperms.

The angiosperms are divided into two subclasses, the Dicotyledones of which there are at least 125,000 species, and the Monocotyledones, with perhaps 75,000 species. The two differ in the following ways:

(1) The embryo in the monocot seed has only one seed-leaf or cotyledon; the dicot embryo has two.

(2) The leaves of monocots have parallel veins and smooth edges; those of dicots have veins which branch and rebranch and, usually, edges which are lobed or indented.

(3) The flower parts of monocots—petals, sepals, stamens and pistils—exist in threes or multiples of three; dicot flower parts usually occur in fours or fives, or in a multiple of these.

(4) Bundles of xylem and phloem vessels are scattered throughout the stem of monocots; in dicots the xylem and phloem vessels occur either as a single, solid mass, running up the center of the stem, or as a ring between the cortex and pith.

There are a great many different families of both monocots and dicots, each of which usually takes its name from some conspicuous member. Some of the monocot families are the grasses, palms, lilies, orchids and irises. Some important dicot families are the buttercup, mustard, rose, maple, cactus, carnation, primrose, phlox, mint, pea, parsley and aster. The rose family, for example, includes, in addition to roses, the apple, pear, plum, cherry, apricot, peach, almond, strawberry, raspberry, hawthorn and other shrubs.

There is some difference of opinion about the evolutionary relationships of the angiosperms, but many botanists regard the monocots as being derived from the dicots, perhaps from the buttercup family.

QUESTIONS

1. Why are the bryophytes called the "amphibians of the plant world"?
2. Discuss the adaptations for land life evident in the bryophytes. What additional adaptations are present in the flowering plants?
3. Why are moss plants restricted to a height of 6 or so inches?
4. How do liverworts differ from mosses?
5. Why are the vascular plants believed to have evolved directly from green algae and not from bryophytes?
6. What are "embryophytes"?
7. How are conifers and flowering plants differentiated?
8. In what ways do ferns resemble the seed

plants? In what ways do they differ from them?

9. What is a seed?
10. How are the following plants classified: (*a*) a ginkgo, (*b*) a mushroom, (*c*) a horsetail, (*d*) a pine tree, (*e*) a club moss, (*f*) an orchid, (*g*) a cactus?

SUPPLEMENTARY READINGS

The evolution of land plants is discussed in Chapter 8 of E. O. Dodson's *A Textbook of Evolution*. A more extended treatment is to be found in Andrew's *Ancient Plants and the World They Lived In.*

Chapter 12

The Evolution of Plant Reproduction

IN THE previous chapters on plants, the methods of reproduction characteristic of each phylum have been described briefly. In plants, much more clearly than in animals, an evolutionary sequence is evident ranging from forms such as the blue-greens and bacteria which reproduce almost exclusively asexually, through other forms with the simplest type of sexual reproduction, to ones with complicated life cycles and highly evolved adaptations to ensure fertilization of the egg and survival of the zygote until it is capable of leading an independent life. Some of the lower forms, such as the puff balls and bracket fungi, which have no reproductive specializations, produce billions of spores so that by chance a few will fall in an environment favorable for germination and survival. The higher plants may produce no more than a few score seeds per plant but each seed has a fairly good chance of growing into a mature plant.

116. ASEXUAL REPRODUCTION

Asexual reproduction is characterized by the presence of a *single* parent, one that splits, buds, fragments, or produces many spores so as to give rise to two or more offspring. Sexual reproduction, in contrast, involves the cooperation of *two* parents, each of which supplies one gamete; the two gametes unite to form the zygote. In self-pollinating plants and in hermaphroditic animals (p. 415), both "parents" may be located in the same plant or animal body, but reproduction is typically sexual nonetheless.

The advantages of sexual over asexual reproduction will be more apparent after the discussion of heredity. Sexual reproduction permits the recombination of characters from two different strains so that the offspring may inherit traits from each parent that make it better fitted for survival than either parent was. In contrast, offspring produced asexually have exactly the same inherited characters as the parent. Evolution by natural selection can occur to any significant extent only when this recombination of traits takes place in each generation.

For most blue-greens and bacteria, asexual reproduction is the only means known by which new individuals are produced. Sexual reproduction has been shown to occur in some bacteria, though very rarely, and it may occur in others. Even in the higher plants reproduction may occur asexually in a variety of ways, and the farmer and florist take advantage of these in producing food and ornamental plants that will be exactly like the parent. Most of the cultivated trees and shrubs are reproduced from the **cuttings** of stems, which sprout roots at their tips when placed in moist ground. Willow stems have an almost unbelievable ability to form roots and grow. The amateur gardener who cuts willow poles to support his beans or tomatoes may be dismayed to find that his willows have taken root and are growing better than his beans! A number of commercial plants—bananas, seedless grapes and navel oranges, to mention just a few—have lost the ability to produce functional seeds and must be propagated entirely by asexual means. Many species of plants which will not propagate in this way naturally can be induced to form roots if treated with a growth hormone (p. 93).

Many plants, such as the strawberry, develop long, horizontal stems called **runners.** These grow several feet along the ground in a single season and may develop new erect plants at every other node. Other plants spread by means of similar stems, called **rootstocks,** which grow underground. Such weeds as witch grass and crab grass are particularly difficult to control because they spread by means of runners or rootstocks. Swollen underground stems or **tubers,** such as the white potato, also serve as a means of reproduction; in fact, some of the cultivated varieties of potato rarely, if ever, produce seed and must be propagated by planting a piece of a tuber containing a bud or "eye."

Grafting, the uniting of the stem of one plant with the stem or root of another, is not a method of reproduction, since it does not result in an increase in the number of individuals. It is widely used commercially to grow the stem of one variety that produces fine fruit on the root of another variety that has hardy, vigorous roots but produces poor fruit. Most Florida sweet oranges, for example, are grown on trees grafted onto the root system of a sour orange variety.

117. THE EVOLUTION OF SEX

No one knows how or when the phenomenon of sexual reproduction first evolved, but within a single living phylum, the green algae, are a variety of species that can be arranged in a series which illustrates how this *might* have happened. The existence of this series does not, of course, *prove* either that sex first evolved in the green algae or that the evolution of sex took place in a series of steps.

The simplest green algae, such as *Protococcus,* exhibit only asexual reproduction by simple cell division. In most other green algae, asexual reproduction occurs by the transformation of a vegetative cell into one or more specialized reproductive cells, called **zoospores,** each of which bears two or more flagella and is well adapted for aiding in the dispersal of the species.

Chlamydomonas, found in pools, lakes and on damp soil, has a vegetative cell that has two flagella and is protected by a heavy cellulose wall. The cell may reproduce asexually by dividing to form two to eight zoospores within the cellulose wall. These are set free by the rupturing of the parental cell wall and swim away, independent plants. Occasionally sexual reproduction occurs: The parent cell divides to form eight to thirty-two smaller cells, **gametes,** which resemble the zoospores and adults but are smaller. Two of these gametes fuse, beginning at the end bearing flagella, to form a **zygote** (Fig. 87). This initially has four flagella, two contributed by each gamete, but in time these are lost. The cell rounds up, secretes a thick cell wall and is capable of surviving long periods in an unfavorable environment. When the environment is again favorable, the zygote undergoes meiosis to form four cells, the cell wall cracks open and the liberated cells develop flagella and become independent plants. *Chlamydomonas* illustrates a very primitive form of sexual reproduction, for the gametes are not specialized cells; they look exactly like the zoospores and adults. In most species of *Chlamydomonas* the two cells uniting are identical in size and structure; this form of reproduction is known as **isogamy.** In a few species, two kinds of gametes occur differing in size but both bearing flagella, and the union of two unequal-sized gametes results in a zygote; this form of reproduction is called **heterogamy.**

Another primitive type of sexual reproduction is found in the pond scum *Spirogyra,* which consists of long filaments of cells arranged end to end. In the fall, when reproduction usually occurs, two filaments come to lie side by side and dome-shaped protuberances appear on the cells lying opposite (Fig. 88). These enlarge, fuse, and form a tube which connects the two cells. The protoplasm of one cell contracts, rounds up, oozes through the tube and joins with the protoplasm of the second. Finally, the nuclei of the two cells unite and fertilization is complete. The resulting cell, or zygote, develops a thick cell wall and is able to survive during the winter. In the spring it divides meiotically to form four haploid nuclei, three of which degenerate. The fourth remains and, after the thick wall breaks, divides mitotically to form a new haploid filament. Sexual reproduction in *Spirogyra* is primitive because it involves unspecialized cells (any cell in the filament can fuse with one from a neighboring filament) and the two fusing cells are similar (isogamy).

Another filamentous green alga, *Ulothrix,* demonstrates what was possibly the next step in the evolution of sex. In this plant, one of the vegetative cells in the filamentous chain undergoes several divisions (Fig. 89) to produce a number of small gametes, each with two flagella. As in *Chlamydomonas,* two of these swimming forms fuse, forming a zygote which initially has four flagella. After swimming for a time, the zygote loses its flagella, secretes a thick cell wall, and is capable of withstanding cold or drying. It later undergoes meiotic division and gives rise to four cells. These are finally liberated from the old zygote wall and develop into new filaments. Thus, in *Ulothrix,* sexual reproduction is isogamous and occurs by the fusion of two identical cells, but these cells are *specialized,* differing from the usual vegetative ones.

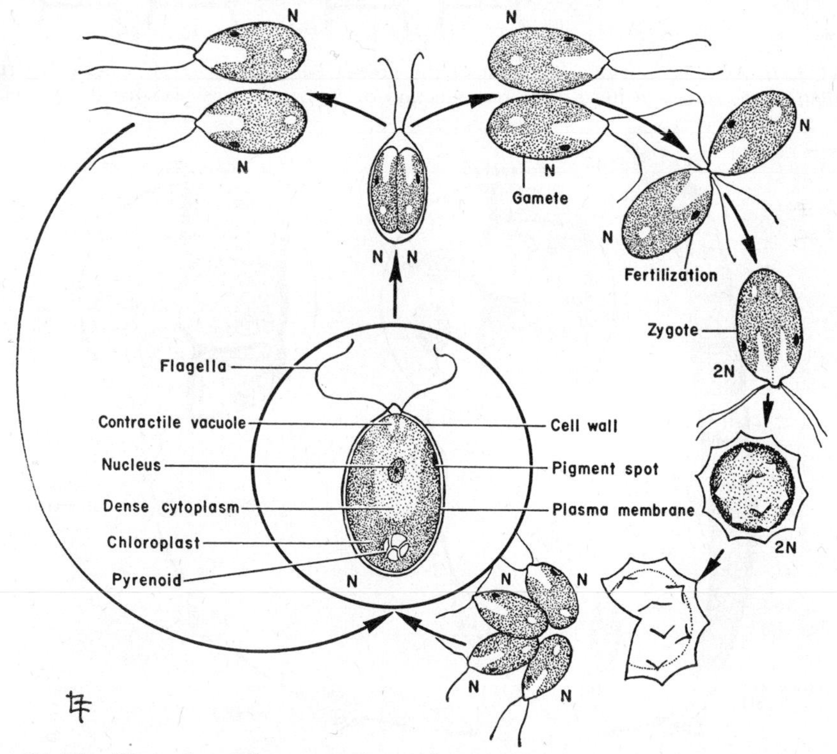

Figure 87. The life cycle of the green alga *Chlamydomonas,* asexual reproduction on the left, stages in sexual reproduction on the right. *Inset:* Enlarged view of a single individual showing body structures.

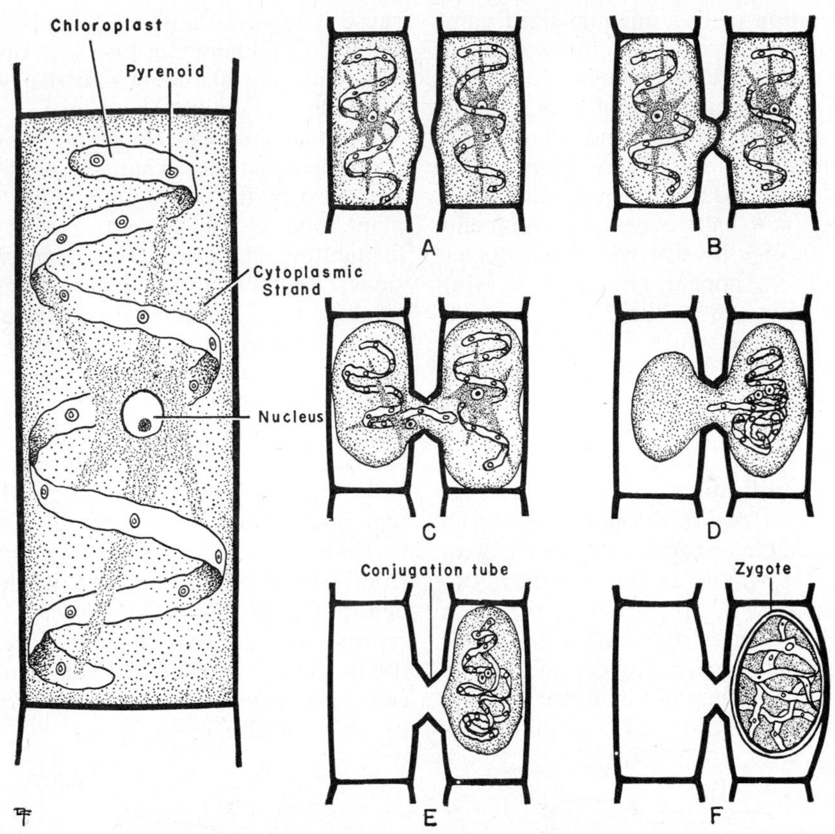

Figure 88. Left, A single cell of a filament of the green alga, *Spirogyra,* showing cell structures. *Right, A–F,* stages in the sexual reproduction of *Spirogyra;* see text for discussion.

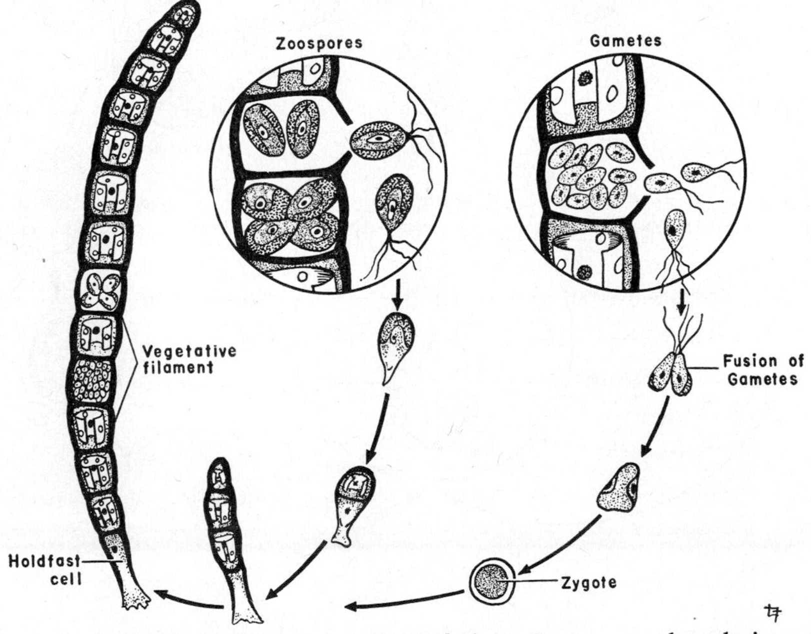

Figure 89. Left, a filament of the green alga, *Ulothrix. Center,* an enlarged view of asexual reproduction by zoospores. *Right,* sexual reproduction by the formation of gametes and the fusion of two gametes to form a zygote.

Another filamentous alga, *Oedogonium,* illustrates a third possible step in the evolution of sexual reproduction. The cells which fuse to form the zygote are *unlike:* One is a large, nonmotile, food-laden egg and the other is a small, motile sperm (Fig. 90). Sexual reproduction by the fusion of unlike gametes is called **heterogamy** and is characteristic of most higher plants. Any vegetative cell can differentiate into either an egg-forming cell or a sperm-forming cell. The egg-forming cell is an enlarged, spherical cell, the protoplasm of which shrinks away from the hard cell wall to form a rounded, nonmotile, food-laden **egg.** The sperm-forming cells are produced when a vegetative cell divides several times to produce a series of short, disc-shaped cells. The protoplasm within each of these cells divides to produce two small sperm, each of which has a circle of flagella at the anterior end (Fig. 90). Both egg and sperm are haploid and their fusion results in a diploid zygote. The sperm swims to the egg, attracted by some chemical substance given off by the egg. It enters the egg-containing cell through a crack and fuses with the egg. The zygote secretes a thick cell wall and in this form can survive periods during which environmental conditions are unfavorable. It eventually undergoes meiosis to form four haploid cells, each of which has a circle of flagella at the anterior end and resembles the asexual reproductive cells, the zoospores. The zoospores, whether formed sexually or asexually, can germinate and divide to form a new *Oedogonium* filament.

The final step in the evolution of sex is exhibited by other algae, such as *Volvox,* and by higher plants and animals, whose specialized gametes are produced only by special cells in the body—the **sex organs**—rather than by any vegetative cell in the organism, as in *Ulothrix* and *Oedogonium. Volvox* is a colonial alga, a hollow ball of cells, each of which bears two flagella and is connected to its neighbors by fine strands of protoplasm. Each colony may contain as many as 40,000 cells, most of which are alike and function only vegetatively. Small sperm cells, which bear two flagella and

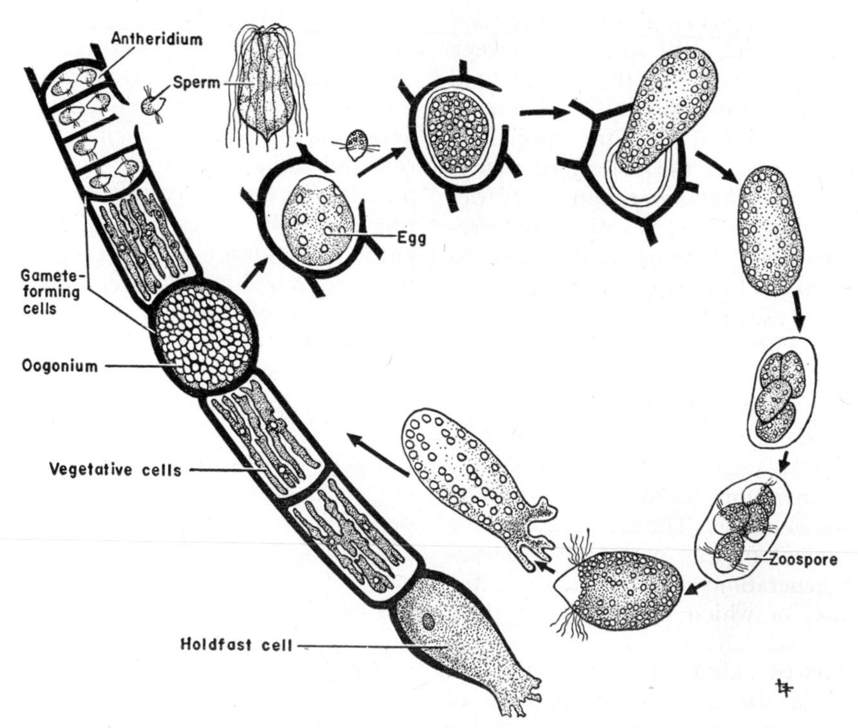

Figure 90. A filament of the green alga, *Oedogonium,* and sexual reproduction involving differentiated eggs and sperm.

are motile, are produced only in special sperm-producing organs or **antheridia** (this term is used generally for sperm-producing organs in higher plants as well). A single, large, nonmotile egg is produced within a special egg-producing organ or **oögonium.** The motile sperm are released and swim to the egg; their union results in a diploid zygote which forms a thick cell wall and can resist unfavorable conditions. During germination, meiosis occurs so that haploid cells are formed. These, by many mitotic divisions, give rise to a new colony. In some species of *Volvox,* a single colony may have both antheridia and oögonia; in other species, a colony has one or the other but not both, and can be said to be male or female. In these forms, sexual reproduction has evolved to a point where there is **sex differentiation.**

The series illustrates several trends in evolution, each of which is toward some sort of specialization. The trend from like gametes (isogamy) to unlike gametes (heterogamy) has obvious advantages for the survival of the species—the large number and motility of the sperm make them effective in seeking out the egg, and the large size and food stores of the egg provide nourishment for the zygote until it can become nutritionally independent. A second trend is toward the specialization of the cells of the colony or many-celled body so that some can carry out only vegetative functions, others only reproductive ones. A third trend is toward differentiation of the sexes. In these primitive plants, asexual and sexual reproduction may occur in the same plant, depending on environmental conditions. In the higher algae and all the higher plants, there is a definite, regular alternation of a generation of plants reproducing sexually with a generation reproducing asexually by means of spores. The trend toward the establishment of this pattern of **alternation of generations** is a fourth one, the beginnings of which occur in the green algae. The sea lettuce, *Ulva,* for example, consists of two kinds of plants which are identical in size and structure. One of these, however, is a diploid sporophyte which, by meiosis, produces haploid zoospores that develop into haploid gametophyte plants. This second type of plant produces gametes which fuse to form a diploid zygote which develops into the diploid sporophyte plant.

118. THE LIFE CYCLE

The **life cycle** of any species may be defined as the biologic processes of development which occur between any given point in an organism's life span, and that same point in the life span of its offspring. For bacteria, blue-greens and *Protococcus,* which reproduce by splitting, the cycle is extremely simple. The filamentous green algae such as *Ulothrix* have a cycle during most of which the colony consists of haploid cells which multiply asexually by mitosis. These divisions result in either new vegetative cells in the filament, haploid gametes, or asexual, haploid zoospores which divide to form new haploid colonies. The only diploid cells are the zygotes, for they divide meiotically to yield haploid vegetative cells. In *Ulva* and the higher plants, the zygote divides mitotically to form a diploid sporophyte plant which in turn produces haploid spores and from them, haploid gametophytes. The resulting life cycle is more complex and involves the marked alternation of generations mentioned previously. Some parasitic plants such as the wheat rust and white pine blister rust have complex life cycles involving two host organisms (p. 148). Many parasitic animals have an even more complex life cycle with three or four different intermediate hosts.

In plants that show an alternation of generations, the generation which reproduces sexually by producing gametes is known as the **gametophyte** and the generation which reproduces asexually by spores is known as the **sporophyte.** The life cycle of such plants consists of the production of haploid gametes by the gametophyte plant, the union of two gametes to form a diploid zygote which develops into a diploid sporophyte, the production of haploid spores by the sporophyte, and the development of these spores into the haploid gametophyte. The relative size and duration of the gametophyte and sporophyte generations vary considerably.

119. THE LIFE CYCLE OF A MOSS

The familiar, small, green, leafy plants called **mosses** are really the gametophyte generation of the whole plant. The gametophyte consists of a single, central stem, surrounded by leaves and held in place in the ground by a number of slender rootlets or **rhizoids,** which absorb water and salts from the soil. The leaf cells produce all the other compounds the plant needs for survival, so that each gametophyte is an independent organism. When the gametophyte has attained full growth, and is ready to reproduce, sex organs develop at the top of the stem, in the middle of a circle of leaves (Fig. 91). In some species the sexes are separate; in others, both male and female organs develop on the same plant. The male organs are sausage-shaped structures (called **antheridia**) which produce a large number of slender, spirally coiled sperm, each of which has two flagella. After a rain or heavy dew the sperm are released and swim through the film of water covering the plant to a neighboring female sex organ, either on the same plant or on another one. The female organ (called an **archegonium**) is shaped like a flask and has one large egg in its broad base. This organ releases a chemical substance which attracts sperm and, guided by this, the sperm swim down

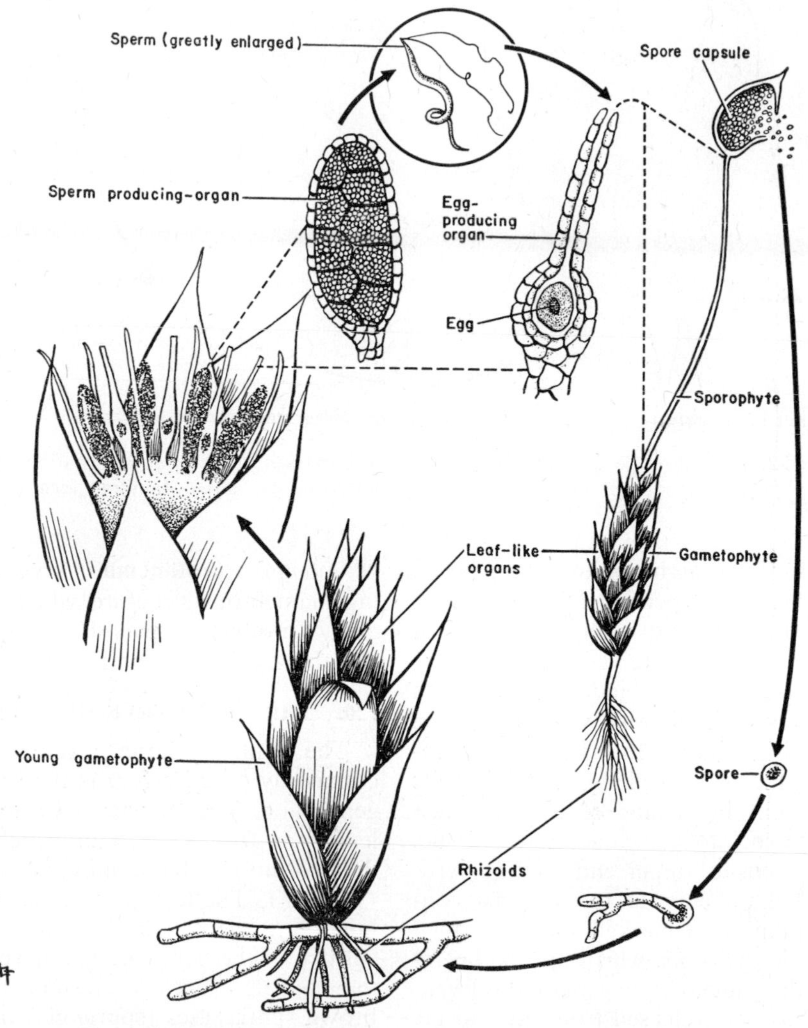

Figure 91. The life cycle of a moss plant. See text for discussion.

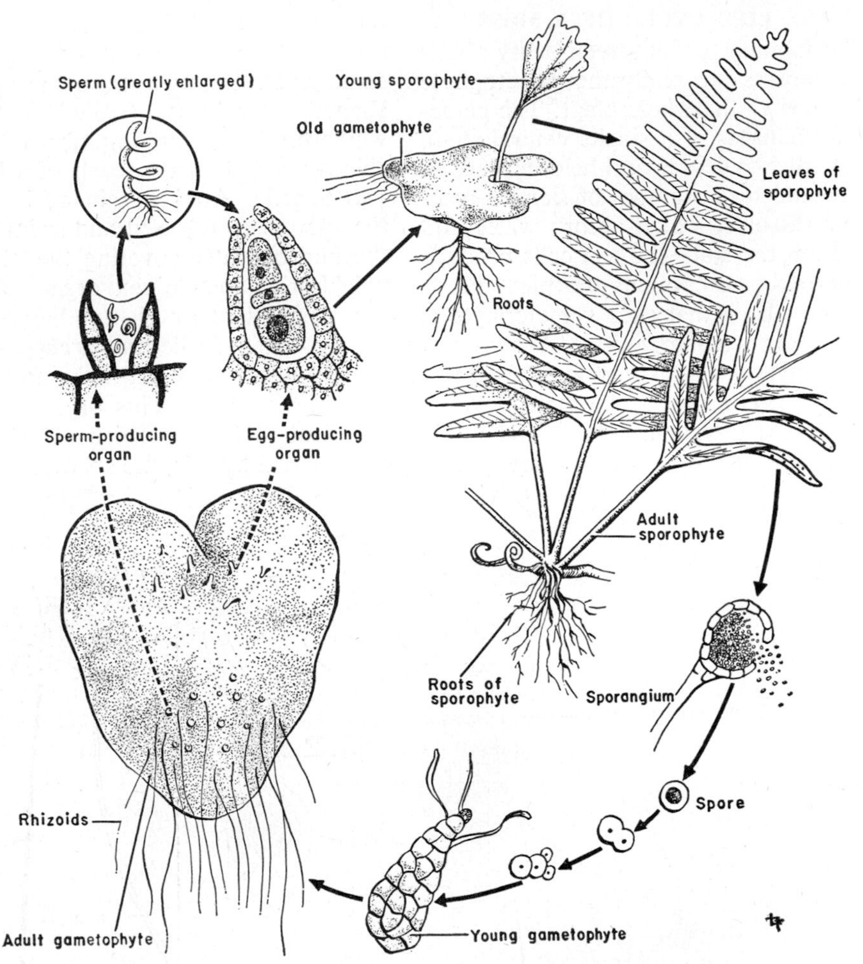

Figure 92. The life cycle of a fern plant. As in the mosses, there is an alternation of a sexual (gametophyte) and an asexual (sporophyte) generation; the large, familiar fern plant is the sporophyte.

the neck of the archegonium and into its base, where one sperm fertilizes the egg. The resulting zygote is the beginning of the diploid sporophyte generation.

In contrast to the independent green gametophyte, the sporophyte is a leafless, brown, single stalk which lives as a parasite on the gametophyte, obtaining its nourishment by means of a foot which grows down into the gametophyte tissue. At the opposite, upper end of the sporophyte stalk, a capsule forms. This contains a number of haploid spores, formed by meiotic divisions, which are the beginning of the gametophyte generation. Eventually these are released from the capsule and drop to the ground, where each develops into a **protonema,** a green, branching filamentous structure which, by budding, produces several gametophytes, thereby completing the life cycle.

120. THE LIFE CYCLE OF A FERN

The relatively large, leafy green plant commonly called a fern is the sporophyte generation. This consists of a number of leaves or fronds, each of which is subdivided into a large number of leaflets (Fig. 92). The leaves are attached to the horizontal stem that lies just under the surface of the soil. The under surfaces of the leaflets develop clusters of small, brown spore cases (**sporangia**) in each of which haploid spores are produced by meiosis. The sporophyte plant may live several years and produce several yearly crops of spores.

The spores are released at the proper time, fall to the ground, and develop into flat, green, heart-shaped gametophytes about the size of a dime. The gametophytes grow in moist, shady places, especially under decaying logs. A number of roots grow from each gametophyte into the soil to absorb water and salts. The male and female sex organs (antheridia and archegonia) develop on the under surface of the gametophyte (Fig. 92). Each archegonium, usually located near the notch of the heart-shaped structure, contains a single egg. The antheridia, located at the other end of the gametophyte, develop a number of flagellated sperm. These are released after a rain and, attracted by a chemical substance, swim through the water on the under surface of the gametophyte to reach the eggs. When an egg has been fertilized it develops into a new large sporophyte to complete the cycle. At first the sporophyte depends on the gametophyte for its nutrients, but it soon develops its own roots, stem and leaves and becomes independent.

The diploid fern sporophyte is fairly well adapted for terrestrial life: it has conducting and supporting tissues and, in contrast to the moss, it is nutritionally independent of the gametophyte. However, the conquest of land by the ferns has remained incomplete, for the gametophyte generation can survive only where there is plenty of moisture and shade, and the union of eggs and sperm in fertilization requires a watery medium.

121. THE LIFE CYCLE OF A GYMNOSPERM

The life cycles of gymnosperms and angiosperms show an even better adaptation to terrestrial life. The gametophyte has been reduced to a few cells enclosed within the tissues of the sporophyte and entirely dependent on it for food. The union of sperm and egg is no longer dependent on the presence of a film of water on the gametophyte, but is brought about by wind- or insect-dispersed pollen and the growth of the pollen tube. The embryo sporophyte is nourished and protected during early growth not by an independent gametophyte plant, as in mosses and ferns, but by a seed with its supply of endosperm and its tough outer coat. The sporophyte of seed plants produces two kinds of spores, larger **megaspores** and smaller **microspores,** and is said to be **heterosporous.** The lycopsid *Selaginella* (p. 155) and a few ferns are also heterosporous.

The reproductive parts of the seed plants were studied and named—stamen, pistil, ovule, etc.—before the stages in the alternation of sporophyte and gametophyte generations were understood and before the essential parallelism of the life cycles of mosses, ferns and seed plants was recognized.

The gymnosperms have life cycles which in certain respects are transitional between those of the ferns and those of the angiosperms (Fig. 93). A pine tree, for example, produces two kinds of cones, **staminate,** which are small, less than an inch long, and **ovulate,** which are the large easily visible cones, as much as 18 inches long in some species. The ovulate cone is composed of many scales and on the surface of each scale are two ovules. Within each **ovule** is a diploid **megaspore mother cell,** which divides by meiosis to form four haploid **megaspores.** Only one of these is functional and grows into a multicellular **megagametophyte.** On each megagametophyte are two or three female sex organs (archegonia) in each of which is a large egg. Each scale of the staminate cone has two **microsporangia** on the underside. Within the microsporangia are many **microspore mother cells,** each of which divides by meiosis to form four **microspores.** While still within the microsporangium or pollen sac, the microspores divide to form a four-celled microgametophyte or **pollen grain.** These are released and carried by the wind. When a pollen grain reaches an ovulate cone it comes to rest near the ovule, in a **pollen chamber.** As much as a year may elapse before one cell of the pollen grain elongates into a pollen tube which grows into the ovule

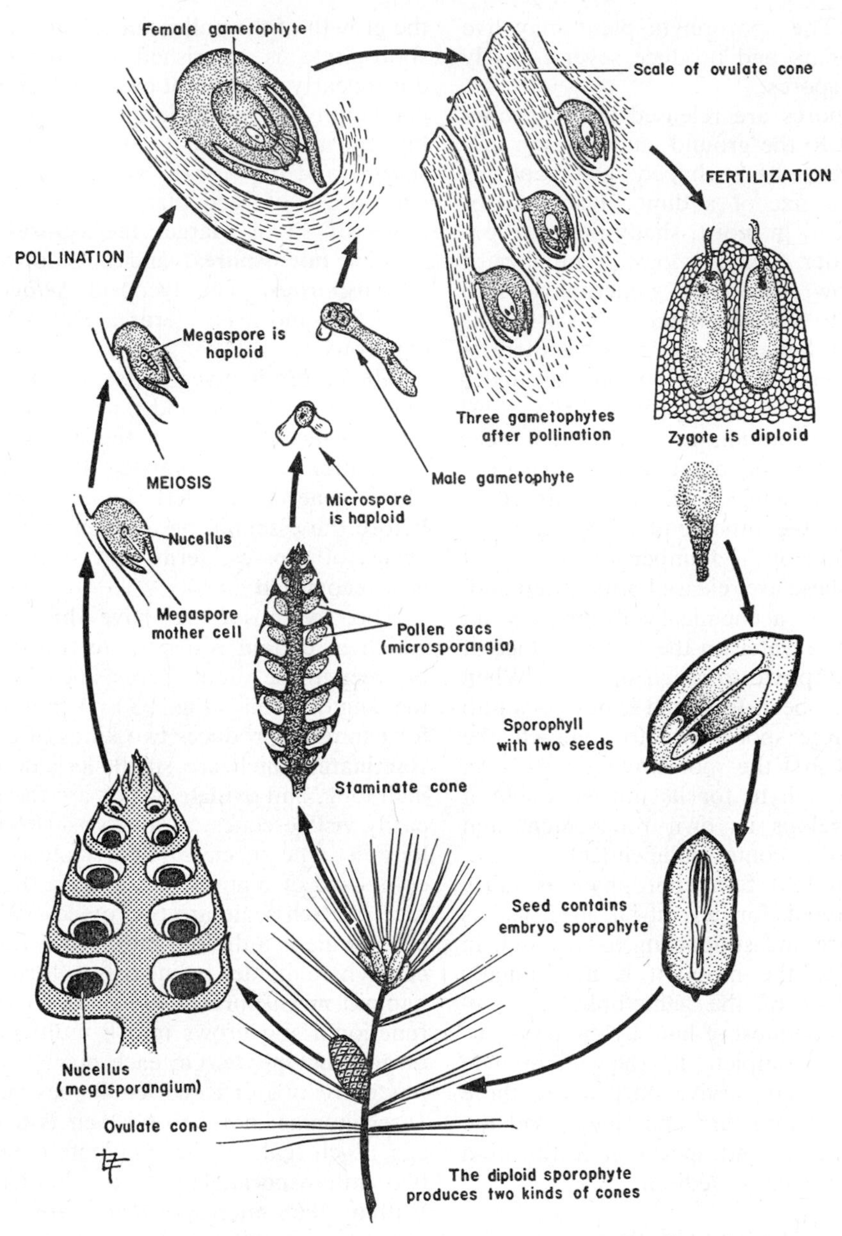

Figure 93. The life cycle of a pine tree. See text for discussion.

and reaches the macrogametophyte. Another cell of the pollen grain divides to form two male **gamete nuclei,** not motile sperm as in lower plants. When the end of the pollen tube reaches the neck of the archegonium and bursts open, the two male nuclei are discharged near the egg. One fuses with the egg nucleus to form the diploid zygote, and the other disintegrates. After fertilization, the zygote divides and differentiates and produces a sporophyte embryo, surrounded by the tissues of the female gametophyte and those of the parent sporophyte. This entire structure is the **seed.** After this short period of growth the embryo remains quiescent until the seed is shed and drops to the ground. When conditions are favorable, it germinates and develops into a mature sporophyte, the pine tree.

122. THE LIFE CYCLE OF AN ANGIOSPERM

In the angiosperms an alternation of generations — gametophyte and sporophyte — still occurs but the gametophyte is reduced to a few cells lying within the tissues of the sporophyte **flower.** The sporophyte is the familiar, visible tree, shrub or herb. Not all flowers are easily recognizable as such: Grasses and some trees have small green flowers which are quite different from the colorful blossoms we usually think of as flowers.

The Flower. The flower of an angiosperm is a modified stem which bears, instead of ordinary foliage leaves, concentric circles of leaves modified for reproduction. A typical flower consists of four concentric rings of parts (Fig. 94) attached to the **receptacle,** the expanded

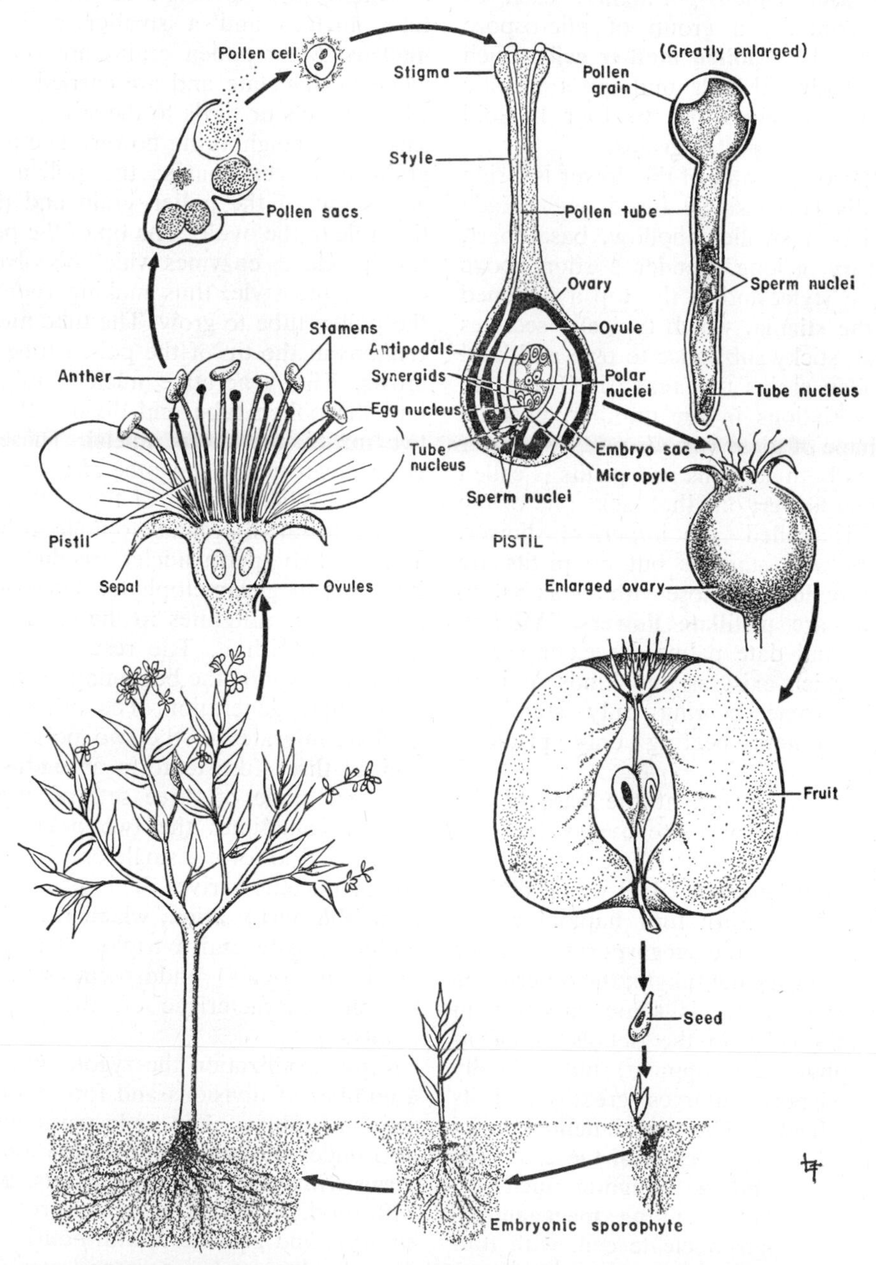

Figure 94. The life cycle of an apple tree. See text for discussion.

end of the flower stem. The outermost parts, usually green and most like ordinary leaves, are called **sepals.** Within the circle of sepals are the **petals,** typically brilliantly colored to attract insects or birds and ensure pollination. Just inside the circle of petals are the **stamens,** the male parts of the flower. Each stamen consists of a slender filament with an **anther** at the tip. The anther is a group of **pollen sacs** (microsporangia), each of which contains a group of microspore mother cells— **pollen mother cells.** Each of these diploid cells undergoes meiotic division and gives rise to four haploid microspores or **pollen grains.**

In the very center of the flower is a ring of **pistils** (or a single fused one). Each pistil has a swollen, hollow, basal part, the **ovary,** a long, slender portion above this, the **style,** and at the top a flattened part, the **stigma,** which typically secretes a moist, sticky substance to trap and hold the pollen grains that reach it. There are great variations in the number, position and shape of these various parts. A flower that has both stamens and pistils is called a **perfect flower;** one that lacks one or the other is called an imperfect flower. Flowers with stamens but no pistils are **staminate flowers;** those with pistils but no stamens are **pistillate flowers.** Willows, poplars and date palms are examples of plants which exist as two kinds of individuals, some bearing only staminate flowers, others bearing only pistillate flowers.

Within the ovary, at the base of the pistils, are one or more **ovules** (macrosporangia). Each ovule typically contains one **megaspore mother cell** which divides by meiosis to form four haploid megaspores. One of the **megaspores** develops into the **megagametophyte;** the other three disintegrate. There is some variation in different species in the details of megagametophyte development, but typically the megaspore enlarges greatly and its nucleus divides. The two daughter nuclei migrate to opposite ends of the cell, then each divides and the daughter nuclei divide again. The resulting **megagametophyte** is an eight-nucleate cell, with four nuclei at each end. One nucleus from each end migrates toward the center; the two come to lie side by side in the center and are known as **polar nuclei** (Fig. 95). One of the three nuclei at one end of the megagametophyte becomes the **egg nucleus,** the other two and the three nuclei at the other end all disintegrate.

The haploid microspore or pollen grain develops into a microgametophyte while still within the pollen sac. The nucleus of the microspore divides into two, a larger **tube nucleus** and a smaller **generative nucleus.** Most pollen grains are released while in this state and are carried by the wind, insects or birds to the stigma of the same or a neighboring flower. The pollen grain then germinates; the pollen tube grows out of the pollen grain and down the style to the ovule. The tip of the pollen tube produces enzymes which dissolve the cells of the style, thus making room for the pollen tube to grow. The tube nucleus remains in the tip of the pollen tube as it grows. The generative nucleus migrates into the pollen tube and divides to form two nuclei, the **sperm nuclei.** These migrate down the pollen tube after the tube nucleus. When the tip of the pollen tube penetrates the megagametophyte, it bursts and the two sperm nuclei are discharged into the megagametophyte. One of the sperm nuclei migrates to the egg nucleus and fuses with it. The resulting diploid cell is the **zygote,** the beginning of the new sporophyte generation. The other sperm nucleus migrates to the two polar nuclei and all three fuse to form an **endosperm nucleus,** made of three sets of chromosomes. Sometimes the two polar nuclei have fused to form a single one before the sperm nucleus arrives. This phenomenon of **double fertilization,** which results in a diploid zygote and a triploid (three sets of chromosomes) endosperm is peculiar to, and characteristic of, the flowering plants.

After fertilization the zygote undergoes a number of divisions and forms a multicellular embryo. The endosperm nucleus also undergoes a number of divisions and forms a mass of endosperm cells, gorged with food, which fill the space around the embryo, and provide it with nourishment. The sepals, petals, stamens, stigma and style usually wither and fall off after fertilization. The ovule with its contained embryo becomes the seed; its walls become thick and form the tough outer coverings of the seed. The seed consists of the embryo plus stored food in the form of the endosperm, all enclosed in a resistant covering derived from the wall of the ovule.

Fruits. The ovary, the basal part of the pistil containing the ovules, enlarges and forms the **fruit.** The fruit thus contains the seed—as many seeds as there were ovules in the ovary. In the strict botanical sense of the word, a fruit is a matured ovary containing seeds—the matured ovules. Although we usually think of only such sweet, pulpy things as grapes, berries, apples, peaches and cherries as fruits, bean and pea pods, corn kernels, tomatoes, cucumbers and watermelons are also fruits, as are nuts, burrs and the

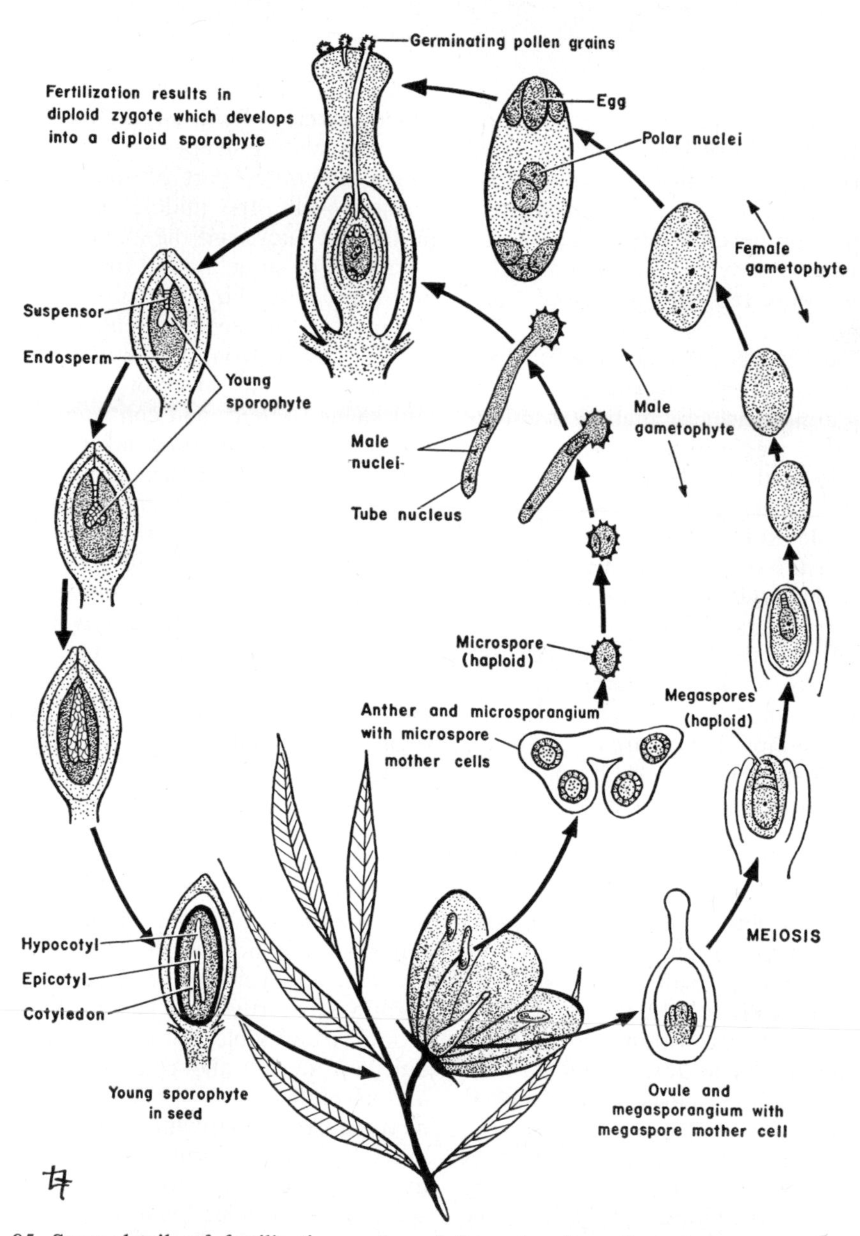

Figure 95. Some details of fertilization and seed formation in a dicot. See text for discussion.

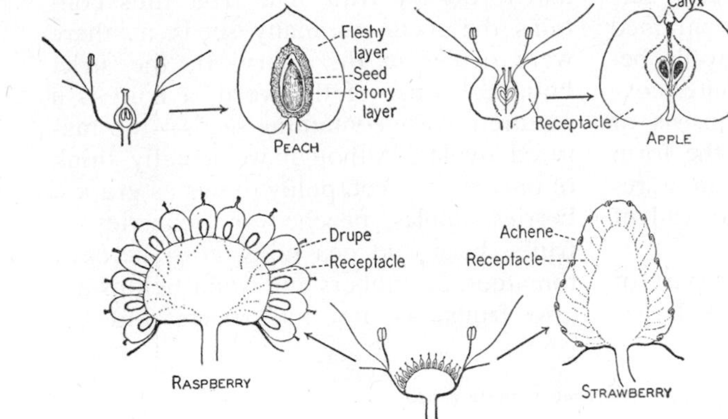

Figure 96. Formation of some fleshy fruits. Raspberries and strawberries are derived from similar flowers, but the pistils of raspberries become fleshy drupes and the pistils of strawberries become dry achenes, the yellow "seed-like" spots scattered over the surface of the fruit. (Weatherwax: Botany.)

winged fruits of maple trees. A **true fruit** is one developed solely from the ovary. If the fruit develops from sepals, petals, or receptacle as well as from the ovary it is known as an **accessory fruit.** The apple fruit consists mostly of an enlarged, fleshy receptacle; only the core is derived from the ovary.

True and accessory fruits are of three general types: simple fruits (e.g., cherries, dates, palms), which mature from a flower with a single pistil; aggregate fruits (raspberries and blackberries), which mature from a flower with several pistils; and multiple fruits (pineapples), derived from a cluster of flowers which unite to form a single fruit. Fruits are also classified as **dry fruits** if the mature fruit is composed of rather hard dry tissues and as **fleshy fruits** if the mature fruit is largely soft and pulpy (Fig. 96). Dry fruits are adapted primarily for being transported by the wind or by being attached to animal bodies by hooks. Fleshy fruits are adapted for dispersal by animals that eat them.

A **nut** is a dry fruit in which the ovary wall develops into a hard shell around the seed. The edible part of a chestnut is the seed within the fruit coat or shell. A Brazil nut is really a seed; there are about 20 such seeds borne within a single fruit. An almond is not a "nut" at all, but the seed or "stone" of a fleshy fruit related to the peach.

Grapes, tomatoes, bananas, oranges and watermelons are examples of fleshy fruits in which the entire wall of the ovary becomes pulpy; such fruits are technically called **berries.** Peaches, plums, cherries and apricots are **drupes** or stone fruits in which the outer part of the ovary wall forms a skin, the middle part becomes fleshy and juicy, and the inner part forms a hard pit or stone around the seed. There are, then, many kinds of fruit, differing in the number of seeds present, in the part of the flower from which they are derived, as well as in color, shape, water and sugar content, and consistency.

Fruits may form, or be induced to form, without the development of seeds. The banana, which has been cultivated for centuries, has only vestigial seeds (the black specks in the fruit) and must be propagated by vegetative means. Plant breeders have been able to develop seedless varieties of grapes, oranges and cucumbers. Many other plants have been induced to form seedless fruits by treatment with plant growth hormones (p. 97).

123. GERMINATION OF THE SEED AND EMBRYONIC DEVELOPMENT

Seeds are said to be "ripe" at the time they are shed from the parent plant, but this does not necessarily mean that they are ready to germinate. A few seeds do germinate shortly after being shed if conditions are suitable, but most seeds remain dormant during the cold or dry season and germinate only with the advent of the next favorable growing season. A prolonged period of dormancy usually occurs only in seeds with thick or waxy seed coats which render them impermeable to water and oxygen. The length of time that a seed will remain viable and capable of germination varies greatly. Willow and poplar seeds must germinate within a few days of being shed or they will not germinate at all. Some seeds remain viable for many years; a long-term experiment in progress at East Lansing, Mich., showed that seeds of the evening primrose and of yellow dock were able to germinate after seventy years. Samples are being tested for germinating ability every ten years; the next ones will be tested in 1959. There are authentic records of lotus seeds germinating 200 years after being shed. The ability of a seed to retain its germinating power depends on the thickness of the seed coat, on a low water content, and on the presence of starch rather than fats as the stored food material. Dormant seeds are alive and do metabolize, though at a very low rate.

Germination is initiated by warmth and moisture and requires oxygen. The embryo and endosperm absorb water, swell and rupture the seed coats. This frees the embryo and enables it to resume development. Most seeds do not need soil nutrients to germinate; they will germinate equally well on moist paper.

The cell divisions that the zygote undergoes following fertilization first produce a filament of cells, called the **suspensor.** Most of the embryo forms from the end cell of this filament, which begins to divide in other planes to form a rounded mass of cells. From this grow (in dicotyledonous plants) two primary leaves or **cotyledons** and a central **axis.** The part of the axis below the point of attachment of the cotyledons is called the **hypocotyl** and the part above it, the **epicotyl** (Fig. 97). The embryo is in about this state of development when the seed becomes dormant.

After germination the hypocotyl elongates and emerges from the seed coat. The primitive root or **radicle** grows out of the hypocotyl and, since it is strongly and positively geotropic, it grows directly downward into the soil. The arching of the hypocotyl in a seed such as the bean pulls the cotyledons and epicotyl out of the seed coat, and the epicotyl, responding negatively to the pull of gravity, grows

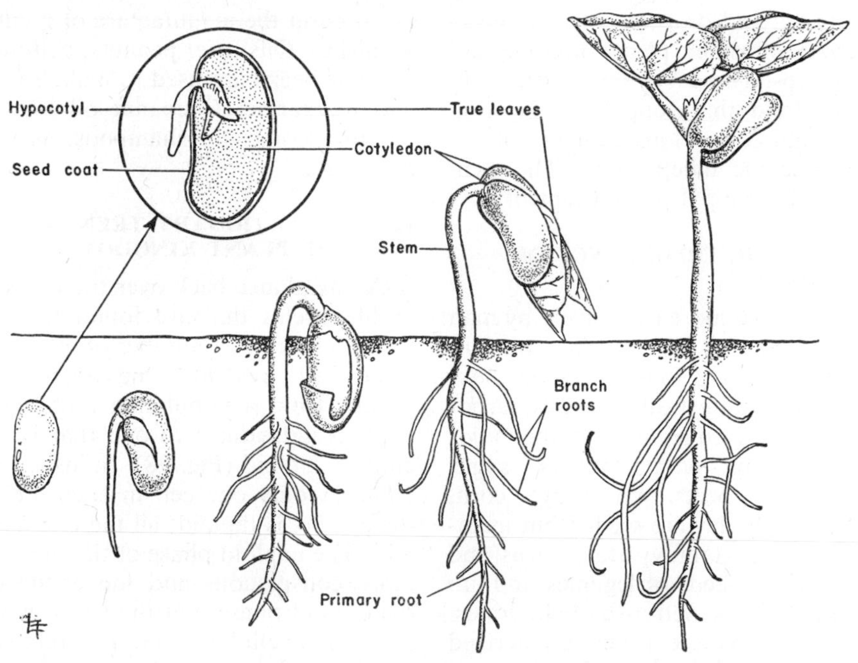

Figure 97. Stages in the germination and development of a bean seed. *Inset:* An enlarged view of an opened seed, showing cotyledons, hypocotyl, from which develops the root, and the epicotyl, from which develop the stem and leaves.

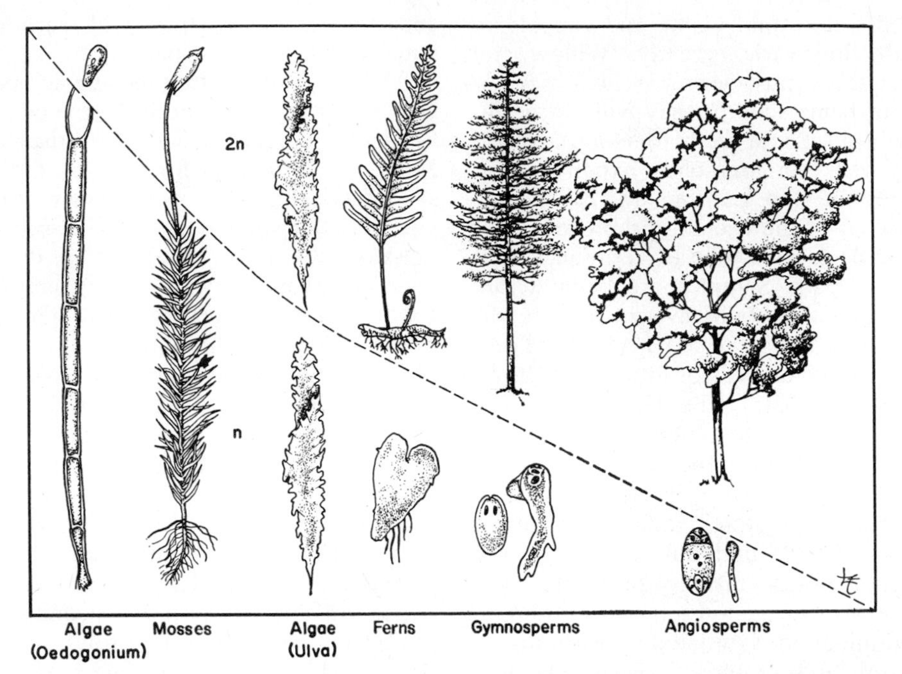

Figure 98. A diagram showing the evolutionary trend toward a greater size and importance of the sporophyte (*2n*) and a reduction in the size of the gametophyte (*n*) generation.

upward. The cotyledons digest, absorb and store food from the endosperm while within the seed. The cotyledons of some plants shrivel and drop off after germination; those of other plants become flat foliage leaves. The cotyledons contain reserves of food that supply the growing seedling until it develops enough chlorophyll to become independent. The stem and leaves develop from the epicotyl.

124. ECONOMIC IMPORTANCE OF SEEDS

Seeds are used more extensively by man than any other part of the plant. They are important sources of food, beverages, textiles and oils. Almost all of man's carbohydrates are derived from seeds, the major exceptions being potato tubers, sugar cane and sugar beets. Wheat, rye, corn, rice, oats and barley are seeds from members of the grass family; beans, peas and peanuts are the seeds of legumes and are rich in proteins as well as carbohydrates. Coffee and cocoa are beverages derived from seeds, and a variety of spices and seasonings such as nutmeg and pepper are made from ground seeds. Cotton fibers are produced as epidermal hairs on the seed coats of the cotton plant. Oils derived from seeds may be important industrially, or as foods. Linseed and tung oils are used in the manufacture of paints and varnishes. Oils from peanuts, cotton seeds and soy beans are used to make salad oils and margarine. Cocoanut oil is used in making soaps and shampoos, as well as margarine.

125. EVOLUTIONARY TRENDS IN THE PLANT KINGDOM

As we glance back over the many types of life cycles that are found from algae to angiosperms, a number of evolutionary trends are evident. One of these is a change from a population that is mostly haploid individuals to one that is almost entirely diploid (Fig. 98). In algae such as *Ulothrix* only one cell in each life cycle, the zygote, is diploid; all the rest are haploid. The haploid phase of the moss is still more conspicuous and longer-lived than the diploid phase, but the latter is a complex, multicellular plant. The relative importance of the two phases is reversed in the ferns: The diploid phase is the obvious, larger plant and the haploid gametophyte, though still an independent plant, is small and inconspicuous. The gymnosperms and angiosperms show progressive reductions of the haploid phase until, in the angiosperms, the male gametophyte consists of three cells and the female gametophyte of eight. The corollary of this trend toward diploidy is the trend toward the reduction of the gametophyte.

There are a number of possible explanations for these evolutionary trends. As long as there was an independent gametophyte generation, the transfer of sperm to the egg required a film of water for the sperm to swim in. This meant that reproduction could not occur unless moisture was present. The evolution of a life cycle in which there is a reduction of the gametophyte to a small group of cells within the sporophyte and sperm are transferred to the egg via a pollen tube permitted reproduction to occur in the absence of moisture. The evolutionary advantages of this are obvious. There may be another, less obvious reason for this; a diploid individual can survive despite the presence of deleterious, recessive genes; a haploid individual would be much more susceptible to the effects of such genes. A third explanation is that, since terrestrial life required the development of conducting and supporting tissues, and since these have appeared only in sporophyte individuals, evolutionary processes on land favored those plants with longer sporophyte and shorter gametophyte generations.

QUESTIONS

1. Distinguish between asexual and sexual reproduction.
2. Discuss the various modes of asexual reproduction exhibited by the higher plants.
3. What trends in the evolution of sexual reproduction are evident among the green algae?
4. What trends in the evolution of the life cycle are evident from mosses to flowering plants?
5. Describe sexual reproduction in the moss. In what ways is this ill-adapted for terrestrial life?
6. What are spores? How are they produced in a moss? In a fern?
7. Describe the life cycle of a pine tree. In what ways does the life cycle of an apple tree differ from this?
8. Draw a diagram of a flower and label the parts. What are the functions of each part?
9. Describe the phenomenon of double fertilization in flowering plants.
10. What is a fruit? Differentiate the several types of fruits.
11. What induces a seed to germinate?
12. Describe the development of a seed.
13. What are the parts of an embryo of a dicot? What does each become in the seedling?
14. Compare the roles of the cotyledons of dicots and monocots.

SUPPLEMENTARY READING

Some interesting aspects of plant reproduction are described in A. W. Naylor's article *Physiology of Reproduction in Plants*. Additional details of the life cycles of plants and of the many different types of fruits and seeds will be found in Fuller and Tippo, Hylander and Stanley, Smith, Gilbert, Bryan, Evans, and Stauffer, or in Weatherwax.

Part Three

The World of Life: Animals

Chapter 13

The Animal Kingdom: Lower Invertebrates

To CATALOGUE the vast array of animals, zoologists use a classification system similar to the one used by botanists, consisting of species, genera, families, orders, classes and phyla. We noted in Chapter 6 that a good classification must be natural, based upon the observation of similarities of structure and development. It is not proper to classify bats with birds because they both happen to fly, or whales with fishes because they both live in the sea. Studies of the anatomy and development of bats and whales show them to have many characteristics in common, and indicate their origin in similar, primitive ratlike or shrewlike mammals. Thus both are members of the class Mammalia. Birds and fishes, which differ as much from bats and whales as from each other, belong to different classes—Aves and Osteichthyes, respectively.

In the study of taxonomy, then, superficial, accidental similarities must be differentiated from the significant, fundamental ones. Those structures of various animals which arise from common rudiments and are similar in basic plan and development are called **homologous.** **Analogous** structures, in contrast, are similar only in function. Accordingly, the arm of a man, the wing of a bird and the pectoral (front) fin of a whale (Fig. 99) are all homologous, with a basically similar structural plan and similar embryonic origins. But the wing of a bird and the wing of a butterfly are simply analogous, enabling their possessors to fly, but having no developmental processes in com-

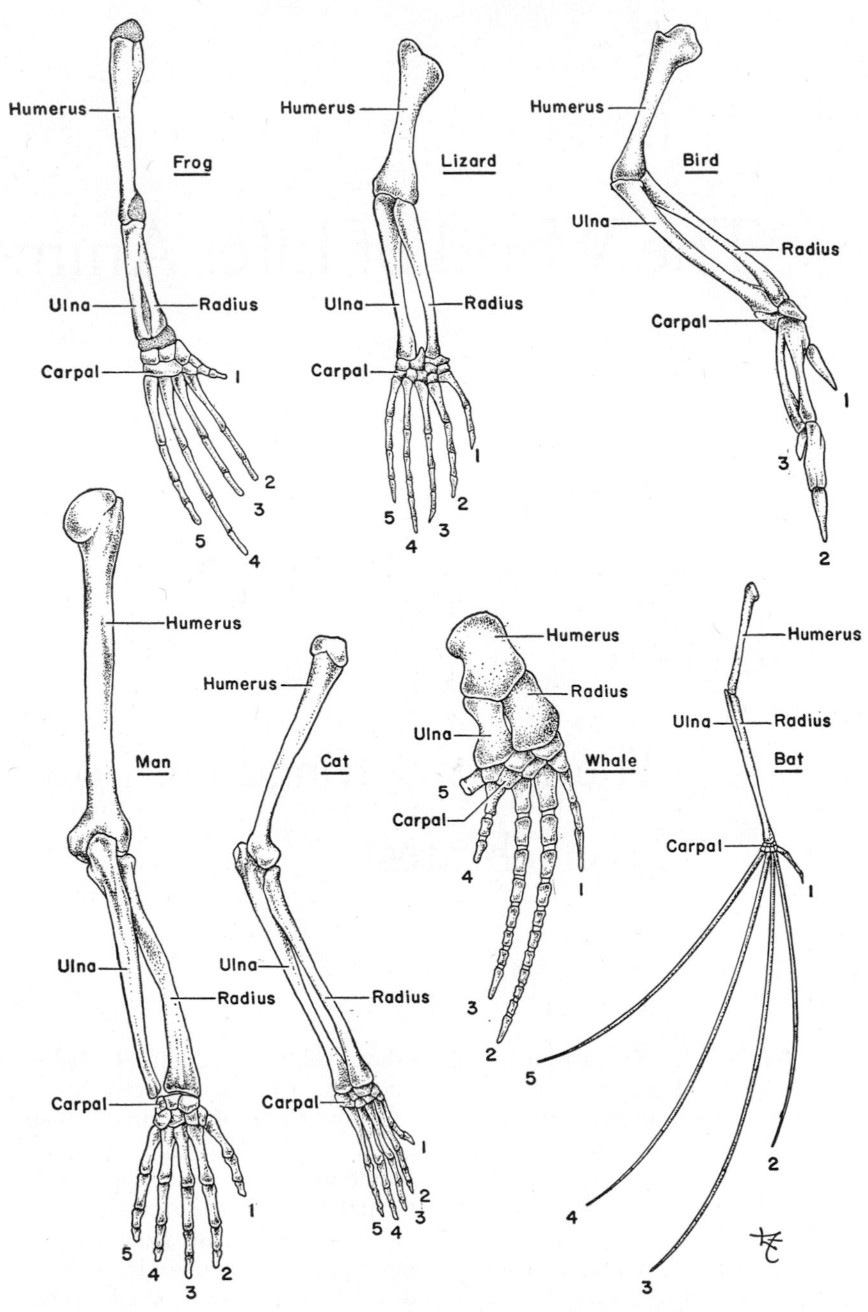

Figure 99. The bones of the forelimbs of a frog, lizard, bird, man, cat, whale, and bat, showing the arrangement of the homologous bones in these superficially different structures.

mon. Structures may be both homologous and analogous; for example, the wings of birds and bats have a similar structural plan and development, as well as the same function. The study of structures and their development to prove homologies often plays an important part in determining evolutionary relationships.

Because all animals have essentially the same problems to solve in order to survive, there is a basic unity of life, so that a discovery about one form has universal application. It is obvious that rats, rabbits and guinea pigs are similar enough to man so that experiments on their digestive, circulatory and excretory systems contribute to our understanding of the corresponding systems in man. But it is not so obvious that much of our knowledge of nerve action can and does come from experiments on squid nerves, that experiments on the heart of the horseshoe crab can give us information about the human heart, and that observations of the toadfish kidney are useful in understanding the human excretory system.

In Chapter 4 we found that the bodies of higher animals, such as man, have elaborate systems composed of organs, and that these, in turn, are made of tissues, which are aggregates of cells made of protoplasm. In the evolution of the higher from the lower forms this organization was built up gradually, and today some of the lower forms illustrate the steps in the process: protozoa represent the **protoplasmic** level of organization, sponges the **cellular** level, coelenterates the **tissue** level, and flatworms the **organ** level. In the proboscis worm we have the most primitive example of the **organ system** level of organization, which reaches a high development in man.

126. THE BASIS FOR ANIMAL CLASSIFICATION

The differences which distinguish related species, such as panthers and leopards, are usually minor and superficial—such as body color, proportion, and so on—but the main divisions of the animal kingdom, the phyla, are differentiated by basic characteristics which usually are not unique for a single phylum, but occur in *unique combinations* in various ones. Some of the factors basic to the determination of an animal's classification are as follows:

(*a*) The presence or absence of cellular differentiation. Animals may be either single-celled (the protozoa) or composed of many kinds of cells, specialized to perform particular tasks in the body's economy. In all the higher animals, collectively known as the **Metazoa,** cells are differentiated and specialized.

(*b*) The type of symmetry present, whether spherical, radial or bilateral (see p. 51).

(*c*) The number of germ layers. Some of the metazoa have only two embryonic cell layers or "germ layers"—an outer **ectoderm,** and an inner **entoderm,** which lines the digestive tract; others have these plus a third, the **mesoderm,** which lies between the ectoderm and entoderm, and comprises the rest of the body.

(*d*) The type of body cavity. In the lower many-celled animals the body is essentially a double-walled sack surrounding a single cavity with only one opening to the outside, the mouth. The higher animals have two cavities, their bodies being constructed on a tube-within-a-tube plan. The inner tube, or alimentary canal, is lined with entoderm, and opens at both ends, the mouth and anus; the outer tube, or body wall, is covered with ectoderm. Between the two tubes the second cavity, or **coelom,** lies within the mesoderm and is lined by it.

(*e*) The presence or absence of segmentation. The members of several phyla are characterized by the fact that their bodies consist of a row of segments, each of which has the same fundamental plan, with or without variation, as the segments in front and behind. In some segmented animals, such as man and most vertebrates, the segmental character of the body is obscured. In man the bones of the spinal column—the vertebrae—are among the few parts of the body still clearly segmental.

(*f*) Unique features. Although there are few structures which belong exclusively to one phylum of the animal world, a few animals do have unique qualities which aid in their classification. For example, the coelenterates alone have stinging cells (nematocysts); the echinoderms have a peculiar "water vascular system" found nowhere else; and, although many kinds of animals have a nervous system, only the chordates (with which man belongs) have a dorsally located, hollow nerve cord.

Animals are not classified according to the environment in which they live, but some of them are found in only one type of habitat; the members of certain phyla always live in the sea, while the members of others are always parasitic, and so on.

127. THE PROTOPLASMIC LEVEL OF ORGANIZATION

The Phylum Protozoa. The Protozoa, the single-celled animals that comprise the first phylum, are functionally complex even though some appear to be relatively simple structurally. There is some division of labor within the single cell, but the protoplasm in general functions as a unit to perform the activities associated with life—digestion, respiration, circulation, excretion, locomotion and reproduction. To carry out these functions, many Protozoa have evolved specialized organelles—cilia or flagella for movement, vacuoles, neurofibrils, eye spots and so on. Almost all protozoa live in water, from small rain puddles to the ocean. Some live in damp soil, in the film of water which surrounds each particle of soil; others live parasitically in the blood and tissue fluids of animals or plants.

Most of the 15,000 or so species of protozoa are microscopic, although a few are big enough to be seen with the naked eye (some of the largest amebas are 1 mm. long). In spite of the fact that many multicellular animals are smaller than the larger protozoa, the latter more closely resemble the primitive ancestor of all animals than do any of the many-celled animals, for that organism undoubtedly was single-celled. The present-day flagellates constitute a bridge between the plant and animal kingdoms, and are believed to be the protozoa nearest the ancestral stem of both. The other classes of protozoa are believed to have evolved from flagellate ancestors. Some flagellates such as *Chlamydomonas* clearly should be called plants, whereas the flagellate blood parasites, trypanosomes, are clearly animals. Others, such as *Euglena,* have a blend of plant and animal characters and cannot reasonably be called one or the other. At this primitive level the distinction between plant and animal disappears.

The body of *Amœba proteus* (Fig. 156) consists of a clear mass of shapeless, naked, gelatinous protoplasm, containing a nucleus and protoplasmic granules, and is often used to study the characteristics of protoplasm. The nucleus of this tiny animal occupies no fixed position in the cell, but is pushed around within the protoplasm as the animal moves. An ameba moves by pushing out temporary protoplasmic projections, called **pseudopods** (false feet), from the surface of the cell. Additional protoplasm flowing into the pseudopods enlarges them until finally all the protoplasm has entered and the animal as a whole has moved. The mechanism responsible for the flow of protoplasm in ameboid motion is unknown. By close observation of a moving ameba one can see that not quite all the protoplasm moves; actually, a stable gel layer at the surface surrounds the central core of liquid, flowing protoplasm. Experimental evidence indicates that protoplasm in the gel state can contract, thus squeezing the contained sol and pushing it forward, a process which is possible because the tip of the pseudopod lacks a gel layer. As the liquid protoplasm reaches the tip, it is pushed to the sides by the protoplasm behind, and converted from a sol to a gel, to form that part of the wall. Meanwhile, at the posterior end of the ameba, the gel walls are converted to a sol to be pushed along. The process of ameboid motion, as well as the chemical reactions supplying the energy for it, is probably fundamentally similar to muscular contraction, which also depends upon reversible changes from sol to gel. The white blood cells of man and other higher animals move by ameboid motion.

The pseudopods are used also to capture food (Fig. 173), two or more of them moving out to surround and engulf a bit of debris, another protozoan, or a small, many-celled animal. The engulfed food is surrounded by a food vacuole, and the encircling protoplasm secretes digestive enzymes to break down the food for absorption. As digestion proceeds, the nutrients and water are absorbed from the food vacuole and the latter gradually shrinks.

Respiration is an uncomplicated process in amebas and other protozoa, since they live in a liquid environment and can take in oxygen and give off carbon dioxide by diffusion. Excretion also occurs by this simple method. Many protozoa, amebas as well as others, have a **contractile vacuole,** a cavity in the protoplasm that regularly fills with water from the surrounding protoplasm, and empties it to the environment. This structure is believed to be not an excretory organ, but simply a pump for the removal of the excess water constantly entering by diffusion. The protoplasm of a fresh-water ameba has a higher concentration of dissolved materials—salts, sugar, and so on —than the surrounding water, so that water tends to pass into it by osmosis. Without a pump to remove excess water, the ameba would soon swell and burst, as our blood cells do when they are placed in fresh water. In contrast, most marine protozoa do not have or need a contractile vacuole, since the concentration of salts in the sea water is about the same as that in their protoplasm.

The *Amœba* belongs to the class **Sarcodina** (fleshlike) containing many other protozoa, all of which move about by means of pseudopods. Some of them, such as the species causing amebic dysentery in man (p. 409), are parasitic. Other, free-living species secrete hard shells around their body, or cement sand grains together into a protective layer around their naked protoplasm. The ocean contains untold trillions of ameboid protozoa called **foraminifera** (Fig. 100), which secrete chalky, many-chambered shells with pores through which the animal extends its pseudopods. The dead foraminifera sink to the bottom of the ocean and form a gray mud which is gradually transformed into chalk. The famous white cliffs of Dover were formed in this way by fossil foraminifera. Because deposits of these fossils are frequently associated with petroleum deposits, they are of great importance to geologists examining rock strata for signs of oil. Still other ameboid protozoa, called **radiolaria,** secrete elaborate and beautiful skeletons made of silica. The skeletons eventually become mud on the ocean floor, and are sometimes converted into siliceous rock such as flint.

A second class of protozoa, the **Ciliata,** is typified by *Paramecium* (Fig. 101). This animal, unlike the ameba, has a definite and permanent shape—clearly rounded in front and pointed in the rear—due to a sturdy, though flexible, outer covering secreted by the cell. The surface of the cell is covered by some 2500 fine, protoplasmic hairs, called **cilia,** which extend through pores in the outer covering and move the animal along by beating in coordinated rhythm. The motion of the cilia is oblique, revolving the animal as it swims, and the coordination is sufficiently good so that the animal can back up and

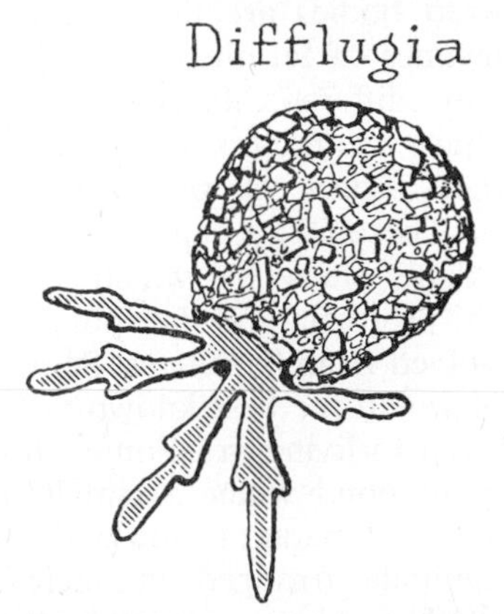

Figure 100. Difflugia, a free-living ameba which builds a protective layer around its naked protoplasm by cementing together grains of sand.

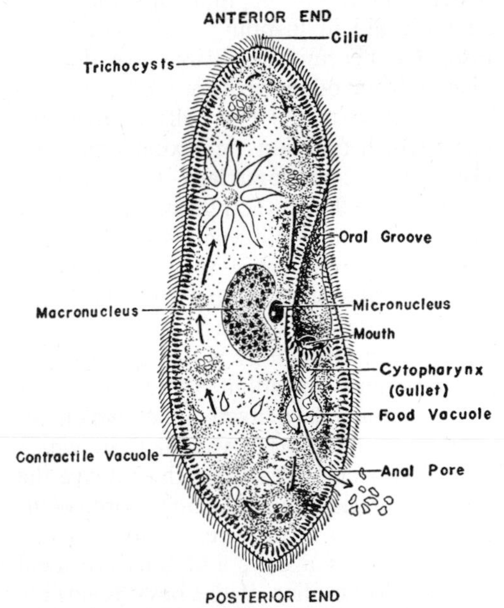

Figure 101. Paramecium, a typical ciliate protozoan. (Hunter and Hunter: College Zoology.)

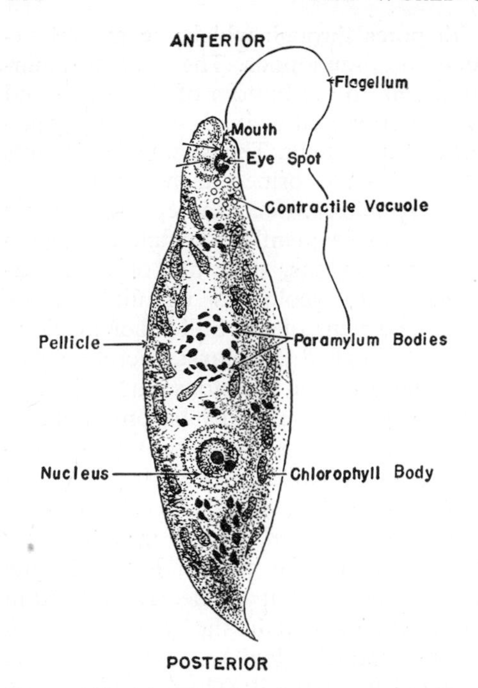

Figure 102. A typical flagellate, *Euglena.* (Hunter and Hunter: College Zoology.)

turn around. All this depends on a system of fine threads that connect rows of granules at the bases of the cilia; if these threads are cut, the coordination is destroyed. Near the surface of the cell are many small bodies, called **trichocysts,** which can discharge filaments that presumably aid in trapping and holding the prey. Ciliates and suctorians are distinguished from other protozoa by the presence of two nuclei per cell: a **micronucleus** which functions in sexual reproduction and a **macronucleus** that governs ordinary cellular activities. Paramecia have two contractile vacuoles, which together can remove a volume of water equal to the total volume of the body within half an hour. In contrast, a man excretes an amount of water equal to his body volume in about three weeks. A paramecium has a fixed gullet which ingests food and forms vacuoles, in which digestion occurs. Waste products leave the body at a fixed point, the anal pore. Well-fed paramecia reproduce by division two or three times a day, and thus are ideal subjects for studies of the laws governing population growth. Such studies have contributed to an understanding of the growth of human populations.

Suctorians constitute a third class of protozoa, very closely related to the ciliates. Young individuals have cilia and swim but the adults are sessile and have no cilia. They are attached to the substrate by a stalk and have delicate protoplasmic tentacles, some of which are pointed to pierce the prey while others are tipped with rounded, adhesive knobs to catch and hold the prey.

A fourth class of protozoa, the **Sporozoa** (spore formers), have no special method of locomotion and are parasitic. Malaria, one of the great plagues of mankind, killing some three million people each year, is transmitted by the bite of a mosquito, but is caused by a sporozoan (p. 415).

The fifth class of protozoa, the **Flagellata,** is named for the long, whiplike protoplasmic projections, called **flagella,** which enable the animals to move. Flagellates are usually more or less oval in shape and have a definite front end from which the flagella project (Fig. 102). Some flagellates engulf food by forming pseudopods, others resemble paramecia and have a mouth and gullet. Although euglenas have chlorophyll, contained in chloroplasts, and synthesize much of their food, no euglena is completely autotrophic. Healthy cultures of euglenas can be maintained only if amino acids are added to the medium. The flagellates with the largest number of flagella and the most specialized bodies are the ones living in the intestines of termites (p. 76). There are many different flagellates, some of which have characters in common with the other orders of protozoa or with sponges. These forms suggest some of the intermediate steps by which the other orders of protozoa and the sponges might have evolved from flagellates. Some flagellates, of which the best known is *Volvox,* have taken to living in colonies; they can be seen in pond water as small, hollow green spheres, made of thousands of individual animals, arranged with their flagella outward. The colony, moving by the beating of the flagella, rolls over and over, but keeps one end directed forward. In such a colony there is a beginning of specialization, for the cells at the front have larger light-sensitive spots than the others, and only those in the rear are capable of reproduction. A few flagellates are parasitic; the organism causing African sleeping sickness, which is transmitted by the tsetse fly, is one of these.

128. THE CELLULAR LEVEL OF ORGANIZATION

The Porifera or Sponges. The bodies of sponges are organized on the cellular level, so that instead of one cell carrying on all the life activities, as in the protozoa, there is a division of labor, with certain cells specialized to perform particular functions: some cells feed the body, while others support or reproduce it. In the sponge there is cellular differentiation, but little or no cellular coordination to form tissues.

Living sponges resemble a piece of raw liver; they are usually drab-colored, slimy to the touch, and have an unpleasant odor. When they are dead and the protoplasm has decayed, the skeleton is revealed. In the bath sponge, the skeleton is made of a protein similar to that of silk and fingernails. Chalk or calcium carbonate forms the skeletal material in some sponges, and silica makes up the skeleton in others called "glass sponges."

These creatures, whose scientific name means "pore-bearer," are living filters. They take in water through certain pores, strain out the microscopic organisms which they use for food, and eject the remaining water through other pores

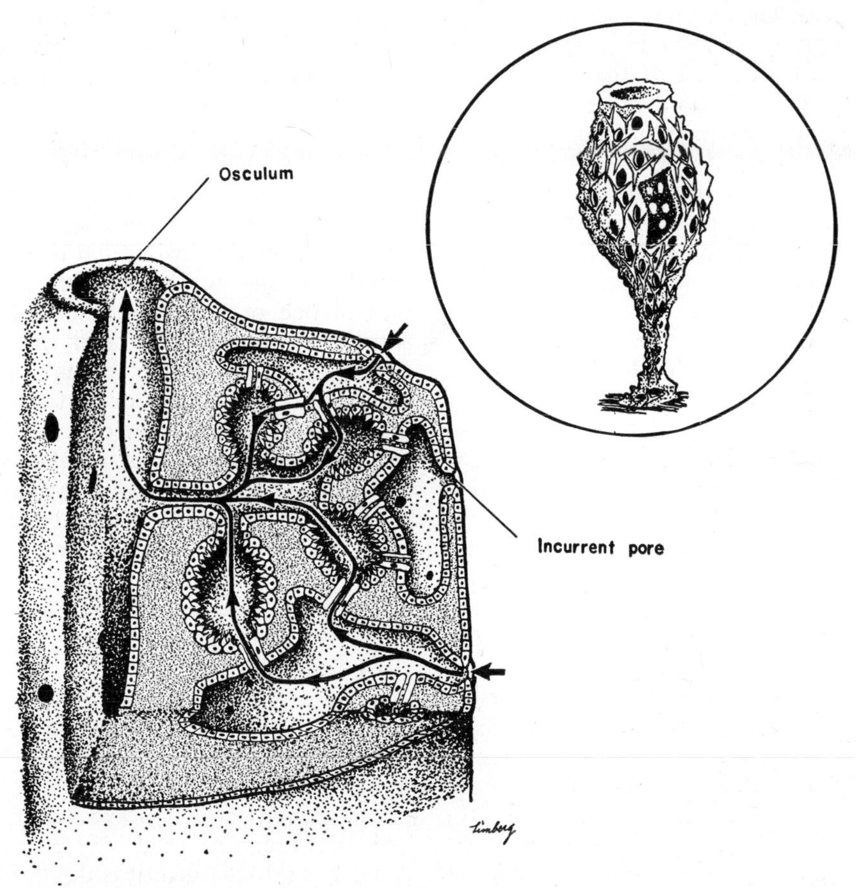

Figure 103. Diagram of a portion of a complex sponge, showing the chambers lined with flagellated cells and the path of the water from the incurrent pore to the osculum. Insert, the simplest type of sponge, consisting of a simple hollow vase, which is a developmental stage of certain calcareous sponges.

(Fig. 103). Certain sponge cells have flagella which beat, creating the current of water necessary for bringing in food and oxygen and for carrying away carbon dioxide and wastes. Wandering through the gelatinous matrix of the sponge body are numerous **amebocytes,** which collect food from the cells lining the pores, secrete the gelatinous matrix and the protein, calcium carbonate or silica of the skeleton, collect wastes, and become converted to epidermal cells as needed. Each cell of the body is irritable and reacts to stimuli, but there are no sense cells or nerve cells to enable the animal to react as a unified whole. Even the effect of a cut is transmitted only a short distance. Sponges often grow fused together in a single large mass, making it difficult to identify the individual animals.

Sponges are believed to be an evolutionary side line which arose from a different group of protozoa from those which gave rise to the rest of the many-celled animals. There is no evidence to suggest that any of the higher animals evolved from the sponges. Because these organisms are undifferentiated, their classification is difficult, but approximately 3000 species have been described. Most of them are marine forms, although a few species occur in fresh water.

129. THE TISSUE LEVEL OF ORGANIZATION

The Coelenterata. The body of these animals is organized typically as a hollow sac, the interior of which is a digestive cavity opening to the outside by a mouth; hence the name, Coelenterata (*coel* = hollow; *enteron* = gut). Coelenterates are believed to have evolved from the same stock as all the higher animals, because, like the latter, they have this central digestive cavity connected to the outside by a mouth (the sponges, in contrast, lack it). The tissues of coelenterates fall into roughly the same categories as do those of higher animals: epithelial, connective, muscular, nervous and reproductive. Although there is considerable division of labor among the cells, they act together in a coordinated way. The cells lining the digestive cavity are the **entoderm;** those covering the outside of the body are the **ectoderm.** In contrast to the higher animals, the coelenterates have no mesoderm cells between these two; the space is filled with **mesoglea,** a gelatinous matrix containing a very few scattered cells.

The body plan of the coelenterates is typified by a tiny animal, *Hydra,* which lives in ponds and looks to the naked eye like a bit of frayed string (Fig. 104). This animal takes its name from the Greek mythologic monster which Hercules fought. You may recall that when the hero cut off one of the Hydra's heads, two more immediately grew in its place. The real hydra has this ability to regenerate, and when it is cut into several pieces, each one grows all the missing parts and becomes a whole animal. The body, which is seldom longer than half an inch, consists of two layers of cells enclosing a central gastrovascular cavity, so called because it performs both digestive and circulatory functions (Fig. 173). The outer ectoderm serves as a protective layer; the inner entoderm is primarily a digestive epithelium. The bases of the cells of both layers are elongated into contractile muscle fibers; those of the ectoderm run lengthwise, and those in the entoderm run circularly. By the contraction of one or the other, the hydra can shorten, lengthen or bend its body. Throughout its life the animal lives attached to a rock, twig or leaf by a disc of cells at its base. At the other end is the mouth, connecting the gastrovascular cavity with the outside, and surrounded by a circlet of tentacles, each of which may be as much as one and a half times as long as the body itself. The tentacles are composed of an outer ectoderm and an inner entoderm and may be hollow or solid.

Coelenterates are unique in producing "thread capsules" or **nematocysts,** which lie in the outer layer and which, when stimulated, can release a coiled, hollow thread containing poison to paralyze the small animals which serve as prey. A nematocyst can be used only once; when it has been discharged, it is discarded and replaced by a new one, produced by special cells. The tentacles encircle the prey and stuff it through the mouth into the gastrovascular cavity, where digestion begins. When the animal has been digested to small fragments, the individual particles are taken up by pseudopods of the entoderm cells, and digestion is completed within food vacuoles in those cells.

Respiration and excretion occur by diffusion, for the body of a hydra is small enough that no cell is far from the surface. The motion of the body, as it stretches and shortens, circulates the contents of the gastrovascular cavity, and some of the entoderm cells have flagella whose beating aids in circulation. The hydra has no other circulatory device.

The first true nerve cells in the animal kingdom are found in the coelenterates. These animals have many nerve cells, which form an irregular network and connect the sensory cells in the body wall with muscle and gland cells. The coordination achieved thereby is of the simplest sort; there is no aggregation of nerve cells to form a "brain" or spinal cord, and an impulse set up in one part of the body goes in all directions more or less equally.

Besides the hydra and hydra-like organisms, the phylum Coelenterata includes such superficially different forms as jellyfish (Fig. 107), corals and sea anem-

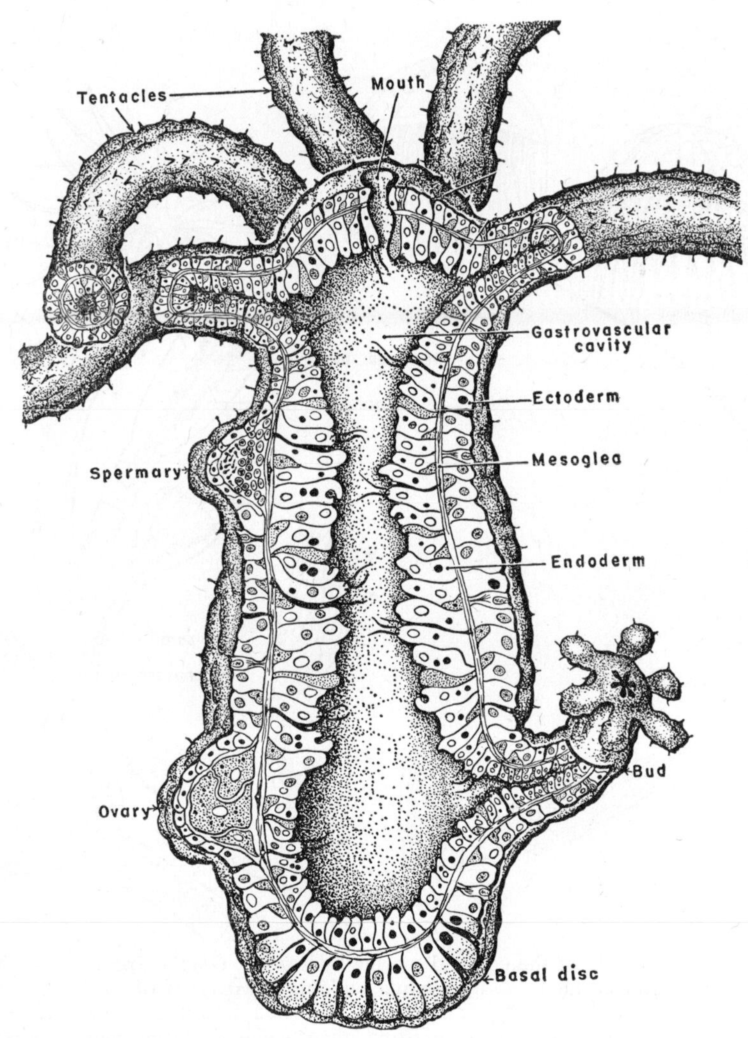

Figure 104. Diagram of a *Hydra* cut longitudinally to reveal its structure. The spermary and ovary for sexual reproduction are represented on the left, and asexual reproduction by budding is represented on the right. The sexes are separate, however; no animal has both an ovary and a spermary. (Modified from Buchsbaum from Hunter and Hunter: College Zoology.)

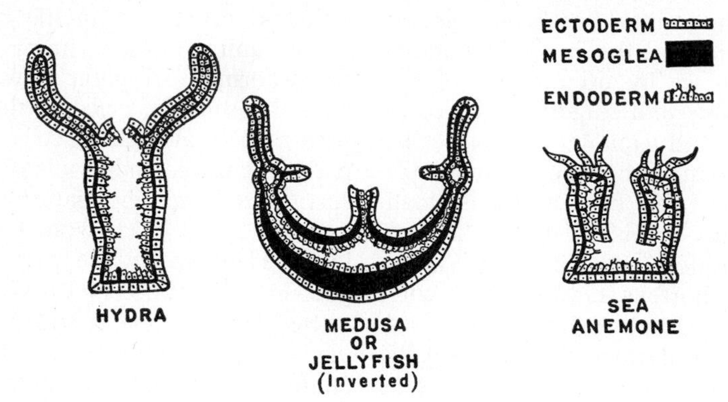

Figure 105. Comparison of a *Hydra,* an inverted jellyfish, and a sea anemone to show their essential structural similarity. (Hunter and Hunter: College Zoology.)

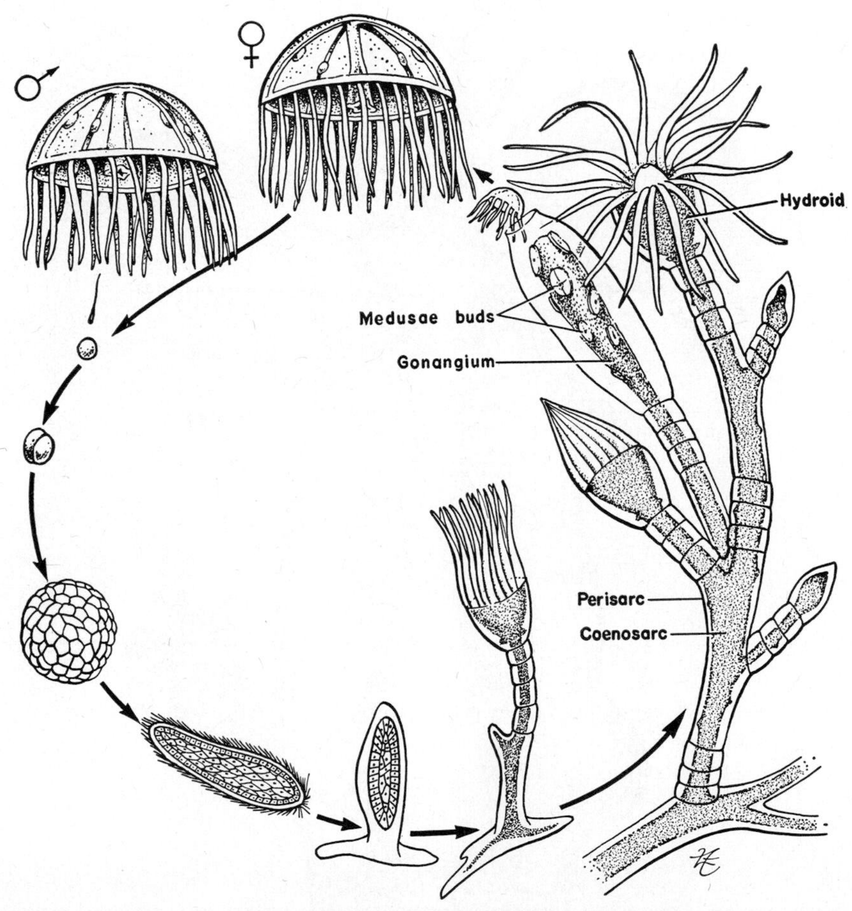

Figure 106. The life cycle of the colonial marine hydrozoan, *Obelia.* Special reproductive members of the colony produce medusae by asexual budding. Two types of medusae, bearing either testes or ovaries, are produced. Eggs and sperm are shed into the water, where fertilization occurs. The zygote divides to produce a blastula, then a planula larva, which swims about and finally becomes attached to some solid surface and develops a new colony of polyps.

ones among its 4500 different species. Both jellyfish and hydroids have bodies composed of an outer ectoderm and inner entoderm, with a nonliving jelly (mesoglea) layer between them; in the hydra, the mesoglea layer is thin, whereas in the jellyfish it is thick and viscous, giving firmness to the body. The fundamental similarity between the two is illustrated in Figure 105. A jellyfish is like a hydra which has been turned upside down and whose mesoglea layer has been greatly increased. The hydra and jellyfish are, then, two ramifications of the same fundamental plan, one adapted for an attached life, the other for a free-swimming life.

Some of the marine coelenterates are remarkable for an alternation of sexual and asexual generations analogous to that in plants. Many species of jellyfish reproduce sexually to give rise to hydra-like animals (**polyps**) which, in turn, reproduce asexually to form new jellyfishes, medusae (Fig. 106). This alternation of generations differs from that of plants in that both sexual and asexual forms are diploid; only sperm and eggs are haploid. Many of the marine coelenterates form colonial organizations of hundreds or thousands of individuals. A colony begins with a single individual that reproduces by budding, but the buds, instead of separating from the parent, remain attached and continue to bud themselves. In time, different types of individuals arise in the same colony, some specialized for feeding, others for reproduction.

The Portuguese man-of-war (*Physalia,* Fig. 107), which superficially looks like a jellyfish, is really a colony of hydroids and jellyfish. The long tentacles of this form are equipped with stinging capsules that can paralyze a large fish and wound a man seriously. The colony is supported by a gas-filled float of vivid, iridescent purplish-green.

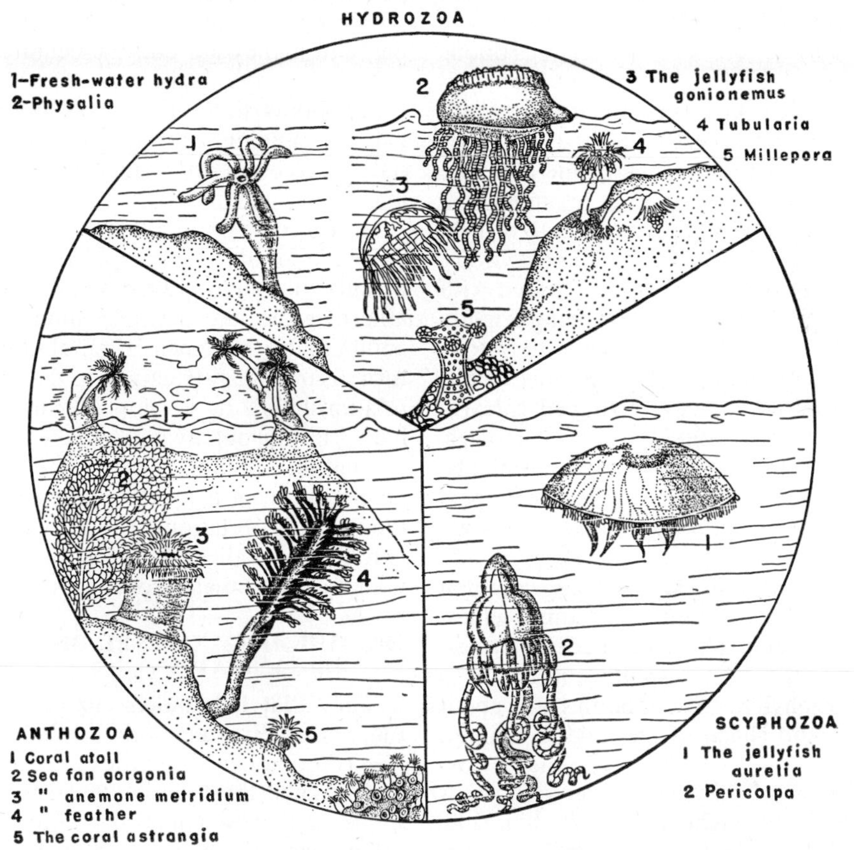

Figure 107. Some common representatives of the phylum Coelenterata, grouped by classes. (Hunter and Hunter: College Zoology.)

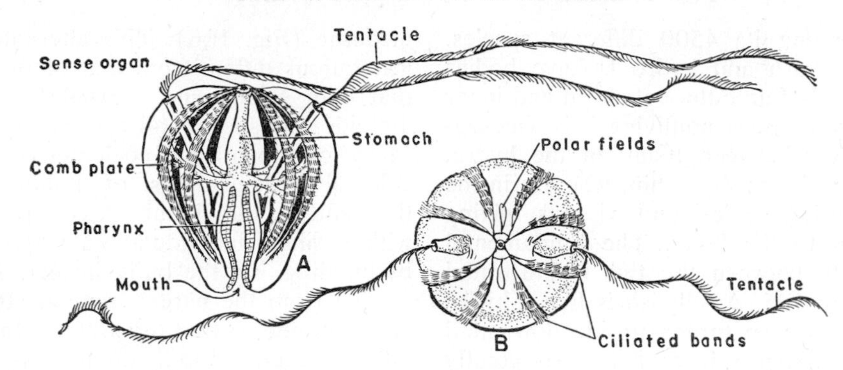

Figure 108. A ctenophore, *Pleurobrachia. A,* Side view; *B,* top view. (Hunter and Hunter: College Zoology.)

The largest known jellyfish, *Cyanea,* may be 12 feet in diameter, and have tentacles 100 feet long. These orange and blue monsters, among the largest of the invertebrate animals, are a real danger to swimmers in the North Atlantic Ocean.

The sea anemones and corals have no jellyfish stage, but occur both as individuals and in colonies. They differ from hydras in that the gastrovascular cavity is divided by a series of vertical partitions into a number of chambers, and the surface ectoderm is turned in at the mouth to line a gullet (Fig. 105). The partitions in the gastrovascular cavity increase the digestive surface, so that an anemone can digest an animal as large as a crab or fish.

In warm shallow seas almost every square foot of the bottom is covered with coral or anemones, most of them brightly colored. The extravagant reefs and atolls of the South Seas are the remains of billions of microscopic, cup-shaped calcareous structures, secreted, during past ages, by coral colonies and by coralline plants. Living colonies occur only in the uppermost part of such reefs, adding their own secretions to the mass.

The **Ctenophores,** whose common names are sea walnuts or comb jellies, are similar in many ways to coelenterates, although they are usually placed in a separate phylum. Their bodies are about the size and shape of an English walnut, and consist of two layers of cells enclosing a mass of jelly; the outer surface is covered with eight rows of cilia, like combs, by which the animal moves through the water (Fig. 108). At the upper pole of the body is a sense organ, containing a mass of limestone particles and balanced on four tufts of cilia connected to sense cells. When the body turns, these particles bear more heavily on the lower cilia, stimulating the sense cells, which cause the cilia to beat faster and bring the body back to its normal position. Nerve fibers running from the sense organ to the cilia control the beating, and if they are cut, the beating of the cilia below the incision is disorganized. Ctenophores differ from coelenterates in lacking stinging capsules and in having only two branched tentacles instead of many.

Both coelenterates and ctenophores have remarkable powers of regeneration; a half, quarter, or even smaller piece is able to grow into a whole animal. These animals also have a marked ability to return disarranged structures to their normal relationships. It is possible to turn a hydra inside out by pulling the base out through the mouth. The hydra, though unable to turn itself inside out, does restore the normal relations of ectoderm and entoderm by the migration of the individual cells to their proper position.

130. THE ORGAN LEVEL OF ORGANIZATION

The Platyhelminthes or Flatworms. Flatworms live in both fresh and salt water, creeping over rocks, debris and leaves. Like the hydra, flatworms have a single gastrovascular cavity (Fig. 173)—sometimes extensively branched—connected to the outside by a single opening, the mouth, which is on the middle of the ventral surface. In addition to an outer ectoderm and an inner entoderm, the flatworm has a third, middle layer, the mesoderm, which comprises most of the body and forms many of the organs. The Platyhelminthes, the flatworms, are the simplest animals that have well developed **organs,** functional units made of two or more kinds of tissue. There are several simple organs—a muscular pharynx for taking in food, eyespots and other sense organs on the head, and complex reproductive organs. The worm is bilaterally symmetrical and has a definite anterior and posterior end. It keeps one surface (the back or dorsal one) always upward as it crawls along. Locomotion is achieved partly by means of cilia on its under or ventral surface, and partly by undulatory muscular contractions, similar to those of the earthworm. Gland cells on the ventral surface secrete a slime which facilitates movement.

The commonest free-living flatworms are the planarians, found in ponds and quiet streams all over the world. The common American planarian is **Dugesia,** about 15 mm. long, with what appear to be crossed eyes and flapping ears (Fig. 109). Planarians are carnivorous and feed on living and dead small animals. Flatworms can survive without food for months, gradually digesting their own tissues and growing smaller as time passes. As in the coelenterates, respiration takes place by diffusion. In order to secrete waste products, the flatworm has an organ not found in lower forms. This structure, which is formed from the mesoderm, consists of a network of fine tubes opening to the surface by pores, and ending in branches known as **flame cells** (Fig. 182). Each of the latter is a single, hollow cell containing a tuft of cilia which, in beating, resembles a flame. The motion of the cilia drives the excreted fluid along the tubes and out the pores. Planarias living in fresh water have the same problem of getting rid of excess water faced by the freshwater protozoa, and the flame cells, like the contractile vacuole, solve it. Some flatworms have an interesting mechanism for defense: from the hydras they eat, they confiscate, intact, the stinging capsules, which become part of their own epithelium and are used in the appropriate manner.

Besides the free-living flatworms like *Planaria,* there are two groups of parasitic ones, the **Trematoda,** or flukes, and the **Cestoda,** or tapeworms.

The flukes are structurally like the free-living flatworms, but differ in having one or more suckers with which to cling to the host, and a thick outer layer, the cuticle, in place of cilia. The organs of digestion, excretion and coordination

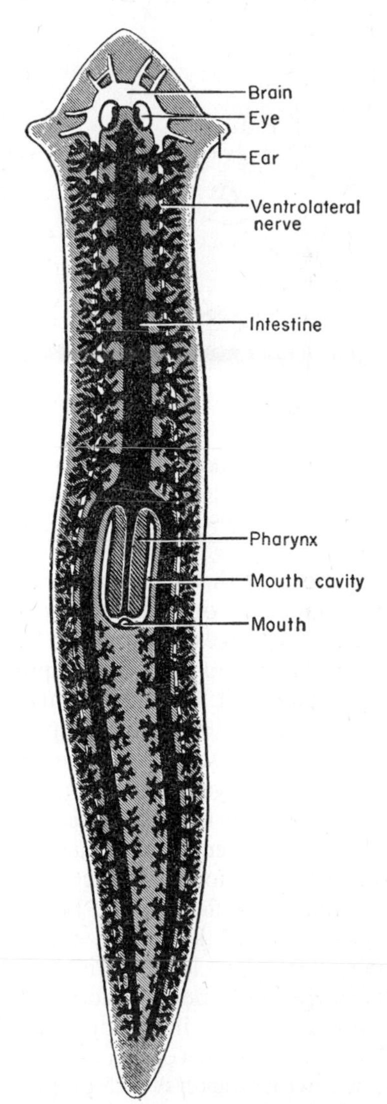

Figure 109. The common American planarian, *Dugesia* (Villee, Walker & Smith: General Zoology).

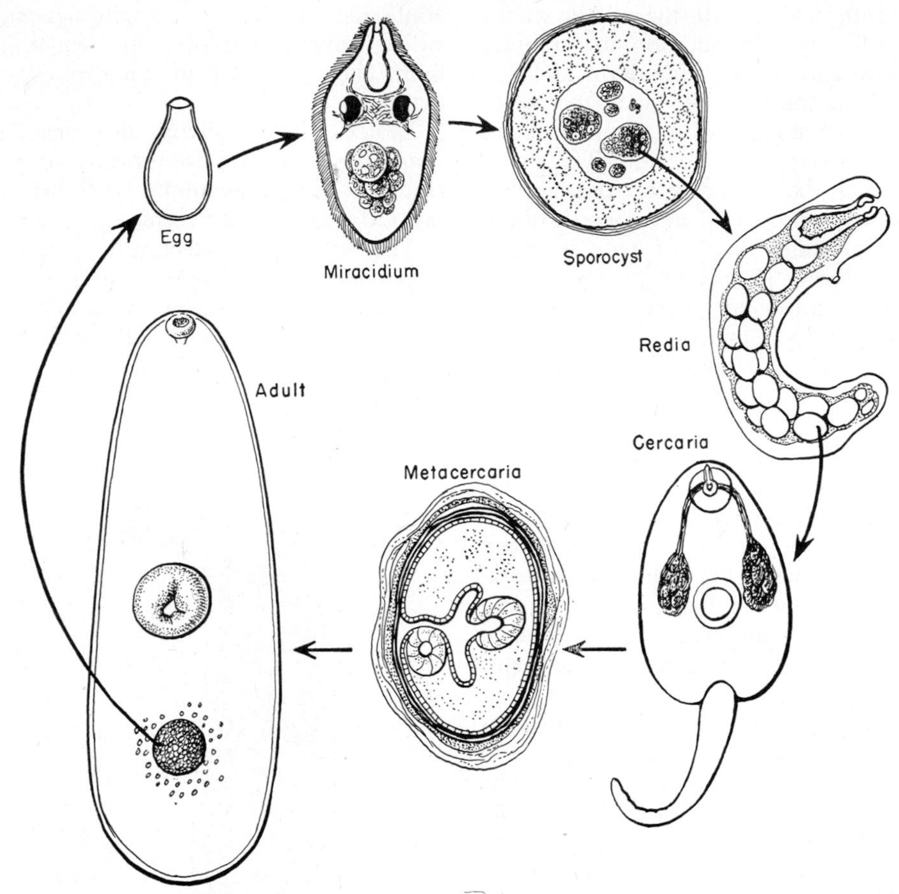

Figure 110. Life cycle of the oriental liver fluke, *Clonorchis sinensis,* one of the parasitic flatworms of the class Trematoda. (Villee, Walker and Smith: General Zoology.)

are like those of the other flatworms, but the reproductive organs are extremely complex. The flukes parasitic on human beings are the blood flukes, widespread in China, Japan and Egypt, and the liver flukes, common in China, Japan and Korea. Both of these parasites go through complicated life cycles, involving a number of different forms, alternation of sexual and asexual generations, and parasitism on one or more intermediate hosts such as snails and fishes (Fig. 110).

Tapeworms are long, flat, ribbon-like animals, some species of which live as adults in the intestines of probably every kind of vertebrate, including man. The head end of a tapeworm (Fig. 226) is equipped with suckers and often has a circle of hooks by means of which it attaches to the lining of the intestine. Behind the head is a growing region which constantly gives rise to new body sections, called **proglottids,** by budding. The rest of the body consists of a series of these sections, which contain little more than a complete set of reproductive organs. Tapeworms have no mouth and no trace of a digestive system; they live by soaking up the digested materials present in the intestine of their host.

131. THE ORGAN SYSTEM LEVEL OF ORGANIZATION

The Nemertea or Proboscis Worms. This group of animals is important to us only as an evolutionary landmark, for the proboscis worms are the simplest living animals which illustrate the organ system level of organization (Fig. 111). None of them is parasitic; hence they are harmless to man and other animals; and none is of economic importance. Almost all of them are marine, although a few inhabit fresh water or damp soil. They have long narrow bodies, either cylindrical or flattened, varying in length from an inch to 60 feet. Some of them are a vivid orange, red or green with black or colored stripes. Their most remarkable organ—the proboscis, from which they take their name—is a long, hollow, muscular tube which they evert from the anterior end of the body and use for seizing food. The proboscis secretes mucus, helpful in catching and retaining the prey. Some species have a hard point at the tip of the proboscis for wounding the enemy, and poison-secreting glands at the base of this point. The proboscis is thrust outward by the pressure of the surrounding muscular walls on the contained fluid; a separate muscle inside the proboscis retracts it.

The important advances displayed by the nemerteans are, first, a complete digestive tract, with a mouth at one end for taking in food, an anus at the other, for eliminating feces, and an esophagus and intestine in between. This is in contrast to the coelenterates and planarians, whose food enters, and wastes leave, by the same opening. In the proboscis worm water and metabolic wastes are eliminated from the body by flame cells, as they are in the flatworms. A second advance exhibited by the nemerteans is the separation of digestive and circulatory functions. These animals are the most primitive organisms to have a separate circulatory system. It is, of course, rudimentary, consisting simply of three muscular tubes—the blood vessels—which run the length of the body and are connected by transverse vessels. Surprisingly, these primitive forms have red blood cells filled with hemoglobin, the same red pigment which transports oxygen in human blood. Nemerteans have no heart, and the blood is circulated through the vessels by the movements of the body and the contractions of the muscular blood vessels. The nervous system is more highly developed than it is in the flatworm; there is a "brain"* at the anterior end of the body, consisting of two groups of nerve cells (ganglia) connected by a ring of nerves extending around the sheath of the proboscis; two nerve cords extend posteriorly from the brain.

The Nematoda. The phylum Nematoda, made up of roundworms, has a great many members (about 8000 species), all of them remarkably similar in general body pattern. Some live in the sea, others in fresh water, in the soil, or in other plants or animals as parasites. About fifty species are human parasites, of which the most detrimental are the hookworm, trichina worm (Fig. 227), ascaris worm (Fig. 112), filaria worm and guinea worm. A microscopic examination of a shovelful of earth from almost anywhere in the world will reveal a number of tiny white worms which thrash around, coiling and uncoiling. Their long, cylindrical bodies, pointed at both ends, are covered

* The word "brain" is loosely applied to the aggregation of nerve cells at the anterior end of the nerve cord, which acts as a reflex center. It should not be inferred that anything like thought processes occurs in any of the lower animals.

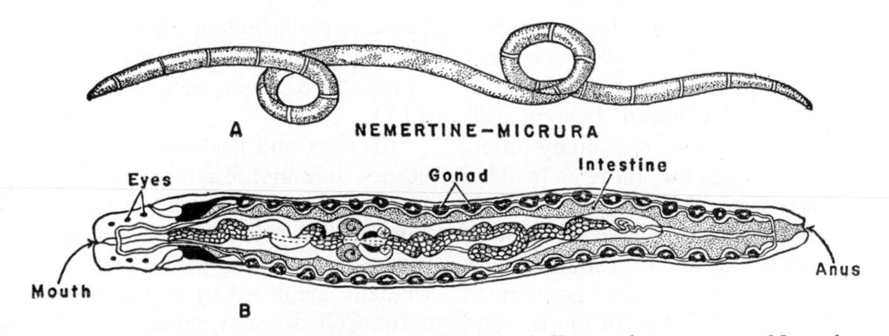

Figure 111. A typical proboscis worm or nemertean. *A,* External appearance. Note the colored bands which are present on many species. *B,* Diagram of the internal anatomy. Note the three longitudinal blood vessels (white) and the digestive system composed of mouth, intestine and anus. (Hunter and Hunter: College Zoology.)

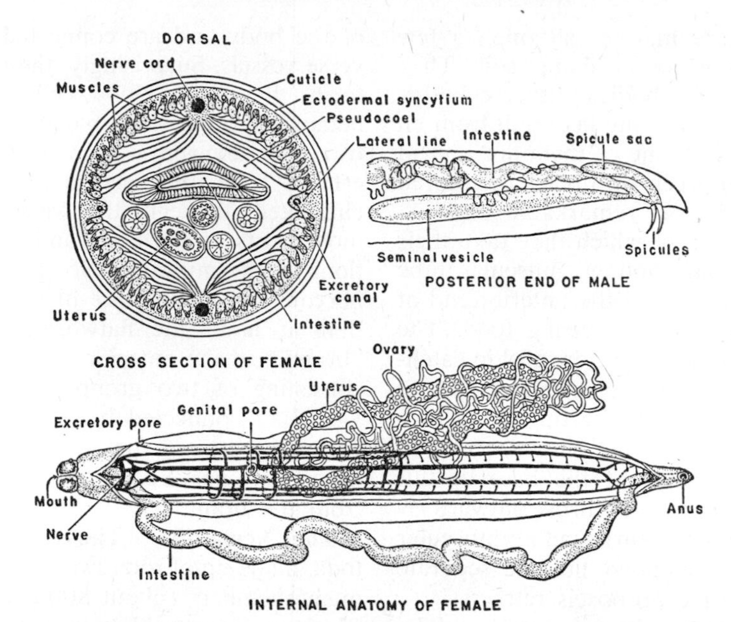

Figure 112. A typical parasitic nematode, *Ascaris lumbricoides.* (Hunter and Hunter: College Zoology.)

with a tough cuticle layer. In contrast to the nemerteans, which have cilia all over the epithelium and the lining of the digestive tract, none of the nematodes has any cilia at all. Nematodes have no circularly arranged muscle fibers, only longitudinal ones. They can only bend, and they swim poorly despite vigorous thrashing movements.

With the evolution of a complete digestive system, a separate circulatory system, and a nervous system composed of a "brain" and nerve cords, as illustrated in the proboscis worm, the essential structures of higher animals were established. The proboscis worm is not believed to be the actual ancestor of the higher forms, but is thought to be like the now extinct ancestor common to both the higher animals and itself. Evolution beyond this point branched out in a great many different directions, and the more advanced animals cannot be arranged in a single series of progressively higher and more complex forms. One main branch of evolution led to the vertebrates, another to the insects and other arthropods; and another to the clams, squids and other molluscs. Besides these familiar forms, there are others which are relatively unknown, about which, since they neither serve nor menace us, we think little. One frequently hears people ask of such animals, "What good are they?" The answer is of course that the various forms of life do not exist to serve man. Each is adapted to survive in a certain habitat and to fill a particular ecologic niche.

The Rotifera. Some of the more obscure animals just mentioned make up the phylum Rotifera (wheel animals). These aquatic, microscopic worms, although no larger than many of the protozoa, have many-celled bodies with a complete digestive tract; an excretory system made up of flame cells and a bladder; a nervous system with a "brain" and sense organs; and a characteristic crown of cilia on the head end which gives the appearance of a spinning wheel (Fig. 113).

Rotifers and gastrotrichs are "cell constant" animals: Each member of a given species is composed of exactly the same number of cells; even each part of the body is made of a precisely fixed number of cells arranged in a characteristic pattern. Cell division ceases with embryonic development and mitosis cannot subsequently be induced; growth and repair are impossible.

The Gastrotricha. This is another phylum of microscopic, aquatic worms which are like rotifers in many respects, but have no crown of cilia. Some of the fresh-water gastrotrichs are peculiar in that they apparently consist entirely of females which reproduce by parthenogenesis; no males of this group have ever been found.

The Bryozoa. The Bryozoa, or moss animals, live in colonies that superficially resemble those of the coelenterates (Fig. 114). The colonies of some species, delicately branched and beautiful, are sometimes mistaken for seaweed; other species form colonies which appear as thin, lacy incrustations on rocks. Each animal secretes about itself a vase-shaped protective case of calcium carbonate or of a horny, protein material, into which it can withdraw when danger threatens. Around the animal's mouth is a circular or horseshoe-shaped ridge, called a **lophophore,** with a set of ciliated tentacles. An adaptation to living in a "vase" is the U-shaped digestive system. In one class the anus is just outside the ring of tentacles; in another it lies immediately inside. New members of the colony arise by budding from the older ones, and new colonies arise as the result of sexual reproduction during certain seasons.

Each bryozoan colony has specialized members, called **avicularia,** which re-

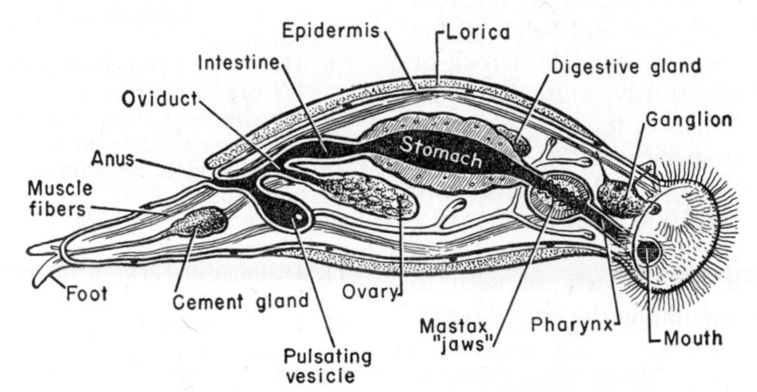

Figure 113. Diagram of the structures of a rotifer.

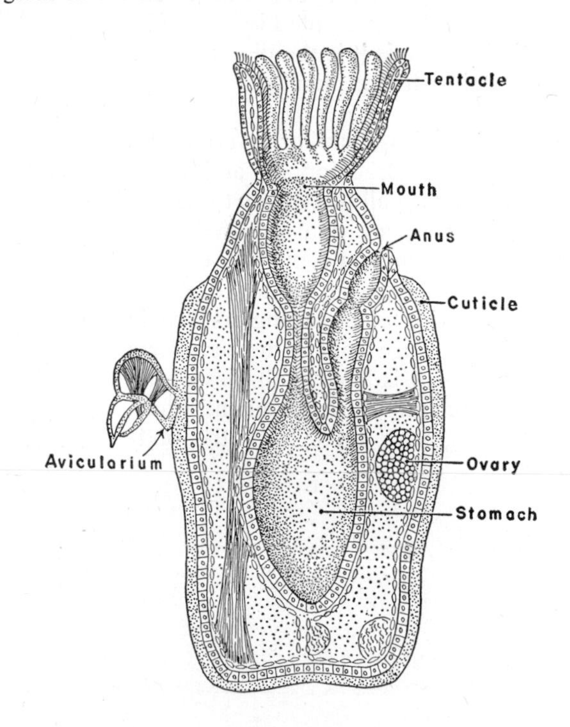

Figure 114. A bryozoan, *Bugula,* cut open longitudinally to reveal its internal structure. (After Curtis and Guthrie, Hunter and Hunter: College Zoology.)

semble the head of a bird. These organisms are in constant motion from side to side, and, as they move, a peculiar organ, shaped like a bird's lower beak and operated by muscles, frequently snaps open and shut. The purpose of the avicularia is not to catch food—other members do that—but to keep small animals from settling on the colony.

The Brachiopoda. This is another phylum characterized by a lophophore. The animals are commonly known as lampshells and superficially resemble a clam. Like the latter, they have two shells, usually calcareous, that can be opened and closed by muscles. But unlike the clam, whose two shells are on the right and left sides of the body, the brachiopod shells lie above and below it, the bottom shell being attached to a rock or other object by a sturdy, muscular stalk. All brachiopods live in the sea. Although only about 200 species of this extremely ancient phylum exist today, there once were more than 3000 species. Because of their great age and well-preserved hard shells, the fossil brachiopods are useful to geologists in determining the age of rocks. Fossils obtained from rocks more than 500,000,000 years old are almost exactly like the brachiopods living today. The genus *Lingula* is represented by both fossil and living forms; it is the oldest known genus that has living members.

Some of the less familiar invertebrates —the Bryozoa, Brachiopoda and Phoronida—as well as all the others to be discussed presently, are characterized by a body cavity or **coelom** lying between the body wall and the wall of the digestive tract. This cavity arises during embryonic development from a split in the mesoderm layer, and hence is lined with mesoderm. The development of the coelom, which freed the digestive tract from the body wall, and permitted the two sets of muscles to contract independently, was an important step toward the development of the higher animals.

QUESTIONS

1. Distinguish between homologous and analogous organs. Why must classifications be based on the former?
2. What is meant by the protoplasmic level of the animal kingdom?
3. What are the outstanding characteristics of the protozoa?
4. How do sponges obtain food and water?
5. Why are coelenterates said to represent the tissue level of the animal kingdom?
6. Compare the body plans of a jellyfish and a hydra and the adaptations to the mode of life present in each.
7. In what ways may the alternation of generations of *Obelia* and a moss plant be compared?
8. How would you distinguish a multicellular organism from a colonial organism?
9. Do you think the sections of a tapeworm constitute a single individual or a colony of individuals comparable to a coelenterate colony?
10. What are flame cells and how do they operate?
11. Distinguish between the organ level and the organ system level of organization.
12. In what ways are a planaria and a hydra similar? In what ways do they differ?
13. An animal is multicellular, has no digestive system, and its body is perforated with pores; what is it?

SUPPLEMENTARY READING

A prime source book for the details of invertebrate structure is Libbie Hyman's *The Invertebrates. Animals without Backbones,* by Ralph Buchsbaum, is a well-written textbook of the invertebrates with outstandingly fine illustrations. Hegner's *Parade of the Animal Kingdom* is a popular book on animals. A large number of the invertebrate animals are marine, and descriptions of them and their modes of life will be found in Ricketts and Calvin, *Between Pacific Tides,* Douglas P. Wilson, *They Live in the Sea,* and MacGinitie's *Natural History of Marine Animals.* More popular accounts of marine life are Carson's *The Sea Around Us* and C. M. Yonge's *A Year on the Great Barrier Reef.*

Chapter 14

The Higher Invertebrates

OF THE four higher invertebrate phyla—the annelids, arthropods, molluscs and echinoderms—only the arthropods are very successful terrestrial animals. It is true that the earthworm is a terrestrial animal, but most annelids are marine; there are a few land snails, but most molluscs live in the sea; all the echinoderms are marine. Of the five classes of arthropods, one, the Crustacea—crabs, lobsters, and so on—is largely marine, but the other four, insects, spiders, centipedes and millipedes, are mostly terrestrial. The fossil record (p. 518) tells us that the first air-breathing land animals were scorpion-like arachnids that came ashore in the Silurian, some 410,000,000 years ago. The first land vertebrates, the amphibians, did not appear until the latter part of the Devonian, some 60,000,000 years later.

132. PROBLEMS OF TERRESTRIAL LIFE

In evolving to become adapted to terrestrial life, animals, like plants (Chap. 11), had certain problems to solve for survival in the absence of a surrounding watery medium. The chief problem that has to be met by all land organisms is that of preventing desiccation. Reproduction provides a second problem: Aquatic forms can shed their gametes into the water and fertilization will occur there; the delicate embryos that result are protected by the surrounding water as they begin development. Land plants, by pollination, transfer the sperm to the egg in the absence of a watery medium, and the developing embryo is protected by the tissues of the parent gametophyte or by seed coverings. Some land animals, such as amphibians, return to the water for reproduction and the young forms—larvae—develop in the water. Earthworms, insects, snails, reptiles, birds and mammals transfer sperm from the body of the male directly to the body of the female by copulation; the sperm are surrounded by a watery medium or semen. The fertilized egg either is covered by some sort of tough, protective shell secreted around it by the female or it develops within the body of the mother. The problem of supporting a body against the pull of gravity in the absence of the buoyant effect of water is not too acute for small animals such as earthworms that burrow in the ground, but the larger ones and the above-ground ones need some sort of skeleton. The arthropods and molluscs evolved one on the outside of the body (called an **exoskeleton**) and the vertebrates have one within the body (an **endoskeleton**). Land forms are subject to much wider variations in temperature than marine organ-

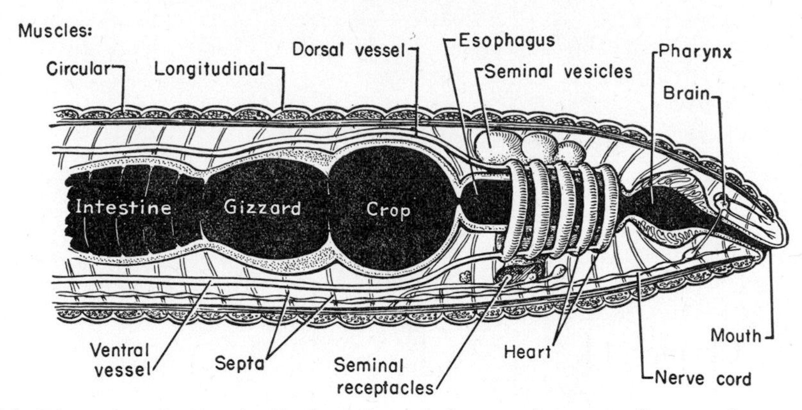

Figure 115. Diagrammatic, longitudinal section of the anterior part of an earthworm, showing internal structures.

isms, for the ocean acts as a great, constant temperature bath. The deep water varies only a few degrees from summer to winter and even a small lake has less drastic temperature fluctuations than air does. Land animals are exposed to both higher and lower temperatures than aquatic ones and had to evolve suitable adaptations to survive.

With all these disadvantages it might seem incredible that any land forms *did* evolve. However, one of the major tendencies in evolution is for organisms to become diversified and to spread into new types of environment. Wherever the environment can support life at all, some form of life, suitably adapted for survival there, will eventually evolve. A land environment is not without some advantages, however. Once land plants had evolved, the land offered the first animals an environment with a plentiful supply of food, no predators, and few competitors.

The tough exoskeleton that evolved in the arthropods and molluscs serves several purposes—it provides stiffening, enabling the body to stand against the pull of gravity; it serves as a point of attachment for muscles, provides protection against desiccation and serves as a coat of armor to protect the animal against predators. Thus the evolution of an exoskeleton solved many of the problems of survival on land.

133. THE ANNELIDA

Among the most familiar invertebrate animals are the earthworms, members of the phylum Annelida (Figs. 115 and 183). This word, which means "ringed," refers to the fact that the body of the worm consists of a series of rings or segments. Both the internal organs and the body wall are segmented, so that each animal is made of about one hundred more or less similar units, each of which contains one or a pair of organs of each system. The segments are separated from each other by transverse, bulkhead-like partitions, the **septa.** The chief evolutionary advance shown by the earthworms over the lower forms is this development of segmentation, for each segment constitutes a subunit of the body that may be specialized to carry on a particular function. The dividing of the body into segments is thus similar, on a larger scale, to the original division of the animal body into cells to provide for local specialization. In the earthworm the individual segments are almost all alike, but in many of the other segmented animals—the arthropods and chordates—the specialization of different segments reaches a point where the segmentation of the body plan is obscured.

The earthworm is protected from desiccation by a thin, transparent cuticle, secreted by the cells of the epidermis or outer layer of the body wall. The glandular cells of the epidermis secrete mucus which forms an additional protective layer over the skin. The body wall contains an outer layer of circular muscles and an inner layer of longitudinal muscles. Each segment except the first bears four pairs of bristles, **chaetae,** supplied with small muscles that can move the chaeta in and out and change its angle. The earthworm moves forward by contracting its circular muscles to elongate the body, grasping the ground or walls of the burrow with its chaetae, and then contracting its longitudinal muscles to draw the posterior end forward; locomotion occurs in waves.

The coelom, or body cavity, of the annelids is large and well developed, and the entire body consists essentially of two tubes, one within the other. The outer tube is the body wall, and the inner tube, the wall of the digestive tract. The coelom is filled with a fluid which bathes the internal organs and is an intermediate, in transporting gases, food and wastes, between the circulatory system and the individual cells of the body.

The digestive system of the earthworm shows several advances over that of the proboscis worm: there is a muscular pharynx for swallowing food, an esophagus, and a stomach of two parts—a thin-walled **crop,** where food is stored, and behind it, a thick-walled muscular **gizzard,** where it is ground to bits. The rest of the digestive system is a long, straight intestine, where digestion and absorption take place, terminating in an anus which opens to the outside at the posterior end.

The circulatory system is also marked by advances in complexity and efficiency over that of the primitive proboscis worm. In the earthworm it consists of two main vessels: one, just dorsal to the digestive tract, collects blood from numerous segmental vessels. It is contractile and pumps the blood anteriorly. The other, in which blood flows posteriorly, lies just below the digestive tract and distributes blood to the various organs. In the region of the esophagus the dorsal and ventral vessels are connected by five pairs of muscular tubes, called "hearts," which propel the blood to the ventral vessel. There are also smaller lateral and ventral distributing vessels, and tiny capillaries in all the organs as well as in the body wall.

The excretory system is composed of paired organs repeated in almost every segment of the body. Each individual organ, called a **nephridium,** consists of a ciliated funnel, opening into the next anterior coelomic cavity, and connected by a tube to the outside of the body (Fig. 183). Wastes are removed from the coelomic cavity partly by the beating of the cilia and partly by currents set up by the contraction of muscles in the body wall. The tube of the excretory organ is surrounded by a capillary network, so that wastes are removed from the blood stream as well as from the coelomic cavity.

The nervous system, too, is more advanced than that of the proboscis worm, for it consists of a large, two-lobed aggregation of nerve cells located just above the pharynx in the third segment, and another ganglion just below the pharynx in the fourth segment. A ring of nerves around the pharynx connects the two ganglia. From the lower ganglion a nerve cord (actually, two closely united cords) extends the entire length of the body, beneath the digestive tract. In each segment there is a swelling of the nerve cord, a **segmental ganglion,** from which nerves extend laterally to the muscles and organs of that segment. The segmental ganglia coordinate the contraction of the muscles of the body wall, so that the worm can creep along. The nerve cord contains a few giant axons which transmit nerve impulses more rapidly than ordinary fibers. These, when danger threatens, stimulate the muscles to contract and draw the worm back into its burrow. Living a subterranean life, the earthworm has no well-developed sense organs, but some of its sea-dwelling relatives, such as the sandworm, *Nereis,* have two pairs of eyes and organs sensitive to touch and to chemicals in the water. The activities of the earthworm are governed by the two ganglia above and below the pharynx, the **brain** and **subpharyngeal ganglion.** Removal of the brain results in *increased* bodily activities and removal of the subpharyngeal ganglion eliminates all spontaneous movements. This is evidence of functional specialization of the nervous system; the brain is in part an **inhibitory center** and the subpharyngeal ganglion is a **stimulatory center.**

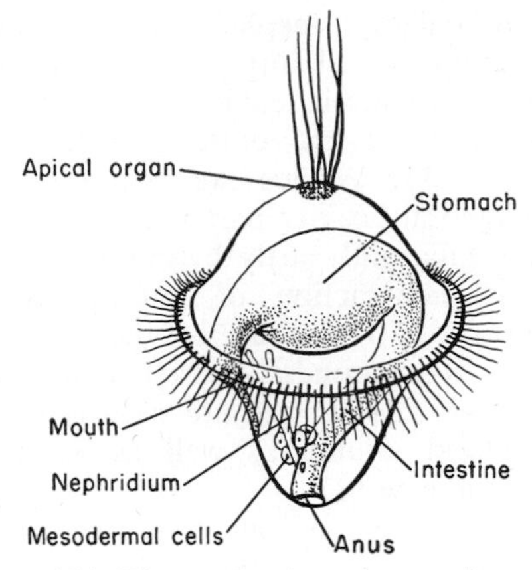

Figure 116. The trochophore larva of a polychaete.

Besides serving as bait and as food for birds, earthworms are of great service to man in turning over the soil. By making it more porous, thus allowing air to penetrate freely and water to drain properly, they facilitate the growth of roots and hence increase the agricultural yield. In their burrowings through the earth the worms swallow bits of soil, grind them to smaller bits, and discharge them above the surface, well mixed with organic matter. The amount of earth moved in this way is remarkable; it is calculated that every ten years earthworms turn over enough soil to form a layer 2 inches thick over the entire land surface of the world.

The phylum Annelida contains some 8000 species divided into four classes. One of these is the **Polychaeta** (many bristles), made up of marine worms which swim freely in the sea, burrow in the sand and mud near shore, or live in tubes formed by secretions from the body wall. Each segment of their bodies has a pair of thickly bristled paddles (called **parapodia**) extending laterally.

The members of the class **Oligochaeta** (to which the earthworm belongs) are characterized, as their name indicates, by having few bristles per segment.

The **Archiannelida** are a small group of simple worms which are not segmented externally and do not have bristles.

The fourth class of annelids, the **Hirudinea,** consists exclusively of leeches. These worms are provided with stout muscular suckers at the anterior and posterior ends for clinging to their prey. Most leeches feed by sucking the blood of vertebrates; they attach themselves by their suckers, bite through the skin of the host, and suck out a quantity of blood, which is stored in pouches in the digestive tract. The leech has glands in its crop which secrete an anticoagulant which insures the leech a full meal of blood. Their meals may be infrequent, but they can store enough food from one meal to last a long time. The so-called "medicinal leech" is a fresh-water worm about 4 inches long which physicians used for blood-letting when the humoral theory of disease was in vogue. The archiannelids and polychaetes appear to represent one branch of annelid evolution and the oligochaetes and leeches another. The development of polychaetes and archiannelids is characterized by a larval form, called a **trochophore** (Fig. 116), which is very similar to the larva of molluscs.

It is generally believed that the annelids and arthropods developed from a common segmented ancestor, a theory substantiated by the existence of a curious animal called **peripatus** (Fig. 117), found in the moist, tropical forests of Africa, Australia, Asia and South America. This caterpillar-like creature, 2 or 3 inches in length, appears to be a connecting link between the two phyla. It is not believed, however, to be the ancestor of the present-day arthropods, but rather a descendent which has not changed much from the original ancestor of both annelids and arthropods, for its anatomy is a mixture of the characteristics of both. It has many pairs of legs, each of which has a pair of claws at the tip. Its excretory, reproductive and nervous systems are similar to those of annelids, but its circulatory system is like that of arthropods, as is its respiratory system, which consists of air tubes (tracheal tubes). Some zoologists classify the peripatus as a separate phylum (Onychophora); others classify it with the annelids; and still others place it with the arthropods.

134. THE ARTHROPODA

The animals which make up this phylum are, without doubt, the most successful, biologically, of all animals, for there are more of them (about 650,000 species are known, of which some 600,000 are insects), they live in a greater variety of habitats, and can eat a greater variety of food than the members of any other phylum. There are six classes of arthropods: the **Trilobita,** ancient marine arthropods that became extinct some 250,000,000 years ago; the **Crustacea** (Fig. 118), including crabs, lobsters, shrimps, barnacles, water fleas and many other marine and fresh water forms; the **Centipedes** (Fig. 119), which are fast-moving carnivorous forms, some of which can inflict a painful bite; the **Millipedes,** which are slower-moving plant-eaters; the **Arachnids,** including spiders, scorpions, ticks and mites; and the **Insects** (Fig. 120).

The word "arthropod" refers to the paired, jointed appendages characteristic of these animals. These are used in a variety of ways: as swimming paddles, walking legs, mouth parts, or accessory reproductive organs for transferring sperm. All arthropods have segmented bodies cov-

Figure 117. Peripatus, a member of the Onychophora, a "missing link" between the Annelida and the Arthropoda. (Dodson: Evolution.)

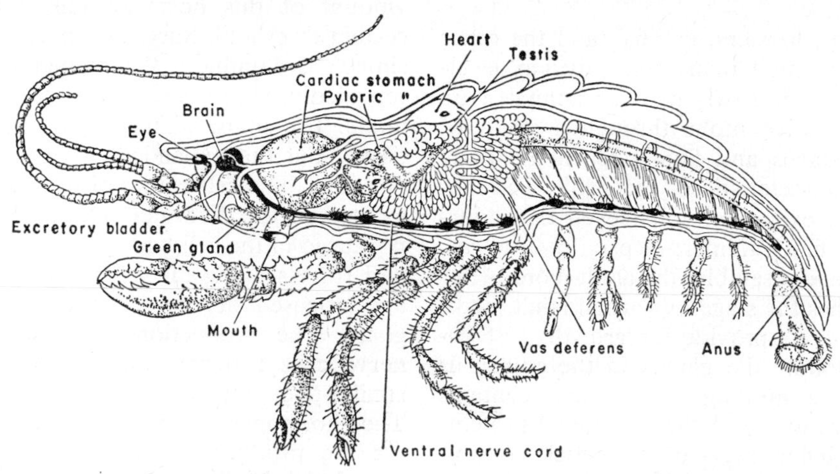

Figure 118. Diagrammatic, longitudinal section of a male crayfish, a fresh-water animal similar to a lobster. (Hunter and Hunter: College Zoology.)

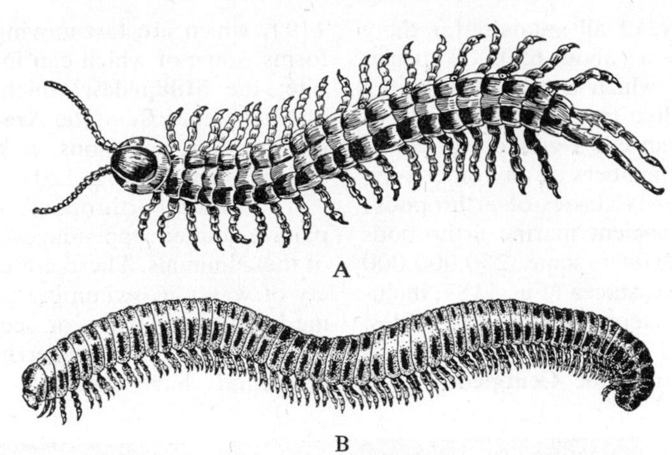

Figure 119. A, Centipede, a member of the class Chilopoda. Centipedes have one pair of appendages per segment. *B,* Millipede, a member of the class Diplopoda. Millipedes have two pairs of appendages per segment. (After Koch; from Hunter and Hunter: College Zoology.)

ered by a hard external coat of cuticle secreted by the underlying epithelium. The cuticle has an outer, waterproof, waxy layer, a rigid, middle layer, and a flexible, inner layer of chitin. The rigid layer is thin in certain regions, such as the joints of the legs and between the body segments, allowing the cuticle to be bent. Such a layer protects the body from excessive loss of moisture and from enemy attack, and gives necessary support to the underlying soft tissues. But it has disadvantages, too: body movement is somewhat restricted, and, in order to grow, the arthropod must shed the outer shell periodically and grow another larger one, a process which leaves him temporarily vulnerable.

Crabs, lobsters, crayfish and the other crustacea molt many times during development. The newly hatched animals pass by successive molts through a series of larval stages and finally reach the body form characteristic of the adult. The lobster, for example, molts seven times during the first summer; at each molt it gets larger and resembles the adult more. After it reaches the stage of a small adult, additional molts provide for growth. Just before molting the glands in the epidermis secrete a **molting fluid** which contains enzymes to digest the chitin and proteins of the inner layers of the cuticle. A soft, flexible, new cuticle is formed under the old one, folded to allow for growth. The digested remains of the old cuticle are absorbed by the body; some substances, e.g., the calcium salts, are stored for reuse. The animal may swallow air or water to aid in swelling up and bursting the old cuticle. It extricates itself from the old cuticle, swells to stretch the folded new cuticle to its full size, and then the epidermis secretes enzymes that harden the cuticle by oxidizing some of the compounds and by adding calcium carbonate to the chitin. Additional layers of cuticle are secreted subsequently.

Molting is under the control of a hormone that is accumulated in the **sinus gland** in the eyestalk. This normally prevents molting; molting occurs when the amount of this hormone falls below a certain threshold. Surgical removal of the sinus gland induces the animal to molt repeatedly, with no resting stage between molts. The sinus "gland" is not glandular tissue, but the expanded tips of bundles of axons. The cell bodies of the axons are located in the **X organ.** The hormone is produced in the cell bodies of the X organ and travels along the axons to be stored and released at their tips in the sinus gland. The production of hormones by nerve cells is termed **neurosecretion.** A similar phenomenon occurs in vertebrates: The hormones vasopressin and oxytocin are produced in the brain and accumulated in the posterior lobe of the pituitary.

135. INSECT METAMORPHOSIS

Some insects develop, like the crustacea, by passing through a series of successive molts. With each molt they get to look more like the adult, but there is no striking change in appearance at any one molt. The grasshopper is an insect that matures in this fashion. In contrast, moths, butterflies, flies, and many other insects pass through successive stages that are quite unlike one another. From the egg hatches a wormlike **larva**—called a caterpillar (moths), maggot or grub (flies, bees)—which crawls about, eats voraciously, and molts several times, each time becoming a larger larva. The last larval molt forms a **pupa,** a form which neither moves nor feeds. Moth and butterfly larvae spin a cocoon and pupate (molt to form a pupa) within that. During pupation all the structures of the larva are broken down and used as raw materials

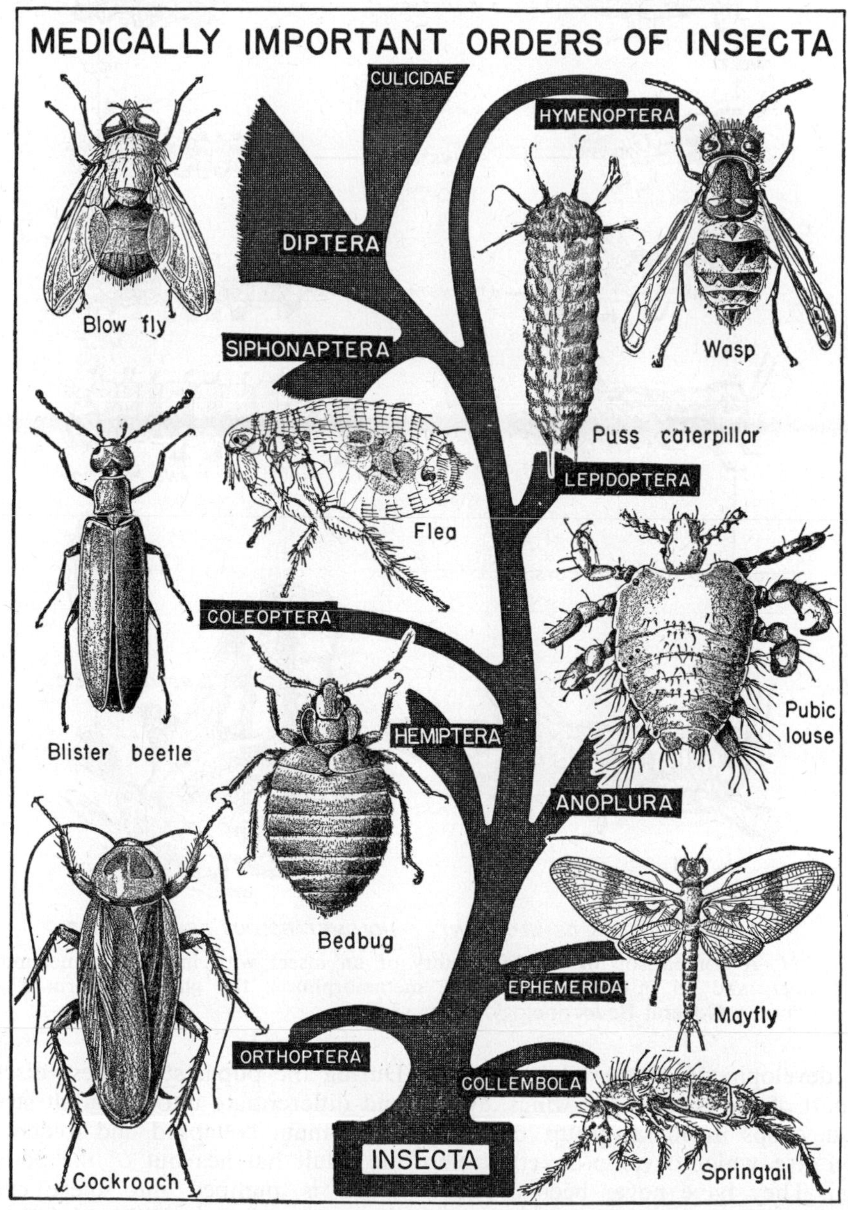

Figure 120. Some medically important orders of the class Insecta. (Mackie, Hunter and Worth: Manual of Tropical Medicine.)

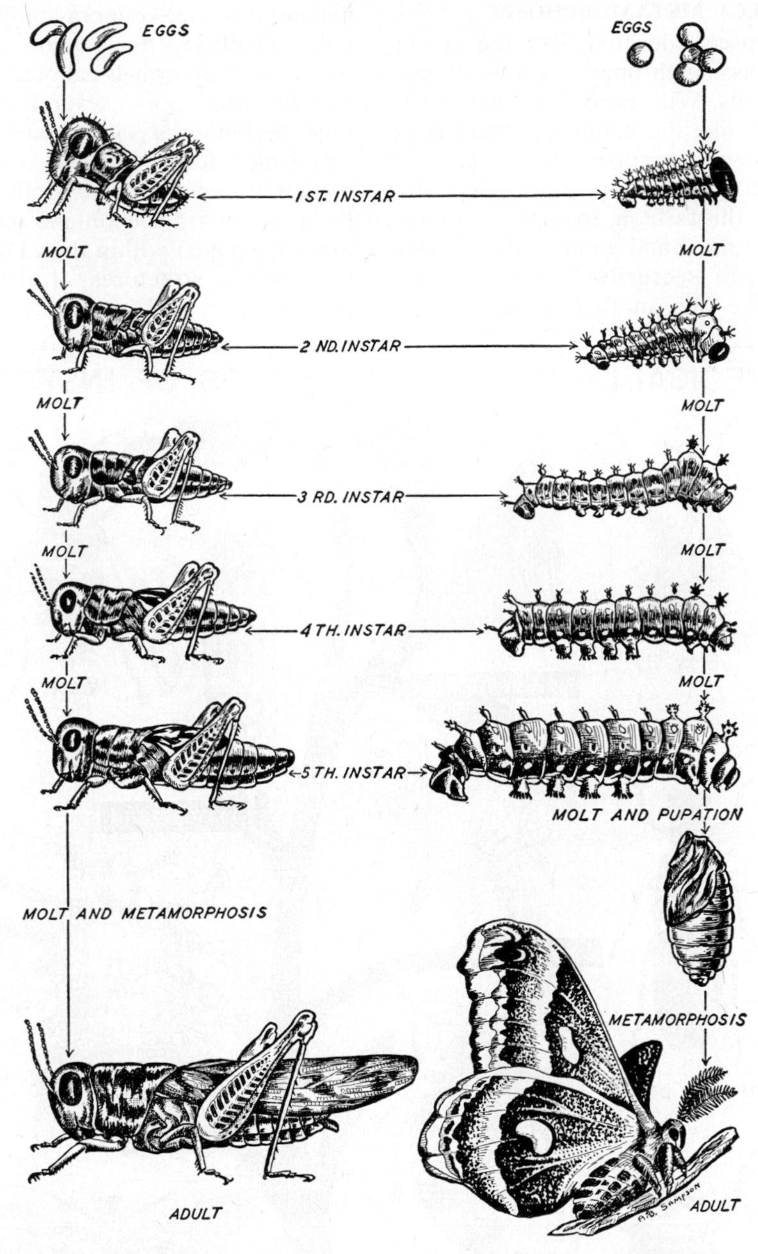

Figure 121. A comparison of the life history of an insect with incomplete metamorphosis, a grasshopper, and an insect with complete metamorphosis, the giant silkworm *Platysamia cecropia.* (Turner: General Endocrinology.)

in the development of the adult animal. Each part of the adult (legs, wings, eyes, etc.) develops from a group of cells called a **disc,** which develop directly from the egg. They have never been a functional part of the larva but remain more or less quiescent during the larval period. During the pupal stage these discs grow and differentiate into the adult structures but remain collapsed and folded. When the adult hatches out of the pupa case, blood is pumped into these collapsed structures, they unfold and inflate, and chitin is deposited to make them hard.

This striking change from larva to adult is called **metamorphosis** (Gr., change of form). The grasshopper is said to exhibit incomplete metamorphosis since it undergoes a gradual change from larva to adult (Fig. 121).

The larva and adult insect not only have a different appearance, they have quite different modes of life. The butterfly larva eats leaves; the adult drinks nectar from flowers. The mosquito larva lives in ponds and eats algae and protozoa; the adult sucks the blood of man and other mammals. In some species, such as the mayfly, the adult lives only a few hours, just long enough to mate and lay eggs.

The hormonal control of insect metamorphosis has been elucidated by the experiments of Wigglesworth in England and Carroll Williams at Harvard (Fig. 122). Williams has worked with the giant Cecropia silkworm, which undergoes a long period of dormancy (diapause) during the pupal stage. When the pupa has been chilled for about six weeks and returned to normal temperature it begins to develop again and completes its metamorphosis in about a month. Williams has found that the chilling stimulates certain gland cells in the insect's brain to secrete prothoracicotropic hormone (PTTH) which in turn stimulates the prothoracic glands to secrete **growth and differentiation** hormone. Karlson and Butenandt in Germany recently isolated pure growth and differentiation hormone from the silkworm *Bombyx*. Its chemical properties suggest that it is a steroid (p. 27) and it has been named **ecdysone.** From 1100 pounds of silkworms they obtained 25 mg. of pure ecdysone! There is a striking

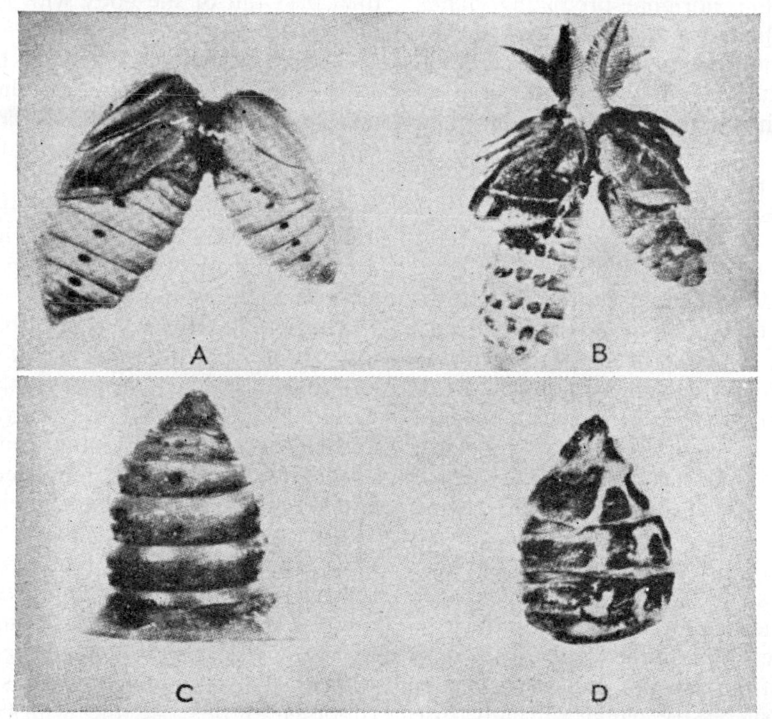

Figure 122. Experiments illustrating the hormonal control of insect metamorphosis. *A,* A brainless, diapausing *Telea polyphemus* pupa (right) is grafted to a chilled *Platysamia cecropia* pupa (left). The chilling stimulates certain brain gland cells to secrete a hormone which in turn stimulates the prothoracic glands to secrete the growth and differentiation hormone. *B,* The growth and differentiation hormone causes metamorphosis of both animals. *C,* The posterior portion of a diapausing *Cecropia* pupa is implanted with both chilled brain glands and prothoracic glands. *D,* The growth and differentiation hormone produced by the implanted prothoracic glands, which were stimulated by the implanted brain glands, brings about the metamorphosis of the pupal abdomen. (After C. M. Williams, from Prosser et al.: Comparative Animal Physiology.)

parallelism between this and the stimulation of production of adrenal hormones in the human adrenal gland induced by ACTH (adrenocorticotropic hormone) secreted by the pituitary (see p. 393). The growth and differentiation hormone then causes the epidermis to secrete molting fluid, thus leading to molt and metamorphosis. The same system of brain hormone and growth and differentiation hormone initiates the change from larva to pupal stage.

An additional hormone, called the **juvenile hormone,** is secreted by glands lying near the brain. This inhibits metamorphosis, but permits molts to occur, and insures that the larva will molt several times and reach a large size before pupating. This hormone is not produced during the last larval stage and so pupation can occur at the ensuing molt. If extra juvenile hormone-producing glands are transplanted into mature larvae they can be induced to undergo additional larval molts and grow to giant size before pupating and eventually forming giant adults. Removal of the glands early in larval life results in premature metamorphosis in some insects.

136. THE ARTHROPOD BODY PLAN

The bodies of most arthropods are divided into three regions: the **head,** always composed of exactly six segments, the **thorax,** and the **abdomen,** both of which are composed of a variable number of segments. In contrast to most annelids, each arthropod has a fixed number of segments which remains the same throughout life. We cannot begin to describe all the variations in body plan and in the shape of the jointed appendages in the numerous species. The nervous system of the more primitive arthropods is like that of the annelids, but in the higher ones the successive ganglia usually fuse together. The arthropods have a variety of well-developed sense organs: complicated eyes, such as the compound eyes of insects; organs located in the antennas, which are sensitive to touch and to chemicals; organs of hearing; and touch cells on the surface of the body. The true coelom is small and is made up chiefly of the cavities of the reproductive system; the large body cavity is not a coelom, but a blood cavity—part of the circulatory system—for the latter includes, besides the enclosed vessels, open spaces throughout the body by means of which the organs are bathed. There is a pumping organ, or "heart," in the dorsal part of the body which stirs the blood around in these spaces. Most of the aquatic arthropods have a system of gills for respiration, whereas the land forms usually have a system of fine, branching air tubes or tracheas, which conduct air to the internal organs. The digestive system, typically, is a simple tube like that of the earthworm, lined in part with a cuticle similar to the outer covering of the body. In insects and some other forms the excretory system consists of tubules which empty into the digestive tube. These metabolic wastes then pass out of the body with the feces, through the anus.

The experiments of Frank A. Brown in 1953 showed that many of the higher marine invertebrates such as crabs and snails have well-marked cyclical periods of activity and rest, correlated with the cycles of high and low tide and of daylight and darkness. These activity cycles are not directly dependent upon these environmental stimuli, for they continue when the animal is taken into the laboratory, away from the influence of the tides and light. In the laboratory the cyclic variations continue, in time with the tidal and light conditions at the beach from which it was removed, regulated apparently by some sort of "built-in clock."

Insects are the only invertebrates to have developed wings (though not all species have done so), but these structures are analogous only, not homologous, to the wings of the vertebrates. Insect wings usually occur in two pairs; the exceptions to this are the wings of flies, which occur as one pair, plus a pair of balancers—the remains of the second original pair. In most insects both sets of wings are functional as flying organs, but in grasshoppers and beetles the anterior pair are simply stiffened protective devices for the functional posterior pair.

The evolutionary relationships of the

several classes of arthropods are not completely clear. The crustaceans and arachnids appear to be lines which evolved independently from the primitive trilobites. Peripatus (p. 202) provides a possible link between the annelids and the terrestrial arthropods—insects, centipedes and millipedes—but not with the trilobite line. Fortunately, in 1930 eleven well preserved fossils from Cambrian deposits were found of a marine, peripatus-like animal that has been named *Aysheaia.* It seems probable that peripatus and *Aysheaia* represent a varied, widespread, ancient group that had developed many of the arthropod characters before the arthropods arose.

137. COLONIAL INSECTS

In a number of species of insects—bees, ants and termites—the population consists not of single individuals, but of colonies or societies made of several different types of individuals, each adapted for some particular function. In this they resemble a coelenterate or bryozoan colony, but differ in that they are not joined together anatomically as are these lower forms; they constitute a **social colony.** A termite colony (Fig. 123), for example, contains "reproductives"—the king and queen—which give rise to all the other members of the colony; "soldiers," which protect the colony from enemies; and "workers," which gather food, build the nest, and care for the young. Both soldiers and workers are sterile, and neither reproductives nor soldiers can feed themselves. Thus the members of the colony are completely dependent on each other. Each year new reproductives develop in a colony as winged forms that leave the group, mate, and form a new colony. A queen termite may lay as many as 6,000 eggs per day every day in the year for years. She is simply a specialized egg-laying machine and must be fed and cared for by the workers.

A honey bee colony consists of a single queen, a few hundred drones or males, and thousands of workers, sterile females. The queen bee mates just once and stores the sperm in a sperm sac in her body. Thereafter she can lay either unfertilized eggs which develop into haploid male drones, or fertilized eggs which develop into diploid females. If the female larvae are fed "royal jelly" for about six days they develop into queens; if fed this for two or three days and then a mixture of nectar and pollen for three days they de-

Figure 123. Model of a royal cell of the termite, *Constrictotermes cavifrons,* from British Guiana. The queen with an enlarged abdomen occupies the center of the chamber with her head toward the right. The king is at the lower left. Most of the individuals are workers. A few soldiers with "squirt gun" heads and reduced mandibles are at the left. (Courtesy of Buffalo Society of Natural Sciences.)

velop into workers. Young, adult workers serve as "nurse bees" to feed the larvae and prepare brood cells. Older adults are "house bees" that stand guard at the entrance of the hive, receive and store nectar, secrete wax for new cells and keep the hive clean. The oldest adults are "field bees" that fly from the hive and forage for water, pollen and nectar.

138. INSECT BEHAVIOR

The behavior of termites and other insects is guided to a remarkable extent by **instincts,** which are unlearned patterns of behavior present from birth. For example, the workers that hatch from the eggs laid by a queen termite build a nest for her exactly like the ones built by other members of the species, although they have no model to copy and, presumably, they cannot learn from the queen how her original nest was constructed. Members of each species of spider spin webs characteristic of the species, again guided only by instinct.

Some very complicated behavior patterns in insects are directed by instincts. The digger wasp female, after being fertilized, digs a tunnel in sandy soil with an enlarged chamber at the end. When she has completed the tunnel she closes the entrance and flies in circles overhead, apparently to fix its location. She then searches for the caterpillar of a sphinx moth, stings it to paralyze it, sucks some of its blood for food, and then takes it to the burrow she has prepared. She opens up the entrance, leaves the caterpillar outside while she goes into the burrow on an inspection tour, then returns and takes the caterpillar down to the enlarged chamber at the end of the tunnel. She then lays a single egg, attaches it to the skin of the caterpillar, closes the burrow and leaves. When she has another mature egg ready to be laid she digs another burrow, finds another caterpillar, and repeats the whole performance. The egg hatches into a larva which eats the caterpillar, pupates, and emerges as an adult wasp. Each female that develops will go through this same procedure in laying eggs. Since the offspring have no contact with the parents there is no possibility of their "learning" from them how to dig a burrow, what kind of caterpillar to sting, and so on. We therefore say that their behavior is governed by unlearned patterns of behavior present from birth—instincts.

The Austrian Karl von Frisch has made intensive studies of the behavior of bees. He was able to show that not all of their behavior was instinctive—they could be conditioned (p. 360). By placing sugar solutions in bowls on colored cards he could train worker bees to go to a bowl on a certain color. In the course of his studies he found that bees can communicate with each other to describe the location and distance from the hive of a food supply. A scout bee that has located food returns to the hive loaded with pollen and does a dance on the wall of the comb. If the food is less than 50 yards away the scout turns about in short circles. If it is 50 to 100 yards away, the dance includes a short straight series of steps between the turns. During this straight part the worker wags her abdomen. The angle of the straight run to the vertical describes the position of the food relative to the sun: If the food is directly toward the sun from the hive, the straight run is vertically upward on the side of the comb, then the bee circles around and repeats the straight run. If the food is directly away from the sun, the straight run is vertically downward; if the food is located 45 degrees to the right of the sun from the hive, the straight run is 45 degrees to the right of vertical, and so on.

What, you may ask, do bees do on a cloudy day? Fortunately, the bee eye is sensitive to ultraviolet light, which penetrates clouds, and they can use the same system.

Some examples of specialization of function in social insect colonies are truly amazing. The workers of honey ants gather a material called "honeydew," a secretion of certain sap-sucking plant lice or aphids. The workers carry this in their crops to specialized ants called "repletes" which have enormously distended crops for storing honeydew. The repletes will give up drops of honeydew to other ants of the colony on demand. Certain other species of ants are slave-makers, capturing workers of other species of ants and forcing them to gather food and build nests for them.

139. THE MOLLUSCA

This phylum, with its 80,000 species, is the second largest of all the animal phyla. It includes the oysters, clams, octopuses, snails, slugs, and the largest of all the invertebrates—the giant squid, which achieves a length of 50 feet, has a circumference of 20 feet, and weighs several tons.

The adult body plan of these animals is quite different from that of any other group of invertebrates, but the more primitive molluscs have a characteristic larval form, called a **trochophore,** similar to the larval form of certain marine annelids. This suggests that the molluscs and worms arose from a common ancestral type; the worms, however, evolved a segmented body plan, while the molluscs evolved a unique body plan without segmentation. The sluggish marine animal, the **chiton** (Fig. 124), that lives by scraping algae off the rocks of the seashore, is probably closest to the ancestral molluscan type. It illustrates the main molluscan characters: a broad, flat, muscular **foot** for creeping across rocks; a **visceral mass** above the foot, containing most of the organs of the body; a **mantle,** or fold of tissue which covers the visceral mass and projects laterally over the edges of the foot; and a hard, calcareous **shell,** secreted by the upper surface of the mantle as eight separate plates. Like the outer covering of the arthropods, this shell gives protection but has the disadvantage of making locomotion difficult.

The molluscan digestive system is a single tube, sometimes coiled, consisting of a mouth, pharynx, esophagus, stomach, intestine and anus. The pharynx characteristically contains a rasplike structure, the **radula,** which, operated by a set of muscles, can drill a hole in another animal's shell, or break off pieces of a plant. The bivalves are the only molluscs that lack a radula; they get their food by straining sea water and would have little use for a radula. The circulatory system consists of a pumping organ which sends blood through a system of branched vessels and open spaces containing the body organs. Two "kidneys," lying just below the heart, extract metabolic wastes from the blood, and discharge them through pores located near the anus. The nervous system consists of two pairs of nerve cords, one going to the foot, another to the mantle. The ganglia of these are connected around the esophagus, at the anterior end of the body, by a ring of nervous tissue, thus forming the "brain." With the exception of squids and octo-

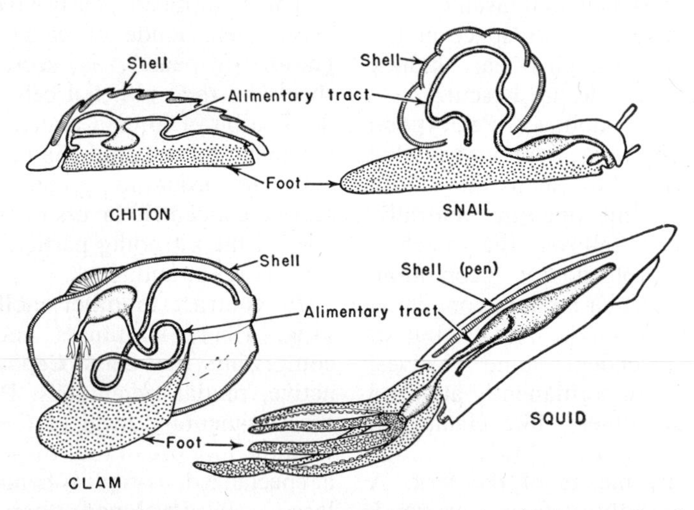

Figure 124. Variations of the basic molluscan body plan in chitons, snails, clams and squids. Note how the shell, foot and alimentary canal have changed their positions in the evolution of the different groups. (Modified after Buchsbaum; Hunter and Hunter: College Zoology.)

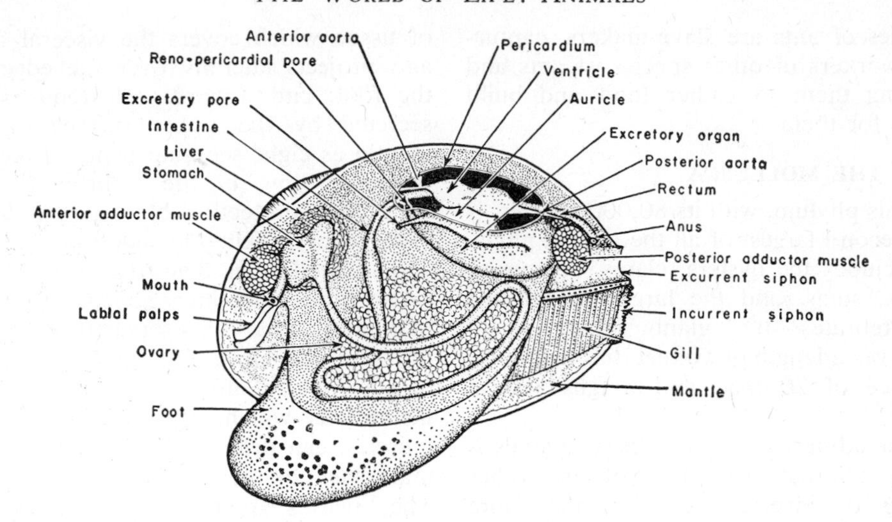

Figure 125. Longitudinal section of the marine clam, *Venus mercenaria,* showing the various organ systems. (Hunter and Hunter: College Zoology.)

puses, the molluscs lack well-developed sense organs.

One usually thinks of snails as having a spirally coiled shell, and many of them do, yet many members of the same class (**Gastropoda**), such as the limpets and abalones, have shells like flattened dunce caps, and others, such as the garden slugs and some marine snails, have no shell at all. At a particular stage in the development of each gastropod there occurs a unique, sudden, permanent twisting of the body so that the anus is brought around and comes to lie above the head. Subsequent growth is dorsal and usually in a spiral coil. The twist limits space in the body and typically the gill, heart, kidney and gonad on one side are absent.

Another class of molluscs, **Pelecypoda** (meaning hatchet-foot), commonly called bivalves, developed two shells, hinged on the dorsal side and opening ventrally. This arrangement allows the hatchet-shaped foot to protrude for locomotion, and the long, muscular siphon, containing two tubes for the intake and output of water, to be extended. Some bivalves, such as oysters, are permanently attached to the substrate; others, like clams and mussels, burrow rather slowly through sand or mud by means of the foot. A third type burrows through rock or wood, seeking protected dwellings (the shipworm, *Teredo,* which damages dock pilings and other marine installations, is just looking for a home). Finally, some bivalves, such as scallops, swim with amazing speed by clapping their two shells together.

Clams (Fig. 125) and oysters obtain food by straining the sea water brought in over their gills by the siphon. The water is kept in motion by the beating of cilia on the surface of the gills, and food particles trapped in the mucus secreted by the gills are carried to the mouth. An average oyster filters about 3 quarts of sea water per hour.

The innermost, pearly layer of the bivalve shell, made of calcium carbonate (mother-of-pearl), is secreted in thin sheets by the epithelial cells of the mantle. If a bit of foreign matter gets between the shell and the epithelium, the epithelial cells, in order to protect the animal, secrete concentric layers of this substance around the intruding particle; in this way, a pearl is formed.

In contrast to other molluscs, squids (Fig. 124), nautiluses and octopuses, comprising the class **Cephalopoda,** are active, predatory animals. By combining the rudimentary head of the chiton type with the molluscan foot they have evolved a specialized, complex head-foot with a large, well-developed "brain" and two big eyes. These are strikingly like vertebrate eyes in structure, but develop quite differently as a folding of the skin, rather than as an outgrowth of the brain. This type of independent evolution of similar structures which carry on similar functions in two different, unrelated animals, is known as **convergent evolution.** The foot of the squid and octopus is divided into ten and eight long tentacles, respectively, covered with suckers for seizing and holding the prey. Besides having a radula, the animals have (in the mouth) two strong, horny beaks for killing the prey and tearing it to bits. The mantle is thick, muscular and fitted with a funnel. By filling the mantle cavity with water and ejecting it through this funnel, the animals attain rapid jet propulsion in a direction opposite to that in which the funnel is pointed.

Cephalopods are equipped with an **ink sac** which produces a thick, black liquid. This is released when the squid or octopus is alarmed. The ink distracts the pursuer; MacGinitie has shown that octopus ink paralyzes the chemoreceptors (organs of smell) of the animals that pursue the octopus.

The shell of a nautilus is a flat, coiled structure, consisting of many chambers, built up year by year; each year the animal lives in the latest and largest chamber of the series. By secreting a gas resembling air into the other chambers the nautilus is enabled to float. The shell of the squid is reduced to a small "pen" in the mantle, and the octopus has no shell at all.

140. THE ECHINODERMATA

The echinoderms (spiny-skinned) include the **sea stars, sea urchins, sea cucumbers, serpent stars** and **sea lilies** (Fig. 126)—a group of animals radically differ-

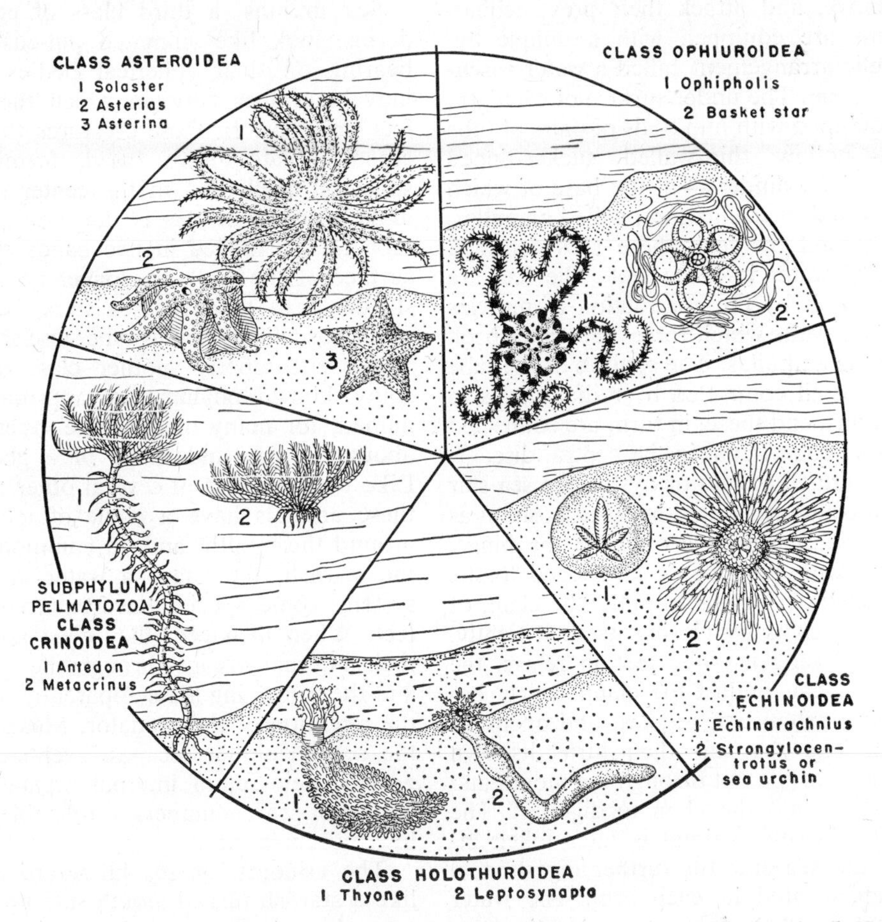

Figure 126. Some representatives of the five classes of Echinodermata. (Hunter and Hunter: College Zoology.)

ent from all other invertebrates. Curiously enough, they appear to be closely related to the vertebrates. The echinoderm larva and those of certain primitive chordates have many features in common. (Fig. 130). All the 6000 or so species of this phylum are marine, and all have radial, rather than bilateral, symmetry.

A starfish or sea star consists of a central disc from which radiate five to twenty or more arms. On the under side of the disc, in the center, is the mouth. The skin of the entire animal is imbedded with tiny, flat bits of calcium carbonate, some of which give rise to spines. A number of these spines are movable. Around the base of some, especially near the delicate skin gills used in respiration, are still tinier, specialized spines, in the form of pincers; operated by muscles, these keep the surface of the animal free of debris. In order to move, and attack their prey, echinoderms are equipped with a unique hydraulic arrangement called a **water vascular system.** The under surface of each arm is equipped with hundreds of pairs of tube feet—hollow, thin-walled, sucker-tipped muscular cylinders—at the base of which are round muscular sacs, called **ampullas.** To extend the feet, these sacs contract, forcing water into the feet. The feet are withdrawn by the contraction of muscles in their walls which forces the water back into the ampullas. The cavities of the tube feet are all connected by radial canals in the arms, and these, in turn, are connected by a circular canal in the central disc.

To attack a clam or oyster, the sea star mounts it, assuming a humped position as it straddles the edge opposite the hinge. Then, with its tube feet attached to the two shells, it begins to pull. The clam, of course, reacts by closing its shell tightly. But by using its tube feet in relays, the sea star can outlast the clam, which, becoming exhausted, relaxes, and opens its shell. The sea star then turns its own stomach inside out through the mouth and inserts it into the clam to digest it. The partly digested animal is later taken inside the sea star for further digestion in glands located in each arm. The water vascular system does not enable the starfish to move rapidly, but since it preys upon slow-moving or stationary clams and oysters, speed of attack is not necessary as it is for most predators.

There are no special respiratory or circulatory systems in these animals; both functions are accomplished by the fluid which fills the large coelomic cavity and bathes the internal organs. Nor is there any special excretory system—wastes pass to the outside by diffusion. The nervous system consists of a ring of nervous tissue encircling the mouth, and a nerve cord extending from this into each arm. There is no aggregation of nerve cells which could be called a brain.

A second class of echinoderms, the brittle stars or serpent stars, also have a central disc but their arms are long and slender, enabling them to move rapidly. The arms are discarded and replaced when injured.

Sea urchins, a third class of echinoderms, look like animated pin-cushions, bearing on their spherical bodies long, movable spines between which the tube feet protrude. In these creatures the calcareous plates have fused, forming a spherical shell, and in the center of the under surface of this is the mouth. The tube feet, arranged in five bands on the surface of the shell, are longer and more slender than those of sea stars, but the water vascular system is otherwise similar.

Sea cucumbers, another class of the spiny-skinned phylum, are appropriately named, for many of them are green and about the size and shape of a gherkin. Like the members of several other phyla, these animals have a circle of tentacles around the mouth, and, in common with the starfish, they have a water vascular system; some species have external tube feet. When menaced, they eject most of their internal organs through the mouth, and the squirming mass apparently diverts the attention of the predator. Meanwhile, the sea cucumber escapes, eventually to grow a new set of internal organs. The bodies of sea cucumbers are flexible, hollow muscular sacs.

The crinoids, or sea lilies, are rather like a starfish turned mouth side up, with a number of arms extending upward and a stalk which attaches the animal to the sea bottom. There are many more fossil than living species.

The foregoing description does not exhaust the great variey of animals. In addition to these phyla, there are other groups of invertebrates sometimes put in phyla of their own, sometimes classified under other phyla which are not important enough to be discussed here.

QUESTIONS

1. What are the fundamental differences between a tapeworm, an earthworm and a roundworm such as *Ascaris?*
2. Why do you suppose echinoderms are able to survive only in the sea?
3. Discuss the various means utilized by land plants and animals to bring about the union of an egg and sperm in the absence of a watery medium.
4. What are the advantages and disadvantages of an exoskeleton? How have some of the disadvantages of an exoskeleton been overcome by the arthropods? By the molluscs?
5. What are the advantages of segmentation?
6. In what ways are the bodies of earthworms and *Nereis* adapted to their habitats?
7. What are the essential features of molluscs? How do they differ in the several classes of molluscs?
8. Describe the peripatus. What is its evolutionary significance? If you were classifying it, would you place it with the annelids, the arthropods, or by itself?
9. What characteristics of arthropods have been important in their evolutionary success?
10. What is meant by metamorphosis? What animals other than arthropods undergo a metamorphosis?
11. Discuss the similarities and differences of a tick and a lobster.
12. What sort of experiment would you devise to demonstrate the production and action of the "juvenile" hormone of insects?
13. Compare and contrast a social colony of insects and a colony of coelenterate coral animals.
14. Describe an experiment to demonstrate that bees can distinguish (*a*) different colors and (*b*) different tastes.
15. Which of the following organisms would be welcome, and which undesirable, in your garden? Why? aphids, centipedes, millipedes, spiders.
16. What is the economic importance of insects? of mites?
17. A recent scientific article described an "artificial clam" used to study the way a starfish attacks a clam. What do you suppose this was like and what investigations would it make possible?
18. What characters are peculiar to echinoderms?
19. What is meant by convergent evolution? Does it result in homologous organs?

SUPPLEMENTARY READING

What we call instinct plays a large role in the control of animal behavior, particularly the behavior of insects. *The Study of Instinct,* by N. Tinbergen, discusses this phenomenon and gives examples of instinctive behavior in a variety of animals. Karl von Frisch's classic studies of bees are presented in *Bees, Their Vision, Chemical Senses, and Language.* The social insects are discussed in A. D. Imms' *Social Behavior in Insects* and W. M. Wheeler's *The Social Insects, Their Origin and Evolution.* Carroll Williams gives a fascinating account of some of his experiments on metamorphosis in his Harvey Lecture, *Morphogenesis and the Metamorphosis of Insects.* Other aspects of insect anatomy and physiology are discussed in W. J. Baerg's *Introduction to Applied Entomology.* An interesting semipopular account of spiders is found in Comstock and Gertsch, *The Spider Book.*

Chapter 15

The Phylum Chordata

THE LAST great phylum of animals, that to which man belongs, is the phylum Chordata, whose members are distinctive in having a notochord, a dorsal, hollow nerve cord and gill slits. The latter are present in all chordate embryos but are not evident in the adult higher vertebrates. In addition to the fishes, amphibia, reptiles, birds and mammals, which make up the classes of the subphylum **Vertebrata**—characterized by a cartilaginous or bony vertebral column—the phylum Chordata includes three subphyla of lowly creatures which show the chordate characteristics to some extent, and are extremely interesting examples of possible connecting links between the vertebrates and invertebrates.

141. ACORN WORMS OR HEMICHORDATES

The title of "chordate-least-like-the-vertebrates" falls to the **acorn worm,** a burrowing marine animal (Fig. 127). On its anterior end is a muscular proboscis used in travelling through the sand; behind this, a collar; and then a long, soft, wormlike body. The mouth is on the lower side of the body at the base of the collar, and just behind the collar, perforated by many gill slits, is the pharynx, through which water passes. As it burrows along, the animal feeds on organic matter in the sand. The nervous system is a diffuse network over most of the body, but concentrated into dorsal and ventral nerve cords in the anterior region. Only the dorsal nerve cord extends into the collar, where it becomes thick and hollow. There is a short, rodlike outgrowth from the anterior end of the digestive tract which extends into the cavity of the proboscis. Although this is called a "notochord," it is not clear whether it really corresponds to the notochord of other chordates. The acorn worm has, then, to a limited degree, the three chordate characteristics. Its larva, however, is much like that of the echinoderms and is often mistaken for it, but of course the later development of the two forms is quite different.

142. SEA SQUIRTS OR TUNICATES

A second subphylum is composed of the **sea squirts** or **Tunicates** (Fig. 128), most of them attached forms unlike the other chordates; indeed, the primitive members are often mistaken for sponges or coelenterates. The larval form of the tunicates, however, is typically chordate, superficially like a tadpole. It has a pharynx with gill slits, and a long tail containing a notochord and dorsal nerve cord. The larva eventually becomes attached to the sea bottom, and loses its tail, noto-

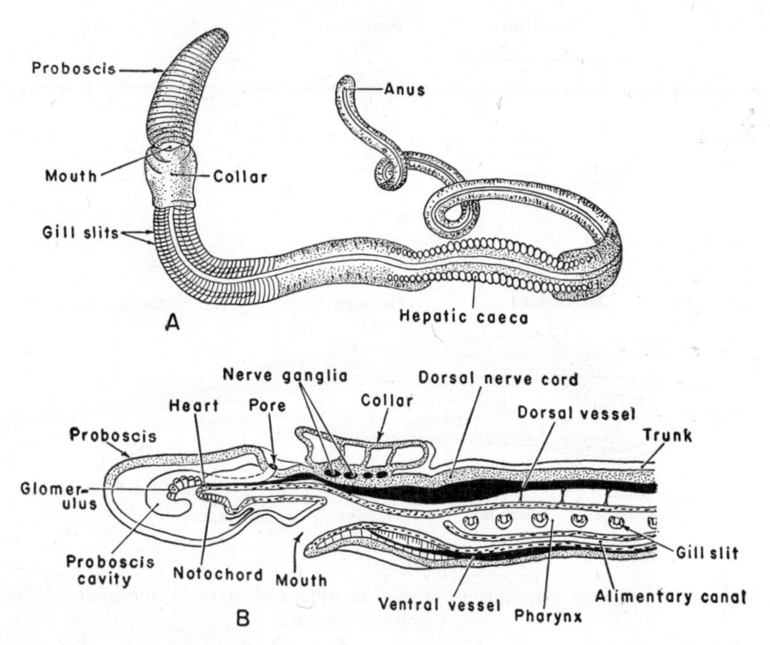

Figure 127. The acorn worm, *Balanoglossus,* a member of the subphylum Hemichordata. *A,* External appearance; *B,* longitudinal section through the anterior end to show the internal structures. (Hunter and Hunter: College Zoology.)

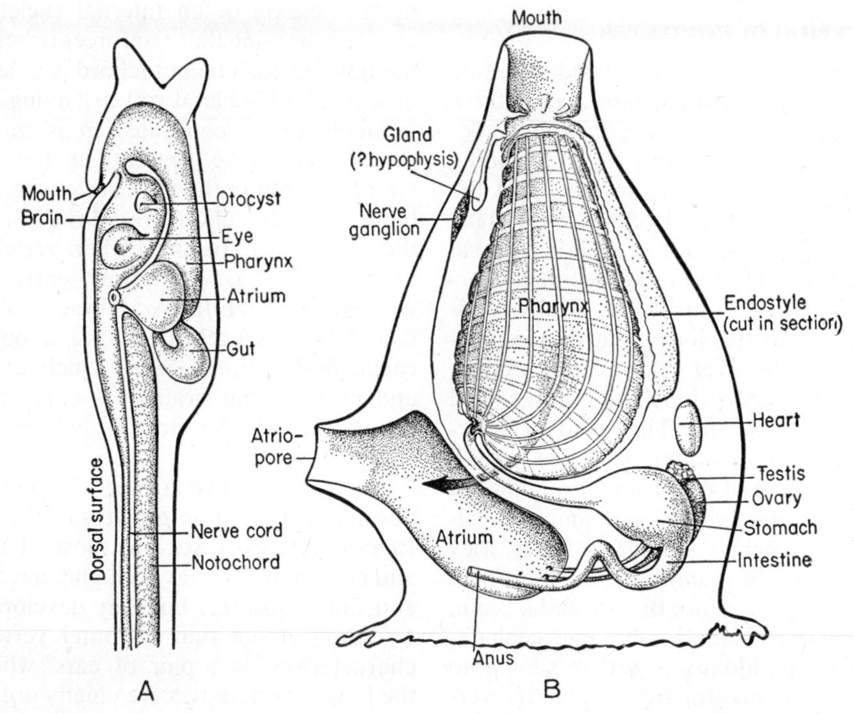

Figure 128. A tunicate, a member of the subphylum Urochordata. *A,* Diagram of the free-swimming larva, anterior end up (the otocyst is an organ of equilibrium). *B,* Diagram of the sessile adult which develops from the anterior part of the larva. (After Delage and Herouard; Romer, A. S.: The Vertebrate Body.)

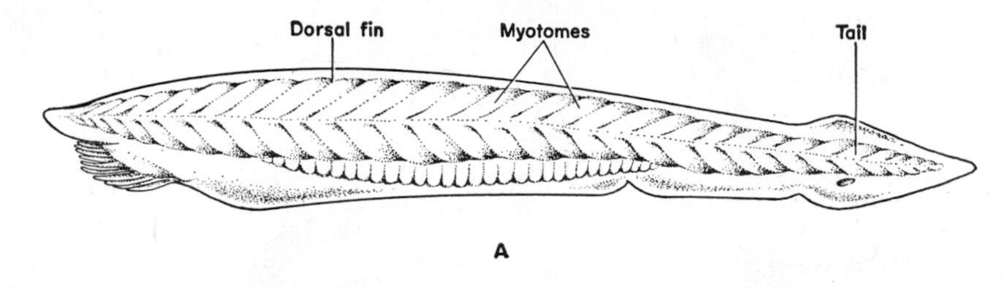

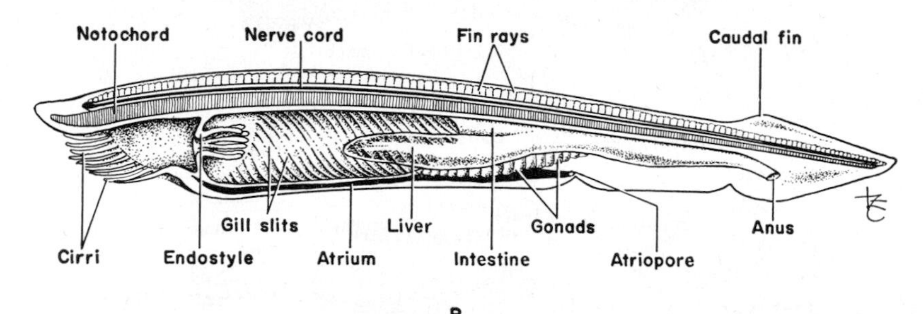

Figure 129. External view and a longitudinal section of Amphioxus, a member of the subphylum Cephalochordata.

chord, and most of its nervous system. In the adult, therefore, only the gill slits suggest that it is a chordate.

143. CEPHALOCHORDATES

In the organisms of the third chordate subphylum, the **Cephalochordata,** all three chordate characteristics are highly developed. There is a notochord extending from the tip of the head to the tip of the tail, a large pharyngeal region, with many pairs of gill slits, and a hollow, dorsal nerve cord (Fig. 129). The cephalochordates are small, translucent, fishlike animals, a few inches long, that live in shallow seas all over the world, either swimming freely or burrowing in the sand near the low tide line. They feed by drawing a current of water into the mouth (by the beating of cilia) and straining out the microscopic plants and animals. Although superficially similar to fishes, they are much more primitive, for they lack paired fins, jaws, sense organs and a brain. It is generally believed that the cephalochordate **Amphioxus** is rather similar to the primitive ancestor from which the vertebrates evolved. In contrast to the invertebrates, the blood of this animal flows anteriorly in the ventral vessel and posteriorly in the dorsal vessel.

144. THE VERTEBRATES

The vertebrates are distinguished from these three types of lower chordates by the possession of an internal skeleton of cartilage or bone that reinforces or replaces the notochord. The notochord is a flexible, unsegmented skeletal rod extending longitudinally in all chordates. It is the only skeletal structure present in the lower chordates, but in the vertebrates segmental bony or cartilaginous **vertebrae** surround the notochord. In the higher vertebrates the notochord is visible only early in development; later the vertebrae replace it completely. Vertebrates have a bony or cartilaginous brain case, which encloses and protects the brain, the enlarged anterior end of the dorsal, hollow nerve cord.

Vertebrates have a pair of eyes which develop as lateral outgrowths of the brain. Invertebrate eyes, such as those of insects and cephalopods, may be highly developed and quite efficient, but they develop from a folding of the skin. Another vertebrate characteristic is a pair of ears, which in the lowest vertebrates are chiefly organs of equilibrium. The **cochlea,** which contains the cells sensitive to sound vibrations, is a later evolutionary development.

The circulatory system of vertebrates is

distinctive in that the blood is confined to blood vessels and is pumped by a ventral, muscular heart. The higher invertebrates such as arthropods and molluscs typically have hearts but they are located on the dorsal side of the body and pump blood into open spaces in the body, called a **hemocoel.** Vertebrates are said to have a **closed circulatory system;** arthropods and molluscs have an open circulatory system, for the blood is not confined solely to tubular blood vessels.

145. THE ORIGIN OF THE CHORDATES

There is no fossil record of the ancestors of the chordates, for, whatever they were, they were undoubtedly small and soft bodied. Theories of the origin of the chordates must depend on other types of evidence. The most widely held theory at present is that the echinoderms and chordates have a common evolutionary origin. This is based on the striking similarity of the **tornaria** larva of the hemichordate and the **bipinnaria** larva of the starfish (Fig. 130) plus the general similarity in the mode of formation of the mesoderm and coelom in the two phyla. Another theory which has some support derives the chordates from primitive nemertean-like worms. This theory postulates that the proboscis sheath and certain cephalic pits of the nemerteans may be comparable to the notochord and gill slits of the chordates. Other theories, which derived the chordates from arthropods or annelids, were based on more superficial resemblances and have now been generally abandoned.

The classes of the subphylum Vertebrata are as follows: the **Agnatha,** the jawless fishes such as the lamprey eels; the **Placodermi,** earliest of the jawed fishes, known only from fossils; the **Chondrichthyes,** the sharks and rays with cartilaginous skeletons; the **Osteichthyes,** the bony fishes, the **Amphibia,** frogs and salamanders; the **Reptilia,** snakes, turtles and alligators plus a host of fossil forms like the dinosaurs; the **Aves,** birds; and the **Mammalia,** the warm-blooded fur-bearing animals that suckle their young.

146. JAWLESS FISHES

The Agnatha, or jawless fishes, includes the **ostracoderms,** which are the earliest known fossil chordates (p. 521) and the living lamprey eels and hagfishes (Fig. 131). These are cylindrical fish, up to 3 feet long, with no jaws or paired fins. Lampreys and hagfishes have a circular sucking disc around the mouth, which is located on the ventral side of the anterior end. They attach themselves by this disc to other fish and, using the horny teeth on the disc and tongue, bore through the skin and get blood and soft tissues to eat. They are the only parasitic vertebrates; hagfishes may bore their way completely through the skin and come to lie within the body of the host. Both are of great economic importance because of their destruction of food fish such as cod, flounder, lake trout, and whitefish. The trout of the Great Lakes have been killed off in great numbers by sea lampreys that apparently came up from the St. Lawrence via the Welland Canal. Lampreys leave the ocean or lakes and swim upstream to spawn.

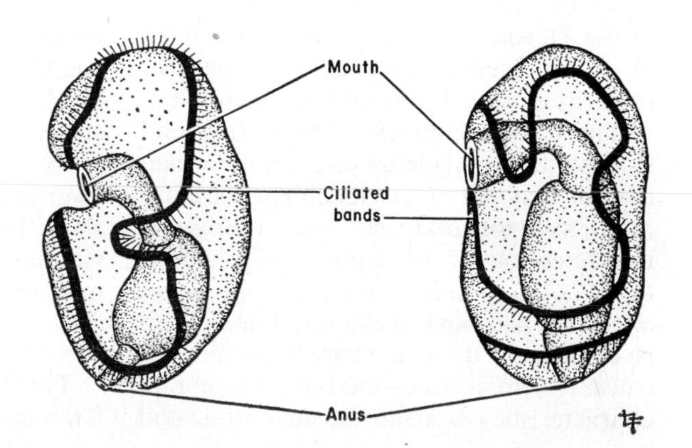

Figure 130. Left, The bipinnaria larva of a starfish and *right,* the tornaria larva of an acorn worm. Note the striking similarities of the two.

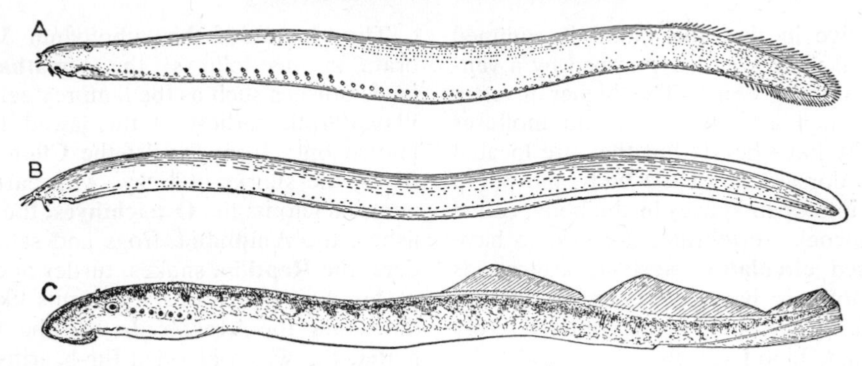

Figure 131. The three main types of cyclostomes. *A,* A slime hag; *B,* a hagfish; *C,* a lamprey. Note the absence of jaws and paired fins. (After Dean, from Romer: The Vertebrate Body.)

They build a nest, a shallow depression in the gravelly bed of the stream, into which eggs and sperm are shed. The fertilized eggs develop into **ammocoetes** larvae in about three weeks. These larvae, which probably resemble the ancestral primitive vertebrate more closely than any living vertebrate does, drift downstream to some pool and live in burrows in the muddy bottom for several years. They then undergo a metamorphosis, become adult lampreys, and migrate back to the ocean or lake.

The placoderms are completely extinct today; the fossil forms will be discussed in Chapter 34. They are probably the ancestors of both the cartilaginous fish and the bony fish. The former probably arose in the sea and the latter in fresh water.

147. CARTILAGINOUS FISHES

The ostracoderms and placoderms were primarily fresh water fish; only a few ventured into the oceans. The cartilaginous fishes evolved as successful marine forms in the Devonian and most have remained as ocean-dwellers; only a few have secondarily returned to a fresh water habitat.

The Chondrichthyes—sharks, rays and skates—have a skeleton of cartilage which may or may not be calcified. The cartilaginous skeleton of these fishes represents the retention of an embryologic condition, not a primitive one, for the adult ancestors had bony skeletons. The dogfish is commonly used in biology classes because it demonstrates the basic vertebrate characteristics in a simple, uncomplicated form. All the Chondrichthyes have paired jaws and two pairs of fins. The skin contains scales composed of an outer enamel and an inner dentine layer. The lining of the mouth contains larger but essentially similar scales which serve as teeth. The teeth of the higher vertebrates are homologous with these shark scales.

All fish, from lampreys to the highest bony fish, have highly vascular gills which have a large surface for the transfer of oxygen and carbon dioxide. In some fish the gills also secrete salts to maintain osmotic equilibrium between the blood and the surrounding water. A current of water enters the mouth, passes over the gills and out the gill slits, constantly providing the fish with a fresh supply of dissolved oxygen.

The whale shark, which reaches a length of 50 feet, is the largest fish known. Sharks are predators, catching and eating other fish. Rays and skates are sluggish, flattened creatures, living partly buried in the sand and feeding on mussels and clams. The sting ray has a whiplike tail with a barbed spine at the tip, which can inflict a painful wound. The electric ray has electric organs on either side of the head; these are modified muscles which can discharge enough electricity to stun fairly large fish. Shark skin is tanned and used in making shoes and handbags, and shark liver oil is an important source of vitamin A. Some sharks and rays are used for food.

148. BONY FISHES

The Osteichthyes include a variety of fresh and salt water fishes, ranging in size from guppies to sturgeons, which may weigh over a ton. Most fish have beautifully streamlined bodies and swim mostly by contracting the body and tail muscles which move the tail back and forth in a sculling motion. The fins are used chiefly for steering. Most bony fish have a **swim bladder,** a gas-filled sac located in the dorsal part of the body cavity (Figs. 132 and 163). By secreting gases into the bladder or by absorbing them from it, the fish can change the density of its body and so hover at a given depth of water. The gills of bony fishes are covered by a hard, bony protective flap called an operculum. The skeleton is composed of bone rather than cartilage, and the head is encased in many bony plates which form a **skull.** Bony fish have protective, overlapping bony scales in the skin which differ from those found in sharks.

Many of the bony fishes, particularly those in tropical waters, are brightly and beautifully colored—red, orange, yellow, green, blue and black. Some fish, such as flounders, can change color and pattern to conform to the color and pattern of the background, and thus render themselves inconspicuous to predators. Fish evolution has led to a tremendous variety of sizes, shapes and colors, and to a number of interesting adaptations—the sea horse male that has a brood pouch in which eggs are carried until they hatch; the deep sea forms that have evolved luminescent structures as lures for their prey; the male stickleback which builds a nest of sticks held together by threads which he secretes, and then guards the eggs in the nest; the true eels which live as adults in streams in North America or Europe but which migrate to the Atlantic near Bermuda to spawn, and so on.

Land vertebrates evolved from the group of bony fishes known as the Crossopterygii or **lobe-finned fishes.** These were

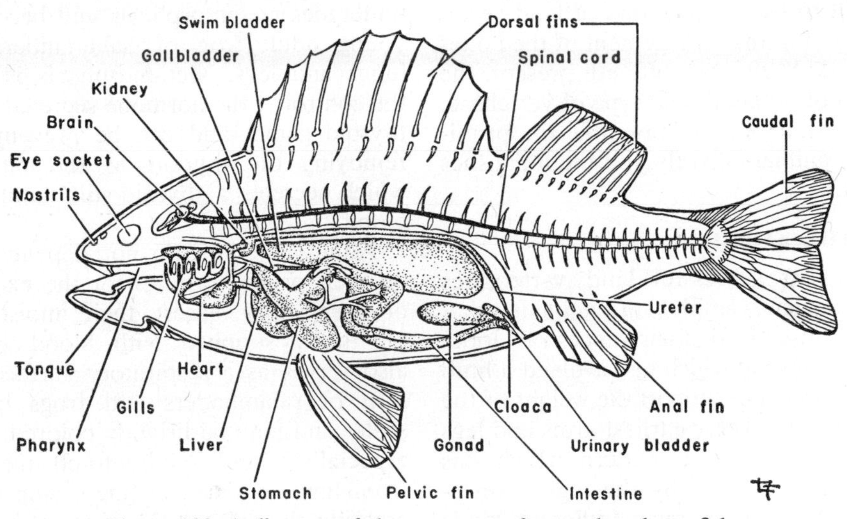

Figure 132. A diagram of the structure of a perch, a bony fish.

Figure 133. A photograph of the 5-foot long coelacanth caught in the Indian Ocean off South Africa in 1952. Note the thick, lobe-shaped fins. (Courtesy LIFE, © TIME, Inc.)

Figure 134. A, Leopard frog adapted to a light background and *B,* one adapted to a dark background. (Turner: General Endocrinology.)

believed to have become extinct in the Mesozoic, but two specimens have been caught in the Indian Ocean off South Africa, one in 1939 and one in 1952 (Fig. 133). The lungfishes (p. 522) were once thought to be the ancestors of land vertebrates but in the arrangement of the bones of the skull, the type of teeth present, the pattern of fin bones and type of vertebrae, the lobe-finned fishes resemble the primitive amphibians closely and the lungfishes do not.

149. THE AMPHIBIA

The first successful land vertebrates were ancient amphibian animals that closely resembled their ancestral lobe-finned fishes but which had evolved a limb strong enough to support the weight of the body on land. These earliest arms and legs were five-fingered, a pattern which has been generally kept by the higher vertebrates. There were many different kinds of ancient amphibia, all of which became extinct in the first part of the Mesozoic. The modern amphibia, the frogs and salamanders, appeared in the latter part of the Mesozoic. The salamanders and water dogs more closely resemble the ancient amphibia; frogs and toads are highly specialized for hopping.

Although some adult amphibia are quite successful as land animals and can live in comparatively dry places, they must return to water to reproduce. Eggs and sperm are generally laid in water and the fertilized eggs, nourished at first by the yolk, develop into larvae or **tadpoles.** These breathe by means of gills and feed on aquatic plants. After a time the larva undergoes metamorphosis and becomes a young adult frog or salamander, with lungs and legs. Metamorphosis is under the control of the hormone secreted by the thyroid gland, and can be prevented by removing the thyroid or the pituitary, which secretes a thyroid-stimulating hormone.

Adult amphibia do not depend solely on their primitive lungs for the exchange of respiratory gases; their moist skin, plentifully supplied with blood vessels, also serves as a respiratory surface. The skin of salamanders and frogs has no scales and may be brightly colored. Frogs especially have the ability to change color, from light to dark, by increasing or decreasing the size of the pigment-containing cells of the skin (Fig. 134). The change in color is controlled by a hormone, called **intermedin,** secreted by the intermediate lobe of the pituitary (p. 389). Salamanders, but not frogs, have a marked ability to regenerate lost legs and tails. In some species of frogs and toads the fertilized eggs do not develop in the water but are kept on the back of the female, in the mouth of the male, or in a string wrapped around the male's hind legs.

A number of frogs, toads and salamanders have skin glands that secrete poisonous substances. These may serve as a means of protection for the species, by discouraging would-be predators.

There is a small group of tropical, legless, wormlike amphibia (the Apoda) which burrow in moist earth.

150. REPTILES

The class Reptilia has many more extinct than living species. They are true land forms and need not return to water to reproduce, as amphibians must. The embryo develops in a watery medium within the protective, leathery egg shell secreted by the female. Since a sperm could not penetrate this shell, fertilization must occur within the body of the female before the shell is added. This in turn necessitated the evolution of some means of transferring sperm from the body of the male to that of the female. Reptiles were the first to evolve a male organ, the **penis,** for this purpose.

The bodies of reptiles are covered with hard, dry, horny scales which protect the animal from desiccation and from predators. They breathe by means of lungs, for the dry scaly skin cannot serve as an organ of respiration. Like fish and amphibia, reptiles do not have a mechanism for regulating body temperature and therefore have the same temperature as their surroundings. In hot weather their body temperature is high, metabolism occurs rapidly, and they can be quite active. In cold weather their body temperature is low, their metabolic rates are low and they are very sluggish. Because of this they are much more successful in warm than in cold climates. The reptiles living today are turtles, alligators, snakes, lizards and the tuatara of New Zealand. The many types of reptiles that flourished in the Mesozoic are discussed in Chapter 34.

151. BIRDS

Birds are characterized by the presence of feathers, which are modified reptilian scales; these decrease the loss of water through the body surface, decrease the loss of body heat, and aid in flying by presenting a plane surface to the air. Birds and mammals are the only animals with a constant body temperature. They are sometimes called **warm-blooded**—other animals are called cold-blooded—but this is inaccurate, for a frog or snake may have a higher body temperature on a hot day than a bird or mammal. Birds and mammals independently evolved mechanisms to keep body temperature constant despite wide fluctuations in the environmental temperature. The constant body temperature permits metabolic processes to proceed at constant rates and enables these animals to remain active in cold climates.

Birds evolved from a group of primitive dinosaurs called **thecodonts.** Like reptiles, birds lay eggs and have internal fertilization. Birds have reptilian scales on their legs, and the earliest birds, known only from fossils, had reptilian teeth. Adaptation to flight has involved the evolution of hollow bones and the presence of **air sacs**—extensions of the lungs that occupy the spaces between the internal organs. Not all birds fly; some, such as penguins, have small, flipper-like wings used in swimming. Others, such as the ostrich and cassowary, have vestigial wings but well-developed legs. Birds have become adapted to a variety of environments, and different species have very different types of beaks, feet, wings and tails.

Men have long been fascinated by the colors, songs and behavior of birds and these have been studied extensively. One of the most fascinating aspects of bird behavior is the annual migration that many birds make. Some birds such as the golden plover and arctic tern fly from Alaska to Patagonia and back each year, flying perhaps 25,000 miles en route. Others migrate only a few hundred miles south each winter and some, such as the bobwhite and great horned owl, do not migrate at all. The stimulus for the northward spring migration of certain birds that winter in California has been shown to be the increasing amount of daylight per day. This in some way stimulates the hypothalamus which in turn stimulates the pituitary gland. This secretes gonadotropic hormones which stimulate the growth of testis or ovary, and the increased amount of sex hormones circulating in the blood initiates the migration.

The services rendered to man by birds include the destruction of harmful rodents, insects and weed seeds and the dispersal of the pollen and seeds of many plants. The guano (excrement) deposited by sea birds in certain regions is a valuable fertilizer. Birds also provide man with food and with feathers that serve a variety of purposes.

152. MAMMALS

The distinguishing features of mammals are the presence of hair, mammary glands, sweat glands, and the differentiation of the teeth into incisors, canines and molars. Mammals have a constant body temperature and the covering of hair serves as insulation to aid in thermoregulation. Mammals evolved from a group of reptiles called **therapsids** (p. 525), probably some time in the Triassic period. The earliest mammals were egg-laying animals called **monotremes.** Only two monotremes have survived to the present —the Australian duck-billed platypus and spiny anteater (Fig. 289). The young, after hatching from the egg, are nourished by milk secreted by mammary glands.

The second subclass of mammals, the **marsupials** or pouched mammals, are also found largely in Australia—kangaroos, koalas and wombats. The opossum is one of the few found elsewhere. Marsupials do not lay eggs; the young are born alive in a very immature state and are transferred to a pouch on the mother's abdomen where they feed on milk secreted by the mammary glands and complete their development.

The third subclass of mammals, the Eutheria or placental mammals, includes all the other mammals, all characterized by the formation of a **placenta** for the nourishment of the developing embryo while within the uterus (womb) of the mother. The placenta is a structure, formed in part from tissues derived from the embryo and in part from maternal tissues, by means of which the embryo receives nutrients and oxygen and eliminates wastes. The young are born alive, in a more advanced state of development than the newborn marsupials. Some of the principal orders of placental mammals are the following:

(1) **Insectivora**—moles, hedgehogs and shrews. These are insect-eating animals, considered to be the most primitive placental mammals and the ones closest to the ancestors of all the placentals. The shrew is the smallest mammal alive; some weigh less than 5 grams.

(2) **Chiroptera**—bats. These mammals are adapted for flying; a fold of skin extends from the elongated fingers to the body and legs, forming a wing. They eat insects and fruit, or suck the blood of other mammals. Bats are guided in flight by a sort of biologic sonar: they emit high frequency squeaks and are guided by the echoes from obstructions. Blood-sucking bats may transmit diseases such as yellow fever and paralytic rabies.

(3) **Carnivora**—cats, dogs, wolves, foxes, bears, otters, mink, weasels, skunks, seals, walruses and sea lions. These are all flesh eaters, with sharp, pointed canine teeth and shearing molars.

(4) **Rodentia**—squirrels, beavers, rats, mice, porcupines, hamsters, chinchillas and guinea pigs. These numerous mammals have sharp, chisel-like incisor teeth.

(5) **Edentata**—sloths, anteaters and armadillos. Mammals with few or no teeth.

(6) **Primates**—lemurs, monkeys, apes and man. These mammals have highly developed brains and eyes, nails instead of claws, opposable great toes or thumbs and eyes directed forward.

(7) **Artiodactyla**—cattle, sheep, pigs, giraffes and deer. Herbivorous hooved animals with an even number of digits per foot.

(8) **Perissodactyla**—horses, zebras, tapirs and rhinoceroses. Herbivorous hooved animals with an odd number of digits per foot.

(9) **Proboscidea**—elephants, mastodons and wooly mammoths. Animals with a long muscular proboscis or trunk, thick, loose skin, and incisors elongated as tusks. These are the largest land animals, weighing as much as 7 tons.

(10) **Sirenia**—sea cows, dugongs and

manatees. These are herbivorous aquatic animals with finlike forelimbs and no hind limbs. They are probably the basis for most tales about mermaids.

(11) **Cetacea**—whales, dolphins and porpoises. These are marine mammals with fish-shaped bodies, finlike forelimbs, no hind limbs, and a thick layer of fat called blubber covering the body. Sulfur bottom whales are the largest animals ever known—up to 100 feet long and 150 tons.

The various members of the animal kingdom cannot be placed on a single scale ranging from lowest to highest, for evolution has occurred in the manner of a branching tree, rather than in a single continuous series (Fig. 135). We cannot say, for example, that the starfish is "higher" or "lower" than the oyster; the two forms are simply representatives of the two main trunks of the evolutionary tree—one of which gave rise to echinoderms and chordates, the other to flatworms, nemerteans, annelids, arthropods and molluscs. Between the two groups are deep-lying differences of structure and development which will be discussed in Chapter 34.

QUESTIONS

1. What is the evidence for believing the vertebrates to be more closely related to echinoderms than to other invertebrates?
2. What are the three chief characteristics of the phylum Chordata? How are these evident in an acorn worm? In man?
3. The acorn worms are placed in a separate phylum by some zoologists. What are

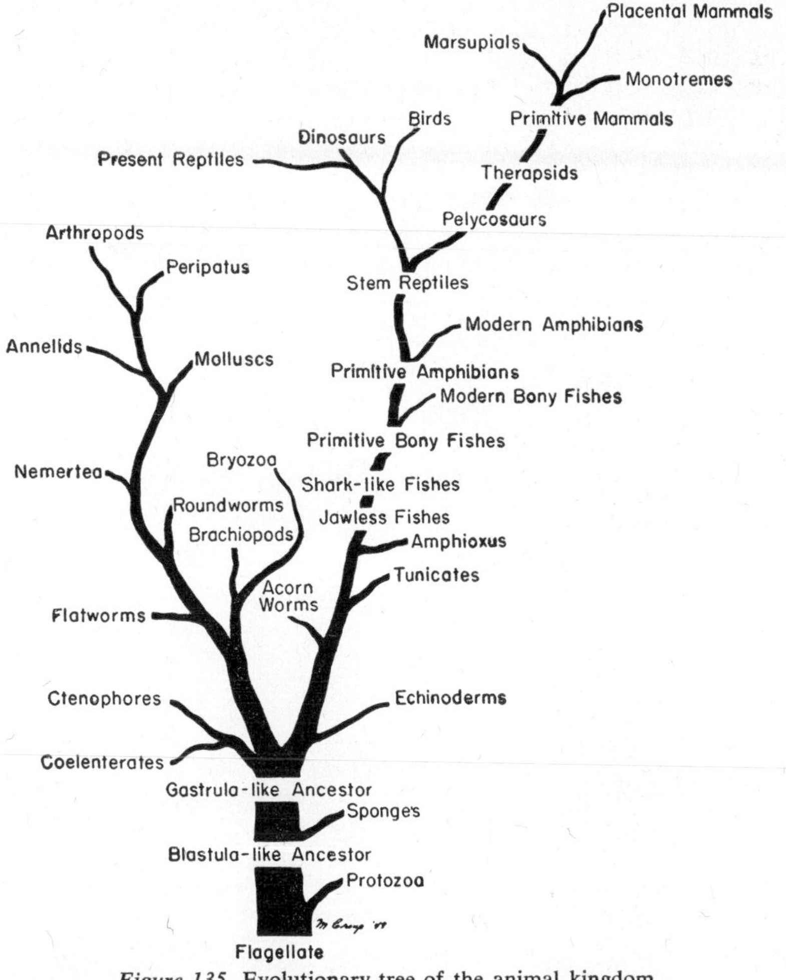

Figure 135. Evolutionary tree of the animal kingdom.

the arguments for and against this proposal?

4. What characteristics distinguish the vertebrates from the rest of the chordates?
5. How do lampreys and hagfishes differ from other fishes? Of what economic importance are they?
6. What are the functions of the gills of fishes?
7. Compare the skins of sharks, frogs, turtles and mammals.
8. Where are the following found and what are their functions: air sacs, swim bladder, placenta?
9. Give the phylum, subphylum and class to which man belongs and name three other animals which each division includes.
10. An animal is multicellular, has a dorsal hollow nerve cord, many pairs of gill slits, a notochord longer than the nerve cord, and no vertebrae. What is it?
11. Why are monotremes considered to be more primitive than marsupials? One group of paleontologists considers them to be therapsid reptiles rather than mammals; give the arguments for and against this theory.
12. Which are more specialized, birds or mammals?

SUPPLEMENTARY READING

For further information on the structure and function of the vertebrates consult A. S. Romer, *The Vertebrate Body,* or Walter and Sayles, *Biology of the Vertebrates.* The role of instincts in determining the behavior of vertebrates is clearly and interestingly presented by Konrad Lorenz in *King Solomon's Ring*. There are many books dealing with the natural history of particular groups of vertebrates, e.g., L. P. Schultz, *The Ways of Fishes,* Thomas Barbour, *Reptiles and Amphibians* and E. A. Armstrong, *Bird Display and Behavior.* Many interesting facts about the various methods used by different animals, vertebrate and invertebrate, to carry on their basic life functions are presented in B. T. Scheer's *Comparative Physiology* and by Prosser, Brown, Bishop, Jahn and Wulff in *Comparative Animal Physiology*.

Part Four

The Organization of the Body

Chapter 16

Blood

THE METABOLIC processes of all cells require a constant supply of food and oxygen and a constant removal of waste products. This is accomplished simply by diffusion in small plants and animals living in a watery environment, but man and all the larger animals have developed some system of internal transport, a circulatory system. The circulatory system of man includes the heart and blood vessels, the lymph vessels, and the blood and lymph. Blood fits our definition of a tissue—it is a group of similar cells specialized to perform certain functions. Many of the studies on the relations of cells to their immediate environment have been done with blood, for this tissue is readily obtained with a hypodermic needle and syringe. The blood cells are not injured when carefully collected and can be studied in their normal state.

The volume of blood in a man depends upon his weight; a person weighing 150 pounds has about 5 quarts of blood (just the quantity of oil in the crankcase of most cars). In addition to transporting food and oxygen to cells and removing wastes from them, blood has the following roles: it transports hormones, the secretions of the endocrine glands; it has a role in regulating the amount of acids, bases and water in cells; it is important in regulating body temperature, cooling organs such as the liver and muscles where an excess of heat is produced and heating the skin where heat loss is greatest; its white cells are a major defense against bacteria and other disease organisms; and its clotting mechanism helps prevent the loss of this valuable fluid.

Although blood appears to be a homogeneous crimson fluid as it pours from a wound, it is composed of a yellowish liquid, called **plasma,** in which float the **formed elements: red blood cells,** which give blood its color, **white blood cells** and **blood platelets.** The latter are small cell fragments, important in initiating the clotting process, which are derived from large cells in the bone marrow. The formed elements make up about 45 per cent of whole blood; the remaining 55 per cent is plasma. The loss of water in profuse sweating may reduce the plasma volume to 50 per cent of the blood, and drinking a lot of water or beer may increase it to 60 per cent. The formed elements, with a specific gravity of 1.09, are heavier than plasma, which has a specific gravity of 1.03, and the two may be separated by centrifuging. Blood is constantly mixed as it circulates in the blood vessels so that the plasma and blood cells do not separate.

153. PLASMA

Plasma is a complex mixture of proteins, amino acids, carbohydrates, fats, salts, hormones, enzymes, antibodies and dissolved gases. It is very slightly alkaline, with a *p*H of 7.4. The two chief constituents are water, 90 to 92 per cent, and proteins, 7 to 8 per cent. The concentrations of glucose (0.1 per cent) and salts (0.9 per cent) are very small but are kept remarkably constant. As blood flows past the body cells, its plasma constantly receives and gives off a wide variety of substances. Yet the composition of blood is relatively constant, for any change in its composition initiates responses on the part of one or more organs of the body to restore the normal equilibrium.

The plasma contains several different kinds of proteins, each with specific properties and functions. Prof. E. J. Cohn, who devised methods for separating these plasma proteins, has shown that these include fibrinogen, alpha, beta and gamma globulins, albumin and lipoproteins. Fibrinogen is one of the proteins involved in the clotting process; albumin and globulins regulate the water content of cells and body fluids. The gamma globulin fraction is rich in antibodies, which provide immunity to certain infectious diseases such as measles and infectious hepatitis. Purified human gamma globulin is used now in the treatment of these diseases. The presence of these proteins makes blood about six times as viscous as water. The plasma protein molecules, which are too large to pass readily through the walls of the blood vessels, exert an osmotic pressure and play an important role in regulating the distribution of water between the plasma and tissue fluids (p. 256).

The normal functioning of nerves, muscles and other tissues requires the presence of the proper balance of sodium, potassium, magnesium and calcium ions. Plasma contains these, together with chloride, bicarbonate and phosphate ions, in a total concentration of about 0.9 per cent in mammals and slightly less in lower animals. The transport of these ions and the regulation of the amount of each present in the tissues is an important role of the blood.

The most important carbohydrate in plasma is glucose, the concentration of which ranges from 0.08 to 0.14 per cent, and averages about 0.1 per cent. It is carried in the blood from the intestines, where it is absorbed into the body, to the liver, where it is stored as glycogen, and eventually to all the cells of the body, where it is metabolized to release energy. The brain cells are especially dependent upon a constant supply of glucose for fuel. If the concentration in the blood falls below 0.04 per cent, the irritability of certain brain cells is greatly increased and muscular twitching and convulsions occur. If the blood glucose concentration remains low, the brain cells become unable to function and coma and death ensue.

When either whole blood or plasma is removed from the blood vessels and allowed to stand, it changes into a sticky, gelatinous mass, called a **clot.** The clot, on standing some time, shrinks and squeezes out a light yellow liquid called

serum. The distinction between plasma and serum is clear: plasma is the liquid part of whole blood; serum is the liquid left after clotting has occurred. The chemical composition of the two is almost identical but plasma contains fibrinogen and serum does not. In the clotting process the soluble fibrinogen is converted to insoluble fibrin, the gel which comprises the firm part of the clot.

154. THE RED CORPUSCLES

The red blood cells, or **erythrocytes,** are biconcave discs 7 to 8 microns in diameter and 1 to 2 microns thick (Fig. 136). Unlike most cells they have no nucleus. An internal elastic framework maintains the disc shape and permits the cell to bend and twist as it passes through blood vessels smaller than its diameter. Erythrocytes cannot move actively but simply float in the blood stream and are moved about by the pumping action of the heart. There are, on the average, about 5,400,000 red blood cells per cubic millimeter of blood in the adult male, and about 5,000,000 per cu. mm. in the adult female. Newborn infants have a larger number, 6 to 7 million per cubic millimeter; this number decreases after birth and the adult number is reached at about three months.

The human body contains about thirty trillion red blood cells. Since the number of red cells per cubic millimeter is an important factor in determining a person's general health, almost all clinical examinations include a **red cell count.** Blood drawn from a patient is first diluted exactly 200 times in a special diluting pipette. The pipette is filled with blood to the 0.5 mark and then with a special diluting fluid to bring the combined blood and fluid to the 101 mark. The blood and diluting solution are carefully mixed in the pipette and the diluted blood is placed on the stage of the counting chamber, called a **hemacytometer.** The stage has two sets of fine lines 50 microns apart, ruled at right angles to each other, which delineate squares 50 microns on a side. On each side of the counting chamber is a ridge of glass exactly 100 microns higher than the surface of the chamber. When a flat glass cover slip is placed on the two ridges, covering the counting chamber, cubes 50 by 50 by 100 microns are established. Each cube thus has a volume of $\frac{1}{4000}$ of a cubic millimeter. The number of red cells in each of several cubes is counted, the numbers are averaged, and then by simple multiplication the number of red cells in $\frac{1}{4000}$ cubic millimeter of diluted blood is converted to the number of red cells in 1 cu. mm. of undiluted blood.

155. HEMOGLOBIN AND THE TRANSPORT OF OXYGEN

Each red corpuscle contains some 265,000,000 molecules of **hemoglobin,** the red pigment which is responsible for the transport of oxygen. Hemoglobin is a protein to which are attached four atoms of iron, each in the center of a complex organic compound called heme. The blood normally contains about 15 to 16 gm. of hemoglobin per 100 ml. Hemoglobin has the unique property of forming a loose chemical union with oxygen, the oxygen

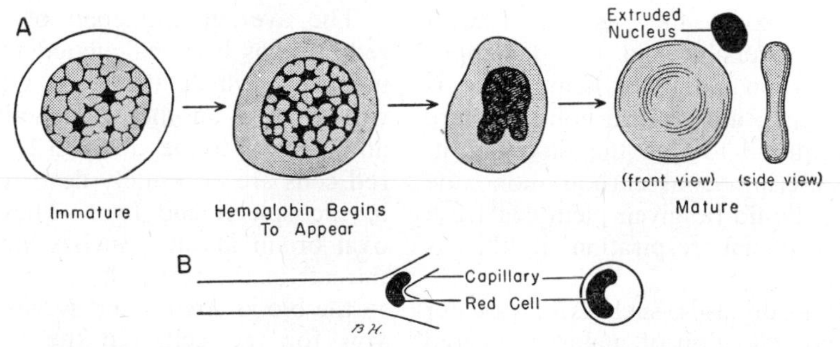

Figure 136. Red blood cells. *A,* Stages in the formation of human red cells; *B,* bending and twisting of red cells as they pass through capillaries. Their supporting elastic framework causes them to resume their normal disc shape after the tension is released.

atoms attaching to the iron atoms of the hemoglobin molecule. Hemoglobin unites with oxygen to form **oxyhemoglobin** in regions where oxygen is abundant, and releases oxygen in regions where oxygen is scarce. Because of the unique properties of the hemoglobin molecule, this substance not only transports oxygen from the lungs (gills in fish) to all the tissues of the body, but it plays a major role in the transport of carbon dioxide from tissues to the lungs and in the prevention of changes in the *p*H of the blood.

The red cells of the fetus contain a slightly different kind of hemoglobin, called **fetal hemoglobin.** This gradually disappears after birth and by twenty weeks has been replaced by the adult type. Cells containing fetal hemoglobin can take up and give off oxygen at lower oxygen tensions than adult cells. This may be important to the fetus, for as it develops in the uterus, it has less oxygen available than the adult has.

Carbon monoxide, present in commercial illuminating gas and in the exhaust gas of automobiles, forms a compound with hemoglobin; in fact, it has a much greater affinity for hemoglobin than oxygen does. Carbon monoxide and oxygen both are attached to hemoglobin by the iron atoms; thus, when hemoglobin has combined with carbon monoxide it is unable to combine with oxygen. When the air breathed in contains only 0.5 per cent carbon monoxide, more than half of the hemoglobin in the blood has combined with this gas and only half is left to transport oxygen. This has the same effect as the sudden loss of half of one's red cells. The union of hemoglobin with carbon monoxide, like its union with oxygen, is reversible but the decomposition of carbon monoxide hemoglobin is a slow process, and several hours in pure air are required to free the blood of it. People suffering from carbon monoxide poisoning should be given plenty of fresh air and artificial respiration if this is necessary.

In certain diseases, such as blackwater fever, a complication of malaria, the red blood cells are destroyed and the hemoglobin is released into the plasma. It is then excreted in the urine and gives the urine a dark color. Other diseases decrease the amount of hemoglobin in the blood, a condition known as anemia.

156. LIFE HISTORY OF RED CELLS

The red cells are constantly being destroyed and new ones are made, yet the total number remains remarkably constant. Red cells originate in the **red bone marrow,** which lies in the central hollow spaces of certain bones. Other bones contain yellow marrow, made of cells modified for the storage of fat. The red marrow consists of a network of connective tissue cells and of thousands of small blood vessels, from the lining of which develop the red cells. Normal cell division can occur, of course, only in nucleated cells, and the precursors of the red blood cells in the blood vessels of the marrow are unspecialized cells with a nucleus and no hemoglobin (Fig. 136). After its last division each cell is gradually transformed into a mature red corpuscle by a process which includes the loss of the nucleus, the manufacture of hemoglobin and the assumption of the biconcave disc shape. During the formation of red cells the blood vessels of the bone marrow are closed, preventing the passage of blood. When the cells are completely formed, the blood vessels open and the new cells are swept out into the circulation. There are thousands of blood vessels in the red marrow; at any given time some are open and others are closed, engaged in red cell production. In certain abnormal conditions, red cells are manufactured too rapidly and are released into the blood stream while still immature and nucleated.

The average life span of the human red cell has been calculated from experiments in which the cells were labeled with one or another radioactive isotope, and proves to be about 127 days. The red cells are eventually destroyed by cells in the spleen and liver. The spleen, an oval organ about 5 inches long, lying to the left of the stomach, is connected only to the blood stream and serves as a reservoir for red cells. In the walls of the blood vessels in the spleen and liver are cells which have the ability to engulf or

phagocytize red cells and thus destroy them. Presumably, only old, worn-out red cells are destroyed, but how these cells distinguish old red cells from new ones is unknown. The hemoglobin molecules of the old red cells are dismantled by the spleen and liver. The iron atoms are recovered and returned to the red bone marrow to be used in the synthesis of new molecules. The heme portion of the molecule undergoes chemical degradation and is excreted by the liver in the bile as **bile pigments.** These substances undergo further reactions by the bacteria in the intestine and pass from the body in the feces. The bile pigments are primarily responsible for the color of the feces; if the bile duct is blocked, by a gallstone, for example, the bile pigments cannot pass into the intestine and the feces are a grayish clay color.

From the total number of red cells in the body, and the average life span, it is simple to calculate that about 15 million red cells are made, and an equal number destroyed, each second throughout the day and night every day. The constancy of the number of red cells provides us with an excellent example of a **dynamic equilibrium.** Under normal circumstances the rates of formation of new cells and of destruction of old ones are just equal and the total number of red cells in the body remains constant. The rate of production of red cells is increased by any factor which decreases the amount of oxygen delivered to the tissues. The loss of red cells by hemorrhage decreases the capacity of the blood to transport oxygen and leads to increased red cell production. The stimulus is not simply the decreased concentration of red cells, for if a person with a normal number of red cells moves to a very high altitude for a few weeks the number of red cells will increase to 6 or 7 million per cubic millimeter. At high altitudes there is less oxygen in the air and consequently less oxygen is delivered to the tissues. The same increase can be produced experimentally at sea level by keeping animals in a chamber the atmosphere of which has a low oxygen content but a total pressure equal to that of air at sea level. The physiologic mechanism by which a decreased oxygen level in tissues leads to an increased rate of red cell production is not clear. It has been shown experimentally that lack of oxygen in the red bone marrow itself does not stimulate red cell production. Presumably, lack of oxygen in some other tissue induces the formation and release of some substance which is carried to the bone marrow and stimulates red cell production.

The synthesis of hemoglobin and the production of red cells are not necessarily correlated. A deficiency of iron, for example, decreases hemoglobin synthesis but red cells are produced at the normal rate, or even at an elevated rate in response to the stimulus of decreased delivery of oxygen to the tissues. The cells produced have less hemoglobin than normal (they are said to be hypochromic) and, of course, are less effective than normal in transporting oxygen.

The cells of the spleen and liver presumably prevent an oversupply of red cells by increasing the rate of red cell destruction. When the bone marrow, stimulated by the decreased number of red cells, has increased the rate of red cell production and has brought the number in the circulating blood back to normal, the stimulus for greater production (i.e., the decreased oxygenation of the tissues) is removed and the red bone marrow gradually returns to its normal rate of red cell production.

157. OXYGEN-CARRYING DEVICES IN OTHER ANIMALS

All other mammals have red cells similar to ours—non-nucleated, biconcave discs containing hemoglobin. Birds, reptiles, amphibians and fishes have oval-shaped red cells which contain hemoglobin but are nucleated. The red cells of the lower vertebrates are typically larger than those of mammals; a frog's red corpuscles are approximately 35 microns on the long axis.

Invertebrates have a variety of devices for oxygen transport. A few worms have blood cells containing hemoglobin, but others, such as the earthworm, have hemoglobin simply dissolved in the

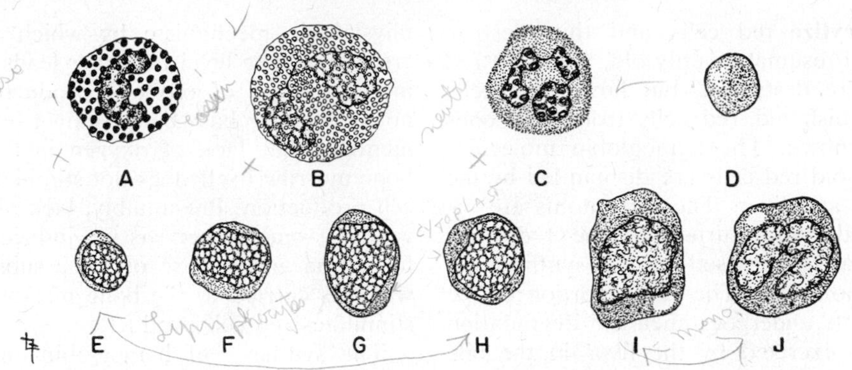

Figure 137. Types of white blood cells. *A,* basophil; *B,* eosinophil; *C,* neutrophil; *E–H,* a variety of lymphocytes; *I* and *J,* monocytes. *D,* is a red blood cell drawn to the same scale.

plasma. Other invertebrates have different blood pigments; crabs and lobsters, for example, have a blue-green blood pigment, **hemocyanin,** which contains copper instead of iron.

The respiratory enzymes of all cells, both plant and animal, the **cytochromes,** which catalyze the transfer of electrons from substrate to oxygen and the concomitant transfer of energy to ATP (p. 59), are hemeproteins closely related chemically to hemoglobin.

158. WHITE CORPUSCLES

There are five types of white blood cells, or **leukocytes,** all of which differ considerably from red cells: they have a nucleus, they have no hemoglobin and hence are colorless, and they move actively, by ameboid movement. Leukocytes can move against the current of the blood stream and even slip through the walls of blood vessels and enter the tissues.

White cells are much less numerous than red ones; there are about 7000 white cells per cubic millimeter on the average, but the number fluctuates from five to nine or ten thousand in different persons and even in the same person at different times of day. The white count is lowest early in the morning and highest in the afternoon. Poorly nourished persons have lower than normal white counts and lower resistance to infection and disease; a drop to 500 or fewer white cells per cubic millimeter is fatal. White cells are counted with the same hemacytometer used in red cell counts. The blood is diluted less (only ten-fold instead of two hundred-fold) and the diluting fluid contains acetic acid to destroy the red cells and gentian violet to stain the white cells and make them more easily visible.

The five types of white cells (Fig. 137) can be distinguished by spreading a drop of blood on a glass slide and staining the smear with Wright's stain or some similar dye.

The **lymphocytes,** the smallest of the white cells, are 8 to 10 microns in diameter. They have a large nucleus, either spherical or slightly indented, which is colored a dark blue-purple by Wright's stain. The small amount of cytoplasm forms a thin layer around the nucleus; it has no granules and stains a light robin's-egg blue. The lymphocytes are less motile than other white cells.

The largest white cells, the **monocytes,** range from 12 to 20 microns in diameter. Their nucleus stains a muddy purple and is considerably indented, with the shape of a bean or a horseshoe. There is a large amount of cytoplasm which is nongranular and stains slate-blue. Monocytes are motile and ingest bacteria and debris particles.

The **neutrophils, eosinophils** and **basophils** all have nuclei divided into from two to five lobes connected by threads of nuclear material, they all have conspicuous granules in the cytoplasm, and they are all between 9 and 12 microns in diameter. The nuclei of all three stain dark purple, but the cytoplasmic granules stain differently. All three types are motile and capable of ingesting bacteria, but neutrophils are more active than the other two. The nuclei of neutrophils have three to five lobes and their cytoplasm contains small granules which stain a pale lavender. The eosinophils have two-lobed or three-lobed nuclei and large granules which are stained bright red by the eosin dye in Wright's stain. The basophil nuclei are usually two-lobed and their cytoplasm contains large granules which stain a deep blue.

The proportion of white cells of each type is determined by a **differential white cell count.** A drop of blood is smeared thinly and evenly on a glass slide, stained with Wright's stain, and examined under the microscope. Several hundred white cells are counted and classified. The average values for a normal person are: 60 to 70 per cent neutrophils, 25 to 30 per cent lymphocytes, 5 to 10 per cent monocytes, 1 to 4 per cent eosinophils and 0.5 per cent basophils.

159. PROTECTIVE FUNCTIONS OF WHITE CELLS

The chief function of the white cells is to protect the body against disease organisms. Neutrophils and monocytes destroy invading bacteria by ingesting them (much as an ameba ingests a particle of food). The engulfing of a particle by a cell is called **phagocytosis.** The phagocytized bacteria are digested by enzymes secreted by the white cell. The white cell continues to ingest particles until it is killed by the accumulated breakdown products. Neutrophils have been observed to phagocytize 5 to 25 bacteria and monocytes as many as 100 bacteria before dying.

When bacteria enter the tissues of the body they destroy cells either by attacking them directly or by producing poisonous chemicals. The blood vessels of the affected region dilate and bring in an increased supply of blood, which causes the characteristic reddening and increased temperature known as **inflammation.** The walls of the blood vessels become more permeable, fluid enters the tissue from the blood stream, and swelling results. The white cells, and the neutrophils in particular, migrate through the walls of the blood vessels and phagocytize the invaders and the remains of any destroyed tissues. The aggregation of dead tissue cells, bacteria, and living and dead white cells forms a thick yellowish fluid called **pus.** The white cells are guided to points of infection by chemicals released by the invading organisms and by the inflamed tissues.

After the bacteria have been destroyed, the lost tissue is replaced. Some tissues have the ability to regenerate by the multiplication of neighboring cells; others have a very limited ability to regenerate and are replaced by connective tissue cells which secrete fibers to form **scar tissue.** The lymphocytes are believed to be active in this process, for they tend to accumulate in areas where healing is occurring. Lymphocytes grown outside the body in sterile media can become connective tissue cells; presumably this same process can occur within the body to facilitate the processes of repair.

The body has a second kind of protection against disease, the production of specific proteins, called **antibodies,** in response to the presence of some foreign substance, called an **antigen,** in the blood or tissues. The antibodies are produced by plasma cells, cells which resemble lymphocytes and which are located in the spleen, lymph nodes and wall of the digestive tract, and perhaps by the lymphocytes themselves.

The eosinophils increase greatly in numbers when the body is infected by an animal parasite such as the *Trichinella* worm which caused trichinosis (p. 409). These leukocytes presumably play some role in counteracting the effects of the parasites. Allergic conditions, such as hay fever, asthma and the response to foreign proteins, also are characterized by a great increase in circulating eosinophils. The injection of the adrenal cortical hormones, cortisone or hydrocortisone, causes within a few minutes a decrease in the number of eosinophils in the blood to half or less of the normal value.

The number of circulating white cells is increased by most infections; the white count may rise to 20,000 per cubic millimeter or more in appendicitis or pneumonia. Inflamed tissues are believed to liberate a substance ("leukocyte promoting factor") which passes via the blood to the bone marrow, where it stimulates the production and release of white cells, especially neutrophils. The number of white cells in the blood is a reflection of the severity of the infection, and successive white cell counts are useful in gauging a patient's recovery. Certain diseases are characterized by an increase in particular kinds of white cells, and a differential white cell count is helpful in diagnosis. The number of lymphocytes is increased in whooping cough and pernicious anemia, by living at high altitudes or in the tropics, by sun tan, and by chronic diseases such as tuberculosis. Typhoid fever and malaria usually effect an increase in the number of monocytes, and pneumonia, appendicitis and other acute bacterial infections typically increase the number of neutrophils. An increase in eosinophils occurs with infections of tapeworm, hookworm and other animal parasites, and in scarlet fever, asthma and some skin diseases.

160. THE LIFE HISTORY OF WHITE CELLS

The several types of white cells originate in different organs. The lymphocytes are formed in the spleen, tonsils and lymph nodes, monocytes are formed in the bone marrow and in the spleen, and the neutrophils, eosinophils and basophils are all formed in the red bone marrow. Although the white cells all contain nuclei, the circulating white cells do not undergo cell division. White cells are not destroyed by any particular organ. Some are killed by bacteria and others wander through the lining of the digestive tract or urinary system and are swept out of the body with the feces or urine. The life span of most white cells is quite short, about two to four days. The blood of a person subjected to intense gamma rays (as from an atomic explosion) loses all of its neutrophils in about three days. The life span of a lymphocyte is estimated to be even shorter, about four hours. This estimate is based on the fact that the number of lymphocytes entering the blood stream from the lymph ducts every 24 hours is several times greater than the total number present at any moment.

161. BLOOD PLATELETS

The blood platelets, a third type of formed element of the blood, are important in initiating the process of blood clotting. The platelets are colorless, spherical, non-nucleated bodies about one third the diameter of a red cell. Most of them, it is believed, originate by the fragmentation of giant cells in the red bone marrow, but recent experiments indicate that some are formed from phagocytic cells in the lungs. Their life span is estimated to be four days.

162. THE CLOTTING OF BLOOD

An elaborate mechanism has evolved to prevent the accidental loss of blood. In some animals, such as the crab, blood loss is prevented by the powerful contraction of muscles in the wall of the blood vessel when the vessel is severed. In man and other vertebrates, and in many invertebrates as well, blood loss is prevented by a series of chemical reactions in which a solid clot is formed to plug the broken vessel. Clotting is essentially a function of the plasma, not of the blood cells, and involves the conversion of soluble **fibrinogen** (one of the plasma proteins) to insoluble **fibrin.**

Blood drawn from a blood vessel into a test tube changes from a liquid to a semisolid gel in about six minutes (the clotting time ranges from four to ten minutes). The fibers of fibrin trap red and white cells, which contribute to the solidity of the clot but are not essential for the clotting process. The clot later shrinks and squeezes out a straw-colored fluid, **serum,** which is similar to plasma in most respects but cannot clot because it lacks fibrinogen.

Blood does not clot, as many people think, because it is exposed to air or because it stops flowing; if it is carefully drawn into a paraffin-lined vessel it will not clot even though it is stationary and exposed to air. The clotting mechanism is actually quite complex, involving many different substances in the plasma which interact in three groups of reactions. Each of the first two reactions produces an enzyme required for the succeeding reaction.

The first reaction in the clotting mechanism involves the disintegration of the blood platelets and the release of a substance called **thromboplastin.** Platelets, when exposed to a rough surface (almost any surface other than the smooth lining of the blood vessels), tend to adhere to it and disintegrate. The disintegration is greatly accelerated by a globulin called "antihemophilic factor" found in normal plasma.

The thromboplastin formed in the first reaction acts enzymatically in the second reaction to convert prothrombin to **thrombin.** This is a complicated reaction which requires calcium ions and at least two additional plasma proteins ("accelerator globulin" and "serum prothrombin conversion accelerator"). **Prothrombin** is a plasma globulin made by the liver. The role of calcium in this reaction is unknown, but if the calcium ions are removed by the addition of citrate or oxalate, clotting is prevented.

Finally, the thrombin formed in the second reaction acts enzymatically to convert fibrinogen to fibrin, a process in which a small part of the fibrinogen molecule is split off and the remainder is polymerized to form the long fibrin threads.

Although this mechanism is quite complex, it is evident that it is admirably adapted to provide for rapid clotting when a blood vessel is injured, yet to prevent clotting in the intact blood vessel. Even normal blood contains a small amount of thromboplastin, for there is a continual breakdown of a few blood platelets. Normal blood also contains a strong anticoagulant called **heparin,** produced in mast cells in the lungs and liver. Heparin prevents the conversion of prothrombin to thrombin and is used clinically to prevent clotting.

Diseases of the liver may lead to defective clotting by interfering with the synthesis of prothrombin. An adequate supply of vitamin K is required for the synthesis of prothrombin. The absorption of vitamin K from the intestine is facilitated by bile; a deficiency of bile may produce a deficiency of prothrombin and lead to defective clotting even though the diet contains an adequate supply of vitamin K. To prevent internal hemorrhages during operations on the liver or bile duct, it is customary to treat the patient previously with injections of vitamin K.

The blood which flows during menstruation does not coagulate, either because its fibrinogen has been removed in the uterus, or because it has already clotted and the fibrin has subsequently been destroyed by proteolytic enzymes.

Some bacteria attack the walls of blood vessels, and the clotting chemicals aid in the repair of such weak spots. The destruction of the cells, and the accumulation and disintegration of the platelets on the roughened wall, liberate thromboplastin and cause the formation of a clot across the weakened area. The clot alone is not very effective, but connective tissue cells from the arterial wall migrate in and secrete a tough, fibrous connective tissue to form a scar and reinforce the clot. This is not without danger, for the clot may completely occlude the blood vessel and prevent the passage of blood. Many organs are supplied by several blood vessels and in these an intravascular clot, called a **thrombus,** is not too dangerous. Other organs, served by a single artery, will have no blood supply at all if that artery is occluded by a thrombus. Sometimes a thrombus formed in one vessel breaks loose and is carried by the blood to some other vessel, which it obstructs. A thrombus or any other particle carried by the blood stream which blocks a blood vessel is called an **embolus.**

The hereditary disease **hemophilia,** a sex-linked trait (p. 478) which affects males primarily and is transmitted to them from their mothers, is characterized by clotting that is so defective that a slight scratch may lead to a fatal hemorrhage. The disease attracted wide attention because it was present in several European royal families, notably the Bourbons and Romanoffs, and was apparently inherited from Queen Victoria of England. Hemophiliacs lack the globulin, "antihemophilic factor," which accelerates the disintegration of platelets and the release of thromboplastin. Their platelets are unusually stable and do not readily disintegrate when blood is shed.

The clotting property of plasma has made it an important tool in tissue culture research. After several unsuccessful attempts to grow tissues outside the body, Dr. Ross Harrison succeeded in 1910 when he used a drop of clotted plasma as a medium for his tissue to grow upon. This discovery provided a technique by means of which much research has been done on the growth, development and fundamental properties of tissues, bacteria and viruses.

The clotting process may be summarized in the following three steps:

of oxygen, and thereby impair the metabolism of all the tissues. In addition, the decreased number of red cells leads to a decreased viscosity of the blood, and this, indirectly, causes the heart to beat faster. The increased work load on the heart is one of the major ill effects of anemia.

Anemia may also result from injury to the bone marrow, liver or spleen. The marrow tissue may be destroyed by tumors, or by substances such as benzol or lead; industrial workers continually exposed to these substances may develop anemia. A fourth cause of anemia is a deficiency of some substance essential to the manufacture of red cells—iron, **cobalamin** (vitamin B_{12}) or folic acid. Pregnant women frequently become anemic because of the demands of the fetus upon

$$\text{(1) Platelets disintegrate} \xrightarrow[\text{antihemophilic factor}]{\text{release}} \text{Thromboplastin}$$

$$\text{(2) Prothrombin} \xrightarrow[\text{calcium, Ac. Glob., SPCA}]{\text{thromboplastin}} \text{Thrombin}$$

$$\text{(3) Fibrinogen} \xrightarrow{\text{thrombin}} \text{Fibrin}$$

163. DISEASES OF THE BLOOD

Anemia. Anemia is not a single, specific disease, but rather a condition which may have many different causes. It is characterized by a decrease in the number of red cells in the blood, by a decrease in the amount of hemoglobin per red cell, or by both. The number of red cells may drop to 4, 3, or even one million per cubic millimeter of blood. Anemia may result from severe loss of blood by hemorrhage or from the destruction of red cells. There are several inherited conditions (e.g., sickle cell anemia) in which the red cells are unusually fragile and are rapidly destroyed. Red cells are destroyed by rattlesnake venom, malaria, burns and certain chemicals. The bone marrow is stimulated to increase red cell production and releases immature, nucleated red cells into the blood stream. Anemias harm the patient by interfering with the transport

the mother's supply of iron and vitamins for red cell manufacture. The fetus must store up enough iron to carry it through the first year or so of life. Milk is quite deficient in iron, and a prolonged diet of milk will produce anemia. Most of the iron of hemoglobin is salvaged and reused; the daily requirement of iron, about 0.01 gm., is small and easily supplied by the average diet.

A severe and formerly fatal disease is **pernicious anemia,** characterized by red cells that are immature, very fragile, and decreased in number. The bone marrow of these patients does not receive enough cobalamin, vitamin B_{12}, to make normal red cells. Cobalamin is present in the diet in adequate amounts but the lining of the stomach of these patients does not secrete the substance called "intrinsic factor" which is necessary for the absorption of cobalamin. In experiments in which dogs were first made anemic by repeated bleeding, and then fed various diets, Dr. Whipple of the University of Rochester noticed that liver was most effective in promoting recovery. Physicians at Harvard then tried feeding liver to patients with pernicious anemia and found that the red cell count began to rise after a few days and reached the normal level in a few weeks. It remained normal as long as the patients ate large amounts of liver. Potent liver extracts were subsequently prepared, and in 1948 the active substance, cobalamin, which is an organic compound containing the element cobalt, was isolated and identified. The primary defect in patients with pernicious anemia is the inability of the lining of the stomach to secrete enough intrinsic factor to absorb cobalamin effectively at the low level at which it is normally present in the diet. However, if cobalamin is injected, or given in the diet in large amounts, red cell production proceeds normally.

Polycythemia. An increase in the number of circulating red cells—the number may reach 11 to 15 million per cubic millimeter—is called **polycythemia.** Diarrhea, by decreasing the fluids of the body, and hence the total blood volume, leads to a temporary increase in the number of red cells *per cubic millimeter,* for the total number of red cells remains the same. True polycythemia results from an overproduction of red cells; the blood becomes very viscous and tends to plug the blood vessels.

Leukemia. The leukemias are diseases of the cells which produce white corpuscles. They multiply too rapidly, and release large numbers of cells into the blood stream, many of which are immature cells. The different types of leukemia are characterized by increases in particular kinds of white cells. Leukemia is a type of cancer, characterized by the abnormally rapid growth of one kind of cell. It is treated, as other cancers are, by x-rays or by the radiation from a radioactive element such as radiophosphorus, or by antivitamins such as **aminopterin,** which is an antagonist of folic acid. The leukemic cells, by filling up the bone marrow and displacing the normal cells which produce red cells, frequently cause anemia.

164. MEDICOLEGAL TESTS FOR BLOOD

As all detective story fans know, "suspicious, brown stains" can be analyzed and proved to be blood, even proved to be human blood of a particular type. In the simplest test, some benzidine (a common organic reagent) and hydrogen peroxide (the familiar bleaching agent) are added to the stain. If hemoglobin is present, the stain turns green. This test merely identifies hemoglobin and does not distinguish human from other blood. A very sensitive test, which can detect one part of blood in the presence of more than a million parts of other materials, uses 3-aminophthalhydrazide. A blue phosphorescent light is produced when this substance, sodium hydroxide, and hydrogen peroxide are added to a blood stain. Even after blood has apparently been removed by repeated washings, this test is effective because of the minute amount of hemoglobin which invariably remains.

Human blood may be distinguished from the blood of other animals by the **precipitin test,** which depends upon an antigen-antibody reaction. An **antigen** is defined as any foreign protein (one not normally present) which when introduced into the blood stream stimulates the production of an **antibody,** a protein which will react specifically with it. To prepare for the precipitin test, human serum, which contains one or more proteins characteristic of man which act as antigens, is injected into a rabbit in divided doses over a period of time until about 30 ml. have been injected. The plasma cells and lymphocytes of the rabbit are stimulated to produce antibodies which will react specifically with human serum. After about five days the rabbit is bled, the serum is separated and kept until needed. To test for human blood, a solution containing the suspected material is made and mixed with the rabbit serum known to contain antibodies for human serum. If the suspected material contains human blood, and therefore human serum antigens, the antibodies and antigens react and cause a precipitation which is evident in one to three hours. If the blood is from an animal closely related to man, a chimpanzee for example, a slight precipitation will result; the blood of more distant relatives causes less precipitation or none at all.

165. BLOOD TYPES AND TRANSFUSIONS

Attempts to replace human blood lost by hemorrhage date back to 1667, when the transfer of animal blood into human veins was tried. Such transfusions were uniformly unsuccessful and evoked severe reactions and often death. The transfusion of blood from one person to another was sometimes successful but occasionally led to the **agglutination** (clumping) of the patient's red cells. Agglutination, caused by an antigen-antibody reaction, must not be confused with coagulation, the clotting of blood caused by the thrombin and fibrinogen reaction.

The mystery of why transfusions were sometimes unsuccessful was solved in 1900 by Landsteiner, who found that the blood of different persons may differ chemically and that agglutination occurs when the bloods of donor and of patient are incompatible because of these chemicals. There are four chief blood groups, designated **O, A, B** and **AB,** and distinguished by the presence of **agglutinogens** (a type of antigen) A or B in the red cells and by **agglutinins** (a type of antibody) a or b in the plasma. The characteristics of the blood groups and the types of transfusions possible are summarized in Table 3. Normally, of course, no blood agglutinates itself, because the corresponding agglutinogen and agglutinin (A and a, for example) are not present together.

In making a transfusion the physician tries to find a donor of the same group as the patient and tests the two bloods by mixing some serum from the patient with some red cells from the prospective donor to make sure that they are compatible. In addition to the four main groups there are two subgroups to be considered. In an emergency, if no donor of the patient's type is available, blood of another type may be used, provided that the plasma of the patient does not agglutinate the red cells of the donor. For example, a type B individual may receive blood from an O donor, because the O red corpuscles have no agglutinogens and will not be clumped by any kind of plasma. The O plasma contains both a and b agglutinins, but in the transfusion they are gradually diluted by the patient's plasma and do not agglutinate the patient's red cells. Type O people are called "universal donors" because their blood can be given to any person, and type AB people are known as "universal recipients," because they can receive blood of any type. The blood groups are inherited (see p. 476), and remain constant throughout life.

A blood sample is typed by mixing it with serum from persons of known blood types. Type A serum contains the b agglutinin, and will cause the clumping of any red cells containing B agglutinogens. A clean glass slide is marked with two circles, labeled A and B; a drop of serum from a person known to have type A blood is placed in one circle and a drop of type B serum in the other. Then drops of blood from the person being tested are added to each kind of serum and mixed. Clumping, if it is to occur, is evident to the naked eye in two to three minutes. By this method it is even possible to detect the type of blood present in a small stain.

Practically all races of man have been typed and some notable differences have been found. No race is characterized by one particular blood type; racial differen-

Table 3. THE HUMAN BLOOD GROUPS

BLOOD GROUP	AGGLUTINOGEN IN RED CELLS	AGGLUTININ IN PLASMA	CAN GIVE BLOOD TO GROUPS	CAN RECEIVE BLOOD FROM GROUPS
O................	None	a and b	O, A, B, AB	O
A................	A	b	A, AB	O, A
B................	B	a	B, AB	O, B
AB...............	A and B	None	AB	O, A, B, AB

ces lie in the relative *frequency* of the different types. Because the blood types are inherited and yet there is no conscious selection of mates for certain blood types, information about them has been useful in anthropology. The proportion of blood types in a population remains constant from one generation to the next as long as there is no intermarriage with other groups. Gypsies, who originally came from India, have lived in Hungary for several hundred years, but because there is little mating between them and the native Hungarians, the proportion of blood types among the Gypsies is similar to that of the Hindus and quite different from that of the Hungarians. The Germans who migrated to Hungary about 1700 have kept, in addition to their German language and customs, a blood type frequency characteristic of the Germans in Germany. With the development of techniques for determining the blood types of mummies, and even of skeletons, the use of blood tests in anthropology has been considerably broadened. Candela tested the blood types of thirty prehistoric Aleutian mummies and found eight of them to be B or AB. Previously the Aleuts had been thought to have been derived from an Eskimo-Indian cross, but both Eskimos and Indians have relatively low frequencies of B and AB groups, whereas B is common in Asia. These findings, indicating that the Aleuts were of Asiatic origin, have been substantiated by other evidence.

The red cells contain, in addition to the A and B agglutinogens, a second pair known as **M** and **N,** which are inherited independently of the A and B pair. These, together with other, less important blood types, provide additional means of identifying blood.

A third set of hereditary factors causes the presence or absence of another agglutinogen, the **Rh factor,** so called because it was first found in the blood of rhesus monkeys. Aoubt 85 per cent of white people are Rh positive, i.e., have the Rh antigen in their red cells, and 15 per cent are Rh negative, with no Rh antigen. If a woman is Rh negative and her husband Rh positive, the fetus may be Rh positive, having inherited the factor from its father. Blood from the fetus may pass through some defect in the placenta into the maternal blood stream and stimulate the formation of antibodies to the Rh factor by the white cells of the mother. Then, when this woman becomes pregnant a second time, some of these antibodies may pass through the placenta into the child's blood stream and cause the clumping of its red cells, a condition known as *erythroblastosis fetalis.* In extreme cases so many red cells are destroyed that the fetus dies before birth; more frequently it is born alive but dies after birth. Newborn children with erythroblastosis fetalis are now given massive blood transfusions to replace essentially all of their red cells.

Blood groups have been demonstrated in a host of mammals and birds: apes, monkeys, sheep, pigs, horses, dogs, cats, rabbits, rats, mice, chickens and pigeons. The O, A, B and AB groups have been demonstrated in chimpanzees, orang-utans and gorillas, which indicates that these blood group substances arose before the primates had finally evolved into different types. Substances similar to, but not identical with, the human A and B substances have been found in the bloods of many mammals and birds. M and N agglutinogens, similar to those of man, have been found in chimpanzees, orang-utans and some monkeys, but not in any other animal. The M and N substances of the chimpanzee are most like those of man, confirming the generally accepted idea that chimpanzees are the most manlike of the anthropoid apes.

The transfusion of whole blood, plasma or plasma fractions is extremely important in saving lives. In transfusing whole blood, after blood has been drawn from a suitable donor it must be treated, usually with sodium citrate (which binds the calcium ions) to prevent coagulation, care must be taken to prevent infection, and the blood must be introduced into the recipient's veins at the proper temperature and rate. The heart will be overloaded if blood is transfused too rapidly. It is now possible to store blood in a "blood bank" for as much as thirty days, by adding citric acid and glucose and keeping the blood cool, at a temperature of 4 to 6°C. By separating the red cells from the plasma, and then suspending the cells in purified albumin, cells can be kept for as much as three months.

The search for substitutes for whole blood to be used in transfusions dates back to 1878, when cow's milk was tried. Plasma and certain plasma fractions, which can be stored much longer than whole blood, are effective substitutes for whole blood in many clinical conditions such as shock. Dried plasma or plasma fractions, prepared by freezing and drying and placed in a sealed sterile container, can be kept even without refrigeration for a long time. To be used, the plasma is mixed with the proper quantity of sterile distilled water and injected. In preparing plasma, bloods from sixteen different people of assorted blood types are pooled so that the different agglutinins are diluted below their effective concentration and will not agglutinate the red cells of the recipient.

Recently, the polysaccharide **dextran,** a glucose polymer made by bacteria, has been widely used as a blood substitute. It can be prepared inexpensively in large amounts, does not cause agglutination of red cells, gives fewer toxic reactions than any other substitute tried, and eliminates the possibility always present in the transfusion of blood or plasma of transmitting the virus of serum hepatitis. Gelatin, pectin, gums and albumin from cow's blood have all been tried as substitutes but none has been very satisfactory.

QUESTIONS

1. Name seven functions of blood.
2. How are these functions carried on in animals which do not possess a circulatory system?
3. Distinguish between plasma, serum, lymph and tissue fluid.
4. How would you measure the volume of blood in a man's body (without draining it out)?
5. What is oxyhemoglobin? What is its significance in the body?
6. Why is carbon monoxide poisonous?
7. Where and how are red cells manufactured? What happens to old red cells?
8. Why does the number of red cells increase at high altitudes? (Note that this "why" refers to the physiologic mechanism involved, not to a teleologic explanation of any possible advantage to the animal of having more red cells at higher altitudes.)
9. What are the chief structural and functional differences between white blood cells and red blood cells?
10. Describe how clotting occurs. What prevents it from occurring at any time within the blood vessels?
11. What is agglutination?
12. What substances may be used in place of whole blood in transfusions?
13. Citric acid forms a tight union with calcium ions so that the calcium is unable to react with other substances. In view of this fact, why is sodium citrate added to whole blood immediately after it is drawn from a blood donor?

SUPPLEMENTARY READING

An interesting account of the discovery of the blood groups and their use in anthropologic research is given in Wiener's *Blood Groups and Transfusions.* An interesting but fairly technical discussion of the chemical constituents of the blood and their uses is given in *Blood Cells and Plasma Proteins* edited by James Tullis.

Chapter 17

The Circulatory System

THE CIRCULATORY system is frequently called the "transportation system" because it carries food and oxygen to all the tissues of the body, removes the waste products of metabolism, carries hormones from endocrine glands to their target organs, and equalizes body temperatures. It includes the heart, the blood vessels and the lymph vessels, in addition to the blood, lymph and tissue fluid. To understand how the system operates as an integrated unit, we must first study the structure and function of each of the organs involved.

166. THE BLOOD VESSELS

There are three types of blood vessels: **arteries, veins** and **capillaries.** Arteries and veins are distinguished from each other by the direction of the flow of blood in them, not by the kind of blood (aerated or nonaerated) that they contain. Arteries carry blood from the heart *to* the tissues of the body; veins return blood *from* the tissues to the heart. Capillaries are microscopic vessels located in the tissues, connecting the arteries and veins. Their chief function is to effect the exchange of food, gases and wastes between the blood and the tissues. They have extremely thin walls consisting of a single layer of cells, the **endothelium,** which is continuous with the endothelial lining of the artery and vein on either side (Fig. 138, *C*). Some capillaries are so small that the red corpuscles are bent in passing through them.

The walls of arteries and veins are too thick to permit diffusion; they have three distinct layers: an **outer** coat of connective tissue, a **middle** coat of smooth muscle cells, and an **inner** coat of endothelium and connective tissue (Fig. 138, *A*). The outer coat contains fibrous tissue which makes the artery tough and resistant to internal pressures, while permitting it to expand and contract each time the heart beats. The smooth muscle in the middle layer regulates the amount of blood going to a particular organ by contracting or relaxing, and so decreasing or increasing the size of the lumen (cavity) of the artery. In addition to the endothelial lining, the inner coat of most arteries contains a strong **internal elastic membrane** to give additional strength to the walls. The arterial walls are supplied with two sets of nerves; impulses carried by one set cause the smooth muscles to contract, those carried by the other set effect a relaxation. The largest artery, the **aorta,** is about an inch in diameter near the heart, and has a wall about 1/8 inch thick.

The walls of veins are much lighter and thinner than those of arteries, but the same three coats are present (Fig. 138, *B*).

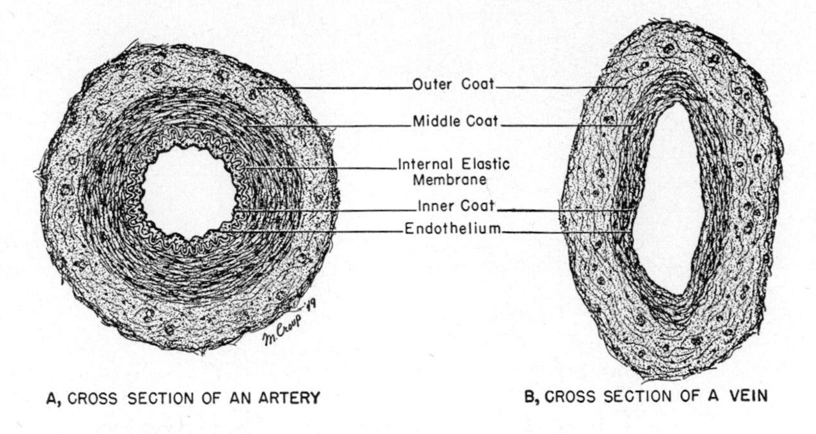

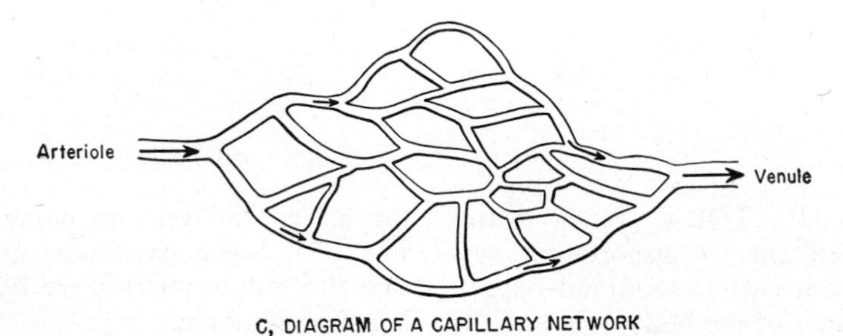

Figure 138. Arteries and veins compared.

The outer one of connective tissue has fewer elastic fibers, the middle muscular layer is thinner than the corresponding arterial layer, and in most veins there is no internal elastic membrane. Veins, but not arteries, are supplied with valves along their length to prevent the back-flow of blood.

The cells of the body are surrounded by and bathed in a liquid called **tissue fluid,** and blood does not come in direct contact with them. Instead, substances must diffuse from the blood, through the wall of a capillary, and across the space filled with tissue fluid to get to the cells (Fig. 139). Professor A. Baird Hastings characterizes this tissue fluid, plus the lymph and blood plasma, as "the sea within us." An adult man has about 10^{15} cells, bathed by only 14 liters of fluid. An equal number of single-celled protozoa living in the sea would require 10,000,000 liters of sea water to provide them with the gases and nutrients they need. The efficient mechanisms—the lungs, liver, intestines and kidneys—which continually replenish the oxygen and food stuffs of these fluids and remove their wastes enable the body to survive even though it has relatively little water.

The fact that the capillaries are small means that each drop of blood passing through a capillary network is exposed to a large surface area through which diffusion may occur. It has been estimated that each cubic centimeter of blood is exposed to 7000 square cm. or about 8 square feet of capillary surface. The number of capillaries in the body is almost beyond calculation. In tissues such as muscle, with a high metabolic rate, they are close together, the distance between adjacent ones being about twice the diameter of the capillary (Fig. 139). One investigator places the number of capillaries in muscle tissue at about 1,500,000 per square inch. Less active tissues are not so well supplied—fatty tissue has few capillaries, and the lens of the eye has none at all. Ordinarily, only a fraction of the capillaries of any organ are filled with blood and functioning, but during periods of intense activity, all, or nearly all, the capillary networks are full.

167. THE HEART

The heart is a powerful muscular organ located in the chest directly under the breast bone. Its walls are made of cardiac muscle tissue held together by strands of connective tissue. Enclosing it is a tough, connective tissue sac (the pericardium). The inner surface of this sac and the outer surface of the heart are covered by a smooth layer of epithelial-like cells, and the cavity contains a fluid which reduces friction to a minimum as the heart beats.

The muscle fibers branch and fuse to form a complex network of protoplasmic connections throughout the heart wall, across which nerve impulses can travel. Because of this, the heart contraction follows the "All-or-None" Law (see p. 337); that is, if a nerve impulse is sufficiently strong to make the heart beat at all, it responds with a maximal contraction. The heart and all the blood vessels are lined with a layer of smooth, thin, flattened cells, the **endothelium,** which prevents blood from clotting within the circulatory system. Any disease or injury of the endothelium which roughens it may cause the formation of a thrombus within the vessel.

The heart of man and other mammals and of birds is divided into four chambers (Fig. 140), the upper right and left

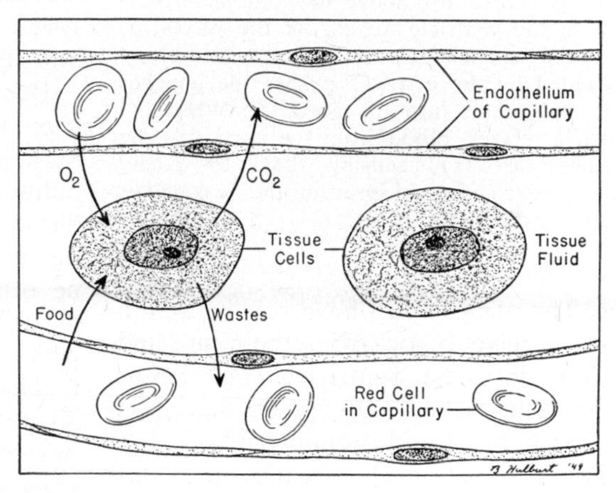

Figure 139. Diagram of the diffusion of materials between capillaries and the cells of the body, by way of the tissue fluid which bathes the cells.

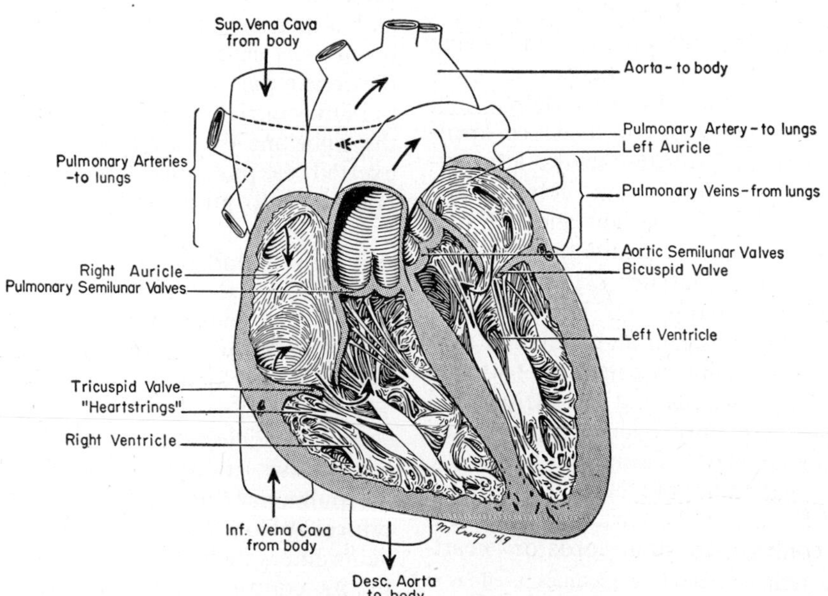

Figure 140. Diagram of the heart, showing chambers, valves and connecting vessels.

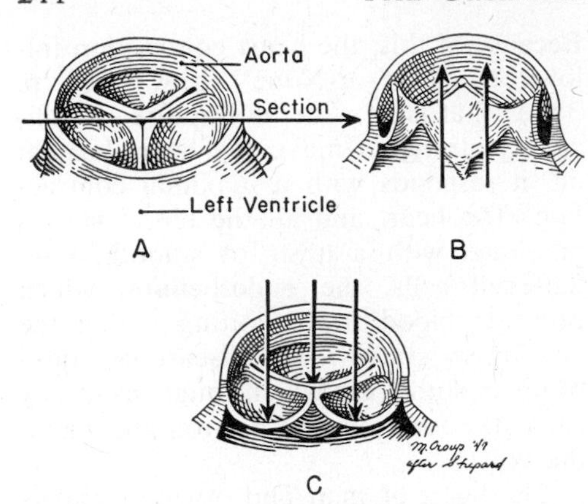

Figure 141. Diagram of the operation of the semilunar valves. *A,* Arrangement of the three pouches of the semilunar valves. The aorta has been cut across just above its point of attachment to the ventricle, to expose the valves. *B,* When the ventricle contracts, the expelled blood (indicated by the arrows) pushes the pouches aside and passes into the aorta. *C,* When the ventricle relaxes, blood from the aorta fills the pouches (arrows), causing them to extend across the cavity and prevent the leakage of blood back into the heart.

atria*, and the lower right and left **ventricles.** The atria, which have relatively thin walls, receive blood from the veins and push it into the ventricles. The latter, with much thicker walls, pump the blood out of the heart and around the body.

For its function as a pump, the heart is supplied with valves that close automatically to prevent blood from flowing in the wrong direction. There is an opening for blood to pass from the right atrium to the right ventricle, and a second opening from the left atrium to the left ventricle, but no connection between the right and left atria or right and left ventricles. The heart is really two separate pumps, then, sometimes called the right heart and the left heart. The valve between the right atrium and ventricle, with three flaps or cusps, is called the **tricuspid** valve. That between the left atrium and ventricle, with only two cusps, is called the **bicuspid** valve. These two valves are held in place and prevented from being pushed back into the atria when the ventricles contract, by stout cords or "heart-strings" attached to the valves and the walls of the ventricles (Fig. 140). At the bases of the two big arteries—the pulmonary and the aorta—which leave the right and left ventricles, respectively, are additional flaps of connective tissue, the two **semilunar** valves, so called because of their half-moon shape. These are really pouches opening away from the heart. When blood passes in the proper direction, the pouches are pushed aside and offer no resistance. But when the ventricles are relaxing and filling with blood, and the blood pressure in the arteries is higher than that in the ventricles, blood fills the pouches, causing them to stretch across the cavity of the pulmonary artery or aorta, and thus prevent a leakage back into the heart (Fig. 141).

* The term auricle is sometimes used as a synonym of atrium, but strictly speaking the auricle is only one part of the atrium, the lateral pouchlike appendage.

Because there is no valve at the openings of the large veins leading into the right atrium, or at the openings of the pulmonary veins into the left atrium, some blood is forced back into the veins when the atria contract. This is less than it might be otherwise because of rings of muscle tissue around the veins in that region, which contract just before the atria do.

The right atrium receives blood from all parts of the body (except the lungs) by way of two large veins: the **superior vena cava,** which drains the head, arms and upper part of the body, and the **inferior vena cava,** which drains the legs and the lower part of the body. The walls of the left ventricle are thicker than those of the right one because more force must be exerted to push the blood through the body than to push it to the lungs.

The path of blood through the heart may be summarized as follows (Fig. 140): Blood from the body enters the right atrium, the contraction of which pushes open the flaps of the tricuspid valve, pumping the blood into the right ventricle. The contraction of the right ventricle then closes the tricuspid valve, opens the semilunar valve, and pushes the blood out via the pulmonary artery to the lungs. Blood returning from the lungs in the pulmonary veins enters the left atrium, and is pumped by its contraction through the bicuspid valve into the left ventricle. The contraction of the left ventricle closes the bicuspid valve, opens the semilunar valve, and sends the blood out the aorta to all parts of the body. Every drop of blood entering the right atrium must go through the lungs before it can get into the left ventricle and be pumped around the body.

The heart muscle is not, as one might assume, nourished by the blood within its chambers, because its walls are too thick for food and oxygen to diffuse through. Instead, it is supplied with blood by arteries which branch off the aorta at the point where that vessel leaves the heart and ramify through the heart muscle. Veins from the heart tissue drain into the right atrium, except for a few small veins which drain from the heart wall directly into the ventricle.

The Heart Beat. Beating is an inherent capacity of the heart, exhibited early in embryonic development and continuing without pause throughout life. Since all tissues need the constant supply of oxygen provided by the circulating blood, unconsciousness results if the heart stops beating for a few seconds, and death ensues if it stops for a few minutes. The heart of a resting human pumps about 5 liters of blood per minute, or about 75 ml. per beat. This means that a quantity of blood equal to the total amount contained in the body passes through the heart each minute. Not all the blood actually goes through the heart once a minute: some of the blood on the shorter circuits will have been through more than once, while that on the longer circuits will not yet have passed through. In man's three score and ten years his heart beats about 2,600,000,000 times, and pumps at least 155,000,000 liters (about 150,000 tons) of blood. Estimates of the work done by the heart indicate that it is enough to raise a weight of 10 tons to a height of 10 miles—a truly remarkable performance for an organ that weighs approximately 11 ounces! During exercise both the number of beats per minute and the amount of blood pumped per beat are greatly increased. Physical training enables the heart to increase its volume per beat, making it possible for the athlete to increase the total blood pumped, without as great an increase in the heart rate as would be necessary in an untrained person.

The fact that the heart continues to beat normally after its nerves have been severed shows that it is not dependent upon stimuli from the brain. If kept in the proper liquid medium, it will beat even when entirely removed from the body. Even a few muscle fibers dissected from the heart retain this ability. The rate of the fundamental, innate contraction is regulated by a number of factors—the **nodal tissue** within the heart, and two sets of nerves from the brain.

Nodal Tissue. Nodal tissue, which is peculiar to the heart, instigates and regulates its beat. It has some of the properties of muscle and some of nervous tissue. In the lower vertebrates, such as fish and frogs, there is a separate chamber of the heart, the **sinus venosus,** into which the veins empty, and which in turn passes blood to the right atrium. In higher forms this has disappeared except for a vestigial mass of nodal tissue called the **sino-atrial node,** or S.A. node, located at the point where the superior vena cava empties into the right atrium (Fig. 142). A second node, lying between the atria just above the ventricles, is known as the **atrioventricular,** or A.V., node. From the latter a bundle of branching fibers passes down and invades all parts of the ventricles. Investigators have demonstrated that the sino-atrial node initiates the heart beat and regulates its rate of contraction. For this reason it is called the **pacemaker.** At regular intervals a wave of contraction spreads from the sino-atrial node through the atrial muscle. When it arrives at the atrioventricular node, the impulse is transmitted to the ventricles by the bundle of nodal tissue. There is no muscular connection between the atria and ventricles; their beating is correlated solely by the specialized nodal tissue, which conducts impulses about ten times as fast as ordinary muscle does (the rates are approximately 5 meters, and 0.5 meter per second, respectively). Conduction by the nodal tissue insures that all parts of the ventricle contract almost simultaneously. If conduction in the ventricles were by

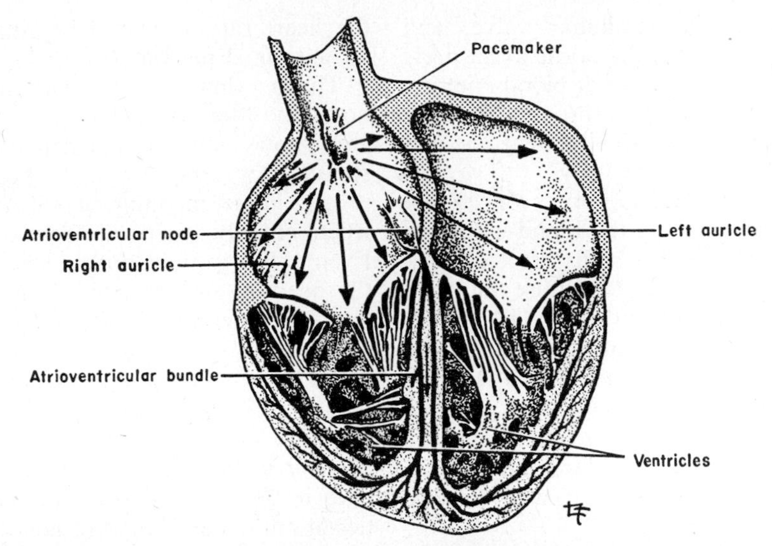

Figure 142. Diagram of the heart showing the location of the pacemaker, the atrioventricular node and the atrioventricular bundle which regulate and coordinate the beating of the parts of the heart.

way of ordinary muscle tissue, the muscles near the base of the ventricles would contract first, causing the uncontracted apex to bulge, and possibly injuring it.

That the sino-atrial node regulates the rate of the heart beat is demonstrated by the fact that warming the node results in an increased rate, whereas cooling it causes a decreased rate. Heat and cold usually have similar effects on other physiologic reactions, but heating or cooling other parts of the heart does not affect the rate of its beat. The acceleration of the beat accompanying fever is caused by the stimulation of the sino-atrial node by the warmer blood. When the sino-atrial node is destroyed by injury or disease, the atrioventricular node takes over its function as pacemaker.

Each heart beat consists of a contraction, or **systole,** of the heart muscle followed by its relaxation, or **diastole.** At the normal rate of seventy beats per minute, each complete beat occupies about 0.85 second. The atria and ventricles do not contract simultaneously: atrial systole occurs first, occupying about 0.15 second, followed by ventricular systole, which takes approximately 0.30 second. During the other 0.40 second all chambers rest in the relaxed state. Because the sino-atrial node, where the wave of contraction originates, is located in the right atrium, this contracts a little before the left atrium does.

The Heart Cycle. The action of the heart in pumping blood follows a cyclic pattern. The successive stages of the cycle, beginning with atrial systole, are as follows:

(*a*) Atrial systole. At this time (Fig. 143, *A*) a wave of contraction, stimulated by the sino-atrial node, spreads over the atria, forcing blood into the ventricles. The ventricles are already partly filled with blood, owing to the fact that the pressure is lower there than in the atria, and the tricuspid and bicuspid valves are open. The conduction of the impulse through the atrioventricular node is slower than in other parts of the nodal tissue, thus accounting for the brief pause after atrial systole before ventricular systole begins.

(*b*) Beginning of ventricular systole (Fig. 143, *B*). The muscles of the ventricular wall, stimulated by the impulse carried by the bundle of nodal tissue from the atrioventricular node, begin to contract, causing a rapid increase of pressure in the ventricles. The bicuspid and tricuspid valves close immediately, producing part of the first heart sound.

(*c*) Period of rising pressure (Fig.

143, *C*). The pressure in the ventricles mounts rapidly, but until it equals the pressure within the arteries, the semilunar valves remain closed and no blood flows into or out of the ventricles.

(*d*) The semilunar valves open (Fig. 143, *D*). When the intraventricular pressure exceeds that in the arteries, the semilunar valves open and blood spurts into the pulmonary artery and aorta. As the ventricles complete their contraction, blood is ejected more slowly until it finally stops. Ventricular diastole then follows.

(*e*) Beginning of ventricular diastole (Fig. 143, *E*). As the ventricles relax, the pressure within them decreases until it is less than the pressure within the arteries, and the semilunar valves snap shut. This causes the second heart sound.

(*f*) Period of falling pressure (Fig. 143, *F*). After the semilunar valves close,

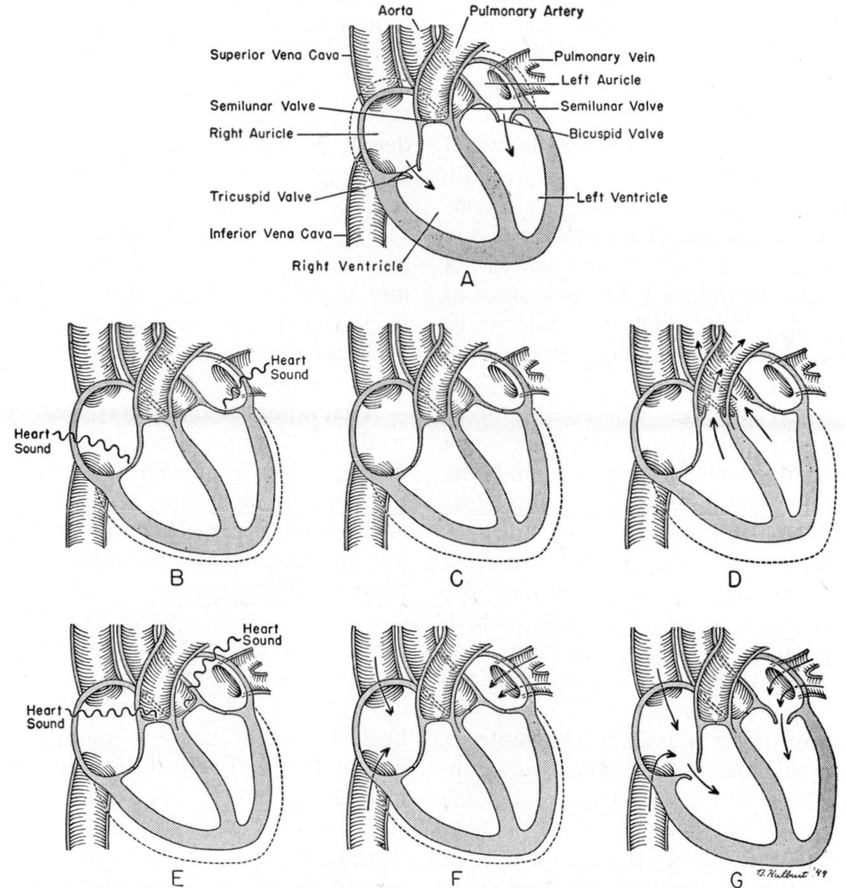

Figure 143. The heart cycle. Arrows indicate the direction of the flow of blood; dotted lines indicate the change in size as contraction occurs. *A,* Atrial systole; atria contract, blood is pushed through the open tricuspid and bicuspid valves into the ventricles. The semilunar valves are closed. *B,* Beginning of ventricular systole; ventricles begin to contract, pressure within ventricles increases and closes the tricuspid and bicuspid valves, causing the first heart sound. *C,* Period of rising pressure. Until the pressure within the ventricles equals that in the arteries, the semilunar valves remain closed, and no blood flows in or out of the ventricles. *D,* The semilunar valves open when the pressure within the ventricles exceeds that in the arteries, and blood spurts into the aorta and pulmonary artery. *E,* Beginning of ventricular diastole. When the pressure in the relaxing ventricles drops below that in the arteries, the semilunar valves snap shut, causing the second heart sound. *F,* Period of falling pressure; ventricles continue to relax, and the pressure within them decreases. The tricuspid and bicuspid valves remain closed because the pressure in the ventricles is still higher than the pressure in the atria. Blood flows from the veins into the relaxed atria. *G,* The tricuspid and biscuspid valves open when the pressure in the ventricles falls below that in the atria, and blood flows into the ventricles.

the ventricular walls continue to relax, and the pressure within the ventricles continues to decrease. The tricuspid and bicuspid valves, which closed during the previous ventricular systole, remain closed because the pressure within the ventricles is still greater than that within the atria. No blood is flowing into or out of the ventricles at this time, although some blood is flowing from the veins into the relaxed atria.

(*g*) Opening of the bicuspid and tricuspid valves (Fig. 143, *G*). The continued relaxation of the ventricular walls finally lowers the intraventricular pressure below the pressure within the atria. At that moment the tricuspid and bicuspid valves open, and blood flows rapidly from the atria into the ventricles. No contraction of the heart causes this; it is due simply to the fact that the pressure within the relaxed ventricle is lower than that of the atria and veins. Half the volume of the ventricles may be filled before the atria undergo systole.

The Heart Sounds. The beating heart produces characteristic sounds which can be heard by placing the ear against the chest or by using a stethoscope, an instrument which magnifies sounds and conducts them to the ear. In most normal persons, two sounds are produced per heart beat, one of which is low-pitched, not very loud, and of long duration. This is caused partly by the closure of the tri- and bicuspid valves and partly by the contraction of the muscle in the ventricle (all muscles make a noise when they contract). The first sound, which marks the beginning of ventricular systole, is followed quickly by a second which is higher pitched, louder, sharper and shorter in duration. It is the result of the closure of the semilunar valves and marks the end of ventricular systole. The two sounds have been described by the syllables "lubb-dup," and their quality indicates to the physician the state of the valves. When the semilunar valves are injured, a soft hissing noise ("lubb-shhh") is heard in place of the second sound. This is known as a heart murmur and may be caused by syphilis, rheumatic fever or any other disease which injures the valves and prevents their closing tightly, so that blood can leak back from the arteries into the ventricles during diastole. Damage to the bicuspid or tricuspid valve affects the quality of the first heart sound.

Electrical Changes in the Heart. When any tissue becomes active—when a muscle contracts, a gland secretes, or a nerve conducts an impulse—it becomes electrically negative with respect to the surrounding tissues; that is, the active tissue acts as the negative pole and the resting tissues as the positive pole of a battery. These slight **action currents** can be detected with sensitive devices. Those accompanying each beat of the heart are detectable even at the surface of the body by means of an **electrocardiograph,** which records them as a complex, curved line (Fig. 144). Since malfunctioning of the heart causes peculiar action currents, the line recording a pathologic condition will be unusual, and a trained person can interpret the curves and diagnose the particular abnormality.

Adaptation of the Heart Beat to Body Activity. Active tissue requires several times as much oxygen and nutrients as the same tissue at rest, and both the heart and blood vessels are active in the adjustments necessary to provide whatever is needed. During periods of intense exercise the heart can pump seven or eight times as much blood as normal, by increasing the number of beats per minute, or by increasing the blood volume per beat. The heart normally pumps about 75 ml. of blood per beat, but it can pump as much as 200 ml. The following stimuli, singly or together, can effect this increase:

(*a*) A rise in the carbon dioxide content of the blood. During exercise, oxidation occurs more rapidly, and the greater amount of carbon dioxide in tissues and blood stimulates the heart to increase the volume per beat.

(*b*) Stretching of the heart muscle. During exercise the pressure in the veins is higher and more blood flows into the heart chambers before they contract, causing the muscular walls to stretch. The contractile power of a muscle, within limits, is increased by the tension it is under when it begins to contract; hence the greater the volume of blood in the heart at the beginning of systole, the greater will be the quantity of blood pumped per heart beat.

An increase in the rate of the heart beat from normal to 170 or 200 beats per minute also is possible during exercise. Again, several factors may be involved:

(*a*) Increased temperature. Enough heat is produced by exercising to raise the body temperature a few degrees. This affects the sino-atrial node (just as fever does), and the heart rate is increased.

(*b*) Hormones. Both epinephrine, which is produced in increased amounts by the adrenal glands during emergencies, and thyroxine, which is produced by the thyroid gland and quickens body metabolism generally, accelerate the heart beat. When thyroxine is injected into an experimental animal, however, acceleration of the heart does not occur until three or four hours have elapsed—a reaction too slow to be effective in bringing about the rapid adjustment which the heart must make continuously, although it can affect the long-term responses of the heart to the general condition of the body.

(*c*) Nerves. In the normal body, adjustments of the heart beat rate are

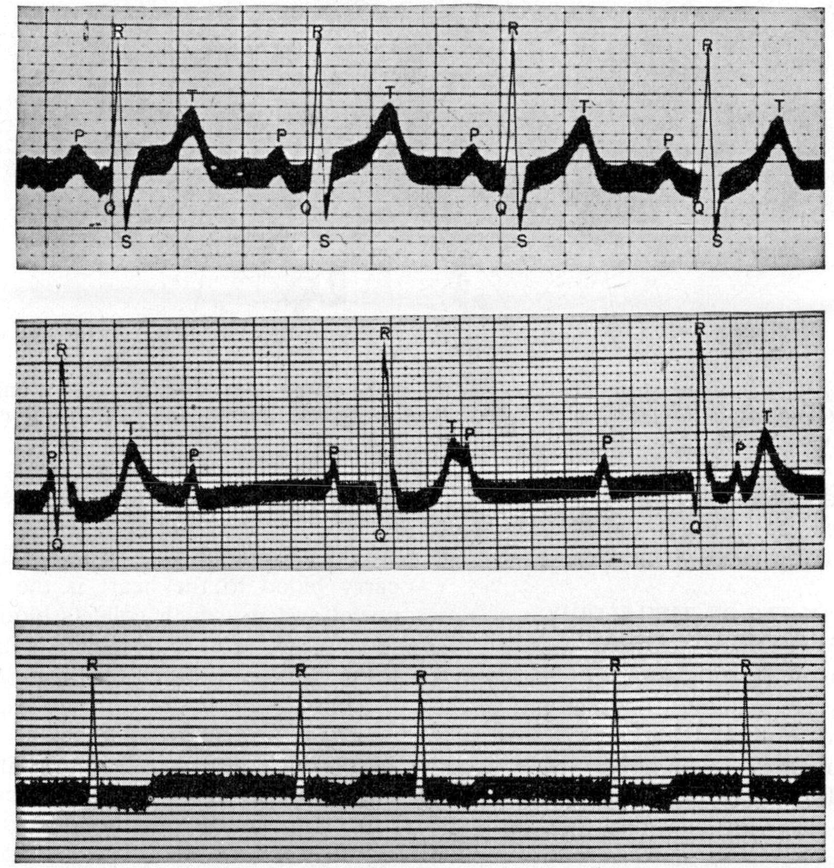

Figure 144. Electrocardiograms. *Above,* Tracing from a normal heart. The P wave corresponds to the contraction of the atria, the QRS complex to the contraction of the ventricle, and the T wave to the relaxation of the ventricle. Each heart cycle is marked by P, QRS, and T waves. *Middle,* Tracing from a man with a complete block of the atrioventricular node, so that the atrium and ventricle are beating completely independently, each at its own rate. Note that P waves appear at regular intervals, and QRS and T waves appear at regular but longer intervals, but that there is no relation between the P and the QRS waves. *Below,* Tracing from a man with atrial fibrillation. Here the individual muscle fibers of the atrium are twitching rapidly and independently, so that there is no regular atrial contraction and hence no P wave. The ventricle beats independently and irregularly, causing the QRS wave to appear at irregular intervals. (Courtesy of Dr. Lewis Dexter and the Peter Bent Brigham Hospital, Boston.)

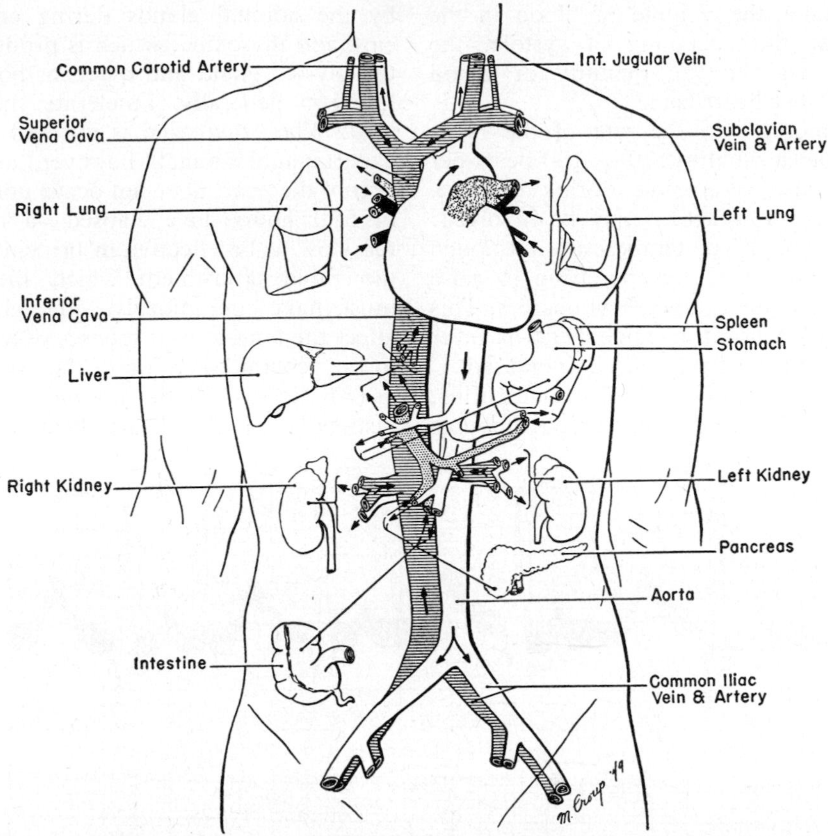

Figure 145. Diagram of the main arteries and veins of the body; arteries (white), pulmonary artery (stippled), veins (horizontal lines), pulmonary veins (black) and hepatic portal system (dotted).

effected chiefly by two sets of nerves, one of which accelerates it, the other of which slows it down (see Chap. 24).

168. ROUTES OF THE BLOOD AROUND THE BODY

To understand how the circulatory system carries material from one part of the body to another, some knowledge of how the blood vessels are connected is necessary (Fig. 145). In any particular vessel, blood flows in one direction only, of course. The head and brain are supplied with blood by the carotid arteries, and are drained by the jugular veins. In addition, the brain is served by a second pair—the vertebral arteries and veins (not shown on the diagram)—running close to the spinal cord. At the base of the brain are interconnections between the carotid and vertebral arteries, so that if one of the vessels is cut or occluded, the brain is still adequately supplied with blood.

One exception to the rule that all veins carry blood to the heart is the **hepatic portal system,** which collects blood from the spleen, stomach, pancreas and intestines and conducts it to the liver. There the portal vein breaks up into capillaries which in turn unite to form the hepatic vein, which drains blood from the liver into the inferior vena cava. Because of this arrangement, all blood from the spleen, stomach, intestines and pancreas must pass through the liver before it reaches the heart. Thus, food absorbed in the intestines is carried directly to the liver for storage.

Fetal Circulation and Changes at Birth. A fetus in its mother's womb, or uterus, cannot obtain food or air directly; consequently its stomach and lungs are nonfunctional. It obtains food and oxygen from the maternal blood by means of blood vessels in the placenta and umbilical cord. There is, however, no *direct* connection between the blood streams of mother and fetus. The blood of the fetus is manufactured within its own body, chiefly in the spleen and liver. Within the placenta the capillaries of mother and fetus come into close contact, and substances pass from one to the other by diffusion; oxygen and food substances diffuse from the maternal to the fetal blood vessels, and carbon dioxide and metabolic wastes diffuse from the fetal to the maternal blood vessels. The developing fetus uses one set of blood vessels while developing a second set which must be ready to take over immediately after birth.

Two **umbilical arteries** grow out of the lower part of the aorta of the fetus and pass to the placenta (Fig. 146). Blood is returned to the child by a single **umbilical vein,** which passes through the liver and empties into the inferior vena cava. The fact that the lungs are small and nonfunctional presents a special problem, for the capillaries in the uninflated lungs can accommodate only a fraction of the blood flowing through the heart; the rest must be passed around the lungs until after birth. This is accomplished by two temporary devices: the **oval window** between

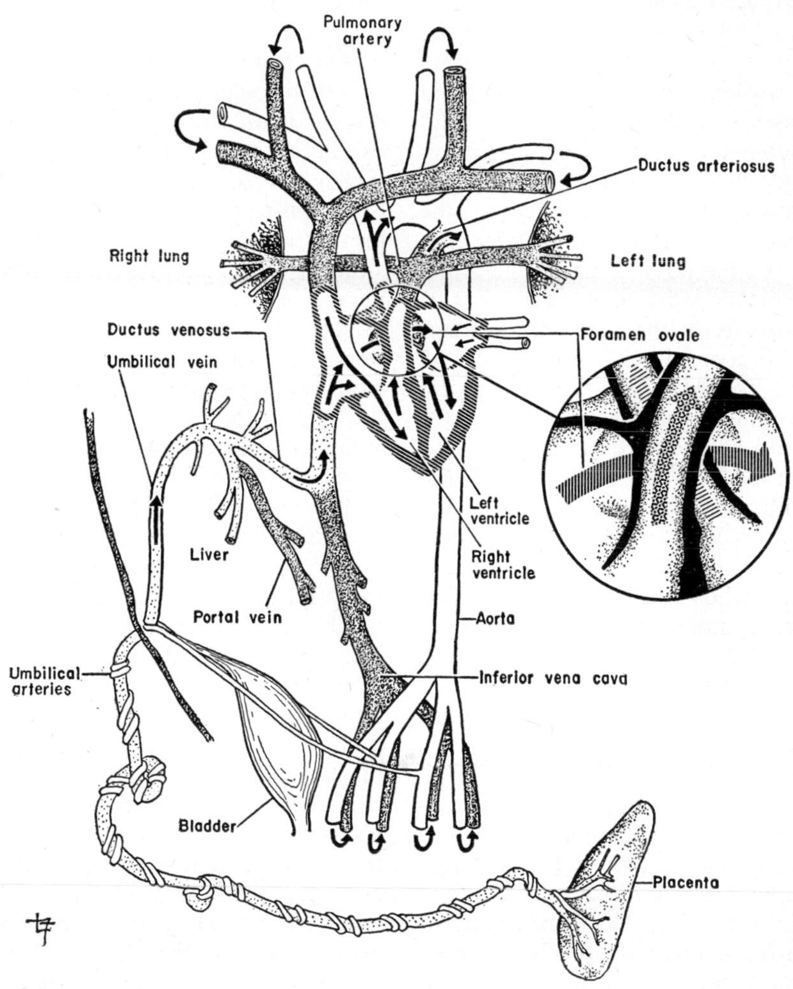

Figure 146. The human circulatory system before birth. The structures peculiar to the fetal circulatory system are the umbilical arteries and veins to the placenta, the *ductus arteriosus* which connects the pulmonary artery and aorta, the *ductus venosus* which connects the umbilical vein and the inferior vena cava, and the oval window (*foramen ovale*) which connects the right and left atria.

the right and left artia, and the **arterial duct** connecting the pulmonary artery and the aorta. The opening of the oval window is covered by a flap of tissue which acts as a valve. The blood from the umbilical vein, rich in oxygen and food, is mixed with the blood of the inferior vena cava and passes to the right atrium. Most of this blood then flows through the oval window (pushing the flap of the valve aside, since the blood pressure is higher in the right than in the left atrium) to the left atrium, whence it flows through the left ventricle to the aorta. From the aorta this blood with a high oxygen content passes chiefly to the arteries of the wall of the heart, the head and the forelimbs. The rest of the blood from the inferior vena cava, together with blood entering the right atrium from the superior vena cava, passes into the right ventricle and out the pulmonary artery. Only a fraction of this blood goes to the lungs; the rest passes across the arterial duct to the aorta. Since the arterial duct joins the aorta beyond the point at which the arteries to the head and arms branch off, this blood passes down the aorta. Only about a third of it goes to the lower part of the trunk and to the legs; the remainder, carried by the umbilical arteries, goes to the placenta, where it receives more oxygen and food. By this arrangement the more rapidly growing head and upper part of the body receive the additional oxygen they need.

At birth, when the infant begins using his lungs, a number of changes must occur quickly. The tying of the umbilical cord deprives the child of that source of oxygen. Carbon dioxide begins to accumulate in the blood stream and stimulates the respiratory center in the brain to bring about the child's first breath. Consequently the lungs and lung capillaries expand, causing more blood to pass through the pulmonary arteries instead of through the arterial duct. This blood, returning to the left atrium, increases the blood pressure in that chamber, and pushes the flap of tissue across the oval window, stopping the flow of blood from the right to the left atrium. This valve eventually grows in place, causing the oval window to be permanently closed. The endothelial lining of the arterial duct grows rapidly to obliterate the cavity of the vessel, and prevent the further passage of blood. Failure of the oval window or the arterial duct to close properly interferes with the normal aeration of the blood; if the condition is extreme, the child may die. If the by-passes close only partially, the hemoglobin is inadequately oxygenated, and a "blue baby" results. The skin capillaries, filled with bluish hemoglobin instead of the normal scarlet oxyhemoglobin, give the child's skin a bluish tinge. The arterial duct, plus parts of the umbilical arteries and veins, remain within the body and are changed into connective tissue to form certain ligaments.

The Rate of Flow of Blood. In its course through the body, blood does not flow at a constant speed. The flow is rapid in the arteries (about 500 mm. or 20 inches per second in the larger ones), a little less rapid in the veins (about 150 mm. per second in the larger ones), and slow in the capillaries (less than 1 mm. per second). The differences in the rate of flow depend upon the total cross-sectional area of the vessels. If a fluid passes from one tube into another of larger size, the rate of flow is less in the larger tube. When the blood flows through a series of tubes of different sizes, connected together end-to-end, its velocity is always inversely proportional to the cross-sectional area of whatever tube it happens to be in.

The circulatory system is constructed in such a way that one large artery (the aorta) branches into many, intermediate-sized arteries. These in turn branch into thousands of small arteries (called **arterioles),** each of which gives rise to many capillaries. Although the branches of the aorta are smaller than the vessel itself, there are so many of them that the *total* cross-sectional area is greater, and the rate of flow correspondingly less. It has been estimated that the total cross-sectional area of all the capillaries of the body is about 800 times that of the aorta. Therefore, the rate of flow in the capillaries is about 1/800th as great as in the aorta. At the other end of the capillary network, the capillaries join to form small veins (venules) which combine to form increasingly larger veins. As this occurs, the total cross-sectional area decreases and the rate of flow increases.

Since the heart pushes blood into the arteries only during ventricular systole, arterial blood moves spasmodically, rapidly when the ventricles contract, slowly at other times. When the semilunar valves are closed, the blood in that part of the aorta nearest the heart is stationary, but the blood in the arteries farther away from the heart does not stop completely between systoles. In the arterioles the alternation in speeds is less marked; in the capillaries the flow of blood is almost constant, so that transfer of materials can occur continuously. This conversion of the intermittent flow in the arteries to the steady flow in the capillaries is made possible by the elasticity of the arterial wall. The force of the contracting ventricles does two things: it pushes the blood forward, and distends and elongates the walls of the arteries (Fig. 147). During diastole the stretched walls contract—as does a stretched rubber band when the stretching force is removed—thus squeezing the blood along. The blood is prevented from flowing backward by the closure of the semilunar valves. The contraction of the arterial wall next to the heart distends the next section of the aorta or pulmonary artery, which, in turn, contracts and distends the next section, and so on. This alternate stretching and contracting passes along the arterial wall at the rate of 7 to 8 meters (25 feet) per second, and is known as the **pulse.** The blood inside the artery flows at a much slower rate, about 20 inches per second.

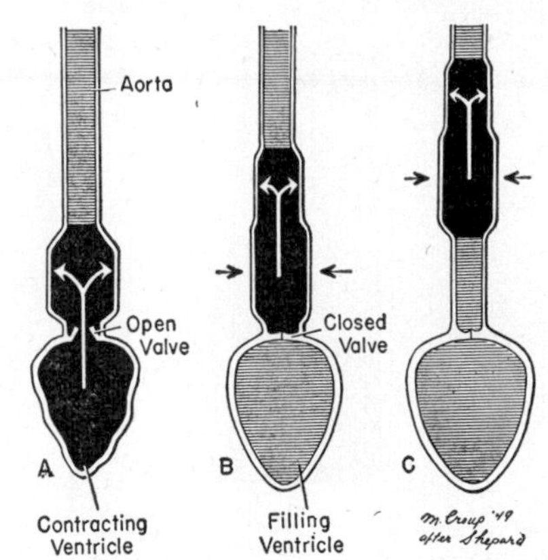

Figure 147. Diagram of the movement of blood from the ventricle through the elastic artery. For simplicity, only one ventricle and artery are shown, and the amount of stretching of the arterial wall is exaggerated. *A,* As the ventricle contracts, blood is forced through the semilunar valves, and the adjacent wall of the aorta is stretched. *B,* As the ventricle relaxes and begins to fill for the next stroke, the semilunar valves close and the expanded part of the aorta contracts, causing the next adjacent part of the aorta to expand. *C,* The pulse wave of expansion and contraction is transmitted to the next section of the aorta.

Two other factors aid the heart in moving blood through the veins—the movements of the skeletal muscles and the motion of the body in breathing. Most of the veins are surrounded by skeletal muscles, which, when they contract, cause the veins to collapse (Fig. 148). As the muscles relax, the collapsed section again fills with blood, which must come from the direction of the capillaries. This mechanism by which muscle contractions "milk" blood along the veins is especially important in returning blood to the heart from the legs against the pull of gravity. If one stands upright quietly for a time, tissue fluid tends to collect in the legs, and swelling (edema) results. In walking, the contractions of the leg muscles force the blood along the veins, and the feet and ankles are less likely to swell. In breathing, the chest muscles and diaphragm contract, increasing the space inside the chest and lowering the pressure within the chest cavity below that outside the body, causing air to flow into the lungs. Since the heart lies within the chest cavity, it too is affected by the breathing movements, and the pressure within the veins in the chest is lowered as one inhales. Blood moves into the chest veins and atria for the same reason that air moves into the lungs.

These two factors are important in enabling the circulatory system to respond to the increased need for blood during exercise. At that time both the "milking" action of the muscles on the veins and the breathing movements are greatly increased, sending more blood to the atria.

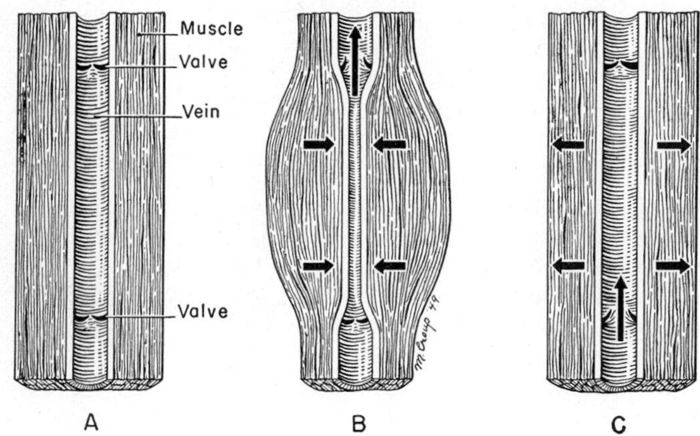

Figure 148. The action of skeletal muscles in moving blood through the veins. *A,* Resting condition. *B,* Muscles contract and bulge, compressing veins and forcing blood toward heart. The lower valve prevents backflow. *C,* Muscles relax, and the vein expands and fills with blood from below; the upper valve prevents backflow.

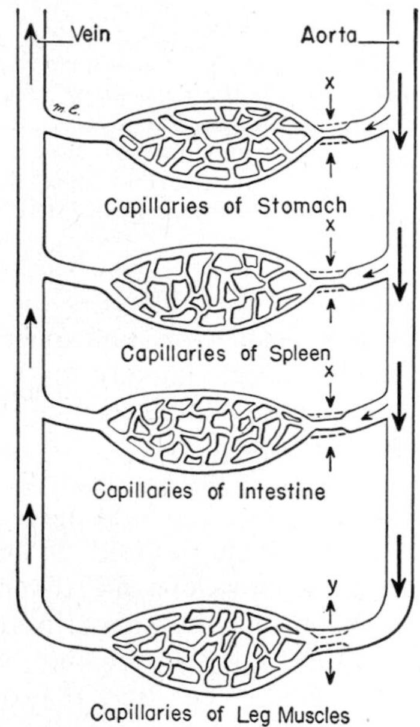

Figure 149. Diagram of the reactions of the circulatory system during exercise. Contraction of smooth muscles in the walls of the arterioles of the stomach, spleen and intestine (*X*) decreases the diameter of the arterioles and the flow of blood to those organs. Relaxation of smooth muscles in walls of arterioles of leg and other muscles (*Y*) increases the diameter of these vessels and the flow of blood to the skeletal muscles.

You will recall that the greater the volume of blood entering the heart, the more the heart muscle is stretched, the more forcibly the heart beats, and the greater the volume of blood ejected per beat. Because of this, the muscular contractions during excitation, which bring about increased requirements for food and oxygen, are partly instrumental in causing the circulatory system to satisfy the increased requirements.

It would be wasteful, however, to have all parts of the body receive an increased supply of blood whenever one part became active. To obviate this, the circulatory system is equipped with devices for regulating the rate at which blood is delivered to any particular part of the body (Fig. 149). The walls of the arterioles contain smooth muscle fibers innervated by two sets of nerves. An increase in the number of nerve impulses in one set causes the muscles to contract, decreasing the size of the arterioles and lessening the supply of blood to that organ or part of the body. An increase in the number of nerve impulses in the other set causes the muscles to relax, increasing both the size of the arterioles and the flow of blood to that organ. These muscles usually are partially contracted because of a balance between the two nervous impulses. This nervous mechanism enables the arterioles to act as valves to regulate the amount of blood each organ in the body receives. The smooth muscles of the walls of the arterioles are affected also by the chemicals—carbon dioxide and epinephrine—which affect the heart output. When a particular organ is metabolizing at a high rate, the greatly increased amount of carbon dioxide acts directly on the smooth muscles, causing them to relax, and thus increases the delivery of blood to the active tissue. Epinephrine causes a relaxation of the walls of the arterioles serving the skeletal muscles, but a contraction of the walls of the arterioles serving the internal organs—the stomach, intestines and liver—which results in a greatly increased flow of blood to the skeletal muscles. These chemical effects are independent of the nerves, and act equally well on normal arterioles and on those with severed nervous connections.

Blood Pressure. The beating of the heart is responsible for the blood pressure in the vessels, which rises with each contraction and falls with each relaxation of the ventricles. The highest pressure, due to systole of the heart, is called **systolic pressure;** the lowest, due to diastole, is called **diastolic pressure** (Fig. 150).

The pressure in an artery can be measured directly by inserting a tube into it and measuring the height to which the column of blood rises. An English clergyman, Stephen Hales, made the first such measurement in 1733. He found the blood pressure in the carotid artery of a mare sufficient to cause a column of blood to rise 9½ feet in a glass tube.

Because it is not practical to puncture an artery in a human being each time the blood pressure is taken, a device known as a **sphygmomanometer** is used. This consists of a cuff, tied around the arm, containing a rubber bag, to which is attached a rubber bulb and a device for measuring pressure (Fig. 151). The pressure of the blood pushing against the sides of the arteries (Fig. 152, *A*) is measured by determining how much pressure in the rubber bag causes the artery to collapse. Systolic pressure corresponds to the pressure in the bag which is just enough to obliterate the wrist pulse (Fig. 152, *B*). To measure diastolic pressure, a stethoscope is placed over the arm artery below the cuff so that the noise produced by the pulsing of the blood can be heard. When the pressure in the cuff is just below systolic pressure (Fig. 152, *C*), a small amount of blood passes the constriction each time the pressure rises at systole. This blood strikes the stationary blood beyond the cuff, producing vibrations audible in the stethoscope. As the pressure in the cuff is gradually lowered, more and more blood passes at each systole, producing an increasingly loud noise (Fig. 152, *D*). When the pressure in the cuff is just above diastolic pressure, the noises are loudest, and as the cuff pressure drops below it, the blood flows continuously and no noises are audible (Fig. 152, *E*).

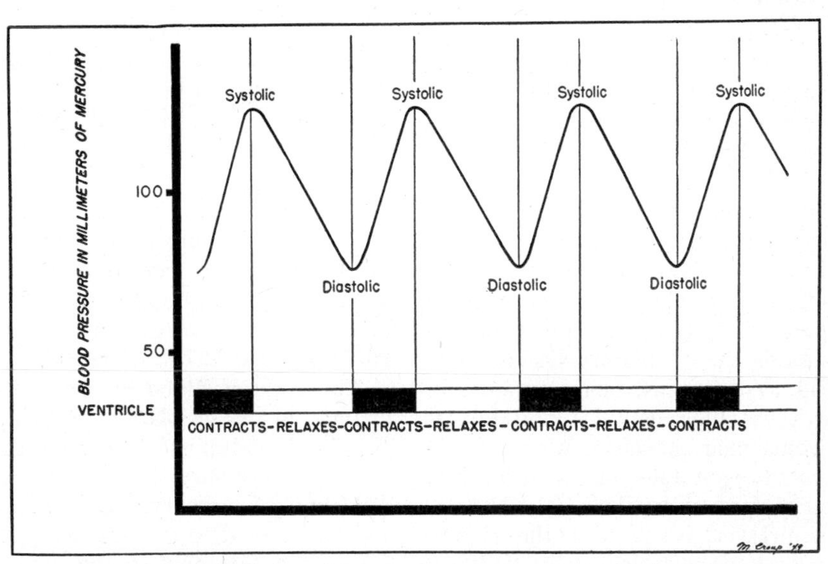

Figure 150. The changes in pressure in an artery as the heart beats.

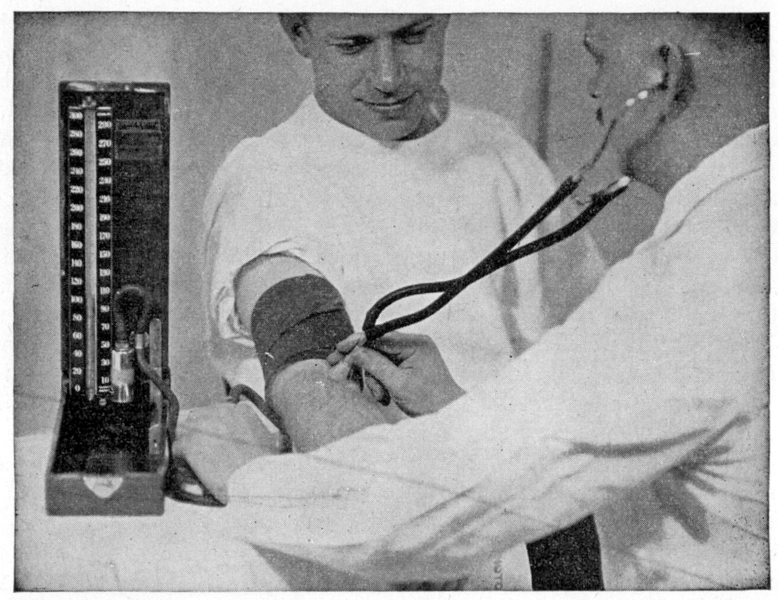

Figure 151. Determining the blood pressure by means of a sphygmomanometer. Using a stethoscope, the physician can detect the increase and decrease in noise of the blood flowing through the artery beyond the rubber cuff, as the pressure in the cuff is raised and lowered. (Crandall, L. A.: Introduction to Human Physiology.)

In man and many mammals, systolic pressure is about 120 mm. of mercury (i.e., equal to the pressure of a column of mercury 120 mm. high), and diastolic pressure about 75 mm. The difference between systolic and diastolic pressures—that is, the change in pressure that occurs each time the heart beats—is called the **pulse pressure.** The arterial pressure, usually measured in the artery of the left arm just above the elbow, is recorded as a fraction, with the systolic pressure as the numerator and the diastolic pressure as the denominator: 120/75. Therefore, when a physician tells you that your blood pressure is 120, he is referring to the pressure in a particular artery in the arm, and does not mean that the blood all over your body is under the same pressure. The blood pressure decreases along the circulatory system from the aorta to the veins, being greatest in the aorta (as high as 140) and lowest in the veins near the atria, where it approaches or even falls below zero, i.e., atmospheric pressure (Fig. 153). The decrease in pressure is caused by the friction of the blood rubbing against the walls of the blood vessels, and is especially marked in the arterioles and capillaries, because those vessels are small and friction is greatest. The gradual decrease in pressure is necessary to keep the blood flowing; if the pressure were the same throughout the circulatory system, blood would not move.

The high pressure in the arteries is necessary to force blood through the capillaries, where resistance due to friction is great, as well as to force blood back to the heart through the veins. This pressure has to be sufficiently great to force blood up to the brain against the pull of gravity, and to return blood from the legs against the same force. High pressure is also necessary to ensure rapid adjustments of the blood flow to certain organs as their needs change; as the arterioles of an organ dilate, high pressure is needed to cause an immediate inrush of blood.

The Role of Blood Pressure in the Exchange of Materials in the Capillaries. As the blood passes through the capillaries, its pressure decreases from about 40 (mm. of mercury) on the side of the capillaries next to the arterioles, to about 10 on the side next to the venules. This enables water and substances dissolved in

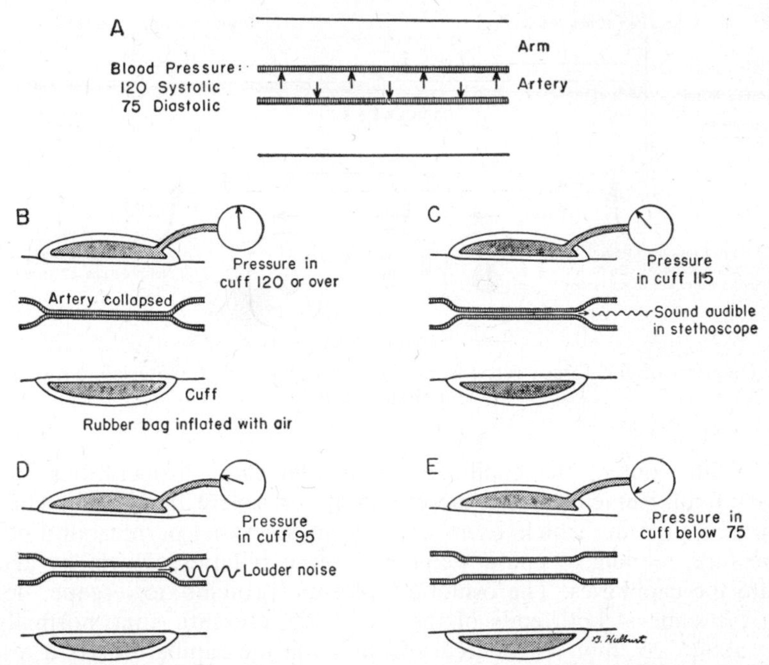

Figure 152. The measurement of blood pressure by the sphygmomanometer. *A,* The blood pushes against the arterial walls with a pressure (systolic) of 120 mm. of mercury when the heart contracts, and a pressure (diastolic) of 75 mm. of mercury when the heart is relaxed. *B,* When the air pressure in the rubber cuff exceeds 120 mm. of mercury, the arm is constricted, the artery is collapsed, and no blood passes. *C,* With the pressure in the cuff just below 120, the artery is collapsed except for a small period during systole, when a small amount of blood squirts through, producing a sound audible in the stethoscope. *D,* With the pressure in the cuff at 95, the artery is open for a longer time during systole, a greater amount of blood passes through, and a louder noise is produced. *E,* When the pressure in the cuff drops below 75, the artery is open continuously, blood passes through continuously, and no noise is heard. Thus systolic pressure is that at which a noise is first heard, as the pressure in the cuff is lowered, and diastolic pressure is that at which the noises cease.

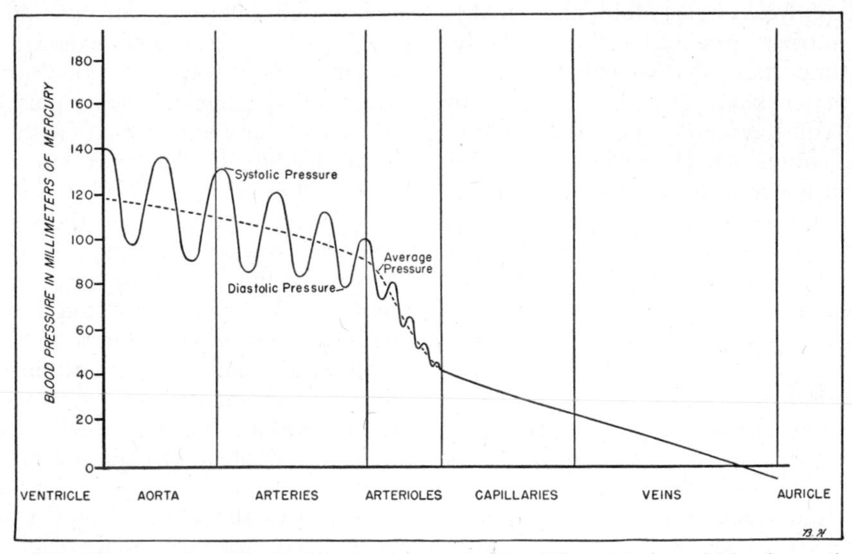

Figure 153. Diagram showing the blood pressure in the different types of blood vessels of the body. The systolic and diastolic variations in arterial pressures are shown. Note that the venous pressure drops below zero (below atmospheric pressure) near the heart.

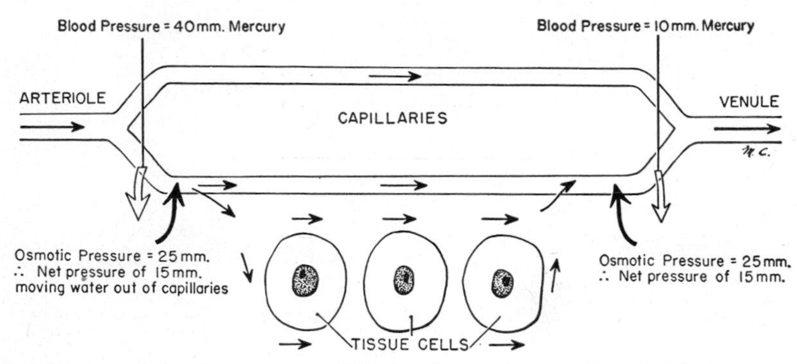

Figure 154. Diagram of the forces responsible for the exchange of materials between capillaries and tissue fluid.

the blood to filter out of the capillaries into the tissue fluid. Remember that blood plasma contains proteins which exert an osmotic pressure, tending to cause water to move into the capillaries. The osmotic pressure is the same at both ends of the capillaries, about 25 mm. of mercury. Thus in the capillaries near the arterioles there is a net pressure of some 15 mm. (40 minus 25) which pushes water out of the capillaries (Fig. 154). At the other end, near the venules, there is a net pressure of about 15 mm. (25 minus 10) pulling water back into the capillaries. Under normal conditions the blood volume remains constant because equal quantities of water pass in and out. This device is important in maintaining a constant stream of fresh tissue fluid and its dissolved nutrients past each cell of the body.

It is important, too, in replenishing the liquid of the blood after hemorrhage. In such circumstances the decrease in blood volume causes a decrease in blood pressure and hence in filtration pressure. But the amount of proteins per cubic centimeter of blood remains the same, so that osmotic pressure is not diminished. This results in a lowered net pressure to move water out of the capillaries at the arterioles, and an increased net pressure to move it into the capillaries at the venule end, thereby increasing the blood volume at the expense of the tissue fluid. This is a temporary measure to prevent heart failure from a lack of blood to be pumped, and true recovery does not occur until later when new blood is manufactured.

"Shock" is a condition frequently present after surgical operations, burns, accidents or severe fright. It is characterized by an increased permeability of the walls of the capillaries, which, by allowing the plasma proteins to escape, lowers the osmotic pressure that normally returns fluids to the capillaries. As a result, fluids escape from the capillaries to the tissues, the blood volume decreases, and, in severe cases, death results. The danger from the shock accompanying severe wounds has been greatly lessened by the use of repeated plasma transfusions which replace the lost plasma proteins and prevent the loss of fluids from the capillaries.

169. HEART AND VESSEL DISORDERS

The heart is subject to several conditions which interfere with its ability to pump. Severe muscular exercise by a person in poor physical condition may strain and damage the heart muscle and reduce its ability to contract. Or the blood vessels leading to the heart muscle may become clogged by a blood clot or some other object which cuts off the supply of food and oxygen, resulting in a heart "attack." If the area of the heart served by the plugged vessel is not too large, the attack will not be fatal; but if a large area is affected, death results within a few minutes. Certain diseases, such as diphtheria, produce poisons which reach the heart via the blood stream and injure the heart muscle. An excess of thyroid hormone causes the atria to beat in an uncoordinated way so that efficiency is impaired. Sometimes the heart valves, damaged by disease organisms such as those of syphilis or rheumatic fever, are unable to close properly, and blood leaks back after the heart contracts. To compensate for this, the heart often enlarges to increase its pumping ability. Rheumatic fever, which is especially prevalent among young people, is dangerous, not only because of the infection of the joints but also because of its effect on the heart.

As one grows older, the walls of the arteries tend to lose their elasticity and become harder and thicker, reducing the size of the cavity. Less blood then passes to the organs, which are unable to function properly. Usually, an increase in blood pressure accompanies the hardening of the arteries, for as they lose their elasticity and are unable to expand and contract each time the heart beats, the latter must pump harder to force the blood along. High blood pressure may be caused by other disorders—kidney disease, for example—and is dangerous because, if excessive, a blood vessel may burst. Usually, only a small vessel bursts, and the loss of blood is not important. But bleeding in a soft structure such as the brain may damage the cells, resulting in paralysis of the muscles served by them, and sometimes death ensues. High blood pressure may bring about heart failure by forcing it to enlarge in order to drive the blood against the increased resistance of the hardened arteries.

The veins sometimes swell, causing large pools to form within the vessels (a condition called **varicose veins**). Varicose veins occur almost always in the legs, and are most frequently the affliction of people who have to stand a great deal. Standing still increases the pressure in the leg veins by depriving them of the "milking" action of the leg muscles, which ordinarily helps to return the blood properly. Varicose veins are more prevalent in women than in men because of the added stresses on the circulatory system of pregnancy and childbirth.

170. THE LYMPH SYSTEM

Besides the blood circulatory system, the body is equipped with a similar, independent group of vessels constituting the **lymph system.** These carry the clear, colorless fluid, **lymph,** which, like tissue fluid, is derived from blood and resembles it closely. It contains much less protein than does blood (because the protein molecules are large and diffuse slowly) and no red cells. It does contain white cells, some of which enter the lymph capillaries from the tissue fluid, others of which are manufactured in the lymph nodes. In other respects lymph is similar to blood.

The lymph system differs from the blood system in that its vessels serve only to *return* fluid to the heart. There are no arteries in this system, only capillaries and veins distributed all over the body. The capillaries are extremely small, with walls only one cell thick. They resemble blood capillaries but are closed at one end (Fig. 155). Lymph diffuses into these capillaries from the surrounding tissue fluid. At the other end the capillaries connect with lymph veins, which resemble blood veins in having valves and thin walls. These in turn empty into successively larger veins, the biggest of which drains into the left shoulder vein of the blood system. It is important to remember that fluids reach the cells of the body by one route only: the arteries, arterioles and capillaries of the blood circulatory system. But there are two possible return routes: the blood capillaries and veins, and the lymph capillaries and veins.

At the points where the lymph vessels join are aggregations of cells known as **lymph nodes,** which manufacture lymphocytes and filter out dust particles and bacteria to prevent their entering the blood stream. The channels by which the lymph passes through the nodes are so minute and tortuous that its flow is sluggish, and invading bacteria can be caught and phagocytized by the white cells. Some bacteria may get past the first lymph node and be caught by the second or third; in a massive infection the bacteria may penetrate all the lymph nodes and invade the blood stream. But even in such cases the lymph nodes are valuable in delaying the spread of the infection, thus giving the body time to accumulate white cells to fight it. The presence of infectious organisms causes the lymph nodes to become swollen and tender—thus the lymph nodes

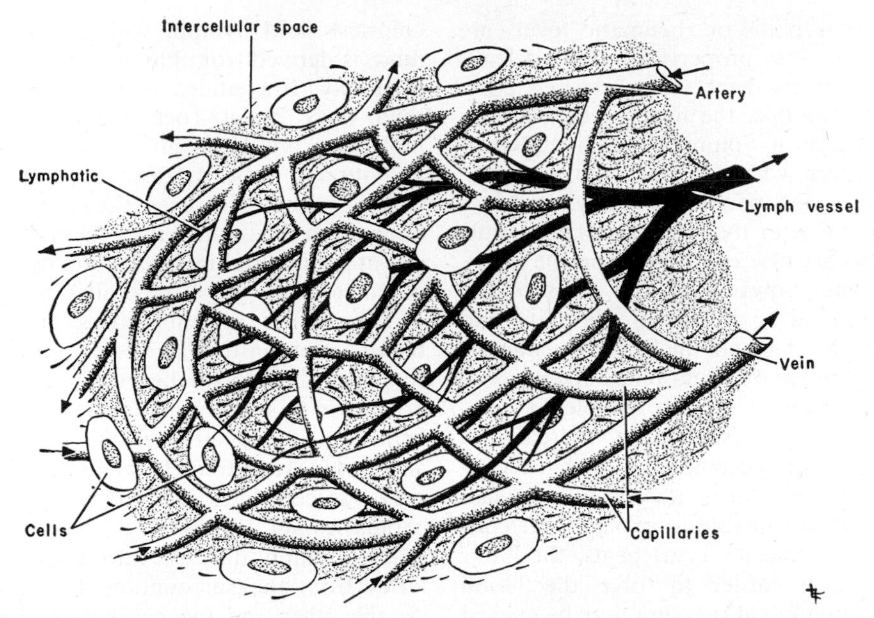

Figure 155. Diagram of the relation of blood and lymph capillaries to the tissue cells. Note that the blood capillaries are connected at both ends, whereas lymph capillaries (black) are "dead end streets." The arrows indicate the direction of flow.

of the neck become noticeably swollen in cases of sore throat. In the lymph systems of city dwellers and heavy smokers the nodes near the lungs become filled with particles of dust and soot and are a dark gray or black. These particles may eventually interfere with the functioning of the lymph nodes, and lower resistance to lung infections, such as tuberculosis.

The Flow of Lymph. The frog has four lymph "hearts" which pulsate and squeeze the lymph along. The hearts are simply thicker-walled, muscular parts of the lymph vessels, without chambers or valves. Human lymph, in contrast, is moved by the movements of the adjacent skeletal muscles, which compress the lymph vessels (the valves prevent its backflow), and the breathing movements of the chest. The flow of lymph from the intestine is aided by the spasmodic contraction and relaxation of the villi, the fingerlike projections from the wall of the intestine into its cavity. Lymph flows much more slowly than blood.

Functions of the Lymph System. The lymph system performs four functions. First, it assists in returning tissue fluids to the blood circulatory system. Since, normally, the walls of the blood capillaries are slightly permeable to the proteins of the plasma, there is a slow, constant leakage of these proteins into the tissue fluid. Without the lymph system the concentration of protein in the tissue fluid would eventually equal the concentration in the blood capillaries. This would interfere with the mechanism that draws tissue fluid into the blood capillaries at their venous end, and so result in swelling (edema). The tremendous swellings of elephantiasis are caused by the blocking of the lymph channels by a parasitic worm (called a filaria) which is injected into the body by the bite of a mosquito.

We have already noted two other functions of the lymph system—the manufacture of lymphocytes, and the filtering of dust and bacteria. A fourth function, the absorption of fats, is accomplished by the lymph vessels that drain the intestines. The products resulting from the digestion of carbohydrates and proteins are absorbed directly into the blood capillaries, but most digested fats are absorbed into the lymph capillaries in the walls of the intestine and carried by the lymph vessels to the left shoulder vein, whence they are emptied into the blood stream.

From the foregoing discussion you can

understand why the lymph system has an important relation to the spread of cancer throughout the body. Cancer is a condition in which certain cells become overactive in growth and cell division, increasing at the expense of the surrounding cells. The stimulus for this may be certain chemicals; x-rays or other rays; the presence of unusual parasites within the cells; and probably other factors. Whatever the cause, the cells proliferate and cause a swelling. They may then break off from the original mass, find their way into a lymph vessel and be carried by it to other parts of the body. If treatment is begun in time, cancer is curable by surgery or radiation, because the lymph nodes prevent the cells from spreading rapidly. In cutting out a cancerous growth, the surgeon must be careful to remove the adjacent lymph nodes, which may contain cancerous cells capable of instigating a recurrence of the condition.

171. CIRCULATION IN OTHER ORGANISMS

All organisms have the same problem of transporting substances from one part of the body to another. The human heart, with its remarkable automatic devices for keeping the blood flowing and for adapting to changing conditions, is the result of a long evolutionary process.

The protozoa have no special system for bringing about circulation of substances; foods, wastes and gases simply diffuse through the cytoplasm and eventually reach all parts of the cell. In most protozoa this process is aided by movements of the cytoplasm. As an ameba moves along, the cytoplasm streams from the rear to the front of the body, distributing substances throughout the cell (Fig. 156, *A*). In other types of protozoa, such as the *Paramecium,* which has a firm, outer shell and does not change shape as it moves, substances are distributed by a rhythmic, circular movement of the cytoplasm in the direction indicated by the arrows in Figure 156, *B*. Food is taken in through a "mouth" and gullet on one side of the animal. Food vacuoles form at the base of this gullet, break off and move through the cell as they digest and give off food. Gases and waste products of metabolism are moved similarly.

The central cavity of the coelenterates (Fig. 173) serves as both a digestive and circulatory organ. When the tentacles have captured their prey they stuff it through the mouth into the cavity, where digestion occurs. The digested food substances then pass to the cells lining the cavity and through them, by diffusion, to the outer layer. The movement of the animal's body, as it alternately stretches and contracts, stirs up the contents of the central cavity and aids circulation.

The flatworm planaria (Fig. 173) re-

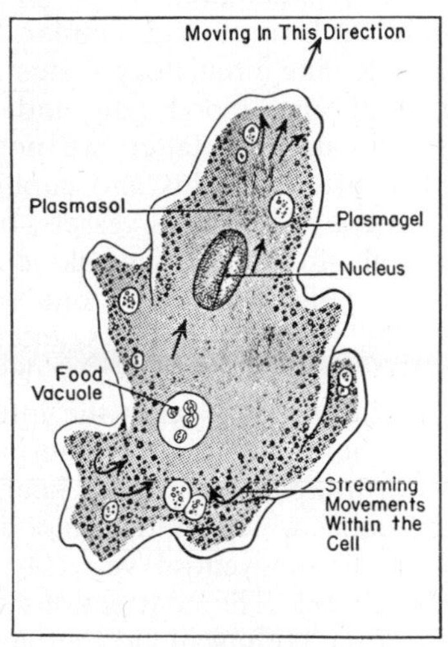

A, AMEBA

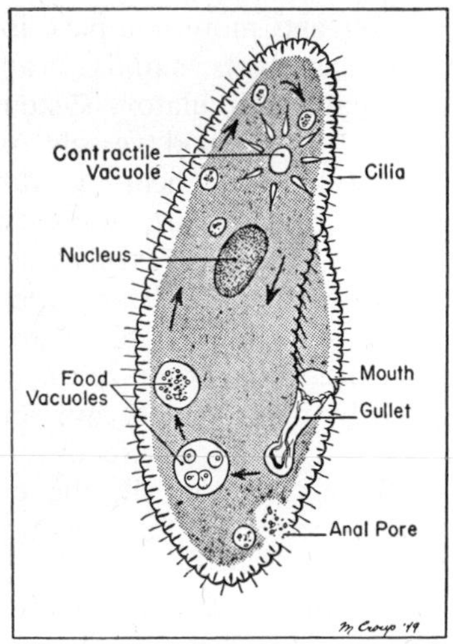

B, PARAMECIUM

Figure 156. Circulation in protozoa by means of streaming movements of the cytoplasm.

sembles the hydra in having a single, central cavity connected to the outside by a single mouth opening. But besides the inner and outer layers of cells found in hydra, the planaria has a third layer of cells, loosely packed between the other two. The spaces between these cells are filled with a tissue fluid somewhat similar to the tissue fluid of human beings. Food enters through the mouth, is digested in the central cavity, diffuses through the inner layer of cells and passes through the tissue fluid to the other cells. As in the coelenterates, circulation is aided by the contraction of the muscles in the body wall which agitate the fluid in the central cavity and the tissue fluid.

In the earthworm and similar forms there is a definite circulatory system, consisting of plasma, blood cells and blood vessels, although the latter are not specialized as arteries, veins and capillaries. There are two main blood vessels, one on the ventral side, in which blood flows posteriorly (Fig. 115), and one on the dorsal side, in which it flows anteriorly. Connecting these in each part of the body are small tubes which serve the intestine, skin and other organs. In the anterior part of the worm are five pairs of "hearts" or pulsating tubes which conduct blood from the dorsal to the ventral vessel to complete the circuit. The contractions of the muscles of the body wall aid the hearts in circulating the blood.

The larger and more complex invertebrates, such as clams, squids, crabs and insects, all have a circulatory system consisting of a heart, blood vessels, plasma and blood cells. The heart is different from a vertebrate heart, consisting in most forms of a single muscular sac. The blood vessels from the heart open into large spaces, enabling the blood to bathe the body cells. From these spaces, blood is collected by other vessels and returned to the heart. The details are different in various animals; but in all, the circulatory system performs the functions of supplying the cells of the body with oxygen and nutrients, and of removing metabolic wastes.

The circulatory systems of all vertebrates are fundamentally the same, from fish and frogs through lizards to birds and man. All have hearts and an aorta, as well as arteries, capillaries and veins, organized on a similar basic plan. Because of this similarity, students can learn much about the human circulatory system from the dissection of a dogfish or frog.

In the evolution of the higher vertebrates, such as man, from the lower, fishlike forms, the principal change in the circulatory system occurred in the heart, and is correlated with the change in the respiratory mechanism from gills to lungs. The fish heart consists of four chambers in a row: sinus venosus, atrium, ventricle and conus (Fig. 157, *A*). Blood from the veins drains into the sinus venosus, while blood from the conus is pumped through the ventral aorta to the gills, where it takes up oxygen. It then goes to the dorsal aorta and is distributed to all parts of the body. Blood passes through the fish heart only once each time it makes a circuit through the body.

In the particular group of fish from which the land vertebrates evolved, a number of changes occurred in the heart and blood vessels which are visible in present-day frogs (Fig. 157, *B*). A partition developed down the middle of the atrium, dividing it into right and left halves. The sinus venosus shifted its connection so that it emptied only into the right atrium. A vein from the lungs emptied into the left atrium, while pulmonary arteries to the lungs grew out of the vessels which originally served the most posterior pair of gills. Thus, in the frog, blood should pass from the veins to the sinus venosus, to the right atrium, to the ventricle, to the aorta, the pulmonary arteries, the lungs, the pulmonary veins, the left atrium, the ventricle, the aorta, and then to the cells of the body. Of course, there is some mixing of aerated and nonaerated blood in the ventricle, and some blood from the sinus venosus may get into the aorta instead of the pulmonary artery, while some from the left atrium may be pumped into the pulmonary artery. There is less mixing than one might suppose, however. Blood from the right atrium enters the ventricle ahead of that from the left atrium, and so lies nearer the exit. As the ventricle contracts, nonaerated blood from the right atrium leaves first and enters the arteries branching off from the aorta—the pulmonary arteries to the lungs. The aerated blood from the left atrium leaves the ventricle toward the end of the contraction, is unable to enter the pulmonary arteries because they are full of other blood, and so passes via the aorta to the body cells. Because aerated and non-aerated blood may mix in the ventricle, blood may pass through the heart once, twice or even more times for each circuit it makes through the body.

In the evolution of reptiles from their ancestral amphibians, one partition developed down the center of the ventricle, and one down the conus (Fig. 157, *C*). Since the ventricular partition is incomplete in all the reptiles except alligators and crocodiles, there is still some mixing of aerated and nonaerated blood, although less than in the frog. The sinus venosus is small, foreshadowing its disappearance in the mammalian heart.

The hearts of birds and mammals (Fig. 157, *D*) show the complete separation of right and left sides. The interventricular partition also is complete, precluding any mixture of blood from the right and left hearts. The conus has split and become the base of the aorta and pulmonary artery. The sinus venosus has disappeared as a separate chamber, although

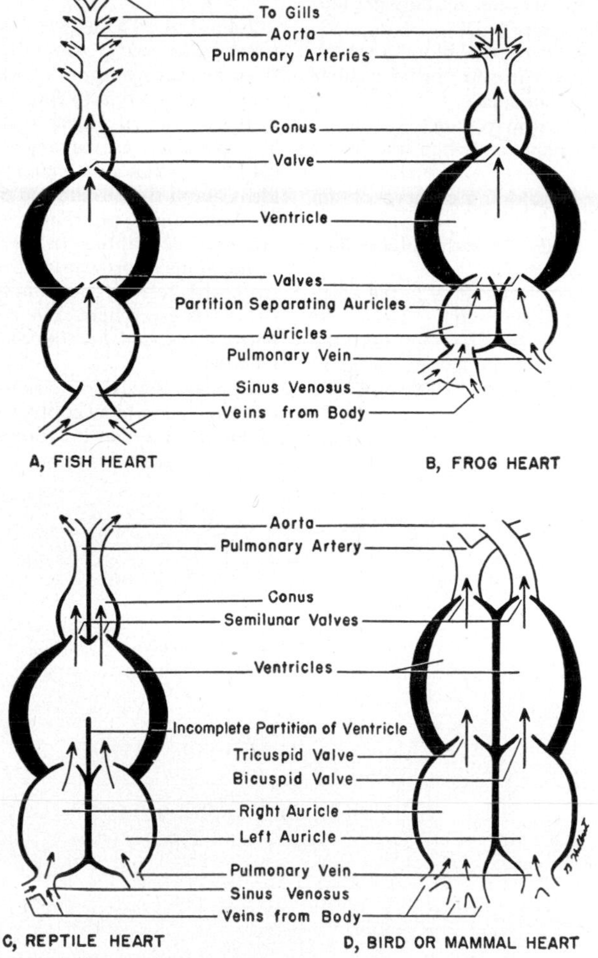

Figure 157. Diagrams illustrating the evolution of the heart in vertebrates.

a vestige remains as the sino-atrial node. The absolute separation of right and left hearts forces the blood to pass through the heart twice each time it makes a tour of the body. As a result, the blood in the aorta of birds and mammals contains more oxygen than that in the aorta of the lower vertebrates. Hence the tissues of the body receive more oxygen, a higher metabolic rate can be maintained, and the warm-blooded condition is possible. Fish, frogs and reptiles are cold-blooded primarily because their blood cannot deliver enough oxygen to the tissues to sustain the high rate of metabolism necessary to maintain a high body temperature in cold surroundings.

QUESTIONS

1. How do food and oxygen actually get into the cells of the body?
2. What is the route of the blood through the heart? What part do the semilunar valves play in its passage?
3. What regulates the heart rate?
4. What actually happens when the heart "beats"?
5. How does the developing embryo obtain food and oxygen?
6. Where in the body does the blood flow most slowly? Why?
7. What is the rate of blood flow in the aorta expressed in miles per hour?
8. How can you explain the fact that the pulse beat passes along an artery more rapidly than the blood passes through the vessel?
9. What is meant by systolic and diastolic pressure?
10. What physiologic changes occur during shock?
11. What is lymph? How does it differ from blood? Where and how is it produced?
12. What are the two ways in man by which fluids may return to the heart?
13. Compare the manner in which food and oxygen are carried about the body in an ameba, a hydra, a planaria, a crab and a frog.
14. What is the principal difference between the circulatory system of a fish and that of man?
15. Trace the path of a red cell from a vein in the leg to the kidney.
16. Trace the path of a molecule of sugar from the capillaries of the intestine to a brain cell, where it is metabolized, and then trace the path of the carbon dioxide formed until it is finally excreted by the lungs.

SUPPLEMENTARY READING

John Fulton's *Selected Readings in the History of Physiology* is a collection of 87 passages from as many authors in which the most significant contributions to physiology, the landmarks in the history of physiologic thought, are given. The passages are arranged chronologically by subject so that one can trace the sequence of ideas about the circulation of the blood in Chapter 2 and about capillaries in Chapter 3. William Harvey's description of the action of the valves of the veins in preventing the flow of blood away from the heart, one chapter of his *De Motu Cordis,* is given here. The entire text of Harvey's book, translated by Chauncey D. Leake, is also available.

Further details of circulation and the many mechanisms which control the rate and volume of blood flow can be found in physiology texts such as Fulton, Starling, or Best and Taylor.

Chapter 18

The Respiratory System

THE ENERGY for all the myriad activities of plants and animals is derived from reactions of **biologic oxidation** (Chapter 5). The essential feature of these reactions is the transfer of hydrogen atoms from one molecule, the hydrogen donor, to another, the hydrogen acceptor. In most animals and plants there is a series of compounds, each of which accepts hydrogen (or its electron) from the preceding and donates it to the subsequent one. The ultimate hydrogen acceptor in the metabolism of most plants and animals is oxygen, which is converted to water. Since only small amounts of oxygen can be stored (as oxyhemoglobin or as the comparable oxymyoglobin in muscle), the continuation of metabolism depends upon an uninterrupted supply of oxygen to each cell. Most cells are rapidly killed if deprived of oxygen; brain cells are especially sensitive and are damaged beyond repair if their supply is cut off for only three or four minutes.

Carbon dioxide is removed from substrate molecules in other reactions. **Decarboxylation,** the removal of carbon dioxide from a larger molecule, can proceed independently of oxygen utilization. Yeasts, for example, can metabolize sugar to alcohol and carbon dioxide without utilizing oxygen. In most animals and plants the utilization of oxygen and the splitting off of carbon dioxide proceed together. The carbon dioxide produced must be removed from the body, for it reacts with water to form carbonic acid, H_2CO_3.

The term **respiration** is used to refer to those processes by which animal and plant cells take in oxygen, give off carbon dioxide, and convert energy into biologically available forms such as ATP (adenosine triphosphate). The term respiration has had three different meanings biologically. It originally was synonymous with breathing, and meant inhaling and exhaling; "artificial respiration" refers to this usage of the term. Then, as it became clear that the important process was the exchange of gases between the cell and its environment, the term respiration was applied to this. Finally, as the details of cellular metabolism became known, the term respiration was used to denote those enzymatic reactions of the cell which are responsible for the utilization of oxygen. Thus the cytochromes are called "respiratory enzymes."

Direct Respiration. The exchange of gases is a fairly simple process in a small, aquatic animal such as a paramecium or a hydra: Dissolved oxygen from the sur-

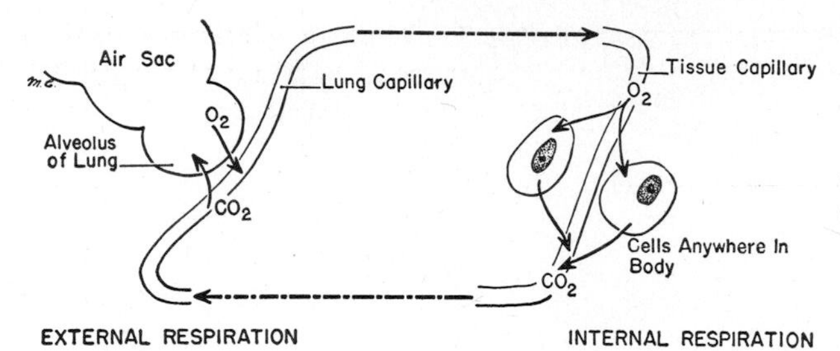

Figure 158. Diagram illustrating the exchanges of gases which occur in external and internal respiration.

rounding pond water diffuses into the cells, the carbon dioxide diffuses out, and no special respiratory system is needed. On this level it is called **direct respiration,** because the cells of the organism exchange oxygen and carbon dioxide directly with the surrounding environment.

Indirect Respiration. As animals evolved into higher, more complex forms, it became impossible for each cell of their bodies to exchange gases directly with the external environment. Some form of **indirect respiration** involving a structure of the body specialized for respiration was necessary. This specialized structure had to be thin-walled (the membrane of the wall must be semipermeable), so that diffusion could occur easily; it had to be kept moist, so that oxygen and carbon dioxide had water to be dissolved in; and it needed a good blood supply. For indirect respiration, fishes, crabs, lobsters and many other animals developed gills; the higher vertebrates, reptiles, birds and mammals developed lungs; the earthworm uses its moist skin; and insects use tracheal tubes, canals running all through the body, connected by pores with the external environment.

In indirect respiration two phases, an external and an internal one, are involved in the exchange of gases between the body cells and the environment. **External respiration** consists of the exchange of gases by diffusion between the external environment and the blood stream, by means of the specialized respiratory organ—for example, the lung in mammals. **Internal respiration** consists of the exchange of gases between the blood stream and the cells of the body (Fig. 158). Between these phases the gases are transported by the circulatory system.

172. STRUCTURE OF THE HUMAN RESPIRATORY SYSTEM

The respiratory system in man and other air-breathing vertebrates includes the lungs and the tubes by which air reaches them. The anatomy of this system may be understood by tracing the path of the oxygen molecules entering the body (Fig. 159). Air enters through the **external nares,** or nostrils, which open into the **nasal chamber,** a large cavity dorsal to the mouth cavity and ventral to the brain. This cavity contains the sense organ of smell, and is lined with mucus-secreting epithelium. Air is cleaned and warmed as it passes through the nasal chamber. When the capillaries in this chamber dilate excessively, causing an oversecretion of mucus, the nose becomes "stopped up" and we have the symptoms of a cold. The air passes to the pharynx via the **internal nares.**

In the **pharynx** the paths of the digestive and respiratory systems cross. Food passes from the pharynx to the stomach by way of the esophagus, and air leaves by way of the larynx and trachea. To prevent food from passing into the larynx and trachea and injuring the delicate membranes lining them, a flap of tissue, the **epiglottis,** folds over the opening of the larynx whenever food is swallowed. Fortunately, this is done automatically so that we do not have to remember to close the epiglottis each time we swallow; occasionally the automatic mechanism fails and food goes down the "wrong throat."

The **larynx,** or voice box (seen on the outside as the Adam's apple), contains the vocal cords, folds of epithelium which vibrate as air passes over them, thereby producing sounds. Muscles adjust the tension of the cords to produce sounds

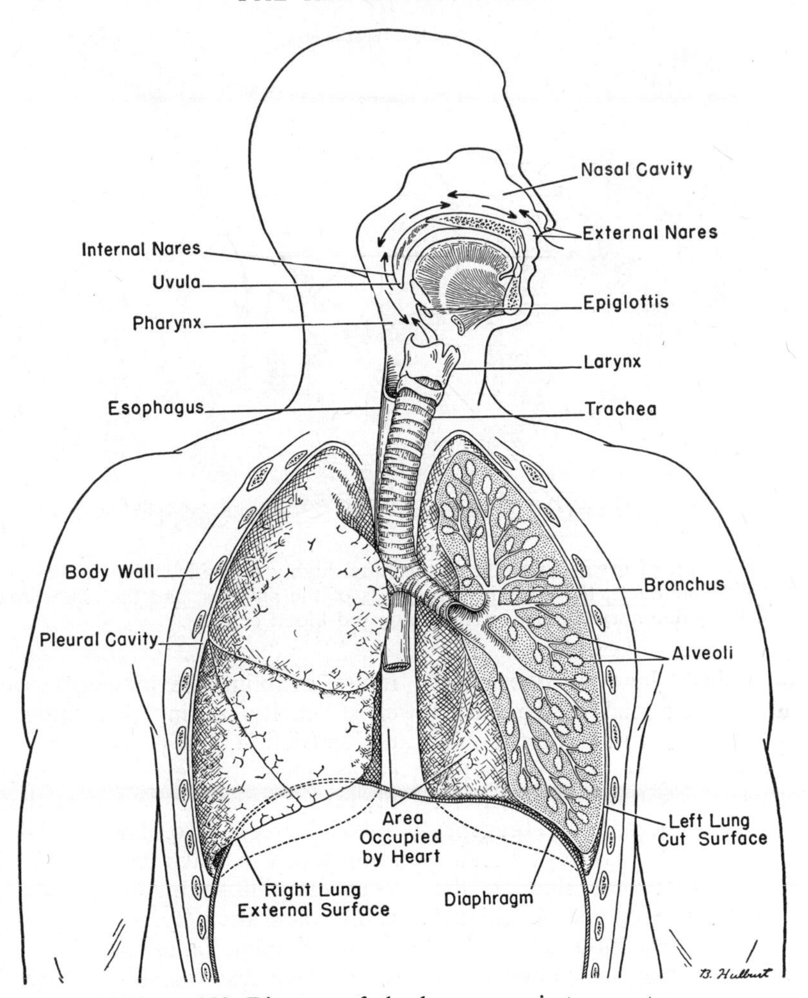

Figure 159. Diagram of the human respiratory system.

of varying pitch. The **trachea,** or windpipe, may be distinguished from the esophagus by the rings of cartilage imbedded in its walls to hold it open. During inspiration the pressure of air in the trachea is less than atmospheric pressure, and without the cartilaginous rings the trachea would collapse.

At the level of the first rib the trachea branches into two cartilaginous **bronchi,** one going to each lung. Inside the lung, each bronchus branches into bronchioles, which in turn branch repeatedly into smaller and smaller tubes leading to the ultimate cavities, the **air sacs.** In the walls of the smaller vessels and the air sacs are minute, cup-shaped cavities, known as **alveoli,** just outside of which are thick networks of blood capillaries (Fig. 160). The walls of the alveoli are thin and moist, enabling gas molecules to pass easily through them into the capillaries. The total alveolar surface, across which gases may diffuse, has been estimated to be greater than 1000 square feet—more than fifty times the area of the skin.

The wall of the trachea or bronchus consists of an inner layer of epithelium, an outer layer of connective tissue, and a middle layer containing the cartilaginous rings and smooth muscle fibers. (In a person suffering from asthma these smooth muscle fibers contract abnormally and reduce the size of the passage, thus making breathing difficult.) The lining epithelium secretes mucus and contains ciliated cells. These cilia beat constantly in one direction, so that when solid parti-

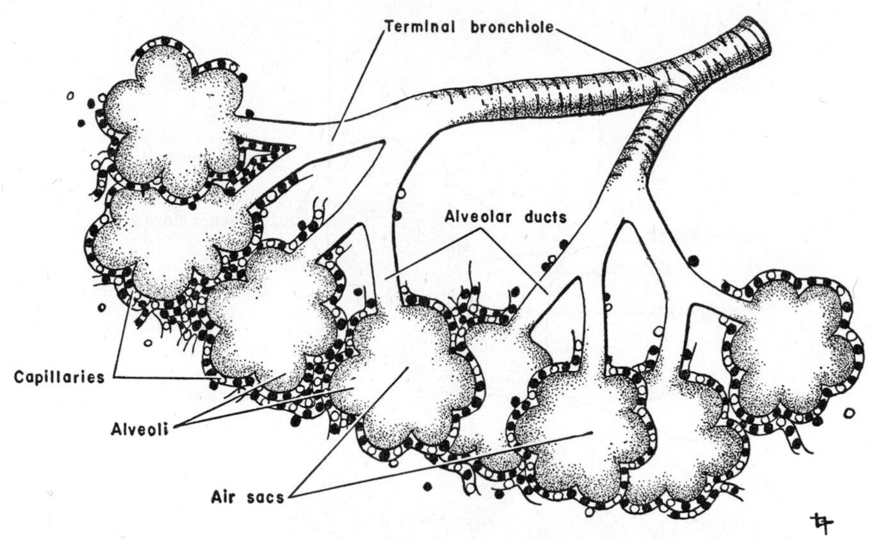

Figure 160. Diagram of a small portion of the lung, highly magnified, showing the air sacs at the end of the alveolar ducts, the alveoli in the walls of the air sacs, and the close proximity of the alveoli and the pulmonary capillaries containing red blood cells.

cles, such as dust, land on the moist surface, they are trapped in the mucus and carried by the beating of the cilia back up to the pharynx. This is an important defense against inhaled bacteria.

As the bronchioles and sub-branches of the respiratory passages become smaller, the walls become thinner, the cartilage layer disappears, and the ciliated cells are replaced by squamous epithelium. The walls of the alveoli are composed of just one layer of these flat epithelial cells, and in some regions even these may be lacking, so that the air in the alveoli is in direct contact with the capillaries (Fig. 160). In between the alveoli and holding them in place are strands of elastic connective tissue, which make the lung pliable. Lungs are so elastic that when they are freshly removed from an animal they may be blown up like a balloon through the trachea. When the pressure is released, the elasticity of the stretched lung deflates it and expels the air that has been blown in. The lung is supplied both with motor nerves, which go to the smooth muscles in the bronchi and bronchioles, and with sensory nerves which ramify throughout it. Each lung, as well as the cavity of the chest in which the lung rests, is covered by a thin sheet of smooth epithelium called the **pleura.** These are both kept moist, enabling the lung to move in the chest cavity during breathing without undue friction. The pressure in the pleural cavity (that between the two sheets of pleura) is generally less than that of the outside atmosphere. The elasticity of the lungs tends to make them pull away from the chest wall, setting up a partial vacuum in the pleural cavity. When these pleural linings become inflamed, they secrete a fluid which accumulates in the cavity between the lung and the chest wall—a condition known as pleurisy. It is sometimes necessary in cases of severe tuberculosis to collapse one lung and give the infected tissues a chance to rest. This is done by puncturing the chest wall and letting sterile air flow into the pleural cavity; the lung then collapses, owing to its own elasticity.

The chest cavity is closed and has no communication with the outside atmosphere or with any other body cavity. It is bounded on the top and sides by the chest wall, containing the ribs, and on the bottom by a strong, dome-shaped sheet of skeletal muscle, the **diaphragm** (Fig. 159).

173. THE MECHANICS OF BREATHING

It is necessary to keep clear the distinction between respiration—the exchange of gases between a cell and its environment (which in man consists of the three phases of external respiration, transportation by the blood stream, and internal respiration) — and **breathing,** which is simply the mechanical process of taking air into the lungs (inspiration), and letting it out again (expiration). Since the lung capillaries are constantly removing oxygen from and putting carbon dioxide into the air in the alveoli, the need for replacing the air in the lungs is obvious. In man the breathing cycle of inspiration followed by expiration is repeated about fifteen to eighteen times a minute.

In human beings and other mammals the ribs, chest muscles and diaphragm are so constructed and arranged as to be easily movable, enabling the volume of the chest cavity to be increased or decreased at will. When it is necessary to increase it, during inspiration, the rib muscles contract, drawing the front ends of the ribs upward and outward, an action made possible by the hingelike connection of the ribs with the backbone (Fig. 161, *right*). At the same time the floor of the chest cavity, the diaphragm, contracts, decreasing its convexity and consequently enlarging the cavity (Fig. 161, *left*). Because the space is closed, this increase in volume results in a lowering of the pressure in the lungs, and when it falls below atmospheric pressure, air from the outside rushes in through the trachea and its branches to the air sacs and alveoli.

Air is expelled from the lungs in expiration by the elasticity of the lungs, and by the weight of the chest wall. During inspiration the lungs are distended as they are filled with air. When the rib muscles relax, the ribs are permitted to return to their original position, and the simultaneous relaxation of the diaphragm permits the abdominal organs to push it back up to its previous convex shape. These factors decrease the chest volume and allow the distended, elastic lungs to contract and expel the air which had been inhaled.

During muscular exercise this passive expiration of the relaxing rib and diaphragm muscles is not rapid enough to expel the air before the next inspiration must start, and the size of the chest cavity is reduced by muscular contraction. Besides the muscles which raise the ribs for inspiration there is another set, with fibers going at right angles to the first, which lowers the front ends of the ribs and thus decreases the thoracic volume. The muscles of the abdominal wall also contract, squeezing the abdominal organs up against the diaphragm, and further hastening the elastic contraction of the lungs. The chest walls never squeeze the lungs or forcibly expel the air; the de-

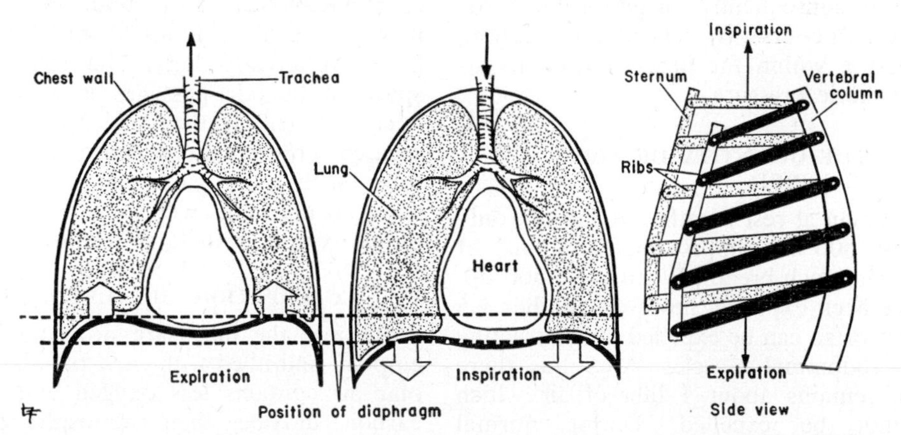

Figure 161. Left, Changes in the position of the diaphragm in expiration and inspiration which result in changes in the volume of the chest cavity. *Right,* Changes in the position of the rib cage in expiration and inspiration. The elevation of the front ends of the ribs by the chest muscles causes an increase in the front-to-back dimension of the chest, and a corresponding increase in the volume of the chest cavity. These two factors, by increasing the volume of the chest cavity, result in the intake of a corresponding amount of air.

crease in the size of the thoracic cavity simply permits the lungs to contract by means of their own elasticity. Coughing and sneezing are types of forced expirations in which a vigorous contraction of the muscles of the abdominal wall push the abdominal viscera against the diaphragm, decreasing the thoracic volume suddenly, and causing a rapid expulsion of air from the lungs.

The trachea, pharynx and other air passages play no active, muscular part in the process of breathing; they function simply as conduction channels. In certain cases of obstruction of the throat it is necessary to make an artificial opening in the neck to carry the air to the trachea; the breathing movements then occur normally.

Air pressure within the lungs changes with each breathing movement. In between breaths the pressure in the lungs is the same as atmospheric pressure, since there is free communication between the outside air and that of the interior. As inspiration starts, there is a slight decrease in the air pressure within the lungs to 1 or 2 mm. of mercury below atmospheric, which causes air to enter the lungs. Toward the end of inspiration the newly entered air has equalized the pressure. As expiration begins, the elasticity of the lungs compresses the air within the lung cavities until the pressure rises to 2 or 3 mm. of mercury above atmospheric; consequently, air passes out of the lungs. Of course, by the end of expiration, pressure within the lungs is back to atmospheric pressure.

174. THE QUANTITY OF AIR RESPIRED

A man at rest breathes in and out only about 500 ml. (approximately a pint) of air with each breath. When these 500 ml. have been expelled, however, another 1.5 liters or so can be expelled by contracting the abdominal muscles. After this there still remains about 1 liter of air which cannot be expelled. During normal breathing, therefore, a reserve of some 2.5 liters of air remains in the lungs, with which the 500 ml. are mixed. After a normal inspiration of 0.5 liter, it is possible, by inspiring deeply, to take in about 3 liters more, and during exercise one can increase the amount of air inspired and expired with each breath, from 0.5 liter to 5 liters. But even in strenuous exercise the full tenfold increase is seldom used; instead, an increased rate of breathing occurs. If one breathes in as deeply as possible and then breathes out as completely as possible into some device for measuring air volume, he will expel about 4500 ml. of air. This total, known as the **vital capacity,** is usually large in trained athletes; in certain heart and lung diseases it may be reduced considerably below normal.

If any air at all has been drawn into the lungs, enough will remain to cause them to float if, at death, they are removed from the body. But a stillborn child, never having drawn a breath, has lungs which sink when placed in water.

Although about 500 ml. of air are breathed in with each inspiration, only some 350 ml. actually reach the alveoli, because the last 150 ml. inhaled remain in the larger air passages, where no exchange of gases between lungs and blood stream can occur. This air is the first to be pushed out with the next expiration. The last 150 ml. expelled from the alveoli with each breath also remain in these tubes, and this air, although laden with carbon dioxide, is the first to be drawn into the alveoli on the next breath. With each breath then, only about 350 ml. of new air reach the lungs to mix with the 2.5 liters already there. The 150 ml. of space in the air passages is known as "dead space." If the dead space is increased (by breathing through a long tube, such as a garden hose) the air going to the lungs will soon be depleted of its oxygen, with fatal results.

175. COMPOSITION OF ALVEOLAR AIR

Because the lungs are not completely emptied and filled with each breath, alveolar air contains less oxygen and more carbon dioxide than atmospheric air (Table 4). A sample of essentially alveolar air can be collected at the end of a maximum expiration. Expired air has had less than one quarter of its oxygen removed and can be breathed over again.

Table 4. PERCENTAGE COMPOSITION OF ATMOSPHERIC AND ALVEOLAR AIR

	ATMOSPHERIC AIR	ALVEOLAR AIR
Nitrogen	79.0	75.3
Oxygen	20.96	13.2
Carbon dioxide	0.04	5.3
Water vapor	Variable	6.2

A poorly ventilated room feels "stuffy," not because of a low oxygen content or high carbon dioxide content, but because of the increased water vapor content and high temperature of air that is breathed over and over.

176. EXCHANGE OF GASES IN THE LUNGS

Oxygen passes from the alveoli to the pulmonary capillaries, and carbon dioxide passes in the reverse direction, simply by the physical process of diffusion; each gas passes from a region of higher concentration to one of lower concentration. The extremely thin alveolar epithelium offers little resistance to the passage of the gases, and since there is normally a greater concentration of oxygen in the lung alveoli than in the blood arriving in the lungs by the pulmonary artery, oxygen diffuses from the alveoli into the capillaries. Similarly, the concentration of carbon dioxide in the blood in the pulmonary artery is normally higher than in the lung alveoli, so that carbon dioxide diffuses from the lung capillaries into the alveoli. In contrast to the cells lining the intestine, which can take a substance from the intestinal cavity and pass it into the blood, where the concentration of that substance may be higher, the alveolar epithelium is unable to move either oxygen or carbon dioxide against a diffusion gradient.

Because the cells of the alveoli cannot force oxygen into the blood, whenever the concentration of that gas in the alveoli drops below a certain value, the blood passing through the lungs cannot take up enough to meet the body's needs and the symptoms of altitude sickness—nausea, headache and delusions—appear. Altitude sickness begins to occur at about 15,000 feet, or even lower with some persons. People can become acclimated to living at high altitudes by increasing the number of red cells in the blood, but no one can live much above 15,000 feet without a supply of oxygen to draw upon. At about 35,000 feet the pressure is so low that even when a man breathes pure oxygen he cannot get enough to supply his body. For this reason airplanes flying at that height must be made airtight and pumps must be provided to maintain the air pressure equal to that at a lower altitude, usually 8000 feet.

In the capillaries of the tissues all over the body, where internal respiration takes place, oxygen moves from the capillaries to the cells, and carbon dioxide from the cells to the capillaries by diffusion. Because of the constant metabolism of glucose and other substances in the cells, there is a continual production of carbon dioxide and utilization of oxygen. Consequently, the concentration of oxygen is always lower, and the concentration of carbon dioxide is always higher, in the cells than in the capillaries.

Throughout the system, from lungs to blood to tissues, oxygen moves from a region of high concentration to one of lower concentration, and is finally used in the cells; carbon dioxide moves from the cells, where it is produced, through the blood to the lungs and out, always toward a region of lower concentration.

177. TRANSPORTATION OF OXYGEN BY THE BLOOD

At rest, the cells of a man's body need about 300 liters of oxygen every twenty-four hours, or 250 ml. per minute. With exercise or work this requirement may increase to as much as ten or fifteen times that amount. If oxygen were delivered to the tissues simply dissolved in plasma, blood would have to circulate through the body at a rate of 180 liters per minute to supply enough oxygen to

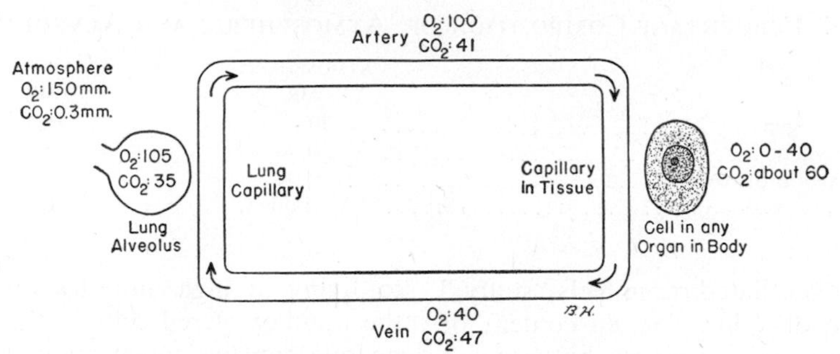

Figure 162. Diagram of the tensions, in millimeters of mercury, of oxygen and carbon dioxide in lungs, vessels and tissues, illustrating the diffusion gradients responsible for the movements of the molecules.

the cells at rest—for oxygen is not very soluble in plasma. Actually, when a man is resting, blood circulates at about 5 liters per minute and supplies all the oxygen the cells need. The difference between 180 and 5 liters per minute is due to the action of hemoglobin.

Hemoglobin is the pigment in red blood cells which carries the burden of transporting nearly all the oxygen and most of the carbon dioxide. Blood in equilibrium with alveolar air can take up in solution only 0.25 ml. of oxygen and 2.7 ml. of carbon dioxide per 100 ml., but by the actions of hemoglobin 100 ml. of blood can carry about 20 ml. of oxygen and 50 to 60 ml. of carbon dioxide.

Approximately 2 per cent of the oxygen in the blood is dissolved in the plasma; the rest is carried in combination with the hemoglobin. After oxygen enters the capillaries in the lungs, it diffuses into the red cells from the plasma and unites with hemoglobin—that is, one molecule of oxygen unites with one molecule of hemoglobin to form a molecule of oxyhemoglobin:

$$\underset{\text{Hemoglobin}}{Hb} + O_2 \rightleftarrows \underset{\text{Oxyhemoglobin}}{HbO_2}$$

The arrows indicate that the reaction is reversible: it can go in either direction, depending on local conditions. Hemoglobin would, of course, be of little value to the body if it could only take up oxygen and not give it off where needed. The reaction goes to the right in the lungs, forming oxyhemoglobin, and to the left in the tissues, releasing oxygen. The difference in color between arterial and venous blood is due to the fact that oxyhemoglobin is a bright scarlet, whereas hemoglobin is purple.

The combination of oxygen with hemoglobin and the breakdown of oxyhemoglobin are controlled by two factors: primarily, the amount of oxygen present, and, to a lesser extent, the amount of carbon dioxide present. In the lungs the concentration of oxygen is relatively high, and oxyhemoglobin is formed. After leaving the lungs the blood passes through the heart and arteries, where there is little change in the oxygen concentration, to the tissues, where the oxyhemoglobin is exposed to an environment with little oxygen. It consequently breaks down, releasing the oxygen to diffuse to the tissue cells. The role of carbon dioxide in controlling this reaction is less evident, and since the chemical details are quite complex, we shall not consider it. The important thing to realize is that the more carbon dioxide the blood contains, the more acid it is, and that the oxygen-carrying capacity of hemoglobin is less in an acid solution.

The factor which actually determines the direction and rate of diffusion is the pressure or "tension" of the particular gas. In a mixture of gases each one exerts, independently of the others, the same pressure it would exert if it were present alone. In the air at sea level, where the total pressure is about 760 mm. of mercury, oxygen exerts one fifth, or 150 mm., of the pressure. The partial pressure or tension of oxygen in the atmosphere is 150 mm. of mercury. Since the alveolar air has less oxygen than atmospheric air, the tension of oxygen in the alveoli is about 105 mm. Blood passes through the lung capillaries too rapidly to become completely equilibrated with the alveolar air, so that the oxygen tension in the arterial blood is about 100 mm. (Fig. 162). The oxygen tension in the tissues varies from 0 to 40 mm., with the result that oxygen diffuses out of the capillaries into the tissues. Not all the oxygen leaves the blood, however; blood passes through the capillaries too rapidly for complete equilibrium to be reached, and the tissues usually have some residual oxygen. The venous blood returning to the lungs has an oxygen tension of about 40 mm. of mercury. At the oxygen tension of arterial blood (100 mm.), each 100 ml. contains about 19 ml. of oxygen. At the oxygen tension of venous blood (40 mm.), each 100 ml. contains 12 ml. of oxygen. The difference of 7 ml. represents the amount of oxygen *delivered to the tissues* by each 100 ml. of blood. Thus the 5 liters of blood in the body can deliver 350 ml. of oxygen on each circuit.

178. TRANSPORTATION OF CARBON DIOXIDE BY THE BLOOD

The transportation of carbon dioxide poses a special problem to the body, because, when carbon dioxide dissolves, it is quickly converted into carbonic acid:

$$CO_2 + H_2O \rightleftarrows H_2CO_3$$

The cells of the body produce, at rest, about 200 ml. of carbon dioxide per minute. If this were simply dissolved in plasma (which can transport in solution only 4.3 ml. of carbon dioxide per liter), blood would have to circulate at a rate of 47 liters per minute instead of 4 or 5. Furthermore, this amount of carbon dioxide would give the blood a *p*H of 4.5, and cells are able to survive only within a narrow range on the alkaline side of neutrality (between about *p*H 7.2 and 7.6). The unique properties of hemoglobin enable each liter of blood to transport about 50 ml. of carbon dioxide from tissues to alveoli, with only a few hundredths of a *p*H unit difference in the acidity of arterial and venous blood. Some of the carbon dioxide is carried in a loose chemical union with hemoglobin, and a small amount is present as carbonic acid, but most of the latter is converted into bicarbonates through neutralization of the carbonic acid by sodium or potassium ions released when oxyhemoglobin is changed into hemoglobin. The chemical details of this process are complicated and beyond the scope of this book, but it is interesting to realize that in the course of evolution, a single chemical, hemoglobin, has been produced, with all the necessary characteristics for assisting respiration: the ability to transport oxygen, to transport carbon dioxide, and to keep the *p*H of the blood constant throughout the transportation process.

Carbon dioxide passes from tissues to blood to lungs because it must diffuse from a region of high tension to one of lower tension. The carbon dioxide tension in the tissues is about 60 mm. of mercury; in the venous blood, about 47 mm.; and in the alevoli, about 35 mm. of mercury. Arterial blood has a carbon dioxide tension of about 41 mm. of mercury, so that the blood contains a great deal of carbon dioxide after it has passed through the lungs. The process of converting carbon dioxide into carbonic acid in the capillaries of the tissues, and of converting the carbonic acid back into carbon dioxide to diffuse out in the lung capillaries, is speeded up 1500 times by a special enzyme called **carbonic anhydrase.**

When the removal of carbon dioxide by the lungs is interfered with, as in pneumonia, its concentration (really bicarbonates and carbonic acid) in the blood increases, and the term **acidosis** is applied to the condition of the blood. This does *not* mean that the blood is actually acid (it is still on the alkaline side of neutrality), but there is a decrease in the alkaline reserves of the blood (chiefly sodium). When the alkaline reserves are exhausted, the blood is no longer able to remain alkaline, but changes its *p*H, and the cells of the tissues die because of exposure to the acid blood. Acidosis also occurs in diabetes. Here, however, the trouble is not due to a failure to remove the carbon dioxide in the lungs, but to an overproduction by the tissues of acids because of impaired carbohydrate metabolism.

179. ASPHYXIA

Asphyxia results whenever there is an interruption in the delivery of oxygen to the tissues or a failure in the utilization of oxygen by the tissues. Therefore, the cause of asphyxia may lie in the lungs, blood or tissues. In drowning, the lung alveoli become filled with water, and in pneumonia they become filled with tissue fluid, bringing about asphyxia from lack of oxygen. In carbon monoxide poisoning, asphyxia results because the hemoglobin of the blood unites with carbon monoxide instead of oxygen and so is unavailable for transporting oxygen to the cells. In cyanide poisoning, asphyxia is caused by the inactivation of one of the enzymes present in each cell (cytochrome oxidase) which is an important link in the chain of enzymes responsible for the utilization of oxygen by the tissues.

Artificial Respiration. In cases of near-drowning or of electric shock the respiratory center may be temporarily impaired, causing the breathing movements to cease, although the person is still alive and his heart beating. A person in such dire straits will soon die unless artificial respiration is substituted for the breathing movements until they can be resumed naturally. A number of mechanical devices called pulmotors have been invented to pump air into the lungs, but they are seldom available when needed. The Schafer and the Holger Nielsen methods of artificial respiration, which require no apparatus, are simple and effective procedures. Victims of drowning accidents sometimes require an hour or more of artificial respiration.

180. THE CONTROL OF BREATHING

Since the needs of the body for oxygen are different at rest and at work, the rate and depth of breathing must change automatically to meet varying conditions. During exercise the oxygen consumption by muscles and other tissues may increase four or five times.

Breathing requires the coordinated contraction of a great many separate muscles, which is achieved by the **respiratory center,** a special group of cells in the part of the brain known as the medulla. From this center, volleys of nervous impulses pass out rhythmically to the diaphragm and rib muscles, resulting in their regular and coordinated contraction every four or five seconds. Under ordinary conditions the breathing movements are automatic and occur without our voluntary control. However, when the nerves to the diaphragm (the phrenic nerves) and the rib muscles are cut or destroyed, as in infantile paralysis, breathing movements stop at once. Of course, we can voluntarily change the rate and depth of breathing. We can even hold our breath for a while, but we cannot hold it long enough to do any serious harm—the automatic mechanism takes over and bring about an inspiration.

The question naturally occurs: Why does the respiratory center give off this volley of impulses periodically? Through a series of experiments it has been determined that if the connections of the respiratory center with all other parts of the brain are cut—that is, if the sensory nerves and those from the higher brain centers are severed—the center sends out a constant stream of impulses, and the breathing muscles contract and remain contracted. The respiratory center, then, if left to its own devices, causes a complete contraction of the breathing muscles. If either the sensory nerves or the nerves from the higher centers of the brain are left intact, however, the breathing movements continue in normal fashion. This means that for normal breathing to occur, the respiratory center must be inhibited periodically, so that it stops sending out impulses which cause contraction of the muscles. Further experiments revealed that the **pneumotaxic center** in the midbrain together with the respiratory center form a "reverberating circuit" which provides for the basic control of the respiratory rate. In addition, the stretching of the walls of the alveoli during inspiration stimulates pressure-sensitive nerve cells in their walls which send impulses to the brain to inhibit the respiratory center and bring about the following expiration.

Many other nervous pathways connected with the respiratory center either stimulate or inhibit it. Severe pain in any part of the body causes a reflex acceleration of breathing. Also, both the larynx and the pharynx have receptors in their linings which, when stimulated, send impulses to the respiratory center to inhibit breathing. These are valuable protective devices. When an irritating gas, such as ammonia or acid fumes, passes down the respiratory tract, it stimulates the receptors in the larynx, which, by sending impulses to the respiratory center to inhibit breathing, bring about an involuntary "catching of the breath." This prevents the harmful substance from entering the lungs. Similarly, when food accidentally passes into the larynx, it stimulates receptors in the lining of that organ to send inhibitory impulses to the respiratory center. This momentarily stops breathing so that the food does not enter the lungs and injure the delicate lining epithelium.

During exercise the rate and depth of breathing must increase to meet the increased needs of the body for oxygen and to prevent the accumulation of carbon dioxide. The concentration of carbon dioxide in the blood is the prime factor controlling respiration. An increased concentration of carbon dioxide in the blood to the brain increases the excitability of both the respiratory center and the pneumotaxic center. Increased activity of the first increases the strength of contraction of the muscles of respiration and increased activity of the second increases the rate of respiration. When the concentration of carbon dioxide returns to normal, these centers are no longer stimulated and the rate and depth of breathing return to normal.

This mechanism also works in reverse. If a person voluntarily takes a series of deep inhalations and exhalations, he reduces the carbon dioxide content of his alveolar air and blood to such a degree that when he stops breathing deeply, all breathing movements cease until the carbon dioxide in the blood again builds up to normal. The first breath of a newborn child is initiated largely by this mechanism. Immediately after a baby is born and separated from the placenta, the carbon dioxide content of its blood increases, stimulating the respiratory center to send nerve impulses to the diaphragm and rib muscles to contract in the first breath. Sometimes, when a newborn infant has difficulty in taking its first breath, air containing 10 per cent carbon dioxide is blown into its lungs, to set off this mechanism.

Experiments have shown that an increase in the carbon dioxide content, rather than a decrease in the oxygen content, of the blood is primarily effective in stimulating the respiratory center. If a man is placed in a small, air-tight chamber so that he breathes and rebreathes the same air, the oxygen in the air gradually decreases. If a chemical is placed in the chamber to absorb the carbon dioxide as fast as it is given off, so that it does not increase in the lungs and blood, the man's breathing accelerates only slightly, even if the experiment is continued until the oxygen content is greatly reduced. If, however, the carbon dioxide is not absorbed, but is allowed to accumulate, the breathing will be greatly accelerated, causing discomfort and a choking sensation in the subject. When he is supplied with air containing the normal amount of oxygen, but with an increased carbon dioxide content, there is again an acceleration of breathing. Obviously, it is primarily the accumulation of carbon dioxide, not an oxygen deficiency, which stimulates the respiratory center.

As additional protection against the failure of the body to respond properly to changes in the carbon dioxide and oxygen content of the blood, still another control has evolved. At the base of each internal carotid artery is a small swelling, called a **carotid sinus,** containing receptors sensitive to changes in the chemicals of the blood. If the carbon dioxide increases, or the oxygen decreases, these receptors are stimulated to send nerve impulses to the respiratory center in the medulla, increasing its activity.

The Effects of Training. The exercises and practice which an athlete performs in training increase his ability to do a certain task. First, the muscles increase in size with use and become stronger (owing to growth in the individual muscle fibers, not to an increase in the number of fibers). Second, as one performs a certain act repeatedly, he learns to coordinate muscles and to contract each one just enough and no more to produce the desired result, with consequent savings in energy. Third, changes occur in the circulatory and respiratory systems. The heart of a trained athlete is somewhat larger than normal and beats more slowly at rest. During exercise it pumps a greater volume of blood, more by increasing the strength of the beat than by increasing its rate. In addition, an athlete breathes more slowly and more deeply than the average person, and during exercise he increases the amount of air breathed by increasing the depth of his breathing, rather than by stepping up the rate. This is the more efficient way of doing it.

181. THE EVOLUTION OF THE HUMAN LUNGS

The lungs of man and other mammals have had a long evolutionary history. The earliest suggestion of a lung is found in certain types of fish. Some ancestral, fossil fish developed an outgrowth of the anterior end of the digestive tract, and in the line of fishes that eventually gave rise to the land vertebrates, this outgrowth became a lung. In other fishes it became a swim bladder—that is, an appendage which facilitates floating, though it may function as a respiratory organ (Fig. 163). The swim bladder is usually single, although it may be paired and exhibit great variation in size and shape. In some fish it has lost its connection with the digestive tract. The cells at the anterior end of the swim bladder have the ability, found nowhere else in the animal kingdom, to secrete oxygen from the blood into the internal cavity. Another set of cells at the posterior end remove oxygen from it for return to the blood stream. By thus "pumping" oxygen in or out of these swim bladders, the fish maintains a certain depth in the water without muscular effort. Several fish are even equipped with a series of bones that connect this organ with the internal ear and presumably act as a depth gauge.

The swim bladder is also an organ for the production of sounds. A few fish, such as the toadfish and sea robin, are able to make a noise, by contraction of a muscle attached to the swim bladder which causes it to vibrate.

Close relatives of the fish that gave rise to the land vertebrates are the lung fish, a few of which have survived to the present time in the headwaters of the Nile and the Amazon, and in certain Australian rivers. These animals live in streams that dry up periodically, and during the dry seasons they remain in the mud of the stream bed, breathing by means of their swim bladders. They are also equipped with gills, by means of which they breathe when swimming. The swim bladders of these fish are simple sacs, single in some forms, paired in others. In contrast to those of other fish, they are equipped with a pulmonary artery. The lungs of the most primitive amphibians, the mud puppies, are two long simple sacs, covered on the outside by capillaries. Frogs and toads have folds on the inside of the lung sac which increase the available respiratory surface (Fig. 163). Since frogs have no diaphragm or rib muscles, their method of breathing is quite different from man's and depends on the action of valves in the nostrils and muscles of the throat. With the nose valves open, the floor of the mouth cavity is drawn down and air is taken into the mouth. Then the nose valves close, and the throat muscles contract, decreasing the size of the mouth cavity and forcing the air back into the lungs. A frog cannot breathe with his mouth open!

The trend in evolution from amphibians on up is toward a greater subdivision of the lung into smaller and smaller sacs, so that the lung structure becomes increasingly complex in reptiles, birds and mammals. The lungs of some lizards—for example, the chameleon—have supplementary air sacs that can be inflated, enabling the animal to swell up—perhaps a protective device to frighten would-be predators. Birds have rather similar sacs at several points throughout the body (Fig. 163) by which air can be drawn all through the lungs, and completely renewed on each inspiration. In addition, when the bird is flying and the chest wall must be held rigid to form an anchor for the flight muscles, the air sacs act as a bellows to move air in and out of the lungs. The air sacs, lying between certain flight muscles, are squeezed and relaxed on each stroke of the wing; thus, the faster a bird flies, the more rapid is the circulation of air through the lungs.

182. RESPIRATORY DEVICES IN OTHER ANIMALS

External respiration in most aquatic animals is carried on by specialized structures called gills. Fish, molluscs (oysters, squids) and many arthropods (shrimps, crabs, spiders, and so forth, but not insects) (Fig. 164) have these organs. Every animal with gills has some arrangement for keeping a current of water flowing over them. In fish, water is taken in through the mouth, passes over the gills, and out the gill clefts. Gills, like the human respiratory organ, have thin walls, and are moist and well supplied with blood capillaries. Oxygen dissolved in the water diffuses through the gill epithelium into the capillaries, and carbon dioxide

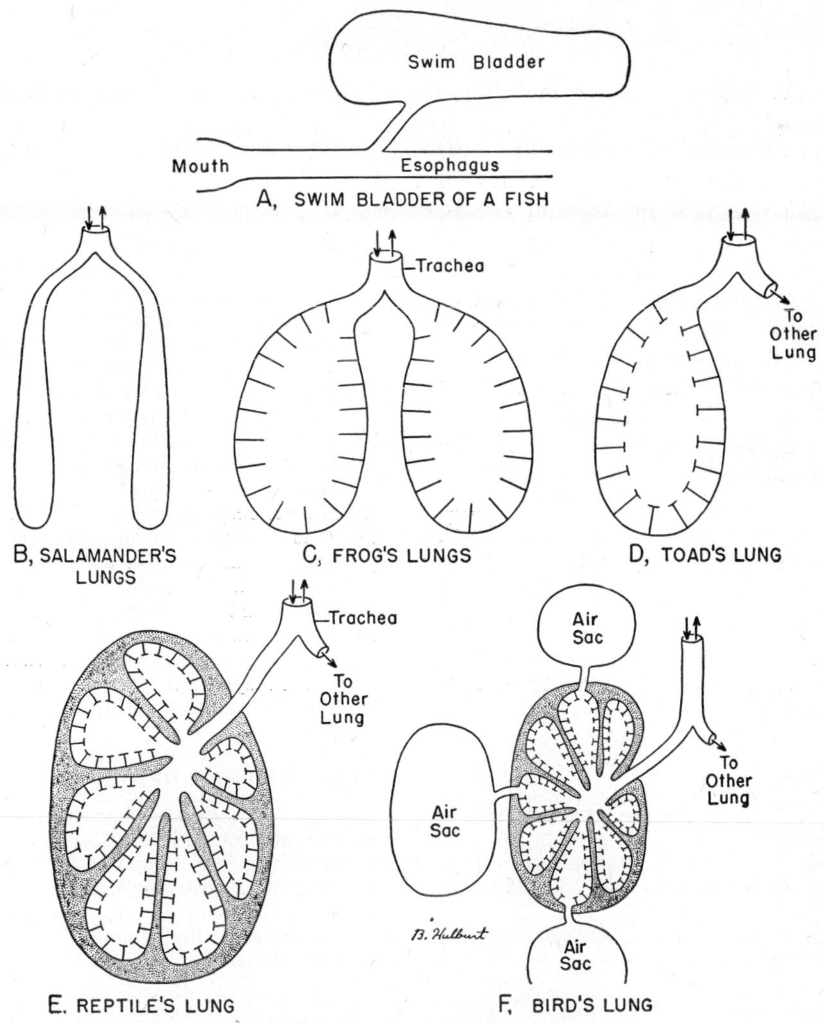

Figure 163. Some stages in the evolution of the lungs.

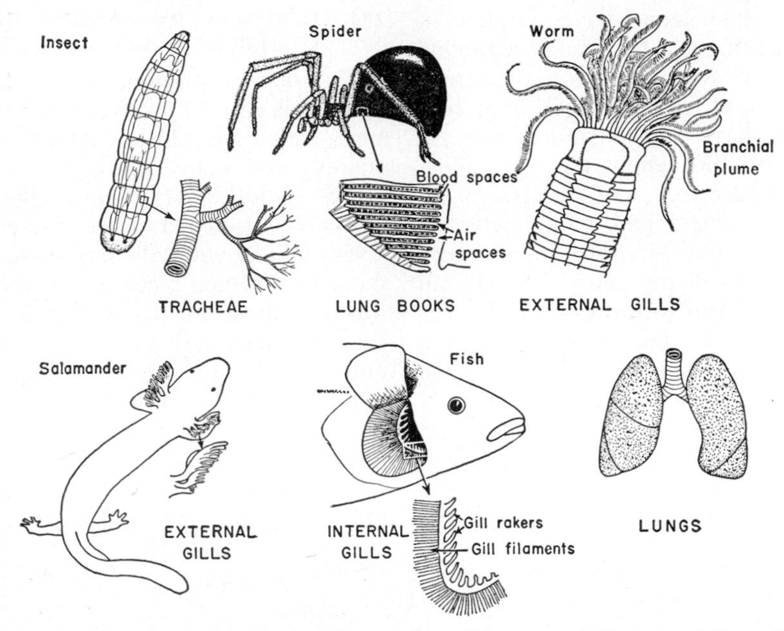

Figure 164. Respiratory devices in certain animals. (Hunter and Hunter: College Zoology.)

diffuses in the reverse direction. Fish suffocate in water lacking sufficient dissolved oxygen, such as that in stagnant ponds.

Insects have quite a different system for getting oxygen to the cells. In each section or segment of the body is a pair of holes, called spiracles, from which a tracheal tube extends into the body, branching and rebranching until it reaches each cell (Fig. 164). The body walls of insects pulsate, drawing air into the trachea when the body expands, and forcing air out when the body contracts. Thus, in contrast to a fish or crab, in which blood is brought to the surface of the body to be aerated in a gill, the tracheal system conducts air deep within the insect body, near enough to each cell so that it can diffuse in through the wall of the tracheal tube.

QUESTIONS

1. What function does oxygen perform in the body?
2. What are the parts of the human respiratory system? How is each adapted for its particular functions?
3. What is the difference between breathing and respiration?
4. What is meant by the "vital capacity" of a person? In what conditions is it increased or decreased?
5. What is the special importance of hemoglobin in respiration?
6. What is meant by oxygen "tension"?
7. What is acidosis? What causes it?
8. What does the respiratory center do and where is it located? Describe briefly the neural control of breathing.
9. Exactly what happens when too much carbon dioxide accumulates in the cells?
10. What has been the major trend in the evolution of the lungs?
11. How do gills operate?
12. How is respiration carried on in insects?
13. Trace the path of an anesthetic such as ether, from the ether cone over the nose to a cell in the brain. List each structure passed and the processes involved.
14. Differentiate between direct and indirect respiration, and between external and internal respiration.
15. Why is it difficult to breathe at high altitudes?

SUPPLEMENTARY READING

Some of the classic experiments on respiration by Hooke, Mayow, Priestley, Lavoisier and Barcroft are quoted in Fulton, *Selected Readings in the History of Physiology,* Chapter 4. Chapter 10 of W. B. Cannon's *The Wisdom of the Body* discusses some of the quantitative aspects of human breathing.

Chapter 19

The Digestive System

ANIMALS MUST provide their constituent cells with a wide variety of substances to be used as raw materials or as sources of energy for the synthesis and maintenance of the many compounds present in protoplasm. Animal cells require carbohydrates, fats, proteins, vitamins, water and minerals, some of which are molecules too large to pass readily into the cells in which they are to be used. The large, complex molecules in the food must be split into smaller ones that can enter the cell. The splitting of these large molecules into smaller ones, which occurs largely by enzymatic reactions, is called **digestion.**

Protozoa, and the simpler animals such as sponges and hydras, take food into **food vacuoles** within the cells in which digestion occurs. In the course of evolution, the higher, more complex animals have developed special organs for obtaining and digesting food. The products of digestion are then transported by the circulatory system to the cells of the body to be utilized. The digestive tract of man (Fig. 165) is essentially a long tube, composed of several separate organs, which carry out ingestion, digestion and absorption. The term **ingestion** refers to the mechanical taking in of food, chewing and swallowing. The passage of substances through the wall of the digestive tract, called **absorption,** can occur only after the food molecules have undergone digestion. The wall of the digestive tract is a semipermeable membrane which will permit only relatively small molecules to pass.

183. THE MOUTH CAVITY

Food is ingested through the mouth into the mouth cavity, supported by jaws and bounded on the sides by the teeth, gums and cheeks, on the bottom by the tongue, and on the top by the **palate.** The last, which separates the mouth cavity from the nasal cavity, has an anterior bony part, the hard palate, and a posterior fleshy part, the soft palate. The latter plays a role during swallowing in preventing food from passing up into the nasal cavity. The tongue, teeth and salivary glands, with ducts emptying into the mouth cavity, are all important in ingestion or digestion. The tongue and teeth in man have assumed an additional function, that of speech.

The Tongue. The tongue consists of several sets of striated muscles oriented in different planes, and by which it can move in or out, up or down, or from side to side. Muscles, of course, can only exert a pull, not a push; so that to stick the tongue out, the muscles running up and down and from side to side contract,

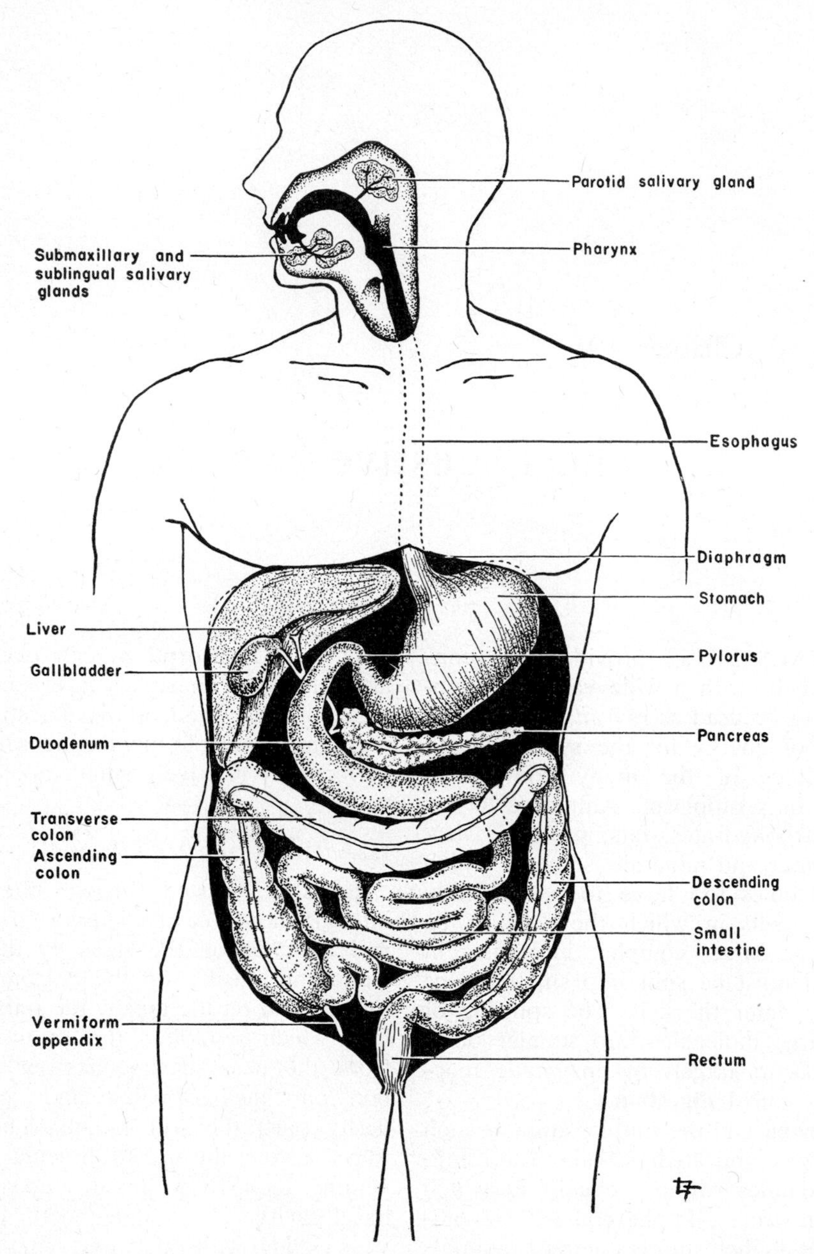

Figure 165. Diagram of the human body showing the parts of the digestive system. The liver, which normally covers part of the stomach and duodenum, has been folded back to reveal these and the gallbladder on its under surface.

while the muscles running from front to back relax. The tongue manipulates the food in such a way that each bit is pushed in turn between the teeth to be chewed. It then shapes the food into a spherical mass, called a **bolus,** to be swallowed. Swallowing is initiated when the tongue pushes a bolus into the pharynx. The tongue of some animals, such as frogs and anteaters, is long and covered with mucus, enabling them to reach out and catch insects in flight. The surface of the tongue contains groups of sensory cells called **taste buds,** which are stimulated by substances in solution and make possible our taste sensations.

The Teeth. The teeth of all vertebrates serve the common function of breaking food up into smaller particles, but they vary in size and shape according to the diet of the particular animal. Although they are superficially quite different, they have a common pattern (Fig. 166). The part of the tooth projecting above the gum is called the **crown,** that surrounded by the gum is called the **neck,** and below the neck is the **root,** embedded in a socket in the jaw bone. Each tooth is composed of several layers, a hard outer one, called **enamel,** an inner layer which is not quite so hard, called **dentin,** and an innermost pulp cavity, filled with blood vessels, nerves and soft tissue, the **pulp.** The enamel covers only the crown and upper part of the neck of the tooth. The tooth is fastened to the jawbone by a substance called **cement.** Dentin resembles bone in its composition and hardness, and consists of about 72 per cent inorganic matter (primarily calcium phosphate) and 28 per cent organic matter. Enamel, the hardest substance in the body, consists of almost 97 per cent inorganic matter.

Unlike the teeth of fish, amphibians and reptiles, which are simple, pointed cones in all parts of the mouth, the teeth of mammals are specialized to perform particular functions. The eight front, chisel-shaped teeth, called **incisors,** are used for biting. These are especially large in gnawing animals such as rats, squirrels and beavers. The four cone-shaped **canine** teeth, one in each front corner of the mouth, are used for tearing food. Flesh eating animals such as wolves and lions have large canine teeth or fangs—in fact, they are called canine or dog teeth because they are so large in dogs. Behind the canine teeth on each side and in each jaw are two **premolars** and three **molars,** with flattened surfaces adapted for crushing and grinding food. In flesh-eating animals the molars are knife-shaped instead of flat and are used in shearing flesh. Vegetarians such as horses and cows have large, flat molars for grinding food and well-developed incisors for cutting off grass. Upper incisors never develop in ruminants; these animals crop grass by pulling it with their tongue and upper lip across the cutting edge of the lower incisors.

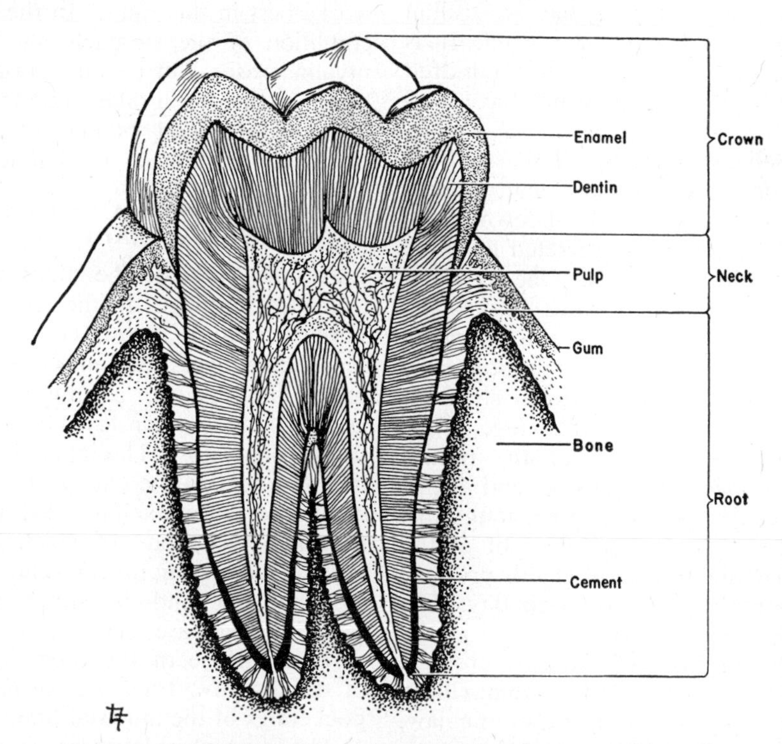

Figure 166. Diagram of a section through a human molar tooth.

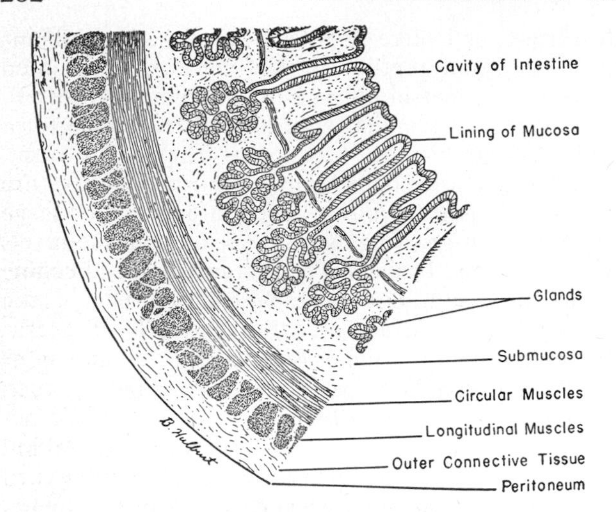

Figure 167. Cross section of the human intestine.

Since man's ancestors were omnivorous for millions of years, human teeth are relatively unspecialized. The last molar, or wisdom tooth, in man frequently fails to erupt, or, if it comes through the gum, it is often crooked and useless. This defect is due to a trend in the evolution of modern man toward a shortening of the jaws, with a resulting crowding of the teeth, which leaves inadequate space for the last molar. In primitive races, such as the Australian aborigines, this has not occurred. It is quite possible that in another hundred thousand years, man may not have wisdom teeth.

The Salivary Glands. To assist the food in moving down the throat, as well as to begin its chemical breakdown, there are two kinds of saliva, secreted by three pairs of salivary glands. One type of saliva is watery, to dissolve dry foods, and the other contains mucus, to lubricate the food as it passes down the esophagus and to make the food particles stick together in a bolus for swallowing. Saliva also protects the lining of the mouth against drying out, cleans it, and facilitates speech by moistening the tongue so that it does not stick to the roof of the mouth. The three pairs of glands produce about 2 quarts of saliva each day. The **parotid glands** (Fig. 165), located in the cheek just in front of the ear, produce only a watery saliva; the **submaxillary glands,** in front of the angle of the jaw, produce a mixture of watery and mucous saliva, as do the **sublingual glands,** located on the floor of the mouth under the tongue. Each glandular mass weighs about an ounce and is connected to the mouth cavity by a duct.

Saliva is one of the digestive juices; it contains the enzyme **ptyalin,** which converts starch into maltose, and a small amount of **maltase,** which splits maltose to glucose. Saliva is normally slightly acid, with a *p*H of 6.5 to 6.8, and ptyalin works best in this range. In the very acid condition of the stomach, the action of ptyalin ceases, but because food is swallowed in masses, it takes time for the acid to penetrate, and ptyalin continues to work in the stomach until that happens.

184. MICROSCOPIC ANATOMY OF THE DIGESTIVE TRACT

All the parts of the digestive system from the esophagus to the rectum have a similar structure and are composed of the same three layers: an inner mucous membrane or mucosa, a muscular middle layer, and an outer layer of connective tissue (Fig. 167). The inner lining of the mucosa, next to the cavity of the tract, is composed of epithelial cells, usually of columnar type, some of which secrete the viscous lubricating mucus. The mucosa in the stomach and intestines is greatly folded to increase the secreting and absorbing surface of the tube. The glands of the digestive tract are formed as outpocketings of the mucosal lining.

The muscular layer is composed of

smooth muscle except in the upper third of the esophagus, where there is striated muscle. Most of the digestive tract has two distinct layers of muscle: an inner one with the fibers arranged circularly, and an outer one with the fibers arranged longitudinally. By contracting these layers alternately or in unison, the digestive organs can perform a variety of movements to churn the food and move it along.

The outermost layer of the digestive tract is made of strong, flexible connective tissue fibers covered by a smooth sheet of the peritoneum. This sheet secretes a fluid to lubricate the surface of the stomach and intestines and reduce friction as the parts of the digestive tract move and rub against each other and the abdominal wall. The esophagus, which lies buried in the muscles of the neck and chest, has no peritoneal covering.

The walls of the digestive tract are richly supplied with nerves to coordinate the actions of the various parts, and with blood and lymph vessels to supply the cells with food and oxygen, to drain away wastes and to carry the absorbed food to a place of storage.

185. THE PHARYNX

Food passes from the mouth cavity into the pharynx, the cavity behind the soft palate where the digestive and respiratory passages cross. It has no less than seven tubes connecting with it: the two internal nares from the nasal cavity, the connection from the mouth, the glottis opening into the trachea, the esophagus going to the stomach, and two eustachian tubes going to the middle ear cavity to equalize pressure on the two sides of the ear drum (see p. 376). An automatic mechanism keeps the substances conducted by these tubes in their proper channels.

Swallowing. When the food is ready to be swallowed, a complicated series of reflexes moves it from the mouth to the stomach. The first part of the swallowing act is under voluntary control: The tongue is raised against the roof of the mouth, and the bolus of food between the tongue and palate is pushed into the pharynx by a wavelike movement of the tongue. When swallowing begins, respiration is stopped momentarily by a reflex mechanism to prevent the passage of food into the larynx or trachea. Once the food is in the pharynx, there are four possible exits for it, only one of which is desirable. Normally, the reflex closing of the other three forces the food down the esophagus when the pharynx contracts. The opening to the nasal cavity is closed by the reflex elevation of the soft palate (Fig. 168), while the tongue is held against the roof of the mouth, preventing the food from

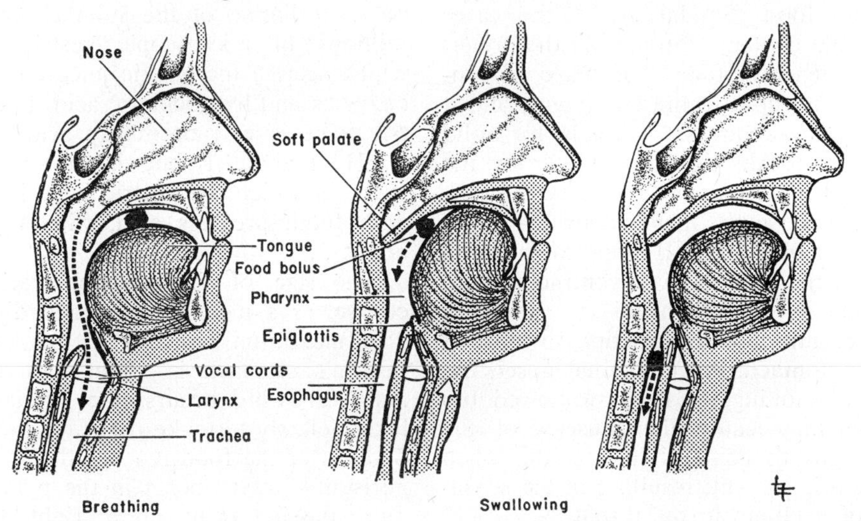

Figure 168. Diagram of the position of the tongue and epiglottis during breathing (left) and swallowing (center and right). Note how a food bolus is pushed from the mouth into the pharynx by the tongue to initiate swallowing (center).

returning. The opening into the larynx is closed by the contraction of muscles which raise the entire larynx, bringing the opening, the glottis, under the fold of tissue called the epiglottis. This action completely closes it and prevents food from going down the trachea, and at the same time enlarges the opening of the esophagus to facilitate the passage of the bolus. The raising of the larynx can be observed in the bobbing up of the "Adam's apple" (which is simply the outside of the larynx) each time one swallows.

186. THE ESOPHAGUS

The esophagus, or gullet, into which food passes next, is a muscular tube leading directly downward from the pharynx to the stomach. It passes between the lungs and behind the heart, penetrating the diaphragm to reach the stomach. It has well-developed muscular walls: the upper third contains striated muscle; the lower two thirds, smooth muscle.

The contraction of the muscles in the wall of the pharynx and the presence of the bolus in the upper part of the esophagus cause a single, powerful, rhythmic wave of muscle contraction in the wall of the esophagus, called **peristalsis,** which pushes the bolus down to the stomach. This wave is preceded by one of relaxation which dilates the tube to make room for the food. Similar peristaltic waves move the contents through all the organs of the digestive tube. The wave of contraction is rapid in the esophagus, and it takes only about six seconds for solid food to reach the stomach from the mouth. Liquids are swallowed even faster, since they simply flow downward under the force of gravity. If some of the food escapes the first wave of contraction and remains in the esophagus, it stimulates another muscular contraction to move it to the stomach. An emotional upset, excessive smoking, or food swallowed too hastily may cause the muscles of the esophagus to contract in a spasm when no food is present, resulting in the sensation of a "lump in the throat."

The opening from the esophagus to the stomach is controlled by a ring of smooth muscle, called a **sphincter.** Normally, this is closed; it opens reflexly only when a wave of contraction in the esophagus has pushed a bolus of food against it. As liquids are swallowed, they fall through the esophagus to the sphincter faster than the accompanying wave of contraction of the esophagus muscles, but the ring of muscle does not open until the peristaltic wave reaches it. Similar sphincters control the movement of food at three other places in the digestive tract: the opening of the stomach into the small intestine, the opening of the small intestine into the large intestine, and the opening at the anus where the digestive tract ends.

187. THE STOMACH

When the food leaves the esophagus, it enters the stomach, a thick-walled, muscular sac on the left side of the body just beneath the lower ribs. This organ is divided into three main regions (Fig. 169): an upper part, nearest the heart, called the **cardiac region;** a deep part, below this, called the **fundus;** and an area extending to the opening of the small intestine, called the **pyloric region.** The muscular layers of the stomach wall are exceptionally thick, being composed of a diagonal layer of fibers in addition to the circular and longitudinal fibers found elsewhere in the digestive tract. The mucosa, or lining, of the stomach contains millions of microscopic gastric glands which secrete the gastric juice containing enzymes and hydrochloric acid. Pure gastric juice is extraordinarily acid, having a *p*H of about 1, but the stomach contents, in which the gastric juice is mixed with food, are less acid, with a *p*H of about 3.

The size of the stomach varies, of course, as a meal is eaten and digested. The maximum capacity of the average person's stomach is about 2½ quarts. As swallowing occurs, the stomach relaxes reflexly to make room for the food. Soon after the food reaches the stomach, peristaltic waves begin in the pyloric region, passing from left to right, toward the opening into the intestine. The rest of the stomach, containing most of the food, remains quiescent at this stage. As digestion proceeds, the waves originate farther and farther to the left, and finally the entire stomach wall, from cardiac to pyloric end, is swept by deep, powerful peristaltic waves which mix the contents and mechanically break the larger bits of food into smaller ones. The food is now the consistency of cream soup, and digestion is well advanced. At intervals, the pyloric sphincter relaxes, and a small amount of **chyme** (as the contents of the stomach and small intestine are called) is pushed into the small intestine by the contraction of the stomach. The opening of the pyloric sphincter at the proper time is regulated by a mechanism whose action is not completely known. The strength of the peristaltic waves in the stomach and the consistency of chyme are important factors in this regulation. In one to four hours, depending upon the amount and kind of food eaten, the stomach is emptied. Carbohydrate foods leave the stomach more rapidly than proteins, and proteins more rapidly than fats. When the stomach is empty it continues to contract, and it is this squeezing of the empty stomach which stimulates nerves in the wall, causing hunger pangs.

Vomiting. Occasionally something may be taken into the stomach which should be ejected. To make this possible, most mammals—with the exception of rabbits, rats and other rodents—have a mechanism called the vomiting reflex. Vomiting may be caused by mechanical irritation of the pharynx (sticking a finger in the throat is used to induce it when some poisonous substance has been swallowed) or by disturbances in the semicircular canals of the ears, as in seasickness. It is controlled by a center in the midbrain, known as the vomiting center, which coordinates the contraction of the stomach and of the muscles of the abdominal wall, the closing of the pyloric sphincter, the opening of the cardiac sphincter and the closing of the glottis.

188. THE SMALL INTESTINE

The small intestine, into which the chyme passes by the force of the peristaltic waves in the stomach, is a long coiled tube, about 22 feet long and approximately 1 inch in diameter. The greater part of digestion and almost all absorption occur in this organ. Only alcohol and a few poisons can be absorbed through the stomach wall.

The length of the intestine is correlated with the type of diet: Plant-eating animals have a long small intestine; meat-eating animals have a short one; and omnivorous

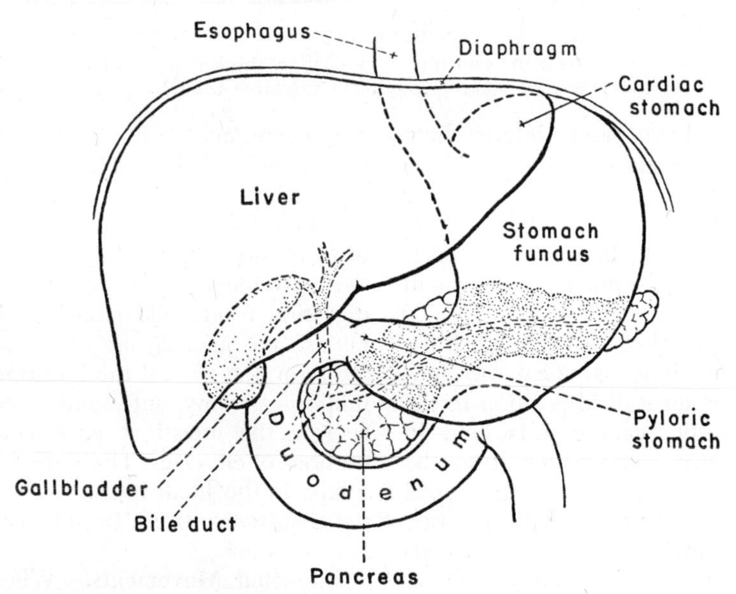

Figure 169. Diagram of the relations of the stomach, liver, pancreas and duodenum.

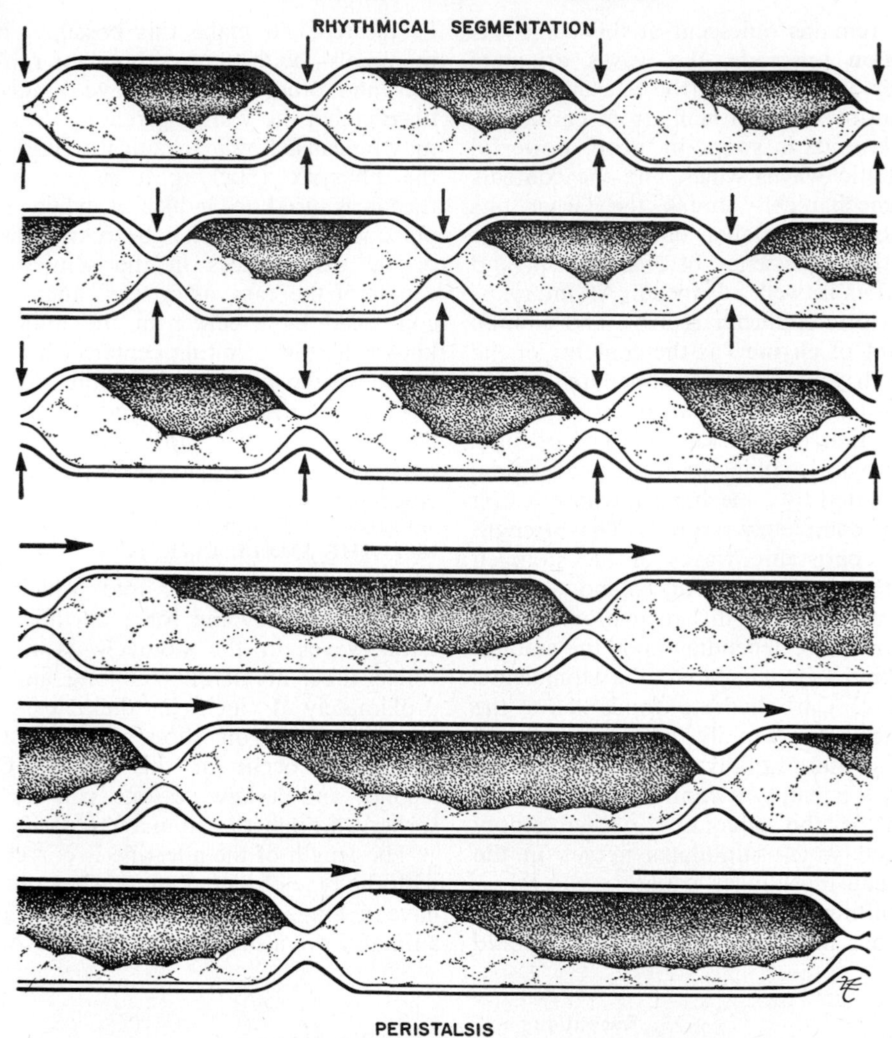

Figure 170. Diagrams to illustrate the churning action of rhythmical segmentation in the intestine, and the movement of food through the digestive tract by peristalsis.

ones, like man, have one of intermediate length. An interesting example of this is to be seen in the frog: The frog larva or tadpole is herbivorous and has a long small intestine, but the adult frog is carnivorous and has a much shorter small intestine.

The first segment of the small intestine, about 10 inches long, is called the **duodenum.** It occupies a fixed position in the abdominal cavity because it is held in place by ligaments connecting it to the liver and stomach, as well as to the dorsal body wall. The rest of the small intestine (and most of the large intestine) is attached only to the dorsal body wall by a thin, transparent membrane called the mesentery, and so is able to move about with considerable freedom. The supporting mesentery provides a means for nerves and blood vessels to pass from the body wall to the intestine. In the duodenum two extremely important juices reach the digesting food: bile from the liver, and pancreatic juice from the pancreas. In addition, the wall of the intestine contains millions of tiny, intestinal glands which secrete the intestinal juice containing a number of enzymes. These three juices are mixed in the small intestine and complete the digestive process begun in the mouth and stomach.

Intestinal Movements. When food is present, the small intestine is more or less constantly in motion, and this intestinal motility is independent of any stimulation from outside, since it continues after all the nerves to the intestine are cut. There are two types of intestinal movements, the **peristaltic contractions,** which move the chyme along, and the **churning movements,** which simply mix the intestinal contents (Fig. 170). A single peristaltic wave does not move far in the intestine; usually after 4 or 5 inches it is dissipated, though occasionally rapid movements, called peristaltic rushes, sweep along for considerable distances. The churning movements are caused by alternate contractions and relaxations of successive segments of the intestine, which are repeated about ten times a minute. These movements complete the mechanical breaking up of the intestinal contents, mix them with the various digestive juices, and ensure that all parts of the intestinal contents will be brought in contact with the intestinal wall so that the digested food may be absorbed into the blood stream. In each part of the intestine these churning movements continue for a time; then a peristaltic wave carries the contents to the next section, and the churning movements begin again. In this way the contents are ultimately carried through the small intestine to the large intestine in about eight hours. By the time the remains of the food pass from the small intestine, digestion has been completed and the digested food particles have been absorbed. The materials passing into the next section of the digestive tract, the large intestine, consist of indigestible matter and large quantities of water derived from the food taken in or from the digestive juices.

189. THE LIVER

Because of its contribution of the digestive juice, **bile,** the liver is vitally important to digestion. It is the largest gland in the body, occupying the whole upper part of the abdominal cavity, just below the diaphragm. The greater part of it lies on the right side of the body, but it extends to the left side and partly covers the stomach.

Bile is formed in all parts of the liver and is collected by a branching system of fine ducts, leading into large ducts, which empty into the **gallbladder** (Fig. 169), where the bile is stored until needed. Here, water and salts are removed from the bile, so that it may be greatly concentrated. Bile is secreted by the liver cells constantly, but passes to the duodenum only after food has been eaten. Its passage down the common bile duct is brought about by the contraction of the muscular wall of the gallbladder. Being alkaline, it helps to neutralize the acid chyme.

Bile contains no digestive enzymes, but provides the **bile salts,** which act as emulsifying agents so that when the churning movements of the intestine occur, a fine emulsion, or suspension, of fat droplets within the chyme is produced. As the fat is broken into small droplets, a greater surface area is created in which the fat-splitting enzyme, **lipase,** may work; consequently an increase in the speed of the digestion of fat occurs. When bile salts are absent from the intestine, as when the bile duct is obstructed, the absorption as well as the digestion of fats is impaired, and much of the fat eaten is lost in the feces. The bile salts themselves are carefully conserved by the body, being reabsorbed from the lower part of the intestine and carried back to the liver to be secreted over again. Another constituent of the bile, cholesterol, plays no part in digestion, but may cause trouble. Cholesterol is not very soluble in water, and under certain circumstances the cholesterol in the gallbladder may be concentrated, by the removal of water, to the point of precipitation, producing the hard little pellets called "gallstones." These may obstruct the bile duct and stop the flow of bile.

The color of bile depends on its pigments. There are two main pigments, a red one and a green one, which are present in different proportions in different vertebrates, so that the color of bile ranges between them; human bile is a deep yellow-orange. In the intestine the bile pigments undergo chemical alterations which change them to a dark brown, and it is to these pigments that the brown color of the feces is due. When the bile duct is obstructed so that bile pigments are not present in the feces, the latter are whitish or clay-colored. In such a condition, and in certain impairments of the functioning of the liver, the bile pigments accumulate in the blood and tissues, giving a yellowish tinge to the skin, a condition known as **jaundice.**

190. THE PANCREAS

The second main digestive gland, the **pancreas,** is an irregular, diffuse mass lying between the stomach and the duodenum (Fig. 169). Its secretion, the pancreatic juice, containing a number of enzymes, passes into the duodenum by way of the pancreatic duct. In addition, certain cells of the pancreas, called the **islets of Langerhans,** secrete into the blood stream the hormone **insulin** (p. 387). These two secretions are entirely separate and unrelated. It just happens that in man, and most vertebrates, the two types of cells occur together in the same gland; in certain types of fish the two types are spatially separated into two different glands.

The pancreatic juice is a clear, watery fluid that is quite alkaline, with a *p*H of about 8.5. It is the main factor in neutralizing the acid chyme. The enzymes secreted by the pancreas and the intestinal wall will not work in an acid medium, hence the need for a neutralizing agent after food has been received from the stomach. Experiments indicate that an average person secretes about 1 or 1.5 liters of pancreatic juice each day. If the pancreatic duct becomes blocked so that the pancreatic enzymes are unable to reach the intestine and act on the food, the person develops a tremendous appetite and eats a great deal. In spite of this he loses weight, which demonstrates the importance of the pancreatic juice in normal digestion.

191. THE ABSORPTION OF FOOD

After food has been digested, it must be absorbed into the body through the lining of the digestive tract. Most of the absorption is done by the small intestine, particularly the lower part of this organ. Water is absorbed by the colon, but almost all organic and inorganic substances are absorbed through the small intestine. The lining of the intestine is greatly folded; this increases the area through which absorption may occur and hence the rate of absorption. In addition, countless small finger-like projections called **villi,** each of which contains a network of blood capillaries and a lymph capillary in its center, cover the inner surface of these folds (Fig. 171).

Absorption is a complex process, involving in part the simple diffusion of substances from the cavity of the intestine through the lining cells and into the blood or lymph capillaries. A number of substances are absorbed despite the fact that there is a greater concentration of the substance in the blood stream than in the intestine. The cells lining the intestine must do work and "pump" these substances into the blood stream against a diffusion gradient. This process is analogous to secretion, in which cells do work to move substances from one region to another. The cells lining the intestine constitute a semipermeable membrane, and as such permit certain substances, such as amino acids and glucose, to pass through while preventing the passage of others, such as intact protein and starch molecules. Glucose and amino acids are absorbed into the blood capillaries and carried by the hepatic portal vein to the liver for storage, later to be distributed to the rest of the body. We shall return to the importance of the liver in storing these substances, changing them into others, and maintaining a relatively constant level of them in the blood.

The products of fat digestion enter the body by a different route. The glycerol and fatty acids, as they pass through the mucosal cells of the villus, are resynthesized into molecules of fat which aggregate as fine globules too large to enter the capillaries of the blood stream. They are taken up by the lymph capillaries; during the absorption of a meal rich in fat the lymphatic vessels of the intestine have a milky color due to this fat emulsion. These intestinal lymph vessels eventually empty into the great thoracic duct and from there into the blood stream, since the great thoracic duct empties into the left shoulder vein. The fat eventually gets into the blood stream and is distributed to the body, but its route is less direct than that of sugars and amino acids.

192. THE LARGE INTESTINE AND RECTUM

The material remaining after the nutrients have been absorbed passes from the small intestine into the large intestine, or **colon,** a U-shaped tube, larger in diameter and with thicker walls than the small intestine. It consists of three parts, the ascending, transverse and descending colons (Fig. 165). The small intestine empties into the side of the colon a short distance from its end, leaving a large blind sac, called the **cecum,** at the tip of which is a small projection about the size of the little finger, the **appendix.** The cecum and appendix were larger in our remote ancestors and active in the digestion of vegetable materials. Herbivores such as rabbits and guinea pigs still have a large functional cecum. As we evolved and changed our food habits, the cecum became superfluous and degenerated. From the junction of the small and large intestines, the ascending colon runs up the right side of the body to the level of the liver, makes a right-angle turn, and, as the transverse colon, runs across the abdominal cavity just below the liver and stomach. When it reaches the left side of the body it makes another right turn, becomes the descending colon, and passes down the left side of the body to the rectum.

The material reaching the colon from the small intestine has had most of the nutrients absorbed from it, but it is still liquid, because, although some water is

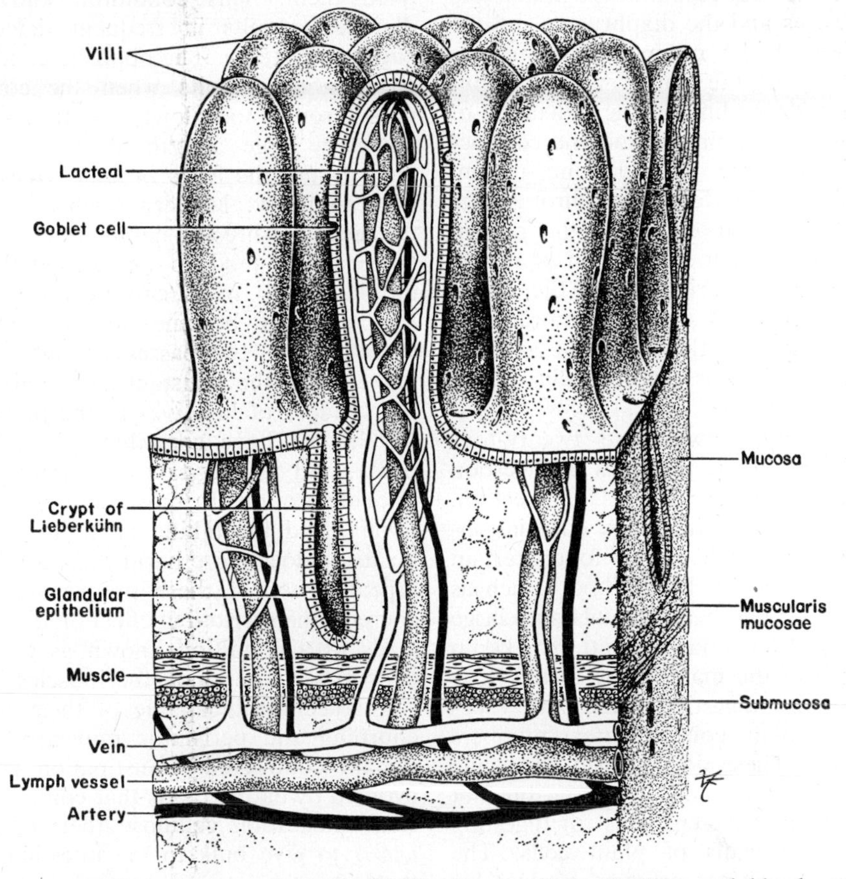

Figure 171. Detailed diagram of intestinal villi, showing their structure and blood and lymph supply.

absorbed in the small intestine, almost as much water is added by the bile and pancreatic juice. The main function of the colon, in addition to transporting the wastes to the rectum to be ejected from the body, is to absorb water and reduce the wastes to a semisolid state. The same two movements occur in the colon as in the small intestine—churning and peristaltic — although both are ordinarily slower and more sluggish than those in the small intestine. Periodically, more vigorous peristaltic movements force the contents along, until they finally reach the rectum. These occur especially after eating, because of a reflex mechanism whereby the filling of the stomach stimulates an emptying of the colon. This mechanism, called the **gastrocolic reflex,** is responsible for defecation usually occurring after a meal.

Defecation is partly voluntary, depending upon the contraction of the abdominal wall muscles and the diaphragm, and the relaxation of the outer ring of muscle (sphincter) of the anus, and partly involuntary, depending on the relaxation of the inner anal sphincter and the contraction of the large intestine and rectum which forces the feces out through the anus. It is the distention of the rectum and the consequent stimulation of nerves in its walls that bring about the desire to defecate. If this signal is ignored, the rectum adapts to the new size and the stimulus diminishes and finally disappears.

It takes from twelve to twenty-four hours for the waste products of digestion to pass through the colon and rectum. The end product, the **feces,** contains indigestible remnants of the food, certain substances secreted by the body, such as bile pigments and heavy metals, and large quantities of bacteria. The latter make up about half of the mass of the feces.

The entire digestive tract, and especially the colon, contains great numbers of bacteria. These do no harm and some are useful to their host. Bacteria, but not most animals, have enzymes for digesting the cellulose walls of plant cells. The cecum of herbivores provides a place for cellulose to be digested by the bacteria present. Bacteria synthesize a variety of vitamins and supply a significant fraction of our daily requirements for many different vitamins. Some bacteria in the colon produce highly poisonous and odorous substances. Only small amounts of these are absorbed and the detoxifying actions of the liver prevent their accumulating in the blood stream in significant quantities.

The headaches and other symptoms which usually accompany constipation are not caused by absorption of "toxic substances" from the feces, but are due to the distention of the rectum. If the rectum is packed with some inert substance such as cotton, the same symptoms appear.

If the lining of the colon is irritated, as in an infection such as dysentery, peristalsis is increased and the intestinal contents are passed through rapidly, with only a small amount of the water removed from them. This condition, known as **diarrhea,** results in frequent defecation and watery feces. The opposite condition, **constipation,** results when the contents pass through too slowly, so that an abnormally large amount of water is removed, and the feces become excessively hard and dry. Modern civilization, and the sedentary life and nervous strain which it produces, tend to decrease intestinal motility and may cause constipation. People often become unduly alarmed when a day or two passes without a bowel movement (the constant exhortations of the vendors of laxatives in the press and on the air have something to do with this!), and usually take something to induce defecation. In most cases nothing is wrong, and the taking of cathartics and laxatives does more harm than good. The repeated use of these preparations, because of their irritating effect on the colon, leads to a condition known as cathartic constipation, in which the muscles in the colon become incapable of their normal churning and peristaltic movements and remain contracted. Constipation can be avoided by eating foods that contain sufficient indigestible cellulose fibers ("roughage") to give bulk to the intestinal contents.

193. DISORDERS OF THE DIGESTIVE TRACT

We have already discussed certain disorders of the digestive system—gallstones, diarrhea and constipation. Another common complaint is "indigestion," which may result from overeating, from the irritation of the stomach lining by an excess of alcohol, or from some disturbance in the functioning of the colon. Since the transverse colon lies just beneath the stomach, it is sometimes difficult to distinguish the source of pain. Any constantly recurring indigestion should receive medical attention. Constant nervous strain may impair the functioning of the digestive system and produce indigestion. Whenever you hear or read an advertisement for some remedy for "acid stomach," remember that the stomach is *normally* acid, and must be acid for pepsin to act, and to stimulate the secretion of secretin, which causes the pancreas to give off pancreatic juice.

About one person in a hundred has a stomach **ulcer,** an area in the stomach where the lining is eroded. This condition is much more frequent in men than in women. The cause of ulcer formation is not clear, but may depend on the secretion of an unusually acid gastric juice, mechanical injury of the wall by some sharp object, or by some interference with the blood supply to the cells lining the stomach so that they are unable to secrete enough protecting mucus. The pain associated with an ulcer is believed to be caused by the contact of the acid gastric juice with the eroded surface, and the danger lies in the fact that the erosion of the wall may increase and reach a large blood vessel, resulting in hemorrhage and death; or the wall may be eroded all the way through so that the stomach contents, which contain bacteria, pass into the peritoneal cavity, causing general inflammation of the peritoneum (peritonitis).

One of the weak spots of the digestive tract is the appendix. The cause of the inflammation of this organ (**appendicitis**) is not known, but again, the danger lies in the possible rupturing of its wall and the release of bacteria into the peritoneal cavity, which may cause peritonitis and death. The surgical removal of the inflamed organ before it bursts is an effective cure.

194. THE CHEMISTRY OF DIGESTION

The chemical reactions which result in the breakdown of food molecules to simple substances which can be absorbed through the intestinal wall are regulated by **enzymes;** these, you will remember, are organic catalysts which enable the body to bring about chemical reactions that can be duplicated in the test tube only at high temperatures and pressures, if at all. Enzymes are *specific;* each will catalyze a certain reaction and no other. For example, a protein-splitting enzyme has no effect on a molecule of fat. The chemistry of digestion is best understood by studying the enzymes involved (Table 5). The names of most enzymes are a combination of *-ase* with the name of the substance worked upon. Thus, lipase means the enzyme acting on fats (lipids). Some of the enzymes, discovered in early days before this convention was agreed upon, have names ending in *-in.* An attempt was made at one time to change all the old names to ones ending in *-ase,* but it was unsuccessful.

Saliva. The first enzyme encountered by food in the digestive tract is **ptyalin,** present in the saliva. Ptyalin attacks cooked starch and other complex carbohydrates and splits them into double sugars, principally maltose, the malt sugar of "malted milk." Although this chemical splitting of starches by ptyalin occurs fairly rapidly, it cannot be completed while the food is in the mouth in the normal time required for chewing. Most of the splitting occur within each bolus of food in the stomach before the gastric juice penetrates and renders the food too acid for ptyalin to work. Different people have saliva containing different amounts of ptyalin, so that the time required to digest a given amount of starch varies. You can test the activity of your own saliva by a simple experiment based on the fact that starch plus iodine gives a blue color, while sugar plus iodine does not. Put 10 ml. of a dilute boiled starch solution in a test tube and add 1 ml. of saliva.

Table 5. ENZYMES IMPORTANT IN DIGESTION

LOCATION	ENZYME	SOURCE	OPTIMUM *p*H	SUBSTANCE ACTED UPON	PRODUCT
Mouth	Ptyalin	Salivary glands	Neutral	Cooked starch	Double sugars
Stomach	Pepsin	Gastric glands	Acid	Proteins	Proteoses and peptones
Stomach	Rennin	Gastric glands	Acid	Milk proteins	Denatured (curdled) protein
Small intestine	Trypsin	Pancreas	Alkaline	Proteins or peptones	Peptones and amino acids
Small intestine	Lipase	Pancreas	Alkaline	Fats	Fatty acids and glycerol
Small intestine	Diastase	Pancreas	Alkaline	Intact or partly digested starch	Double sugars
Small intestine	Peptidases	Intestinal glands	Alkaline	Peptones	Amino acids
Small intestine	Maltase	Intestinal glands	Alkaline	Maltose (double sugar)	Glucose (simple sugar)
Small intestine	Sucrase	Intestinal glands	Alkaline	Sucrose	Simple sugars
Small intestine	Lactase	Intestinal glands	Alkaline	Lactose (milk sugar)	Simple sugars
Small intestine	Enterokinase	Intestinal glands	Alkaline	Trypsinogen (inactive)	Trypsin (active)

Mix thoroughly and at the end of each minute after adding the saliva, remove a drop and add it to a drop of iodine solution. At first the solution will be blue; later samples will be violet or red, and finally the samples will be colorless, indicating that the starch has been digested to maltose. The red color is due to the intermediate substances formed in the breakdown of starch to sugar. All enzymes are proteins and "denatured" (changed) by heating, so that they no longer function. You can prove this to be an enzyme-controlled reaction by repeating the test, using saliva that has been boiled. Saliva also contains some maltase, which breaks maltose down to form glucose.

Gastric Juice. The second enzyme to act on the food in its passage through the digestive tract is **pepsin,** secreted by certain cells of the gastric glands in the fundus of the stomach. Other cells in the fundus secrete hydrochloric acid. The glands in the pyloric region secrete mucus, and the entire lining of the stomach contains other mucus-secreting cells.

Pepsin splits the large molecules of protein into smaller, soluble molecules known as proteoses and peptones, but is unable to complete the breakdown of these to their constituent amino acids. This is done by other enzymes in the intestine. Pepsin is active only in an acid medium and any acid will speed up its action, but hydrochloric acid is most effective. In contrast, the protein-splitters of the intestine, trypsin and the peptidases, operate only in an alkaline medium. All the body cells have intracellular protein-splitting enzymes (called **cathepsins**) which act in an essentially neutral medium.* If a piece of tissue is removed from the body and kept moist and warm, its cathepsins will eventually dissolve it. Pepsin is secreted by the gastric glands in an inactive form, called pepsinogen; this is converted into pepsin when it comes in contact with acid.

Gastric juice also contains an enzyme known as **rennin,** which acts on a specific protein present in milk, casein, changing

* Plant cells also have protein-splitting enzymes, which may be present in large quantities. For example, pineapple contains so much protein-splitting enzyme that one cannot prepare a gelatin containing fresh pineapple. After pineapple has been heated, as in canning, the enzyme is destroyed, so that canned pineapple can be used in gelatin salads.

it from a soluble to an insoluble form, thereby curdling the milk. This is necessary to keep milk in the stomach long enough for pepsin to act upon it, since, if milk remained liquid, it would pass through as rapidly as water. For centuries rennin has been extracted from the stomachs of calves and used in making certain types of cheese, and the milk pudding "junket" is made with rennin.

Since the stomach wall is made of protein, you may wonder why it is not digested by the pepsin and hydrochloric acid it secretes. The stomach has a number of safeguards against this, but when they fail, the wall is digested and an ulcer results. The entire stomach wall is normally lined with a thick coat of mucus which protects it from the active enzyme. This coat is constantly renewed by the mucus-secreting cells lining the stomach. When the stomach is empty, little gastric juice is secreted, and that present has a low acidity, so that the pepsin is not very active. At the end of a meal, a little of the alkaline intestinal juice passes through the pylorus into the stomach and neutralizes the acid there. All these factors serve to protect the lining of the stomach from the action of pepsin.

Pancreatic Juice. In the small intestine the enzymes of the pancreatic juice continue the digestion of carbohydrates and proteins begun by the saliva and gastric juice and perform almost all the digestion of fats. The three main enzymes secreted by the pancreas are **trypsin, diastase** and **lipase.** Trypsin acts on intact or partially digested protein molecules, breaking them up into simple groups of several amino acids. Since trypsin can digest intact proteins, some mechanism is necessary to protect the pancreas itself from digestion. This enzyme is not secreted in an active form, but as an inactive substance called trypsinogen. The trypsinogen passes down the pancreatic duct and is acted upon in the intestine by another enzyme, **enterokinase,** secreted by glands in the wall of the intestine, which converts it into active trypsin. If active trypsin does get into the pancreas accidentally, it will digest it and cause death. The second pancreatic enzyme, diastase, supplements the action of ptyalin and digests cooked or uncooked starches to maltose. It is more important than ptyalin in the digestion of starches and has a longer time to work on them. Most animals do not have ptyalin in their saliva, but rely on pancreatic diastase to digest the starches eaten. Lipase, the third pancreatic enzyme, breaks molecules of fat into glycerol and fatty acids. Since the pancreas is the only gland secreting appreciable amounts of a fat-splitting enzyme, there are more marked disturbances in fat digestion than in protein or carbohydrate digestion when the flow of the pancreatic juice is interfered with. In such cases, fat passes through the digestive tract essentially unchanged and is lost in the feces.

Intestinal Juice. The walls of the intestine contain many small glands which secrete additional enzymes to complete the digestion of proteins and carbohydrates. It is difficult to measure the amount of intestinal juice secreted, but it is estimated to be about 3 liters daily. The intestinal glands secrete a number of different **peptidases,** which split small groups of amino acids (peptides) into individual ones which can be absorbed. **Maltase, sucrase** and **lactase** are enzymes that act on malt sugar, cane or beet sugar and milk sugar to split them into simple sugars such as glucose. The intestinal glands secrete enterokinase, the enzyme which transforms trypsinogen into trypsin, and large amounts of mucus to lubricate the passage of food.

By the time the chyme reaches the lower part of the small intestine, digestion is complete and absorption occurs rapidly; the proteins have been split into their component amino acids, carbohydrates into simple sugars, and fats into glycerol and fatty acids.

In addition to the enzymes secreted by the digestive tract, other digestive enzymes are secreted by bacteria in the colon. In man and the carnivores bacterial enzymes are not very important, but the cellulose-digesting enzymes secreted by bacteria are vital to herbivorous animals.

195. METHODS OF STIMULATING THE DIGESTIVE GLANDS

Each of the glands secreting enzymes must be stimulated to release its product at the proper time. It would be wasteful and even harmful if the glands secreted constantly. The coordination of the flow of the digestive juices with the presence of food is achieved in two ways—by the nervous system and by hormones. Hormones are chemical substances secreted in one part of the body and carried by the blood stream to another part where they produce a specific effect.

The salivary glands are controlled entirely by the nervous system. Either smelling or tasting food stimulates nerve cells in the nose or mouth to send impulses to the salivation center in the medulla of the brain; these are relayed to the salivary glands, causing them to secrete saliva. The mere presence in the mouth of tasteless, odorless objects, such as pebbles, stimulates other cells in the lining of the mouth which act similarly to cause salivation. Or impulses may come from the higher centers of the brain —simply seeing or thinking of food can bring about salivation. The salivary glands will respond, then, to chemical, mechanical or psychic stimuli. It is necessary to have the flow of saliva under rapid nervous control, since the food remains in the mouth such a short time.

Much of our knowledge of the mechanism controlling the secretion of gastric juice we owe to the Russian physiologist, Pavlov, who devised many experimental techniques (such as the Pavlov pouch) and performed many critical experiments. One of these was to sever the esophagus of a dog, bringing the two cut ends to the surface of the neck so that when the dog was fed, the food, instead of going to the stomach, went out through the hole in the neck. Although the food was not used, such a "sham-feeding" caused a flow of gastric juice, about one quarter of the normal amount (the normal amount of gastric juice in man is 400 to 800 ml. per meal). This quarter of the normal flow is stimulated by nerve impulses originating in the taste buds, or in the eye, and passing to the brain, whence they go to the stomach. The flow is completely abolished when the nerves to the stomach are cut. By putting food into the cut end of the esophagus leading to the stomach, and preventing the dog from seeing, smelling or tasting it, about half the normal flow was stimulated when the food reached the stomach. This flow occurs even when the nerves to the stomach are cut, although in reduced amount. It therefore depends in part on nervous stimulation of the gastric glands by impulses from cells in the stomach lining, and in part on a hormone called **gastrin.**

The cells of the mucosal layer of the pyloric part of the stomach secrete gastrin into the blood stream whenever partly digested food comes into contact with these cells. If extracts of the cells are injected into an animal, the gastric glands begin to secrete within a short time. The final proof of the existence and action of this hormone was given by cross circulation experiments, in which the blood system of one dog was connected by rubber tubes with that of another. When food was placed in the pyloric region of one dog, the gastric glands of the other began to secrete. Since there were no nerve connections between the two dogs, the secretion of gastric juice by the second dog must have been caused by the substance carried by the blood, the hormone gastrin.

Some secretion of gastric juice is caused by the presence of food in the intestine. Perhaps amino acids, absorbed into the blood from the small intestine, are responsible for this, or it may be caused by some as yet unrecognized reflex or hormone. The operation of so many different mechanisms enables the stomach to provide the proper amount of gastric juice for the amount and type of food eaten. A meal high in protein content causes a copious amount of gastric juice to flow; a meal with little protein and much carbohydrate causes a moderate flow; while one with a high fat content causes a small flow.

The stimulation of the pancreas depends upon a hormone, **secretin,** given off by cells in the lining of the upper part of the small intestine. These in turn are stimulated by the acidity of the food entering the intestine from the stomach; they can be stimulated experimentally by introducing acid into the cavity of the small intestine. Normally, as the acid chyme enters the small intestine, it stimulates the cells in the wall to give off secretin into the blood vessels of the intestine. The hormone, carried by the blood stream all over the body, finally reaches the pancreas and causes it to secrete the enzymes it has prepared (Fig. 172). Secretin also causes an increased production of bile by the liver.

When the nerves to the pancreas are stimulated, little secretion results; and when they are cut, there is little or no suppression of the flow, indicating that its secretions are almost entirely dependent on secretin.

196. COMPARISON OF DIGESTIVE SYSTEMS OF OTHER ANIMALS

The chemistry of digestion and the enzymes involved are much the same in man as in the ameba, but the protozoa, whose bodies consist of single cells, do not, of course, have any digestive system. Digestion in protozoa is intracellular, occurring within the single cell which is the entire animal. An ameba engulfs a bit of food by oozing around it and forming a food vacuole in which the food is shaped into a ball and surrounded by a membrane (Fig. 173, *A*). Digestion occurs within this vacuole as it circulates in the cytoplasm. The digestive enzymes manufactured by the cytoplasm are poured into the vacuole and digest the food inside. The broken down food molecules are then absorbed through the vacuole wall into the cytoplasm, where they are assimilated or used for energy.

The hydra, although multicellular, has no differentiated organs and therefore can have no digestive system. The body consists of two layers of cells (Fig. 173, *B*), an outer layer of ectoderm for protection and an inner entoderm. Only the cells of the entoderm are capable of digestion. Food, consisting of smaller animals and plants caught by the tentacles, enters the central cavity via the one central opening

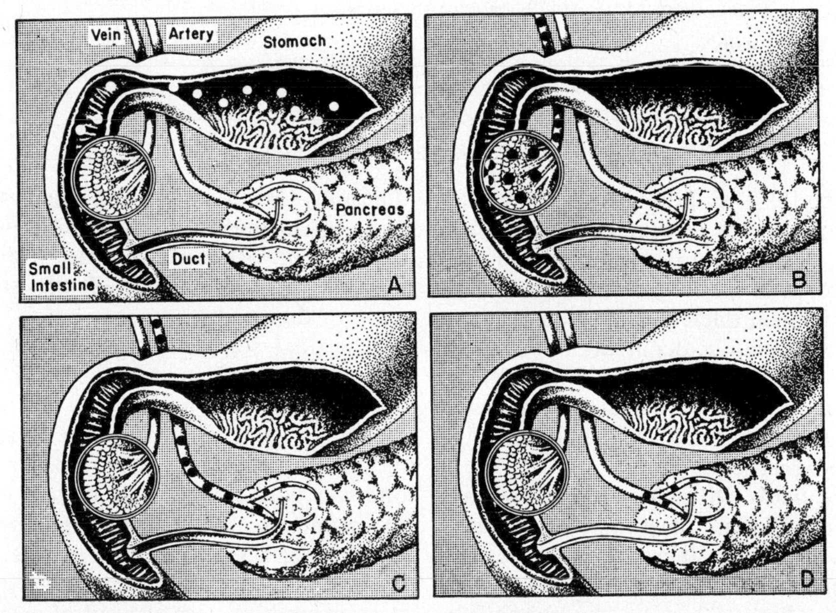

Figure 172. The control of the release of pancreatic juice by the hormone secretin. *A,* Hydrochloric acid (white dots) is secreted by the glands in the wall of the stomach and passes through the pylorus to the duodenum. *B,* Some of the hydrochloric acid diffuses into the wall of the duodenum and causes cells there to secrete the hormone secretin (black dots) which passes into the adjacent capillaries. *C,* Secretin is distributed by the blood vessels to all parts of the body and some of it is carried via the pancreatic artery to the pancreas. *D,* The secretin stimulates the pancreas to secrete pancreatic juice, which is visible in the pancreatic duct. The duct carries the pancreatic juice to the small intestine where its enzymes are important in digestion.

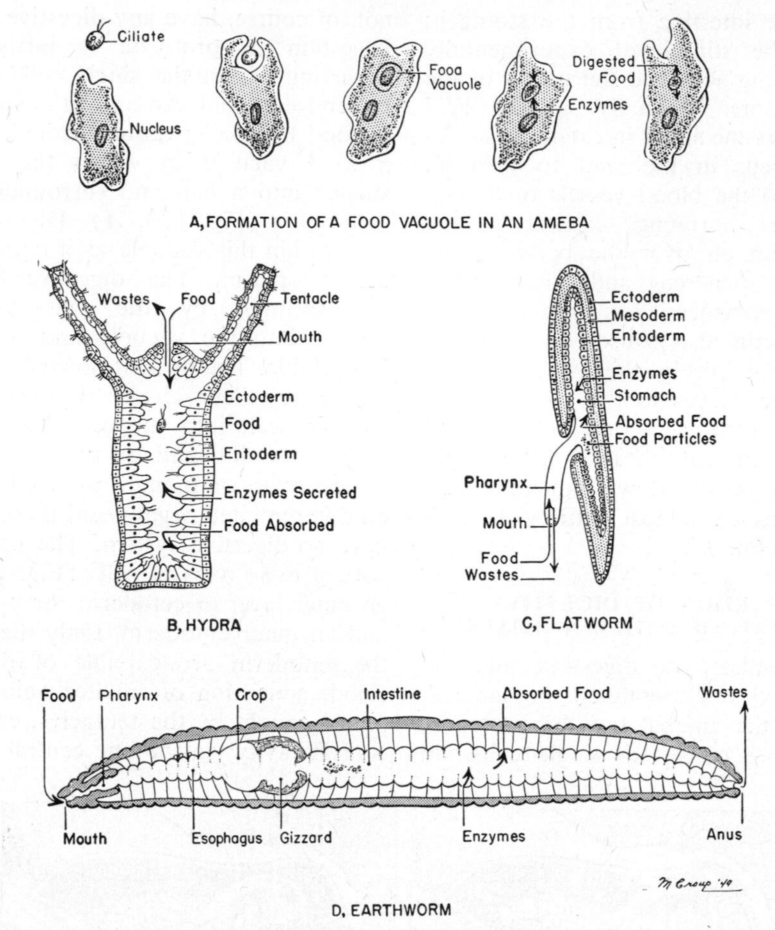

Figure 173. Digestion in some of the lower animals.

of the body. The entoderm cells secrete enzymes into this cavity, where the food is broken down into small pieces and then absorbed into the entoderm cells, where digestion is completed. Digestion is, then, partly extracellular, occurring in the space outside the cells, and partly intracellular. There is no anal aperture; undigested wastes are excreted via the opening through which food enters.

In the flatworm, planaria, the digestive organs are the mouth, pharynx and stomach; as in hydra, there is no anus, and food enters, and wastes leave, by the same aperture (Fig. 173, *C*). The stomach is greatly branched and extends throughout most of the body, facilitating the distribution of digested and absorbed food. Digestion is both extracellular and intracellular and is carried on by the entoderm cells. These animals have the peculiar ability of digesting their own organs when starved, and so can go for many months without eating. The organs of the body slip, cell by cell, into the cavity of the stomach and are digested.

The next stage in the evolution of the digestive system is demonstrated in the earthworm. This animal has a complete digestive system with two apertures, mouth and anus, so that food travels a one-way traffic road. The system consists of a mouth, a muscular pharynx, an esophagus, a soft-walled crop where food is stored, a hard, muscular gizzard where it is ground up with the aid of small pebbles taken in with the food, a long straight intestine where extra-cellular digestion occurs, and an anus through which undigested wastes pass (Fig. 173, *D*).

As the higher animals such as the vertebrates evolved, this system was gradually elaborated and organs added, resulting in the complex human mechanism we have just discussed. The digestive systems of the vertebrates from fish to man are similar, and in all animals, from lowest to most complex, the chemistry of digestion and the enzymes involved are much alike.

QUESTIONS

1. What are three functions of the digestive system?
2. Name the organs of the digestive system in order and state the functions performed by each.
3. What keeps the food moving through the digestive tract?
4. Why is bile important in digestion? Where is it manufactured and how does it reach the food?
5. What are gallstones and how are they formed?
6. What are the islets of Langerhans? What is the relationship of insulin to the pancreatic juice?
7. What part do enzymes play in digestion?
8. Discuss the absorption of glucose and amino acids. How does this differ from the absorption of glycerol and fatty acids?
9. In a certain abnormal condition the stomach does not secrete hydrochloric acid. What effect would this have on digestion?
10. What prevents the stomach from being digested by its own secretions?
11. What controls the secretion of the digestive enzymes?
12. Discuss the way digestion occurs in a paramecium, a hydra, a flatworm and an earthworm.
13. How does digestion in plants differ from digestion in animals?
14. Describe the path taken by a molecule of sugar from the mouth to the liver.

SUPPLEMENTARY READING

William Beaumont was a military physician stationed in upper Michigan. On June 6, 1822, a trapper, Alexis St. Martin, received a gunshot wound that opened a hole from his stomach to the outside. Beaumont treated St. Martin and made unique observations on the movement of the stomach during digestion, the stimuli effective in evoking gastric secretion, the normal appearance of the stomach mucosa, and so on. His book, *Experiments and Observations on the Gastric Juice and the Physiology of Digestion,* describing these observations was reprinted in 1929 and is a beautiful example of the contribution to basic science possible by careful clinical observation. Part of this, plus experiments by Reaumur, Spallanzani, Prout, Pavlov, and Bayliss and Starling's discovery of secretin, is given in Fulton, *Selected Readings,* Chapter 5.

Chapter 20

Metabolism and Nutrition

FOOD MAY be defined as any substance taken into the body which can be utilized for the release of energy, for the building and repair of tissues, or for the regulation of body processes. This broad classification includes carbohydrates, fats, proteins, water, mineral salts and vitamins. The first three are energy producers; the latter three, though not energy producers, are equally essential to life.

After being taken into the body, foods participate in a variety of chemical reactions, and these, plus all the other chemical activities of the organism, go under the name of **metabolism.** The presence of metabolic processes, you will recall, is one of the outstanding characteristics of living things. After foods are absorbed from the intestine they are either built into new tissue or oxidized to provide energy. Some of this energy is used in the building of new tissue, some in the functioning of the cells (transmission of nerve impulses, contraction of muscles, secretion of enzymes, and so forth), and some appears as heat.

There are many ways of subdividing the general field of metabolism. We may study the metabolism of a single tissue—e.g., liver metabolism—or we may study the chemical reactions undergone by a particular kind of food. Carbohydrate metabolism, for example, includes all the chemical reactions which starches and sugars undergo from the time they are taken in, until, after digestion and absorption, they are stored, converted into something else, or oxidized for energy and leave the body as carbon dioxide and water.

Human beings, unlike many other animals, can adapt to a variety of diets. We are able to operate quite well on one made up chiefly of protein with only small amounts of fats and carbohydrates, on one composed mainly of carbohydrate, with small amounts of proteins and fats, or even on one that is primarily fat with small amounts of the others. The diet of Eskimos is an example of the last.

Although the body can obtain the greater part of its energy requirements from any of the three types of fuels, they are not equally effective; the oxidation of a gram of carbohydrate or protein in the body yields about 4 Calories, but the oxidation of a gram of fat yields about 9 Calories. That is why such foods as whipped cream, mayonnaise and butter are more fattening than fruits, meat or bread. In the average American diet about one half of the energy used daily comes from carbohydrates, one third from fats and one sixth from proteins.

197. THE BASAL METABOLIC RATE

The daily expenditure of energy varies widely from person to person, depending on activity, age, sex, weight and body proportions. In order that doctors may make comparisons between the metabolic rates of different people and detect abnormally low or high rates, the conditions under which metabolism is measured must be standardized. If a person is reclining at complete rest, at least twelve hours after his last meal, the energy used from moment to moment keeps the heart beating and the breathing movements going, but most of it is used to maintain the body temperature. The amount of energy expended by the body just to keep alive, when no food is being digested and no muscular work is being done, is called the **basal metabolic rate.** The basal metabolic rate for young adult men is about 1600 Calories per day; for women it is about 5 per cent less. In other words, if a young adult remained in bed for twenty-four hours without eating or moving he would expend about 1600 Calories in keeping alive. From thousands of determinations of basal metabolic rates in different people, tables have been established giving the normal basal metabolic rate for a given age, sex and total body area. The rate is proportional to the surface area of the body, which can be calculated from the height and weight. A normal young adult uses 40 Calories per square meter of body surface per hour.

Since chemical reactions occur more rapidly at higher temperature, the basal metabolic rate increases about 5 per cent for each degree of rise in body temperature. This is the primary reason why weight is lost during feverish illnesses.

The basal metabolic rate can be determined directly by measuring the heat given off. The subject is placed in an insulated chamber surrounded by water, and the increase in the temperature of the air in the chamber and the surrounding water is measured. The simpler, clinical method is to determine the person's oxygen consumption over a short period of time. Since the release of energy and the production of heat depend on the uniting of oxygen with glucose and other foods, the amount of heat produced can be calculated from the amount of oxygen consumed.

Energy Requirements. If a person remains in bed for twenty-four hours and receives nourishment, he will expend about 1800 Calories, the extra 200 being required for movements of the muscles of the digestive tract and for secretion of the digestive juices. A person living a sedentary life uses about 2500 Calories per day, the additional 700 supplying energy for contraction of the body muscles and the additional work of the heart, since it beats faster. With heavy physical work, the daily expenditure of energy may reach 6000 Calories.

When the number of calories taken in as food just balances the energy requirement, the body weight does not change. Most people achieve a balance between the intake and outgo of calories, and their weight remains remarkably constant for years. There is a tendency for middle-aged people to gain weight because physical activity, but not appetite, decreases with age.

When the caloric intake is less than the daily energy requirement, the body must draw on its stored materials. The first to be used are the carbohydrates, stored as glycogen in the liver and muscles. Next, fat is withdrawn from storage in the fat deposits of the body and used to supply energy. The average adult male has about 9 kilograms and the average adult woman about 11 kilograms of stored fat (about 15 per cent and 21 per cent, respectively, of the total weight). The calories from the stored fat will supply energy for five to seven weeks of life. Finally, when the carbohydrate and fat reserves are used up, the body begins to consume its own structural materials, the proteins, first of the less essential organs, then of the essential ones (the brain, heart and other internal organs), until death occurs.

198. THE FUELS

Carbohydrates. Sugars and starches are the principle sources of energy in the ordinary human diet, but they are not essential to the body in any way; it could obtain energy as well from a mixture of protein and fat. Carbohydrates are the cheapest foods commercially, and the fact that the average diet contains large amounts of them is due primarily to this economic factor. In addition to sugars and starches, substances such as the citric acid of citrus fruits and the malic acid of apples and tomatoes are carbohydrates and serve as sources of energy.

Fats. Fats and oils are the most concentrated foods, since they not only supply more than twice as many calories per gram as carbohydrates and proteins, but also contain less water than these substances. They are digested and absorbed more slowly than other foods, so that one does not become hungry after a meal rich in fats as soon as after one of proteins and carbohydrates.

Fats are broken down into glycerol and fatty acids. The body is able to synthesize most, but not all, of the fatty acids, and the few which cannot be synthesized must be taken in with the diet; hence they are called "essential." The amount of essential fatty acids required is small, and provided by almost any diet. The fact that they are essential was discovered only when animals were raised on highly purified diets from which fat had been removed chemically. Fats and oils are important also because they contain fat-soluble vitamins.

Proteins. Proteins are the most expensive of the three types of fuels, and the amount of protein in the diet largely depends on the person's income. Because the body is constantly tearing down and replacing the protein in protoplasm, there must be a constant supply of protein in the food, even for adults whose growth has ceased and in whom there is no increase in the total amount of protoplasm. In growing children, pregnant women and people recovering from wasting diseases—in all people, that is, whose cells are manufacturing more protoplasm than is being destroyed each day—the protein intake must exceed the output. It is difficult to say just how much is necessary in the diet to maintain health, since that depends on the kind eaten and the amount of other substances in the diet. You will remember that proteins are composed of smaller units known as amino acids, of which there are some twenty-five different kinds. Proteins differ widely in the number and kind of amino acids they contain. When the body cells are synthesizing a particular type of protein, all the specific amino acids which make it up must be available. If they are not, or are present in inadequate amounts, the protein cannot be made. Animal cells can manufacture certain amino acids, but not all of them, and the ones they cannot synthesize, called "essential amino acids," must be supplied in the diet. There are ten of these essential amino acids, and proteins which contain all of them in adequate amounts are called "adequate proteins." Milk, meat and eggs contain biologically adequate proteins, but the protein in corn (zein) lacks two essential amino acids, so that an animal raised on a diet of which corn is the sole source of protein loses weight and dies. If the two amino acids, tryptophan and lysine, are fed in addition to zein, normal growth takes place. Gelatin lacks three of the essential amino acids. By eating a mixture of different proteins, one avoids any danger of suffering from a deficiency of any of the essential amino acids. But since the diet must contain an adequate quantity of all the essential ones, the total amount of protein necessary to maintain health is that which provides enough of even the least plentiful. Experiments have shown that it is possible for an adult to maintain health on as little as 25 gm. of protein per day, but larger amounts, from 70 to 100 gm., are desirable. If an excess of protein is ingested it is converted into fat or carbohydrate to be stored, or used for energy.

The element nitrogen is found in proteins, but not in carbohydrates or fats. By measuring the amount of nitrogen in the urine and feces and in the food ingested, it is possible to determine whether the body is maintaining nitrogen and therefore protein "balance." If analyses show that the daily intake of nitrogen is less than the output in urine and feces, the person is said to be in "negative nitrogen balance."

Soon after protein is eaten, the metabolic rate increases temporarily to as much as 30 per cent above the basal level, which is believed to represent the energy necessary to change some of the amino acids into carbohydrate or fat. This phenomenon is called the **specific dynamic action** of proteins. It depends upon some action of the liver on the amino acids, for if protein is fed to an experimental animal whose liver has been removed, the metabolic rate does not increase.

199. THE METABOLISM OF CARBOHYDRATES, FATS AND PROTEINS

In the previous chapter we traced the history of foods from the mouth to their absorption through the wall of the small intestine—proteins and carbohydrates being absorbed into the capillaries of the villi, and fats by way of the lymph vessels of the villi. After absorption the amino acids and simple sugars are carried to the liver by way of the hepatic portal vein. Perhaps, originally, the liver was important in digestion only, but during the evolutionary process it assumed many other functions and is now a chemical jack-of-all-trades. It protects the body by detoxifying certain harmful substances; it is active in the storage and interconversion of carbohydrates, fats and proteins; it is important in the metabolism of hemoglobin; it stores certain vitamins; it manufactures substances necessary for the coagulation of the blood; and it converts some of the harmful waste products produced by the metabolism of other body cells to less harmful, more soluble ones which can be excreted by the kidneys.

Carbohydrate Metabolism. Three kinds of simple sugars—glucose, fructose, and galactose—are derived from the breakdown of different kinds of double sugars and are absorbed from the digestive tract. They pass to the liver, which converts the other simple sugars to glucose and stores them all as **glycogen.** Glycogen is nothing more than a collection of glucose molecules bound together chemically for storage.

The role of the liver in storing carbohydrates was discovered by the French physiologist, Claude Bernard. He analyzed the glucose content of blood entering and leaving the liver just after a meal and found a much higher concentration of sugar in the blood entering the liver than in that leaving it. Analysis of the liver showed that new glycogen appeared simultaneously. Later the liver reconverts the glycogen to glucose with a resulting higher concentration of glucose in the blood leaving the liver than in that entering it. In this way Bernard discovered that the liver keeps the glucose concentration of the blood more or less constant, throughout the day.

The liver can store enough glycogen to supply glucose for about twelve to twenty-four hours; after that it must maintain the normal glucose level of the blood by converting other substances, principally amino acids, into glucose. If the supply of protein is sufficient, up to 60 per cent of the amino acids ingested can be converted by the liver into glucose.

Since glucose is the primary source of energy for all cells, its concentration in the blood must be maintained above a certain minimum level, about 60 mg. per 100 ml. of blood. The brain is the first organ to suffer when the concentration falls below this, because in contrast to most other cells of the body, the brain cells are unable to store any appreciable quantity of glucose and they cannot use fats or amino acids as sources of energy. When the glucose level is low, the diffusion of this substance from the blood stream into the cells where it is oxidized is not fast enough to supply the brain with adequate fuel. This results in symptoms similar to those accompanying a lack of oxygen—mental confusion, convulsions, unconsciousness and death. Whenever the brain cells (or any other cells) are deprived of either glucose or oxygen, they cannot carry on the metabolic processes which yield energy for their normal functioning. Other tissues normally get glucose for energy from the blood, but they can use other substances when necessary.

Muscle cells also can change glucose into glycogen for storage, but muscle glycogen serves only as a local fuel deposit, available for muscular work, and is not available for regulating the blood glucose level.

Besides being stored as glycogen or oxidized for energy, glucose may be transformed into fat for storage. Whenever the supply of glucose exceeds the immediate needs, the liver converts it into fat, to be used for energy at some later time. It has been known for many years that eating large amounts of starches or sugars is fattening; thus starch in the form of corn or wheat eaten by cattle and pigs is converted into the fat of butter and bacon. But only recently, with the use of radioactive or stable isotopes, has it been possible to demonstrate that a particular carbon or hydrogen atom put into the body as carbohydrate can be recovered in the fat of adipose tissue and liver.

The functioning of the liver in carbohydrate metabolism is regulated by a complex interaction of four hormones—one from the pancreas, one from the adrenal medulla, one from the adrenal cortex and one from the pituitary.

Fat Metabolism. Each species of animal or plant deposits fat containing a certain proportion of the different kinds of fatty acids. When we eat beef fat or olive oil, it must be changed into the type of fat characteristic of human beings. This is done by the liver, which also converts the absorbed fat into a form which can be stored in adipose tissue. The fat in such tissue, besides being available as a source of energy when needed, serves as a supporting cushion for certain internal organs and as an insulating layer under the skin, preventing too rapid heat loss. The liver also partially breaks down fats so that they can be oxidized by other tissues. There is evidence that unless glucose is being metabolized at the same time, the oxidation of fats does not proceed properly.

Diabetics, whose carbohydrate metabolism is interfered with, have abnormal fat metabolism also, and certain injurious intermediate products (called acetone bodies) resulting from the breakdown of fats tend to accumulate in their blood. In addition, large amounts of fat collect in the liver, a symptom which occurs in certain other abnormalities of liver function.

Fats, as well as proteins, are structural components of protoplasm, especially of the nuclear and plasma membranes, to which they give the property of differential permeability.

Fat metabolism is controlled partly by hormones from the pituitary and adrenals, and partly by sex hormones, but the details of this regulation are not clear. Any severe disturbance of liver function results in the almost complete absence of fat from the usual adipose tissues, indicating that fat must be acted upon in some way by the liver before the body can use it for storage or as a source of energy.

Protein Metabolism. Most of the amino acids entering the liver from the hepatic portal vein are removed from the blood and stored temporarily. Later, some of them are returned to the blood and carried to other cells to be incorporated into new protoplasm. Recent experiments, using amino acids labeled with the isotope N^{15}, or "heavy" nitrogen, have shown that the body proteins are constantly and rapidly being torn down and rebuilt.

If there are more amino acids in the diet than are necessary to maintain the protoplasm, enzymes in the liver remove the amino group from amino acids, a process called **deamination.** Other enzymes, by combining the split-off amino groups with carbon dioxide, convert them into a waste product, **urea,** which is carried by the blood stream to the kidneys and eliminated in the urine. The parts of the amino acids remaining after deamination are simple organic acids, converted by the liver either into glucose and glycogen to be used for energy, or into stored fat. There is little or no storage of proteins as such in the body; the proteins the body draws upon when carbohydrates and fats are exhausted are not stored proteins, but are the actual substance of the protoplasm of the cells.

The hormonal control of protein metabolism is even more obscure than that of lipids. Since growth is essentially the deposition of new protein in protoplasm, the growth hormone of the pituitary plays some part in it, but its nature is unknown. Insulin, the sex hormones, and one or more of the hormones from the adrenal cortex are also involved in the control of protein metabolism.

200. OTHER COMPONENTS OF THE DIET

Minerals. Some fifteen elements are known to be essential as mineral salts in the diet, of which a few are needed only in traces. The daily requirements of some of these are as follows: sodium chloride, 2–10 gm.; potassium, 1–2 gm.; magnesium, 0.3 gm.; phosphorus, 1.5 gm.; calcium, 0.8 gm. (more during growth or pregnancy); iron, 0.012 gm.; copper, 0.001 gm.; manganese, 0.0003 gm.; iodine, 0.00003 gm. The constant loss of mineral salts from the body (about 30 gm. per day) in the urine, sweat and feces must be balanced by the intake of equal amounts in the food. A diet that contains no minerals is more rapidly fatal than no food at all, because the excretion of wastes from the metabolism of carbohydrates, fats and proteins requires the simultaneous excretion of a certain amount of salt (to keep the *p*H of the blood constant). Thus, a salt-free diet actually exhausts the body reserve of salts. Deficiencies of minerals are rather rare, since meat, cheese, eggs, milk and vegetables are rich sources. But deficiencies of iron, calcium and iodine do occur.

Since blood and other body fluids are about 0.9 per cent salt, and most of this is sodium chloride (common table salt), sodium and chlorine are important in maintaining osmotic balance in the body. These elements are important components of the secretions of the digestive tract—the hydrochloric acid of the stomach, and the pancreatic and intestinal juices. The salts in these secretions are reabsorbed and used over again, so that the loss of salts via the digestive tract is negligible. The daily requirement for sodium chloride varies widely and depends on the amount lost in perspiration. Men doing heavy work in hot places—for example, "sand hogs" digging tunnels—must drink salt water instead of plain water, or the amount of salt in the blood will decrease, resulting in muscular cramps and heat exhaustion.

Potassium and magnesium are necessary for muscle contraction, as well as for the functioning of many enzymes.

Calcium and phosphorus are the chief constituents of bones and teeth, and a deficiency in childhood of either one (or of vitamin D; see p. 306) produces rickets. Phosphorus is extremely important in carbohydrate and protein metabolism; both types of substances must be changed into intermediate compounds containing phosphorus before they can be utilized as sources of energy.

Iodine is a constituent of the hormone of the thyroid gland, and if the diet is deficient in this substance, the gland is unable to make its hormone and enlarges to form a **goiter** (p. 383). Iodine is abundant in sea water and sea foods, but is rare elsewhere, and formerly it was common for people living in inland regions to suffer from goiter. Now, most table salt is fortified with small amounts of potassium iodide to prevent this.

Iron is a constituent of hemoglobin and of the cytochromes. This iron is used over and over, and as long as there is no loss of blood the amount of iron needed daily in the diet is negligible. Because women lose a considerable amount of blood each month by menstruation, they are more likely to become anemic due to an iron deficiency than are men.

Small amounts of copper are necessary in the diet to bring about the proper utilization of iron, and for normal growth. Traces of manganese, zinc and cobalt are also required for normal growth and as activators of certain enzymes. Zinc is a constituent of the enzyme carbonic anhydrase. Traces of fluorine in the drinking water are remarkably effective in preventing dental decay.

Water. Water makes up about two thirds of the human body and is an essential component of every cell. It is the fluid part of blood and lymph, and is the medium in which the other chemicals are dissolved and in which all chemical reactions occur. It is indispensable for digestion, since the splitting of carbohydrates, proteins and fats requires a molecule of water for each pair of sugar molecules or amino acid molecules separated. Water removes metabolic wastes, distributes and regulates the body heat, and, as perspiration, cools the body surface. The amount of water lost daily averages about 2 liters, although it varies with individual activities and the climate. The loss must be replaced promptly; men can live weeks without food, but only a few days without water.

All foods contain some water, and some, such as green vegetables and fruits, may contain as much as 95 per cent water. Certain desert animals live indefinitely without drinking water, obtaining it from the foods they eat.

Condiments and Roughage. Pepper and other spices, collectively known as condiments, have little or no food value themselves, but are important for making foods more palatable. By stimulating the appetite they help insure that a sufficient quantity of food is eaten.

We have already discussed the importance of "bulk" or undigested matter in stimulating the movements of the intestines and preventing constipation. The diet should contain some indigestible matter for this purpose, such as the cellulose material from vegetables and fruits.

201. VITAMINS

One of the most notable biochemical achievements since the turn of the century has been the discovery of vitamins and the analysis of their properties. Vitamins are relatively simple organic compounds which, though present in such scanty amounts that they cannot be used as sources of energy, are absolutely essential to life. The various vitamins are quite different chemically, but are similar in that they cannot be manufactured by the animal body and therefore must be present in the diet in small amounts for metabolism to occur normally. When an adequate amount of any one of them is not present, a specific pathologic condition or deficiency disease occurs, curable only by administration of the specific vitamin; for example, only vitamin C is effective in curing scurvy.

In 1912, investigators found that animals could not survive on a diet of purified carbohydrates, proteins and fats, but that accessory growth factors or vitamins were necessary. Since at first the chemical structure of these substances was unknown, they were referred to as vitamins A, B and C, which prevented night blindness and rickets, beriberi and scurvy, respectively. Today, the chemical structure of nearly all of them is known, and most of them have been made synthetically. The original vitamins A and B are complexes of several vitamins, A being subdivided into A, D and E, while vitamin B consists of almost a dozen different ones. The vitamins of known chemical structure are usually referred to by their chemical names, e.g., thiamine rather than vitamin B.

The distinction between vitamins and such things as essential amino or fatty acids is not clear cut; the latter are also simple organic substances, essential for life, which cannot be made by the animal body and must be taken in with the food. They are needed in much larger amounts, however; the term vitamin is reserved for the substances required by the body in very small quantities.

The average adult eating a normal varied diet has no need to take vitamin pills; he will obtain the necessary kinds and amounts of vitamins from his food. Infants and younger children, whose diets are more restricted, may need supplementary amounts of certain vitamins, especially A and D. The vitamin requirements of different animals are not the same; most animals do not require vitamin C, since it is made in their own bodies; only man, monkeys and guinea pigs need it in their diet. Insects require only cholesterol and the B complex vitamins in their diet.

Thus, what is a vitamin for one animal is not a vitamin for another, although it is probable that all animals and plants require all or nearly all the known vitamins. The function of a number of vitamins has been discovered; each one has been found to serve as an integral part of a coenzyme for one or more of the fundamental enzyme reactions common to all protoplasm. Clear evidence that plants require the same vitamins as animals, but that they normally synthesize all the vitamins they require has been provided by the experiments of Beadle and Tatum and their collaborators with the mold *Neurospora.* The ordinary "wild" strain of this mold requires in its culture medium only a single vitamin, biotin, in addition to salts and a sugar of some sort. By exposing mold organisms to x-rays or ultraviolet light, thus causing a gene mutation which interfered with some step in the synthesis of a vitamin, these investigators produced mutant strains which would grow only when one additional vitamin was added to the culture medium. Now there are strains which require each of the vitamins necessary for growth in animals. This is evidence that the mold cells (and probably other plant cells) need these substances for growth just as much as animal cells do, but that ordinarily they synthesize all the vitamins they need except biotin.

It is quite likely that the ancestors of animals also had the ability to synthesize all the substances we now call vitamins, but that over the intervening millions of years, mutations occurred which interfered with the synthesis of these substances, so that to survive, the organisms had to obtain the compounds in their diet. And since, in different evolutionary lines, a variety of mutations has occurred, the species living today have different vitamin requirements. If such mutations occurred in a green plant, the plant would die, since it would have no way of obtaining the vitamin.

Vitamin A. Vitamin A occurs only in animal products such as butter, eggs and fish liver oils, but plants contain a yellowish substance, called **carotene** or provitamin A, which is easily changed into vitamin A in animal cells. Vitamin A itself is fat-soluble and can be stored in the body. The daily requirement for adults is about 1.5 mg. (5000 International Units), for a child under three about 0.6 mg., and for older children an intermediate amount.

This vitamin is necessary for the maintenance of the epithelial cells of the skin, eye, digestive and respiratory tracts. In vitamin A deficiency these cells become flat, brittle and less resistant to infection than normal (vitamin A is sometimes called the "anti-infection vitamin"). In advanced cases of the deficiency the eye epithelium forms a dry and horny film over the cornea, resulting in a characteristic type of blindness called **xerophthalmia** (Fig. 174). Vitamin A is necessary also for the maintenance of normal nerve tissue and for the growth of bone and the enamel of teeth. It is involved in the chemistry of vision, so that **night blindness,** inability to see in a dim light, may result from vitamin A deficiency. The retina of the eye contains a substance called **visual purple,** made up of vitamin A and a protein; this is broken down when light strikes it, stimulating the receptor cells which send impulses to the brain, resulting in the sensation of sight. Ordinarily, this substance is quickly rebuilt, but when insufficient vitamin A is present, the reconstruction of visual purple is retarded and night blindness results. Deficiencies severe enough to produce xerophthalmia are rare in the United

A

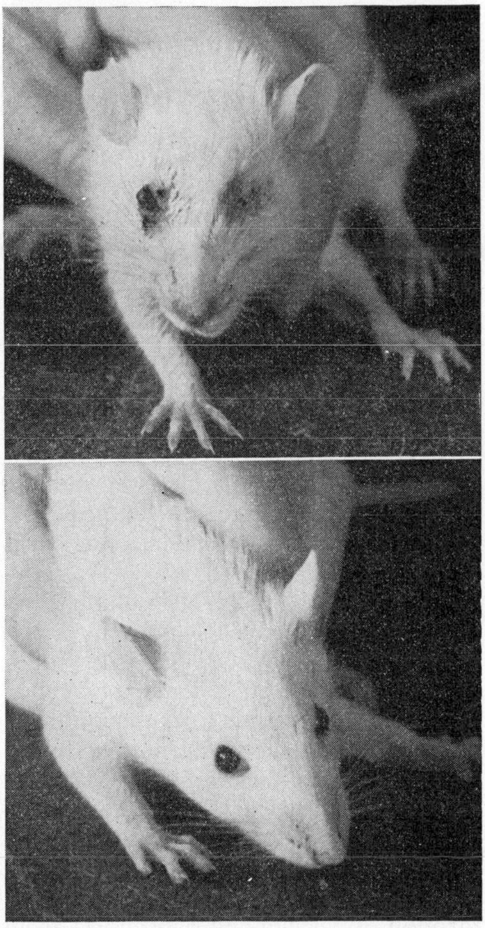

B

Figure 174. A, Typical eye condition produced by lack of vitamin A. *B,* The eyes restored to normal by the feeding of 3 units (about 0.001 mg.) of vitamin A daily. (Courtesy of E. R. Squibb and Sons.)

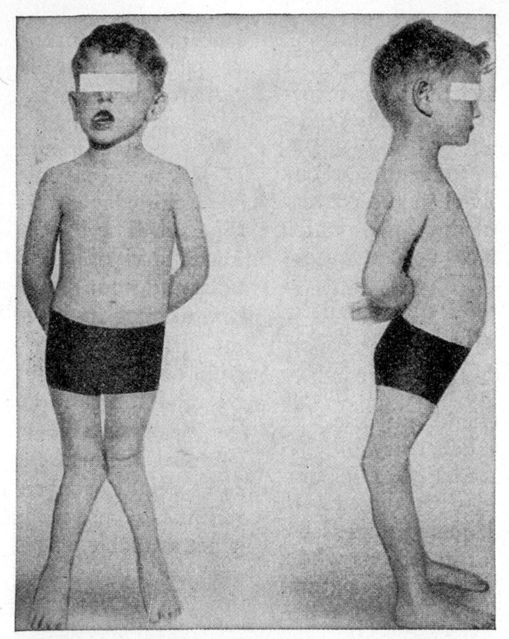

Figure 175. A child with rickets. A deficiency of vitamin D decreases the ability of the body to absorb and use calcium and phosphorus in bone formation, resulting in soft, malformed bones. (Cooper, Barber and Mitchell: Nutrition in Health and Disease for Nurses, J. B. Lippincott Co.)

States, but night blindness is more common. During World War II, the pilots of night fighter planes were fed diets particularly rich in vitamin A to prevent this. Toxic symptoms due to an overdose of vitamin A have been observed in man; some of the first cases occurred in people who had eaten polar bear liver, which is very rich in vitamin A!

Vitamin D. Another fat-soluble vitamin, D, or **calciferol,** is unique in that it can be made in the body under the influence of sunlight (it is sometimes called the "sunshine" vitamin) from a substance called ergosterol normally present in the skin. Calciferol is also found in liver oils, butter, eggs and milk, and any excess manufactured in the skin during the summer months is stored in the liver. This substance is necessary for the normal absorption of calcium and phosphorus from the intestine; about 0.02 mg. per day is recommended for children and adults. When there is a deficiency of calciferol, calcium and phosphorus are not absorbed in normal amounts, so the formation of bones and teeth is retarded for lack of raw material. This results in the disease known as **rickets** (Fig. 175), characterized by soft, weak bones, enlarged ankles, knees and wrists, bowing of the legs, beaded ribs, and defective tooth development. There is danger in overdosing a person with vitamin D, however, because it may lead to calcification of the soft tissues and death.

Vitamin C. The deficiency disease **scurvy,** resulting from a lack of vitamin C, is one of the principal noninfectious plagues of history, characterized by bleeding gums, bruised skin, painful swollen joints, and general weakness. It occurs whenever people are deprived of fresh fruits, vegetables and meat for long periods of time, as they are on extended sailing voyages or during the long northern winters. Modern methods of preserving and shipping foods have practically eliminated this disease, but slight deficiencies, characterized by laziness, gloom and irritability, still occur.

The earliest report of a cure for scurvy is found in the records of Jacques Cartier's expedition to Canada in 1536. His crew suffered severely from scurvy and were cured by an extract of fir needles prescribed by the Indians. The scurvy-preventing vitamin was isolated in 1933 and proved to be ascorbic acid (or hexuronic acid), a substance that had been known for many years, but whose antiscorbutic properties had not been suspected. Ascorbic acid is rather unstable and is rapidly destroyed by cooking, so that the best sources of it are fresh fruits and juices, although modern canning processes preserve most of the ascorbic acid content of food. The exact role of ascorbic acid in metabolism is unknown, but it plays some part in cellular oxidations, particularly in the oxidation of one of the amino acids, tyrosine. It is necessary for the maintenance of normal connective tissue. In its absence the capillaries become exceedingly fragile and easily ruptured, resulting in hemorrhages under the skin and in the joints. The development of bones and teeth also is abnormal. Normal human adults require between 75 and 100 mg. of ascorbic acid daily, an amount supplied by an 8-ounce glass of orange juice.

Vitamin E (Alpha-tocopherol). Experimental studies on rats, chicks and ducks have shown the existence of vitamin E, or alpha-tocopherol, which is necessary to prevent sterility. If it is absent from the diet, male animals become sterile, owing to degenerative changes in the testes, and females are unable to complete pregnancy successfully, since the embryos die and are resorbed. Eggs from vitamin E-deficient hens fail to hatch. So far it has not been shown that a deficiency of vitamin E is responsible for human sterility, but this is possible. No figure can be given for the daily human requirement of the substance, but it is so widespread in both vegetable and animal oils that a deficiency in any normal diet is almost impossible.

Experiments have shown that a deficiency of vitamin E produces progressive deterioration of the muscles and paralysis, presumably by degeneration of the nerves (just as the destruction of nerves in infantile paralysis leads to the wasting of muscles and paralysis). Certain types of paralysis in human beings have been treated with vitamin E preparations, with beneficial results.

Vitamin K. Normal coagulation of blood, which depends upon the manufacture of prothrombin by the liver, is connected with the specific action of a number of different chemicals referred to as vitamin K. These chemicals with the same effect occur in a variety of foods and are manufactured by the bacteria in the human intestine, so that vitamin K deficiency in man usually results from some abnormality in its absorption rather than the lack of it in the diet. Since it can be absorbed only in the presence of bile salts (vitamins A, D and E also require bile salts to be absorbed), an obstruction of the bile duct results in vitamin K deficiency, no matter how much is present in the diet or made by the intestinal bacteria. Patients with vitamin K deficiencies are poor surgical risks because of the likelihood of hemorrhages after the operation; the administration of vitamin K (and bile salts, if necessary) before operation removes this danger and has saved many lives. Newborn infants, before they acquire their quota of intestinal bacteria, are likely to be deficient in vitamin K, and administration of this substance to the mother in the last few days of pregnancy helps prevent the hemorrhages that often occur in infants after delivery. No estimate can be given of the daily requirement of this vitamin, but in vitamin K deficiencies, 1 to 5 mg., administered daily, brings the clotting time of the blood back to normal.

The Vitamin B Complex. The original vitamin B was characterized as the anti-beriberi factor, but from the same extracts of liver, yeast or rice hulls which yield an anti-beriberi substance, nine other materials with specific biologic effects have been separated. Some of these substances were once given separate alphabetical designations: riboflavin was called vitamin G, and biotin was called vitamin H, but they are now all grouped together as members of the B complex, not because they are similar chemically or in their effects, but because they tend to occur together. Individual members of the B complex are referred to by their chemical names.

Thiamine (Vitamin B_1). This substance, the first to be separated from the rest of the complex, prevents beriberi. It is a white, crystalline material with a yeastlike odor, found in small quantities in a wide variety of foods. Yeast, liver, nuts, pork and whole grain cereals are the best sources of all the vitamin B complex. Since the average American diet is somewhat deficient in thiamine, flour, bread and breakfast cereals are now enriched with it. The daily requirement varies with the body weight, the number of calories eaten, and the proportion of carbohydrate in the diet (the more carbohydrate, the more thiamine needed), but the amount needed daily by the average person is from 2 to 3 mg. It is now prepared synthetically in large quantities, and the price has dropped from about $1000 a gram (1935) to $0.16 a gram (1957). Thiamine and the other B complex vitamins are not stored in the body to any great extent, and evidence of a deficiency appears within a few weeks. Most diets

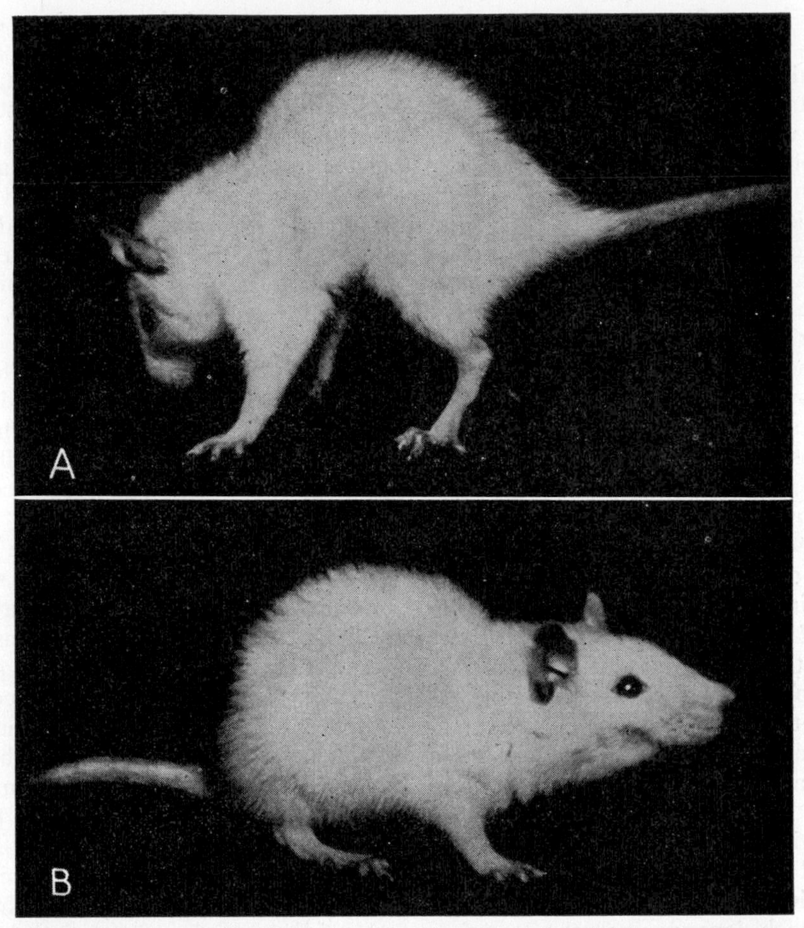

Figure 176. A, Polyneuritis (beriberi) in a rat raised on a diet deficient in thiamine. Note that the back is arched and the hind legs are stretched and far apart. Such animals have a peculiar halting gait and are particularly awkward in turning, readily losing their balance. When rotated, they have great difficulty in regaining equilibrium, probably because of degeneration of the nerves to the semicircular canals. *B,* The same rat eight hours after receiving an adequate dose of thiamine: the back and hind legs are normal, and the animal readily regains equilibrium when spun. (Courtesy of the Upjohn Company.)

contain enough thiamine to prevent the appearance of beriberi, but may not contain enough for maximum health.

The function of thiamine in the body is to form the active part or "coenzyme" of certain enzymes involved in the metabolism of carbohydrates, particularly of pyruvic acid. When a thiamine deficiency interferes with carbohydrate metabolism, a number of characteristic symptoms appear: in mild deficiencies there is fatigue, loss of appetite, weakness and muscular cramps; in more marked deficiencies these symptoms are accentuated, and there is also a painful degeneration of the nerves and a secondary wasting of the muscles resulting in paralysis. This condition is known as **beriberi** (Fig. 176). The symptoms disappear rapidly when thiamine is given. Any diet deficient in thiamine is likely to be deficient in the other B complex vitamins as well, so that cases of thiamine deficiency alone are rare.

Riboflavin (Vitamin B_2 or G). Riboflavin is a yellow pigment found in both plant and animal tissues; it occurs most abundantly in foods rich in thiamine: yeast, liver, wheat germ, meat, eggs and cheese. Riboflavin forms part of coenzymes necessary for the functioning of enzymes involved in the metabolism of glucose, amino acids and certain cellular oxidative processes. One to 2 mg. of riboflavin per day are required to maintain health in man. A deficiency of riboflavin is marked by the appearance of cracks in the corners of the mouth, a characteristic purplish-red color of the tongue, and stunted growth. In experimental riboflavin deficiencies in rats there is failure of growth, loss of hair, cataract, inflammation of the eyes, and death (Fig. 177).

Niacin or Nicotinic Acid. Niacin is a component of two different coenzymes which are the active parts of many different enzymes in cellular metabolism. Niacin was known as an organic compound for over fifty years before its function as a vitamin was recognized. It is found in yeast, fresh vegetables, meat and beer. Corn meal has an unusually low niacin content, and wherever this food forms a large part of the diet, the deficiency disease **pellagra** is fairly prevalent. Pellagra is characterized by dermatitis (reddened inflammation of the skin, especially in those parts of the body exposed to light), diarrhea and dementia. The normal functions of niacin are, then, the maintenance of the epithelia of the skin and digestive tract and of normal nerve functioning — processes which depend

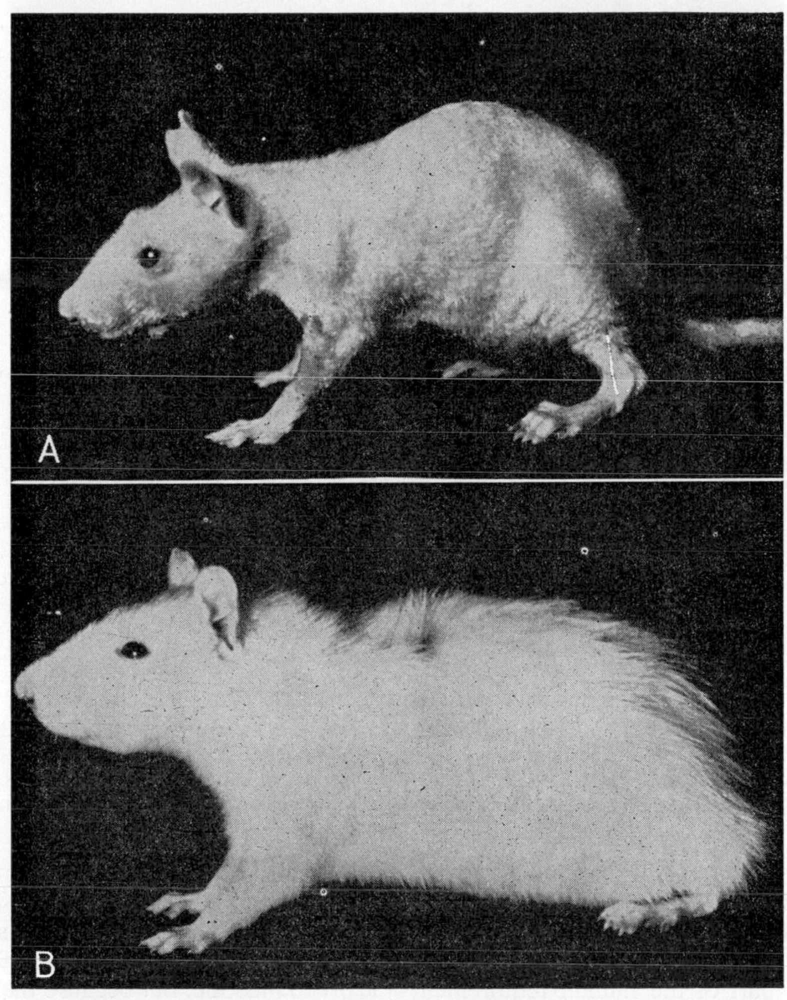

Figure 177. A, Riboflavin-deficient rat, with stunted growth, general inflammation of the skin (note the open sore on the left front leg), scanty hair, and inflammation of the eyes. *B,* The same rat after two months of treatment with riboflavin: no signs of the deficiency are visible; growth has been resumed, and the lesions of the skin and eyes are cured. (Courtesy of the Upjohn Company.)

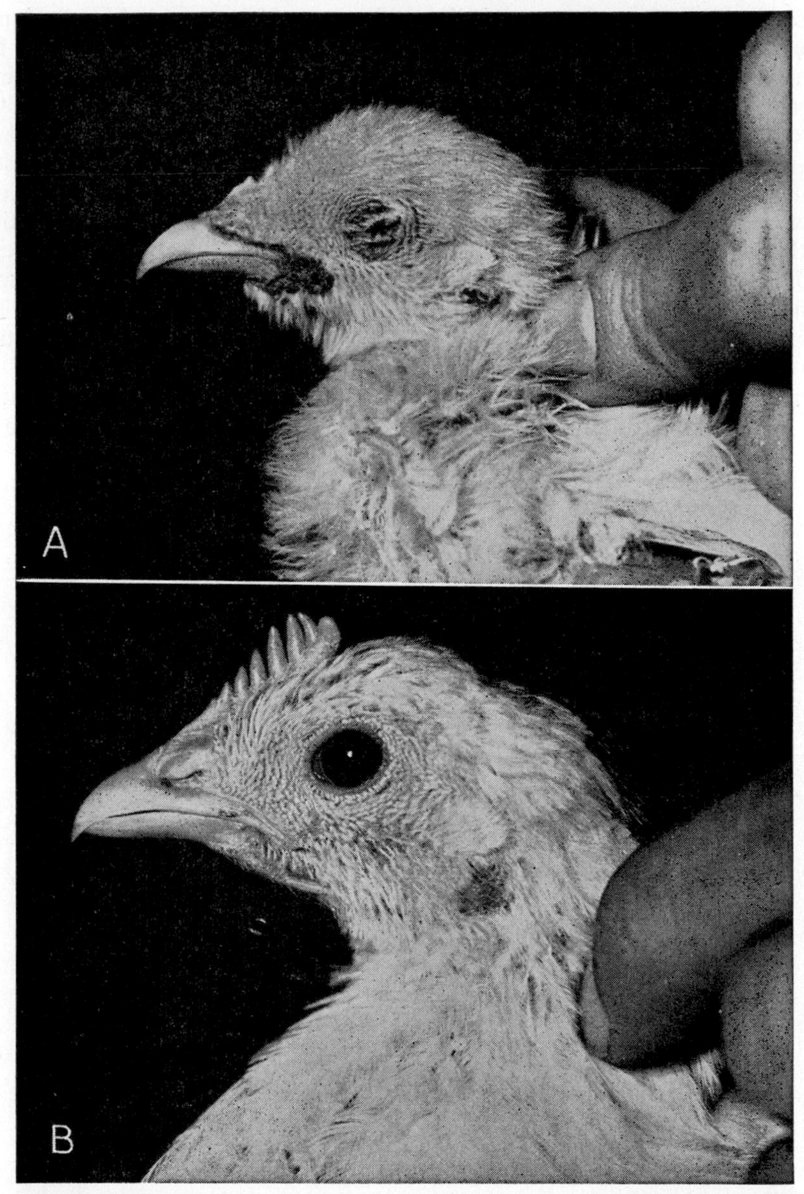

Figure 178. A, Chick after being fed a diet deficient in pantothenic acid (one of the B complex vitamins). The eyelids, corners of the mouth, and adjacent skin are inflamed. The growth of feathers is retarded, and the feathers are rough. *B,* The same chick after three weeks on a diet with pantothenic acid: the lesions are completely cured. (Courtesy of the Upjohn Company.)

upon its action as the coenzyme of one or another of many different enzymes. The recommended daily allowance of niacin is about 20 to 25 mg., but a considerable part of the human requirement is synthesized by the intestinal bacteria. When a person is treated with sulfa drugs for some infection, the intestinal bacteria are killed and deficiencies of a number of vitamins, including niacin, may occur.

The original name for this substance was nicotinic acid, since it is related chemically to nicotine, but because the two may be confused, the name niacin has come into use.

Pyridoxine (Vitamin B_6). It has been

known for some time that pyridoxine is necessary in the diet of rats, and it is now believed that human beings also require it. The vitamin occurs in a wide variety of foods—meat, eggs, nuts, whole grain cereals, and beans—so that a clear-cut deficiency of pyridoxine in man has not been found. A derivative of it, pyridoxal phosphate, is the coenzyme for a number of enzyme reactions in the metabolism of amino acids. Experimental animals fed a diet deficient in pyridoxine fail to grow, become anemic and have atrophied lymph tissue, resulting in a lack of white blood cells and antibodies, with a consequent lowered resistance to infection. The administration of pyridoxine has been tried in a number of human disorders, but without consistent results. The daily requirement is about 1 to 2 mg., but it varies with the amount of protein in the diet, since the more protein eaten, the more pyridoxine is needed to stimulate growth.

Pantothenic Acid. This vitamin is necessary for the maintenance of normal nerves and skin, and experimental deficiencies of it cause the failure of growth, dermatitis, gray hair and damage to the adrenal gland (Fig. 178). The "burning foot" syndrome suffered by some prisoners in Japanese prison camps during the war responded to treatment with pantothenic acid. Almost any normal diet will provide the 20 mg. required daily by human beings. Especially rich sources of it are eggs, meat, sweet potatoes and peanuts. It forms part of another coenzyme, "coenzyme A," important in a number of steps in the metabolism of carbohydrates, fats and proteins and in the transfer of energy.

Biotin. This was first discovered as a factor indispensable for the growth of yeast. It has since been shown to be necessary in the diet of mammals, though only in extremely small amounts. Some rich sources of it are molasses, egg yolk and liver. Egg white contains a protein called **avidin,** which combines with biotin in the intestine and prevents its absorption. Avidin is destroyed by heat, however, so that cooked egg white does not interfere with the absorption of biotin, and much more than the amount in an eggnog or two is needed to cause biotin deficiency in experimental animals (Fig. 179). One of the few cases of human biotin deficiency occurred in a man who lived almost entirely on raw eggs and wine and suffered an inflammation of the skin; it cleared up when biotin was administered. Biotin is believed to form a coenzyme involved in carbon dioxide fixation.

Folic Acid, Vitamin B_{12}, Choline, Inositol and Para-aminobenzoic Acid. Folic acid and vitamin B_{12} are necessary to prevent anemia and are used in conjunction with liver extract in treating pernicious anemia. They are active as coenzymes in the metabolism of certain substances involved in the synthesis of amino acids and nucleic acids. Their role in preventing anemia apparently is that of facilitating the nucleic acid synthesis involved in the production of red blood cells.

Choline is a growth factor the absence of which causes hemorrhages in the kidneys and a bone deformity in chicks called perosis. It is important in the metabolism of fats and proteins, not as a coenzyme as many other B vitamins are, but as a source of methyl groups to be used in building up certain essential substances. An adult requires about 2000 mg. of choline daily.

Inositol and para-aminobenzoic acid have been reported as important in preventing the loss of hair and the graying of hair, respectively. Both are necessary for normal growth of rats and presumably of other animals, including man. These five B vitamins are also synthesized by the intestinal bacteria.

202. ANTIMETABOLITES

D. D. Woods found in 1940 that para-aminobenzoic acid reversed the action of the "sulfa drug," sulfanilamide, on bacteria. Sulfanilamide is bacteriostatic—it prevents the multiplication of bacteria and thus aids the body defenses in dealing with invading bacteria. This observation suggested the theory that sulfanilamide interferes with bacterial growth by acting as a competitive inhibitor of some bacterial enzyme for which para-aminoben-

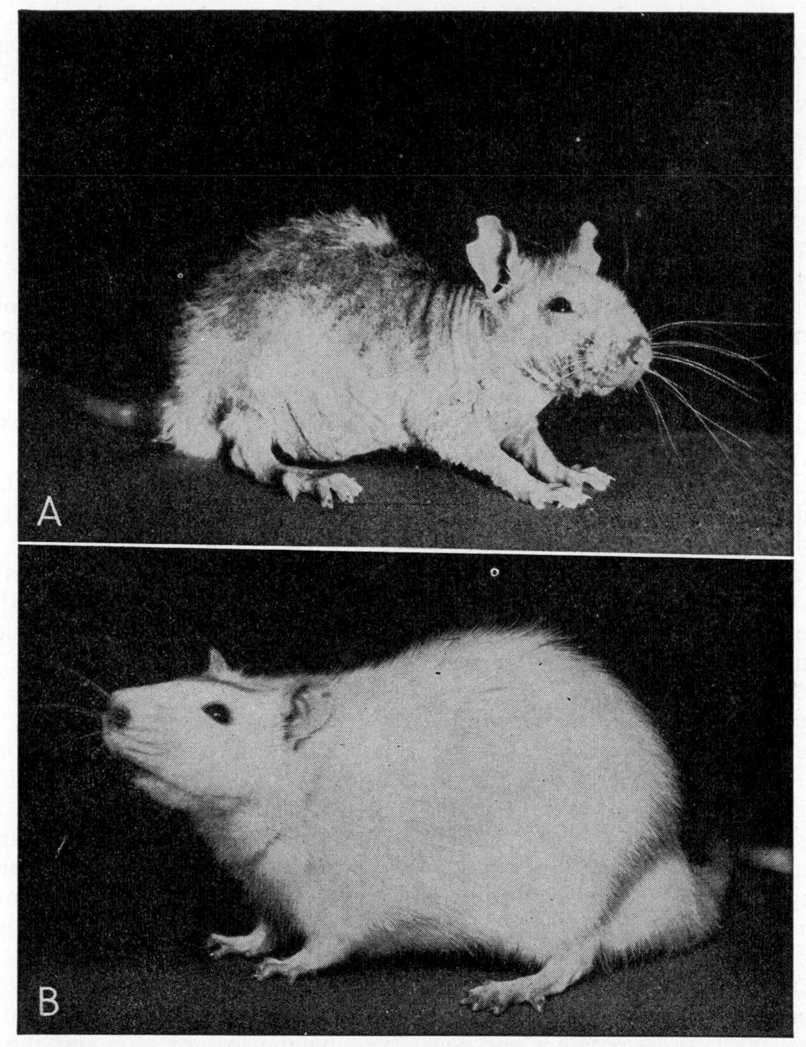

Figure 179. A, Rat after being fed a diet deficient in biotin, to which raw egg white was added. Growth has been retarded, and there is generalized inflammation of the skin. *B,* The same rat after three months on a diet containing adequate amounts of biotin: growth is normal and the skin lesions are completely healed. (Courtesy of the Upjohn Company.)

zoic acid is the normal coenzyme or forms an integral part of the coenzyme. Sulfanilamide is quite similar in chemical structure to para-aminobenzoic acid, similar enough to fool the enzyme and be taken into the enzymatic mechanism, but different enough so that the enzymatic mechanism becomes jammed. This theory set off a search for other substances (called **antimetabolites**) like, but slightly different from, ordinary vitamins to serve as inhibitors of bacteria or of the growth of cancer cells. Aminopterin, an antimetabolite of folic acid, has been successful in alleviating certain kinds of leukemia.

203. DIET

The most important nutritional problem in the United States at present is **obesity.** Surveys show that some 15 per cent of the population are overweight. Obesity predisposes to a number of diseases such as diabetes and materially decreases life expectancy.

Dr. Clive McCay of Cornell University has shown that a low calorie diet, particularly during the early part of life, will double the lives of rats, hamsters and dogs. The animals fed diets restricted in calories are healthier, spryer and more fertile than the control animals fed *ad libitum.* The statistics on these experiments prove what animal breeders and trainers have long known—that an animal looks and behaves better when it is slightly underfed. There is every indication that what is true for rats and dogs is true for human beings. Many physiologists, by performing experiments on themselves, have found that a restricted diet produces beneficial psychologic effects: they felt more alert, happier, and able to stick to tedious, problem-solving tasks much longer.

An adequate diet must supply water, salts and vitamins; sufficient calories to balance the daily expenditure of energy (unless one wishes to lose weight); and enough fats and proteins for tissue repair. Except where there is extreme poverty, ignorance, idiosyncrasy of taste or some accidental factor which prevents the eating of a normal diet, there is little danger of anyone's failing to obtain the specific, essential substances. A daily diet including two or three glasses of milk, an egg, one or two servings of lean meat, plenty of green and yellow vegetables, fresh fruit, whole grain cereals and bread, with moderate servings of butter or margarine, and a glass of citrus fruit juice, provides adequate amounts of all the necessary foods. The more fresh—and especially raw—vegetables, and the fewer fried foods, pastries, rich sauces, whipped cream and candy, the better. The latest research in the field of dental health indicates that sweets of all kinds are an important factor in producing caries. In children particularly, there is an unmistakable correlation between high sugar consumption and a large number of cavities. Table 6 gives the nutritional content and caloric value of many common foods.

QUESTIONS

1. What are three functions of foods?
2. Specify the six main types of food. Which of these yield energy?
3. Define a Calorie.
4. Why is a square of butter more fattening than an equal amount of bread or meat?
5. What is meant by the basal metabolic rate? What are two pathologic conditions which increase it?
6. What are "adequate" proteins?
7. What is meant by the specific dynamic action of proteins?
8. What is glycogen? What role does it play in cell metabolism?
9. Why is the liver of great importance in nutrition?
10. What is meant by deamination?
11. Name the minerals most essential to the body, and give the function of each in cell metabolism.
12. What are vitamins? State the diseases which may result from vitamin deficiencies, and which vitamin cures each condition.
13. What vitamins are found in milk? Eggs? Green vegetables? Steak?
14. Utilizing the data of Table 6, make up a diet for one day which would be adequate for an active young man. How should the diets of a pregnant woman, a ten year old boy and a sixty-five year old man differ from this?

SUPPLEMENTARY READING

Quotations from papers by F. G. Hopkins and by Casimir Funk on the discovery of vitamins are found together with a description by James Cook of the measures taken to prevent scurvy in his crew, in Chapter 8 of Fulton's *Selected Readings.* Hopkins' discovery of the accessory food factors is described in Needham and Baldwin's *Hopkins and Biochemistry.*

Table 6. ANALYSES OF COMMON FOODS
(From Bowes and Church)

FOOD	WT. (GM.)	APPROXIMATE MEASURE	PROTEIN (GM.)	FAT (GM.)	CARBOHYDRATES (GM.)	CALORIES	CALCIUM (MG.)	PHOSPHORUS (MG.)	IRON (MG.)	VITAMINS					
										A (I. U.)	THIAMINE (MCG.)	RIBOFLAVIN (MCG.)	NIACIN (MG.)	ASCORBIC ACID (MG.)	D (I. U.)
Breads:															
Bread, white, enriched	25	1 slice, average	2.1	0.5	13.1	65	14	25	0.5		60	37	0.55		
Bread, whole wheat, 60%	28	1 slice, average	2.5	0.8	13.8	72	14	42	0.6		84	49	0.90		
Doughnut, yeast	30	1 small, 3″ diam.	2.0	6.5	14.0	123	13	24	0.4	51	67	56	0.50		1
Griddle cakes, white	50	1 medium, 4″ diam.	2.4	2.4	11.1	76	32	38	0.4	110	62	78	0.42		2
Roll, white, hard	35	1 av., no milk or butter	2.9	2.1	18.9	106	20	35	0.6		84	53	0.77		
Cereals:															
Corn flakes, Kellogg's	28.35	1 oz., 1⅓ cups	2.2	0.1	24.5	107	1	11	0.5		120		0.60		
Oats, Quaker	30	⅔ cup, cooked	4.9	2.0	19.4	105	15	131	1.2		204	39	0.33		
Rice, white	30	¾ cup, cooked	2.3	0.1	23.8	105	3	28	0.2		15	9	0.42		
Spaghetti, tomato sauce	220	1 serving	6.4	11.3	35.8	271	23	80	1.0	1,336	125	76	1.57	19	
Wheaties	28.35	1 oz., 1 cup	3.1	0.5	22.5	107	9	100	1.3		150	37	1.50		
Crackers and Similar Products:															
Crackers, saltines	4	1 cracker, 2″ square	0.4	0.5	2.8	17	1	4	0.06		2				
Pretzel	20	1 medium	1.8	0.6	14.9	72	7	26	0.3						
Ry-Krisp	6.5	1 piece	0.8	0.08	4.9	24	4	25	0.3		21	15	0.15		
Dairy Products:															
Butter	10	1 pat, average	0.06	8.1	0.04	73	2	2	0.02	330		1	0.01		4
Cheese, American	28.35	1 oz., average serving	6.8	9.2	0.5	112	247	173	0.2	493	11	142	0.06		
Cheese, cottage	100	½ cup	19.2	0.8	4.3	101	82	263	0.5	30	20	290	0.10		
Ice cream, plain	100	⅙ quart, av. serving	4.0	12.3	20.8	210	132	104	0.1	540	40	190	0.10	Trace	
Milk, whole	180	6 oz., 1 medium glass	6.1	7.0	8.8	123	212	167	0.13	288	72	306	0.18	2	4
Egg, boiled	50	1 medium	6.4	5.8	0.4	79	27	105	1.4	570	52	161	0.05		45

Fish:															
Halibut steak	100	1 serving	18.6	5.2		121	11	209	0.9	10	90	65	3.00		
Oysters, with liquor	100	4 to 6 medium	6.0	1.2	3.7	50	68	172	7.1		180	230	1.20	2	5
Salmon, canned, red	100	1 cup	20.5	6.2		138	194	289	1.3	325	30	225	6.50		800
Fruits and Fruit Juices:															
Apple, fresh	150	1 large	0.5	0.6	22.5	97	9	15	0.5	135	60	30	0.30	8	
Banana, fresh	100	1 medium	1.2	0.2	23.0	99	8	28	0.6	430	90	60	0.60	10	
Orange juice, fresh	100	½ cup			10.1	40	33	23	0.4	190	80	30	0.20	49	
Prunes, stewed	100	4 to 5 medium	1.1	0.3	40.5	169	27	43	1.9	945	50	80	0.85	1	
Tomato juice	100	½ cup	1.0	0.2	4.3	23	7	15	0.4	1,050	50	30	0.70	16	
Watermelon	600	1 slice, 6″ x 1½″	3.0	1.2	41.4	188	42	72	1.2	3,540	300	300	1.20	36	
Meats:															
Bacon, cooked	30	3 strips, 5″ long, crisp	4.2	8.1	0.5	93	6	48	0.3		180	30	0.87		
Beef steak	115	1 pc., 4″ x 2″ x 1″	20.3	25.0		306	12	218	3.0		84	125	4.42		
Lamb chop	105	1 chop, 1″ thick	17.9	21.0		261	12	192	2.6		180	224	4.96		
Liver, beef	41	1 slice	9.9	3.3	3.0	82	6	187	6.1	9,600	115	1,190	6.85	?	23
Pork chop, fried	70	1 chop, medium	16.4	18.3		230	10	180	2.5		876	170	3.76		
Veal cutlet, breaded	122	1 chop	24.8	18.0	5.8	284	23	272	4.1	114	168	350	7.15		9
Poultry:															
Chicken, fried	95	½ medium	20.6	10.4		176	12	218	1.9	138	93	153	7.3		
Turkey	30	1 piece, lean	10.7	1.5		56	6	112	1.6		49	65	2.12		
Nuts:															
Almonds, salted	15	10 to 12 nuts	2.7	8.0	2.8	94	38	71	0.7		38	101	0.69	Trace	
Peanut butter	15	1 tablespoon	3.9	7.2	3.2	93	11	59	0.3		30	24	2.43		
Walnuts, English	15	8 to 15 halves	2.3	9.7	2.3	106	12	57	0.3	9	77	20	0.18	Trace	
Sweets:															
Chocolate creams	13	1 average piece	0.5	1.8	9.4	56	?	10	0.06			7			
Hershey's milk chocolate	43	1 bar, 5 cents	3.2	14.5	23.8	238	86	89	1.1	64	44	219	0.14		43
Assorted jams, commercial	20	1 tablespoon	0.1	0.06	14.2	58	4	3	0.06	2	2	4	0.03	Trace	

Table 6. ANALYSES OF COMMON FOODS *(continued)*

FOOD	WT. (GM.)	APPROXIMATE MEASURE	PRO-TEIN (GM.)	FAT (GM.)	CARBO-HYDRATES (GM.)	CALO-RIES	CAL-CIUM (MG.)	PHOS-PHORUS (MG.)	IRON (MG.)	VITAMINS					
										A (I. U.)	THIA-MINE (MCG.)	RIBO-FLAVIN (MCG.)	NIA-CIN (MG.)	ASCORBIC ACID (MG.)	D (I. U.)
Vegetables:															
Beans, string, fresh	100	½ cup, cooked	2.4	0.2	7.7	42	65	44	1.1	630	80	100	0.60	19	
Beets, fresh, raw	100	½ to ⅝ cup, diced	1.6	0.1	9.6	46	27	43	1.0	20	30	50	0.40	10	
Carrots, raw	100	1 large	1.2	0.3	9.3	45	39	37	0.8	12,000	70	60	0.50	6	
Corn, sweet, canned	100	½ cup	2.0	0.5	16.1	77	4	51	0.5	200	20	50	0.80	5	
Potatoes, white, baked	139	1 medium large	3.0	0.2	28.7	129	17	84	1.1	30	153	60	1.76	22	
Spinach, fresh	100	⅔ cup, cooked	2.3	0.3	3.2	25		55	3.0	9,420	120	240	0.70	59	
Tomatoes, fresh	100	1 medium, ripe	1.0	0.3	4.0	23	11	27	0.6	1,100	60	40	0.60	23	
Beverages:															
Chocolate milk shake	342	1 regular	11.1	17.7	57.7	435	362	323	0.4	708	120	534	0.37	2	9
Coca-Cola	170	1 bottle			20.4	82									
Coffee, roasted	100	3½ oz.									900	70	8.69		
Beer, lager, American	240	1 large glass	1.2		8.9	103					10	72	1.92		
Desserts:															
Cake, chocolate, 2 layers	100	1 pc., vanilla icing	4.3	8.3	68.7	368	43	80	0.6	229	25	110	0.23		8
Pudding, bread, with raisins		1 serving	6.1	7.0	29.3	205	99	119	1.1	349	116	193	0.98		16
Pie, apple		⅙ medium pie	2.6	13.3	34.0	266	11	36	0.6	116	53	30	0.42	1	
Miscellaneous:															
Salad, gelatin with fruit	155	1 serving with lettuce	1.9	9.0	20.8	172	16	18	0.3	331	23	32	0.18	2	3
Sandwich, bacon, lettuce, tomato		1, white toast	7.0	20.2	24.1	306	38	97	1.5	776	232	115	1.78	12	7
Stew, oyster	325	1½ cups	14.3	18.7	15.5	287	353	397	7.3	714	248	575	1.30		12
Soup, vegetable	140	¾ cup	4.6	4.2	11.6	103	31	45	0.6	1,612	54	35	0.57	7	

Chapter 21

The Excretory System

THE NORMAL processes of cellular metabolism and the constant building up and breaking down of proteins and nucleic acids result in the production of waste products such as urea, uric acid, creatinine and ammonia. These nitrogenous wastes are not only useless but toxic. When kidney function is impaired by disease, these substances rapidly accumulate in the blood and tissues and cause death. In a normal person the concentration of these substances in the blood remains constant at a low level, for the kidneys remove them from the blood as rapidly as the tissues produce them.

The terms **defecation, excretion** and **secretion** are sometimes confused. Defecation refers to the elimination of wastes and undigested food, collectively called feces, from the anus. Undigested food never enters any of the body cells and so cannot take part in cellular metabolism; hence these are not metabolic wastes. Excretion refers to the removal of substances which have no further use in the body from the cells and blood stream via the urine and perspiration. The excretion of wastes by the kidneys involves an expenditure of energy by the cells, but the act of defecation requires no such effort of the cells lining the intestine. Secretion is the giving off by a cell of some substance that is used elsewhere in some body process—for example, the salivary glands secrete saliva, used in the mouth and stomach for digestion. Secretion also involves cellular activity and requires the expenditure of energy by the secreting cell. It can be shown, for example, that the consumption of oxygen and glucose increases when cells begin secreting.

The excretory system of the human body includes more than the kidneys and their ducts; the skin, lungs and digestive tract have excretory functions too. We have discussed previously the elimination of carbon dioxide, one of the most important metabolic wastes, by the lungs; the excretion of bile pigments, the breakdown products of hemoglobin, by the liver; and the excretion of certain metals, such as calcium, by the colon. The sweat glands of the skin are primarily concerned with the regulation of body temperature, but they also serve the purpose of excreting 5 to 10 per cent of all metabolic wastes. Sweat contains the same substances (salts, urea and other organic compounds) as urine, but is much more dilute, having only about one eighth as much solid matter. The volume of perspiration varies from about 500 ml. on a cool day to as much as 2 or 3 liters on a hot one. From tests made by the Army during World War II, it was found that, while doing a maximum of work at high

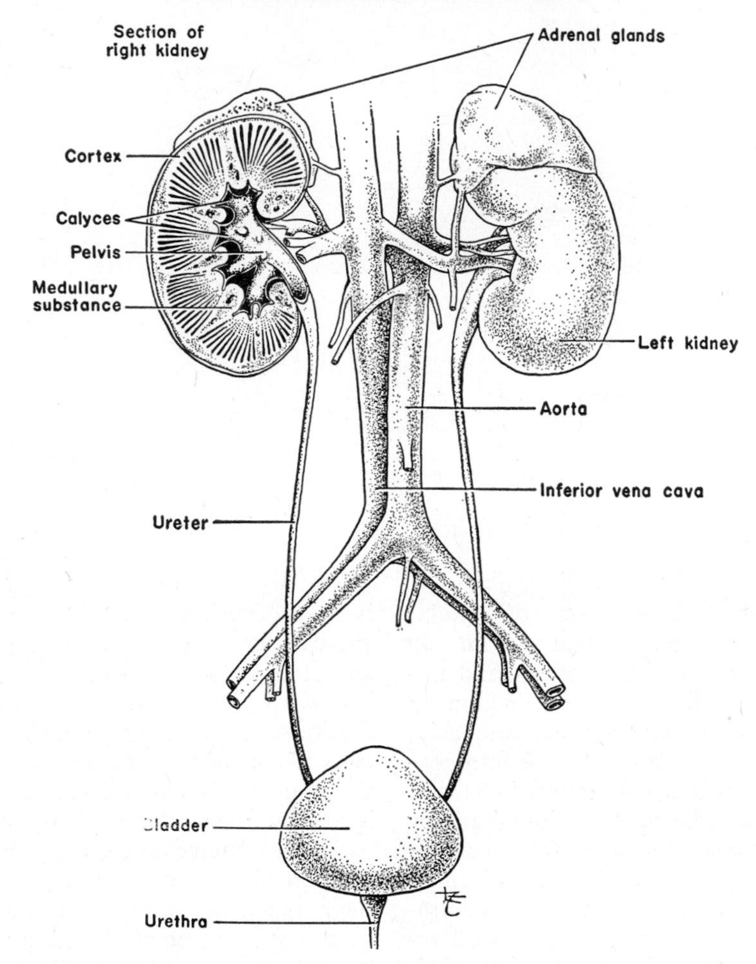

Figure 180. The human urinary system, seen from the ventral side. The right kidney is cut open to reveal the internal structures.

temperatures, some men would excrete from 3 to 4 liters of sweat in an hour!

204. THE KIDNEY AND ITS DUCTS

Although the kidneys are the most important excretory organs of mammals, performing approximately 75 per cent of the work of excretion, they have a number of other important functions as well. They regulate the concentration of various substances dissolved in the blood, maintain the balance between acids and bases, and keep the blood volume constant. Since the concentration of substances in all body fluids is determined largely by their concentration in the blood, the kidneys indirectly regulate the composition of all body fluids.

The kidneys are a pair of bean-shaped structures about 4 inches long, one of which is located on each side of the mid-dorsal line of the abdominal cavity, just below the level of the stomach (Fig. 180). On the medial, concave side of each kidney is a funnel-shaped chamber called the **pelvis.** The urine, excreted by the kidney in a continuous trickle, collects in the pelvis and passes down the **ureters** by peristaltic waves of contraction of the ureter walls to the **urinary bladder,** a hollow, muscular organ located in the lower, ventral part of the abdominal cavity (Fig. 180). The muscular walls of the bladder relax and distend to make room for the urine as it accumulates. Valves at the openings of the ureters into the urinary bladder prevent the backflow of urine, and keep any bacteria which may be in the bladder from ascending to the kidney. As the volume of urine in the bladder increases, the distention of the muscular walls stimulates nerve endings located there to send impulses to the brain, producing the sensation of fullness. To make urination possible, impulses originating in the brain cause a contraction of the bladder and a relaxation of the sphincter guarding the opening from the bladder to the urethra.

The kidney consists of masses of microscopic **tubules** arranged in an inner core, the **medulla,** and an outer layer, the **cortex.** The tubules are richly supplied with blood vessels and supported by a fine network of connective tissue fibers. The kidney has solved the problem of removing from the blood the useless, toxic waste products of metabolism without losing any of the substances that the body needs. The individual tubule is the structural and functional unit of the kidney; each one excretes a small part of the daily urine output. A branch of the aorta, the kidney artery, enters each kidney and ramifies to all parts of the organ. Each ultimate arteriole goes to the end of a tubule and branches into a spherical tuft of capillaries, called a **glomerulus** (Fig. 181). The end of each kidney tubule is a double-walled, hollow sac of cells, called **Bowman's capsule,** which surrounds the glomerulus. The inner wall of Bowman's capsule consists of flat epithelial cells which adhere closely to the capillaries of the glomerulus, so that substances may diffuse easily from the capillaries into the cavity of Bowman's capsule.

205. THE FORMATION OF URINE

The combination of the three processes of **filtration, reabsorption** and **augmentation** enables the kidney to remove wastes but conserve the useful components of the blood. Filtration occurs at the junction between the glomerular capillaries and the wall of Bowman's capsule. The blood is "filtered" as it passes through the capillaries so that water, salts, sugar, urea and all the substances in the blood, except the blood cells and large molecules such as the plasma proteins, pass across this junction into the cavity of Bowman's capsule to become the **glomerular filtrate.** The total blood flow through the kidneys is about 1200 ml. per minute, or one-quarter of all the blood pumped by the heart! The plasma going through the glomerulus loses about 20 per cent of its volume to the glomerular filtrate; the rest passes from the glomerulus in the outgoing blood vessel. The mechanism underlying this process is the purely physical one of filtration, and depends on the fact that the small artery entering the glomerulus is larger than the vessel leaving it. Consequently the blood pressure in the glomerular capillaries is relatively high, and a fraction of the plasma filters across into the capsule. By introducing a fine glass syringe into the Bowman's capsule of a frog's kidney and actually collecting and analyzing some of the glomerular filtrate, Dr. A. N. Richards of the University of Pennsylvania showed that it has the same concentration of urea, salts, glucose, and so forth, as the plasma, but lacks its proteins. The cells of Bowman's capsule are thin and unable to move materials from the capillaries; the work of pushing the filtrate from the plasma into the capsule is done by the heart. It can be shown experimentally that as the blood pressure, and consequently the filtration pressure, rise and fall, the quantity of glomerular filtrate varies accordingly. The amount filtered is also regulated by the constriction or dilation of the arterioles leading to and from the glomerulus. The amount filtered is increased by the constriction of the efferent arterioles and dilation of the afferent arterioles.

If the composition of the urine excreted were like that of the glomerular filtrate, excretion would be a wasteful process, and a great deal of water, glucose, amino acids and other useful substances would be lost. However, the kind and quantity of substances present in the urine are quite different from those in the plasma and glomerular filtrate. From each Bowman's capsule, located in the cortex, the filtrate passes first through a **proximal convoluted tubule** (also in the cortex), then through a long loop going into the medulla and back into the cortex, then through a second region in the cor-

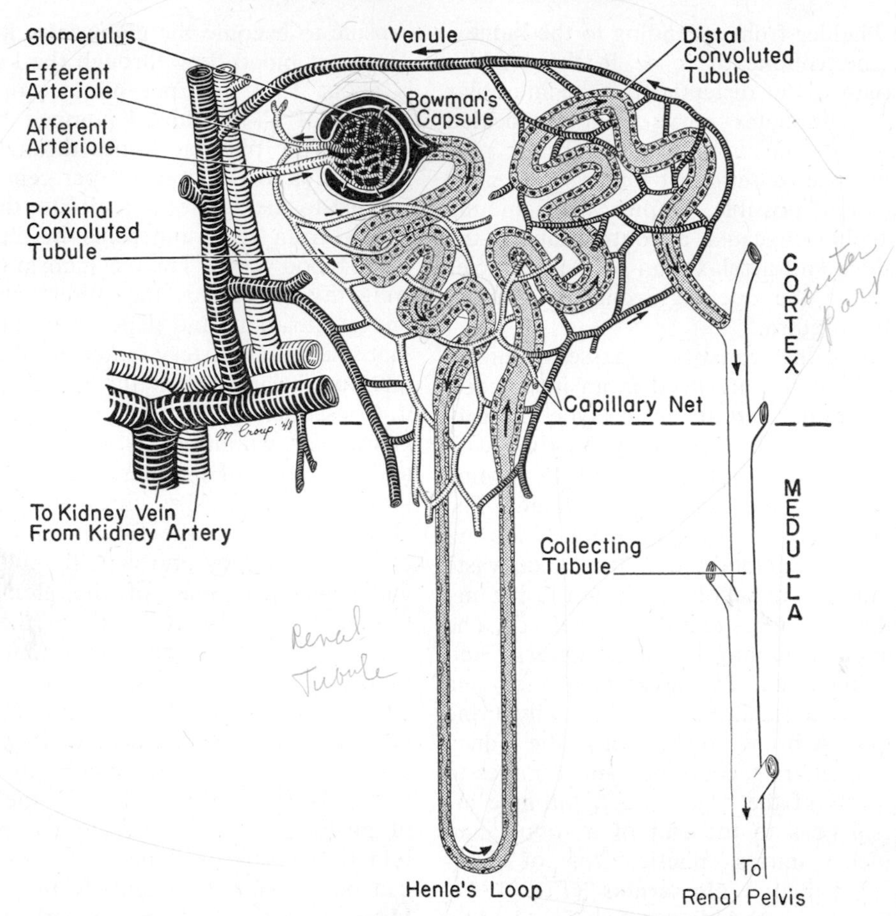

Figure 181. Diagram of a single kidney tubule and its blood vessels.

tex, the **distal convoluted tubule,** to empty at last into a collecting tubule, through which it passes to the pelvis (Fig. 181). There is no later change in the urine as it passes from the pelvis of the kidney through the ureters, bladder and urethra; the changes in concentration occur when the materials pass from the Bowman's capsules through the long coiled tubules to the collecting tubule.

The walls of the kidney tubules are made of a single layer of flat or cuboidal epithelial cells. As the filtrate passes through, these remove or reabsorb much of the water and virtually all the glucose, amino acids and other substances needed by the body, and secrete them back into the blood stream. This is possible because the arteriole leaving the glomerulus does not go directly to a vein, but connects with a second network of capillaries around the proximal and distal convoluted tubules (Fig. 181). Thus the route of blood in the kidney differs from that in every other organ—for blood to pass from a kidney artery to a kidney vein, it must pass through two sets of capillaries. The ability of the kidney to excrete urine and regulate the composition of the blood depends upon this fact. Substances are reabsorbed into the blood stream selectively, according to the momentary needs of the body; if there is already too much glucose in the blood, as in a patient suffering from diabetes, glucose is not reabsorbed, but passes on in the urine. The cells lining the tubules must do work to secrete these substances back into the blood stream, usually against a diffusion gradient. In fact, a given amount of kidney tissue consumes a larger amount of oxygen per hour than an equivalent weight of heart muscle, indicating that the kidneys work harder than the heart. They obtain the energy for this work from the oxidation of carbohydrates; when the kidney is deprived of oxygen, reabsorption (though not filtration) ceases.

The human kidney produces about 125 liters of filtrate for every liter of urine formed; the other 124 liters of water are reabsorbed. In this way the waste products, such as urea, are greatly concentrated as the filtrate passes down the tubules. The concentration of urea in the urine is about sixty-five times that in the glomerular filtrate and would be even higher but for the fact that a small amount of urea is reabsorbed by the tubules. The quantity of water reabsorbed depends also on the body's current needs for it. If a large quantity of water or beer is drunk, less water is reabsorbed and a copious, dilute urine is excreted. If water intake is restricted, a maximum amount of water is reabsorbed by the cells of the tubules, conserving water for the body, and a scanty but concentrated urine is excreted.

The cells of the kidney tubules not only remove substances from the filtrate and put them back into the blood stream, but also excrete additional waste materials from the blood stream into the filtrate. This process, called **augmentation,** probably plays only a minor role in human kidney function, but in animals like the toadfish, whose kidneys lack glomeruli and Bowman's capsules, excretion by the tubules is the only method used. When the blood pressure, and consequently the filtration pressure, drop below a certain point, filtration ceases in man, although urine is still formed by tubular excretion. Dyes injected into experimental animals can be seen to pass from the blood stream into the urine through the cells lining the tubules. Experiments have revealed that drugs such as penicillin and Atabrine are removed from the blood and excreted by the process of augmentation. There is no doubt, then, that augmentation can occur in man and other animals, but just how large a role it normally plays in the process of excretion is unknown.

When the fluid reaches the end of the distal convoluted tubule, and some substances have been reabsorbed and others added, the glomerular filtrate has become urine.

206. THE REGULATORY FUNCTION OF THE KIDNEY

By the selective excretion of certain substances and the reabsorption of others, the kidney plays an extremely important role in regulating the composition of the blood and other body fluids. Any excess of acid or base released during metabolism is excreted by the kidneys, thereby maintaining the proper *p*H of the blood. The kidneys regulate the osmotic pressure of the body fluids bathing the cells by regulating the concentration of salts in the blood. This is important, because if the concentration of salts in the body fluids exceeds that within the cells, water passes out of the cells and they shrivel and die. Or if the concentration of salts in the body fluids falls below that within the cells, water passes into the cells, causing them to swell and burst.

Although there is glucose in the glomerular filtrate, there normally is none in the urine because it is reabsorbed by the cells of the tubules. But if there is so much glucose present in the blood, and consequently in the glomerular filtrate (as there is in diabetics) that not all of it can be reabsorbed in the time the filtrate passes through the tubules, some glucose will appear in the urine. When the concentration of glucose in the blood reaches this point, it is said to have reached the "kidney threshold," which, for glucose, is about 150 mg. per 100 ml. of blood. Many other substances have "kidney thresholds," but the concentration at which the substance begins to appear in the urine is different for them.

The kidney regulates the total volume of blood as well as the concentration of the substances dissolved in it. When the total amount of blood is decreased after a hemorrhage, the blood pressure is lessened. Since the filtration pressure depends on the blood pressure, it is decreased correspondingly and less liquid is filtered through the glomeruli into the Bowman's capsules. The kidney produces a smaller volume of urine, and body fluids are conserved. When a great deal of liquid is taken in, the blood volume, blood pressure and filtration pressure are raised, a larger volume of urine is formed, and blood volume is restored to normal.

The quantity of urine excreted depends not only on how much liquid is consumed, but also on the amount of salts and other solids to be excreted from the blood. When food is unusually salty, the kidney must excrete a proportionately larger amount of salt to maintain the proper osmotic pressure of the blood, and the urine volume is increased. Since the solids excreted are dissolved in the urine, an increased amount requires more water to pass them. That is why **diabetes,*** of which one of the chief symptoms is the presence of sugar in the urine, is also characterized by a copious flow of urine. An increased amount of dissolved solids in the glomerular filtrate increases its osmotic pressure and consequently decreases the rate of reabsorption of water, resulting in the increase in urine volume. This mechanism enables the kidney to respond to increases in the concentration of urea and other wastes in the blood, since the higher concentration itself stimulates the flow of urine and the excretion of wastes.

Another factor controlling urine volume is a hormone (known as "antidiuretic hormone") produced by the posterior lobe of the pituitary which controls the rate at which water is reabsorbed by the kidney tubules. In a comparatively rare disease, **diabetes insipidus** (insipid indicating a lack of sugar), due to a deficiency of this hormone, the output of urine may reach 30 or 40 liters per day instead of the normal 1.2 to 1.5 liters, and the patient suffers from an insatiable thirst.

* The word "diabetes" simply means "to go through." In diabetes mellitus (mellitus means sweet, indicating the presence of sugar) a pathologic condition of the pancreas decreases the amount of insulin formed, resulting in impaired carbohydrate metabolism and increasing the concentration of glucose in the blood and urine.

207. SUBSTANCES PRESENT IN THE URINE

Since the kidneys excrete, in addition to the normal products of metabolism, most of the abnormal substances that may enter or be formed in the body, analyzing the urine is a reliable method for determining the general metabolic condition of the body. Normal urine is about 96 per cent water, 1.5 per cent salts, and 2.5 per cent organic wastes, chiefly urea. The salts in the urine are the same as those in the blood and protoplasm, chiefly sodium chloride, with small amounts of potassium, calcium, magnesium and ammonium sulfate, phosphate and carbonate. The amount and kinds of salts excreted in the urine are controlled by aldosterone, a hormone secreted by the adrenal cortex (p. 389).

The 1200 to 1500 ml. of urine excreted daily contain about 60 gm. of solids; half of this is **urea,** which comes from the deamination of proteins. Other normal constituents of urine are **uric acid,** a breakdown product of nucleic acids, and **creatinine,** which comes from muscle metabolism.

The yellow color of urine is not due to any of the substances mentioned, but depends on the presence of a pigment called **urochrome,** a breakdown product of hemoglobin related to the bile pigments.

208. DISEASES OF THE KIDNEYS

A painful and often fatal disease, **nephritis,** is caused by bacterial infection of the kidney cells. They usually attack the glomeruli, causing them to become more permeable than normal, so that proteins and even intact blood cells pass through into the urine. Because of the constant loss of proteins from the blood, the ability of the plasma to reabsorb water from the tissue fluid around the capillaries is lowered, and a watery swelling, known as **edema,** occurs, especially in the lower parts of the legs. In the last stages of this disease there is a marked decrease in the amount of urine formed, and the waste products normally excreted by the kidneys accumulate in the blood (a condition called uremia). Because some of these substances are toxic, death occurs unless the condition is relieved immediately.

Impairment of the kidneys is often associated with high blood pressure, because the abnormal pressure tends to injure the glomeruli, and the decreased supply of blood to the kidney cells, resulting from the constriction of the arteries, may cause damage. Recent experiments have shown that injured kidneys give off a substance called **renin,** which causes constriction of the blood vessels, thus raising the blood pressure and completing the vicious cycle.

Some of the solids normally present in the urine, uric acid and calcium phosphate, are not very soluble, and when the amount present in the urine is increased, they may be precipitated in the urinary passages as "kidney stones." If these become large enough to block the passage of urine, they must be removed surgically, since once they form, it is impossible to redissolve them.

209. EXCRETORY DEVICES IN OTHER ANIMALS

Every organism has had to solve the problem of getting rid of metabolic wastes. In protozoa, such as amebas and paramecia, the wastes simply diffuse through the cell wall into the outside environment where the concentration is lower. Protozoa living in fresh water have a special problem of getting rid of water, because their protoplasm, being hypertonic to pond water, tends to absorb it continuously. To control this situation, which would otherwise result in the swelling and bursting of the cell, they have a **contractile vacuole,** a small vesicle in the protoplasm which empties water from the interior of the cell as fast as it is taken in. Hydras and other coelenterates also have direct excretion of wastes by diffusion through the cell membrane.

In larger animals the elimination of wastes by diffusion is not adequate to prevent toxic accumulations, and various excretory devices are necessary. Flatworms have specialized **flame cells** (Fig. 182), single cells that absorb fluid from the surrounding spaces and excrete it into the excretory tubules. The beating of a tuft of cilia on the flame cell propels the fluid down the tubules. The excretory tubules from several flame cells join and eventually open to the exterior through an excretory pore. The beating of the cilia

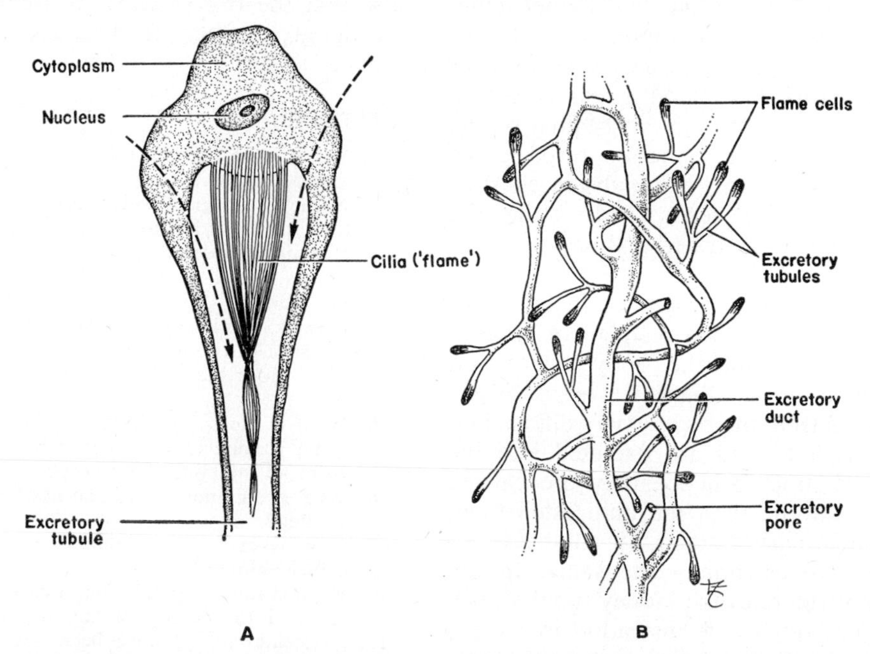

Figure 182. The excretory organs of the flatworm. *A,* A single flame cell. *B,* Flame cells are connected by excretory tubules and ducts to the pore which opens to the outside.

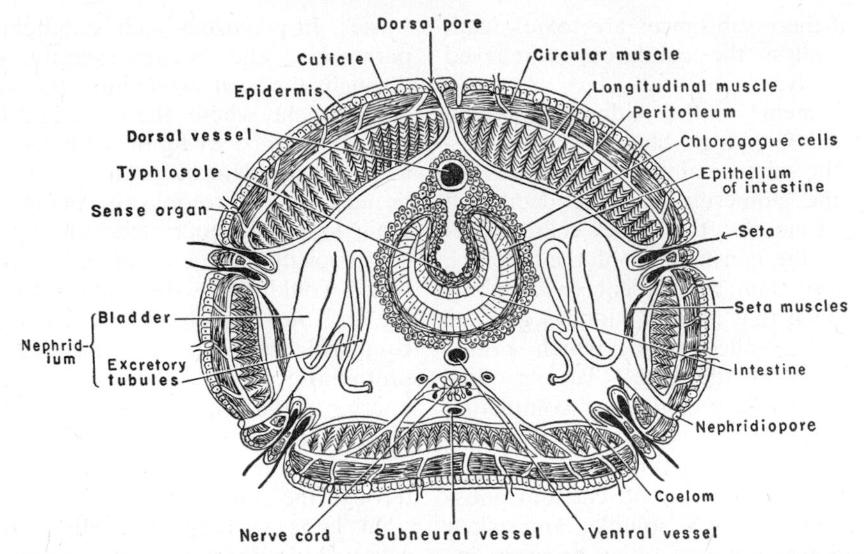

Figure 183. Cross section of the common earthworm, *Lumbricus terrestris,* showing the excretory organs, the paired nephridia. Each consists of a ciliated funnel opening into the coelomic cavity, a coiled tubule, and a pore opening to the outside. (Hunter and Hunter: College Zoology.)

resembles the flickering of candlelight, hence the name.

Earthworms have in each segment of their bodies a pair of specialized organs, called **nephridia,** which function in excretion. The nephridium, in contrast to the flame cells of flatworms, is a tubule open at both ends, the inner end connecting to the coelom by a ciliated funnel (Fig. 183). Around each tubule is a coil of capillaries, which permits the removal of wastes from the blood stream. As the body fluid, moved by the beating of the cilia in the funnel, passes through the nephridium, water and substances such as glucose are reabsorbed, while the wastes are concentrated and passed out of the body.

The excretory system of insects consists of organs called **malpighian tubules,** which lie within the body cavity and empty into the digestive tract. Waste products from the body cavity diffuse into these tubules and are excreted into the digestive tract, whence they are carried to the exterior with the undigested food.

The urinary systems of all the vertebrates are essentially the same. In the lower vertebrates the kidney tubules open into the body cavity instead of into Bowman's capsules, and thus represent a type of excretory organ intermediate between the nephridia found in the earthworm and the kidneys of higher vertebrates. The evolution of the urinary system is complicated by the fact that in many animals the reproductive system has come to share some of the structures of the urinary system, so that several organs play a dual role. This relationship is so close that the two systems are frequently considered together as the "urogenital" system.

QUESTIONS

1. What is the difference between excretion and defecation?
2. Name the organs which perform excretory functions in the body.
3. Explain in detail how the metabolic wastes are eliminated from the body by the kidney.
4. How do the kidneys regulate the osmotic pressure of the body fluids?
5. What does the term "kidney threshold" mean?
6. What is the composition of urine?
7. What is edema? Why may it result from from an injury to the kidney?
8. What are kidney stones and how are they formed?
9. How is excretion accomplished in one-celled animals?
10. Explain how the nephridia of earthworms carry on the process of excretion.
11. Artificial kidneys have been devised for patients critically ill with kidney disease. How do you suppose they work? How

would you go about designing an artificial kidney?

SUPPLEMENTARY READING

A discussion of excretion and the regulation of the internal environment in other animals is found in Chapter 7 of Prosser's *Comparative Animal Physiology*, or in B. T. Scheer's *Comparative Physiology*. The evolutionary significance of the kind of nitrogenous waste materials excreted by members of the different classes of vertebrates is discussed in Baldwin's *An Introduction to Comparative Biochemistry*.

An interesting discussion of the role of the kidney in maintaining a constant water and salt content of the body is given in Chapters 4 and 5 of Cannon's *The Wisdom of the Body*.

Chapter 22

The Integumentary and Skeletal Systems

THE SKIN, which covers the body, and the bony framework which supports it, are both organ systems, groups of organs that act together to perform one of the primary life functions. The integumentary and skeletal systems function independently of each other, but since both act as protective devices for the body, and because together (with the muscles) they determine the shape and symmetry of the body, we shall consider them in the same chapter.

210. THE SKIN

All multicellular animals are covered externally by a skin or **integument,** consisting of one or many layers of cells. The skin is much more than merely an outer wrapping for the animal; it is one of the important body organs and performs many diverse functions. Perhaps the most obvious and vital of these is to protect the body against a variety of external agents and to help maintain a constant internal environment. Being tough and pliable, it shields the underlying cells from mechanical injuries caused by pressure, friction or blows. The skin is practically germ-proof as long as it is unbroken, and it protects the body against disease-producing organisms. Its waterproof quality protects the body from excessive loss of moisture, or, in aquatic animals, from the excessive intake of water. Important, too, is the protection it affords the underlying cells from the harmful ultraviolet rays of the sun, by virtue of the pigment it can produce (sun tan).

The skin also functions as a thermostatically controlled radiator, regulating the elimination of heat from the body. Since heat is constantly being produced by the metabolic processes of the cells and distributed by the blood stream, a certain amount of heat must be lost all the time to maintain a constant temperature within the body. Some heat leaves the body in the expired breath and some in the feces and urine, but approximately 90 per cent of the total lost is given off by the skin. When the external temperature is low, temperature-sensitive nerve endings in the skin are stimulated, and the capillaries in the skin are reflexly contracted, thereby decreasing the flow of blood through the skin and the rate of heat loss. In a warm environment the reverse occurs: the capillaries expand, and the increased flow of blood causes the skin to appear flushed and results in an increased heat loss. In very warm environments this mechanism

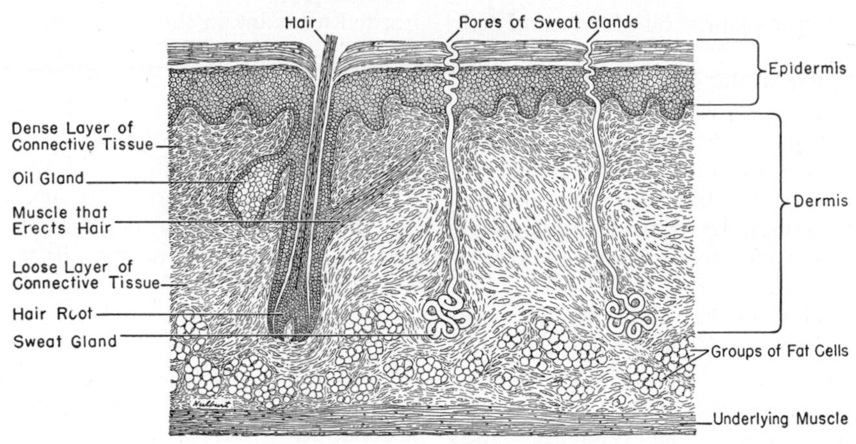

Figure 184. Microscopic section through the human skin, showing dermis, epidermis, hair, and oil and sweat glands.

is not sufficient to get rid of the necessary amount of heat, so that the sweat glands of the skin give off an unusually great amount of perspiration. The evaporation of sweat from the surface of the skin lowers the body temperature by removing from the body the heat necessary to convert the liquid sweat into water vapor—540 Calories are required to convert a liter of water to water vapor.

The skin contains a number of different sense receptors which are responsible for our ability to feel pressure, temperature and pain, and to discriminate between the various objects touched (Fig. 205). Specialized glands are located in the skin: some 2½ million sweat glands occur all over the body, but are most numerous on the palms of the hands, the soles of the feet, in the arm pits and on the forehead; oil glands, too, are found all over the body, but are especially numerous on the face and scalp. They secrete a film of oil to keep the hair moist and pliable and to prevent the skin from drying and cracking. The mammary glands of mammals are also derivatives of the skin, specialized for the secretion of milk.

Parts of the Skin. The skin is composed of two main parts: a comparatively thin, outer layer, the **epidermis,** free of blood vessels; and an inner, thicker layer, the **dermis,** packed with blood vessels and nerve endings (Fig. 184). The epidermis is really made up of several layers of different kinds of cells, which vary in number in different parts of the body. The thickness of the skin varies considerably from one part of the body to another. It is thickest on the soles of the feet and the palms of the hand, where the epidermal surface is thrown into countless tiny ridges which form the fingerprint patterns. These patterns are unique in each person and remain constant throughout life. The layer of the epidermis lying next to the dermis is made up of columnar cells which undergo frequent cell division to give rise to the layers above. The outer layers of the skin are constantly sloughing off and being replaced by cells from beneath. As each cell is pushed outward from the bottom layer, it is compressed into a flat epithelial cell, which is lifeless and scale-like. Dandruff consists of the flaky particles of dead, outer epidermal cells of the scalp.

The dermis is much thicker than the epidermis and is composed largely of connective tissue fibers and cells. The outer part, made of thickly matted connective tissue fibers, is the portion which is tanned to make leather. Below this, and connected with the underlying muscles, is a layer composed of many fat cells and a more loosely woven network of fibers. This part of the dermis is one of the principal depots of body fat. The fat helps prevent excessive loss of heat and acts as a cushion against mechanical injury. The dermis is richly supplied with blood and lymph vessels, nerves, sense organs, sweat glands, oil glands and hair follicles.

The color of the skin depends on three factors: the yellowish tinge of the epidermal cells, their translucent quality which allows the pink of the underlying blood vessels to show through, and the kind and amount of pigment—red, yellow or brown—contained in the inner layer of epidermal cells.

Outgrowths of the Skin. The hair and nails of man and the feathers, scales, claws, hoofs and horns of other vertebrates are derivatives of the skin. The entire skin, except the palms of the hands and soles of the feet, is equipped with countless hair follicles—inpocketings of cells from the inner layer of the epidermis. These cells undergo division and give rise to the hair cells, just as the inner layer of the epidermis gives rise to the outer layers. But the hair cells die while still in the follicle, and the hair visible above the surface of the skin consists of tightly packed masses of their remains. Hair grows from the bottom of the follicle, not from the tip. Its color, and that of feathers and fur (which is a form of hair), depend on the amount and kind of pigment present, on the number of air bubbles, and on the nature of the surface of the hair, which may be smooth or rough.

Fingernails and toenails also develop from inpocketings of cells from the inner layer of the epidermis, and the growth of nails is similar to that of the hair. Nails are composed of densely packed dead cells which are translucent, allowing the underlying capillaries to show through and giving the nails their normal pink color.

Oil and sweat glands are derived from the inner layer of the epidermis by inpocketings which go deep into the dermis. Each hair follicle is associated with an oil gland (Fig. 184).

211. THE SKELETON

The first and most obvious function of the skeleton is to give support and "shape" to the body. In order that an animal may rise off the ground and move around, some hard, durable substance is needed to maintain the soft tissues against the pull of gravity and act as a firm base for the attachment of muscles. These requirements are met by the bones. The skeleton also protects the delicate underlying organs, such as the brain and lungs, from injury. The marrow tissue, within the cavity of the bones, performs the special task of manufacturing all red corpuscles and some kinds of white ones.

The skeletal system is not composed solely of bones; connective tissue fibers are important in helping to maintain body form by holding the organs together. Two specialized kinds of connective tissue fibers, ligaments and tendons, attach bones to bones, and muscles to bones, respectively, thereby playing an indispensable role in locomotion.

Types of Skeletons. The skeleton of an animal may be located on the outside of the body (an **exoskeleton**) or inside the body (an **endoskeleton**). The hard shells of lobsters and crabs, and the shells of oysters, clams and snails, are examples of exoskeletons. The advantage of an exoskeleton as a protective device is obvious, but a serious disadvantage is the attendant difficulty of growth. Snails and clams meet the difficulty by secreting additions to their shells as they grow, and lobsters and crabs have evolved a complicated solution whereby the outer shell is first softened by the removal of some of its salts, so that it can be split down the back. The animal then crawls out of the old shell, grows rapidly for a short time, and produces a new, larger shell which hardens by the redeposition of mineral salts. During this process the animal, being weak and barely able to move, easily falls prey to its enemies.

Man and all the other vertebrates characteristically have an endoskeleton. The skeleton of sharks and rays is made of cartilage, but in the higher fishes and other vertebrates most of the cartilage has been transformed to bone. The human skeleton consists of approximately 200 bones; the exact number varies at different periods of life, as some of the bones which at first are distinct, gradually become fused. Most of the bones are hollow and contain the bone marrow cells, which manufacture red and white blood cells. Bony scales and plates are present in the dermis of many vertebrates; some of these become associated with, and an integral part of, the skull and pectoral girdle.

Parts of the Skeleton. The vertebrate skeleton may be divided into the **axial skeleton** (the bones and cartilages in the middle or axis of the body), and the **appendicular skeleton** (the bones and cartilages of the fins or limbs) (Fig. 185). The axial skeleton includes the **skull, backbone** (vertebrae), **ribs** and **breast bone** (sternum).

The skull is made up of a number of bones fused together: the **cranium** or bony case immediately around the brain, and the bones of the face. The lower vertebrates have gill arches, cartilages or bones supporting the gill pouches. In the higher vertebrates these have disappeared or have been converted into other structures, such as the small bones in the middle ear—the hammer, anvil and stirrup—which transmit sound waves from the ear drum to the inner ear. Certain gill arches of fishes have been transformed, in the higher forms, to parts of the epiglottis and larynx.

The backbone is made of thirty-three separate **vertebrae** which differ in size and shape at different points along the spine. A typical vertebra (Fig. 186) consists of a basal portion, the **centrum,** to which is attached, dorsally, a ring of bone, called the **neural arch,** which surrounds and protects the delicate spinal cord. Different vertebrae have different projections for the attachment of ribs and muscles and for articulating (joining) with neighboring vertebrae. The skull articulates with the first vertebra (called the atlas), which has rounded depressions on its upper surface into which fit two projections from the base of the skull.

The rib basket is composed of a series of flat bones which support the chest wall and keep it from collapsing as the diaphragm contracts. The ribs are attached dorsally to the vertebrae, each pair of ribs being attached to a separate vertebra. Of the twelve pairs of ribs in man, the first seven are attached ventrally to the breastbone, the next three are attached indirectly by cartilages, and the last two have no attachments to the breast bone, and so are called "floating ribs."

The bones of the appendages, or arms

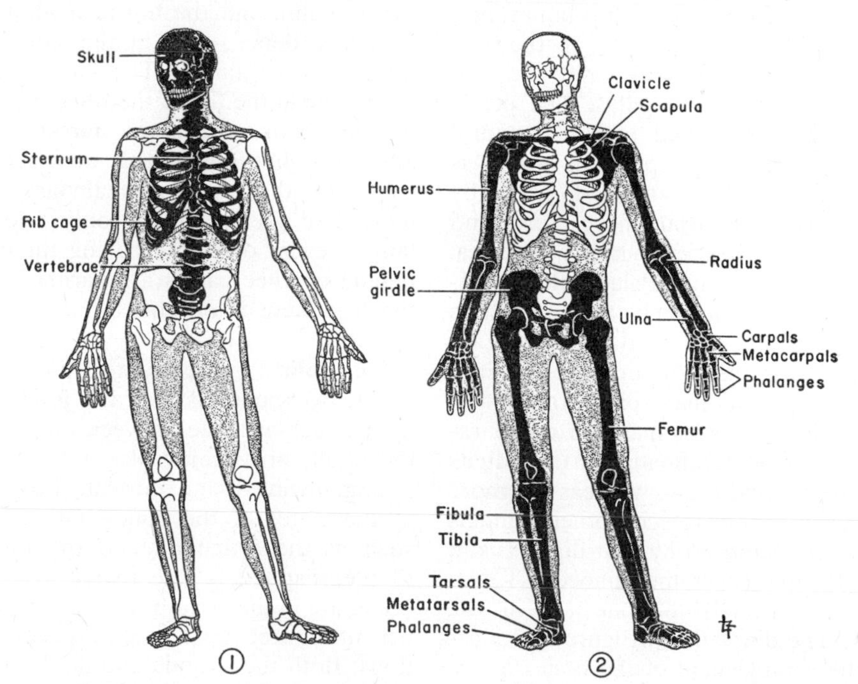

Figure 185. Diagrams of the human body showing *1,* the bones of the axial skeleton and *2,* the bones of the appendicular skeleton.

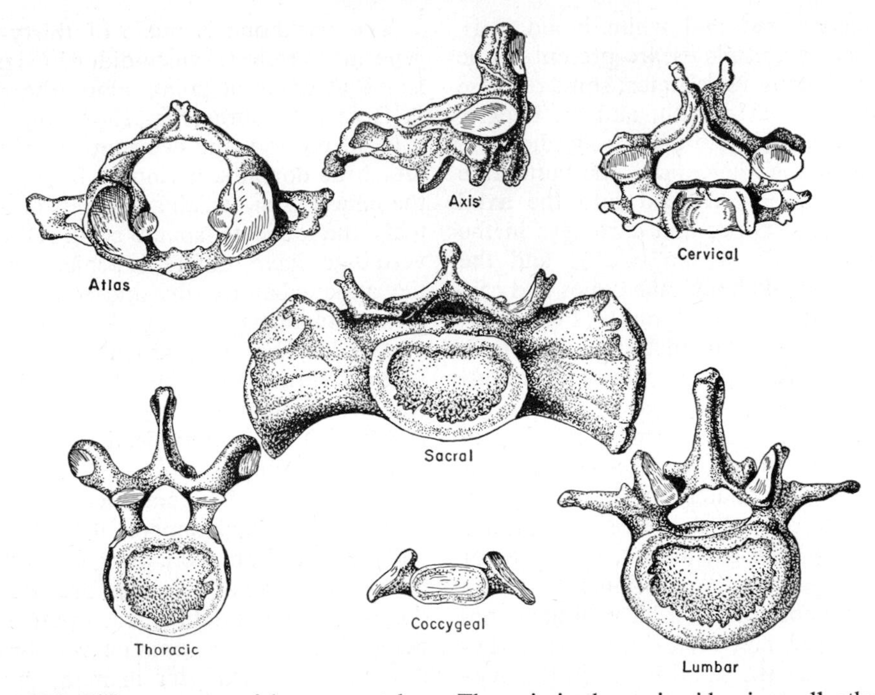

Figure 186. Different types of human vertebrae. The axis is shown in side view; all others are seen from above. (Hunter and Hunter: College Zoology.)

and legs, and the **girdles** which attach them to the rest of the body, make up the appendicular skeleton. The **pelvic girdle** consists of three fused hip bones, and the **pectoral girdle** consists of the two collar bones, or **clavicles,** and the two shoulder blades, or **scapulas.** The pelvic girdle is securely fused to the vertebral column, whereas the pectoral girdle is loosely and flexibly attached to it by means of muscles. Both the pectoral and pelvic fins of fishes are simple structures, adapted for paddling. Paleontologic evidence shows that such fins as these evolved into limbs adapted for moving on land, and that these in turn evolved into wings, hoofs, and the flippers of whales.

The appendages of man are comparatively primitive, terminating in five digits —the fingers and toes—whereas the more specialized appendages of other animals may be characterized by four digits (as in the pig), three (as in the rhinoceros), two (as in the camel), or one (as in the horse). The diagrams in Figure 187 illustrate the arrangement of the bones in the appendages.

The bones of the arm are the **humerus** of the upper arm, the **radius** and **ulna** of the lower arm, the eight tiny **carpals** of the wrist, the five slender **metacarpals** of the palm, and the fourteen **phalanges,** or finger bones—two in the thumb and three in each finger. The leg bones are the **femur** in the thigh, the **tibia** and **fibula** in the shank, the seven **tarsals** in the ankle, the five **metatarsals** across the instep, and the fourteen **phalanges** in the toes. The great toe has only two phalanges, every other toe having three. The **patella** or knee cap is a separate bone of the leg; there is no counterpart for it in the arm.

The Joints. The point of junction between two bones is called a **joint.** Some joints, such as those between the bones of the skull, are immovable and extremely strong, owing to an intricate dovetailing of the edges of the bones. Other joints, such as the articulation of the humerus to the scapula, or the femur to the hip bone, are in the form of a ball-and-socket, permitting free movement in several directions. Both the scapula and hip bone contain rounded, concave depressions to accommodate the rounded, convex heads of the humerus and femur, respectively. Between these two extremes are joints with moderate freedom of motion, such as the hinge joint at the knee, which is restricted to movement in a single plane, or the pivot joints at the wrists and ankles, which are intermediate in freedom of movement between the hinge and the ball-and-socket types (Fig. 188).

Wherever two bones move on one another, their ends do not touch directly, but are covered with smooth, slippery **cartilage** which reduces friction. The bearing surfaces are completely enclosed in a liquid-tight capsule made of ligaments, and the joint cavity is filled with a liquid, secreted by the membrane lining the cavity, which acts as a lubricant. This liquid is similar to lymph or tissue fluid, but contains a small amount of mucus. During youth and early maturity the lubricant is replaced as needed, but in middle and old age the supply is often decreased, with a resulting stiffness of joints and difficulty of movement.

Types of Locomotion. Animals differ as to the part of the foot they put on the ground in walking and running. Men and bears walk flat on the palm of the foot, a type of locomotion adapted for a comparatively slow gait, called **plantigrade.** To increase the effective limb length, and thus the running speed, some animals, such as dogs and cats, have become adapted to running on their digits, a type of locotion called **digitigrade.** Speed is increased still further in the hoofed animals, horses and deer, by the lengthening of

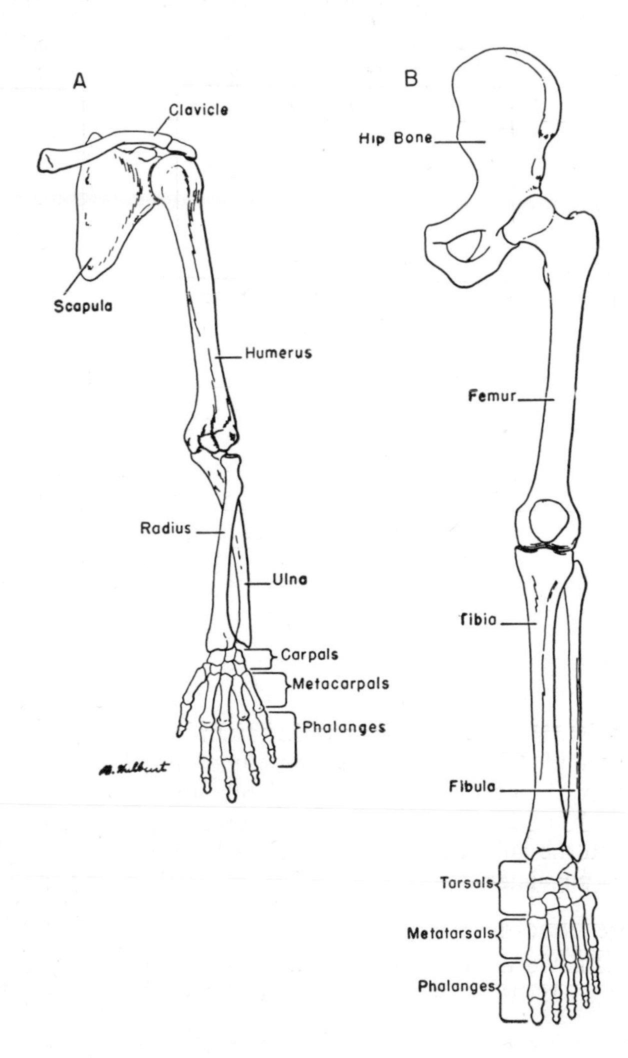

Figure 187. Diagrams of the bones of the left arm (*A*) and left leg (*B*), as seen from the front. Note the fundamental similarity of pattern in the arrangement of the bones in the two limbs. Human limbs have the primitive pentadactyl (five-fingered) arrangement.

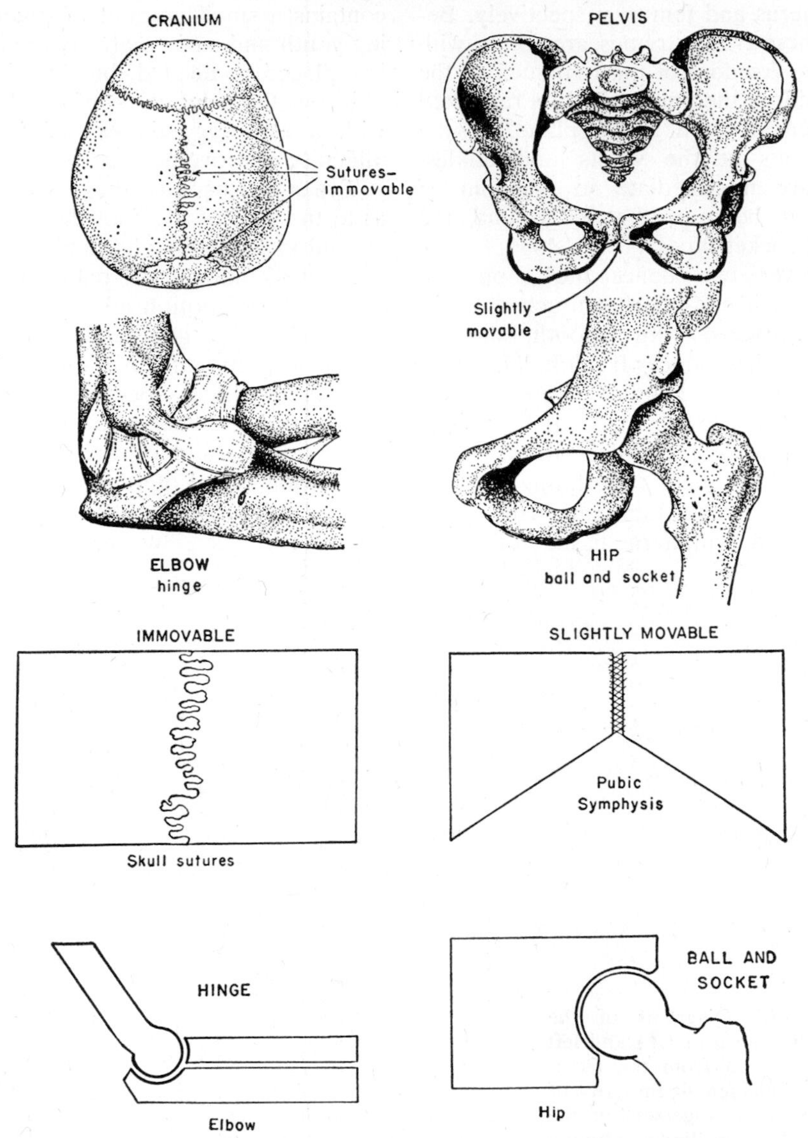

Figure 188. Above, Drawings of the types of joints found in the human body and *below,* diagrams which illustrate the principles underlying the mode of operation of these joints. (Hunter and Hunter: College Zoology.)

the limb bones and the raising of the wrist and ankle still further from the ground, so that the animal runs upon the tips of one or two digits of each limb. This is known as **unguligrade** locomotion, and the animals which have it are known as **ungulates.**

QUESTIONS

1. What is the primary function of the skin? Name four other functions it performs.
2. Draw a diagram of a cross section of the skin, labeling all parts.
3. How are hairs formed? What is their microscopic structure? What structures are associated with each hair follicle?
4. What three types of tissue are included in the skeletal system? How is each adapted for its role in supporting the body?
5. Differentiate between the axial and appendicular skeletons. What is the difference between an exoskeleton and an endoskeleton? What vertebrates have an exoskeleton?

6. How do bones grow?
7. What characteristics would enable an expert to determine whether a skeleton was from a man or a woman? What characteristics would enable him to determine the age of the person at death?
8. What is a joint? Why may they become stiff in old age?
9. How is the skeleton of a horse adapted for running?
10. Compare the mode of walking of a bear, a cat and a deer.

SUPPLEMENTARY READING

The anatomy and evolution of the vertebrate skeleton is discussed in detail in Romer, *The Vertebrate Body*. D'Arcy Thompson's interesting *Growth and Form* discusses the skeleton as an example of the application of engineering principles.

Chapter 23

The Muscular System

IN MAN and most animals the ability to move depends upon a group of specialized contractile cells, the muscle fibers. Man, and indeed most vertebrates, are quite muscular animals; almost half of the mass of the human body consists of muscle tissue. In the vertebrates three types of muscle fibers have evolved to perform various kinds of movements: **skeletal muscle,** which is attached to and moves the bones of the skeleton; **cardiac** or heart muscle, which enables the heart to beat and move the blood through the circulatory system; and **smooth** muscle, which makes up the walls of the digestive tract and certain other internal organs, and moves material through the internal hollow organs. All three types of muscle have the ability to shorten when stimulated, and ordinarily this stimulation reaches the muscle fibers by a nerve. Both cardiac and smooth muscle can contract in the absence of nervous stimulation, and both the heart and the digestive tract function almost normally even when all the nerves leading to them have been cut. In contrast, when the nerves to skeletal muscle are severed or blocked, the muscle is completely paralyzed. For a few weeks it will respond to artificial stimulation, such as an electric shock applied to the overlying skin, but even this ability is gradually lost.

The drug **curare,** the chief ingredient of the arrow poison of the South American Indians, blocks the junction between nerve and muscle so that impulses can no longer pass. This produces the same effect as the cutting of the nerves to all the muscles of the body. The muscles of a curarized animal will still respond to direct electric stimulation, however, demonstrating that muscle is "independently irritable" and need not receive its stimulation through a nerve.

212. THE SKELETAL MUSCLES

A typical skeletal muscle is an elongated mass of tissue composed of millions of individual muscle fibers bound together by connective tissue fibers. The entire structure is surrounded by a tough, smooth sheet of connective tissue so that it can move over adjacent muscles and other structures with a minimum of friction. The two ends of the muscle are usually attached to two different bones, although a few muscles pass from a bone to the skin, or even, as in the case of the muscles of the face used in speech and expression, from one part of the skin to another. The end of the muscle which remains relatively fixed when the muscle contracts is known as the **origin;** the end that moves is called the **insertion;** and the thick part between the two is called the

belly (Fig. 189). The origin of the biceps muscle is at the shoulder, and its insertion is on the radius bone of the forearm; when the biceps contracts, the shoulder remains fixed and the elbow is bent.

Muscles never contract singly, but always in groups. No matter how hard you try, you cannot contract the biceps muscle alone—you can only bend the elbow, which involves the contraction of a number of other muscles besides the biceps. Furthermore, muscles can exert only a pull, not a push. Hence they usually are paired as antagonists, one pulling a bone one way, another pulling it the opposite way. The names **flexor** and **extensor** are applied to muscles to indicate the type of movement they effect. Thus the biceps, which bends or flexes the arm, is called a flexor, and its opposing muscle, the triceps, with its origin on the shoulder and upper arm, and its insertion on the ulna, straightens or extends the forearm and is called an extensor. Similar pairs of opposing flexors and extensors are found at the wrist, knee, ankle and other joints. Whenever a flexor contracts, the opposing extensor must relax to permit the bone to move, and this requires the proper coordination of the nerve impulses going to both sets of muscles. Some of the muscles of the ventral side of the body are shown in Figure 190.

When muscles are not contracting to effect a movement, they are not completely relaxed. As long as a person is conscious, all the muscles are contracted slightly, a phenomenon called **tonus.** Posture is maintained by the partial contraction of the muscles of the back and neck, and of the flexors and extensors of the legs. When a person is standing, both the flexors and extensors of the thigh must contract simultaneously so that the body sways neither forward nor backward on the legs, and the simultaneous contraction

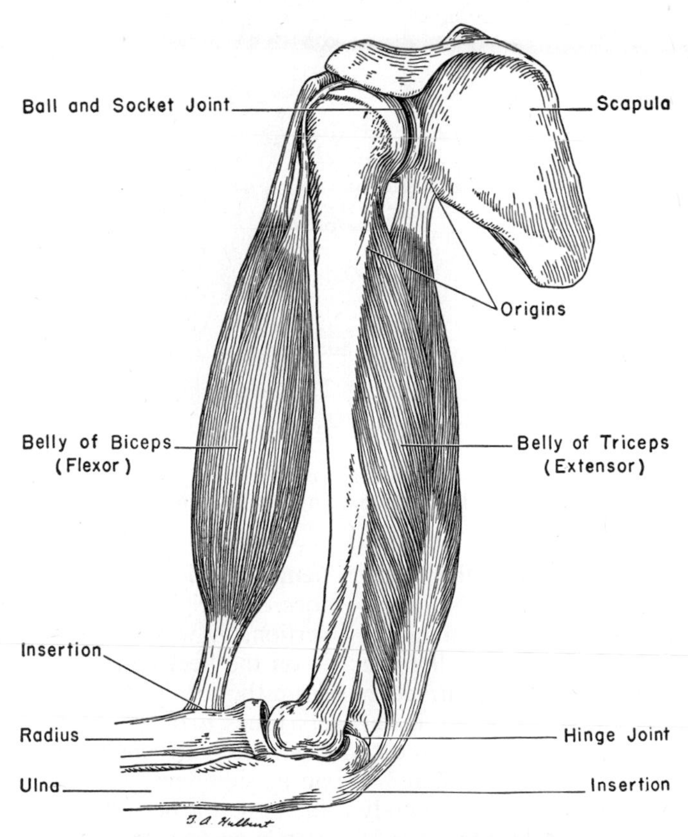

Figure 189. Muscles and bones of the forearm, showing origin, insertion and belly of a muscle, and the antagonistic arrangement of the biceps and triceps.

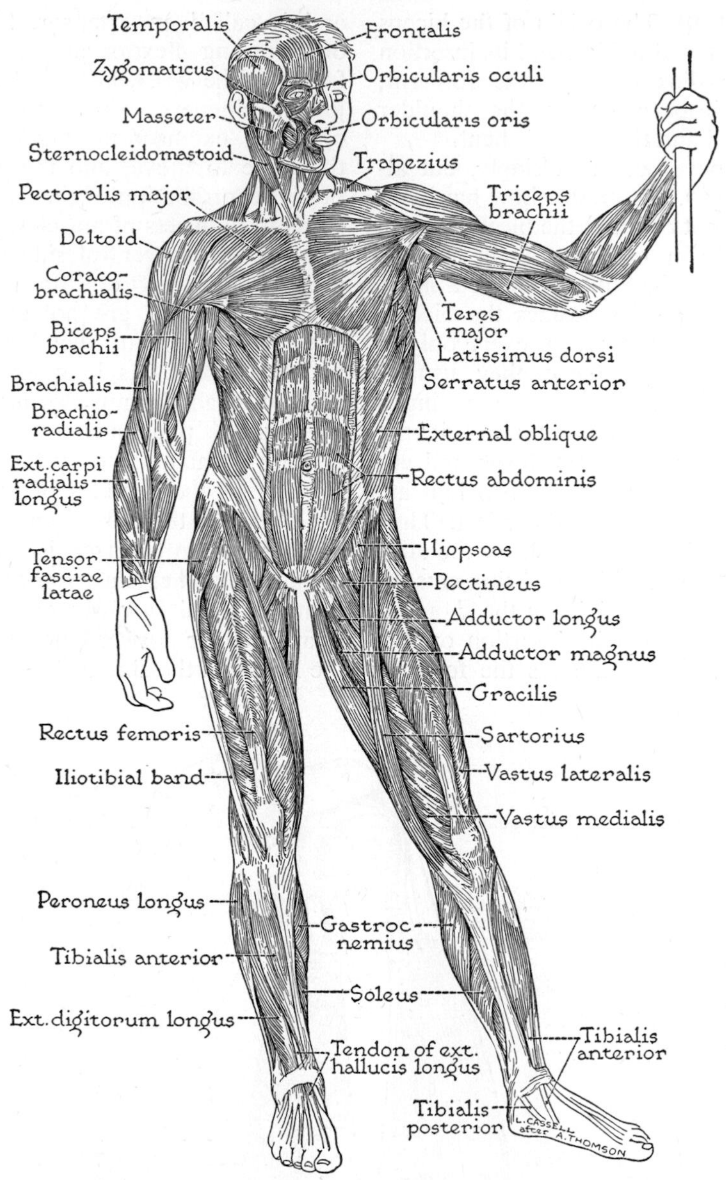

Figure 190. A ventral view of the superficial muscles of the human body. (Millard, King and Showers: Human Anatomy and Physiology.)

of the flexors and extensors of the shank locks the knee in place and holds the leg rigid to support the body. When movement is added to posture, as in walking, a complex coordination of the contraction and relaxation of the leg muscles is required. It is not surprising that the process of learning to walk is long and tedious.

Some of the larger muscles of the body are extremely strong. Consider the muscle of the calf of the leg, called the **gastrocnemius,** which is used in rising on one's toes. Its origin is at the knee, and its insertion, by way of the tendon of Achilles, is on the heel bone; because the distance from the toes to the ankle joint is at least six times that from the ankle joint to the heel, the gastrocnemius is working against an adverse lever ratio of 6:1. This means that when a person weighing 150 pounds stands on one leg and rises on his toes, the one gastrocnemius muscle is exerting a force of 900 pounds; and if a man were to hold another person in his arms and perform this action, the muscle would be exerting a force of nearly a ton.

Because of nervous coordination, no normal person can cause his muscles to contract maximally, but in certain diseases in which the nervous control is absent, muscles do contract forcefully enough to rip tendons and break bones.

213. KINDS OF CONTRACTION

To understand how muscles contract, it is necessary to discriminate carefully between the contraction of a whole muscle and the contraction of its individual fibers. For although a muscle cannot contract maximally, a *single* fiber of it can respond *only* maximally, or not at all. This phenomenon, described as the "All or None Law," can be demonstrated experimentally by dissecting out a muscle fiber and giving it repeated stimuli of increasing intensity, beginning with ones too weak to cause contraction. As the strength of the stimulus is increased, there will be no response until a certain level is reached, at which time the fiber will contract completely. No stimuli of greater intensity can cause any greater contraction. Since a whole muscle is made up of thousands of individual muscle fibers, however, the nature and strength of *its* contraction depend upon the number of its constituent fibers which are contracting, and upon whether the fibers are contracting simultaneously or alternately.

In studying the different kinds of contraction, advantage is taken of the fact that a muscle will retain its ability to contract after it has been removed from the body. The muscle usually used for experimental purposes is the calf muscle of a frog, and if care is taken to keep it

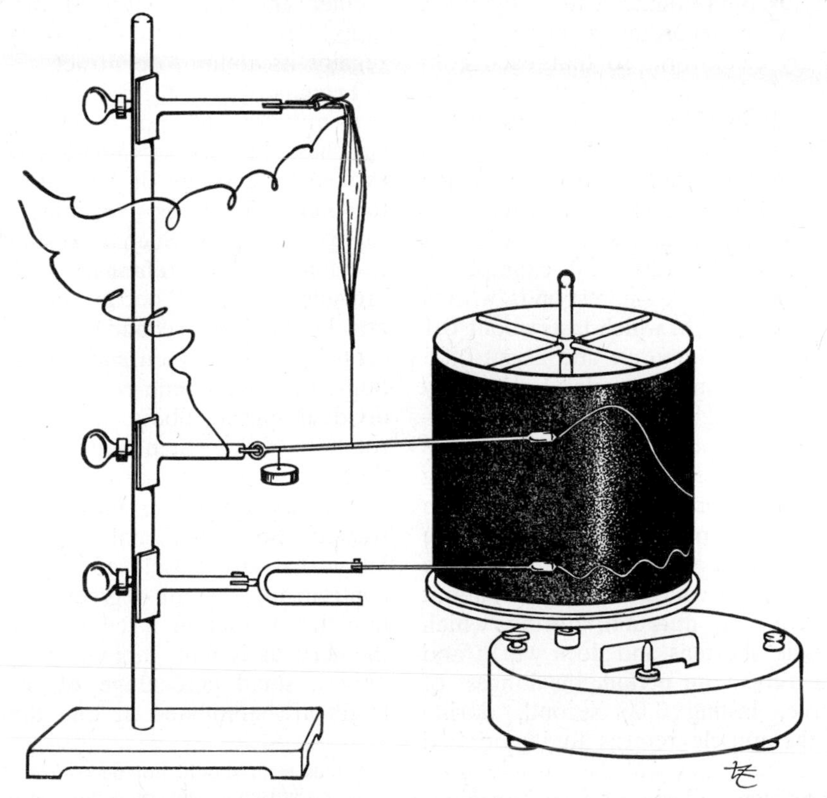

Figure 191. The apparatus used to study the contraction of an isolated muscle. The stylus attached to the insertion of the muscle writes a record of the contraction on the rotating cylinder, the kymograph. The contraction is timed by the vibrating tuning fork, whose stylus draws a wavy line on the kymograph record.

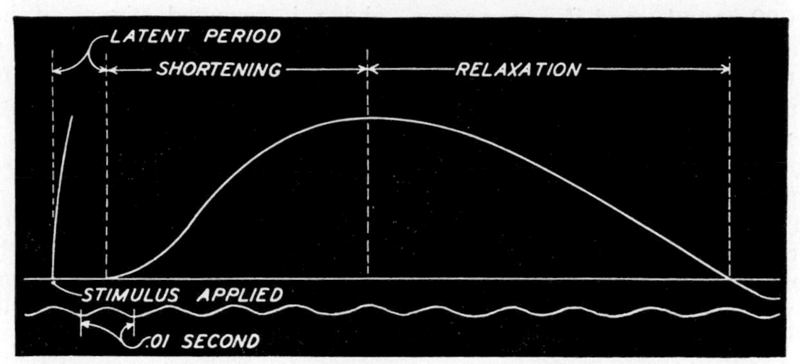

Figure 192. The recording of a single twitch of a frog muscle made on a rapidly moving smoked plate, mounted on a kymograph. The wavy line at the bottom was made by a tuning fork vibrating 100 times a second. (Carlson and Johnson: The Machinery of the Body, published by University of Chicago Press.)

moist, it will contract for hours. To make a record of these contractions, the muscle is mounted with its origin attached to a fixed hook and its insertion connected, by means of another hook, to a lever with a pointed stylus at its tip (Fig. 191). This stylus is in contact with a cylinder, covered with recording paper and revolved by clockwork, so that each contraction of the muscle raises the stylus and records its vigor and duration. Additional styluses can be used to record an appropriate time scale and to mark when the stimulus is given to the muscle.

The Single Twitch. When a muscle is given a single stimulus—for example, a single electric shock—it responds with a single, quick twitch, which lasts about 0.1 second in a frog's muscle and about 0.05 second in a human muscle. Laboratory records of a single twitch (Fig. 192) indicate that it consists of three separate phases: (1) the **latent period,** lasting about 0.01 second, an interval between the application of the stimulus and the beginning of the visible shortening of the muscle; (2) the **contraction period,** about 0.04 second in duration, during which the muscle shortens and does work; and (3) the **relaxation period,** the longest of the three, lasting 0.05 second, during which the muscle returns to its original length. After a twitch the muscle consumes oxygen and gives off carbon dioxide and heat at a rate greater than during rest, marking a **recovery period** in which the muscle is restored to its original condition. This recovery period lasts for several seconds, and if a muscle is stimulated repeatedly so that successive contractions occur before the muscle has recovered from the previous ones, the muscle becomes fatigued and the twitches grow feebler and finally stop. If the fatigued muscle is allowed to rest for a time, it regains its ability to contract.

Tetanus. The normal contractions of the muscles do not occur as single twitches, but as sustained contractions evoked by a volley of separate stimuli—the nerve impulses—reaching them in rapid succession. Such a sustained contraction is called **tetanus,*** and while it prevails, the stimuli occur so rapidly (several hundred per second) that relaxation cannot occur between successive contractions. In most tetanic contractions the individual muscle fibers are stimulated in rotation, rather than simultaneously, so that although they contract and relax, the muscle as a whole remains partly contracted. From personal experience you know that any muscle of your body can contract to different degrees. This gradation of contraction is controlled through the nervous system: in a weak contraction only a small percentage of the muscle fibers are stimulated at one time; for a

* This term should not be confused with tetany, the muscular spasms occurring in deficiencies of the parathyroid hormone, or with the disease tetanus (lockjaw), characterized by abnormal muscular contractions and caused by the tetanus bacillus.

stronger contraction a larger percentage of muscle fibers contract simultaneously.

Tonus. The term **tonus** or "tone" refers to the state of sustained partial contraction present in all normal skeletal muscles as long as the nerves to the muscle are intact. Cardiac and smooth muscles exhibit tonus even after their nerves are cut. Each muscle is normally stimulated by a continuous series of nerve impulses which cause a constant, slight contraction or tonus. Severing the nerve to a skeletal muscle eliminates tonus immediately. Tonus is a mild state of tetanus, present at all times and involving only a small fraction of the fibers of a muscle at any moment. It is believed that the individual fibers contract in turn, working in relays, so that each fiber has a chance to recover completely, while other fibers are contracting, before it is called upon to contract again.

214. THE CHEMISTRY OF MUSCLE CONTRACTION

A steam engine can convert only about 10 per cent of the heat energy of its fuel into useful work; the rest is wasted as heat. But muscles are able to use between 20 and 40 per cent of the chemical energy of the food molecules, such as glucose, in contraction. The remainder is converted into heat, but is not wholly wasted, since it is used to maintain the body temperature. If one refrains from contracting the muscles, the heat produced elsewhere in the body is insufficient to keep it warm in a cold place. In these circumstances the muscles contract involuntarily (one "shivers"), and heat is produced thereby to restore and maintain normal body temperature.

Physiologists and biochemists have been attempting for many years to solve the problem of how protoplasm can exert a pull, but the actual chemical and physical events that occur in muscle contraction are still a matter of conjecture rather than established fact. Chemical analysis reveals that muscle is about 80 per cent water, the rest being mostly protein, with small amounts of fat and glycogen, and two phosphorus-containing substances, **phosphocreatine** and **adenosine triphosphate.** The actual contractile part of a muscle fiber is believed to be a protein chain which shortens by folding of the links or by an expulsion of water from within the "interstices" of the protein molecule. Two proteins are involved in this, **myosin** and **actin,** neither of which is capable of contracting alone. But when they are mixed together in a test tube, and potassium and adenosine triphosphate are added, contraction does occur. This demonstration of contraction in a test tube, made by Dr. Albert Szent-Györgyi, is one of the most exciting discoveries yet made in biochemistry.

The first step in unraveling the mystery of contraction consists in making analyses to determine what substances are used up in the process. Glycogen, oxygen, phosphocreatine and adenosine triphosphate decrease in amount during contraction, and carbon dioxide, lactic acid and inorganic phosphate increase. The fact that oxygen is used up and carbon dioxide is formed suggests that muscular contraction is an oxidative process. But this oxidation is not essential, for a muscle can twitch a good many times even when completely deprived of oxygen—for example, when it is removed from the body and placed in an atmosphere of nitrogen. Such a muscle becomes fatigued, however, sooner than one contracting in an atmosphere of oxygen. Furthermore, although we breathe more rapidly during muscular exertion, the accelerated breathing continues for some time after the physical work has ceased. This suggests that oxidation is involved, not in muscular contraction, but in the process of recovery from contraction.

The disappearance of glycogen and the formation of lactic acid are related, for in the absence of oxygen, the amount of lactic acid formed is just equivalent to the glycogen that disappears. Since the breakdown of glycogen to lactic acid requires no oxygen, and since it liberates energy rapidly, it was once thought that this reaction is directly responsible for muscle contraction. When oxygen is present, the muscle oxidizes about one fifth of the lactic acid to carbon dioxide and water, and the energy released by this oxidation is used to reconvert the other four fifths of the lactic acid to glycogen. This explains why lactic acid does not accumulate as long as muscle has sufficient oxygen, and why a muscle becomes fatigued more rapidly (uses up its glycogen and accumulates lactic acid) when it contracts in the absence of oxygen.

About 1930 it was found that a muscle poisoned with iodoacetate, which inhibits the chemical reactions by which glycogen breaks down to lactic acid, can still contract, although it is capable of twitching only sixty to seventy times instead of the 200 or more times achieved by a muscle deprived of oxygen. But the fact that it can twitch at all when the breakdown of glycogen is prevented shows that this is not the primary source of energy for contraction.

The other change that can be detected chemically during contraction is a splitting off of inorganic phosphate from phosphocreatine and adenosine triphosphate, accompanied by the release of energy. It is now believed that this is the immediate source of energy for contraction. The bond between phosphate and the organic compounds creatine and adenylic acid is different from that of most phosphate compounds and has been called a "high-energy" or "energy-rich" phosphate bond because energy is released when the bond is broken. The major part, and perhaps all, of the energy of foodstuffs must be converted to these energy-rich phosphate compounds to be available for use. The reactions whereby glucose and other substances are metabolized to yield energy-rich phosphate compounds such as adenosine triphosphate were described in Chapter 5. After a muscle has contracted, the breakdown of glycogen to lactic acid and the oxidation of lactic acid in the Krebs citric acid cycle provide energy for the resynthesis of adenosine triphosphate and phosphocreatine.

In summary, the chemistry of muscle contraction involves the following chemical reactions:

$$\text{Organic phosphates} \rightleftharpoons \text{Phosphate} + \text{Organic compounds} + \text{Energy}$$
(used in actual contraction)

$$\text{Glycogen} \rightleftharpoons \text{Intermediates} \rightleftharpoons \text{Lactic Acid} + \text{Energy}$$
(used in resynthesis of organic phosphates)

$$\text{Part of lactic acid} + O_2 \longrightarrow CO_2 + H_2O + \text{Energy}$$
(used in resynthesis of rest of lactic acid to glycogen and in resynthesis of organic phosphates)

It is estimated that the energy from organic phosphates alone could sustain maximal muscular contraction for only a few seconds. A man might run a 50 yard dash with this. By calling upon all the sources of energy available in the absence of oxygen a man might be able to continue maximal contractions for 30 to 60 seconds.

The Oxygen Debt. The fact that the actual contraction, and part of the recovery from contraction, occur without oxygen is extremely important. Our muscles are often called upon to do great spurts of work, and although both the rate of breathing and the heart rate increase during exertion, oxygen cannot be supplied in sufficient quantities to permit these exertions. During violent exercise, such as running the 100 yard dash, glycogen breaks down to lactic acid faster than the lactic acid can be oxidized, so that the latter accumulates. In such circumstances the muscle is said to have incurred an **oxygen debt,** which is afterward repaid by our rapidly breathing enough extra oxygen to oxidize part of the lactic acid, which furnishes energy for resynthesizing the rest to glycogen. In other words, during short spurts of extreme muscular activity, muscles use energy from sources that do not require the utilization of oxygen. After the activity has ceased, the muscles and other tissues pay off the "oxygen debt" by utilizing an extra amount of oxygen to restore the energy-rich phosphate compounds and glycogen to their normal condition. During a long race a runner may reach an equilibrium in which he gets a "second wind," and, because of the increase in breathing and heart rate, takes in enough oxygen to oxidize the lactic acid formed at that moment, so that the oxygen debt is not increased.

Fatigue. A muscle that has contracted many times, exhausted its stores of organic phosphates and glycogen, and accumulated lactic acid, is unable to contract any more and is said to be **fatigued.** Fatigue is primarily induced by this accumulation of lactic acid, although animals feel fatigue before the muscle reaches the exhausted condition.

The exact spot most susceptible to fatigue can be demonstrated experimentally if a muscle and its attached nerve are dissected out and the nerve stimulated repeatedly by electric shocks until the muscle no longer contracts (Fig. 193). If the muscle is then stimulated directly, by placing the electrodes on the muscle tissue, it will respond vigorously. With the proper apparatus for detecting the passage of nerve currents, it can be shown that the nerve leading to the muscle is not fatigued; it is still capable of conduction. The point of fatigue, then, is the *junction* between the nerve and the muscle, where nerve impulses instigate muscle contraction.

The Nature of Contraction. Electron micrographs show that muscle fibrils are made of longitudinal filaments of actin molecules which extend through both dark and light bands. These are surrounded in the light bands by aggregations of myosin molecules which are close to, but not combined with the actin; the two are held apart by electrostatic charges. Contraction appears to occur almost entirely in the light bands, with the dark bands serving as "tendons" between adjacent light bands. According to one theory, the electrical stimulation of the muscle fiber produces changes in the ionic pattern around the actin and myosin which permit them to come together, form actomyosin, and contract.

At present there are two hypotheses of the nature of muscle contraction. One states that the energy for contraction is released from the organic phosphate to the muscle fiber at the moment of contraction, and that after this energy has been used in the physical shortening of the fiber, the fiber automatically relaxes. The other states that contraction resembles the releasing of a stretched spring, the muscle before contraction being analogous to a spring that has been stretched and held by a kind of trigger mechanism. When this trigger is tripped by the nerve impulse, the spring contracts; energy is then required to restretch the spring for the next contraction. When a muscle is stimulated repeatedly and becomes fatigued, it does not suddenly lose its ability to contract; instead, its relaxations become slower and more labored, and when fully fatigued, it can no longer relax, but must remain contracted. This condition we call "cramps." After death, also, the muscles lose their power to relax and go into a state of contraction (the so-called *rigor mortis*) which lasts until the muscle fibers begin to distintegrate.

When a muscle contracts, it becomes shorter and fatter, but there is no change in its total volume. This has been shown experimentally by dissecting out a muscle, placing it in a glass vessel with a narrow

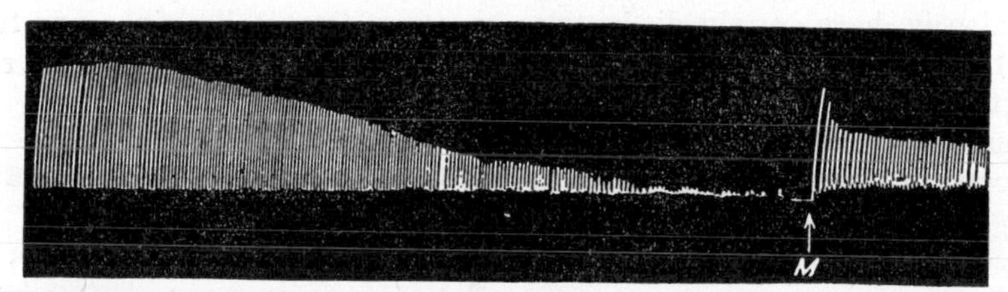

Figure 193. Fatigue. Contractions of a frog muscle (recorded by means of a kymograph) induced by rapidly repeated stimuli applied to the nerve of the muscle. When the muscle no longer responded to nerve stimulation, stimuli applied directly to the muscle (beginning at *M*) caused definite contractions, demonstrating that the fatigue was not primarily in the muscle itself. (Carlson and Johnson: The Machinery of the Body, published by University of Chicago Press.)

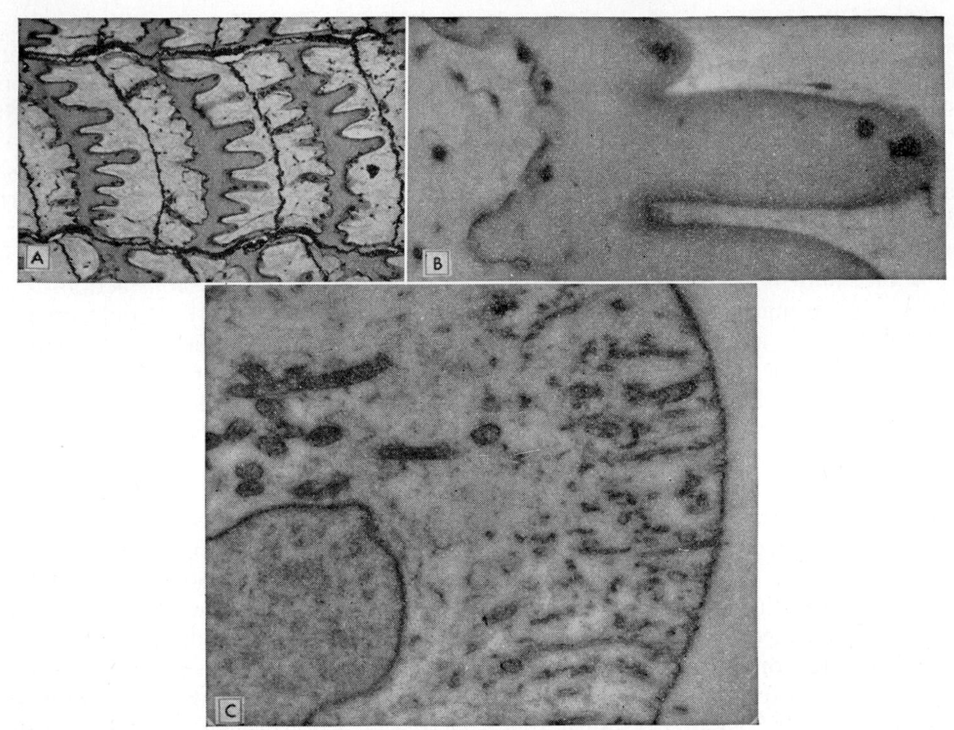

Figure 194. The electric organ of the eel. *A,* Photomicrograph of a small portion of the electric organ showing 6 to 8 electroplaxes (cells specialized for the generation of electricity) within their connective tissue compartments. Magnified 250 ×. *B,* Photomicrograph of a single electroplax, magnified 1500 ×. *C,* Electron micrograph of the noninnervated surface of an electroplax, magnified 25,000 ×. A nucleus (the round structure in the lower left corner) and mitochondria (the round or elongate dark structures near the nucleus) are shown. The plasma membrane (right) is invaginated by numerous tubules which dip into the interior of the cell. (Photographs courtesy of Dr. John Luft.)

neck, and filling the vessel with water. The muscle is then stimulated electrically, and as it contracts and relaxes, there is no change in the water level in the neck of the vessel.

215. CARDIAC AND SMOOTH MUSCLE

The muscles of the heart and internal organs, though resembling skeletal muscle in a general way, have certain distinctive characteristics. They are both much slower to contract than skeletal muscle: while skeletal muscle fibers contract and relax in 0.1 second, cardiac muscle requires from one to five seconds, and smooth muscle needs from three to 180 seconds. All the phases of contraction are prolonged.

Smooth muscle exhibits wide variations in tonus; it may remain almost relaxed or tightly contracted. And it also, apparently, can maintain the shortened condition of tonus without the expenditure of energy, perhaps owing to a reorganization of the protein chains making up the fibers.

Each beat of the heart represents a single twitch. Cardiac muscle has a long **refractory period,** the period following one stimulus when it is unable to respond to any other. Consequently, it is unable to contract tetanically, since one twitch cannot follow another quickly enough to maintain a contracted state.

216. THE MUSCLES OF LOWER ANIMALS

The muscles of all animals from the flatworm to man are similar in that they are all made of long cylindrical or spindle-shaped fibers which are contractile because of their protein chains. Even the coelenterates, which lack true muscle fibers, have cells which can contract. Differences do exist, however; for example, most of the invertebrates have only smooth muscle, whereas arthropods have only striated muscle. Electrical phenomena are associated with all types of muscle contraction, (cf. the electrocardiograph; p. 248), but in some animals, such as the electric eel, specialized muscle cells have evolved in which contraction is at a minimum and the production of electricity is at a maximum (Fig. 194). The "electric organ" of the eel may produce a potential of 400 volts or more, sufficient to stun or kill the fish on which it preys, and to give quite a jolt to a man.

QUESTIONS

1. What are the three types of muscle fibers and what are their distinguishing characteristics?
2. Describe a typical skeletal muscle.
3. What is tonus?
4. State the "All-or-None" Law. Does the muscle as a whole operate according to this principle? If not, what does control the degree of contraction of a muscle?
5. What is tetanus?
6. What causes the phenomenon of shivering?
7. What is meant by the term "oxygen debt"?
8. What happens when a muscle contracts? When it relaxes?
9. Explain in detail how fatigue is caused.
10. What is responsible for the contractile quality of a muscle fiber?
11. Describe the muscle mechanism involved in the maintenance of an erect posture.
12. Which muscles contract and which relax when you hold your right arm out at the side? When you throw a ball?

SUPPLEMENTARY READING

Two classic books on muscular exercise, the effects of exertion and of training are A. V. Hill's *Muscular Movement in Man* and Bainbridge, Bock and Dill's *The Physiology of Muscular Exercise.* Albert Szent-Györgi's latest experiments and theories on the details of muscle contraction are given in his *Chemical Physiology of Contraction in Body and Heart Muscle.* Quotations of some of the classic papers on muscle physiology are found in Chapter 6 of Fulton, *Selected Readings in the History of Physiology.*

Chapter 24

The Nervous System

To keep the various parts of anything as complicated as the human body functioning properly requires a correspondingly complex coordinating device; the nervous system, which integrates the activities of all the parts of the body, is undoubtedly the most complex of all the body systems. The muscles and glands of an animal are collectively called **effectors;** the eyes, ears and other sense organs are known as the **receptors.** The nervous system, composed of brain, spinal cord and nerve trunks, connects receptors with effectors and conducts impulses or "messages" from one to the other. It is able to do all this in such a way that when a given receptor is stimulated the proper effector responds appropriately. The two chief functions of the nervous system are **conduction** and **integration.**

217. THE NEURONS

Although their interrelations are exceedingly complex, the cells which comprise the nervous system, called **neurons,** are all fundamentally similar, consisting of one **axon,** one or more **dendrites** (see p. 47), and, in between, a **cell body,** containing the nucleus. Neurons differ widely in the shape of the cell body, and in the length, number and amount of branching of the axons and dendrites. They are subdivided into **sensory, motor** and **connector** neurons on the basis of their connections. Sensory neurons have dendrites connected to receptors, and axons connected to other neurons; motor neurons have dendrites connected to other neurons, and axons to some effector; connector neurons have both dendrites and axons connected to other neurons. The simplest pathway that a nerve impulse can follow consists of one sensory, one connector and one motor neuron.

The **nerve trunks** or **nerves** of the body consist of large numbers of axons and dendrites bound together in a common connective tissue sheath. The cell bodies of the neurons are not scattered at random along these nerve trunks, but occur in aggregations, called **ganglia** if they occur outside the brain and spinal cord, and **nerve centers** if they occur within the brain or spinal cord.

In addition to and surrounding the cell membrane of the axon and dendrite may be one or two nerve sheaths, an outer **neurilemma** and an inner **myelin** sheath (Fig. 195). The neurilemma sheath is cellular. The myelin sheath is composed of noncellular, fatty substances which give a white appearance to nerves which have it. Nerve fibers in the spinal cord and brain have *only* a myelin sheath. Those going to the viscera have a neurilemma sheath and a thin myelin sheath, so that they are gray rather than white. Nerves to the skin and skeletal muscles have both sheaths. The work of Dr. Betty Uzman shows that the myelin sheath is formed by the folding of the walls of the neurilemma cells (Fig. 196).

The position of the myelin sheath suggests that it may act as insulation, preventing the nerve impulse from passing from one fiber to another and thus causing the wrong effector to be stimulated, but there is no positive evidence for this suggestion. One might suppose that the myelin sheath provides a reserve supply of nourishment for the fiber, but the evidence available indicates that the fiber receives its nourishment solely from the cell body. Another theory of the function of the myelin sheath is that it increases the speed of conduction of the nerve impulse. In man and other mammals a nerve impulse is conducted about 300 feet per second in a myelinated nerve, and 25 to 50 feet per second in the nerves going to the internal organs, which have only a thin myelin sheath. But there is no proof that the greater speed of conduction of the former is due to the myelin sheath and not to some other property of the nerve.

The neurilemma sheath seems to play some role in the regeneration of a cut nerve fiber (axon or dendrite). When fibers are severed, the parts on the side of the cut away from the cell bodies degenerate and disappear in a few weeks, leaving the hollow tubes of the neurilemma sheaths. If the cut ends of the whole nerve trunk are then clamped or sewed together, the nerve fibers may grow from the cut ends out through the tubes formed by the neurilemma sheaths that surrounded the degenerated fibers, to the structures innervated by the original fibers. In this way the sensations and muscular control lost when the nerve was cut may be regained. The length of time required for regeneration depends on how far the nerve has to grow; as much as two years may be necessary. When a cut occurs in the brain or spinal cord, where the fibers have no neurilemma sheath, regeneration does not occur. Whether the neurilemma simply directs the course of the regenerating fiber, aids in its nourishment, or forms the myelin sheath is not known.

218. THE NERVE IMPULSE

The study of the nature of the nerve impulse has been fraught with special difficulties because nothing visible occurs when an impulse passes along a nerve. Only recently, with the development of microchemical techniques, has it been possible to show that the nerve fiber must expend more energy, consume more oxygen and give off more carbon dioxide and heat while transmitting an impulse than it does in a resting state. This indicates that oxidative processes are involved in the conduction of an impulse, or in the recovery of the nerve after conduction, or both.

With the discovery, about a century ago, that the nerve impulse is accompanied by certain electrical changes, the belief arose that the impulse itself is an electric current. At that time the speed of electricity was known to be fast, and it was predicted that the speed of the nerve impulse was too great to be measurable. Ten years later Helmholz measured the speed of impulse conduction in nerves by stimulating the nerve to a muscle at different distances from the muscle, and measuring the elapsed time between stimulus and contraction. In this way he demonstrated that the nerve impulse travels much more slowly than electricity—at about 100 feet per second in frog nerves. This, of course, was evidence that the nerve impulse is not an electric current like that in a copper wire. Furthermore, dead or crushed nerves still conduct elec-

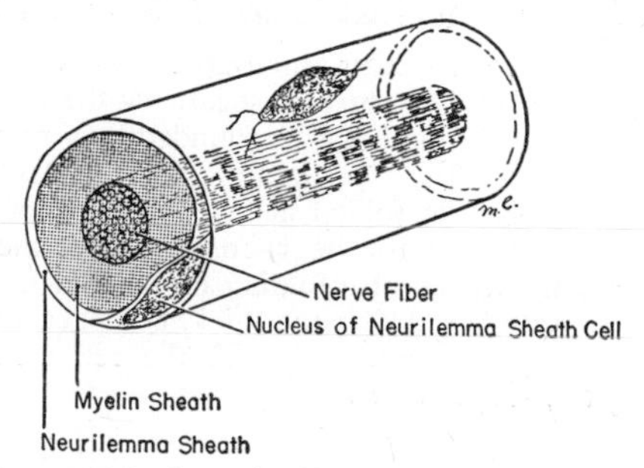

Figure 195. A typical nerve fiber and its surrounding sheaths.

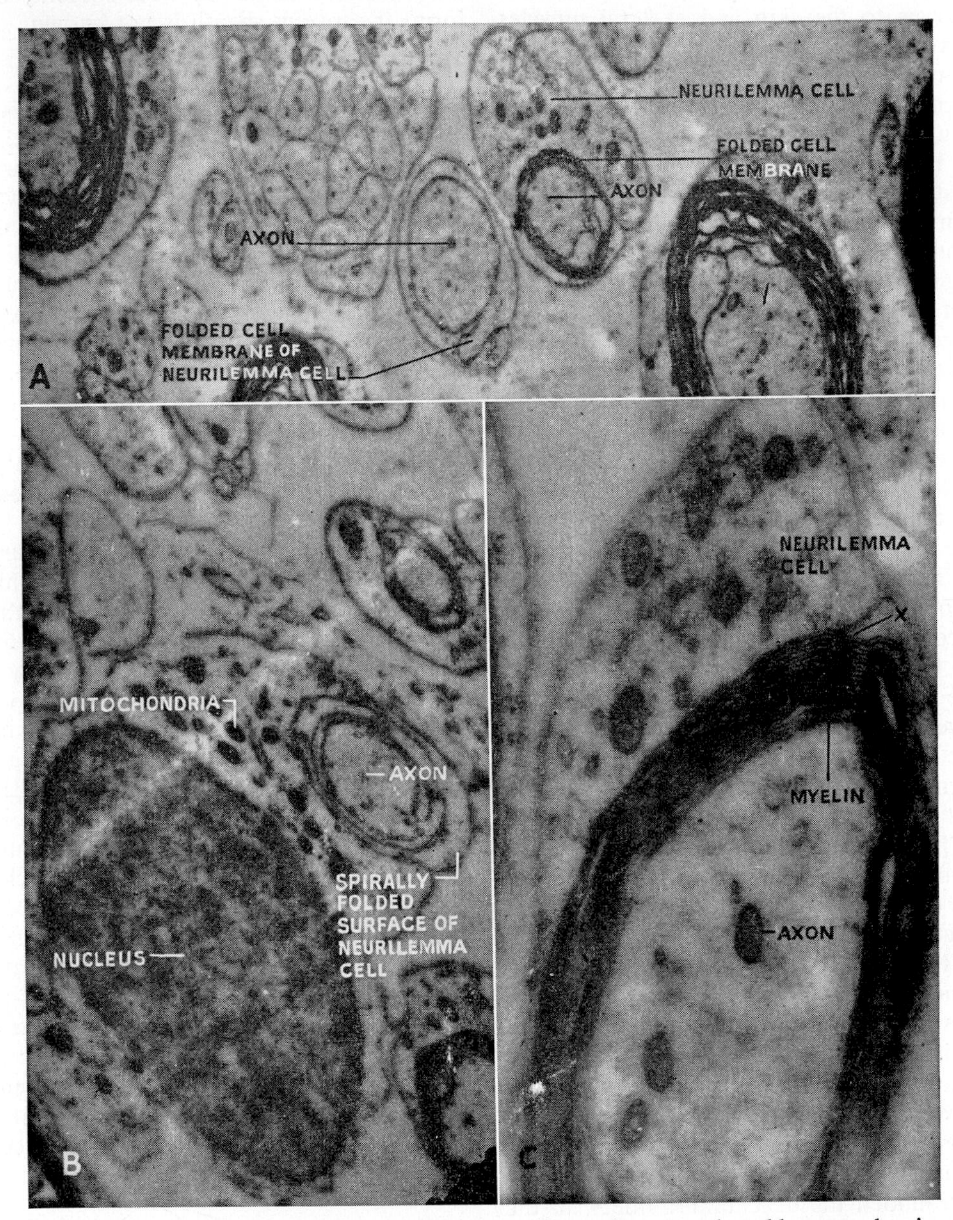

Figure 196. Electron micrographs of sciatic nerve fibers of a seven day old mouse showing the development of the myelin sheath by the folding of the cell walls of the neurilemma. *A,* Left, an early stage with an axon surrounded by a neurilemma cell; only a small portion of the neurilemma cell is visible. Right, a slightly later stage with much more of the neurilemma cell visible and with a thicker layer of folded neurilemma cell membrane. × 26,000. *B,* The large, dark oval structure is the nucleus of the neurilemma cell. Some mitochondria are evident between the nucleus and the axon. The spiral infolding of the cell membrane of the neurilemma cell is visible. × 28,000. *C,* Later stage showing the thick, compact, many-layered myelin sheath. Note at X that the layers of myelin are continuous with the cell membrane of the neurilemma. × 83,000. (Courtesy of Dr. Betty G. Uzman.)

tricity, but not nerve impulses, and whether a nerve is stimulated electrically or by touch, heat or chemicals, the resulting impulse travels at the same rate of speed. We conclude that the nerve impulse is not an electric current, but an **electrochemical disturbance** in the nerve fiber. A stimulus instigating this disturbance in one section of a nerve fiber causes a similar disturbance in the adjacent section, and so on until the impulse reaches the end of the fiber. The transmission of an impulse is thus like the burning of a fuse: the heat given off by the burning of one section causes the next section to catch fire, and so on. Of course, in the nerve, it is not the heat but the electrical changes produced by one section which stimulates the next.

The transmission of the nerve impulse shows further similarities to the burning of a fuse. The rate at which a fuse burns is independent of the amount of heat applied to light it, provided enough heat is applied to light it at all. Furthermore, the method of setting it off is unimportant. The same thing is true of a nerve. It will not respond unless a stimulus of a certain minimum intensity is applied, but increasing the strength of the stimulus beyond this minimum does not cause the impulse to travel any faster. This is due to the fact that energy for conducting the impulse is derived from the nerve itself and not from the stimulus. The phenomenon is expressed in the **All-or-None Law:** The nerve impulse is independent of the nature or strength of the stimulus setting it off, if the stimulus is strong enough to set it off at all. Although the rate of conduction is independent of the strength of the stimulus, it does depend on the state of the nerve fiber, and drugs can retard or prevent the transmission of an impulse.

A burned fuse cannot be reused but a nerve fiber is capable of recovering and transmitting other impulses. It does not do this immediately, however; a definite period of time elapses after the passage of one impulse before the fiber can transmit another. This interval, known as the **refractory period,** lasts from 0.001 to 0.005 second. During this time, chemical and physical changes occur which restore the fiber to its original condition.

So far as we know, the impulses transmitted by all types of neurons—motor, sensory and connector—are essentially alike. The fact that one impulse results in a sensation of sight, another in a sensation of sound, another in muscle contraction, and another in glandular secretion, is due entirely to the nature of the structures to which the impulses travel, and not to any peculiarity of the impulses themselves.

Although a nerve fiber can be stimulated anywhere along its length, stimulation normally occurs at one end only, from which the impulse passes along the fiber to the other end.* The junction between consecutive neurons is called a **synapse.** A nerve impulse is transmitted from the tip of the axon of one neuron to the dendrite of the next across the synaptic junction by means of a chemical secretion at the tip of the axon. This initiates a nerve impulse in the dendrite of the next neuron. Transmission across the synapse is considerably slower than transmission along the nerve. Impulses normally pass in one direction only: Those in sensory neurons pass from the sense organs to the spinal cord and brain; those in motor neurons, from the brain and spinal cord to the muscles and glands. The synapse controls this, because only the tip of the axon is capable of secreting the chemical which stimulates the next neuron. Any individual nerve fiber can conduct an impulse in either direction; if it is stimulated electrically in the middle, two impulses will be set up, one going in one direction and one in the other (these can be detected by appropriate electrical devices), but only the one going *toward* the *axon* can stimulate the next neuron in line. The one going toward the dendrite will stop when it arrives at the tip.

The chemical and electrical processes

* The only nerve that is occasionally stimulated at a point along its length, as well as at the end, is the "crazy bone" nerve, or ulnar nerve, which lies close to the surface of the skin at the elbow. This nerve is made up of fibers for pain, touch, cold, pressure, and so on, which is why one receives the tingling, "mixed" sensation when it is bumped.

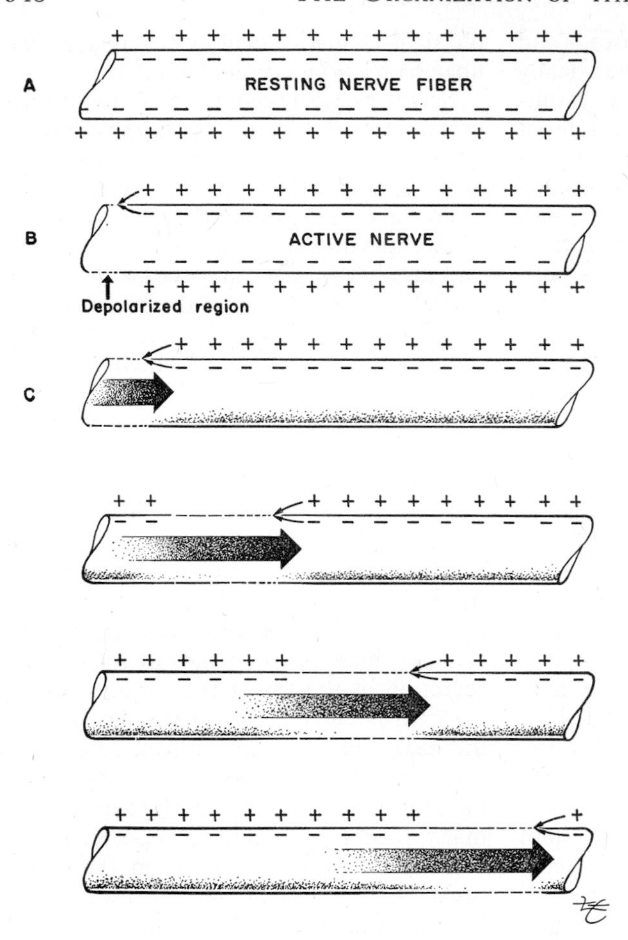

Figure 197. Diagram illustrating the membrane theory of nerve transmission. *A,* Resting nerve, showing the polarization of the membrane with positive charges on the outside and negative ones inside. *B,* Nerve conducting an impulse, showing, from left to right, the depolarized region where the impulse is, and the polarized region ahead of the impulse. *C,* Stages in the passage of the impulse along the nerve.

involved in the transmission of a nerve impulse are similar in many ways to those involved in muscle contraction. But a transmitting nerve expends little energy compared to a contracting muscle; the heat produced by 1 gm. of nerve, stimulated for one minute, is equivalent to the energy liberated by the oxidation of 0.000001 gm. of glycogen. This means that if a nerve contained only 1 per cent glycogen to serve as fuel, it could be stimulated continuously for a week without exhausting the supply. Nerve fibers are practically incapable of being fatigued as long as an adequate supply of oxygen is available. Whatever mental fatigue may be, it is not a real fatigue of the nerve fibers.

The generally accepted theory of the nature of the nerve impulse is the **membrane theory.** According to this, the membrane surrounding each nerve fiber is semipermeable, allowing certain ions, but not all, to penetrate it. The metabolic activities of the nerve maintain this membrane in a polarized state; that is, there is an extra number of positive ions (positively charged chemical particles) on its outer surface, and an equal excess of negative ions on its inner surface, so that the resting nerve resembles a charged condenser (Fig. 197*A*). The positive and negative ions are kept from mixing, and neutralizing each other, by the membrane, which is impermeable to them. The potential across the membrane due to the excess of positive ions outside and negative ions inside is from 0.03 to 0.06 volt.

When a nerve is stimulated at one point, the stimulation depolarizes the membrane and increases its permeability, so that the ions from the adjacent, not-yet-activated region pass through the depolarized region and neutralize each other

(Fig. 197*B*). This depolarizes the adjacent region and makes it permeable to the migration of ions from the next region, and so on. In this way the nerve impulse moves along the surface of the nerve fiber as a wave of depolarization of the membrane. It seems probable that some chemical reactions must occur in the depolarized membrane to make it permeable, and other reactions must occur during the refractory period to recharge the membrane and prepare it to be depolarized by the next impulse. Although there are many other details, the transmission of the nerve impulse rests on this rather simple electrochemical mechanism. All the complex phenomena of reflexes, learning and creative thinking depend on these extremely feeble currents and small chemical changes.

This theory, which is supported by all the experimental evidence, affords an explanation of the refractory period as the time necessary to repolarize the membrane. And it accounts also for the all-or-none phenomenon, since no matter what the strength of the stimulus, the depolarization can go only as far as zero.

219. TRANSMISSION AT THE SYNAPSE

The process whereby a nerve impulse is transmitted across a synapse from one neuron to the next is not perfectly understood. It was once believed that an impulse reaching the end of one neuron simply excites the next one as though it were a continuation of the first, and this may actually happen at certain synapses. But at others a completely different mechanism operates, one which depends on the secretion of a chemical substance, called a neurohumor, by the tip of the axon. This chemical then diffuses to the dendrite of the next neuron or to the muscle or gland innervated, and stimulates it.

The existence of these chemicals was discovered in experiments on the heart, an organ supplied with two sets of nerves, one which speeds it up, another which slows it down. When a frog's heart is isolated from the body, and the vagus (inhibitory) nerve to it is stimulated, it stops beating. If the heart is then washed out with salt solution, and the fluid is placed in a second heart, the latter also stops! Therefore, the vagus nerve must release some chemical which inhibits the heart beat. Further research has shown this substance to be a relatively simple chemical called **acetylcholine;** pure, synthetic acetylcholine has the same effect. Here the chemical transmission is from nerve to muscle, but the same mechanism is involved in at least some types of synapses between two neurons. By similar experiments it has been shown that the nerves speeding up the heart also release a chemical substance, called **sympathin,** which is similar to or identical with the hormone epinephrine. You may wonder why, if the tip of the axon secretes a substance such as acetylcholine, that substance does not stimulate the next dendrite or the muscle continuously. The answer is that blood and most tissues contain a potent enzyme, called **cholinesterase,** which specifically destroys acetylcholine, thereby preventing its continued effect. A different enzyme oxidizes sympathin.

It is not yet definitely known whether transmission across the synapses in the brain and spinal cord is by means of acetylcholine, sympathin or some other chemical, or by an electrical mechanism.

The synaptic junction is a point of resistance to the flow of impulses in the nervous system, and not every impulse reaching the synapse is transmitted to the next neuron. The resistance varies in different synapses, so that they are important in determining the route of impulses through the nervous system and the response of the organism to a specific stimulus.

Functionally, the entire nervous system is a unit, and an impulse arising in any receptor can be transmitted to every effector in the body. Consider, for example, the effects of burning a finger: The muscles of the arm contract to pull the finger away from the heat, a sensation of pain is produced in the brain, a cry may be emitted, the heart beat, digestion and breathing may be interrupted—in fact, it is conceivable that every muscle and gland in the body may be affected

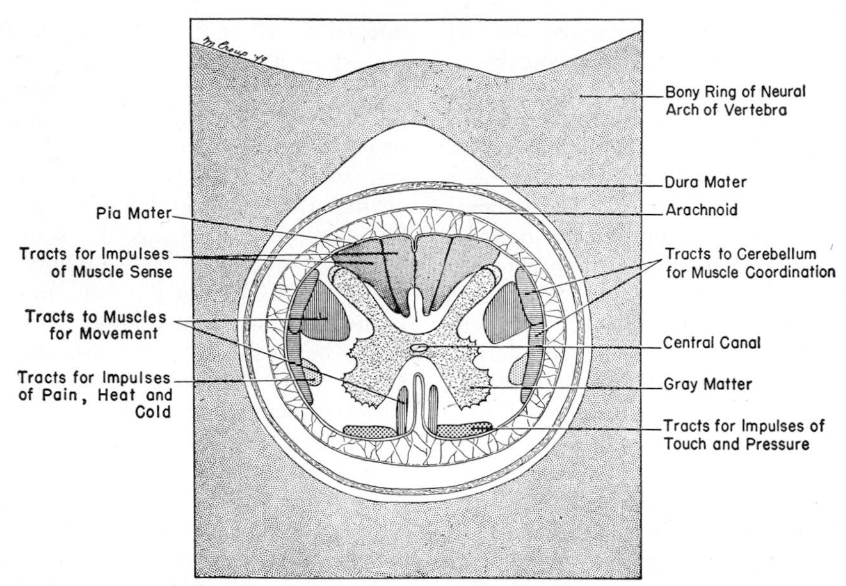

Figure 198. Cross section of the spinal cord surrounded by the bony vertebra, showing the meninges (*dura mater, pia mater* and *arachnoid*), the gray matter, and some of the important nerve tracts of the white matter.

temporarily. Our sense organs receive a constant stream of stimuli, but selective resistance at the synaptic junction prevents the uncontrolled, continuous contraction of muscles and secretion by glands. The drug **strychnine** decreases synaptic resistance; the slightest stimulus to a person suffering from strychnine poisoning sets off the secretion of all glands and the convulsive contraction of all the muscles of the body.

The amount of synaptic resistance can be modified by nerve impulses, so that one impulse cancels out the effect of another. This is known as **inhibition.** The opposite condition, whereby one impulse strengthens another, is called **reinforcement.** These two processes are of prime importance in effecting integration of body activities. We have seen that all the muscles of the body are in a state of constant, slight contraction, known as tonus, due to a constant volley of nerve impulses reaching them. But when one muscle, such as the triceps, is to contract, its antagonist, the biceps, must relax. This is achieved by the simultaneous operation of impulses which inhibit the volley of impulses to the biceps, and of impulses which reinforce the volley to the triceps. Inhibition and reinforcement can occur only at the synapse, since once an impulse starts along a neuron, it can be neither stopped nor accelerated. Whether or not a given impulse crosses a synapse depends on whether it is inhibited or reinforced by other impulses.

220. THE CENTRAL NERVOUS SYSTEM

The ten billion or so neurons which make up the nervous system are divided into two main parts: those belonging to the **central nervous system,** which make up the brain and spinal cord, and those belonging to the **peripheral nervous system,** which make up the cranial and spinal nerves.

The Spinal Cord. The tubular spinal cord, which is surrounded and protected by the neural arches of the vertebrae, has two important functions: to transmit impulses to and from the brain, and to act as a reflex center. In cross section it is seen to consist of two regions, an inner, butterfly-shaped mass of **gray matter,** made up of nerve cell bodies, and an outer mass of **white matter** made up of bundles of axons and dendrites (Fig. 198). The whiteness of these bundles is due to the myelin sheaths of the axons and dendrites; the ends of the axons and dendrites, present in the central gray matter, have no myelin sheath. The "wings" of the gray matter are divided into two dorsal horns and two ventral horns. The latter contain the cell bodies of motor neurons, whose axons pass out through the spinal nerves to the muscles; all the other neurons in the spinal cord are connector neurons.

The axons and dendrites of the white matter are segregated into bundles with similar functions: the **ascending tracts,** which carry impulses to the brain, and the **descending tracts,** which carry impulses from the brain to the effectors. Neurologists have patiently collected data on persons with injured spinal cords, and by carefully noting the symptoms, and correlating these with the particular tracts found to be destroyed when the patient's nervous system was examined after death, they have been able to map out the location and functions of the various tracts (Fig. 198). For example, the dorsal columns of the white matter transmit impulses originating in the sense organs of muscles, tendons and joints, by means of which we are aware of the position of the parts of the body. In advanced syphilis these columns may be destroyed, so that the patient cannot tell where his arms and legs are unless he looks at them, and he must watch his feet in order to walk.

One curious fact, still not satisfactorily explained, emerged from these studies of the location and function of the fiber tracts. All the fibers in the spinal cord *cross over* from one side of the body to the other somewhere along their path from sense organ to brain, or from brain to muscle. Thus the right side of the brain controls the left side of the body and receives impressions from the sense organs of the left side. Some fibers cross in the spinal cord itself; others cross in the brain.

In the center of the gray matter is a small canal, running the entire length of the spinal cord, filled with **cerebrospinal fluid,** which is similar to plasma. The spinal cord and brain are wrapped in three sheets of connective tissue, known as **meninges.** Meningitis is a disease in which these wrappings become infected and inflamed. One of these sheets (dura mater) is fastened against the bony neural arches of the vertebrae; another (pia mater) is located on the surface of the spinal cord, and the third (arachnoid) lies between. The spaces between the meninges are filled with more cerebrospinal fluid, so that the spinal cord (and the brain) floats in this liquid and is protected from bouncing against the bone of the vertebrae (or skull) with every movement.

The Brain. There is a general correlation between the size of the brain and what we call intelligence. Modern man has a brain volume of about 1500 ml., while the higher apes have a brain size of only 600 ml. Our immediate evolutionary ancestors had intermediate-sized brains: Java man, 940 ml.; Peking man, 1000 ml.; Solo man, 1300 ml.; and Neanderthal man, 1550 ml. The important factor is not absolute, but relative, size, however, since a whale with a 3000 ml. brain is not so intelligent as a rat with a 20 ml. brain. And even the ratio of brain weight to total body weight is not completely valid, for some vertebrates, notably certain monkeys, have a higher such ratio than man. Intelligence can best be correlated with the ratio of brain weight to spinal cord weight, which indicates the extent of the control of the higher centers over the lower. In frogs and fishes this ratio is less than one; the brain weighs less than the spinal cord. In lower mammals it is between two and four; in apes, fifteen; and in man the brain weighs fifty-five times as much as the spinal cord.

The brain is the enlarged, anterior end of the spinal cord. In man the enlargement is so great that much of the resemblance to the spinal cord is obscured, but in the lower animals the relationship of brain to spinal cord is clear. The detailed anatomy of the brain is exceedingly complex, and we shall consider only the six main regions: medulla, pons, cerebellum, midbrain, thalamus and cerebrum (Fig. 199).

The most posterior part of the brain, lying next to the spinal cord, is the **medulla.** Here the central canal of the spinal cord enlarges to form a large cavity called the **fourth ventricle** (three other ventricles lie farther up in the brain). The roof of

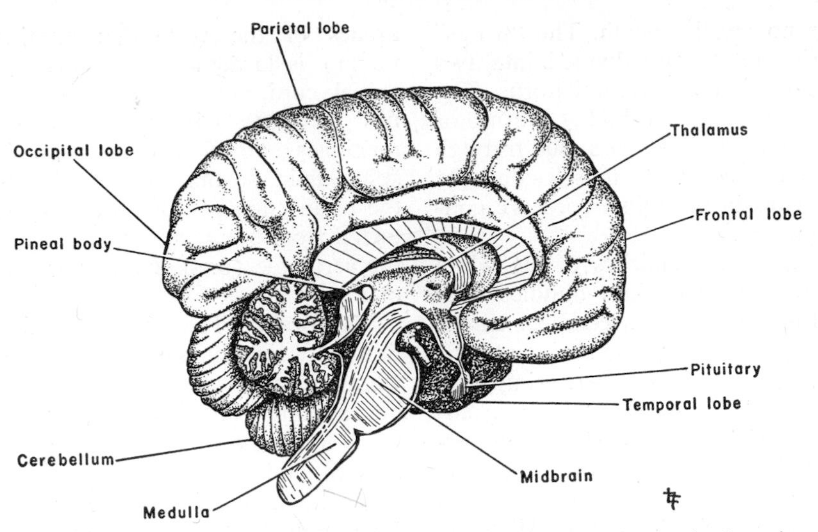

Figure 199. The parts of the human brain seen in a median sagittal section.

the fourth ventricle is thin and contains a cluster of blood vessels which secrete part of the cerebrospinal fluid; the rest of this fluid is secreted by similar clusters of blood vessels in the other ventricles. In the roof of the fourth ventricle are three tiny pores through which cerebrospinal fluid escapes into the meningeal spaces. The walls of the medulla are thick and made up largely of nerve tracts connecting with the higher parts of the brain. The medulla also contains a number of clusters of nerve cell bodies, the **nerve centers,** which are reflex centers controlling many of the fundamental physiologic processes of the body—respiration, heart rate, the dilatation and constriction of blood vessels, swallowing and vomiting.

Above the medulla is the **cerebellum,** consisting of a central part, and two hemispheres extending laterally, which resemble pine cones in shape. Its gray surface is made up of nerve cell bodies, and beneath is a mass of white tissue composed of fiber tracts connecting with the medulla and with the higher parts of the brain. The size of the cerebellum in different animals is roughly correlated with the amount of their muscular activity. Since it regulates and coordinates muscle contraction, it is proportionately large in extremely active animals, such as birds. Removal or injury of the cerebellum is not accompanied by paralysis, but by impairment of muscle coordination. When the cerebellum of a bird is removed surgically, it is unable to fly and the wings thrash about jerkily. When the human cerebellum is injured by a blow or disease, all muscular movements are uncoordinated, and any activity requiring delicate coordination, such as threading a needle, is impossible.

Running crosswise on the ventral side of the brain just below the cerebellum is a thick bundle of fibers, known as the **pons** or bridge, which carries impulses from one hemisphere of the cerebellum to the other, thus coordinating muscle movements on the two sides of the body.

In front of the cerebellum and pons lies the **midbrain,** which has thick walls and a small central canal connecting the fourth ventricle of the medulla with the third ventricle in the thalamus. The thick walls of the midbrain contain certain reflex centers and the main fiber tracts leading to the thalamus and cerebrum. On the upper side of the midbrain are four low, rounded protuberances called **optic lobes,** in which are centers for certain visual and auditory reflexes. The reflex constriction of the pupil when light shines on the eye is controlled by a center in the optic lobes, as is the pricking up of a dog's ears in response to a sound.

The midbrain also contains a cluster of nerve cells regulating muscle tonus and posture.

In front of the midbrain the central canal again widens and becomes the third ventricle, the roof of which contains another cluster of blood vessels secreting cerebrospinal fluid. The thick walls of the third ventricle are called the **thalamus.** This is a relay center for sensory impulses; fibers from the spinal cord and lower parts of the brain synapse here with other neurons going to the various sensory areas of the cerebrum. In the floor of the third ventricle are centers regulating body temperature, appetite, water balance, carbohydrate and fat metabolism, blood pressure and sleep. Curiously, the front part of the thalamus prevents a rise in body temperature and the rear part prevents a fall. The thalamus also appears to regulate and coordinate the external manifestations of emotions; thus, by stimulating the thalamus, a sham rage can be produced in a cat—the hair stands on end, the claws protrude, the back becomes humped, and other signs of anger are evinced. But as soon as the stimulation stops, the appearance of rage ceases.

The parts of the brain considered so far have to do with unlearned, automatic behavior which is determined by the fundamental structure of these parts, a structure which is essentially the same from fish to man. The **cerebrum,** the most anterior and the largest single part of the human brain, has a basically different function, that of controlling learned behavior. The complex psychologic phenomena of consciousness, intelligence, memory, insight and the interpretation of sensations have their physiologic basis in the activities of the neurons of the cerebrum. The importance of the cerebrum to different animals can be investigated by removing it surgically. A cerebrumless frog behaves almost exactly as a normal one, and a pigeon whose cerebral cortex has been removed can fly and balance on a perch, but tends to remain quiet for hours. When stimulated, it moves about, though in a random, purposeless way, and, since it fails to eat when given food, will starve unless fed artificially. A dog whose cerebral cortex has been removed can walk and will swallow food if it is placed in the mouth, but shows no signs of fear or excitement. Human infants occasionally are born whose cerebral cortex fails to develop, and although they can carry out the vegetative functions of breathing and swallowing, they are incapable of learning and make no voluntary movements. Such creatures usually die soon after birth.

The cerebrum contains slightly more than half of the ten billion neurons of the human nervous system. These are arranged in two **cerebral hemispheres,** which develop as outgrowths of the anterior end of the brain. In man and other mammals they grow back over the rest of the brain and hide it from view. Each hemisphere contains a cavity, the first and second ventricles, each of which is connected to the third ventricle of the thalamus by a canal. These ventricles, like the others, contain clusters of blood vessels which secrete the cerebrospinal fluid. The cerebrum is made of both gray and white matter; the latter, composed of tracts of nerve fibers, is on the inside of the cerebrum, while the gray matter, made of nerve cell bodies, lies on the surface or **cortex** of the cerebrum. Deep in the substance of the cerebral hemispheres lie other masses of gray matter, nerve centers which act as relay stations to and from the cortex. The lower vertebrates, with little gray matter, have smooth cerebral cortices, but in man and other mammals the surface of the cerebral hemispheres is convoluted. Ridges separated by furrows increase the amount of space available for the cortical gray matter. The pattern of these convolutions is quite constant even in men of widely different degrees of intelligence, and the geography of the cerebral cortex has been carefully studied (Fig. 200). The idea that certain parts of the brain have special functions is an old one, the "science" of phrenology having been based on the premise that functions were localized in the brain, so that if a person were specially gifted, a particular area would be enlarged and cause a bump on the head. It was believed that an analysis of such

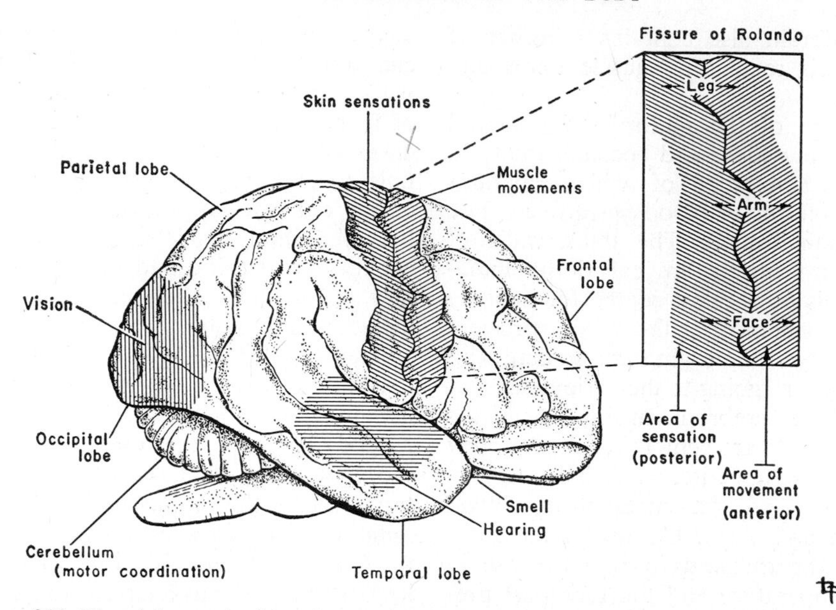

Figure 200. The right cerebral hemisphere of the human brain, seen from the side. The stippled areas are regions of special function; the light areas are "association areas." *Inset:* Enlarged view of the sensory and motor areas adjacent to the fissure of Rolando, showing the location of the nerve cells supplying the various parts of the body.

bumps would indicate what a man was best fitted for. How a tissue as soft as the brain could cause a bump in the hard skull was apparently overlooked.

Experimental evidence has established that there is a considerable amount of localization of function in the cortex. By surgical removal of particular regions of the cortex from experimental animals it has been possible to localize many functions exactly; and by observing the paralysis or loss of sensation in a man with a brain injury or tumor, and then examining the brain after death to see where the injury was located, it has been possible to map the human brain. During operations on the brain, surgeons have electrically stimulated small regions and observed which muscles contracted, and, since brain surgery can be carried on under local anesthesia, the patient could be asked what sensations he felt when the particular regions were stimulated. Curiously, the brain itself has no nerve endings of pain, so that stimulation of the cortex is not painful. The most recent method of studying brain activity is that of measuring and recording the electrical potentials or "brain waves" given off by various parts of the brain when active (see p. 355).

By combining the data obtained in several ways, investigators have been able to locate many functions of the brain (Fig. 200). The back part contains the visual center; removal of it causes blindness, and stimulation of it, even by a blow on the back of the head, causes the sensation of light. Removal of the region from one side of the brain causes blindness in half of each eye, for the nerves from each eye split, and half go to each side of the brain. The center for hearing is located on the side of the brain, above the ear. Stimulation of it by a blow causes a sensation of noise; although removal of both auditory areas causes deafness, removal of one does not cause deafness in one ear, but a decrease in the auditory acuity of both ears.

Running down the side of the cerebral cortex is an easily recognizable, deep furrow, called the fissure of Rolando, which separates the motor area, controlling the skeletal muscles, from the area just behind the furrow which is responsible for the sensations of heat, cold, touch and pressure from stimulation of sense organs in the skin. In both there is a further specialization along the furrow from the top of the brain to the side—neurons at the top of the cortex control the muscles of the feet; then next in line control those of the shank, thigh, abdomen, and so on; and the neurons farthest around to the side control the muscles of the face. The size of the motor area in the brain for any given part of the body is proportional not to the amount of muscle but to the elaborateness and intricacies of movement; thus there are large areas for the control of the hand and face. There is a similar relationship between the parts of the sensory area and the region of the skin from which it receives impulses. Thus, in the connections between the body and the brain, there is not only a twisting of the fibers so that one side of the brain controls the opposite side of the body, but a further "reversal" which makes the uppermost part of the cortex control the lower extremities of the body.

When all the areas of known function are plotted, they cover almost all of the rat's cortex, a large part of the dog's, a moderate amount of the monkey's, but only a small part of the total surface of man's cortex. The rest, known as **association areas,** is made up of neurons that are not directly connected to sense organs or muscles, but which supply interconnections between the other areas. These regions are responsible for the higher intellectual faculties of memory, reasoning, learning, imagination, and for personality. In some way, the association regions integrate all the diverse impulses constantly reaching the brain into a meaningful unit, so that the proper response is made. They interpret and manipulate the symbols and words by means of which our thought processes are carried on. When disease or accident destroys the functioning of one or more association areas, the condition known as **aphasia** results, in which the ability to recognize certain kinds of symbols is lost. In one type of aphasia the patient is unable to understand written words, although he comprehends spoken words perfectly. In another type of aphasia the names of objects are forgotten, although their functions are remembered and understood.

221. BRAIN WAVES

Metabolism is invariably accompanied by electrical changes, and the electrical activity of the brain can be recorded by a device known as the **electroencephalograph.** To obtain recordings, electrodes are fastened to different parts of the scalp by adhesive tape, and the activity of the underlying parts of the cortex is measured. The electroencephalograph has shown that the brain is continuously active, even when no thinking is going on, and that the most regular manifestations, called alpha waves, come from the visual areas at the back of the brain when the subject is resting quietly with his eyes shut. These waves occur rhythmically at the rate of nine or ten per second and have a potential of about 45 microvolts (Fig. 201). When the eyes are opened, the waves disappear and are replaced by more rapid, irregular waves. That the latter are produced by objects seen can be demonstrated by presenting the eyes with some regular stimulus, such as a light blinking at regular intervals, and observing that brain waves of a similar rhythm appear. Sleep is the only normal condition in which the brain waves are drastically altered. During sleep the waves become slower and larger (have a greater potential) as the subject falls into deeper and deeper unconsciousness. The dreams of a sleeping subject are mirrored in flurries of irregular waves.

Certain brain diseases alter the character of the waves; epileptics, for example, exhibit a peculiar and readily recognizable wave pattern, and even people who have never had an epileptic attack, but might in certain conditions, show similar abnormalities. The location of brain tumors can be detected by noting the part of the brain showing abnormal waves.

222. SLEEP

The neural mechanisms involved in sleep are unknown, and investigators are still trying to discover why sleep is necessary and why it has a recuperative effect.

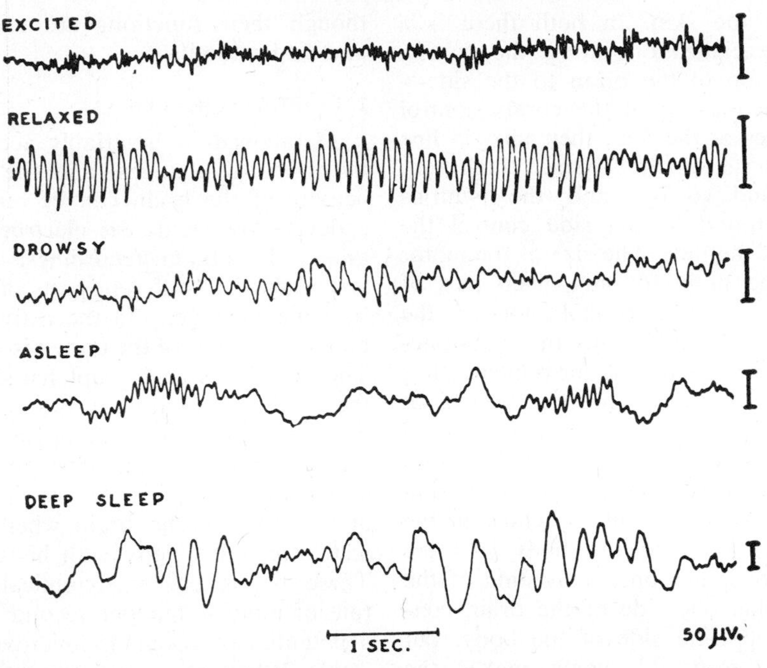

Figure 201. Electroencephalograms made while the subject was excited, relaxed, and in various stages of sleep. Recordings made during excitement show brain waves which are rapid and of small amplitude, whereas in sleep the waves are much slower and of greater amplitude. The regular waves characteristic of the relaxed state are called alpha waves. (From Jasper; in Epilepsy and Cerebral Localization, by Penfield and Erickson.)

Only the higher vertebrates with fairly well-developed cerebral cortices sleep, and those with larger hemispheres seem to require more sleep than others. Fatigue is popularly considered the cause of sleep, but there is no experimental evidence to verify this belief. An important sleep-inducing factor is the absence of stimuli; it is easy to go to sleep, even when one is not particularly tired, if there is nothing interesting to occupy the mind. But although we tend to be wakeful in the presence of attention-holding stimuli, there is a limit beyond which sleep is inevitable. For all the higher animals, life is characterized by a basic rhythm of sleep alternating with wakefulness, a pattern regulated by the hypothalamus. There is a sleep center in the anterior part, and a wakefulness center in the posterior part of the hypothalamus. It is believed that the change from wakefulness to sleep and back is controlled by "feed-back" circuits involving these two centers. The human habit of sleeping for eight hours of the twenty-four is definitely learned; the basic, natural rhythm is one in which sleep and wakefulness alternate every three or four hours, as exemplified by the behavior of infants.

223. INSANITY AND NEUROSES

Certain types of brain derangements have an easily understood basis in damage to the brain tissue produced by disease or a wound. If the pores in the roof of the fourth ventricle become clogged, trapping the cerebrospinal fluid within the ventricles, the pressure of this fluid inside the brain will gradually destroy the tissue there. Or a blood vessel in the meninges covering the brain may rupture and the pressure of the accumulated blood destroy parts of the brain. Tumors and infectious diseases, such as syphilis, can damage the tissue; the actual symptoms—paralysis, loss of sensation or other functions—depend on which part of the brain is affected.

The cause of other types of disorder, the so-called functional disorders—neuroses and psychoses—is more baffling, for they occur without any structural or chemical change in the brain which pathologists have so far been able to detect. Typically, these involve emotional disturbances rather than changes in intelligence.

Neuroses are comparatively mild and common disorders with a great variety of symptoms: anxiety, fear, shyness and oversensitiveness. The emotional upsets may actually produce organic disorders, such as irregular heart beat or digestive disturbances. The cause of this type of mental disorder is not positively known, and there is evidence for believing that it differs from person to person, and is complex in every instance. One theory states that neuroses are due to deep-seated conflicts, and in some cases this would seem to be a reasonable explanation. Usually, however, heredity, present environment, past experience and general health all play a part in causing the condition. In any event, the patient is often completely unaware of the cause or causes of his unhappiness. There is no single cure for the various neuroses; many of them respond to psychiatric treatment, by means of which the reason for the sense of guilt, conflict or fear is brought to the patient's attention. Other neuroses disappear gradually for no apparent reason, others become increasingly worse, and a few develop into the more serious psychoses.

Psychoses are the severe mental diseases which usually require hospitalization. There are three main types of psychoses, each of which represents an exaggeration of normal tendencies. The **manic-depressive** psychosis is characterized by an alternation of excessive elation and depression, sometimes accompanied by delusions and hallucinations. Most manic-depressives are normal for most of their lives, but suffer recurrent episodes of insanity. **Paranoia** is a psychosis characterized by delusions, typically of grandeur or persecution. **Dementia praecox** or **schizophrenia** is marked by a withdrawal from the everyday world into a world of daydreams which becomes the world of reality.

Psychoses are much more difficult to cure than neuroses, and most of them are not permanently curable. One of the most drastic therapeutic methods is the shock treatment, based on the theory that psychotics actually can be jolted back into sanity. Violent fits are produced by the injection of insulin or Metrazol or by the application of electric currents. The drawbacks to such treatments are many, and the neural mechanism underlying the results are not clearly understood, but a number of cases have been cured by some variation of the shock treatment. Treatment with certain new drugs such as chlorpromazine has been successful in many cases and is gradually supplanting the shock treatment.

224. THE PERIPHERAL NERVOUS SYSTEM

Emerging from the brain and spinal cord and connecting them with every receptor and effector in the body are the paired cranial and spinal nerves; these make up the peripheral nervous system. Cranial and spinal nerves are made of bundles of nerve fibers, the axons and dendrites. The only nerve cell bodies present in the peripheral nervous system are those of the sensory neurons, aggregated into clusters known as **ganglia,** near the brain or spinal cord, and of certain motor neurons of the autonomic system which will be discussed later.

Cranial Nerves. Twelve pairs of nerves originate in different parts of the brain and innervate primarily the sense organs, muscles and glands of the head. The same twelve pairs, innervating similar structures, are found in all the higher vertebrates—reptiles, birds and mammals; fish and amphibia have only the first ten. Like all nerves, these are composed of neurons, although some of them have only sensory neurons (nerves I, II and VIII), some are composed almost completely of motor neurons (III, IV, VI, XI and XII), and the others are made up of both sensory and motor neurons. The names and structures innervated by the cranial nerves are given in Table 7. One of the most important cranial nerves is the **vagus,** which forms part of the autonomic system and innervates the internal organs of the chest and the upper abdomen.

Spinal Nerves. All the spinal nerves

Table 7. THE CRANIAL NERVES OF MAN

NUMBER	NAME	ORIGIN OF SENSORY FIBERS	EFFECTOR INNERVATED BY MOTOR FIBERS
I	Olfactory	Olfactory mucosa of nose (smell)	None
II	Optic	Retina of eye (vision)	None
III	Oculomotor	Proprioceptors of eyeball muscles (muscle sense)	Muscles which move eyeball (with IV and VI); muscles which change shape of lens; muscles which constrict pupil
IV	Trochlear	Proprioceptors of eyeball muscles (muscle sense)	Other muscles which move eyeball
V	Trigeminal	Teeth and skin of face	Some of muscles used in chewing
VI	Abducens	Proprioceptors of eyeball muscles (muscle sense)	Other muscles which move eyeball
VII	Facial	Taste buds of anterior part of tongue	Muscles of the face; submaxillary and sublingual glands
VIII	Auditory	Cochlea (hearing) and semicircular canals (senses of movement, balance and rotation)	None
IX	Glossopharyngeal	Taste buds of posterior third of tongue, lining of pharynx	Parotid gland; muscles of pharynx used in swallowing
X	Vagus	Nerve endings in many of the internal organs—lungs, stomach, aorta, larynx	Parasympathetic fibers to heart, stomach, small intestine, larynx, esophagus
XI	Spinal accessory	Muscles of shoulder (muscle sense)	Muscles of shoulder
XII	Hypoglossal	Muscles of tongue (muscle sense)	Muscles of tongue

are mixed nerves, having motor and sensory components in roughly equal amounts. In man they originate from the spinal cord in thirty-one symmetrical pairs, each of which innervates the receptors and effectors of one region of the body. Each nerve emerges from the spinal cord as two strands or roots which unite shortly to form the spinal nerve. All the sensory nerves *enter* the cord through the *dorsal root* and all motor fibers *leave* the cord through the *ventral root* (Fig. 202). If the dorsal root is severed, the part of the body innervated by that nerve suffers complete loss of sensation without any paralysis of the muscles. If the ventral root is cut, there is complete paralysis of the muscles innervated by that nerve, but the senses of touch, pressure, temperature, kinesthesis and pain are unimpaired. The size of each spinal nerve is related to the size of the body area it innervates; the largest in man is one of the pairs supplying the legs. Each spinal nerve, shortly beyond the point of junction of the dorsal and ventral root, divides into three branches: the dorsal branch, serving the skin and muscles of the back; the ventral branch, serving the skin and muscles of the sides and belly; and the autonomic branch, serving the viscera (Fig. 202).

225. REFLEXES AND REFLEX ARCS

A reflex is an inborn, automatic response to a given stimulus depending only on the anatomic relationships of the neurons involved. The neurons which conduct the impulse form a **reflex arc,** the simplest of which is made of one sensory, one connector, and one motor neuron (Fig. 203). Reflexes are the functional units of the nervous system, and most of our activities are the result of them. We have already seen how important reflexes are in controlling heart rate, blood pressure, breathing, salivation, movements of the digestive tract, and so forth. When we step on something sharp or come in contact with something hot, we do not wait until the pain is experienced by the brain and then, after deliberation, decide what to do; our responses are immediate and **reflex.** We have begun to withdraw the foot or hand by reflex action *before* the pain is experienced by the brain. Many of the more complicated activities of our daily lives, such as walking, are regulated to a large extent by reflexes. Those present at birth, and common to all men, are called **inherited reflexes;** others, acquired later as the result of experience, are called **conditioned reflexes.**

Reflexes are classified according to the number of nerve pathways involved. A **simple reflex,** in which stimulation of a receptor produces contraction of a single muscle, is typified by the "knee jerk." When the tendon of the knee cap is tapped, and thereby stretched, receptors in the tendon are stimulated, an impulse travels over the reflex arc up to the spinal cord and down again, and the muscle attached to the tendon contracts, resulting in a sudden straightening of the leg. An important, though fairly simple, reflex is the **flexion** or withdrawal reflex. When an arm or leg is stimulated by anything injurious, the flexor muscles are stimulated and the injured member is pulled away before further damage occurs. Many reflexes are more complicated than these examples, but even the most complicated of the inherited ones do not require the operation of the higher thought processes. An experiment demonstrating this consists in removing the brain of a frog, while leaving the spinal cord intact, and then applying a piece of acid-soaked paper to the animal's back. No matter how many times the piece of paper is placed on the skin, one leg will invariably come up and flick it away. This response, involving many muscles working in a coordinated fashion, is purely reflex, and clearly demonstrates one of the chief characteristics of a reflex: fidelity of repetition. A frog with a brain might make the response two or three times, but eventually it would do something else—perhaps hop away. Most reflexes have some survival value to the animal; the anatomic configuration responsible for the

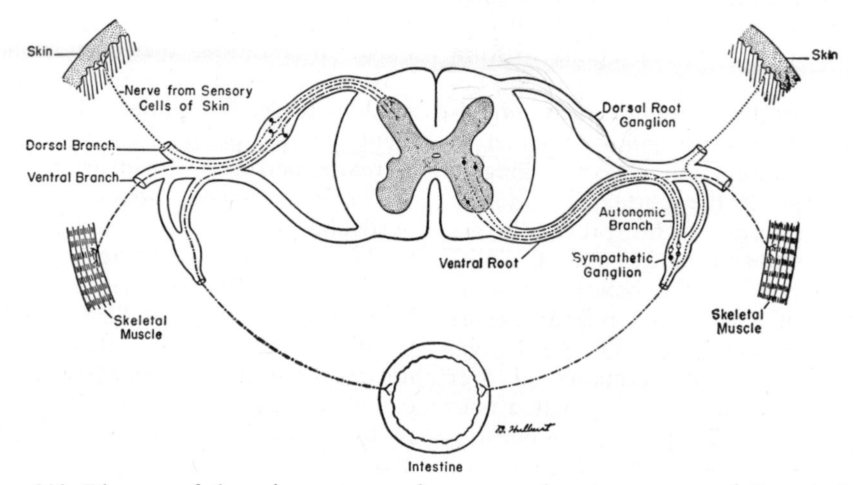

Figure 202. Diagram of the primary types of sensory and motor neurons of the spinal nerves, and their connections with the spinal cord. For convenience, the sensory neurons are shown on the left and the motor neurons on the right, though both kinds are found on each side of the body.

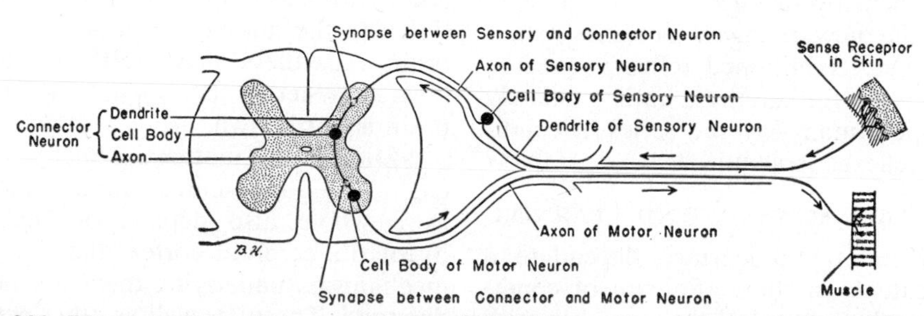

***Figure 203.* Diagram of a reflex arc, showing the pathway of an impulse, indicated by arrows.**

reflex was selected in evolution because of this survival value.

The behavior of a newborn child is determined largely by its inborn, inherited reflexes; but as he grows older, the relationship between stimulus and response may be altered, and a new stimulus substituted for the old one in eliciting a response. The relationship of new stimulus and old response is known as a **conditioned reflex.** These are usually complicated, indirect and difficult to investigate. The classic investigations of the Russian physiologist, Pavlov, on conditioned reflexes associated with salivation have made a beginning toward an explanation of the mechanisms involved, and have shown how much learned behavior is built upon inborn behavior.

Normally, saliva is secreted reflexly only when food is in the mouth, but if another stimulus, such as the ringing of a bell, occurs every time food is eaten, a conditioned reflex is established and saliva flows when the bell is rung, whether or not food is present. The ability to form conditioned reflexes is related to the development of the cerebral cortex and is greater in mammals and birds than in lower vertebrates, and almost lacking in invertebrates. If the cortex is removed, the animal loses all his previously acquired conditioned reflexes and the ability to form new ones. It is believed that the formation of conditioned reflexes occurs in the association areas of the cortex. Here, impulses from receptors all over the body converge and are shunted onto motor neurons. The passage of one impulse across a synapse tends to lower the synaptic resistance, so that the impulse from the new stimulus follows the path of the impulse in the previously established reflex; then, by facilitation through repetition, the new pathway becomes established. One conditioned reflex may serve as the basis for another, until ultimately a complex, many-layered series of conditioned reflexes is acquired.

226. HABIT, MEMORY AND LEARNING

Both habit and learning depend to a large extent on the setting up of conditioned reflexes, and these, together with memory, depend upon the fact that a series of impulses passing across the synapses of a given chain of neurons results in a lasting increase in the chain's ability to conduct subsequent impulses. This phenomenon, known as **facilitation,** increases as the synapses continue to be used, until a maximum is reached. The amount of repetition necessary to reach the maximum—really the ability of the animal to "learn"—varies in different species and in different members of the same species. Facilitation is gradually lost if the particular synaptic pathway is not used for a long time; this is the basis for forgetting and loss of habit.

Learning is a complex and poorly understood process that depends on cerebral activities other than the formation of conditioned reflexes. Certain experiments, such as those in which animals learn to run a maze or labyrinth to get food or to avoid electric shocks, have emphasized the role of "trial and error" in learning. The higher mammals, apes and men, show, in addition, the phenomenon of "insight" or getting the idea. After a few random trials, the subject of the experiment sees the point of the problem, and thereafter performs with invariable success.

The experiments demonstrating that particular functions are localized in particular regions of the cortex suggest that at least a few kinds of learning are localized. But the work of Karl Lashley on learning in rats has indicated that this is not so. Learning, it seems, is a function of the cortex as a whole. By removing parts of the brain and then testing the ability of the rat to learn, he found that the important thing was not which particular areas were left, but the *total amount* of the cortex remaining. The ability to learn appeared to be proportional to the amount of cortex left in the brain, regardless of where the cortex was located. Whether this is true of the human brain is not known.

Moods and Emotions. Such phenomena as moods, emotions and personality as a whole also depend on the activity of the cerebral cortex, but the neural mechanisms underlying them are not understood. These, as well as other activities of the higher centers of the brain, are influenced greatly by the conditions of the body; the state of mind can be markedly affected by the state of the stomach. The secretions of several endocrine glands also affect the functioning of the brain; for example, many women have periods of mental depression just before and during menstruation, and it is quite common for the menopause (the period from forty to fifty years of age when the recurring menstrual cycle ceases) to be accompanied by profound emotional and mental disturbances.

227. THE AUTONOMIC NERVOUS SYSTEM

The heart, lungs, digestive tract and other internal organs are innervated by a special set of peripheral nerves, collectively called the **autonomic nervous system.** This system in turn is composed of two parts: the **sympathetic** and the **parasympathetic** nerves.

The autonomic system as a whole contains both sensory and motor nerves, but is distinguished from the rest of the nervous system by several features. There is no willful control by the cerebrum over these nerves; we cannot voluntarily speed up or slow down the heart beat or the action of the muscles of the stomach or intestines. Secondly, there is less direct connection between the sensory fibers and the cerebrum, so that the normal stimulation of these fibers does not result in sensations. Another important characteristic of the autonomic system is that each internal organ receives a *double* set of fibers, one set coming via the sympathetic nerves and one set via the parasympathetic nerves. Impulses from the sympathetic and parasympathetic nerves always have antagonistic effects on the organ innervated. Thus, if one speeds up an activity, the other decreases it. These effects are summarized in Table 8.

Still another peculiarity of the autonomic system is that the motor impulses reach the effector organ from the brain or spinal cord, not by a single neuron, as do those to all other parts of the body, but by a *relay* of two or more neurons. The cell body of the first neuron in the set, or chain, is located in the brain or spinal cord; that of the second neuron is located in a ganglion somewhere outside the central nervous system (Fig. 202). The ganglia of the sympathetic nerves are close to the spinal cord; those of the parasympathetic nerves are close to, or actually within, the walls of the organs they innervate.

The Sympathetic System. The sympathetic system consists of nerve fibers whose cell bodies are located in the lateral portions of the gray matter of the spinal cord. Their axons pass out through the ventral roots of the spinal nerves, in company with the motor neurons to the skeletal muscles, and then separate from these and become the autonomic branch of the spinal nerve going to the sympathetic ganglion. These ganglia are paired, and there is a chain of eighteen of them on each side of the spinal cord from the neck to the abdomen (Fig. 204). In each ganglion the axon of the first neuron synapses with the dendrite of the second neuron. The cell body of this second neu-

Table 8. ACTIONS OF THE AUTONOMIC SYSTEM

ORGAN INNERVATED	ACTION OF SYMPATHETIC SYSTEM	ACTION OF PARASYMPATHETIC SYSTEM
Heart	Strengthens and accelerates heart beat	Weakens and slows heart beat
Arteries	Constricts arteries and raises blood pressure	Dilates arteries and lowers blood pressure
Digestive tract	Slows peristalsis, decreases activity	Speeds peristalsis, increases activity
Urinary bladder	Relaxes bladder	Constricts bladder
Muscles in bronchi	Dilates passages, makes for easier breathing	Constricts passages
Muscles of iris	Dilates pupil	Constricts pupil
Muscles attached to hair	Causes erection of hair	Causes hair to lie flat
Sweat glands	Increases secretion	Decreases secretion

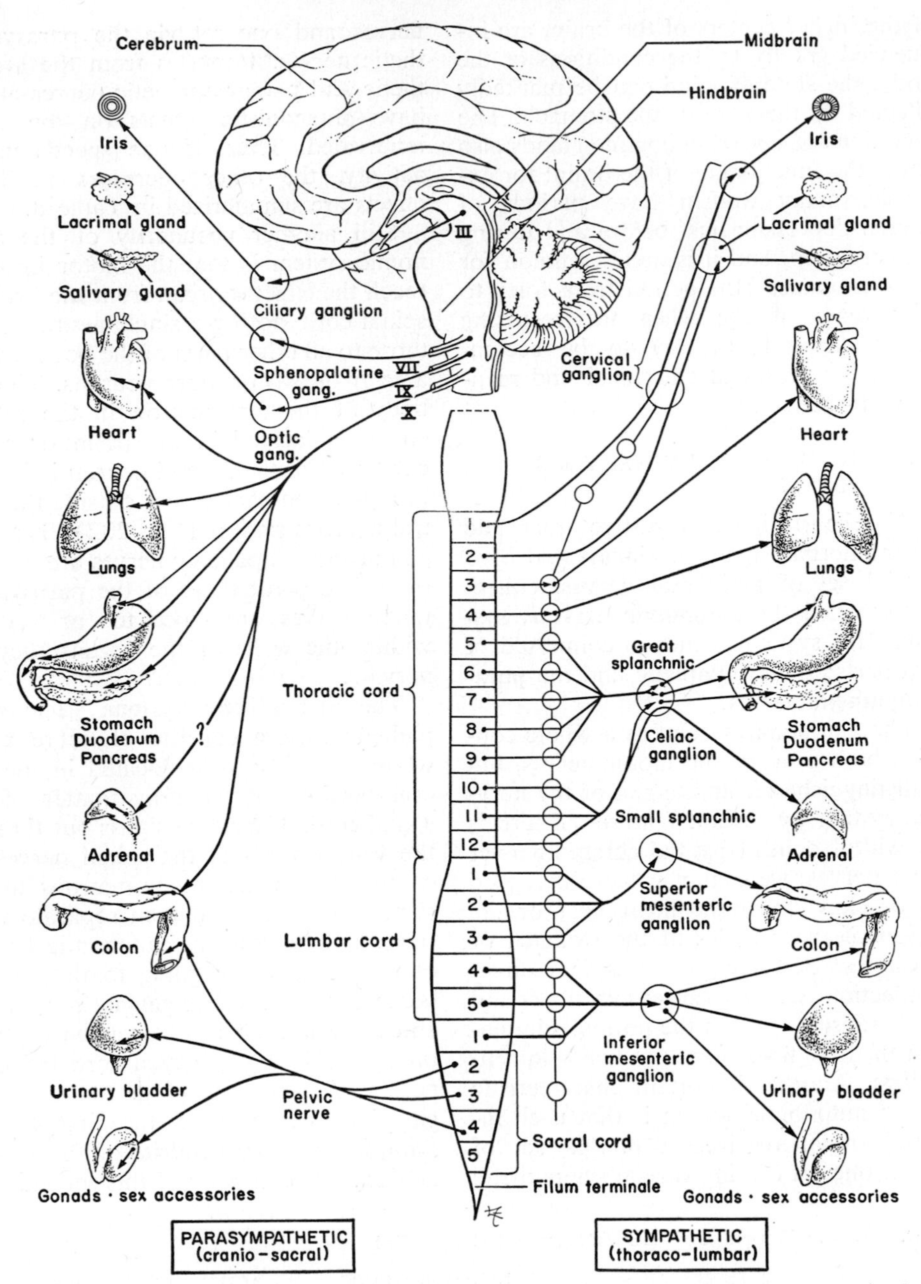

Figure 204. Diagram of the autonomic nervous system. The parasympathetic system is shown on the left, the sympathetic system on the right. Roman numerals refer to the numbers of the cranial nerves.

ron is located within the ganglion, and its axon passes to the organ innervated. In addition to the fibers going from each spinal nerve to each ganglion, there are fibers passing from one ganglion to the next. The axons of some of the second neurons pass from the sympathetic ganglion back to the spinal nerve and through it to innervate sweat glands, the muscles that make the hair stand erect, and the muscles in the walls of blood vessels. The axons of other second neurons pass from the sympathetic ganglia of the neck up to the salivary glands and the iris of the eye. The sensory neurons (fibers) of the sympathetic system are located within the same nerve trunks as the motor neurons, but enter the spinal cord by way of the dorsal root, together with other sensory nerves of the nonautonomic system.

The Parasympathetic System. The parasympathetic system consists of fibers originating in the brain and emerging via the third, seventh, ninth and especially the tenth or vagus nerves, and of fibers originating in the pelvic region of the spinal cord and emerging by way of the spinal nerves in that region (Fig. 204). The vagus nerve arises from the medulla and passes down the neck to the chest and abdomen, innervating the heart, respiratory system and digestive tract as far as the small intestine. The large intestine and the urinary and reproductive systems are innervated by parasympathetic nerves from the pelvic spinal nerves. The iris of the eye, the sublingual and submaxillary glands and the parotid gland are innervated by the third, seventh and ninth cranial nerves, respectively. These nerves all contain the axons of first neurons in the chain; the ganglia of the parasympathetic system are located in or near the organs innervated, so that the axons of the second neurons are all relatively short.

228. THE NERVOUS SYSTEMS OF LOWER ANIMALS

The unicellular animals, such as the ameba and paramecium, have no neurons, since their entire body consists of but one cell. The ameba does very well without any structure for integration, but the paramecium, whose body is covered with thousands of tiny cilia, needs some mechanism to coordinate the beating of these cilia so that it can move. This is accomplished by a system of tiny fibers, called **neuromotor fibers,** which stretch from the anterior end of the animal to all the cilia. These fibers can be severed by a surgical operation performed under the microscope; thereafter the cilia no longer beat in a coordinated fashion, but at random.

The lowest multicellular animals, the sponges, have no nervous system either; specialized nerve cells are first found in the hydra and other coelenterates. The nerve cells of these animals are not separated from each other by synapses, and associated to form a true system, but are arranged either as separate branched cells or as a nerve net, composed of many cells whose branched processes connect, so that an impulse starting in one part of the body can spread in all directions to every part. There is no differentiation of the neurons of a hydra into sensory, connector and motor types, but some of the branches of the nerve net go to receptor cells, while others go to contractile cells. The amount of response in a hydra depends on the strength of the stimulus: a faint needle prick to a tentacle may cause that single tentacle to roll up, but a stronger prick will cause all the tentacles and the trunk to contract as the animal draws itself into a ball.

Most of the invertebrates have a highly centralized nervous system. This is especially true of the arthropods (the insects, spiders, crabs and lobsters), the molluscs (squids and snails), and the annelids, or earthworms. The earthworm has a true nervous system with axons and dendrites arranged in definite nerve cords and fibers. There is a differentiation into central and peripheral systems with separate sensory, connector and motor neurons joined by synapses so that nerve impulses can travel in one direction only. This permits the central nervous system to act as an integrator, selecting certain incoming sensory impulses and passing them on to effectors, while inhibiting or suppressing others. The earthworm has a central nerve cord which runs the entire length of the body, enabling its separate segments to move in a coordinated fashion. The most anterior part of the cord is an enlargement sometimes referred to as the "brain," which sends impulses down the cord to coordinate movements. After removal of this "brain," the animal can move almost as well as before, but it persists in futile efforts to go ahead, instead of turning aside, when it comes to some obstacle. The "brain," therefore, is necessary for adaptive movements. The nervous systems of the other higher invertebrates are rather similar to that of the earthworm. In all, the nerve cord is ventral to the digestive system and is solid.

In the vertebrates the nerve cord is dorsal to the digestive system and is hollow, with a central cavity. The nervous systems of all the vertebrates are fundamentally alike; differences are primarily those of development of the various brain regions, and in the size of the brain relative to the spinal cord.

QUESTIONS

1. What are the primary functions of the nervous system?
2. Differentiate between neurons and nerves.
3. How many kinds of neurons are there, and what functions does each type perform?
4. What is the evidence for the theory that the neurilemma sheath plays some role in the regeneration of cut nerves?
5. Discuss the nature of the nerve impulse.
6. Is the nerve impulse stimulated in the eye by a beam of light the same as that stimulated in the skin by a hot poker?
7. In what ways does transmission at the synapse differ from transmission along the neuron?
8. What are nerve centers and where are they located?
9. Make a diagram of the brain, labeling the principal parts, and list the functions carried on by each.
10. What is a reflex arc?
11. Give an example of a conditioned reflex from your own experience.
12. Explain the term "facilitation."
13. What is the autonomic nervous system? To which of the larger divisions of the entire nervous system does it belong?
14. Pilocarpine is a drug which stimulates the nerve endings of parasympathetic nerves. What effect would you expect this drug to have on the digestive tract, on the iris of the eye, and on the heart rate?
15. Atropine blocks the action of the parasympathetic system and thus produces the equivalent of a stimulation of the sympathetic system. What effects would you expect atropine to have on the digestive tract, on the iris of the eye, and on the heart rate?

SUPPLEMENTARY READING

The role of the nervous system in maintaining constant body conditions is discussed in Chapters 15 and 16 of W. B. Cannon's *The Wisdom of the Body, and* in more detail in Sherrington's *Integrative Action of the Nervous System.* A comprehensive account of Pavlov's experiments is given in his *Conditioned Reflexes.* The development of psychology and psychiatry is presented in Zilboorg's *A History of Medical Psychology.* Some of the classic experiments in the field are described in *Great Experiments in Psychology* by H. E. Garrett. The details of the structure of the brain and nerves are found in Ranson and Clark, *The Anatomy of the Nervous System.* Stanley Cobb's *Foundations of Neuropsychiatry* is an interesting discussion, at not too technical a level, of the anatomic and physiologic bases of normal and abnormal mental activity. The analogy between the regulation of body activities by the nervous system and the regulation of industrial processes by "feed-back" mechanisms is developed by Norbert Wiener in his *Cybernetics.* The biologic basis of memory is discussed by J. Z. Young in *Doubt and Certainty in Science.* Some classic experiments in neurophysiology are given in Chapters 6 and 7 of Fulton's *Selected Readings.*

Chapter 25

The Sense Organs

THE PROTOPLASM of the ameba and other single-celled animals is sensitive to many different stimuli, as evidenced by the fact that the protozoan will move away from bright lights, certain chemicals, electric currents and so forth. But on a higher, more complex level of existence, where the activities of searching for food, attracting a mate, and escaping enemies are correspondingly more complex and hazardous, the animal needs specialized cells sensitive to one or a few types of stimuli, to help him in his struggle for life. In the evolutionary process such receptors have been developed; we call them **sense organs.** The receptors in these sense organs are remarkably sensitive to the appropriate stimulus; the eye is stimulated by an extremely faint beam of light, whereas a strong light is required to stimulate the optic nerve directly. The negligible amount of vinegar that can be tasted, or the amount of vanilla that can be smelled, would have no effect at all if applied to the nerve fibers directly.

Traditionally, man is supposed to have five "senses" (i.e., of touch, smell, taste, sight and hearing), but this is misleading, for some of the five can be divided into several completely different senses. Thus, touch, pain, pressure, cold and heat are all included under the sense of touch. In addition, there are more vague and generalized, but none the less important, senses for determining internal states of the body. The receptors for such senses are located in the viscera, the throat, and other places.

The sense organs of some animals are sensitive to stimuli that are quite ineffective in man. Dogs and cats can hear high-pitched whistles inaudible to us. Bats in flight make very high-pitched, short noises. They are guided in flight by the echoes that rebound from objects in their path; they even catch insects by the echoes from their small prey!

Stimulation of any sense organ initiates what might be considered a coded message, transmitted by the nerve fibers and decoded in the brain. In the transmission of the impulse there may be (1) differences in the number of fibers transmitting, (2) differences in which particular fibers are transmitting, (3) differences in the total number of impulses passing over a given fiber, (4) differences in the frequency of the impulses passing over a given fiber, and (5) differences in the time relations between impulses in different fibers. These, then, are the possibilities in the "code" sent along the nerve fiber; how the sense organ initiates different codes and how the brain analyzes and interprets them to produce various sensations are still unknown.

229. THE STIMULUS-RECEIVING PROCESS

For all types of sense organs the actual excitation of the sensitive cells is either mechanical or chemical. So far as we know, no organism has receptors sensitive to cosmic or radio waves or to electric currents. The stimulation of the touch and pressure receptors of the skin depends on the mechanical stresses transmitted through the surrounding capsules; the proprioceptors (kinesthetic receptors) respond to the mechanical pressure exerted on them when the surrounding muscle cells or tendon fibers are stretched or compressed, and the sensitive cells in the organs of hearing and balance are believed to be excited by ripples or waves in the fluids bathing them. In contrast, the olfactory cells of the nose and the taste buds of the tongue are stimulated chemically by the molecules that come in contact with them. The stimulation of the pain receptors probably depends upon the action of chemical substances released by damaged cells, since they respond to strong stimuli of any sort. The receptors for heat and cold respond to chemical changes induced in them by changes of temperature, and the cells of the retina respond to the chemical reactions that occur when light falls on them.

230. THE PERCEPTION OF SENSATIONS

No organism receives a perfectly true and complete picture of his world, or even of himself. In the first place, there are undoubtedly many types of phenomena, a few of which have already been mentioned, for which he has no effective receptors. In the second place, even the stimuli for which he is sensitive undergo some "distortion" in the act of being apprehended. How much of our perception is a product of the sense organs and the brain, and how much is representative of the "real" world, is an interesting philosophical question which we cannot go into here. Most biologists assume that our perceptions do give us a fairly valid picture of our environment.

Whatever awareness we do have is made possible by a "hook-up" between the sense organs and the brain. It is most important to remember that *all nerve impulses are qualitatively the same*. This means that the impulse instigated by the ringing of a bell is exactly the same as the impulse initiated by the pressure of a pin against the skin, or any other of the countless possible impulses emitted by any sense organ. This being so, it follows that the qualitative differentiation of stimuli must depend upon the sense organ itself, upon the brain, or both. In fact, it depends on both. Primarily, our ability to discriminate red from green, hot from cold, or red from cold, is due to the fact that particular sense organs and their individual sensitive cells are connected to particular parts of the brain.

Since only those nerve impulses which reach the brain can result in sensations, any blocking of the impulse along the nerve fibers by an anesthetic has the same effect as removing the original stimulus entirely. The sense organs, of course, will continue to initiate impulses, which can be detected by the proper electrical apparatus, but the anesthetic prevents them from reaching their destination.

231. THE LOCALIZATION OF STIMULI

The localization of stimuli, like the ability to distinguish different qualities of stimuli, depends on the specific connections between the sense organ and the brain. The infant early learns that two sensations, though identical in quality, are associated with stimuli coming from different sides of the body. He is able to localize a bright light in his left eye or a sharp stab in his right side, simply because the nerve fibers from the left eye and the right side have different destinations in the brain from those of the right eye and the left side. It is more difficult to localize odors and still more difficult to localize sounds, since the stimuli for these are more general and diffuse.

The importance of the brain in making sensations possible is emphasized by the phenomenon of "misreference," which occurs occasionally in connection with pain. A well-known example of this is the experience of people suffering from gallbladder diseases, who complain of pain in the right shoulder. Actually, of course, the stimulus originates in the gallbladder, but in some as yet not understood way, the nerve impulse ends up in the same part of the brain as impulses genuinely originating in the right shoulder.

In contrast to the quality and localization of stimuli, which depend to such a large extent upon the parts of the brain involved, the intensity of sensation depends almost entirely upon the sense organ initiating the original impulse. For almost all sensations, not one but many impulses are sent along the nerve fiber, and it is the number of these which determines how strong the sensation will be. An intense blow will initiate more impulses per second than a light tap, and the greater the area covered by the blow, the stronger the ultimate sensation, since more receptors will be stimulated and more impulses sent.

232. THE TACTILE SENSES

The skin contains several different types of simple sense organs, some of them merely the free ends of dendrites, while others consist of the ends of dendrites enclosed in special cellular capsules (Fig. 205). By making a careful survey of a small area of skin, point by point, using a stiff bristle to test for touch, a hot or cold metal stylus to test for temperature, and a needle to test for pain, it has been found that receptors for each of these sensations are located at different spots. By comparing the distribution of the different types of sense organs and the types of sensations produced, it has been found that the free nerve endings are responsible for pain perception, and that particular types of encapsulated nerve endings are responsible for the others.

Kinesthesis. Each muscle, tendon and joint is equipped with nerve endings, called **proprioceptors,** which resemble certain sense organs of the skin. These are sensitive to changes in the tension of the muscle or tendon, and initiate impulses to the brain which are responsible for our awareness of the position and movement of the various parts of the body, the sense referred to as **kinesthesis.** The existence of this sense enables us, even with our eyes closed, to perform manual acts, such as dressing or tying knots. Impulses from the proprioceptors are also extremely important in insuring the harmonious contraction of different muscles involved in a single movement; without them, complicated, skillful acts

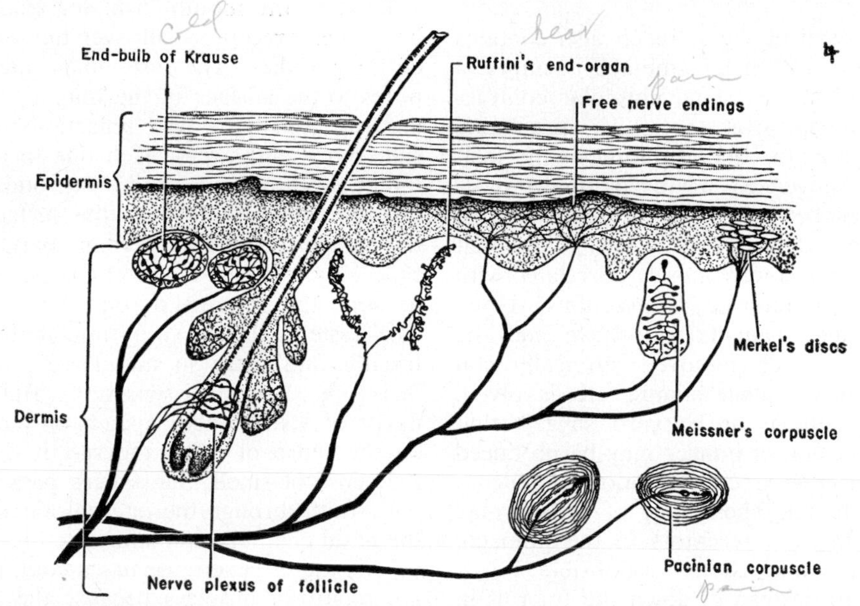

Figure 205. Diagrammatic section through the skin showing the types of sense organs present. The sense organs respond to the following stimuli: cold—end-bulbs of Krause; warmth—Ruffini's end-organs; touch—Meissner's corpuscles and Merkel's discs; deep pressure—Pacinian corpuscles, and pain—free nerve endings.

would be impossible. These impulses are also important in the maintenance of balance. The proprioceptors are probably more numerous and more continuously active than any of the other senses, although we are less aware of them than of any others; in fact, the existence of this sense was discovered only about a hundred years ago. One obtains some idea of what life without proprioceptors would be like when a leg or arm "goes to sleep" —the feeling of numbness results from the lack of proprioceptive impulses.

Visceral Sensitivity. The sensations associated with the receptors located in the internal organs, which are extremely important in regulating the activities of the viscera, seldom reach the level of consciousness. They bring about reflex control of the functioning of the internal organs by way of reflex centers in the medulla, midbrain or thalamus. A few of the impulses from these receptors do get to the cerebrum, however, and give rise to sensations such as thirst, hunger and nausea. The sensation of thirst originates in receptors in the lining of the throat; when this lining becomes dry, the receptors send impulses to the brain which we interpret as a feeling described as "being thirsty."

The wall of the stomach also contains receptors. When the stomach is empty, a series of strong, slow, muscular contractions sweeps over the walls, stimulating these receptors and resulting in the feeling of hunger. By having a subject swallow a rubber balloon connected by a tube to a recording device, it was found that hunger pangs were closely correlated with these characteristic contractions. However, since patients who have had the entire stomach removed surgically still feel hunger, other stimuli are involved. Recent studies at Harvard suggest that the sensation of hunger may be produced by a decreased concentration of glucose in the blood. The feeling of nausea may originate from receptors in the stomach, but the contractions responsible for it move up, instead of down the tract as in normal peristalsis.

Still other receptors lie in the mesenteries holding the internal organs in place. When these mesenteries become inflamed or stretched by unusual movements of the organs to which they are attached, sensations of pain result. There are other nerve endings for pain in the linings of the organs themselves.

The sense of fullness and the urge to defecate and urinate depend upon receptors in the walls of the rectum and urinary bladder, respectively, which are stimulated by the tension resulting when those hollow organs are distended by their contents.

Many other, less well-defined visceral sensations are felt during sexual activity, illness or emotional crises.

233. THE CHEMICAL SENSES: TASTE AND SMELL

The sensations of taste and smell result from the stimulation of **chemoreceptor cells** in the tongue and nose by specific chemical substances. Embedded in the mucous membranes of the tongue and soft palate are special sense organs known as **taste buds,** each of which consists of a few sensitive cells connected with sensory neurons and surrounded by supporting cells (Fig. 206 *B*). In other vertebrates the taste buds are not restricted to the mouth; fish, for example, have chemoreceptors all over the outside of their bodies. The taste buds open by pores to the surface of the tongue. There are only four basic taste sensations: sour, salt, bitter and sweet, each due to a different kind of taste bud. The buds are distributed unevenly over the surface of the tongue, so that certain parts are especially sensitive to sweet things, others to sour things, and so on (Fig. 206). The taste buds can be stimulated electrically, and when so stimulated produce their characteristic sensation. But the flavor of a substance depends only partly on the sense of taste; the rest is due to the sense of smell. Substances pass from the mouth through the internal nares into the nasal chamber and stimulate the sense organs there. When one has a cold, foods are relatively tasteless because the sense of smell is partly or wholly lost.

There are inherited differences in the sense of taste. Some people find that the chemical phenylthiocarbamide is bitter; others find it tasteless. Ability to taste the chemical is inherited through a single pair of genes.

The sense organs of smell are located in the epithelial lining of the upper part of the nasal cavity in a region not ordinarily washed by the incoming air. Particles entering the nostrils reach them by diffusion and dissolve in the mucus covering the sensitive cells. The sensory cells for smell occur singly and are distinguished from ordinary epithelial cells by hairs which project into the mucous layer (Fig. 206 *A*).

In contrast to the sensations of taste, the various odors cannot be classified into definite types; each substance has its own distinctive smell. The olfactory organs respond to remarkably small amounts of a substance. The synthetic substitute for the odor of violets, ionone, can be detected by most people when it is present to the extent of one part in more than thirty billion parts of air. The sense of smell is rapidly fatigued, and air originally having a powerful stimulus may seem odorless after a few minutes. This fatigue is specific for the particular substance producing it; thus receptors that have become insensitive to one substance will react to another quite normally. This suggests that there are many different kinds of sense cells, each specific for a particular chemical. Some people either completely lack a sense of smell, or are able to smell some substances, but not others.

234. VISION

Light-sensitive cells exist in almost all living matter; even protozoa which lack any special organ respond to changes in light intensity, usually moving away from the source of light. Most plants orient their leaves and flowers toward the sun, although they have no special light-sensitive structures. In most of the higher animals this light-sensitivity is localized in certain cells and is highly developed. The human eye is an excellent example of an extremely sensitive, specialized organ for perceiving light. Selig Hecht and his collaborators at Columbia University have shown that a well dark-adapted eye can

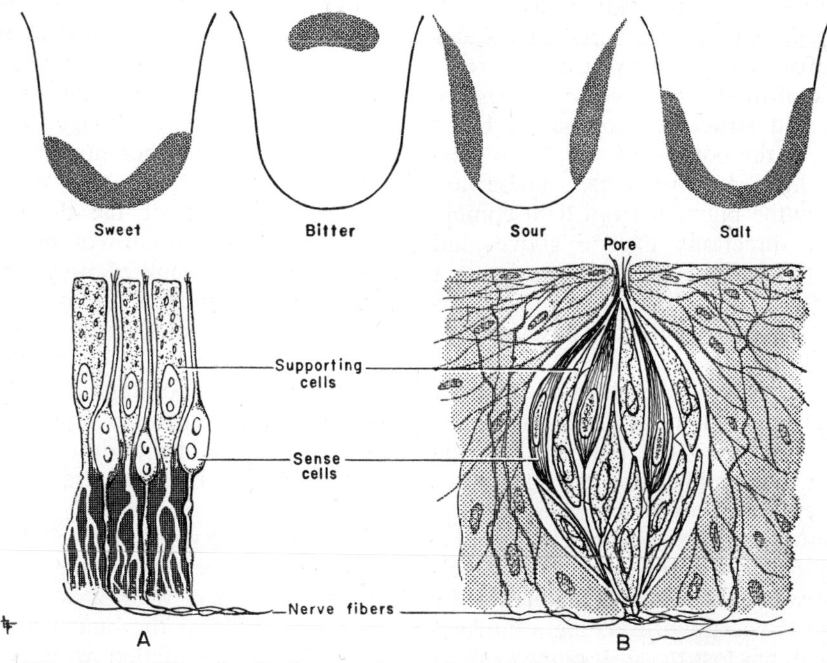

Figure 206. Above, The distribution on the surface of the tongue of taste buds sensitive to sweet, bitter, sour and salt. *Below, A,* Cells of the olfactory epithelium of the human nose. *B,* Cells of a taste bud in the epithelium of the tongue. The free ends of the sensory cells are hairlike projections which are stimulated by chemicals to produce the sensation of odors and tastes.

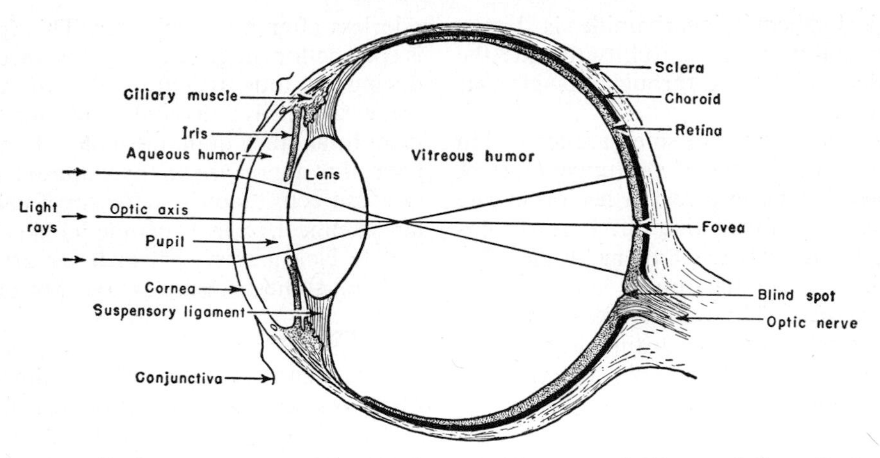

Figure 207. Horizontal section of the human eye. (Hunter and Hunter: College Zoology.)

detect as little as 6 to 10 quanta of light. Just as matter consists of tiny particles known as atoms, light consists of small corpuscles called photons, and, by definition, the energy of 1 photon is 1 quantum. The light reaching the eye from a candle fourteen miles away is just at the limit of visibility of a normal eye and is about 6 or 7 quanta of light.

Some protozoa have "eye spots" which are more sensitive to light than the rest of the cell, but the most primitive light-sensitive organs in the evolutionary scale are those of the flatworms. Their eyes are bowl-shaped structures containing black pigment, at the bottom of which are clusters of light-sensitive cells. These are shaded by the pigment from light coming from all directions except above and slightly to the front. This arrangement enables the planarian to detect the direction of the source of light. These eyes, of course, do not register images as do our eyes, but are merely sensitive to different light intensities. Planaria have other light-sensitive cells over the entire body surface, for they still react to light after their eyes have been removed, though more slowly and less accurately.

Many marine worms have well-developed eyes and other sense organs on the head, but the earthworm, being a burrowing animal, has lost these. It emerges from its burrow at night, however, and the light-sensitive cells on its dorsal surface enable it to recognize the coming of dawn. In addition, the earthworm has cells sensitive to touch and temperature, and to substances placed on its skin and in its mouth and pharynx. Earthworms have definite tastes in food, preferring carrot leaves to celery, and celery to cabbage.

The eyes of insects and crabs are "mosaic" eyes, quite different from the camera eyes of vertebrates. Mosaic eyes are composed of many, sometimes thousands, of visual units, each with a small bundle of light-sensitive cells and a fixed, immovable lens. Such an eye does not give a single, sharp picture but produces a mosaic, to which each unit of the eye contributes a separate image. The effect is imagined to be rather similar to a poor newspaper photograph. Arthropods, then, are not so aware of the details in an image as of the *motion* of objects. Because any movement of prey or enemy is immediately picked up by one of the eye units, this type of organ is peculiarly suited to the arthropod's way of life. Among the invertebrates, only the squid and octopus have camera eyes with lenses that can focus for far and near vision and reproduce reasonably accurate images of the animals' environment.

The Human Eye. The squid or octopus eye is rather like a simple Brownie camera, equipped with slow, black and white film, whereas the human eye is like a de luxe Leica loaded with extremely sensitive color film.

The analogy between the human eye and a camera is complete: the eye (Fig. 207) has a **lens** which can be focused for different distances, a diaphragm (the **iris**) which regulates the size of the light opening (the **pupil**), and a light-sensitive **retina** located at the rear of the eye, corresponding to the film of the camera. Next to the retina is a sheet of cells filled with black pigment which absorbs extra light and prevents internally reflected light from blurring the image (cameras are also painted black on the inside). This sheet, called the **choroid coat,** also contains the blood vessels which nourish the retina.

The outer coat of the eyeball, called the **sclera,** is a tough, opaque, curved sheet of connective tissue which protects the inner structures and helps maintain the rigidity of the eyeball. On the front surface of the eye this sheet becomes the thinner, transparent **cornea,** through which light enters.

Immediately behind the iris is a transparent, elastic ball, the **lens,** which bends the light rays coming in, bringing them to a focus on the retina. It is aided by the curved surface of the cornea and by the refractive properties of the liquids inside the eyeball. The cavity between the cornea and the lens is filled with a watery substance, the **aqueous humor;** the larger chamber between the lens and the retina is filled with a more viscous fluid, the **vitreous humor;** both fluids are important in maintaining the shape of the eyeball. They are secreted by the **ciliary body,** a doughnut-shaped structure which attaches the ligament holding the lens to the eyeball.

The eye accommodates, or changes focus for near or far vision, by changing the curvature of the lens. This is made possible by the stretching and relaxing of the lens by the **ciliary ligament,** which attaches the lens to the ciliary body. Because of the pressure of the fluids within, the eyeball is under tension, which is transmitted by the ciliary ligament to the lens. When the lens is stretched, it flattens and focuses the eye for far vision, the condition of the eye at rest. Just in front of the ciliary body, and attached to the ciliary ligament, are ciliary muscles which, when contracted, take up the strain on the ligament and lens, leaving the latter free to assume the more spherical shape for near vision.

As people grow older, the lens becomes less elastic and thereby less able to accommodate for near vision. When this occurs, spectacles with one part ground for far vision and one for near vision are worn to accomplish what the eye can no longer do.

The amount of light entering the eye is regulated by the **iris,** a ring of muscle which appears as blue, green or brown, depending on the amount and nature of pigment present. The structure is composed of two sets of muscle fibers, one arranged circularly, which contracts to decrease the size of the pupil, and one arranged radially, which contracts to increase the size of the pupil. The response of these muscles to changes in light intensity is not instantaneous, but requires ten to thirty seconds; thus when one steps from a light to a dark area, some time is needed for the eyes to adapt to the dark, and when one steps from a dark room to a brightly lighted street, the eyes are dazzled until the size of the pupil is decreased.

Each eye has six muscles stretching from the surface of the eyeball to various points in the bony socket which enable the eye as a whole to move and be oriented in a given direction. These muscles are innervated in such a way that the eyes normally move together and focus on the same area.

The only part of the human eye which is light-sensitive is the **retina,** a hemisphere made up of an abundance of receptor cells, called, according to their shape, **rods** and **cones.** There are about 125,000,000 rods and 6,500,000 cones. In addition, the retina contains many sensory and connector neurons and their axons. Curiously enough, the sensitive cells are at the *back* of the retina; to reach them, light must pass through several layers of neurons. There is no logical basis for this, but as the eye develops as an outgrowth of the brain, it folds in such a way that the sensitive cells eventually lie on the farthermost side of the retina (Fig. 208). At a point in the back of the eye, the individual axons of the sensory

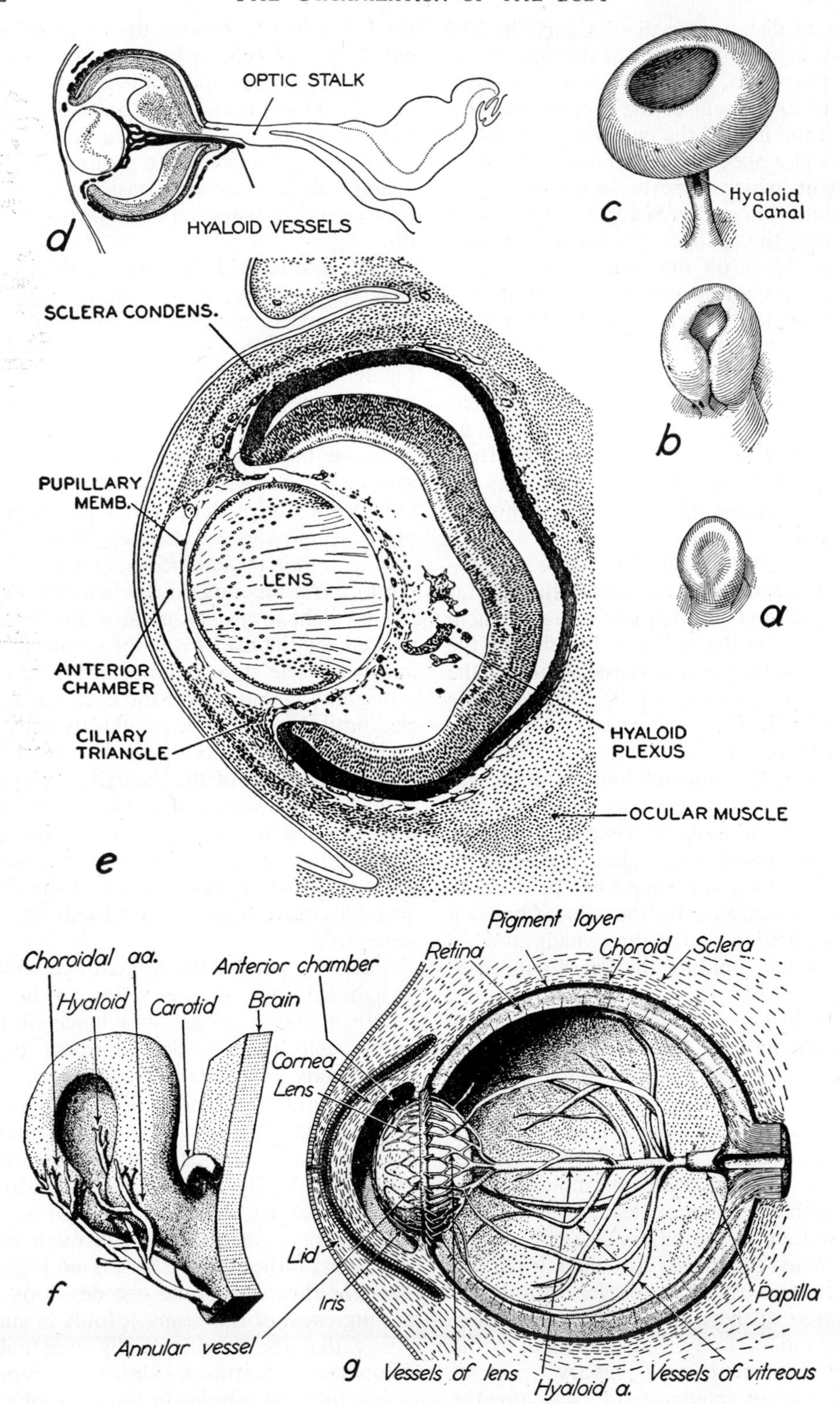

Figure 208. Eye development in man. *a–c,* Development of the optic cup. *d,* Section through eye and optic stalk; *e,* Late stage. *a–d,* × 25; *e,* × 60 (Streeter 1951, Carnegie Contrib. 230). *f, g,* Vascularization of the eye (after Mann from Walls 1942). (Witschi: Development of Vertebrates.)

nerves unite to form the optic nerve and pass through the eyeball. Here there are no rods and cones. This area is called the "blind spot," since images falling on it cannot be perceived. Its existence can be demonstrated by closing the left eye and focusing the right one on the + in Figure 209. Starting with the page about 5 inches from the eye, move it away until the circle disappears. At that position the image of the circle is falling on the blind spot and so is not perceived.

In the center of the retina, directly in line with the center of the cornea and lens, is the region of keenest vision, a small depressed area called the **fovea.** Here the light-sensitive cones, responsible for bright light vision, for the perception of detail and for color vision, are concentrated.

The other light-sensitive cells, the rods, are more numerous in the periphery of the retina, away from the fovea. These function in twilight or dim light and are insensitive to colors. One is not ordinarily aware that color can be perceived only in those objects which are more or less directly in front of the eyes, but the fact can be demonstrated by a simple experiment. Close one eye and focus the other on some point straight ahead. As a colored object is gradually brought into view from the side, you will be aware of its presence and of its size and shape before you are aware of its color. Only when the object is brought closer to the direct line of vision, so that its image falls on a part of the retina containing cones, can its color be determined. The rods are actually more sensitive in dim light than are the cones. Since the rods are located not in the center, but in the periphery of the retina, it is a curious fact that you can see an object better in dim light if you do not look at it directly (for then its image will fall on the cones in the center of the retina), but slightly to one side of it, so that its image falls on the rods in the periphery of the retina.

Every rod contains a light-sensitive chemical known as **visual purple,** a compound of vitamin A and a protein. When light falls on the rods, the light energy causes a chemical breakdown of the visual purple to a different chemical known as visual yellow. Each quantum of light falling on the retina causes the breakdown of one molecule of visual purple, which stimulates one rod cell. For a light to be perceived, about six to ten rods have to be excited at once. In the dark, the visual yellow is resynthesized into visual purple, and the rod is ready to be stimulated again. Since visual purple contains vitamin A, a certain amount of that substance is necessary for this synthesis; hence people whose diets are deficient in vitamin A are unable to resynthesize visual purple at an adequate rate and become unable to see in dim light. After thirty to sixty minutes in the dark, the eye becomes adapted and is much more sensitive—perhaps a thousand times more so—than in the light. Although the rods do not function in bright light, some of the visual purple is constantly being broken down as long as the eyes are exposed to it; but when one remains in the dark, all the visual yellow is converted to visual purple, and the eye attains its maximum sensitivity. For this reason, aviators frequently remain in the dark or wear thick red glasses to dark-adapt their eyes before flying at night.

Less is understood about the chemistry of the cones, but they are known to contain a substance called **visual violet,** also made of vitamin A plus a protein. A number of theories, none of which is completely satisfactory, have been proposed to explain color vision. It seems likely that there are at least three different kinds of cones, sensitive to three different light

Figure 209. Demonstration of the blind spot on the retina. See text for details.

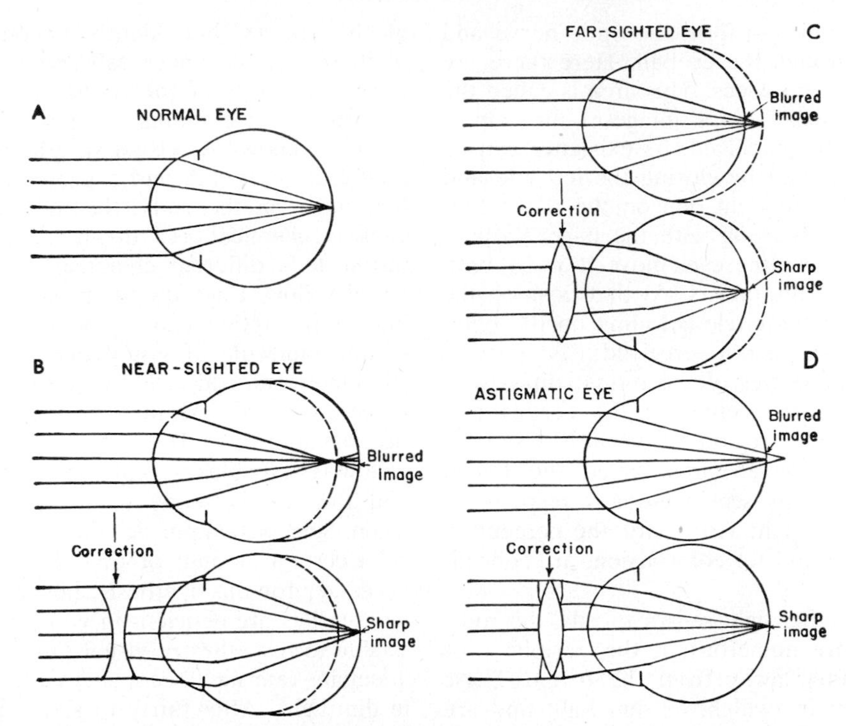

Figure 210. Diagram illustrating common abnormalities of the eye. *A,* Normal eye, in which parallel light rays coming from a point in space are focused as a point on the retina. *B,* Near-sighted eye, in which the eyeball is elongated so that parallel light rays are brought to a focus in front of the retina (on dotted line, which represents the position of the retina in the normal eye) and so form a blurred image on the retina. This situation is corrected by placing a concave lens in front of the eye. This diverges the light rays, making it possible for the eye to focus these rays on the retina. *C,* Far-sighted eye, in which the eyeball is shortened and light rays are focused behind the retina. A convex lens converges the light rays so that the eye focuses them on the retina. *D,* Astigmatic eye, in which light rays passing through one part of the eye are focused on the retina, while light rays passing through another area of the lens are not focused on the retina, owing to unequal curvature of the lens or cornea. A cylindrical lens will correct this by bending light rays going through only certain parts of the eye. (Hunter and Hunter: College Zoology.)

wavelengths, which produce the sensations of redness, blueness and greenness. When all are stimulated equally, the sensation of whiteness results. Impressions of intermediate colors, such as orange, result from the simultaneous but unequal stimulation of two or more types.

Defects in Vision. The defects of the eye most common in man are near-sightedness (myopia), far-sightedness (hypermetropia) and astigmatism. In the normal eye (Fig. 210 *A*) the shape of the eyeball is such that the retina is the proper distance behind the lens for the light rays to converge in the fovea. In a near-sighted eye (Fig. 210 *B*) the eyeball is too long, and the retina is too far from the lens, so that the light rays converge at a point in front of the retina, and are again diverging when they reach it, resulting in a blurred image. In a far-sighted eye (Fig. 210 *C*), the eyeball is too short and the retina too close to the lens, causing the light rays to strike the retina before they have converged, again resulting in a blurred image. Concave lenses correct for the near-sighted condition by bringing the light rays to a focus at a point farther back, and convex lenses correct for the far-sighted condition by causing the light rays to converge farther forward.

Astigmatism is a condition in which the cornea is curved unequally in different planes, so that the light rays in one plane are focused at a different point from those in another plane. To correct for astigmatism, lenses must be ground unequally to compensate for the unequal curvature of the cornea.

In old age the lens may lose its transparency, become opaque, and interfere with the transmission of light to the retina, causing blindness. The only cure for this is surgical removal of the lens. This restores sight, but removes the ability to focus, so that a lensless person must wear special spectacles as a substitute for the lens.

The position of the eyes in the head of man and certain other higher vertebrates permits both of them to be focused on the same object. This **binocular vision** is an important factor in judging distance and depth. To focus on a near object, the eyes must converge (become slightly cross-eyed). In the eye muscles causing this convergence are proprioceptors stimulated by this contraction to send impulses to the brain; hence part of our judgment of distance and depth depends upon impulses which result when the sensory fibers in those muscles are stimulated. In addition, the eyes, being a little over 2 inches apart, see things from slightly different angles and thus get slightly different views of a close object.

235. THE EAR

The organs of two different senses, hearing and equilibrium, are located in the ear. These organs are buried deep in the bone of the skull, and a number of accessory structures are needed to transmit sound waves from the outside to the deep-lying sensory cells. The ear may be divided into outer, middle and inner parts; by reference to the diagram (Fig. 211) the path of the sound waves may be followed.

The outer ear consists of two parts, the skin-covered cartilaginous flap or **pinna,** and the **auditory canal** leading from it to the middle ear.

The pinnas, or visible ears, are of some slight use in man for directing sound waves into the canal, but in other animals such as the cat, the larger, movable pinnas are very important. At the junction of the auditory canal and the middle ear is stretched a thin, connective tissue membrane, the **ear drum,** which the sound waves set vibrating.

The middle ear is a small chamber containing three tiny bones connected in a

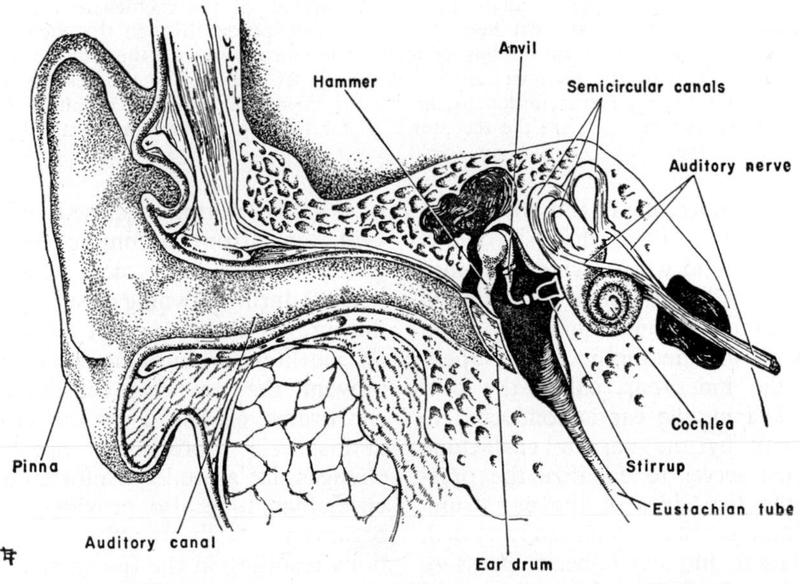

Figure 211. Structure of the human right ear, cut open to show schematically the outer, middle and inner ear.

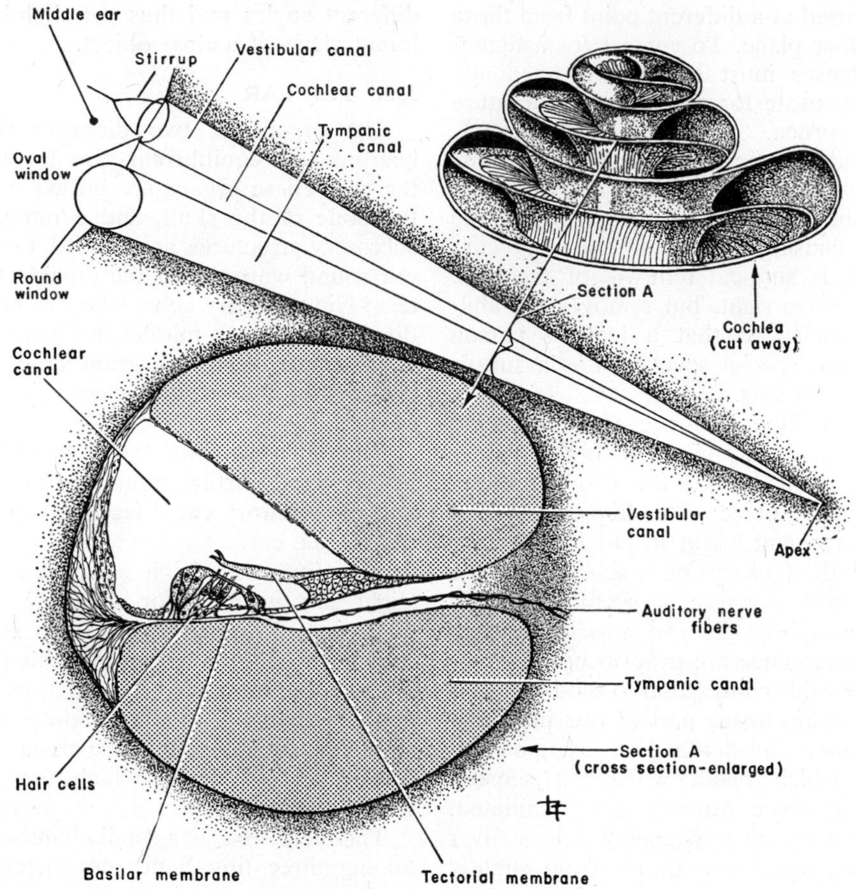

Figure 212. Upper right, The coiled cochlea shown dissected out of the skull and cut open to show the vestibular and tympanic canals. *Center,* A diagram of the cochlea as though it were uncoiled and drawn out in a straight line. *Below,* A cross section through the cochlea at *A* to show the organ of Corti resting on the basilar membrane and covered by the tectorial membrane. Vibrations transmitted by the hammer, anvil and stirrup set the fluid in the vestibular canal in motion; these vibrations are transmitted to the basilar membrane and the organ of Corti. The hair cells of the organ of Corti are the receptor cells for hearing and are innervated by branches of the auditory nerve.

series, the **hammer, anvil** and **stirrup** (so called because of their shapes), which transmit the sound waves across the middle ear cavity. The hammer is in contact with the ear drum, and the stirrup is in contact with the membrane of the opening into the inner ear, called the **oval window.** The middle ear is connected to the pharynx by the narrow **eustachian tube,** which serves to equalize the pressure on the two sides of the ear drum. If the middle ear were completely closed, any variation in atmospheric pressure would cause a pronounced and painful bulging or caving in of the ear drum. At the pharyngeal end of the eustachian tube is a valve, normally closed, which prevents one from becoming unpleasantly aware of his own voice. This valve is opened during yawning or swallowing, and during an abrupt ascent or descent in an elevator or airplane such acts help prevent the cracking sensation of the ear drums produced by the changes in atmospheric pressure accompanying changes in altitude. Unfortunately, the eustachian tube also provides a path for organisms which sometimes cause infections resulting in the fusing of the middle ear bones and loss of hearing.

The inner ear consists of a complicated group of interconnected canals and sacs, often referred to, most appropriately, as the **labyrinth.** The part of the labyrinth concerned with hearing is a spirally coiled tube of two and a half turns, resembling a snail's shell, called the **cochlea.** If the cochlea were uncoiled, as in Figure 212 it could be seen to consist of three canals separated from each other by thin membranes and coming almost to a point at the apex. The oval window is connected to the base of one of these tubes, the **vestibular canal.** At the base of the **tympanic canal** is another opening covered by a membrane, the **round window,** which also leads to the middle ear. These two canals are connected with each other at the apex of the cochlea and are filled with a fluid known as the **perilymph.** Between the two lies a third, the **cochlear canal,** filled with a fluid called **endolymph** and containing the actual organ of hearing, the **organ of Corti.** This structure consists of five rows of cells with projecting hairs which extend the entire length of the coiled cochlea. Each organ of Corti contains about 24,000 of them. These cells rest upon the **basilar membrane,** which separates the cochlear from the tympanic canal. Overhanging the hair cells is another membrane, the **roof** (or tectorial) **membrane,** attached along one edge to the membrane on which the hair cells rest, and with its other edge free. The hair cells initiate impulses in the fibers of the auditory nerve.

For a sound to be heard, sound waves must first pass down the auditory canal and set the ear drum vibrating. These vibrations are transmitted across the middle ear by the hammer, anvil and stirrup, which are so arranged that they decrease the amplitude, but increase the force, of the vibrations. The stirrup transmits the vibrations, via the oval window, to the fluid in the vestibular canal. Since fluids are incompressible, the oval window could not cause a movement of the fluid in the vestibular canal unless there were an escape valve for the pressure. This is provided by the round window at the end of the tympanic canal. The pressure wave presses upon the membranes separating the three canals, is transmitted to the tympanic canal, and causes a bulging of the round window. The movements of the basilar membrane produced by these pulsations are believed to rub the hair cells of the organ of Corti against the overlying roof membrane, thus stimulating them and initiating nerve impulses in the dendrites of the auditory nerve, lying at the base of each hair cell.

Since sounds differ in pitch, intensity and quality, any theory of hearing must account for the ability to discriminate such differences. Microscopic examination of the organ of Corti reveals that the fibers of the basilar membrane are of different lengths along the coiled cochlea, being longer at the apex and shorter at the base of the coil, thus resembling a harp or piano. It is believed that specific fibers in the basilar membrane are tuned to and set into vibration by sounds of specific pitches. When the ear is subjected to intense, continuous sound, the organ of Corti is injured. This was demonstrated by an experiment on guinea pigs in which the animals were exposed to continuous pure tones for a period of several weeks. When their cochleas were examined microscopically after death, it was found that the guinea pigs subjected to high-pitched tones suffered injury only in the lower part of the cochlea, while those subjected to low-pitched tones suffered injury only in the upper part of the cochlea. Workers, such as boilermakers, subjected to loud, high-pitched noises over a period of years, frequently become deaf to high tones, because of injury of the cells toward the base of the organ of Corti. Recent research indicates that the nerve impulses produced by particular sounds have the same frequency as those sounds, so that the brain may recognize particular pitches by the frequency of the nerve impulses reaching it, as well as by the identity of the nerve fibers conducting the impulses.

The auditory nerves send two kinds of nerve impulses: ordinary nerve impulses like those of any other nerve, and a different type called **microphonic.** The energy for the latter is not derived from the metabolism of the nerve fiber, as is the energy for the former; instead, the cochlea acts as a microphone to convert the mechanical energy of the sound vibrations into electrical energy. For this reason the wave form of the electrical potential from the cochlea closely resembles that of the stimulating sound wave. In fact, Wever and Bray placed electrodes on the auditory nerve of a decerebrated cat, and then, listening with a telephone receiver to the amplified signals of the nerve, were able to hear not only musical tones, but actual words spoken to the cat. The hair cells of the organ of Corti are believed to be responsible for this conversion of mechanical to electrical energy, the upper and lower ends of the cochlea responding to low and high tones, respectively. It is still a disputed question, however, whether these microphonics have anything to do with the actual sensation of hearing in the normal animal.

The intensity or loudness of a tone depends upon the number of hair cells stimulated. A gentle sound wave does not produce as intense a sensation as a stronger one because it fails to create as great a vibration of the basilar membrane and hence of the individual hair cells.

Variations in the quality of sound, such as are produced when an oboe, a cornet and a violin play the same note, depend upon the number and kinds of **overtones** or **harmonics** present, which stimulate different hair cells in addition to the main stimulation common to all three; thus, differences in quality are recognized by the *pattern* of the hair cells stimulated. Careful histologic work has shown that the nerve fibers from each particular part of the cochlea are connected to particular parts of the auditory area of the brain, so that certain brain cells are responsible for the perception of sensations of high tones, others for low tones.

The human ear is equipped to register sounds of frequencies between about 20 and 20,000 cycles per second, although there are great individual differences. Some animals—dogs, for example—can hear sounds of much higher frequencies. The human ear is more sensitive to sounds between 1000 and 2000 cycles per second than to higher or lower ones. Within this range the ear is extremely sensitive; in fact, when compared with the power of light waves necessary to produce a sensation, the ear is ten times more sensitive than the eye.

The normal human ear is just about as efficient a hearing device as anything could possibly be, for, like the eye, it has evolved to the point where any further increase in sensitivity would be useless. If it were more sensitive, it would pick up the random movement of the air molecules, which would result in a constant hiss or buzzing. If the eye were more sensitive, a steady light would appear to flicker because the eye would be sensitive to the individual photons (light particles) impinging on it.

There is little fatigue connected with hearing. Even though it is constantly assailed by noises, the ear retains its acuity and fatigue disappears after a few minutes. When one ear is stimulated for some time by a loud noise, the other ear also shows fatigue—loses acuity—indicating, not unexpectedly, that some of the fatigue is in the brain rather than in the ear itself.

Deafness may be caused by injuries or malformations of either the sound-transmitting mechanisms of the outer, middle or inner ears, or of the sound-perceiving mechanism of the latter. The external ear may become obstructed by wax secreted by the glands in its wall; the middle ear bones may become fused after an infection; or, more rarely, the inner ear or auditory nerve may be injured by a local inflammation or the fever accompanying some disease.

Relatively few animals have a sense of hearing. The vertebrate ear began as an organ of equilibrium, the cochlea being a later evolutionary outgrowth of the saccule which reaches full development only in mammals. The human ear is indeed a curious evolutionary hodgepodge: the cells sensitive to sound are apparently adaptations of cells sensitive to the motion of liquids; the middle ear and eustachian tube were originally part of the respiratory apparatus of fish; the stirrup was originally a structure which attached the jaws of primitive fishes to the cranium, and the hammer and anvil are the remnants of the lower and upper jaws, respectively, of our ancestral fish. In the jawless fish ancestral to these the structures were part of the support for the gills. Thus, respiratory organs became, first, eating organs, and then organs for hearing. This is an example of one of the fundamental patterns of evolution—the reshaping of old organs to perform new functions, rather than the setting up of completely new structures.

236. EQUILIBRIUM

Besides the cochlea, the labyrinth of the inner ear consists of two small sacs—the **saccule** and **utricle**—and three **semicircular canals** (Fig. 213). These structures are filled with endolymph and float in a pool of perilymph. Destruction of them causes a considerable loss of the sense of equilibrium, and a pigeon whose organs have been destroyed is unable to fly. In time he can relearn to maintain equilibrium using visual stimuli.

Equilibrium in man depends upon the sense of vision, stimuli from the proprioceptors, and stimuli from cells sensitive to pressure in the soles of the feet, as well as upon stimuli from these organs in the inner ear. In certain types of deafness the equilibrium organs of the inner ear as well as the cochlea are inoperative, yet the sense of equilibrium remains unimpaired.

The utricle and saccule are small, hollow sacs lined with sensitive hair cells and containing small ear stones or **otoliths** made of calcium carbonate. Normally, the pull of gravity causes the otoliths to press against particular hair cells, stimulating them to initiate impulses to the brain via sensory nerve fibers at their bases. When the head is tipped, the otoliths press upon the hairs of other cells and stimulate them.

Many invertebrates such as the crayfish and lobster have similar organs. An ingenious experiment was performed to demonstrate the action of these organs in the crayfish; it depended on the fact that as the crayfish molts—sheds its skin and grows another, larger one—it also develops new organs of equilibrium and supplies them with grains of sand picked up from the environment. By supplying the molting crayfish with particles of iron, the experimenters caused the animals to respond to a magnet. When the magnet was placed directly over the animal, pulling the iron filing against the hair cells on the top of its equilibrium organ, the crayfish thought that "up" was "down" and responded by turning over and swimming on its back.

The labyrinth of each ear has three semicircular canals, each of which consists of a semicircular tube connected at both ends to the utricle. The canals are so arranged that each is at right angles to the

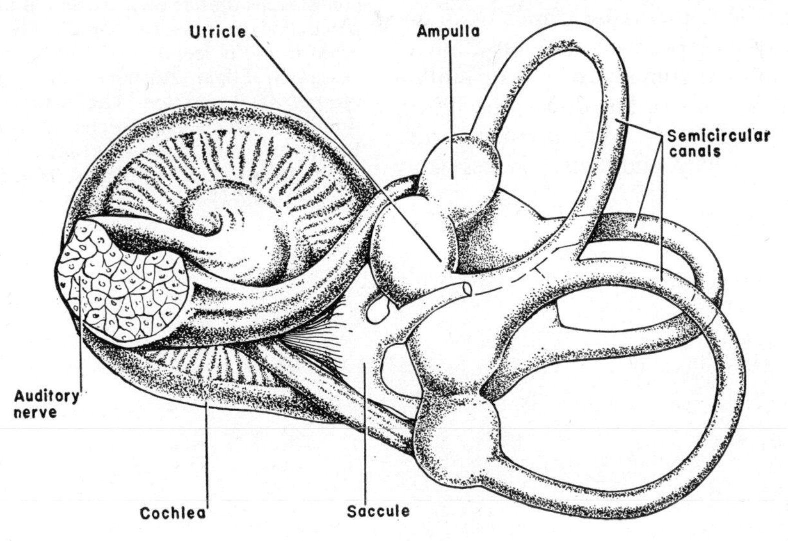

Figure 213. The right semicircular canals and cochlea of an adult man, shown dissected free of surrounding bone and enlarged about five times, seen from the inner and posterior side. Note that the plane of each semicircular canal is perpendicular to the other two.

other two. At one of the openings of each canal into the utricle is a small, bulblike enlargement containing a clump of hair cells similar to those in the utricle and saccule, but lacking otoliths. These cells are stimulated by movements of the fluid (endolymph) which fills the canals. When the head is turned, there is a lag in the movement of the fluid within the canals, so that the hair cells in effect move in relation to the fluid, and are stimulated by its flow. This stimulation produces not only the consciousness of rotation, but certain reflex movements in response to it, movements of the eyes and head in a direction opposite to the original rotation. Since the three canals are located in three different planes, a movement of the head in any direction will stimulate the movement of the fluid in at least one of the canals. By irrigating the canal of the outer ear with warm or cold water, convection currents can be set up causing movements in the fluid of the canals without movement of the head. Sensations of rotation and dizziness result. Man has become used to movements in the horizontal plane, which stimulate certain semicircular canals, but he is unused to vertical movements parallel to the long axis of the body. Such movements—the motion of an elevator or of a ship pitching in a rough sea—stimulate the semicircular canals in an unusual way and produce the sensation of nausea and the vomiting of sea or motion sickness. When one lies down, the movement stimulates the semicircular canals in a different way, and nausea is less likely to occur.

QUESTIONS

1. Define proprioceptors and explain their function.
2. How do anesthetics act to eliminate pain?
3. Draw a diagram of the eye, labeling all parts, and explain the mechanics of the stimulation of the photoreceptors by light.
4. What is the role of visual purple in sight?
5. What produces myopia? Hypermetropia? Astigmatism?
6. Explain how the eye is regulated for far and near vision and for seeing in bright and dim lights.
7. Draw a diagram of the ear, labeling all parts. Explain the mechanics of the stimulation of the sensory cells of the ear by sound waves.
8. What are otoliths and what is their role in maintaining equilibrium?
9. The philosopher Berkeley, as well as many others, believed that there is no material world independent of our sense perceptions, but that what we call perceptions of the external world are really ideas transferred from God's mind to our own. Do you think there are any facts known today which absolutely disprove this theory?
10. Do you think there is any justification for the oft-repeated statement that "one person's taste is as good as another's"? What do you think is or should be the basis for esthetic judgment? Could there be esthetic standards completely independent of the human race? Of any conscious, perceptive organism?

SUPPLEMENTARY READING

Dennis' *Readings in the History of Psychology* contains a number of the original papers on the perception of sensory stimuli. The role of the brain in the perception of sensory stimuli is discussed in Edgar Adrian's *The Physical Background of Perception.* The structure and functioning of the eye is presented in *Vision,* by S. H. Bartley, and the sense of hearing is discussed in detail in Stevens and Davis, *Hearing: Its Psychology and Physiology.*

Chapter 26

The Endocrine System

THE human body is equipped with two main sets of controls which regulate the activities of the whole organism. One of these is the nervous system; the other is a group of hormone-secreting glands, collectively called the **endocrine system.** The basic difference between nervous and hormonal control is in the speed with which they effect changes. By means of the nervous system an organ can undergo extremely rapid adaptation to changes in the environment. In contrast, hormonal regulation is much slower, but longer lasting. You are already familiar with a typical example of its effects, the stimulation of the flow of pancreatic juice from the pancreas by the hormone **secretin,** manufactured by cells in the lining of the intestine (see p. 294).

The endocrine glands are distinguished by the fact that they secrete substances into the blood stream, rather than into a duct leading to the outside of the body or to one of the internal organs. For this reason they are sometimes referred to as ductless glands or glands of internal secretion. Some glands—the thyroid, parathyroids, pituitary and adrenals—function only in the secretion of hormones and are strictly ductless glands. Others, such as the pancreas, ovaries and testes, have both external secretions, via ducts, and internal secretions, carried by the blood stream. The pancreas, which is really two functionally separate organs combined in the same structure, produces digestive enzymes as well as hormones. In some of the lower animals the two parts of the pancreas are anatomically separate.

Hormones cannot be defined as belonging to any particular class of chemicals. All hormones are organic substances, but some are proteins, while others are simpler chemicals, such as amino acids or steroids. The distinguishing factor about them all is that they are secreted by cells in one part of the body and carried by the blood stream to some other part where they are effective in very low concentrations in regulating and coordinating the activities of the cells. All the hormones, and of course the glands secreting them, are necessary for the normal functioning of the body; a deficiency or an oversecretion of any one produces a characteristic pathologic condition. These conditions are frequently called **functional diseases** to distinguish them from deficiency diseases caused by the lack of vitamins, and from infectious diseases, caused by the presence of some infective agent such as bacteria.

Practical knowledge of endocrinology—exemplified by the castration of both men

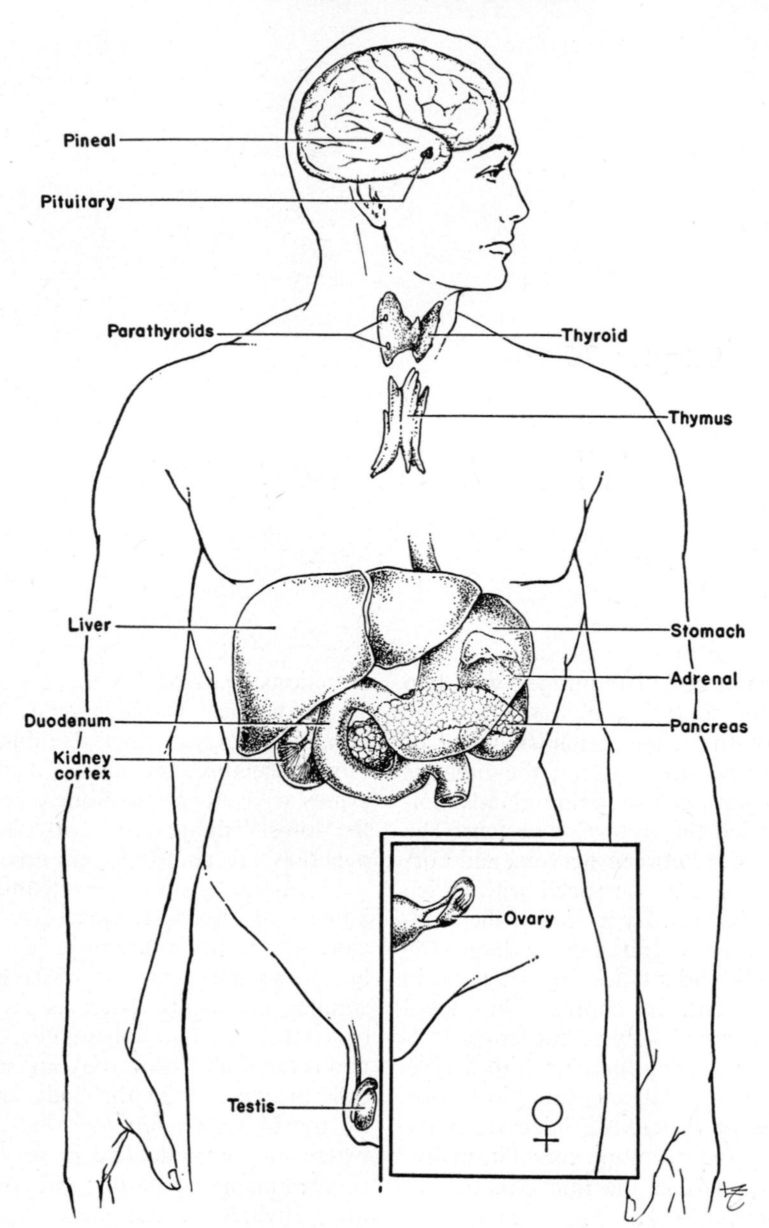

Figure 214. The approximate locations of the endocrine glands in man. Although the pineal body, thymus and stomach are shown, they are not definitely known to secrete hormones.

and animals—has existed for several thousand years, but modern endocrinology usually is said to date from 1849. From the results of experiments in which he transplanted testes from one bird to another, Berthold postulated that these male sex glands secrete some substance carried by the blood stream which is essential for the differentiation of the male secondary sex characters. The first attempt at endocrine therapy was made in 1889, when the French physiologist, Brown-Séquard, injected himself with testicular extracts and claimed that the injections were beneficial and rejuvenating. It is uncertain whether the extracts he used had any effect other than a psychologic one, but his claims stimulated a great deal of research. Since then, experiments on the nature and function of hormones have shown that to a large extent what we are physically, mentally and emotionally depends upon the action of our endocrine glands. Figure 214 shows the location of the glands of the human body known to secrete hormones.

Because extremely small quantities of a hormone produce marked effects, only small amounts are secreted at any one time. This makes the isolation of a pure hormone difficult. In order to obtain a few milligrams of pure female sex hormone, an extract had to be made of over two tons of pig ovaries!

Hormones are gradually inactivated and eliminated from the body, and must be continually replaced by the appropriate endocrine gland. Both the synthesis and the inactivation and destruction of hormone molecules are enzymatic processes.

237. THE THYROID GLANDS

All vertebrates, from the lowest fish to man, have a pair of glands called the **thyroids,** located in the neck. In man the two glands are joined by a narrow isthmus of tissue which passes across the front of the trachea, immediately below the larynx. The thyroids develop as outgrowths of the floor of the pharynx, but by the time a human embryo is ¼ inch long, this connection has been lost, and the glands exist as independent structures. The thyroids consist of groups of cuboidal epithelial cells, arranged in hollow spheres one cell thick. Each cavity of the spheres contains the secretion of the cells, and serves as a storehouse.

The hormone secreted by the cells is a protein called **thyroglobulin.** The active part of thyroglobulin is the amino acid thyroxine, which contains four atoms of iodine per molecule. It was synthesized artificially in 1927, and much of the material administered to patients today is synthetic thyroxine. If two of its four iodine atoms are removed, the activity of the substance is greatly decreased, and if all four iodine atoms are removed, activity is completely lost.

The fundamental effect of thyroxine is to speed up the oxidative, energy-releasing processes in all the body tissues. When an extra amount is given, the body uses more oxygen, produces more metabolic wastes, and gives off more heat than it normally does. When the supply of thyroxine is inadequate, the basal metabolic rate falls to as little as 600 to 900 Calories per day, or between 30 to 50 per cent of the normal amount. Individual tissue slices from an animal with thyroid deficiency also show a metabolic rate lower than normal when incubated *in vitro*. By its effect on metabolism, thyroxine has a marked effect on growth and differentiation. Removal of the thyroids of a young animal produces decreased body growth, retarded mental development, and delayed or decreased growth of the genitalia.

Deficiency in the amount of thyroxine secreted in an adult results in a condition known as **myxedema,*** a disease characterized by a low metabolic rate which results in decreased heat production. The body temperature may drop to several degrees below normal, so that the patient constantly feels cold. In addition, the pulse is slow and the patient is physically and mentally lethargic. The appetite usually remains normal, however, and since the food consumed is not used up at the normal rate, there is a tendency towards obesity. The skin becomes waxy and puffy, owing to the deposition of mucus fluid in the subcutaneous tissues, and usually the hair falls out (Fig. 215). Myxedema responds well to the administration of thyroxine or dried thyroid gland. Since thyroxine is not digested appreciably by the digestive juices, it can be given by mouth.

Myxedema is caused by an underactivity or degeneration of the thyroid gland itself. Another type of hypothyroidism results when the diet contains insufficient iodine for the synthesis of thyroxine. In such a case the gland itself tends to compensate for the insufficiency by increasing in size. The resulting enlargement, known as a simple **goiter,** may be a small swelling barely detectable by touching the neck, or a large, disfiguring mass, weighing several pounds (Fig. 216). The symptoms

* Any deficiency of a gland secretion is indicated by the prefix "hypo," while an oversecretion is designated by the prefix "hyper." Hence myxedema is a type of hypothyroidism.

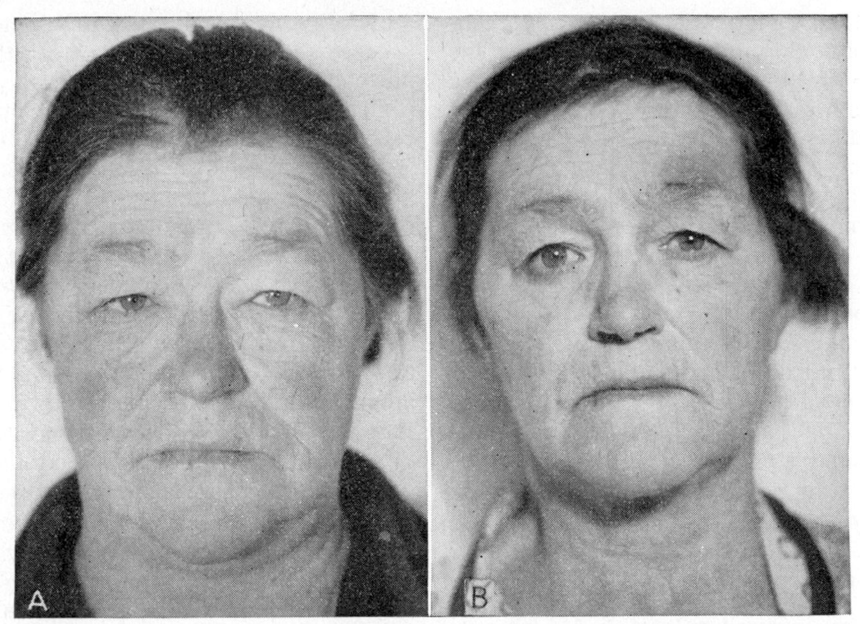

Figure 215. Myxedema. *A,* Woman who had had myxedema within the previous seven years. Note the puffy features and coarse, dry hair. She was rather obese (had gained 40 pounds in the seven years) and moved slowly and clumsily. *B,* After six months of treatment with dried thyroid substance the patient had lost 23 pounds without dieting; her skin and hair were less dry. (Courtesy of Drs. H. Lisser and R. F. Escamilla.)

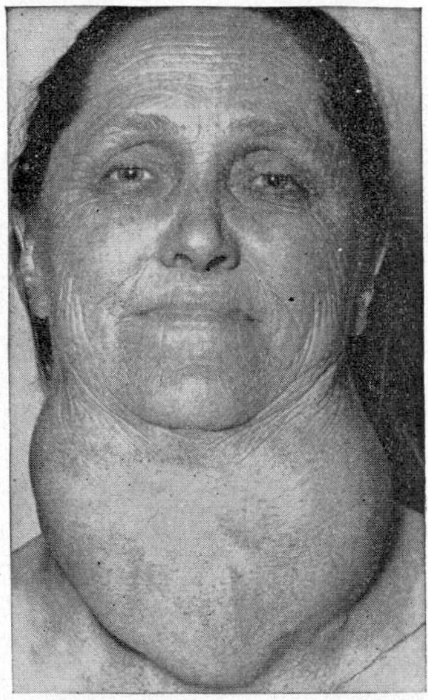

Figure 216. A simple goiter, caused by insufficient iodine in the diet. (Crile, G., Jr.: Practical Aspects of Thyroid Disease.)

accompanying the goiter resemble those of myxedema, but are much milder. This type of goiter occurs in areas where the soil lacks iodine, or in regions remote from the sea where seafood (which is rich in iodine) is unobtainable. The incidence of this type of goiter has been greatly reduced by modern packing and shipping methods which permit the transport of seafood to all parts of the country, and by the addition of iodine (as potassium iodide) to table salt. The efficacy of this preventive measure was demonstrated in Detroit, where in seven years the incidence of goiter in school children was reduced from 36 to 2 per cent.

Hypothyroidism which is present from birth is known as **cretinism.** Children suffering from the disease are dwarfs of low intelligence who never mature sexually (Fig. 217). If treatment is begun early, normal growth and mental development can be effected.

Hyperthyroidism results either from the overactivity of a normal-sized gland or from an increase in the size of the gland itself. In both cases the basal metabolic rate increases to as much as twice the normal amount. The excessively rapid heat production causes the hyperthyroid to feel uncomfortably warm, and to perspire profusely. Because the food he eats is used up quickly, he tends to lose weight even on a high caloric diet. High blood pressure, nervous tension and irritability, muscular weakness and tremors are symptomatic of the condition. But probably the most characteristic symptom is the protrusion of the eyeballs, called **exophthalmos,** which gives the patient a wild, staring expression (Fig. 218). The swelling of the gland as the result of hyperthyroidism is known as **exophthalmic goiter,** to distinguish it from the simple goiter caused by insufficient iodine. Identical symptoms can be caused by feeding thyroid substance or thyroxine to normal people.

Hyperthyroidism can be treated by removing some of the thyroid gland, usually by surgery, although radium and x-rays, which kill the individual cells, have also been used with success. Recently the disease has been treated by administration of the drug thiouracil, which inhibits the synthesis of thyroxine. Another method of treatment is the injection of radioactive iodine. Since the thyroid accumulates iodine, the rays given off by it are concentrated in the tissue of the gland and destroy the cells.

In those vertebrates which undergo a change from larval to adult form the thyroid hormone controls the alteration. In frogs, for example, the metamorphosis of the tadpole into an adult frog is regulated by the thyroid, and if the gland is removed surgically, the young tadpole never develops into an adult. But young tadpoles which are placed in pond-water containing small amounts of thyroxine undergo a precocious metamorphosis and develop into completely formed adult frogs, the size of a large fly.

238. THE PARATHYROID GLANDS

The parathyroid glands are masses of tissue, about the size of a small pea, which, in man, are attached to or embedded in the substance of the thyroid gland. There are usually four parathyroids, two in the upper part and two in the lower part of the thyroid, but there may be fewer or

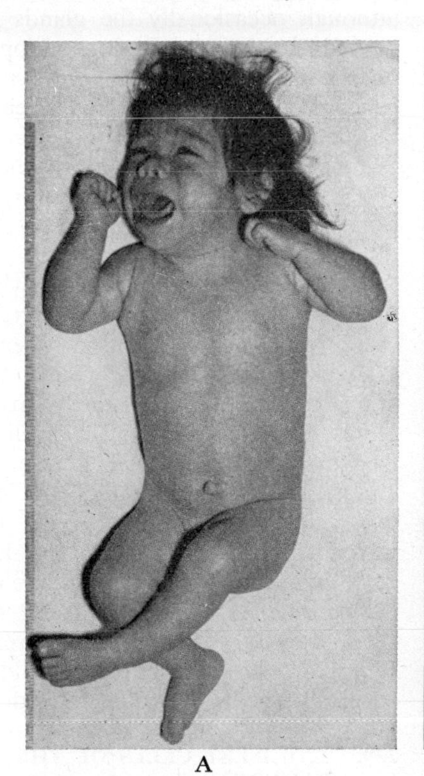

A

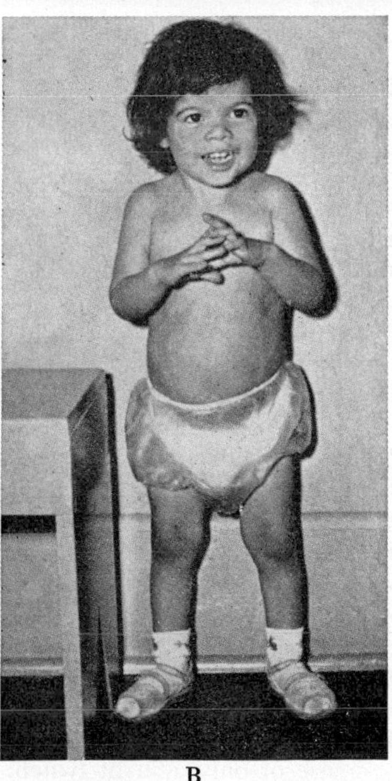

B

Figure 217. A, Photograph of a cretin, aged 3 years, 10 months. No previous treatment with thyroid hormone had been administered. *B,* The same child after 14 months of treatment with thyroxine. (Williams, B. H., and Cramm, C. J., in Med. Clin. N. Am. July, 1955.)

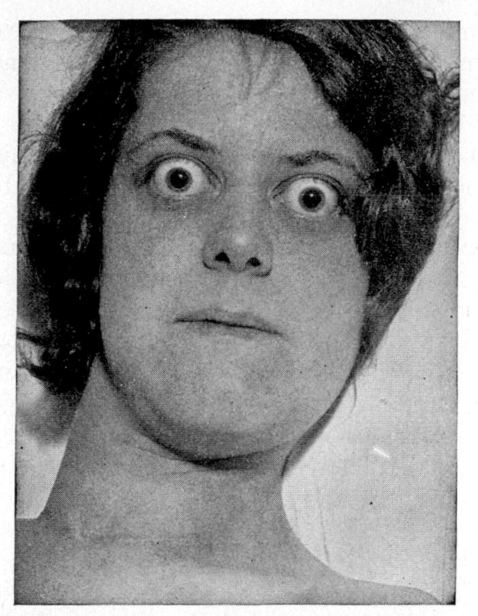

Figure 218. A woman with exophthalmic goiter. Note the marked staring expression of the eyes. (Courtesy of Surgery, Gynecology and Obstetrics.)

more than this number. Although the parathyroids are located on or within the thyroids, they have a completely independent function and a different microscopic appearance. The parathyroid cells are in a compact mass quite different from the hollow, spherical arrangement of cells in the thyroid. Like that gland, they originate embryologically from outgrowths of the pharynx and are evolutionary remnants of the gill pouches of fish.

The secretion of the parathyroids, known as **parathormone,** is essential for life, and an animal whose glands have been removed will die in about a week. In some of the early operations on human goiter the parathyroids were removed along with the thyroid, and death invariably followed shortly after. At first it was believed that death was due to the removal of the thyroid, but it is now known that removal of the thyroid is not fatal.

When the parathyroids are removed, the animal suffers muscular tremors, cramps and convulsions in response to stimuli which, in a normal animal, produce no response or only a slight twitch. This condition, called **tetany,** is due to an increased irritability of muscles and nerves, caused by a decrease in the calcium content of the blood and tissue fluids. The calcium content of the blood of a parathyroidectomized animal falls to about half the normal amount. If a solution of a calcium salt is injected into a vein of an animal in tetanic convulsions, the convulsions cease within a minute, and further convulsions can be prevented by the repeated injection or feeding of calcium. The amount of phosphorus in the blood increases as the calcium decreases and is decreased by the injection of parathormone. The basic function, then, of the parathyroid hormone is to regulate the calcium and phosphorus content of the blood and tissue fluids. Recent experiments indicate that the parathyroids secrete two hormones, both proteins; one of these regulates the excretion of phosphorus by the kidney and the other regulates the deposition of calcium in tissues. The parathyroid hormones are inactivated by digestive enzymes and cannot be administered orally.

Parathyroid deficiency is rare in man, although occasionally the glands are removed accidentally during an operation on the thyroid, and sometimes they degenerate as the result of an infection. If the deficiency is mild, tetany occurs only during excessive strain, as in childbirth, and is treated, like the more severe condition, by giving parathormone or calcium, or both.

Hyperfunction of the parathyroids is brought about by tumors or enlargements of the glands, and is characterized by a high blood calcium level. Since the calcium comes, at least partly, from the bones, hyperparathyroidism is characterized by soft bones which are easily bent and fractured. The muscles are less irritable than normally, and may become atrophied and painful. As the level of calcium in the blood increases, the mineral is deposited in abnormal places, such as the kidney. The disease can be treated by removing the excess parathyroid tissue surgically, or by destroying it with x-rays.

239. THE ISLET CELLS OF THE PANCREAS

In addition to the cells which secrete the digestive enzymes, the pancreas contains clusters or islets of cells which look and stain differently and have no associated ducts. The existence of these cells was recognized for a long time before their function was understood.

In 1886 two German investigators, Minkowski and von Mehring, were studying the role of the pancreas in digestion by removing the gland surgically from dogs and noting the digestive disturbances which occurred. The caretaker in charge of the animals noticed that their urine attracted swarms of flies to the cages. Upon analysis, large amounts of sugar were found in the urine, and the resemblance to diabetes was recognized. This disease had been known since the first century A.D., but its cause was unknown, and no treatment was effective. A short time before Minkowski and von Mehring performed their experiments, the cure for myxedema—the feeding of thyroid gland—had been discovered, and it was hoped that a similar feeding of pancreas, or the injection of pancreatic extracts, would cure diabetes. After 1892, when the discoveries were published, many scientists tried to prepare effective extracts of pancreas, but none of the preparations was very good, and many were toxic. The digestive enzymes of the pancreas destroyed the hormone before it could be extracted and purified. Finally, in 1922, two Canadians, Banting and Best, obtained an active substance by making extracts from pancreases of which the ducts had been tied for several weeks, so that the enzyme-producing cells had degenerated. Since the islet cells develop before the enzyme-producing cells in embryonic animals, they were also able to obtain active extracts from that source. Since January, 1922, when the extracts were first used in the treatment of human diabetes, **insulin,** the name they gave to the substance extracted from the islet cells, has added years of life to millions of diabetics. Since the work of Banting and Best, methods have been worked out for extracting insulin from ordinary beef and hog pancreas, obtainable at the slaughterhouse. The hormone was first obtained as pure crystals in 1927. The brilliant work of F. Sanger in England has revealed the exact sequence of the amino acids in each of the two peptide chains which comprise the insulin molecule. One chain contains 21 and the other 30 amino acids.

Because insulin is a protein, it is destroyed by the digestive juices if taken orally, and so must be injected. Ordinary insulin must be taken several times a day to maintain the proper level in the blood. **Protamine zinc insulin** is absorbed from the site of injection more slowly, and for most patients one injection a day is sufficient.

Most commercial preparations of insulin were found to contain a second hormone, which increases blood sugar concentration instead of decreasing it as insulin does. This hormone, called **glucagon,** has been isolated and found to be a protein. It is secreted by the alpha cells and insulin by the beta cells of the pancreas; alpha and beta cells have different staining properties. Glucagon stimulates an enzyme, called phosphorylase, which synthesizes and breaks down glycogen.

The primary function of insulin is to control the metabolism of glucose. It is necessary for the storage of glucose as glycogen in the liver, and for the oxidation of glucose to yield energy. In persons suffering from diabetes mellitus these functions are interfered with, and a host of other metabolic processes are affected secondarily. A great deal of urine is excreted, leaving the patient dehydrated and thirsty. A high concentration of sugar occurs in the blood and urine, and there is a steady loss of weight with increasing weakness, until at last the patient becomes comatose and dies. In spite of the excess glucose in the blood, the amount of glycogen in the liver decreases, because the tissues, unable to utilize glucose normally, are forced to use fats and proteins instead. The increased oxidation of fats causes an accumulation of incompletely oxidized fatty acids, known as **ketone bodies,** which are toxic. These substances are volatile and have a sweetish smell, which gives to the breath of diabetics its peculiar, characteristic odor. The amino acids from the proteins of the body are converted to glucose, and this also is excreted. If carbohydrates are eliminated from the diet, the patient still excretes glucose, which comes from this breakdown and transformation of proteins.

Diabetes is fatal because of the toxicity of the accumulated ketone bodies, and the continuous loss of weight. The injection of insulin alleviates all the diabetic symptoms: the patient is enabled to utilize carbohydrates normally, and the other symptoms disappear. The action of insulin persists for a short time only, however, which is the reason why repeated injections are necessary. Insulin does not "cure" diabetes, since the pancreas does not begin secreting its hormone again, but continued injections of it prevent the appearance of the symptoms and enable the diabetic to lead a normal life. The fundamental reason why the pancreas stops secreting adequate amounts of insulin is not known, but there appears to be a hereditary basis for it.

The secretion of insulin is controlled by the level of glucose in the blood. When the blood glucose level rises, after a meal, for example, the secretion of insulin is stimulated and it restores the glucose level to normal. When the glucose concentration has been lowered, the stimulus for insulin secretion is removed and it decreases or stops.

If, in the treatment of diabetes, too much insulin is injected, the blood sugar level falls drastically and shock results. The nerve cells of the brain require a certain level of glucose for their normal functioning; if this level is not maintained, they become overirritable, and convulsions, unconsciousness and death may follow. There are rare cases of enlarged pancreases due to tumors, which produce so much insulin that the patients suffer from recurring attacks of convulsions and unconsciousness. These can be relieved by eating candy, but the condition is cured only by the surgical removal of part of the pancreas.

240. THE ADRENAL GLANDS

Over the upper end of each kidney lies one of a pair of small glands, the **adrenals.** The two glands together weigh less than half an ounce, but they are richly supplied with blood vessels. Actually, the adrenal is a combination of two entirely independent glands: a dark reddish brown central core, the **medulla,** and a pale yellowish pink outer coat, the **cortex.** The two parts have quite different embryonic origins and cellular structures, and secrete dissimilar hormones. It just happens that, in man and other mammals, the two glands lie together; in some of the lower vertebrates the cells corresponding to the medulla and cortex are anatomically separate.

The Adrenal Medulla. The cells of the central core of the adrenal gland are derived embryologically from the same source as the nervous system, and in structure resemble nerve cells. The secretion of the medulla, **epinephrine,** is similar in its effects to sympathin, secreted by the tips of certain nerve cells (p. 349). Epinephrine, also known as adrenin and adrenaline, is probably the simplest hormone chemically (its formula is $C_9H_{13}O_3N$), and is now made synthetically.

The physiologic effects of an injection of epinephrine are marked and varied. The hormone causes a rise in blood pressure, an increase in the heart rate, an increase in the glucose content of the blood, a decrease in the glycogen content of the liver, and an increase in the rate of blood coagulation. In addition, the skin becomes pale because of the contraction of the arteries serving it, the pupils dilate, and the muscles which erect the hairs contract, producing gooseflesh. All these effects can be produced by stimulating the sympathetic nervous system.

The adrenal medulla is not essential for life; it can be removed surgically without interfering in any way with the normal functioning of the body. Ordinarily, the gland secretes a small amount of epinephrine continuously, the rate being controlled by the nervous system. It is the only hormone whose secretion is affected by nervous stimulation.

Some investigators believe that epinephrine is secreted during emergencies to coordinate the activities of the various parts of the body for fighting or escaping. Indeed, the various effects of epinephrine would enable the animal to fight more effectively; but whether the medulla actually secretes enough of the hormone during an emergency to produce this result is still open to question. Adrenaline is widely used as a drug in treating asthma, in increasing blood pressure, and in starting a heart that has stopped beating.

The Adrenal Cortex. The adrenal cortex is a more complex gland than the medulla, both structurally, since it is made of three distinct layers, and functionally, since it secretes several physiologically important hormones. It develops from mesoderm adjacent to the mesoderm that forms the mesonephric kidneys.

When the adrenals are removed surgically from an experimental animal, the blood undergoes profound changes: the concentration of sodium and chloride decreases, the potassium concentration rises, and there is a loss of water resulting in a diminished blood volume, and hence a lowering of the blood pressure. The amount of sugar in the blood and of glycogen in the tissues decreases. The tissues apparently lose the ability to convert proteins into carbohydrates for oxidation. After a complete adrenalectomy, death ensues within a few days.

The human disease resulting from insufficient secretion of the adrenal cortex was first described in 1855 by the English physician Addison. **Addison's disease** is usually caused by a tubercular or syphilitic infection of the cortex which destroys its cells. It is characterized by low blood pressure, muscular weakness, digestive upsets, and a peculiar bronzing of the skin caused by the deposition of the pigment **melanin.**

The cortical hormones are present in the cortex in small amounts only, and great quantities of cortical tissue must be extracted to obtain a few milligrams of the active substances. The first extracts to be successful in relieving adrenalectomized animals, and people suffering from Addison's disease, were made in 1927. Since that time some twenty-eight hormones have been isolated from the adrenal cortex; all of them are steroids.

There are three types of cortical hormones, with different physiologic effects and different chemical structures. The glucocorticoids stimulate the conversion of proteins to carbohydrates; the most potent glucocorticoid is **hydrocortisone** (Compound F). The mineralocorticoids regulate sodium and potassium metabolism. **Aldosterone,** discovered in 1953, is the most potent mineralocorticoid; **desoxycorticosterone** (DOCA) is an effective regulator of salt and water metabolism and is widely used clinically. **Cortisone,** which has some salt- and some carbohydrate-regulating effects, is most famous for its effects in the treatment of arthritis, leukemia and certain skin conditions.

The adrenal cortex of both men and women produces **androsterone**, a steroid with male sex hormone activity. Hyperfunction of the adrenal cortex in male children leads to precocious sexual maturity, with the muscular development, hair distribution and voice of a mature man. Cortical hyperfunction in females causes masculinization—growth of a beard, deep voice, regression of the ovaries, uterus and vagina, and the development of the clitoris to resemble a penis.

From the foregoing we can conclude that the various cortical hormones perform three major functions in the body: they regulate the concentration of sodium, potassium and chloride in the blood and body tissues; they control the metabolism of carbohydrates, and especially the formation of carbohydrates from proteins; and they regulate the development of the sex organs, in a manner similar to that of the male sex hormones.

241. THE PITUITARY

In a small depression on the floor of the skull, just below the hypothalamus of the brain, to which it is attached by a narrow stalk, lies a small gland about the size of a pea, the **pituitary**. This too is a double gland, of which one part, the anterior lobe, forms in the embryo as an outgrowth of the roof of the mouth, while the other (posterior) lobe grows down from the floor of the brain. When the two parts meet, the anterior lobe grows partly around the posterior one, but the two lobes are distinguishable microscopically, even in the adult. The anterior lobe loses its connection with the mouth, but the posterior lobe retains its connection with the brain. The pituitary, like the adrenal, consists of two parts with quite different functions, which simply happen to be located together. In the whale the two parts are anatomically separate.

The Posterior Lobe. Two potent extracts have been prepared from the posterior lobe of the pituitary. One, known as **oxytocin,** causes increased strength of contraction of the uterine muscles, and is sometimes injected after childbirth to contract the uterus. Whether the posterior lobe normally secretes oxytocin to aid in childbirth is unknown, but animals can give birth to their young quite easily after the posterior lobe has been removed. The other secretion, **vasopressin**, causes a constriction of the small arteries of the body and thus a general increase in blood pressure. It also regulates the reabsorption of water from the kidney tubes and is sometimes referred to as the **"antidiuretic" hormone,** since in its absence the individual excretes large amounts of urine. The brilliant work of Vincent du Vigneaud, for which he received the Nobel Prize in 1955, led to the isolation of these two hormones, the determination of their molecular structure, and their synthesis. Each is a peptide containing nine amino acids, seven of which are identical in the two. It is of considerable interest that these two substances, with quite different physiologic properties, differ chemically in only two amino acids. Most investigators agree that oxytocin and vasopressin are not produced in the pituitary but in certain cells in the brain. They pass along certain nerve tracts to the pituitary, where they are stored and subsequently released.

Injury of the posterior lobe itself, or of the nerve tracts leading to it, may result in a hormonal deficiency causing **diabetes insipidus.** This disease, characterized by the failure of the kidney to concentrate urine, results in the patient's excreting as much as 10 gallons of urine daily, and hence suffering from excessive thirst. Injection of vasopressin does not cure the disease, just as the injection of insulin does not cure diabetes mellitus, but it does relieve all the symptoms, and by repeated injections of vasopressin the patient can live a normal life.

The Anterior Lobe. A number of different types of cells are distinguishable in the anterior lobe of the pituitary, each of which is believed to produce a different hormone. Six clearly defined fractions are known to be secreted, and there is evidence that half a dozen more may exist. All the fractions separated so far are proteins.

The importance of the pituitary in the body's economy is demonstrated by the symptoms which result when the gland is removed experimentally (Fig. 219). Young animals whose pituitary is removed immediately stop growing and never reach sexual maturity. If adults are operated upon, both males and females show a regression of the reproductive organs, with the accompanying atrophy of both thyroid glands and the adrenal cortex. When pituitary extracts are injected into normal young animals, growth is stimulated and they become giants, reaching sexual maturity at an early age. The adrenal cortex, the thyroid and the sex organs respond by growing abnormally large and oversecreting.

The Growth Hormone. Giants have been known since the beginning of history, but it was not until 1860 that excessive growth was correlated with an enlargement of the pituitary. The first hormone from this gland to be discovered was the **growth-stimulating hormone,** finally isolated as a pure protein from extracts of beef pituitary in 1944. This hormone controls general body growth, and especially the growth of the long bones. Consequently, when the pituitary is overactive during the growth period, there is a general acceleration of the process, resulting in a very tall, though fairly well-proportioned person (Fig. 220). Most circus giants are of this type. If the pituitary is underactive during the growth period, a small, normally proportioned person, known as a **midget**, is the result. Oversecretion of the growth hormone after normal growth has been completed, however, produces a condition known as **acromegaly**. Since by this time most of the

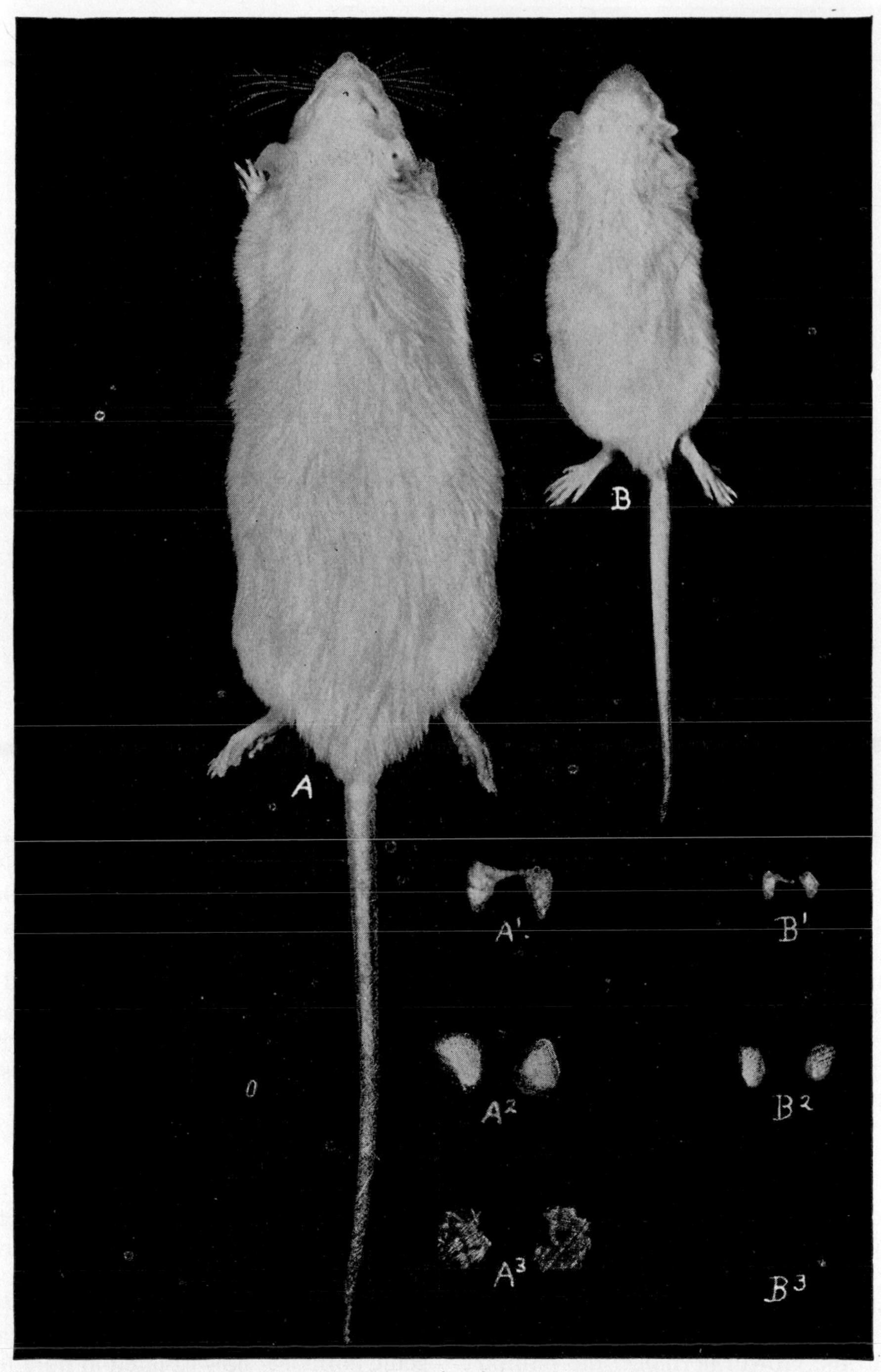

Figure 219. The effects of hypophysectomy in the rat. *A,* Normal control. *B,* Hypophysectomized rat from same litter, hypophysectomized when 36 days old, when it weighed 70 gm. These photographs were made at 144 days of age when control animal weighed 264 gm. and the hypophysectomized rat weighed 80 gm. A^1, A^2 and A^3 are the thyroids, adrenals and ovaries of the normal rat and B^1, B^2 and B^3 are the thyroids, adrenals and ovaries of the hypophysectomized rat. (Turner: General Endocrinology.)

Figure 220. A hyperpituitary giant posed with two men of normal height. The giant, approximately 7 feet, 3 inches tall, had a large pituitary tumor, visible by x-ray. He also exhibits some of the characteristics of acromegaly—note the enlarged lower jaw and hands. (Courtesy of Dr. E. Perry McCullagh.)

parts of the body have lost their capacity for growth, only the hands, feet and bones of the face develop. The hands and feet become grossly enlarged, the jaws grow abnormally long and broad, and the bony ridges over the eyes and the cheekbones enlarge, giving the acromegalic a heavy, unpleasant facial appearance (Fig. 221).

In 1930 a strain of hereditarily dwarf mice was discovered in which the dwarfness was due to the absence in the pituitary of the type of cells believed to secrete the growth hormone. When pituitaries from normal mice were grafted into the dwarf ones, they attained normal size.

The Gonadotropic Hormones. In addition to the growth hormone, the pituitary produces two different protein hormones which stimulate the primary sex organs of both sexes, the ovaries of the female and the testes of the male. Both are necessary for the achievement of sexual maturity and for maintenance of the sexual functions in mature animals. One of them, called **follicle-stimulating hormone,** or **FSH,** stimulates the development of graafian follicles in the ovaries, and the formation of sperm by the seminiferous tubules of the testes. The other, called **luteinizing hormone**, or **LH**, is necessary for the release of ripe eggs from the follicle, the formation of corpora lutea, the production and release of estrogens and progesterone, and the formation of male sex hormone by the interstitial cells of the testes. The two hormones together regulate the menstrual cycle in women and the estrus cycles of other animals.

The Lactogenic Hormone. The lactogenic hormone, now isolated in pure form, maintains the secretion of estrogens and progesterone by the ovary and initiates the secretion of milk by the mammary glands. It is effective only after the breast has been stimulated by the proper amounts of estrogen and progesterone. If the pituitary is removed immediately after childbirth, the flow of milk from the mammary glands ceases. The lactogenic hormone is responsible also for much of the maternal instinct, and male animals injected with it develop an interest in the young, even of another species, making attempts to protect and care for them. If a male is treated first with female sex hormones and then with lactogenic hormone, his mammary glands can be made to secrete milk.

Adenocorticotropic Hormone. The **adrenocorticotropic hormone,** or **ACTH**, has become famous in recent years because of its use in the treatment of such diseases as arthritis and certain allergic conditions. Its normal function is to stimulate the adrenal cortex to produce its battery of hormones. An injection of ACTH produces, within a few minutes, a marked increase in the amount of hydrocortisone in the blood. ACTH is necessary for the development and maintenance of the normal structure and functioning of the adrenal cortex; in its absence the adrenal cortex regresses. ACTH probably acts only by stimulating the adrenal cortex, for the adrenal cortical hormone cortisone has about the same effects in the treatment of arthritis and allergies.

Other Pituitary Hormones. The anterior lobe of the pituitary also secretes a **thyrotropic** hormone which is necessary for the development and normal functioning of the thyroid gland. Extracts of the pituitary have a number of other effects but it is difficult to determine whether these are mediated by separate hormones or are side effects of one of the well-known pituitary hormones. The anti-insulin effect of the pituitary, for example, is probably caused by the growth hormone. All of the hormones are proteins and it is difficult to separate and purify them.

The release of each of these tropic hormones is controlled in part by the amount of the target hormone in the blood. The release of ACTH is inhibited by hydrocortisone, the release of thyrotropin is inhibited by thyroxine, and so on. This provides a cut-off mechanism so that the secretions of the pituitary and its target glands are kept in balance. The pituitary is also controlled by the **hypothalamus.** A considerable body of experimental evidence indicates that certain cells in the hypothalamus secrete a "neurohumor" which is carried by blood vessels from the hypothalamus to the pituitary located directly beneath, and there stimulates the

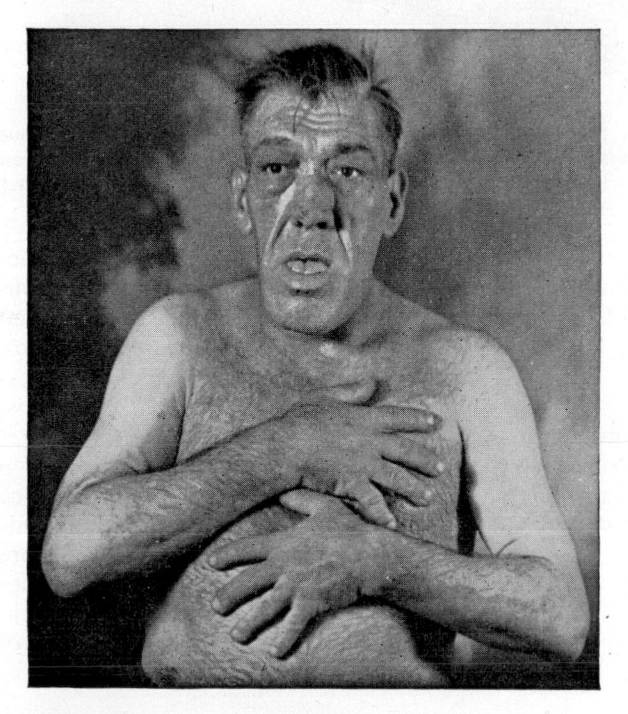

Figure 221. A case of acromegaly. The oversection of the growth hormone of the pituitary in the adult results in an overgrowth of those parts of the skeleton which are still able to respond. Note the enlargement of the lower jaw and hands and the thickening of the nose and ridges above the eyes. (Turner: General Endocrinology.)

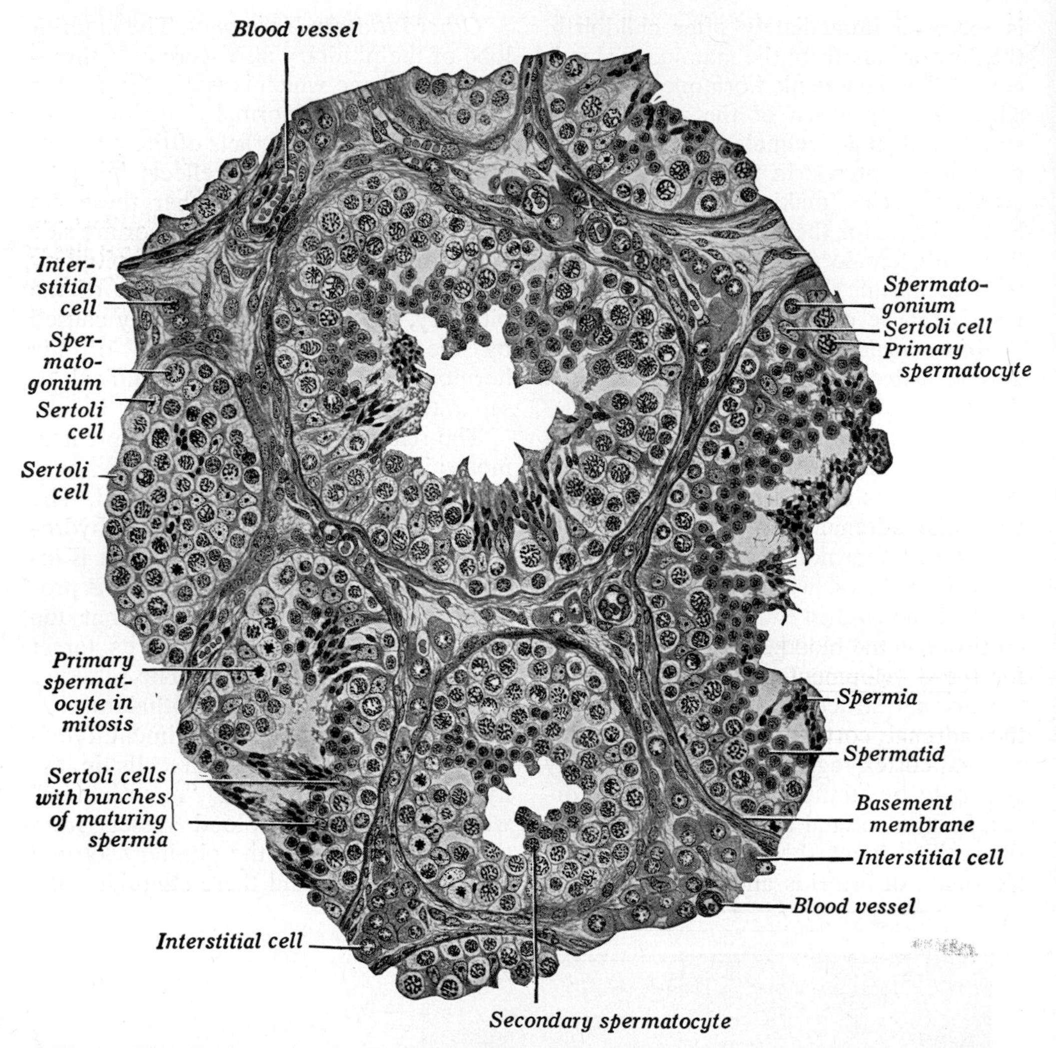

Figure 222. Section of a human testis obtained from an operation. Note the seminiferous tubules seen in cross section, containing cells in various stages of spermatogenesis, and, between the tubules, the interstitial cells which secrete the male sex hormone. (Maximow and Bloom.)

release of ACTH. The release of the other hormones of the anterior pituitary may also be controlled by secretions from the hypothalamus.

242. THE TESTES

In addition to cells which manufacture sperm, the testes contain endocrine cells which produce the male sex hormone, **testosterone.** These hormone-secreting cells, known as **interstitial cells**, are located between the seminiferous tubules which produce sperm (Fig. 222). Testosterone stimulates the development of the so-called secondary male sex characters—the beard, the growth and distribution of hair on the body, the deepened voice, the enlarged and stronger muscles, and the accessory sex structures, the prostate gland, seminal vesicles and penis. Castration, the removal of the testes, prevents this stimulation and results in a eunuch, a man with a high-pitched voice and beardless face. Since castration tends to make an animal larger, fatter and more placid, it has been practiced for centuries on domestic animals. The effects of castration can be reversed by the injection of testicular extracts or by the grafting on of testicular tissue. Testosterone injected into a female causes many of the secondary male sex characteristics to develop.

Testosterone belongs to the group of chemicals known as the steroids (its formula is $C_{19}H_{28}O_2$). The activity of this hormone is commonly measured by its ability to induce growth of the comb in a castrate chicken or capon. Pure testosterone is extremely potent: a pound of it would cause growth of the comb in 500,000 capons!

Evidence of the source of the male sex hormone came from a study of the condition in which the testes fail to descend into the scrotal sac, but remain within the abdominal cavity. When this happens, the man is completely sterile—unable to produce sperm—although he usually has normal secondary sex characteristics. When an undescended testis is examined microscopically, the cells of the sperm-forming tubules are found to have degenerated, but the interstitial cells lying between them are normal. For a long time it was not understood why the sperm-forming cells of the testis degenerated when the gland remained in the body cavity, since it received a normal blood and nerve supply. Then it was discovered that the sperm-forming cells are particularly susceptible to heat and require the approximately 3 degrees cooler temperature of the scrotal sac. The slightly higher temperature of the abdominal cavity destroys the sperm-forming cells, but not the interstitial, hormone-secreting cells. It is probable that an attack of fever temporarily sterilizes a man.

The male sex hormone also determines sexual behavior to some extent, and is partly responsible for the sex urge. Many animals, even when castrated before reaching sexual maturity, will pursue females and exhibit mating behavior for months after castration, although eventually sexual behavior decreases.

243. THE OVARIES

Like the testes, the ovaries are endocrine glands and secrete sex hormones as well as produce eggs for reproduction.

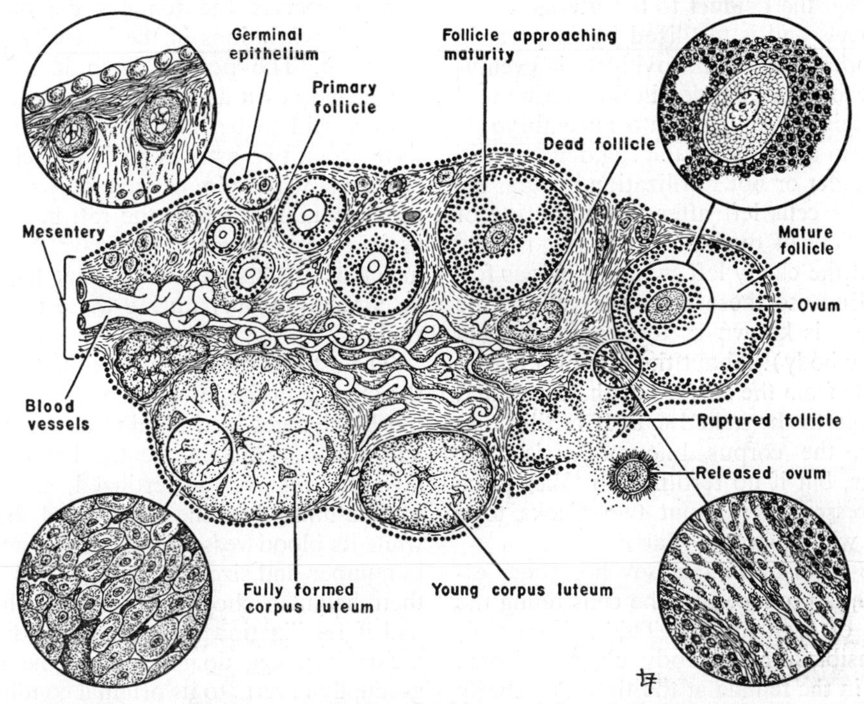

Figure 223. Diagram of the stages in the development of an egg, follicle and corpus luteum in a mammalian ovary. Successive stages are depicted clockwise, beginning at the mesentery. The insets show the cellular structure of the successive stages.

Both ovaries and testes develop from mesoderm, from the genital ridge on the ventral side of the mesonephric kidneys. The two ovaries are bean-shaped structures, about 1½ inches long, supported in the back part of the abdominal cavity by mesenteries. The outer layer of each gland is the **germinal epithelium**, from which the eggs develop, while the central part is composed of connective tissue and blood vessels. Just under the germinal epithelium is a thick layer of spherical groups of cells, or follicles, each enclosing an egg (Fig. 223). At birth, several hundred thousand of these have already originated from the germinal epithelium. Some degenerate, but the majority remain quiescent until the age of puberty, when the process of their growth and development begins.

Each month, one or more of the follicles begins to enlarge and becomes distended with follicular fluid, until it finally protrudes above the surface of the ovary and bursts, releasing the egg cell contained in it. This process is known as **ovulation**. The released egg passes by way of a channel called the **oviduct** to the **uterus**. If the egg meets and is fertilized by a sperm in the upper part of the oviduct, it eventually becomes embedded in the uterine wall and begins to develop into an embryo. If no sperm are present, the egg degenerates.

Whether or not fertilization occurs, the follicular cells left after the rupturing of the follicle in ovulation multiply rapidly and fill the cavity left by the previous follicle. Because these cells are yellow, the structure is known as the **corpus luteum** (yellow body). About the size of a pea, it projects from the surface of the ovary and is visible to the naked eye. If fertilization occurs, the corpus luteum persists for months, but if no fertilization takes place it regresses after about two weeks to a small, white patch of scar tissue.

The primary female sex hormone, **estradiol**, is produced by the cells lining the cavity of each follicle. This substance is responsible for the body changes which occur in the female at the time of puberty or sexual maturity: the broadening of the pelvis, development of the breasts, growth of the uterus and vagina, growth of the pubic hair, change in the voice quality and the onset of the menstrual cycle.

The second female sex hormone, **progesterone**, is produced by the cells of the corpus luteum. It is necessary for the completion of each menstrual cycle, because it completes the alterations in the uterus begun by estradiol. It also makes possible implantation of the fertilized egg in the uterine wall, and causes development of the breasts during the latter months of pregnancy. Progesterone is related chemically to the adrenal cortical hormones and is believed to be an intermediate in their synthesis, as well as an intermediate in the synthesis of estradiol and testosterone.

Both male and female sex hormones are produced by both sexes; in fact, one of the richest sources of female sex hormone—one used commercially—is the urine of stallions. Recently, female sex hormones have been found in palm nut oil and pussy willows! What they are doing there no one knows.

The Estrous Cycle. In most mammalian species the females demonstrate rhythmic variations in the intensity of the sex urge. The period when it is at its height is known as **estrus**, and the animal is then said to be in heat. Cats and dogs have about two estrous periods each year, and most wild animals have only one, but some animals, such as the rat, have them as frequently as every five days. Most females of any species will accept the male in copulation only during the estrous period.

The estrous cycle is marked not only by the change in the intensity of the sex urge, but by changes in the lining of the vagina and uterus, which make the latter better able to receive a fertilized egg. The uterine lining becomes softer and thicker, while its blood vessels and glands increase in number and size. These processes reach their height a short time after ovulation, and if fertilization and the embedding of a fertilized egg do not occur, the lining gradually reverts to its original condition.

The Menstrual Cycle. Human and anthropoid ape females do not experience any distinct period of estrus; instead, the cycle is marked by periods of bleeding, known as **menstruation**, which occur about every twenty-eight days and last about four days. The menstrual flow consists of pieces of the ruptured uterine lining, and blood from its vessels.

Since the lining of the uterus is almost completely destroyed by each menstruation, it is thinnest just after the menstrual flow. At that time, under the influence of the follicle-stimulating hormone, or FSH, secreted by the pituitary, one or more of the follicles in the ovary begin to enlarge rapidly, while the follicular cells secrete estradiol. This hormone causes the uterine lining to grow, and by the end of the first week, it is about half as thick as it will become without pregnancy (Fig. 224).

Stimulated by the proper mixture of FSH and LH from the pituitary, ovulation occurs about fifteen days after the beginning of the previous menstrual period. The ripe egg is released from the ovary by the rupturing of the follicle. The follicular cells are then transformed into a corpus luteum, which, under the stimulation of LH secreted by the pituitary, secretes the second hormone involved in the menstrual cycle, progesterone. This hormone completes the development of the uterine lining, preparing it to receive a fertilized egg. It also promotes the growth of the mammary glands, and prevents the development of any additional follicles and eggs.

If the egg has not been fertilized, about twenty-seven days after the beginning of the previous menstrual period the corpus luteum undergoes regression, thus terminating the secretion of progesterone. Since the uterine lining depends on the hormone for maintenance, this marks the beginning of its breakdown, and the menstrual flow begins. In the ovary, another follicle, released from the inhibition of progesterone, begins to ripen, and the next menstrual cycle starts.

If pregnancy occurs, the corpus luteum remains and secretes progesterone almost until the time of birth. The continued secretion of the hormone during the early months of pregnancy is necessary for its

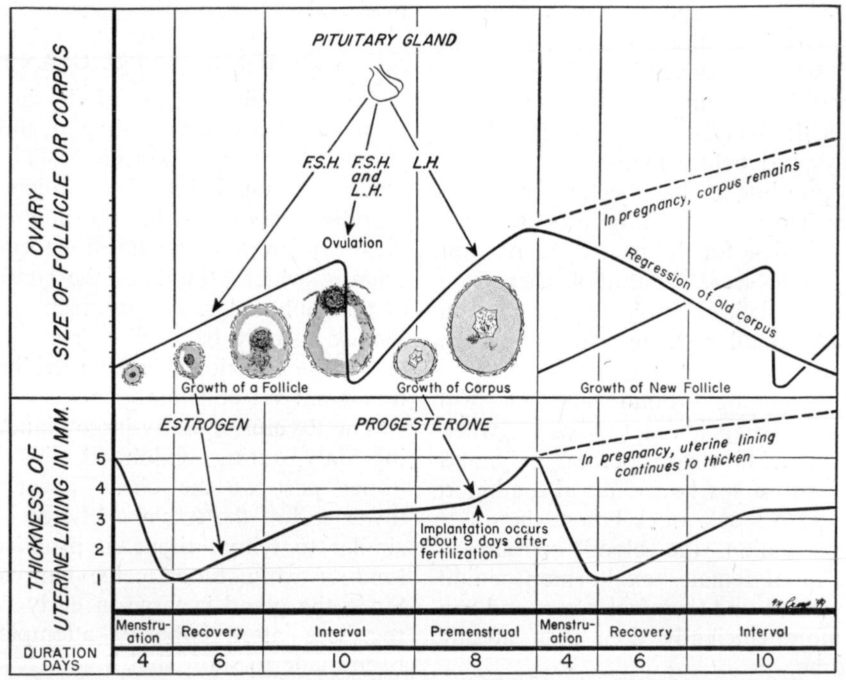

Figure 224. The menstrual cycle in the human female. The solid lines indicate the course of events if the egg is not fertilized; the dotted lines indicate the course of events when pregnancy occurs. The actions of the hormones of the pituitary and ovary in regulating the cycle are indicated by arrows.

continuation, and if the corpus luteum is removed, pregnancy ceases at once with an abortion. In some animals the placenta produces enough progesterone so that loss of the corpus luteum does not result in abortion. The corpus luteum hormone also stimulates the growth of the mammary glands during the latter months of pregnancy and prepares them for the action of the lactogenic hormone of the pituitary, which is secreted shortly after birth and causes the flow of milk.

After the egg has been released from the ovary and is passing down the oviduct, it can be fertilized only within a short time, probably about twenty-four hours. When the sperm are deposited, through intercourse, in the female reproductive system, they quickly lose their ability to fertilize an egg, within forty-eight hours at the most. The period of maximum fertility in human beings, then, narrows down to the time of ovulation, about midway between successive menstrual periods.

The reason why women are subject to nervous upsets and fits of depression just before and during the menstrual period is now fairly well understood. The maintenance of emotional equanimity in women seems to depend on the presence of a certain amount of female sex hormone in the blood and tissues. But for a few days just before menstruation, when the corpus luteum has begun to regress and has stopped secreting progesterone, and the follicle for the next cycle has not yet begun to secrete estradiol, there is a measurable deficiency of sex hormone in the blood. Similar changes in disposition frequently occur at the menopause, a period beginning around the forty-fifth year, and lasting about two years, when the menstrual flow stops permanently, and the amount of sex hormones in the blood decreases. About a week before the onset of menstruation, the blood contains a maximum of female sex hormones, and many women find they feel better and can work more efficiently then than at any other time.

244. THE PLACENTA

Although the placenta is primarily an organ for the support and nourishment of the developing embryo (see p. 423), it is also an endocrine gland. It secretes estradiol, progesterone, adrenal corticoids and a substance similar to the luteinizing hormone of the pituitary. During pregnancy, large amounts of this hormone are produced, and the excess passes into the urine; this is the basis for several pregnancy tests. When small amounts of a pregnant woman's urine are injected into a female animal—a nonpregnant mouse, rabbit or rat—significant effects on the animal's ovaries are produced. The test is accurate and makes diagnosis of pregnancy possible as early as two weeks after conception, long before any other signs are apparent.

In some animals, such as the rabbit, the placenta is a significant source of **relaxin.** This protein hormone, also produced by the ovary, relaxes the ligaments of the pelvis to facilitate the birth of the young. Relaxin is effective only after the connective tissue of the pubic symphysis has been sensitized by estradiol. Relaxin also inhibits the motility of the uterine muscles and has been used clinically to prevent premature birth.

245. OTHER ENDOCRINE GLANDS

Certain other organs of the body, although not usually considered endocrine glands, do produce hormones. These include the small intestine, which yields **secretin,** responsible for the flow of pancreatic juice, and the hormone **cholecystokinin**, which stimulates the contraction of the gallbladder. The stomach, liver and kidney also are believed by some to have endocrine functions, but the evidence for this is not yet conclusive.

The **thymus**, a fairly large gland, existing only during childhood, lies in the upper part of the chest, covering the lower end of the trachea. Microscopically similar to lymph tissue, it produces one kind of white blood cells, lymphocytes. Since the gland is large in early life and regresses after puberty, attempts have been made to show that it secretes a hormone affecting sexual maturity, but clear evidence of this has not been obtained.

The **pineal gland**, a small, round structure on the upper surface of the thalamus, between the two halves of the cerebral cortex, has long been suspected of endocrine activity, largely because it has no other known function, but there is no evidence that it secretes a hormone. Neither its removal nor injection of extracts from it has any effect on experimental animals.

246. INTERRELATIONSHIPS BETWEEN THE ENDOCRINE GLANDS

For the sake of simplicity we have had to consider each gland and its effects separately. Recent research shows, however, that nearly every gland affects the functioning of almost every other one. We have already discussed the extensive control which the anterior pituitary exerts over the sex glands, the thyroid, the adrenal cortex, the pancreas, and probably others as well. Consider, for example how the pituitary affects the ovary, stimulating the production first of estrogen and then progesterone. These hormones, in turn, affect the secretion of hormones by the pituitary itself. Thus progesterone inhibits the pituitary from secreting FSH to start another menstrual cycle, until the previous one is completed or pregnancy terminates. The rate of cell metabolism and the relative rates of utilization of carbohydrates, fats and proteins are controlled by the complex interplay of thyroxine, insulin, epinephrine, glucagon, growth hormone, hydrocortisone, estradiol and testosterone. Normal growth requires not only growth hormone and thyroxine but insulin, androgens and other hormones.

Hans Selye of the University of Montreal has done much in recent years to investigate the role of hormones in "stress." Stresses such as operations, burns, fractures or cold stimulate the adrenal medulla to release epinephrine. The pituitary is stimulated by the epinephrine to secrete ACTH and this in turn stimulates the adrenal cortex to produce cortisone and other hormones. The adrenal hormones produce changes in mineral and carbohydrate metabolism which assist in adapting the animal to the conditions of stress. Long continued stress may eventually overcome the body's adaptive ability and cause exhaustion or shock. The intimate functional association of adrenal and pituitary in this and other situations has led to the concept of the "adrenal-pituitary axis" as the primary center for the control of the body's adaptation to the environment.

QUESTIONS

1. In what ways are the endocrine and nervous systems comparable? In what ways do they differ?
2. Define a hormone. How would you go about proving that a particular gland secretes a particular hormone?
3. Where is the thyroid gland located? What does it secrete?
4. Differentiate between myxedema, simple goiter, and cretinism.
5. What is the function of the parathyroid?
6. What is the importance of insulin in the body? Explain why the feeding of thyroid gland cures myxedema, while the feeding of pancreas does not cure diabetes.
7. What are the parts of the adrenal gland? What functions does each part perform?
8. What is Addison's disease? What reasons can you advance to explain why Addison's disease is not cured by eating adrenal glands?
9. What are oxytocin and vasopressin and what effects do they have on the body?
10. Why is the pituitary sometimes called the "master gland"?
11. In view of the newer evidence of the source of the hormones of the posterior lobe of the pituitary, what function might be ascribed to the pineal body? Describe experiments which would provide evidence for your hypothesis.
12. What is ovulation? When does it occur in the human?
13. What is the function of progesterone? Of luteinizing hormone?
14. How can you account for the fact that one of the best sources of estrogens is the urine of a stallion?
15. Make a list of all the structures in the body known to be endocrine glands and the hormones secreted by each.
16. To what extent do you think an individual's personality is affected by his endocrine glands?
17. Discuss ways in which a knowledge of physiology and psychology can help society establish fair laws regarding behavior.

SUPPLEMENTARY READING

The biologic aspects of endocrinology are treated in C. D. Turner's *General Endocrinol-*

ogy, while the more clinical aspects of endocrinology are discussed in great detail in R. H. Williams' *Textbook of Endocrinology* and in Hans Selye's *Textbook of Endocrinology*. Selye's theory of the role of stress in inducing endocrine imbalances is presented there. The role of hormones in controlling behavior in the several classes of vertebrates is surveyed in Frank Beach's *Hormones and Behavior*. Endocrine mechanisms in other animals, particularly the invertebrates, are described by Frank Brown in Prosser's *Comparative Animal Physiology*.

Chapter 27

Infectious Diseases and the Body's Defenses

THE HUMAN body may be the host of a wide variety of viruses, rickettsias, bacteria, fungi, protozoa, worms, insects and other forms, many of which cause no ill effects. Others of them produce marked upsets of one sort or another in the body's economy. The upsets produced by a certain kind of organism are fairly regular and recognizable as the symptoms of a certain disease. A disease caused by the presence of some parasitic organism, one that is transmitted from one person to another by the transfer of the organism, is known as an **infectious disease.**

It is rather startling to think that a hundred years ago man knew little more about the nature of contagious diseases than was known by the early Greeks. Through all the centuries of recurring plagues, which swept countries and continents and killed literally millions of people, there was apparently no inkling of what caused and spread them. Malaria was believed to be caused by "bad air" (the word malaria means bad air). The humoural theory of disease was in vogue from the time of Hippocrates (460 B.C.) until about 1870. According to this there were four humours of the body—blood, phlegm, black bile and yellow bile—normally present in certain proportions. Disease resulted when there was too much or too little of one of these humours. Since blood was the easiest "humour" to reach, treatment for all sorts of diseases consisted of blood letting, by cutting a vein or by attaching leeches.

The germ theory of disease was established by the work of Pasteur, Koch, and their students. This simply states that contagious diseases are caused by "germs"—bacteria, viruses, etc.—which are transmitted from person to person by contact, drinking water, food or the bite of an insect. Pasteur first studied the "diseases" of souring wine and beer, then a disease of silkworms before turning his attention to anthrax, which was threatening to wipe out all the sheep and cattle in France. The dramatic experiment in which he showed the value of inoculation for the prevention of anthrax is well told in Vallery-Radot's *Life of Pasteur*. In 1885 he demonstrated the value of inoculation in saving human life from the hitherto inevitably fatal disease rabies, transmitted by the bite of a mad dog.

The word "disease" means any malfunctioning of the plant or animal body, and includes those caused by the lack of some nutrient (deficiency diseases), those caused by some abnormal functioning of the cells, such as cancer or the endocrine diseases (clinically called metabolic diseases), and mental ailments, as well as the contagious ones discussed here.

247. HOW MICROORGANISMS CAUSE DISEASE

Three factors make it possible for these pathogenic organisms, tiny as they are, to be effective in incapacitating and killing organisms as large as man. The first is their prodigious rate of multiplication, which enables them to increase their numbers to countless billions within a few hours, and to place a tremendous material burden on the tissues. The second is their ability to destroy tissues of the body, thus interfering with the function of certain organs. But perhaps more deadly than either of these is their third method of attack—the production of poisonous substances, known as **toxins.** A toxin usually affects one particular organ or organ system—the central nervous system or the red blood cells, and so forth—rather than the body as a whole, thus producing a characteristic set of **symptoms** by means of which the physician can diagnose the disease and the organism responsible.

These chemical substances which bacteria produce are of two types, called **exotoxins** and **endotoxins.** An exotoxin is an extremely potent poison secreted by the bacterial cell to the *outside* environment. A mere 0.002 ml. of diphtheria toxin is lethal for a guinea pig, as is 0.0005 ml. of tetanus (lockjaw) toxin, and 0.0001 ml. of botulinus (food poisoning) toxin. Why these proteins are so poisonous is unknown, for they do not seem to differ in any essential way from completely harmless ones, such as egg albumin. It may be that they block some fundamental chemical process of the cell, or they may function as enzymes to destroy some essential part of it. Whatever the explanation, they are easily destroyed by enzymes (the botulinus poison is an exception to this, so that it alone of the exotoxins is effective when eaten, the others being destroyed by the digestive enzymes) and by heat.

The other type of toxins, endotoxins—not all of which are proteins—are made and kept *within* the bacterial body, and released only after the bacterium dies and dissolves. These are not so powerful as exotoxins, but neither are they so easily destroyed by heat and enzymes.

Each kind of pathogenic bacterium both invades tissues and produces a toxin to some extent, and they can be arranged in a series, from the botulinus organism, which is very toxic, but least invasive,* through tetanus, diphtheria, streptococcus and staphylococcus bacteria, to the tuberculosis bacillus, which is not very toxic, but invades and destroys tissues extensively.

Bacteria are not unique in producing toxins. Substances similar to exotoxins are found in the seeds of certain plants, such as the castor oil bean, and in the secretions of certain animals, such as the rattlesnake and cobra, the scorpion and black widow spider.

248. BODY DEFENSES AGAINST DISEASE

The body's first line of defense against pathogenic organisms is the skin, mucous membranes and digestive tract lining. Bacteria cannot ordinarily penetrate the skin, unless a break is provided by a wound. Sometimes they enter by way of the sweat gland ducts or hair follicles, but they are not often able to reach deeper tissues by these routes. Furthermore, the skin can actively destroy many kinds of bacteria, although its ability to do so is reduced by the presence of dirt, and ceases completely within fifteen minutes after death. The sticky mucous secretions of the digestive tract, nose and lungs trap bacteria and prevent their spread, and the mucous membranes themselves serve as mechanical barriers to their entrance. Still another barrier lies in the ciliated cells of the upper passages of the respiratory tract, whose constant beating moves the bacteria

* Usually this anaerobic organism does not itself invade the body. Its toxin—the most deadly of all—is taken into the body via improperly canned food, and is lethal almost immediately. From this it is obvious that botulism is not a contagious disease, like the others discussed in this chapter.

up towards the pharynx, where they are swallowed. In the stomach the majority of organisms are destroyed by the gastric juice, but a few, protected perhaps by a particle of food or a tough capsule, survive and multiply in the intestine. These are constantly being eliminated with the feces. Both male and female urethras are normally quite free of bacteria, owing to the constant flushing by the slightly acid urine. The normally acid vaginal secretions are lethal for most species of bacteria.

If disease organisms are successful in penetrating these first lines of defense, other mechanisms come into play to control them. One of these is the process of **inflammation,** effective in localizing an infection. The presence of foreign bodies stimulates a dilation of the capillaries (hence the reddening of the area), and the increased blood supply brings blood-clotting elements and white cells to combat the invaders. The formation of a clot traps them so that they cannot spread, while connective tissue fibers grow around the periphery of the clot and wall them off. The pus of an infection such as a pimple or boil consists simply of masses of bacteria in the process of being engulfed by white corpuscles. How the lymph nodes function in preventing the spread of infection has already been discussed (Chap. 17). When the body is in poor condition, owing to anemia, lack of vitamins, previous infectious diseases, or any other cause, it is less effective against invasion by disease organisms. This is why some people seem to get one thing after another, while others seem invulnerable to disease.

Experiments at the University of Notre Dame with rats raised from birth under completely sterile conditions indicate that the body's resistance to bacteria is gained gradually in response to constant exposure. When these rats which have never been exposed to bacteria are removed from their sterile quarters they rapidly die of infection. They apparently have no resistance to bacteria and within three hours after being removed from their sterile quarters, every tissue of the body is infected with a variety of bacteria.

Acquired Immunity. In the eighteenth century many people, feeling sure that they would contract smallpox at some time in their lives, exposed themselves deliberately in order to have it at their convenience and get it over with. Even today some parents expose their children to childhood diseases from the same motive, because they know that certain diseases do not occur twice in a lifetime. This type of resistance—the body's last line of defense—is known as **acquired immunity.** But a person immune to smallpox, from having been a victim of it, is no more immune to measles or any other disease than someone who has never had smallpox, and for this reason we say that immunity is *specific.*

Actively acquired immunity depends upon the body's production of specific proteins, called **antibodies,** and the release of these substances into the blood and tissue fluids, after the invasion of some foreign protein, called an **antigen.** These two chemicals—antigen and antibody—react upon each other with the result that the body is protected from injury. For example, if some egg albumin (a protein) is injected into a rabbit, the animal's cells respond by producing antibodies specific for albumin. Similarly, the body can produce a particular kind of antibody, known as an **antitoxin,** in response to the presence of a toxin (usually protein) released by a bacterium. When enough antitoxin has been produced, the body is no longer affected by that particular toxin.

Antibodies react with the antigen in one of several different ways. They may combine with the toxin, neutralizing its poisonous qualities; they may dissolve the bacterial cells; or they may sensitize the bacteria and make them more vulnerable to the white corpuscles. Some agglutinate the organisms, thus preventing their spread and making their entrapment by the lymph nodes more certain.

Another, more pleasant way in which immunity may be acquired is through **inoculation** with a **vaccine.** A vaccine is the commercially produced antigen of a particular disease, strong enough to stimulate the body to make antibodies, but not sufficiently strong to cause the disease's harmful effects. Toxicity of the antigen is reduced in various ways. In some vaccines only a small amount of the toxin is present. In others both the toxin and the antitoxin are combined, so that while the antigen stimulates the body to produce more antibodies, the antibodies of the vaccine act as protection for the cells. Still other toxins are heated or treated chemically to destroy their deleterious properties, without destroying their ability to stimulate antibody production; such a vaccine is known as a **toxoid.** Another method is to attenuate the cultures of bacteria by growing them for long periods of time in test tubes, until they have lost some of their toxicity. Rabies vaccine is attenuated by drying, others by injection into a series of laboratory animals. Typhoid vaccine may be prepared with dead typhoid bacteria.

The vaccination technique was discovered late in the eighteenth century by the English physician Edward Jenner, when he noticed that dairy workers handling cows with cowpox never had smallpox. When he tried scratching some serum from the pustules on a cow's udder into a human being, a mild disease resulted, with a single localized pox at the point of injection. Men so vaccinated never acquired smallpox. Cowpox and smallpox are caused by two distinct but closely related viruses; inoculation with cowpox virus (vaccinia) stimulates the production of antibodies that will also react with the closely related smallpox (variola) virus. Theoretically, it should be possible to immunize against all diseases by inoculations, but the means for doing so have not been developed for many important diseases, among them tuberculosis, influenza and syphilis.

When a person has contracted a disease and needs antibodies immediately to combat its antigens, his body may not be able to manufacture them quickly enough. In such cases, antibodies produced by some other animal—usually a horse—are injected to tide him over until his own body can produce enough to protect him. The injection of prepared antibodies (called a **serum**) is the only method of *passively* acquiring immunity; although immediately effective, the immunity disappears entirely after a few weeks.

To prepare a serum, bacteria are first cultured in test tubes until they have produced a large amount of toxin. This is then injected, in increasing doses, into a horse, which responds by gradually building up a tremendous amount of antitoxin in its blood. Finally, blood is withdrawn from the animal at intervals and processed to remove the blood cells and concentrate the antitoxin.

Natural Immunity. For all animals and plants, immunity to certain diseases is a part of their heritage and does not have to be acquired. There is evidence that the various races of man differ hereditarily in their resistance to such diseases as tuberculosis, diphtheria and influenza. Hereditary invulnerability to disease is called **natural immunity,** and is passed on to each new generation.

One type of natural immunity is that built up in a population which has been exposed to a particular disease for many generations. Many diseases, such as measles, which are relatively mild in civilized nations today, were extremely virulent among the American Indians and South Sea Islanders when they first came into contact with them. Syphilis, too, is a much milder disease now than at the time of its introduction into Europe, when it often was fatal in less than a month. And many tropical diseases, such as malaria and sleeping sickness, are more severe in white men than they are in the natives. Other diseases, originally widespread, have become rare with the passage of time—leprosy, for example, was extremely frequent in biblical times.

This progressive change is usually interpreted as being the result of a gradual process of "natural selection." Thus people who survived in the early days of the disease passed on their "survival ability" to their offspring, and so on. It is possible, also, that in some diseases the microorganism itself has undergone an adaptation, resulting in a decrease of virulence.

249. ALLERGY

Similar to the antigen-antibody reaction of immunity, is another, responsible for the type of hypersensitivity called **allergy,** which is expressed in such conditions as asthma, hay fever, skin rashes, and so on. This is due to the fact that any protein (antigen) which gets into the body not only stimulates the production of antibodies specific for it, but changes the reactivity of the cells. Then, if it enters the body again (at any time after about two weeks), its presence provokes any one or several of a multitude of symptoms: nausea, weakness, lowered temperature, convulsions and even death. Such symptoms are the outward signs of the body's reaction to the antigen. An allergy, like immunity, is highly specific for a particular protein. Sometimes the unpleasant symptoms disappear after the reaction to the second invasion, but often they recur for years or even throughout life, whenever the antigen is present. Desensitization may be accomplished by successive injections of minute amounts of the antigen.

Perhaps the most common of all allergies is that involving the pollen of certain plants—roses, goldenrod, ragweed, and so on—known as hay fever, rose fever, and the like. The protein antigen of the pollen enters the body through the mucous membranes of the nose or mouth and causes a sensitization. When, later, more pollen comes into contact with these membranes, they become swollen and irritated, and all the symptoms of a cold result.

Allergy to a particular kind of food is initiated when some undigested protein passes through the lining of the digestive tract (via a cut or lesion in the wall), and on into the blood stream. After sensitization to the protein has occurred, the next time the person eats the food he may become nauseated, or have hives, or some other reaction.

To test for an allergy, a small amount of sterile solution of a particular protein is injected into the patient's skin. If he is allergic to the protein, a large, localized, inflamed area appears, owing to the antigen-antibody reaction. The greatest care must be exercised that the test solution does not get under the skin, where it might come into contact with the blood stream and sensitize the person to the substance if he is not already sensitized.

It should hardly be necessary to say that, although allergies are a form of disease (asthma is an example of how devastating they may be), they are not infectious, since they are not caused by organisms. We include them in this chapter because the mechanism responsible for them is similar to that of immunity.

250. ANTIBIOTICS

In any natural environment there are many species of bacteria and fungi, some of them antagonistic to others. As early as 1879 it was known that, when microorganisms are grown together in the laboratory, one usually overcomes and kills the others. This phenomenon was given the name "antibiosis" and was first explained as being due to the organisms' competition for the same nutrients. Investigators later realized that certain organisms manufacture substances which are detrimental and even lethal to others, and these they called **antibiotics.** The earliest antibiotic to be discovered (pyocyanin) was isolated from pus in 1860, even before the bacterium which produces it was known. Many antibiotics have since been isolated from bacteria, and from a wide variety of plants, such as tomatoes and onions.

Unlike bacteriophages, antibiotic chemicals are successfully used against pathogenic bacteria. Today, the most important ones are isolated from molds. Of these, **penicillin,** which comes from a fungus closely related to those of Roquefort and Camembert cheese, is the most efficacious. This was first discovered by Fleming in 1929, but its possibilities were not fully appreciated until 1940. There are three different forms of this antibiotic, only one of which is highly active against bacteria. In contrast to many other antibiotics, it is not toxic when injected into man and other animals. Bacteria exposed to the action of penicillin swell and cannot divide, so that they are easily destroyed by the body's white blood cells.

Not all bacteria are vulnerable to penicillin, of course; some are susceptible to **streptomycin** which is extracted from Actinomyces—organisms intermediate between bacteria and fungi. Streptomycin is rather toxic and is now used primarily in the treatment of tuberculosis, for it is the only antibiotic effective against the tubercle bacillus.* Aureomycin, chloromycetin and Terramycin, derived from other Actinomyces, are effective against a number of viruses, rickettsias and bacteria. The antibiotics are of diverse chemical natures: some are protein-like, some are fatlike, and some are complicated organic chemicals of other types. A continual problem in the field of antibiotics is the development of strains of organisms that are resistant to the antibiotic. The constant hunt by the drug companies for new antibiotics has managed to keep us one "cure" ahead of the germs.

251. THE SPREAD OF MICROORGANISMS

The number of deaths each year due to infectious diseases has decreased steadily and markedly since the discoveries of Pasteur and Koch. The explanation of this lies partly in the discovery of better methods of treatment—the use of antitoxins, sulfa drugs, penicillin, streptomycin, Aureomycin and so on—but even more in our increasing efforts to prevent the *spread* of microorganisms. To infect a new host there must be some way for organisms to get from one person to another, and since most of them can exist for only short periods of time outside the body, usually the best way to eliminate contagious diseases is to eliminate the means whereby the causative agents are transmitted. The Federal Government, the individual states, and many cities and counties have departments of Public Health, whose task is the prevention of infectious diseases by the enforcement of quarantines, inspection of food supplies, elimination of insect pests, testing of dairy herds, supervision of milk pasteurization, filtration and chemical treatment of water supplies, and sanitary disposal of sewage. These measures, which were possible only after scientists had learned enough about microorganisms to determine how they could be attacked, now help protect us from the terrible epidemics and plagues of former ages. Some public health measures, however, were begun before there was any understanding of microorganisms. For example, the ancient Jewish soldier had to carry an implement to dispose of his excreta: "And thou shalt have a paddle upon thy weapon; and it shall be, when thou wilt ease thyself abroad, thou shalt dig therewith and shalt turn back and cover that which cometh from thee" (Deuteronomy 23:13).

Many microbes can pass from man to man only by the immediate contact of the two. In such cases isolation or quarantine of diseased persons is an effective way of preventing the spread of the disease. Quarantines were used long before the nature of disease was understood, for man early discovered that some illnesses were contagious. But even the strictest quarantine measures fail to prevent some spread of disease because many people are what is known as **immune carriers,** infected persons who have antibodies protecting them from the harmful effects of the organisms, and thus the appearance of disease symptoms. In any disease epidemic several times as many people have a mild case, and perhaps never know that they have had it, as those who suffer the severe, recognizable form. Immune carriers can, of course, transmit the microorganisms to others who are susceptible, and so unknowingly infect them. Typhoid, diphtheria, pneumonia and infantile paralysis are spread to a large extent by immune carriers.

Today, the number of people suffering from typhoid is less than a hundredth of what it was in 1890, owing chiefly to the use of chlorine and other disinfectants in drinking water. The drinking water of American cities is so pure that inoculation for typhoid is unnecessary except in emergencies, such as floods, when the water is likely to become contaminated.

Most of the milk sold in American cities has been pasteurized, or heated to a

* Isoniazid and *p*-aminosalicylic acid, which are effective against tubercle bacilli, are not antibiotics; they are not produced by the metabolic processes of some organism.

temperature of 145° F. for thirty minutes. This treatment kills all the harmful bacteria known to be transmitted by milk: the germs of diphtheria die at 129°, those of typhoid at 136° and those of tuberculosis at 138° F. But some harmless bacteria survive the treatment, so that milk is not sterile and will turn sour if left in a warm place. And of course any harmful organisms which get into it after pasteurization multiply rapidly, for milk is an excellent culture medium for most bacteria.

Perhaps the greatest single reason for the decrease in contagious diseases is simply the increasing tendency toward general cleanliness—the increase in bathing, and the more careful handling of food, disposal of garbage and sewage, and the like, are all measures destructive of bacteria which otherwise might get into the body.

252. SOME COMMON INFECTIOUS DISEASES

Poliomyelitis. The causative agent of poliomyelitis ("infantile paralysis") is a filtrable virus, which enters the body either through the nasal membrane to the olfactory nerves, whence it passes to the brain and spinal cord, or by the digestive tract, whence it passes to the nervous system via the blood stream. It attacks and destroys the motor nerves, thus causing paralysis of certain parts of the body. Withering of the limbs, however, is caused only indirectly; because they are paralyzed and hence unusable, the muscles atrophy from lack of exercise. Diagnosis in the early stages is difficult because many of its symptoms resemble those of a bad cold. There is a complement-fixation test which is useful in identifying the disease. There is no specific cure for polio; present treatment consists in keeping the patient in bed and well nourished so that a maximum of his energy can be used for combating the disease.

Polio is much more frequent in countries with high standards of sanitation and health. It appears that the virus is universally distributed and that in many countries infants are exposed to it, and develop immune antibodies to it, almost from birth. In countries with good sanitation, a child may not be exposed to the virus until later years; then the virus, unopposed by antibodies, rapidly produces severe symptoms. Probably most adults have been exposed to the virus many times and have developed protective antibodies. Antibodies to polio can be demonstrated in the gamma globulin fraction of the plasma of many people. The Salk vaccine developed in the last few years greatly increases an individual's resistance to the virus and has decreased both the total number of cases of polio and the fraction of patients who become paralyzed.

Typhus. A rickettsia transmitted to man by rat fleas, or lice infecting either rats or humans, is the cause of one of the great plague diseases, typhus. Usually associated with filth and overcrowding, epidemics often follow the disruptions of war, and after World War I, 25,000,000 cases occurred in Russia alone. That such an epidemic did not accompany World War II was primarily due to the extermination of fleas and lice by liberal use of the insecticide DDT. The disease begins with a violent headache and fever, followed by prostration and the eruption of red spots. Frequently it is lethal, the fatality rate varying from 5 to 70 per cent during different epidemics. There is a diagnostic test for it (the Weil-Felix reaction) which depends on an antigen-antibody reaction, but there is no satisfactory treatment.

Syphilis. This disease was not recognized in Europe before 1493, and its origin has always been a matter of dispute: some have maintained that it was brought back to Europe from the New World by Columbus' crew; others, that it was already present in Europe. At any rate, some of Columbus' crew accompanied King Charles VIII of France on his invasion of Italy in 1494, and the epidemic which started in Italy at that time spread over Europe as the troops scattered at the end of the war. The French called it the Italian disease, and, naturally, the Italians called it the French disease. It derives its present name from the principal character in a poem by

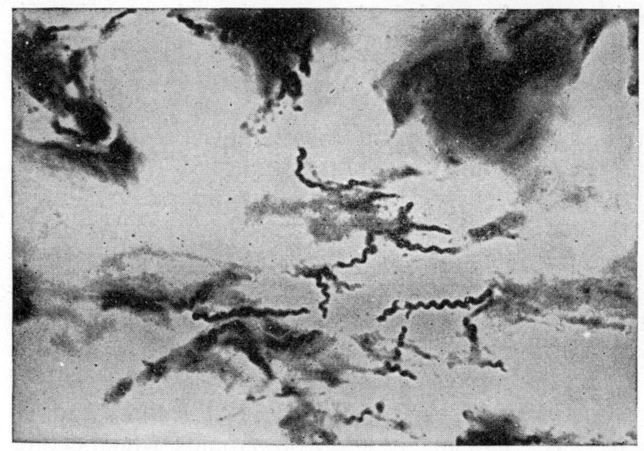

Figure 225. The spirochete, *Treponema pallidum,* which causes syphilis. The typical cork-screw shape is evident. Magnified 2000 times. (Kral, in Burrows, W.: Jordan-Burrows Textbook of Bacteriology.)

Fracastorius, written in 1530, which described the then current methods of treatment with sulfur and mercury.

Its causative agent is a long, tightly coiled bacterium, called a spirochete (Fig. 225), which usually is brought into the body by the contact of the mucous membranes of the penis and vagina during sexual intercourse. It may possibly be contracted occasionally by indirect means, if a person puts to his mouth something which has just been in contact with the mouth of a syphilitic. Or syphilis may be present from birth, transmitted by an infected mother to her child either before or at the time of birth.

The infection proceeds through three stages. In the *primary stage* a superficial lesion or sore appears on the external genitalia, or point of contact, ten to thirty days after infection. This is often small and painless and may be overlooked. The first stage lasts a few weeks only, and is followed by a long *secondary stage,* lasting months or years. During this period the spirochetes spread through the blood stream to other parts of the body, causing additional sores and lesions. But a long time may pass during which there is no external evidence of infection, so that a patient may think he is cured when he is not. If he is not cured, the *tertiary stage* occurs after several years and usually is fatal. All the organs of the body, and especially the central nervous system, may be attacked, resulting in paralysis and insanity due to the destruction of the nerve cells.

As you might expect, diagnosis varies according to the stage of the disease. In the first stage, demonstration with the aid of a darkfield microscope of the presence of spirochetes in the chancre or sore is a positive test. In the secondary and tertiary stages the famous Wassermann test, or one of its refinements, is used. The Wassermann test, which is based on a complement* fixation reaction, cannot be used in the primary stage because the antibodies do not develop until two or three weeks after infection. For many years, treatment of syphilis depended upon repeated injections of the chemical salvarsan, or one of its derivatives, containing arsenic. The discovery of this chemical, which is poisonous enough to kill the spirochetes, but not sufficiently toxic to kill the patient, was the result of a long process of trial and error, carried on by the German physician Ehrlich. Salvarsan has now been superseded by penicillin in treatment of syphilis.

The prevalence of syphilis in the United States varies widely in different parts of the country and among different groups of people. In one random series of autopsies, it was found in about 5.5 per cent of the total 150,000 cases. Of the first two million men drafted in World War II, 2.3 per cent of the whites, and 27.2 per cent of the Negroes, were infected. Since spirochetes cannot live outside the body, syph-

* Complement is a substance normally present in plasma which is removed as an antigen-complement-antibody complex when certain antigens and their corresponding antibodies are present. Somewhat similar complement fixation reactions are used in testing for a variety of virus-induced antibodies.

ilis could be abolished by the discovery and treatment of every infected person—a program which has been remarkably successful in some of the European countries, notably Denmark and Sweden. The difficulty lies in getting everyone to take a Wassermann, and in persuading those infected to continue treatment until they can no longer infect others.

Athlete's Foot. This widespread, irritating, but usually harmless disease is transmitted from person to person in public showers, or similar damp places. Spores of the fungi from an infected person are deposited on the moist floor and picked up by the next one to step there. Diagnosis can be made by microscopic examination of scrapings from the infected parts, but ordinarily this is not necessary, for the symptoms are common and easily recognizable. The fungus usually grows between the toes, where the skin is soft and moist, but sometimes it appears on the soles of the feet, the hands, or even in the scalp or beard. The first signs of it are itching, and tiny blisters which break, exposing the raw, pink underskin. Athlete's foot can be cured by the application of weak acids, and can be prevented by the use of fungicidal baths in locker and shower rooms, and by thorough drying of all parts of the body after bathing.

Amebic Dysentery. Amebic dysentery is well known as a "traveler's disease," because it is often picked up by the unwary foreigner in countries where human feces are used for fertilizer. It is also common around army bivouacs and on board merchant marine vessels, and it even occurs from time to time in the United States. In short, people suffer from it whenever conditions of sanitation are poor.

The agent causing amebic dysentery is the ameba, *Endamoeba histolytica* (*histo,* tissue; *lytic,* dissolving), which gets into food in the form of cysts. In the stomach, the cysts dissolve and the amebas are released. They burrow into the tissues of the large intestine, producing ulcers which interfere with the functioning of the colon, and consequently cause diarrhea. Diagnosis of the disease is made certain when microscopic examination of the feces reveals the amebas and their cysts. It is treated by the administration of drugs to kill the ameba. Sanitation, including thorough sewage disposal and cleanliness in the preparation of food, is its only preventive. Other important protozoan-induced diseases are malaria (see p. 415) and African sleeping sickness.

Tapeworm Infections. The tapeworm, which has become a kind of joke, but is not at all amusing to those who have one, owes its name to its shape, which resembles a measuring tape. The adult worm, which may be 6 to 20 or more feet long, lives only in the digestive tract of man. It lays eggs which it fertilizes itself (the worm is hermaphroditic), and which pass out of the body with the feces. The eggs are then eaten by an intermediate host—a hog or cow—in whose digestive tract the eggs grow into larvae. These bore through the wall of the intestine, enter the blood stream and are carried to the muscles, where they develop into small, cyst-like "bladderworms" (Fig. 226). A person becomes infected when he eats meat from an infected hog or cow that has not been cooked enough to kill the bladder worms. The bladderworm passes into the intestine, rapidly matures into an adult, and fastens onto the wall of the intestine by hooks in its head. Here it remains, "soaking up" digested food from its host (for it has no mouth and no digestive tract of its own) and consequently causing cramps, indigestion and malnutrition. Diagnosis of the disease depends upon the demonstration, by a trained laboratory technician, of tapeworm eggs in the feces. Treatment consists in administering drugs to kill or anesthetize the worms, followed by purges to flush them out. Prevention is effected by thorough cooking of all flesh eaten (other tapeworms are acquired from raw fish), by the disposal of sewage to prevent the infection of intermediate hosts, and by inspection of all carcasses destined for the market.

Trichinosis. The small roundworm causing this disease belongs to a different phylum from the tapeworm and is structurally quite different. The larvae, which are minute, coiled worms, 1 mm. long, lie within the muscles of a hog, enclosed

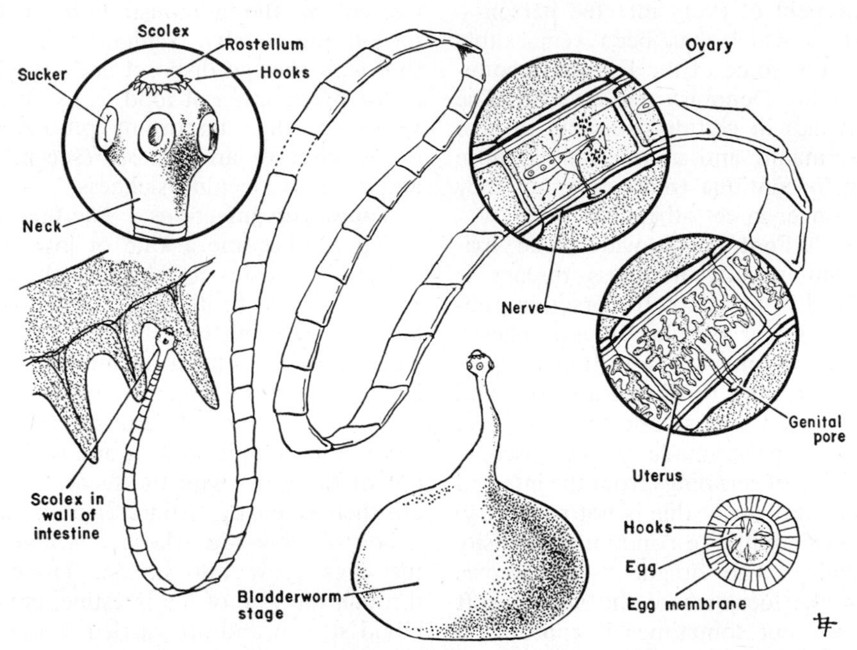

Figure 226. The pork tapeworm, *Taenia solium.* Insets show the head, an immature and a mature section of the body.

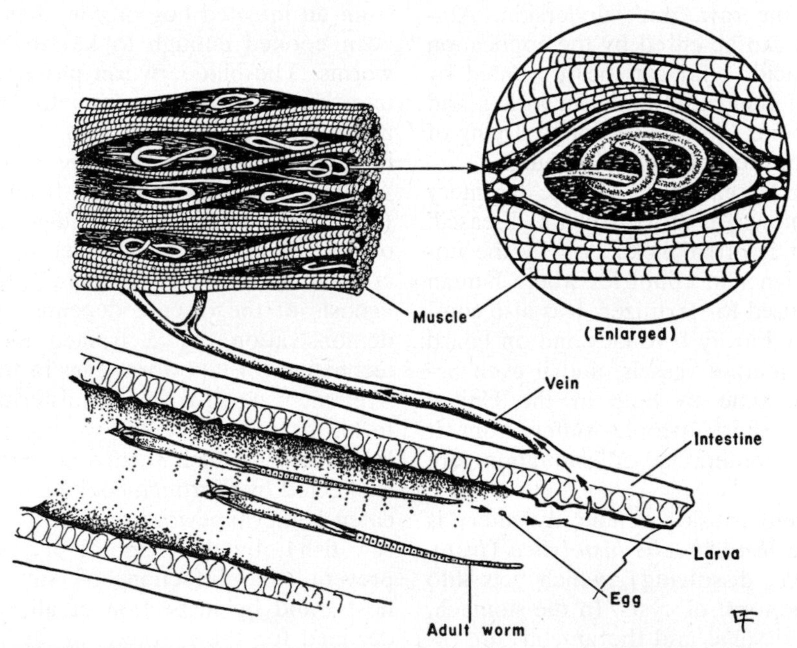

Figure 227. The roundworm, *Trichinella spiralis,* which causes trichinosis. When a piece of pork infected with *Trichinella* is eaten, the larvae are released and grow rapidly to maturity in the intestine. After fertilization, the females produce tiny larvae which burrow into the blood vessels and are carried to the muscles, where they encyst (inset, upper right).

by lemon-shaped cysts (Fig. 227). When improperly cooked pork is eaten, the cysts are digested away in the stomach and the worms enter the small intestine and mature in about two days. After a week or so, when the females' eggs have been fertilized by the males' sperm, the females deposit tiny larvae in the intestinal lining. The larvae pass through the blood stream from the intestine to the muscles, where they form cysts. There are, then, two distinct phases of the disease: the first, when the intestine is infected with the adult worms, which involves severe gastrointestinal disturbances and is sometimes fatal, and the second, which follows invasion of the muscles by the larvae and is characterized by muscular soreness.

Trichinosis is not directly transmissible from person to person (or could be so only through cannibalism), and the hogs themselves become infected by eating pork scraps, or rats which have eaten pork scraps. The disease can be diagnosed by an antigen-antibody skin test, but unfortunately there is no known cure. Government inspection of pork does not include inspection for the trichinosis roundworm, which would be impractical, but the disease can be prevented by the thorough cooking of all pork products. Further preventive measures include outlawing the feeding of raw garbage to pigs and the elimination of rats from pig farms. It has been found recently that quick-freezing pork is an effective method of killing encysted *Trichinella* larvae.

QUESTIONS

1. What are three different types of disease? What is the cause of each?
2. What is meant by the germ theory of disease?
3. What characteristics enable microorganisms to be such successful parasites?
4. What are antibodies? How do they combat disease?
5. Differentiate between exotoxins and endotoxins; antigens and antibodies; vaccines and serums.
6. What means does the body have to defend itself against the invasion of disease organisms?
7. Explain how serums are prepared.
8. What is the physiologic explanation of allergies?
9. What measures have been effective in decreasing the incidence of typhoid? of syphilis?
10. What insects are parasites of man?
11. What problems are involved in the use of antibiotics in the treatment of disease?

SUPPLEMENTARY READING

The details of the transmission, cure and prevention of infectious diseases can be found in textbooks of bacteriology such as the ones by Frobisher, Grant or Krueger. Jenner's description of his method of vaccination as a prevention of smallpox is printed in Holmes Boynton's *The Beginnings of Modern Science*. The great influence of plagues on the course of history is discussed in Hans Zinsser's *Rats, Lice and History* and in Smith's *Plague on Us*. The relation of geography to disease is demonstrated in J. S. Simmons' *Global Epidemiology*.

Part Five

The Reproductive Process

Chapter 28

Reproduction

THE SURVIVAL of each species of plant or animal requires that its individual members multiply, that they produce new individuals to replace the ones killed by predators, parasites or old age. The actual process involved varies tremendously from one kind of animal to another, but there are two basic types of reproduction, asexual and sexual. **Asexual reproduction** involves only a single parent, which splits, buds or fragments to give rise to two or more offspring which have hereditary traits identical to those of the parent. **Sexual reproduction** involves two parents, each of which contributes a specialized cell or **gamete** (eggs and sperm) which fuse to form the **zygote** or fertilized egg. The egg generally is nonmotile and large, with a store of yolk to supply nutrients for the embryo which will result if the egg is fertilized; sperm are usually small and motile, adapted to swim actively to the egg by the development of a tail. The biologic advantage of sexual reproduction is that it permits the recombination of the best inherited characteristics of the two parents, thus giving rise to offspring better able to survive than either parent.

The evolution of sexual reproduction in plants was described in Chapter 12; no similar clear series exists in animals and we can only guess as to how and when sexual reproduction began. Even the highest animals may reproduce asexually; the production of identical twins from the splitting of a single fertilized egg is a kind of asexual reproduction.

253. ASEXUAL REPRODUCTION

Living things can give rise to a new generation asexually in a number of ways. When the body of a parent simply splits into two more or less equal daughter parts, which become new organisms, the

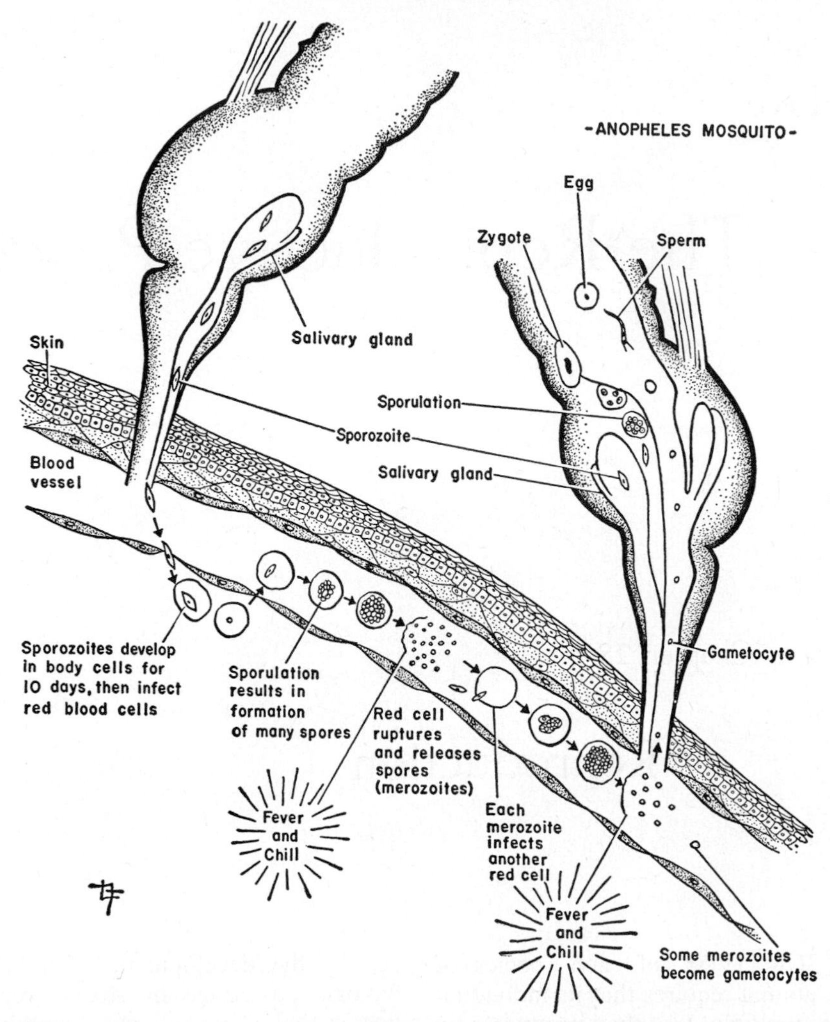

Figure 228. A diagram of the life cycle of the malaria parasite, *Plasmodium.* An infected mosquito (*left*) bites a man and injects some *Plasmodium* sporozoites into his blood stream. These reproduce asexually by sporulation within the red blood cells of the host. The infected red cells rupture and the new crop of merozoites released then infects other red cells. The bursting of the red cells releases toxic substances which cause the periodic fever and chill. In time some merozoites become gametocytes which can infect a mosquito if one bites the man. The gametocytes develop into eggs and sperm (*right*) and undergo sexual reproduction in the mosquito and the zygote, by sporulation, produces sporozoites which migrate to the salivary glands.

process is called **fission.** This method occurs predominantly among the single-celled animals and plants such as amebas, paramecia and algae. The cellular division involved in the splitting process takes place by mitosis.

Hydras and yeasts reproduce by **budding,** in which a small part of the parent's body separates from the rest and develops into a new individual, eventually either taking up an independent existence or becoming a more or less independent member of the colony.

Many higher animals—salamanders, lizards, starfish and lobsters—can grow a new leg, tail or other organ if one is lost. When this ability is carried to an extreme, it becomes a method of reproduction, called **fragmentation.** By this method the entire body of the parent may break into pieces—as many as several hundred—each of which develops into a new animal.

The process is particularly common among flatworms, but starfish also have the ability to a high degree—a single arm can regenerate a whole new starfish. At one time, when oystermen caught starfish feeding in oyster and mussel beds, they would cut them in half and throw the pieces back into the water, believing that they had killed them. But each half would quickly regenerate, making twice as many predators for the oysters as before. Now oystermen know that starfish must be killed by being kept out of water.

Some animals and most plants reproduce asexually by means of spores, which are special cells with resistant coverings, adapted to withstand unfavorable environmental conditions such as excessive heat, cold or desiccation. One of the most interesting examples of reproduction by spore formation occurs in the single-celled animal *Plasmodium,* which causes malaria fever. The organism has a complex life cycle involving both man and a certain mosquito, *Anopheles* (Fig. 228). Through the bite of the mosquito, the malaria organisms infect a person's blood stream and enter the red blood cells. Inside the red cell each *Plasmodium* divides into from twelve to twenty-four spores, each of which enters the blood stream when the cell, after swelling, disintegrates. The spore forms released infect new cells and the process is repeated. The simultaneous breakdown of billions of red cells causes the malarial chill, followed by fever as the toxic substances released penetrate to other organs of the body. If a second, uninfected mosquito bites the infected man, it becomes infected by sucking up some of his blood. Inside the mosquito's stomach a complicated process of sexual reproduction results in the formation of thousands of new spores, some of which migrate into the mosquito's salivary glands, ready to infect the next man bitten.

For hundreds of years man has used an extract of the bark of the cinchona tree, *quinine,* to treat malaria. Since long treatments and large amounts are required and toxic side effects are common, an intensive search was made during World War II for a more effective antimalarial. Chloroquine, Paludrine, Atabrine, Daraprim and Primaquine were found among the thousands of compounds synthesized and tested, and these have now replaced quinine to a large extent.

254. SEXUAL REPRODUCTION IN ANIMALS

A few animals such as the coelenterates have alternate sexual and asexual generations reminiscent of the life cycles of plants, but most animals reproduce solely by sexual means and have permanent sex organs.

Some of the protozoa, such as paramecia, have a complicated process of sexual reproduction in which two individuals come together on their oral surfaces, fuse, and exchange micronuclear material. The original micronucleus of each divides several times before one of the resulting nuclei migrates across to the other animal and fuses with one of its nuclei. After that the animals separate and there is further nuclear and cytoplasmic division, eventually resulting in four individuals. It was once thought that sexual reproduction is occasionally necessary to rejuvenate the strain, but Woodruff of Yale maintained a strain of paramecia for twenty-five years without sexual reproduction and with no loss of vitality. Although paramecia are not differentiated into sexes, T. M. Sonneborn has shown that there are definite mating groups, and that individuals of one group will mate only with those of another. The mating groups are determined by inheritance.

Hermaphroditism. Many of the lower animals are **hermaphroditic;** both ovaries and testes are present in the same individual and it produces both eggs and sperm. Some hermaphroditic animals, the parasitic tapeworms, for example, are capable of self-fertilization. Since a host may be infected with but one parasite, this ability is an important adaptation for the survival of the species. Most hermaphrodites, however, do not reproduce by self-fertilization; rather, two animals come together in copulation, each inseminating the other. This method is used by earthworms. In other species, self-fertilization is prevented by a development of the testes and ovaries at different times. Oysters, for example, are first male, then female.

Parthenogenesis. A rather rare modification of sexual reproduction, common among honeybees, wasps and certain other arthropods, is **parthenogenesis,** the development of an unfertilized egg into an adult animal. Some species of arthropods consist entirely of females which reproduce in this way. More commonly, parthenogenesis occurs for several generations only, after which males develop and fertilize a generation of eggs. Often, fertilization occurs in the fall, and the fertilized eggs are quiescent during the winter. Among honeybees the female or queen bee is fertilized by a male just once in her entire life time. The sperm she receives are stored in a little pouch connected with her genital tract, and closed off by a muscular valve. As the queen lays eggs, she can either open this valve, permitting the sperm to escape and fertilize them, or keep the valve closed, so that the eggs develop without fertilization. The fertilized eggs become females—queens and workers; the unfertilized eggs becomes males (drones). Some species of wasps give birth alternately to a parthenogenetic generation and one from fertilized eggs.

Eggs from species which do not normally undergo parthenogenetic development may be stimulated artificially to develop without fertilization. Frog eggs can be so stimulated by pricking them with a fine needle; other eggs can be made to divide by shaking, by adding chemicals to the water in which they are lying or (as in marine eggs) by changing the concentration of salt in the water. The animals resulting from artificial parthenogenesis are often weaker and smaller than normal, and may be unable to complete development. Adult frogs and rabbits have been produced by this means in the laboratory of Gregory Pincus.

Many species of worms, and most of the other animals farther along the evolutionary scale, have permanent structures for sexual reproduction. The gonads, their ducts and accessory structures may be single, paired or multiple, perhaps present in several segments of the body. Among the vertebrates a wide assortment of accessory structures has evolved to facilitate the union of egg and sperm, and to ensure the development of the embryo.

Types of Fertilization. Most aquatic animals simply liberate their sperm and eggs into the water, so that their union is haphazard. No accessory structures are needed, except the ducts which transport the cells to the outside of their bodies. This, the most rudimentary and uncertain method of uniting the gametes, is called, for obvious reasons, **external fertilization.**

Other animals, especially those living on land, have accessory sex organs for transferring the sperm from the body of the male to that of the female, so that fertilization occurs within the latter. This method, called **internal fertilization,** requires the cooperation of the sexes, and many species have evolved elaborate patterns of mating behavior to insure that it takes place.

An unusual one is that exhibited by the salamander: the male clasps the female and rubs her nose with his chin; he then dismounts in front of her, and deposits a spermatophore, a sac containing a large number of sperm. The female picks up the sac and places it in her cloaca. Here the sac breaks, releasing the sperm to fertilize her eggs.

Many breeding habits are dangerous or even fatal to the animals performing them; for instance, salmon swim hundreds of miles upstream to spawn and die; and the male spider is frequently eaten by the female after he has performed the necessary act of inseminating her. Yet the survival of the species requires that the individual sacrifice his own interests in the performance of these acts.

The evolution of instincts for the care of the young has accompanied the evolution of more efficient methods for bringing about fertilization. Fish and amphibia in general take no care of the developing eggs, and great quantities must be laid each year in order that by chance a few will develop. The eggs of reptiles are usually laid in the sand or mud, where they develop without parental care, warmed only by the sun. Birds lay their eggs in nests, and incubate them by sitting on them. The newly hatched birds are quite helpless and require parental attention for several weeks.

In contrast to these eggs which develop more or less at the mercy of the environment, the mammalian egg (with the exception of those of the monotremes, which are laid and develop outside the body) develops within the uterus of the famale, where it is safe from predators and from environmental changes until it is able to cope with them.

The young of many vertebrates—primarily the amphibians—do not resemble their parents at all, but go through an early larval form. At the proper time the larva, such as a tadpole, undergoes a relatively rapid change, or **metamorphosis,** to become the adult form. In general, the existence of the larval form is correlated with a relatively small amount of yolk in the egg, so that the young must become self-supporting at an early time. The larvae of many insects are specially adapted to feed and store food, while the adult animal eats nothing and lives only long enough to reproduce.

In the evolution of the vertebrates from fish to man, the trend has been towards the production of fewer eggs, and the development of instincts for parental care of the young. Thus the codfish produces millions of eggs a year, few of which ever develop into adult fish, while mammals have few offspring, but take such good care of them that the majority reach adulthood.

In the vertebrates a number of accessory structures have developed to facilitate the transfer of sperm from the male to the female reproductive tract, and to provide a place for the development of the fertilized egg. These structures have evolved either from or in close association with the urinary system, and the two systems together are frequently referred to as the **urogenital system.**

255. HUMAN REPRODUCTION

Human reproduction, in common with that of most animals and plants, is accomplished sexually, by the union of specialized cells called gametes—ova or eggs produced by the female, and sperm produced by the male. All the parts of the complicated reproductive system in both sexes, as well as the various physiologic and psychologic phenomena associated with sex, have just one purpose: to insure the successful union of the egg and sperm, and the subsequent development of the fertilized egg into a new individual.

The Male Reproductive Organs. The testis, discussed previously as an endocrine gland, performs an equally important function as the source of the male sex cells. A pair of these glands develops within the abdominal cavity of all vertebrates, but in man and some other mammals, they descend shortly before or after birth into the **scrotal sac,** a loose pouch of skin which is an outpocketing of the body wall. The cavity within the scrotal sac is part of the abdominal cavity, and during development is connected to it by a passageway, the **inguinal canal.** After the testes have descended through it, this canal is usually closed by the growth of connective tissue, so that the abdominal and scrotal cavities are separate. Occasionally, the inguinal canal reopens or fails to close, resulting in an inguinal hernia. In such cases a loop of the intestine may slip down into the scrotal sac and be caught there by the contraction of the abdominal muscles, which cuts off its blood supply. Unless an operation is quickly performed, the tissue of that part of the intestine dies. The normal descent of the testes into the scrotal sac is necessary for the production of sperm. If the testes remain in the abdominal cavity, as occasionally they do, the slightly higher temperature there prevents the formation of sperm.

Each testis consists of about one thousand highly coiled **seminiferous tubules,** which actually produce the sperm, and, lying between them, masses of interstitial cells which produce the male sex hormones (Fig. 222). The seminiferous tubules are lined with a germinal epithelium made up of rounded cells with large nuclei. These cells undergo division to form cells which develop into the sperm, with compact heads, containing the nucleus, and a long, whiplike tail for

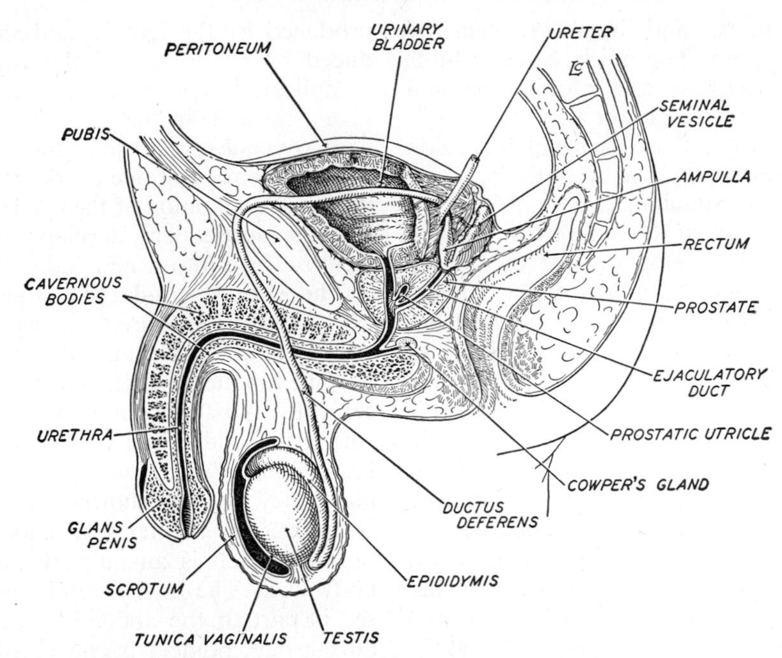

Figure 229. Schematic sagittal section of the male pelvic region, showing the genital organs and their relations to the urinary bladder and urethra. (Turner, C. D.: General Endocrinology.)

locomotion. In addition, the seminiferous tubules contain "nurse" cells which nourish the sperm as they develop from rounded cells into the mature, tailed form. The formation of sperm proceeds in waves along the tubules, so that the sperm which are not yet motile are moved along largely by the pressure of other sperm behind them.

At one end of each seminiferous tubule is a fine tube called the **vas efferens,** which connects it to a single, complexly coiled tube, the **epididymis,** where the sperm are stored. Each of the two epididymides lies close to the base of the testis to which it is attached. From each epididymis a duct, the **vas deferens,** passes from the scrotum through the remains of the inguinal canal, into the abdominal cavity, and over the urinary bladder to the lower part of the abdominal cavity, where it joins the urethra (Fig. 229).

As soon as the sperm are produced they are suspended in a liquid, the **seminal fluid,** which protects them as they are transferred to the female reproductive tract. This fluid is produced by three different glands. A short distance before the vas deferens joins the urethra, a pair of glands called the **seminal vesicles** empty into it by ducts. Farther on, around the urethra, at its source from the urinary bladder, are the **prostate glands** (in man these are fused into a single gland), which secrete their part of the seminal fluid into the urethra by two sets of fine, short ducts. Finally, farther along the urethra, at the base of the spongy tissue of the penis, lies a third pair of small glands, called **Cowper's glands,** which contribute the last component to the seminal fluid. The seminal fluid contains buffers that protect the sperm from the acids normally present in the urethra and the female reproductive tract, glucose and fructose as substrates for the metabolism of the sperm, and substances which lubricate the passages through which the sperm travel.

The **urethra** is a tube leading from the urinary bladder to the outside of the body. In the male the last part of it runs through the **penis,** the external reproductive organ, just above and in front of the scrotal sac. Within the penis the urethra is flanked by three columns of erectile tissue, which is spongy and capable of being filled with blood. Ordinarily, the spaces of the spongy tissue, though containing blood, are not distended, but during sexual excitement the arteries leading to these blood spaces dilate so that the spaces are filled with blood under pressure. This causes the penis to change from its usual flaccid state to become hard and erect, and thus able to enter the vagina. Ejaculation of the **semen**—the sperm in their seminal fluid—occurs when the penis is further stimulated by friction against the walls of the vagina, and is due to reflex waves of contraction of the muscles of the vas deferens, the urethra and the associated glands.

The Female Reproductive Organs. The egg-producing organs of the female—the **ovaries**—are held in place by ligaments within the lower part of the abdominal cavity, between the hips. Each of the pair is about the size and shape of a shelled almond. When a girl reaches puberty there are some 30,000 eggs in each of her ovaries. Apparently no new ones are ever produced thereafter. Since normally a woman ovulates thirteen times a year for approximately thirty years, and but a single egg ripens each month, only 400 or so of these eggs ever reach maturity and escape from the ovary; the rest degenerate and are absorbed. The ovaries alternate in releasing their eggs, but the alternation is irregular and unpredictable.

In contrast to the male reproductive tract, the female tract does not consist of a series of connected ducts. Instead, the egg is released into the abdominal cavity at the time of ovulation (p. 396), whence it passes into one of two tubes, called the **oviducts** or **fallopian tubes** (Fig. 230). Each of this pair of tubes ends in a funnel-shaped enlargement which normally covers a part of the surface of the ovary. When the egg has been discharged into the body cavity, it normally is directed into one of these funnels by currents set up by the beating of cilia which line the oviducts. Rarely, having failed to enter the oviduct, an egg is fertilized within the abdominal cavity, where, attached to some organ such as the liver, it begins to develop. Usually, when this occurs, development cannot be completed and the embryo must be removed from its position in the abdominal cavity; there have been instances, however, of such embryos completing development and being "born" by means of a surgical operation on the mother.

The two oviducts empty directly into the upper corners of a pear-shaped organ, the **uterus** or **womb,** the primary function of which is to house the developing embryo until the time of birth. This organ

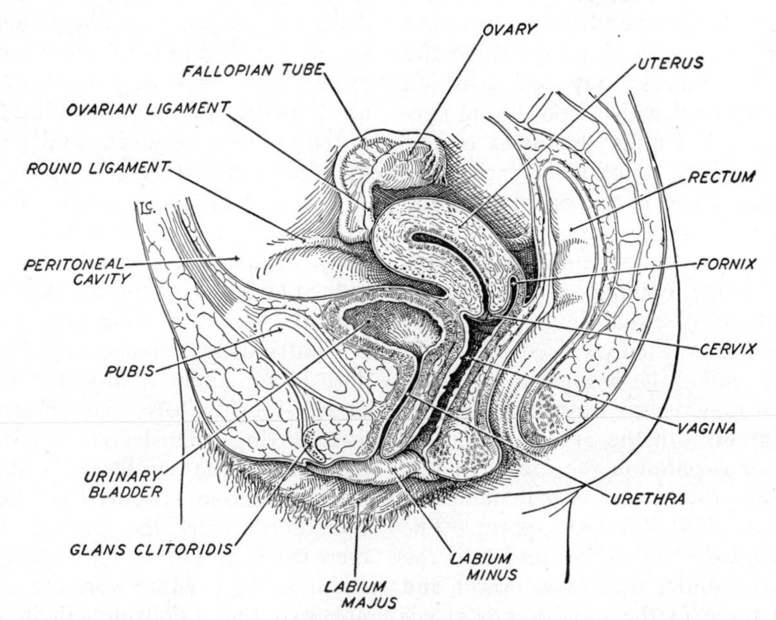

Figure 230. Schematic sagittal section of the female pelvic region, showing the genital organs and their relations to the urinary bladder and urethra. (Turner, C. D.: General Endocrinology.)

lies in the middle of the lower part of the abdominal cavity (Fig. 230) just behind the urinary bladder. It is about the size of a clenched fist, and has thick muscular walls and a mucous lining richly supplied with blood vessels. From the center of its lower end a single muscular tube, the **vagina,** passes to the outside of the body. The vagina is larger than any other tube in the reproductive tract; it serves both as receptacle for the sperm, and so must accommodate the penis, and as birth canal when the prenatal development is complete. The uterus terminates in a muscular ring, the **cervix,** which projects a short distance into the vagina.

The external female sex organs are collectively known as the **vulva.** At the lowest extremity of the torso, in front, is a mass of fatty tissue forming a slight elevation known as the **mons veneris.** From this point, two folds of skin, covered with hair, the **labia majora,** extend back and down, to enclose the opening of the vagina, and grow together into a single tissue behind it. Inside the labia majora are two other folds of skin, free of hair, the **labia minora.** At the junction of these two in front is a sensitive, erectile organ, about the size and shape of a small pea. This is the **clitoris,** an organ homologous to the penis of the male; that is, it develops from the same embryonic structure from which the penis develops. Like the penis, it contains spongy tissue which becomes engorged with blood during sexual excitement. Unlike the male organ, however, its only function is excitatory. Behind the clitoris is the opening of the urethra, which in the human female has only a urinary function, and behind that the much larger opening of the vagina. Before the first sexual intercourse, the latter opening is usually closed by a thin membrane, called the **hymen,** but disease or accident may destroy this in childhood.

Fertilization. In the act of sexual intercourse or **copulation** the erect penis is inserted into the vagina, where it ejaculates about 200,000,000 sperm. The sperm travel up the vagina and into the uterus partly under their own power and partly by force of the muscular contractions of the walls of these organs. Most of the sperm become lost on the journey, but a few find their way to the openings of the oviducts and swim up them. Sperm can swim against a current, and the same current which draws the egg from the abdominal cavity into the oviduct probably assists them in finding their way. If ovulation has occurred shortly after or before copulation, the egg which passes into the oviduct probably will be fertilized by one of the sperm. Fertilization usually occurs in the upper third of the oviduct; by the time the egg reaches the lower part of the oviduct it has lost the capacity to be fertilized. Only one of the millions of sperm deposited at each ejaculation fertilizes a single egg.

Each human egg is surrounded by a layer of cells (derived from the follicle) called the *corona radiata,* which must be pierced before a sperm is able to unite with the egg. The individual cells of the layer are held together by a complex organic substance known as **hyaluronic acid,** which is acted upon by the enzyme **hyaluronidase.** Since each sperm contains only a small amount of hyaluronidase, it requires the combined supply of several hundred thousand sperm to break down the hyaluronic acid and separate some *corona radiata* cells, in order that one sperm may enter and fertilize the egg. As soon as the egg has united with a sperm, it develops a membrane which prevents the entrance of others. The unused sperm and the unfertilized eggs die in the oviducts or uterus and are removed by white corpuscles.

Considering how many factors are working against it, it seems remarkable that fertilization ever does occur, and, indeed, man is a comparatively infertile animal. The best evidence available at present indicates that sperm remain alive and retain their ability to fertilize for twenty-four to forty-eight hours at most after having been deposited in the female tract, while the egg loses its ability to be fertilized about twenty-four hours after ovulation. Thus the period during the menstrual month when fertilization can occur is brief. Moreover, the sperm cells are extremely delicate; their cytoplasm contains small resources of food, and, being sensitive to heat, they die quickly at body temperature. Still another hazard lies in the leukocytes of the vaginal epithelium, which always engulf countless millions of sperm. It is only because copulation is so frequent and so many sperm are deposited at each ejaculation, that the human race is able to maintain itself as well as any other species.

After fertilization has occurred, the zygote, while passing down the oviduct to the uterus, begins to divide (Fig. 231). Eight to ten days elapse from the time the egg is fertilized until it is implanted in the uterine wall. Until it is firmly implanted, the developing embryo is nourished by a secretion of the uterine glands known as "uterine milk." By the time of implantation the embryo consists of a cluster of several hundred cells derived by division from the single, original, fertilized egg.

Implantation. The implantation of the developing embryo in the lining of the uterus is a process that involves activity on the part of both embryo and uterine lining. The embryo secretes substances which destroy a few of the cells of the lining and then penetrates at that point. This stimulates the uterine tissues to grow and surround the embryo. It is possible to get this imbedding reaction by pricking the lining of the uterus with a glass needle; when a rat is treated in this way, the uterus develops just as though an embryo were present and the rat undergoes a short pseudopregnancy.

Nutrition of the Embryo. After implantation in the uterine lining, the embryo continues to develop, at first obtaining its nourishment by enzymatically breaking down the cells of the uterine wall immediately around it, and later by extracting the nutritional essentials from the blood stream of the mother via the blood vessels of the placenta.

The new human being develops only from the cells which lie along one side of the hollow ball originally implanted in the uterus; the other cells form membranes which nourish and protect the developing child and eventually form part of the "afterbirth." The problem of supplying the embryo with food during development has been solved in different ways by the several groups of vertebrates.

Fish and amphibia produce relatively large eggs, containing yolk to supply the necessary proteins, fats and carbohydrates; these eggs are laid and develop in

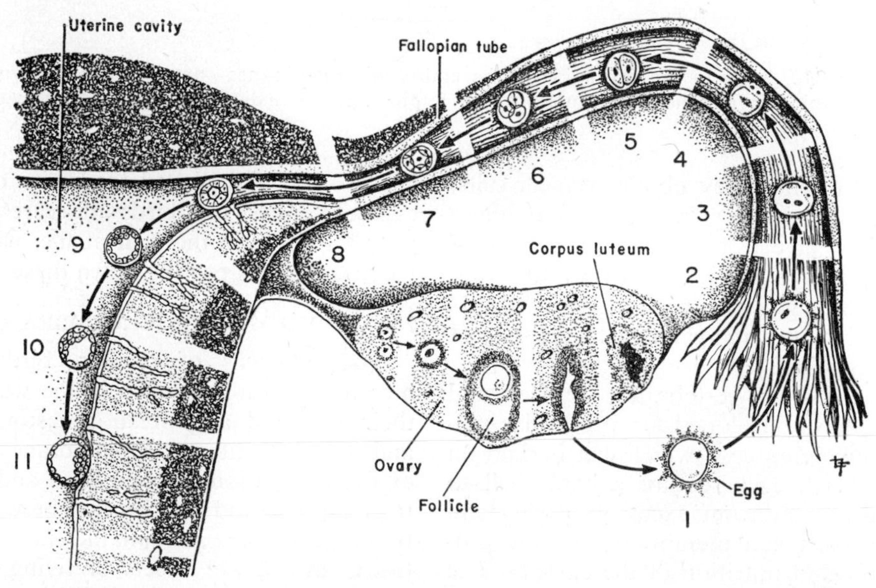

Figure 231. Schematic diagram of the maturation of an egg in a follicle in the ovary, its release (ovulation) (*1*), fertilization in the upper part of the oviduct (*2*), cleavage of the egg as it descends the oviduct or fallopian tube (*3–7*), stages in the development of the embryo in the uterus before implantation (*8–10*), and implantation of the embryo in the wall of the uterus (*11*).

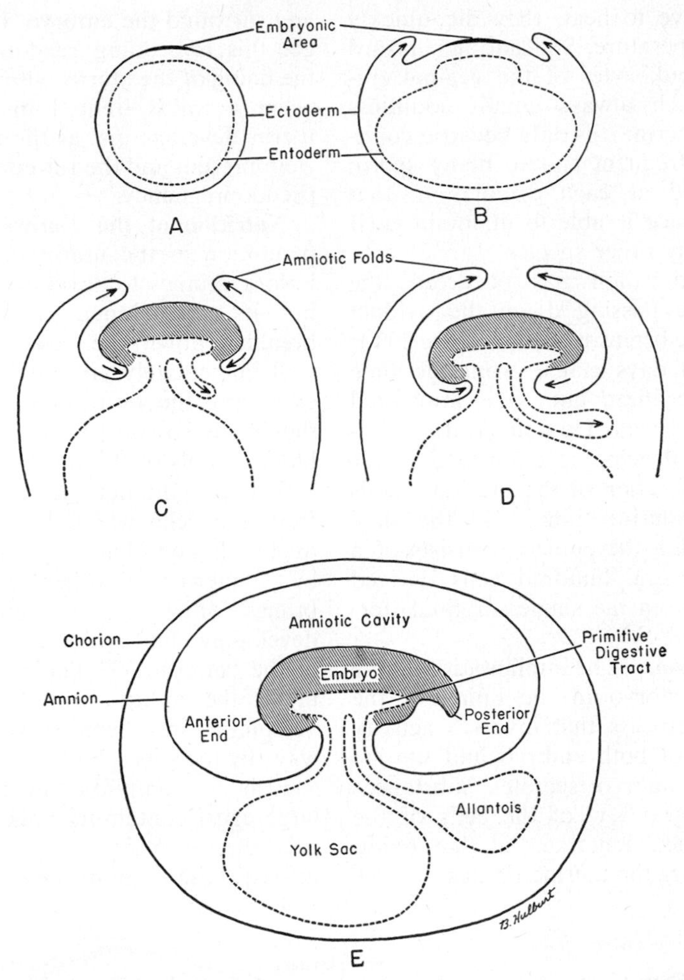

Figure 232. Steps in the formation of the embryonic membranes—amnion, chorion, yolk sac and allantois—in a typical mammal such as a pig. Arrows indicate direction of growth and folding.

water, whence they obtain oxygen, salts, and water itself. The embryos of these animals have a pouchlike outgrowth of the digestive tract, known as the **yolk sac,** which grows around the yolk, digests it, and makes it available to the rest of the organism.

The higher vertebrates, reptiles and birds, have developed a reproductive system involving eggs which can be laid on land. Such eggs require a hard shell to protect them from excessive drying and have additional membranes for the protection and nutrition of the embryo. The familiar "white" of the hen's egg is an extra store of protein and water to carry the embryo through to the time of hatching. Both the shell and the white of the eggs of reptiles and birds are secreted by glands located in the wall of the oviducts while the egg is passing down these tubes.

256. THE EMBRYONIC MEMBRANES

The embryos of reptiles, birds and mammals do not develop in the water as their fish and amphibian ancestors did, and several embryonic membranes have evolved to enfold the embryo and protect, support and nourish it. These membranes, called the **amnion,** the **chorion** and the **allantois,** are sheets of living tissue growing out of the embryo itself—the amnion and chorion out of the body wall to enfold the embryo (Fig. 232), and the allantois out of the digestive tract to function in the absorption of food.

The formation of the amnion is a complex process, differing in detail in different species, but in all it is essentially an outfolding of the body wall of the embryo which grows around the embryo to meet and fuse above it (Fig. 232). The space created between the embryo and the amnion, known as the **amniotic cavity,** becomes filled with a clear, watery fluid secreted by both the embryo and the membrane. Thus embryos of the higher vertebrates reach the birth stage enclosed in a small pool within the shell or uterus. The amniotic fluid acts as a protective cushion which absorbs shocks and prevents the amniotic membrane from sticking to the developing embryo, while permitting the organism a certain freedom of motion. During the birth process of human beings, pressure of the amniotic fluid helps to dilate the neck of the uterus; later, the amnion normally ruptures, releasing, shortly before the fetus is born, about a quart of amniotic fluid, the so-called "waters." Sometimes the amnion fails to burst, and the child is born with it still enveloping its head. It is then popularly known as a "caul," and is the source of many odd superstitions.

The amnion develops from the inner part of the original fold from the body wall; the outer part forms a second membrane, the **chorion.** This membrane, in the eggs of reptiles and birds, rests in contact with the inner surface of the shell, and in mammals it is established next to the cells of the uterine wall.

The allantois, like the yolk sac, is an outgrowth of the digestive tract. It grows between the amnion and chorion and, in animals like the chick, where it is a large and functional membrane, fills almost all the space between these two. The allantois of the chick fuses with the chorion to form a compound membrane, full of blood vessels, by means of which the embryo takes in oxygen, gives off carbon dioxide, and excretes waste products. Since the embryo "breathes" through the shell, it will suffocate if the shell is coated with wax. In the human, the allantois is small and nonfunctional, except for furnishing blood vessels to the placenta, and the yolk sac is completely nonfunctional. When the chick hatches from the shell, or when the child is born, most of the allantois and all the other membranes are discarded. But the base of the allantois, the part connected to the digestive tract originally, remains within the body and is converted into part of the urinary bladder.

As the human embryo grows, the region on the ventral side from which the folds of the amnion, the yolk sac and the allantois grew, becomes relatively smaller, and the edges of the amniotic folds come together to form a tube which encloses the other membranes. This tube is the **umbilical cord,** which contains, in addition to the yolk sac and allantois, the large blood vessels through which the embryo obtains nourishment from the wall of the uterus. The umbilical cord is composed chiefly of a peculiar jelly-like material found nowhere else. It is about half an inch in diameter and about 2 feet long. It is usually twisted spirally, and in its contortions before birth, the fetus* sometimes passes through a loop of the cord and actually ties a knot in it.

257. THE PLACENTA

The outer surface of the chorion, in man and the higher mammals, is thin over most of its surface, but at the outer extremity of the umbilical cord it develops a number of finger-like projections, known as **villi,** which grow into the tissue of the uterus. These villi, plus the tissues of the uterine wall in which they are embedded, make up the organ known as the **placenta,** by means of which the developing embryo obtains food and oxygen and gets rid of carbon dioxide and metabolic wastes (Fig. 233). These processes can be accomplished because of the many capillaries in the villi which receive blood from the embryo by way of one of the two

* As soon as the zygote or fertilized egg begins to divide, it is called an embryo. After the embryo has begun to resemble a human being (some two months after fertilization), it is referred to as a fetus until the time of birth.

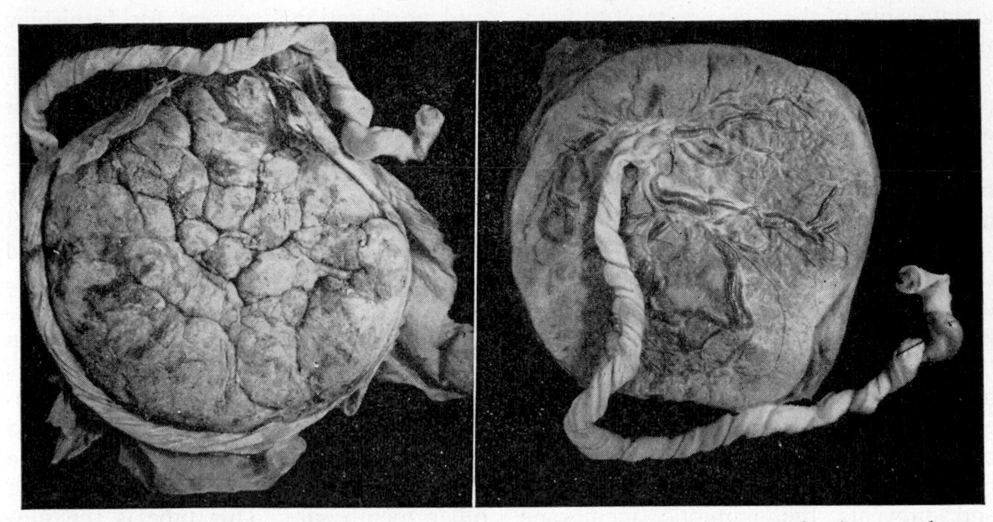

Figure 233. Photographs of the maternal (left) and fetal (right) surfaces of a human placenta at the end of a normal pregnancy. (Greenhill: Obstetrics.)

umbilical arteries, and return it to the embryo by way of the umbilical vein. The part of the uterine wall which makes up the placenta is a mass of spongy tissue filled with blood which originates in the mother's blood stream. *The bloods of the mother and fetus do not mix at all in the placenta or any other place;* the blood of the fetus in the capillaries of the chorionic villi comes in close contact with the mother's blood in the tissues between the villi, but they are always separated by a membrane, through which substances must diffuse. As the embryo develops, the placenta necessarily grows too. At the time of birth, it is a thick, circular disc, about 7 inches in diameter, 1 inch thick, and weighing about a pound. Besides serving as the nutritive, respiratory and excretory organ of the fetus, the placenta is an important endocrine gland (see p. 398).

The uterus also increases in size as the fetus grows, and by the end of nine months its mass is twenty-four times as great as at the beginning of pregnancy. After six months of fetal development the upper end of the uterus is on a level with the navel; by eight months it is as high as the lower edge of the breastbone. Within the uterus the fetus assumes a characteristic position with elbows, hips and knees bent, arms and legs crossed, back curved, and the head bowed and turned to one side. At birth the fetus usually is turned head downward so that its head emerges first, but occasionally the buttocks or feet are presented first, making delivery more difficult.

258. BIRTH

The factors which actually initiate the process of birth or **parturition** after the period of pregnancy is complete, are unknown. Childbirth begins with a long series of involuntary contractions of the uterus, experienced as "labor pains." Labor may be divided into three periods. During the first one, which lasts about twelve hours, the contractions of the uterus move the fetus down toward the cervix, and cause the latter to dilate, enabling the fetus to pass through. At the end of this period the amnion usually ruptures, releasing the amniotic fluid, which flows out through the vagina. In the second period, which normally lasts between twenty minutes and an hour, the fetus passes through the cervix and vagina and is born or "delivered" (Fig. 234). At this time, before the umbilical cord is cut, the contractions of the uterus squeeze much of the fetal blood from the placenta back into the infant. After a few minutes the pulsations in the umbilical cord cease, and the cord is ready to be tied and cut, severing the child from the mother. The stump of the cord gradually shrivels until nothing remains but the depressed scar, the **navel.** During the last stage of labor, which lasts ten or fifteen minutes after the birth of the child, the placenta and the fetal membranes are loosened from the lining of the uterus by another series of contractions and expelled. At this stage they are called collectively the "afterbirth." In man and certain other mammals this is accompanied by some loss of blood, for part of the lining of the uterus is torn away with the afterbirth. In other mammals, where the connection between the fetal membranes and uterine wall is not so close, the placenta can pull away from the uterine wall without causing bleeding. After birth the size of the uterus decreases, while its lining is rapidly restored.

In general, women of civilized countries have more difficulty during childbirth than do the women of primitive

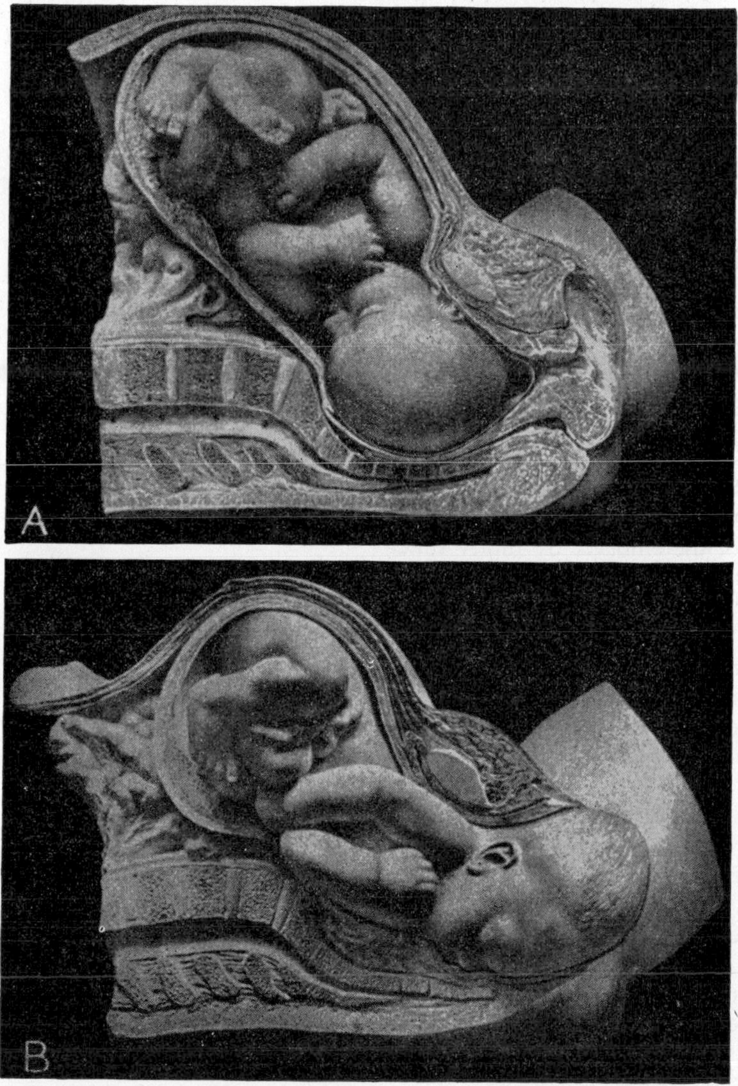

Figure 234. Models showing the process of birth. (From the Dickinson-Belski series, the Maternity Center Association.) *A,* Head passing through the dilated cervix of the uterus into the upper part of the vagina. *B,* Head passing through opening of vagina. (Patten, B. M.: Human Embryology, The Blakiston Co.)

tribes, and they frequently require artificial aid in being delivered. The physician present at birth, called an **obstetrician,** sometimes administers drugs, such as the hormone of the posterior lobe of the pituitary, to increase the contractions of the uterus, or he may have to assist the expulsion of the child by using a pair of forceps to draw it out. In some women the aperture between the pelvic bones, through which the vagina passes, is too small to permit the passage of the baby. In such cases the child must be delivered by an operation in which the abdominal wall and uterus are cut open from the front. This operation is now called a "cesarean operation." It seems unlikely that Julius Caesar was born this way (his mother was still alive when he was a grown man). The term is probably derived from the latin *caedere,* to cut, or from the roman law, *lex caesarea,* that governed the cutting open of a dying woman in an attempt to save her unborn child.

In a little more than 20 per cent of all known pregnancies the infant is born before it is prepared to carry on an independent existence. When this occurs, the birth is called an **abortion** or **miscarriage.** Such births are caused by improper implantation of the embryo, by faulty functioning of the placenta, or by injury to the mother.

Among mammalian species there are great differences as to the condition of the young at birth. The newborn of some species, such as the rat, are blind, hairless and helpless, while others, such as the guinea pig, are well developed and able to walk and eat solid food. There is also great variation in the weight of the newborn in comparison to the mother's weight: a newborn polar bear weighs 0.1 per cent of its mother's weight; a newborn human weighs about 5 per cent as much as its mother; and a newborn bat may weigh as much as 33 per cent of its mother's weight.

259. NUTRITION OF THE INFANT

The enlargement of the breasts during pregnancy and the onset of milk secretion occur under the stimulus of hormones from the ovary and the anterior pituitary, but the continuation of the milk secretion depends upon the presence of a suckling child. If the newborn infant is not breast-fed, the breasts stop secreting milk after a few days and decrease in size. Otherwise, they continue to yield milk for six to nine months or even longer. Formerly, it was not unusual for women to nurse their children for as long as a year and a half.

As everyone knows, milk is an excellent food, containing proteins, fats and carbohydrates. But it is deficient in some things, especially iron and vitamins C and D, so that at the present time an infant's diet is usually supplemented after the first month or so with eggs to supply iron, and orange juice and cod liver oil to supply the vitamins. The milk of various species differs in its content, and often it is difficult to raise a human infant on cow's milk, because the latter has more protein and less sugar than human milk. Usually, by diluting cow's milk and adding sugar to approximate the content of human milk, an acceptable formula can be made up.

QUESTIONS

1. Define the following: hermaphroditism, parthenogenesis.
2. Compare the devices used by different vertebrates to ensure fertilization.
3. What trends are discernible in the evolution of reproduction in animals?
4. Are humans capable of regeneration? If so, to what extent?
5. How are eggs and sperm adapted structurally for their respective functions?
6. Distinguish between copulation and fertilization.
7. How is the seminal fluid produced?
8. Describe the movements of the egg from the time it is released from the ovary until it is implanted in the wall of the uterus.
9. Trace the path of a sperm from the testis to its union with the egg.
10. Discuss the nutrition of the human embryo from the time of fertilization until birth.
11. What are the amnion, chorion, and allantois? What functions have they?
12. What is the amniotic cavity?
13. Describe the formation of the umbilical cord.
14. What is the "afterbirth"?

SUPPLEMENTARY READING

W. P. Pycraft's *The Courtship of Animals* gives an interesting account of mating instincts in animals. *The Hormones in Human Reproduction*, by George W. Corner, gives a clear picture of the role of the pituitary, ovary and testis in controlling secondary sex characters, the menstrual cycle and gestation. The details of reproduction in other animals are given in Hegner and Stiles' *College Zoology*.

Chapter 29

Embryonic Development

THE DIVISION, growth and differentiation of a fertilized egg into the remarkably complex and interdependent system of organs which is the adult animal is certainly one of the most fascinating of all biologic phenomena. Not only are the organs complicated and reproduced in each new individual with extreme fidelity of pattern, but many of the organs begin to function while still developing. The fetal heart, for instance, begins to beat during the fourth week, long before its development is complete.

The pattern of cleavage, blastula formation and gastrulation is seen, with various modifications, in all multicellular animals. The details of later development are quite different in animals of different phyla but are similar in more closely related forms. The main outlines of human development can be discerned by studying the embryos of rats or pigs, or even chicks or frogs. The development of man and other primates is characterized by the very early development of the embryonic membranes, the amnion, chorion and allantois.

260. TYPES OF EGGS

One important factor in determining the type of development is the amount of yolk or stored food present in the egg, since the yolk, being nonliving and inert, will modify the processes of development in the surrounding cytoplasm.

Probably the most primitive kind of egg is that with a small amount of yolk distributed more or less evenly throughout the cytoplasm. Such eggs, known as **isolecithal** (meaning "equal yolk"), are found in such animals as the sea urchin and the primitive marine chordate, Amphioxus.

Another type of egg, produced by fish, amphibia, reptiles and birds, has a large amount of yolk massed toward the lower or so-called **vegetal pole** of the egg, while the cytoplasm is concentrated toward the upper or **animal pole.** Such eggs, known as **telolecithal** (meaning "end yolk"), differ in the amount of yolk present. The frog's egg is about half yolk, whereas the hen's egg is more than 95 per cent yolk, with its cytoplasm restricted to a small disc at the animal pole.

The insect egg is an example of a third type, known as **centrolecithal,** in which the yolk is concentrated in the center of the egg and surrounded by a thin layer of cytoplasm.

The duck-billed platypus (p. 535) and other egg-laying mammals have small yolk-filled eggs, comparable in size to those of a small lizard, which are laid and develop outside the mother's body. Other mammalian eggs are small and relatively free from yolk (Fig. 235). They superficially resemble isolecithal eggs but develop more nearly like telolecithal ones, probably because mammals descended from reptiles which laid the latter kind. The loss of the yolk from mammalian eggs is a result of the fact that the yolk is no longer needed to nourish the developing embryo.

261. CLEAVAGE AND GASTRULATION

The process by which any single, fertilized egg develops into a many-celled embryo is called **cleavage**, indicating that it actually splits or divides. This division, called **mitosis**, is accompanied by a complicated series of processes *within* the nucleus and cytoplasm of the cells, which will be considered in Chapter 30. For the present we shall be concerned only with the larger, *external* changes brought about by cleavage.

When a simple, isolecithal egg is fertilized, it divides by mitosis to form a two-celled embryo. The line of this first division passes through both the animal and vegetal poles of the egg, splitting it down the center into two equal cells (Fig. 236). This is followed by a second cleavage division, also passing through both poles of the cells (now called an **embryo**), but at right angles to the first, which divides the two cells into four. The third division is a horizontal one, at right angles to the other two divisions, which splits the four cells into eight—four above and four below the line of cleavage. Further divisions divide these eight cells into thirty-two, sixty-four, 128 and so on, but the later divisions are

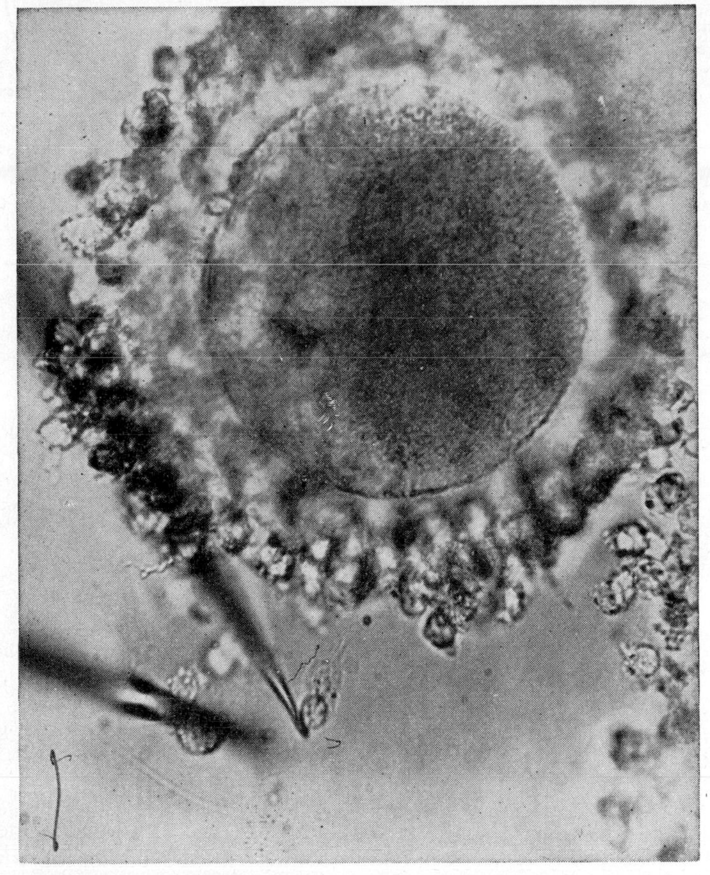

Figure 235. A human egg recovered from an ovarian follicle. The large egg cell is surrounded by the many small cells which make up the "corona radiata"—cells which were originally part of the follicle. The photograph was made in the course of an experiment in which microneedles were used to dissect away the corona radiata cells and to dissect the egg itself. The two needles can be seen in the lower corner of the picture, dissecting the nucleus out of one of the corona radiata cells. (Magnified 600 times.) (Courtesy of Dr. William R. Duryee.)

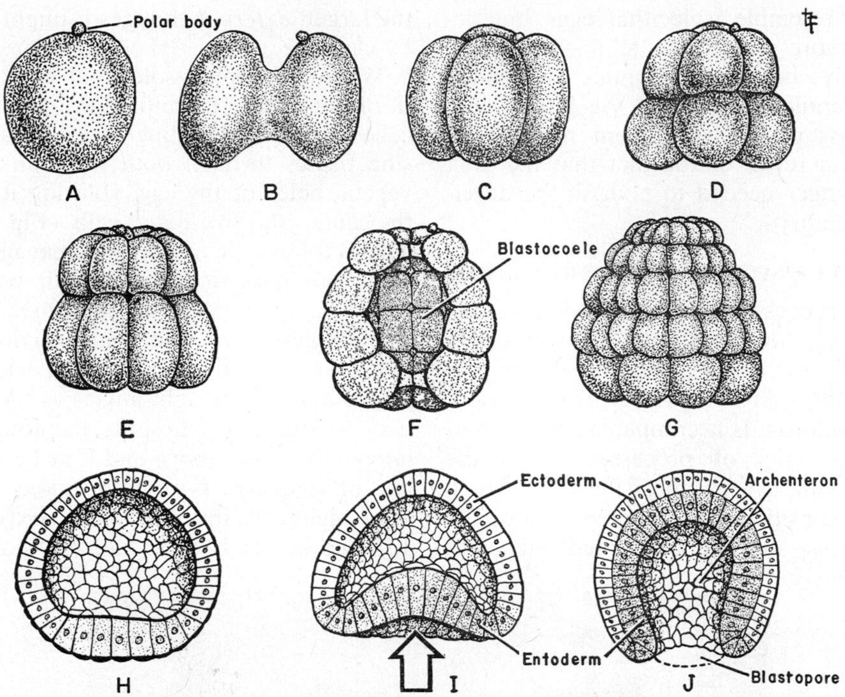

Figure 236. Isolecithal cleavage and gastrulation in Amphioxus viewed from the side. *A,* Mature egg with polar body. *B—E,* Two-, four-, eight- and sixteen-cell stages. *F,* Thirty-two-cell stage cut open to show the blastocoele. *G,* Blastula, and *H,* blastula cut open. *I,* Early gastrula showing beginning of invagination at vegetal pole. *J,* Late gastrula, invagination completed and blastopore formed.

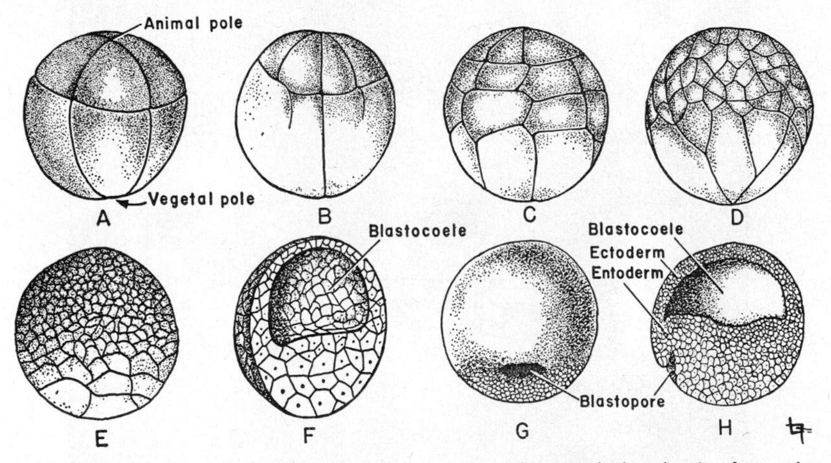

Figure 237. Successive stages in telolecithal cleavage and gastrulation in the frog, viewed from the side. *A–D,* cleavage; *E,* blastula; *F,* blastula cut open; *G,* early gastrula; *H,* early gastrula cut open.

usually somewhat irregular. Each of the cells formed by cleavage is exceedingly small, and the total mass of all the cells is less than the mass of the original fertilized egg, because some of the stored food is used up in the process.

Blastula Formation. As the cells undergo further division, a spherical mass is formed, in the center of which a cavity appears. With additional cleavages the cavity increases in size, until finally the embryo consists of several hundred cells, arranged in the form of a hollow ball. At this stage the embryo is called a **blastula**

(Fig. 236), and the cavity in the center, filled with fluid, is known as the **blastocoele**. The wall of the blastula consists of a *single* layer of cells.

Gastrulation. Almost as soon as the single-layered blastula is formed, it begins to change into a double-layered form called a **gastrula**. In simple, isolecithal eggs, gastrulation occurs by the pushing in (invagination) of a section of one wall of the blastula (Fig. 236, *I*). The pushed-in wall eventually meets the opposite wall, so that the original blastocoele is obliterated. The new cavity of the gastrula is known as the **archenteron** (meaning primitive gut), because it forms the rudiment of what is to become the digestive system. This opens to the outside by the **blastopore**, which marks the place where the indentation for gastrulation began. The formation of the two-layered embryo is accompanied by rapid growth and division of the cells, and the resulting gastrula has about the same diameter and shape as the blastula from which it came.

The outer of the two walls of the gastrula is called the **ectoderm** (outer skin); it eventually gives rise to the skin and nervous system. The inner wall, lining the archenteron, is known as the **entoderm** (inner skin); it finally becomes the digestive tract and its outgrowths—the liver, lungs and pancreas.

Cleavage and Gastrulation in Frog and Chick Eggs. The processes of cleavage and gastrulation are markedly modified by the presence of large amounts of yolk: the cleavage divisions of the cells originating from the lower part of the frog's egg are slowed up by the presence of the inert yolk, with the result that the blastula consists of many small cells at the animal

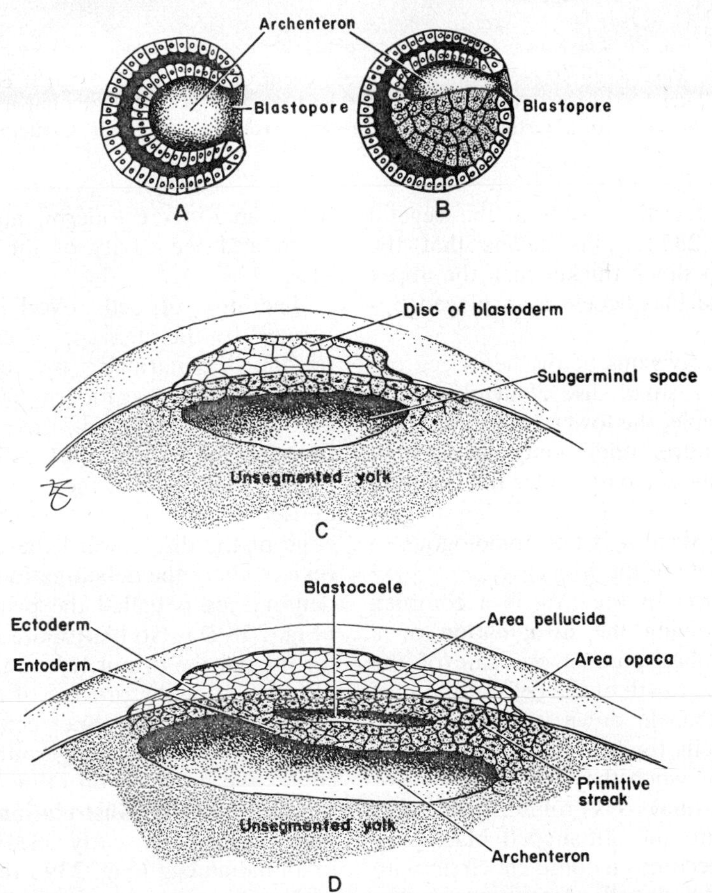

Figure 238. Gastrulation in Amphioxus (*A*), the frog (*B*) and the chick (*C* and *D*). Note that there is no blastopore in the chick.

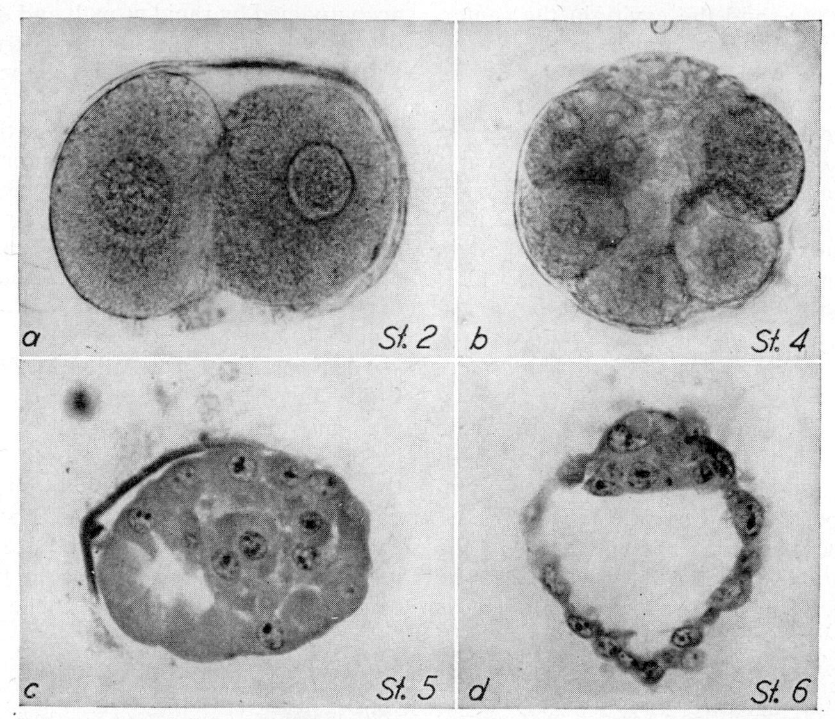

Figure 239. Very early stages of human development. *a,* Two cells; *b,* eight cells; *c,* 58 cell blastocyst; *d,* later blastocyst showing the chorion and the ball of cells at the upper margin which gives rise to the embryo. (Hertig, Rock, Adams and Mulligan: Carnegie Contributions to Embryology.)

pole and a few large cells at the vegetal pole (Fig. 237). This means that the lower wall is much thicker than the upper wall and the blastocoele is displaced upward.

Cleavage divisions of the hen's egg occur only in a small disc of cytoplasm at the animal pole, the lower, yolk-filled part of the egg never undergoing cleavage at all. The shallow cavity under the dividing cells (Fig. 238 *C*) is called the **subgerminal space** since it is not homologous to the blastocoele of the frog egg.

Gastrulation in the frog is a complex process involving the **invagination** of a slit-shaped blastopore at one side of the blastula, the growth of the cells of the roof of the blastocoele down over the lower, yolk-filled cells **(epiboly)**, and a rolling in of these cells when they reach the blastopore **(involution)**. The rolling in of cells continues and the slit-shaped blastopore eventually becomes a complete circle with some yolk-filled cells remaining in the center as a **yolk plug.** This process establishes an outer ectoderm, an inner entoderm and the cavity of the archenteron (Fig. 238 *B*).

The disc of cells overlying the yolk formed by the cleavage of the bird's egg becomes separated by cell division into an upper ectoderm and a lower entoderm; the small, irregular space between the two may be called the blastocoele. The cells of the ectoderm become denser in an elongated pattern from one edge of the disc toward the center in the region where the delamination began. This denser area is called the **primitive streak** (Fig. 238 *D*). No blastopore opening ever forms, but the primitive streak is its homologue. After the process of delamination has established an upper ectoderm and a lower entoderm the subgerminal space becomes the archenteron (Fig. 238 *D*).

Cleavage and Gastrulation in the Human Egg. The early cleavage of the mammalian egg (Fig. 239) resembles that of Amphioxus in forming a cluster of cells called a **blastocyst.** After this the mammalian egg differs in that the mass of cells divides into two parts—an outer, hollow sphere of cells and, attached to one side of this, an inner, solid ball of cells. The outer sphere is the fetal membrane, the chorion, which forms at this early stage in development and unites with the maternal tissue of the uterus as implantation occurs. The embryo and the other embryonic membranes develop from the inner, solid ball of cells.

The inner ball proceeds to form a gastrula, consisting of ectoderm and entoderm, but the process is quite different from that in the lower vertebrates. Within the inner cell mass, two cavities form simultaneously (Fig. 240 *B*). The upper one is the cavity of the amnion, lined with ectoderm. The lower one is the cavity of both the yolk sac and the primitive gut, lined with entoderm. Between the two cavities the cells spread out in the shape of a flat, two-layered plate, from which the embryo develops. At what is to become the posterior end of the embryo this plate connects with the outer chorion by a group of cells known as the **body stalk** (Fig. 240 *D*). Into this body stalk grows the (nonfunctional) allantois, which has developed as a tube from the rear end of the entodermal (yolk) sac. Thus we find, after about two weeks of development, that the human embryo consists of a flat, two-layered disc, about 0.01 inch across, and a stalk which connects the disc with the outer chorion.

262. MESODERM FORMATION

In all animals (except sponges and coelenterates) a third layer of cells, the **mesoderm,** develops between the ectoderm and entoderm. The method of forming this layer is quite different in various species. In Amphioxus it forms as a series

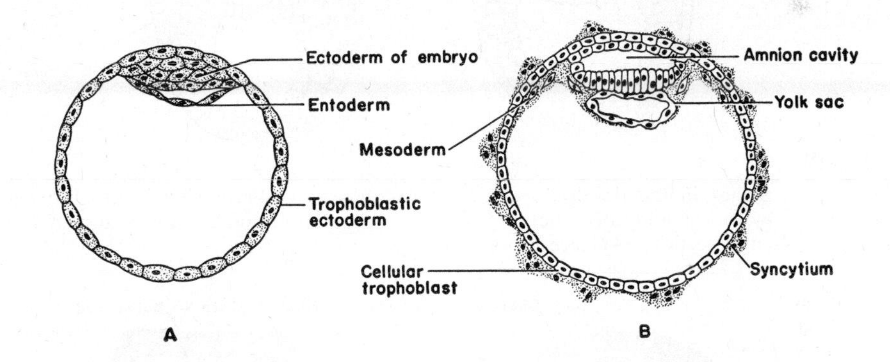

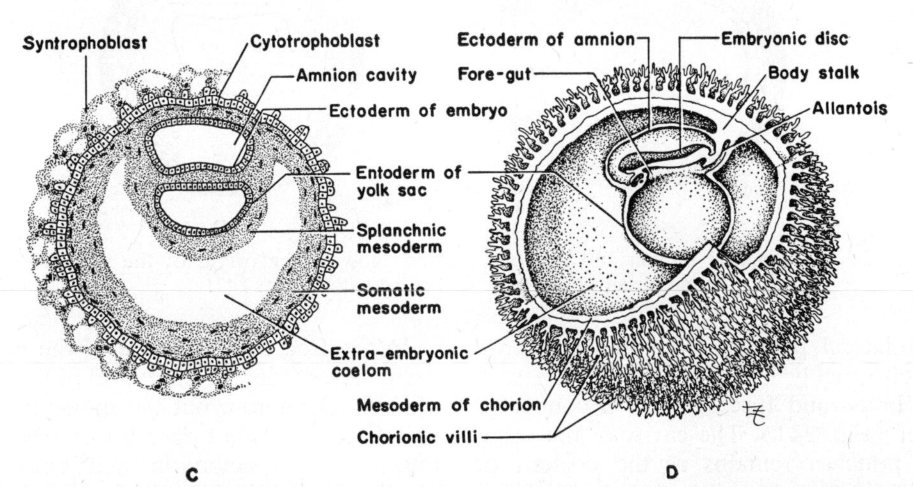

Figure 240. Diagrams of human embryos ten (*A*) to twenty (*D*) days old showing the formation of the amniotic and yolk sac cavities and the origin of the embryonic disc.

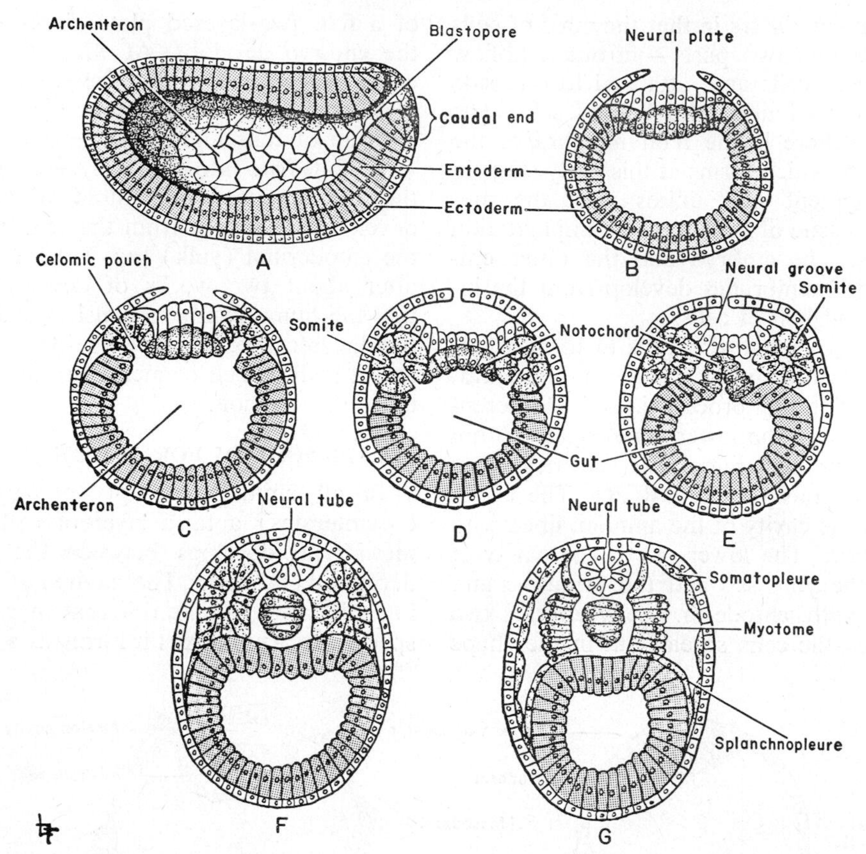

Figure 241. Stages in the development of an Amphioxus embryo, showing the formation of the mesoderm by the budding of pouches from the archenteron, and the formation of the neural tube. *A* is a sagittal section, *B–G* are cross sections.

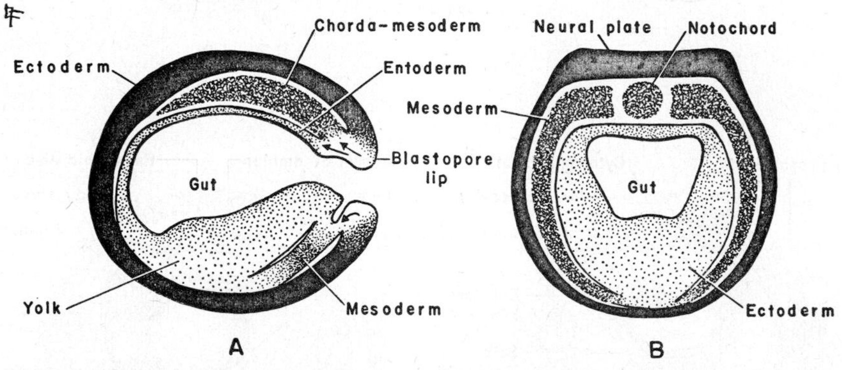

Figure 242. A, Sagittal section and *B,* cross section showing the origin of the notochord and mesoderm in the frog. Compare with Figure 241.

of bilateral pouches from the entoderm, which quickly lose their connection with the latter and fuse to form a connected layer (Fig. 241). The cavity of the original pouches remains as the coelom or body cavity.

In the frog embryo some of the mesoderm forms from the entoderm of the roof of the archenteron, but the major part of it is formed at the upper lip of the blastopore, where ectoderm and entoderm meet (Fig. 242). This region is one of especially active proliferation of cells and also gives rise to some additional ectoderm and entoderm.

In the chick embryo, where the presence of so much yolk prevents invagination and the formation of a true blastopore, and in mammals, where no blastopore forms, mesoderm formation occurs at the primitive streak. The rapid growth of cells in the upper layer of the chick embryo results in the movement of cells to the primitive streak. Here they sink in and move laterally between the ectoderm and entoderm to establish the mesoderm. In the human embryo the primitive streak (Fig. 244) is a thickened band of ectoderm and entoderm extending forward from the body stalk, marking the longitudinal axis of the body. When cells of this band proliferate, some migrate out laterally between the ectoderm and entoderm layers to form the mesoderm. The primitive streak is a center of growth for ectoderm and entoderm as well as for mesoderm, and apparently in some way determines the position and order of appearance of the various parts of the embryo. In front of the anterior end of the streak, the head and trunk develop. The streak itself does not increase in length, but gradually decreases as the rest of the embryo grows, until it finally disappears in the tail region.

Regardless of how the mesoderm originates, it splits into two sheets which grow laterally and anteriorly between the ectoderm and entoderm; one of these sheets attaches itself to the ectoderm, the other to the entoderm. The cavity between the two mesodermal sheets becomes the coelom, and when the digestive tract becomes separated as a tube from the yolk sac cavity, the mesoderm grows around it (Fig. 244). The layer of mesoderm associated with the ectoderm forms the muscles of the body wall, and the mesodermal sheet associated with the entoderm forms the muscles of the digestive tract. The entoderm itself forms only the *inner lining* of the digestive system.

The Notochord. The notochord is a flexible, unsegmented, skeletal rod which extends longitudinally along the dorsal midline of all chordate embryos and is formed at the same time as the mesoderm. Its exact method of formation differs from species to species, as does that of the mesoderm. In Amphioxus it is formed from the entoderm as a bud, like those which produce the mesodermal pouches (Fig. 236). It lies dorsal to the gut and just below the developing nervous system.

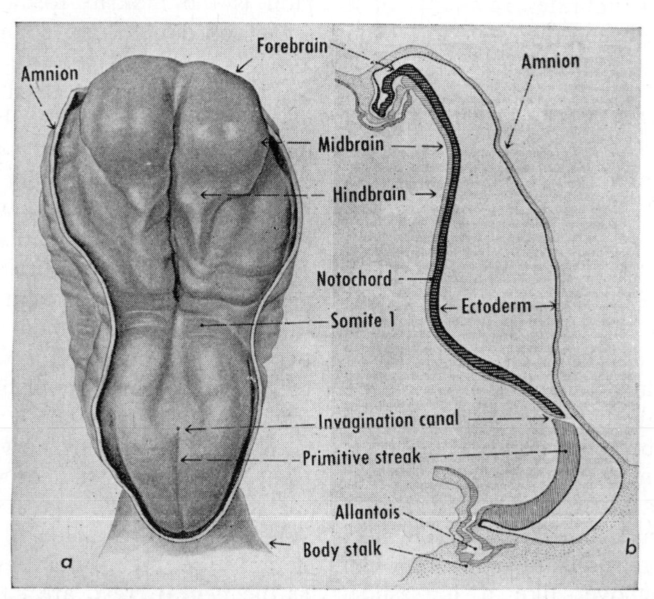

Figure 243. A human embryo undergoing neurulation. Left, Dorsal view of whole embryo; the head end is uppermost. Right, A sagittal section through the plane of the primitive streak. × 32. (Ludwig 1928.)

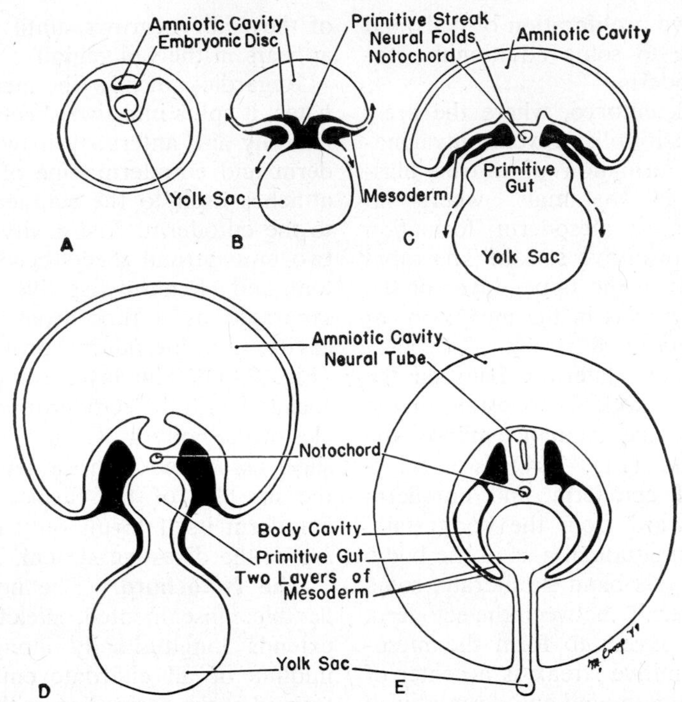

Figure 244. Stages in the origin of body form and the growth of mesoderm around the digestive tract in man.

In the chick and mammals it forms as an anterior outgrowth of the primitive streak (Fig. 243). In all vertebrates the notochord is a short-lived structure, eventually replaced by the vertebral column, but in some lower vertebrates remnants of it can be found between the vertebrae even in adult animals.

263. DEVELOPMENT OF THE NERVOUS SYSTEM

Although the two week old human embryo is a simple flat disc, the two-month embryo has nearly all its structures in rudimentary form. The brain and spinal cord are among the earliest organs to appear. At about the third week the ectoderm just over the notochord in front of the primitive streak develops a thickened plate of cell called the **neural plate.** The center of this becomes depressed, and is known as the **neural groove** (Fig. 245), while the outer edges of the plate rise in two longitudinal **neural folds** which meet at the anterior end and appear, when viewed from above, like a horseshoe. Gradually, these folds come together at the top, forming a hollow neural tube. The cavity of the anterior part of this neural tube becomes the ventricles of the brain, while the cavity of the posterior part becomes the neural canal, running the length of the spinal cord. The brain region is the first to form, and the long spinal cord develops slightly later (Fig. 246).

The anterior part of the neural tube, which gives rise to the brain, is much larger than the posterior part, and continues to grow so rapidly that the head region comes to bend down at the front end of the embryonic disc. All the regions of the brain—forebrain, midbrain and hindbrain—are established by the fifth week of development, and a week or two later the outgrowths which will form the large cerebral hemispheres begin to grow.

The various motor nerves of the body grow out of the brain or spinal cord but the sensory nerves have a separate origin. When the neural folds fuse to form the neural tube, bits of nervous tissue, known as the **neural crest,** are left over on each side of the tube (Fig. 245). These migrate downward from their original position and form the dorsal root ganglia of the spinal nerves and the sympathetic ganglia. From cells in the dorsal root ganglia grow sensory nerves sending dendrites out to the sense organs, and axons in to the spinal cord.

264. DEVELOPMENT OF BODY FORM

The conversion of the two-week-old flat disc into a roughly cylindrical embryo is accomplished by three processes: (*a*) the growth of the embryonic disc, which is more rapid than the growth of the surrounding tissue; (*b*) the underfolding of the embryonic disc, especially at the front and hind ends; and (*c*) the constriction of the ventral body wall to form the future umbilical cord.

Growth is rapid at the front end of the embryonic disc, and soon the head region bulges forward from the original embryonic area. The tail, which even human embryos have at this stage, bulges to a lesser extent over the posterior end. The sides of the disc grow downward, eventually to form the sides of the body. The embryo becomes elongated, because growth is more rapid at the head and tail ends than laterally. The enlarging of the embryo has been compared to the increase of size in a soap bubble blown from a pipe, which, as it grows, swells out in all directions above the mouth of the pipe (the yolk sac). What is to become the mouth and heart originally lies in front of the embryonic disc, and as the disc

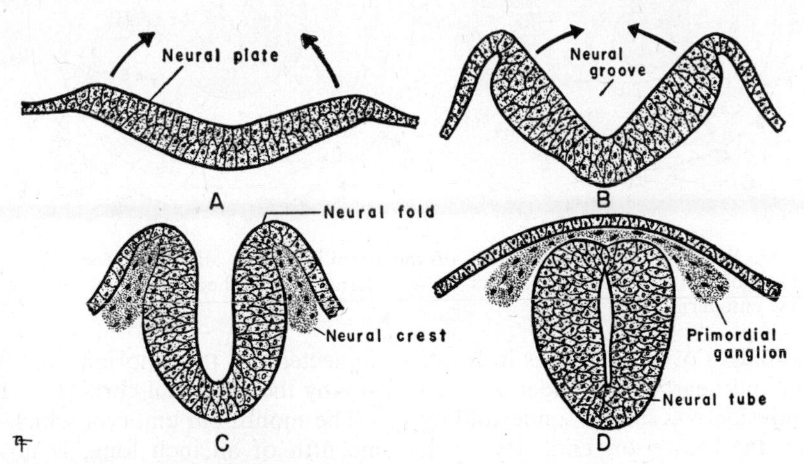

Figure 245. Cross sections of the ectoderm of human embryos at successively later stages to illustrate the origin of the neural tube and the neural crest, which forms the dorsal root ganglia and the sympathetic nerve ganglia.

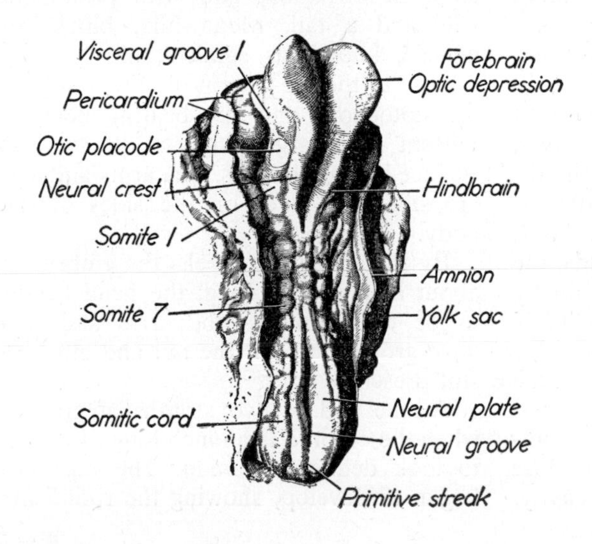

Figure 246. A dorsal view of a human embryo showing the closure of the neural tube between somites 4 and 7 and the open neural folds anterior and posterior to this. (Witschi: Development of Vertebrates.)

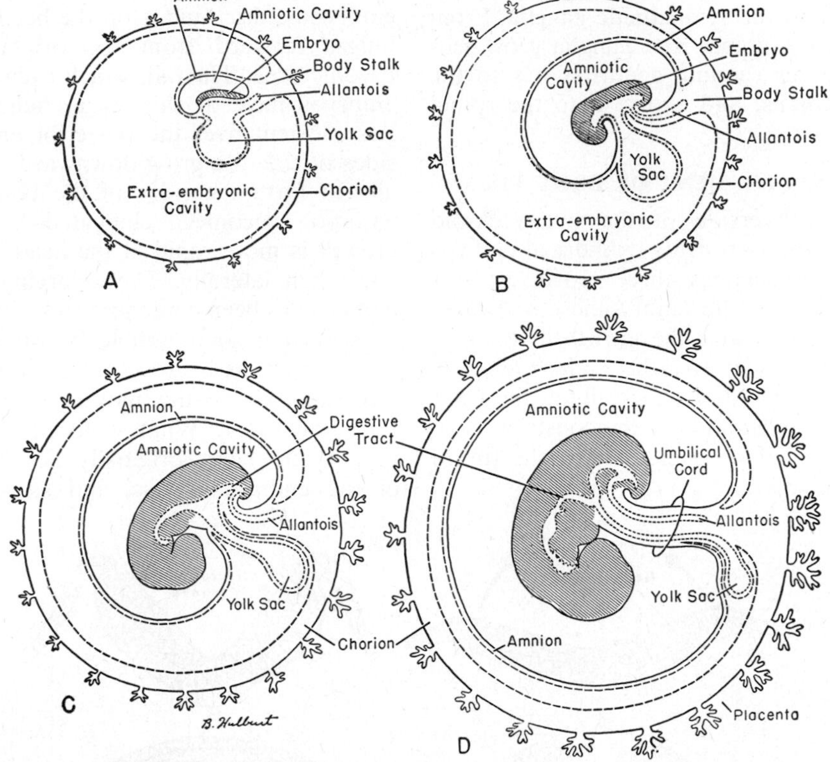

Figure 247. Stages in the development of the umbilical cord and body form in the human embryo. The solid lines represent layers of ectoderm; the dashed lines, mesoderm; and the dotted lines, entoderm.

grows and bulges over the tissues in front, the mouth and heart swing underneath to the ventral side. A similar underfolding occurs at the posterior end. By such growth and underfoldings the lateral and eventually the ventral walls of the body are formed, and the embryo becomes more or less cylindrical in shape.

While the embryo is still a simple disc, its entire under surface is open to the yolk cavity. As the body wall folds, a foregut and a hindgut (which form, respectively, the anterior and posterior parts of the digestive tract) are cut off from the yolk sac but remain connected to it by the yolk stalk. As the embryo grows and folds, the amnion also grows to enclose it, finally constricting the yolk stalk and the body stalk (with its allantois and blood vessels) together into a single cylindrical tube, the **umbilical cord.** This takes place about four weeks after development is commenced, and allows the embryo to float free in the liquid-filled amniotic cavity, connected to the chorion and placenta only by the umbilical cord (Fig. 247).

The month-old embryo, which is about one fifth of an inch long, is now recognizable as a vertebrate of some kind. It has become cylindrical, with a relatively large head region, and with prominent gills and a tail. Meanwhile, blocks of muscle, known as **somites,** are forming rapidly in the mesoderm on either side of the notochord and the beating heart is present as a large bulge on the ventral surface behind the gills. The arms and legs are still mere buds on the sides of the body.

By the end of six weeks the embryo is about half an inch long; the head begins to be differentiated; the arms and legs have grown out, but the tail and gills are still present.

At the end of two months of growth, when the embryo is an inch long, it begins to look definitely human. The face has begun to develop, showing the rudiments of eye, ear and nose. The arms and legs have developed, at first resembling tiny paddles, but by this stage the beginnings of fingers and toes are evident (Fig. 248). The tail, which was prominent during the fifth week of development, has begun to shorten and to be concealed by the growing buttocks. As the heart moves posteriorly on the ventral side, and the gill pouches become less conspicuous, a neck region appears. Now most of the internal organs are well laid out so that development in the remaining seven months consists mostly of an increase in size and the completion of some of the minor details of organ formation (Fig. 249).

The embryo is about 3 inches long after three months of development, 10 inches long after five months, and 20 inches long after nine months. During the third month the nails begin forming and the sex of the fetus can be distinguished. By four months the face looks really human; by five months, hair appears on the body and head. During the sixth month, eyebrows and eyelashes appear. After seven months the fetus resembles an old person with red and wrinkled skin. During the eighth and

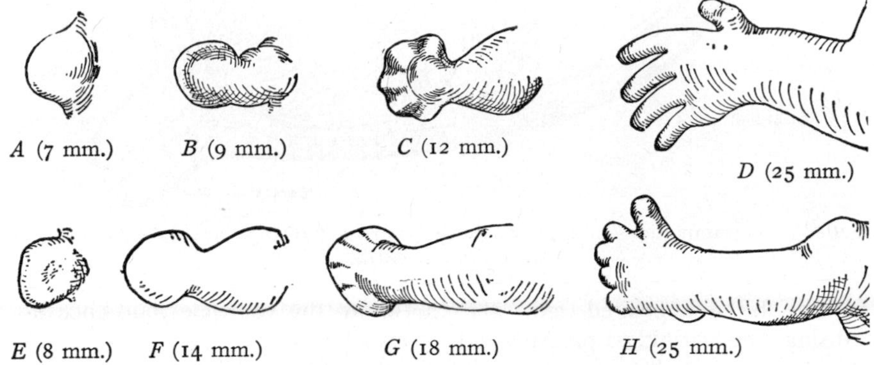

Figure 248. Stages in the development of the human arm (upper row) and leg (lower row), between the fifth and the eighth weeks. (Arey.)

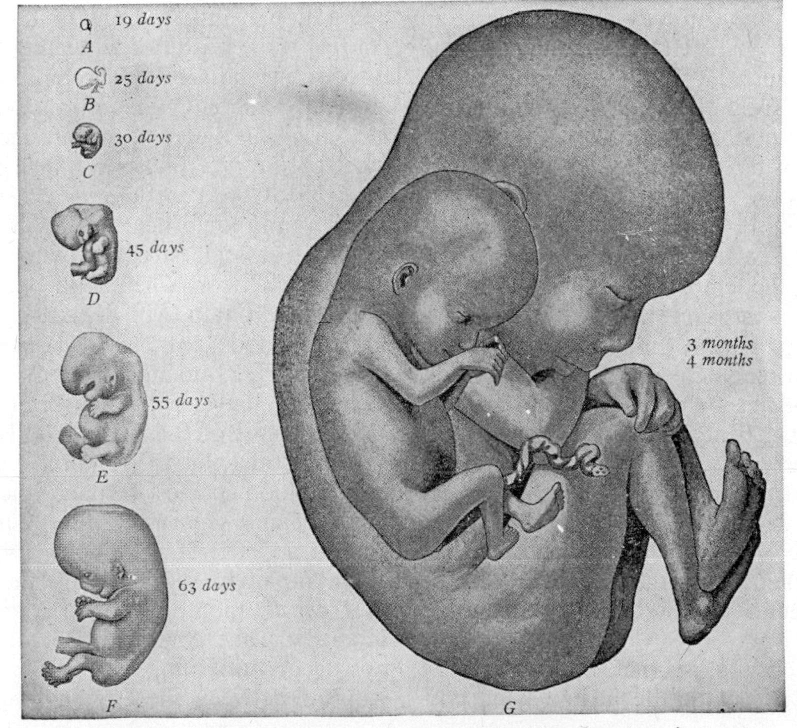

Figure 249. A graded series of human embryos. (Arey.)

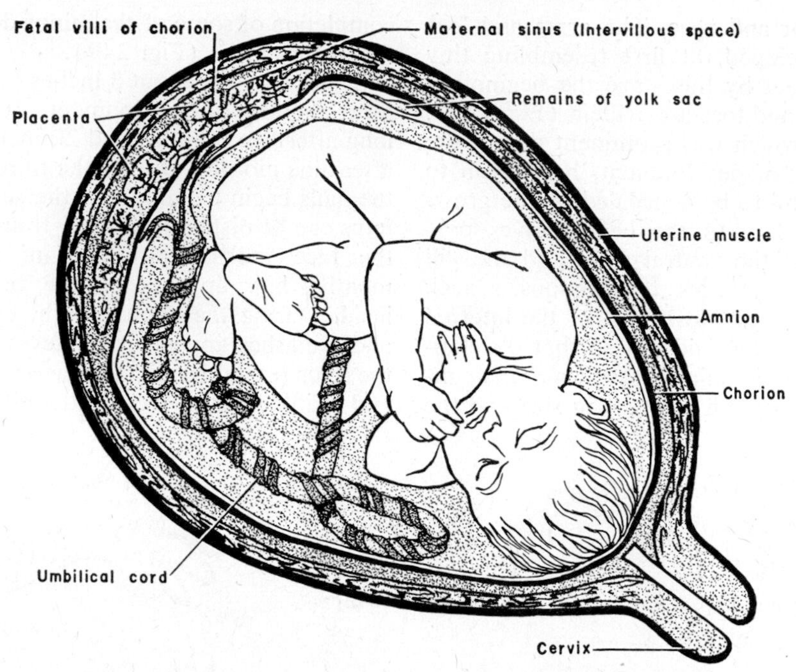

Figure 250. A diagrammatic section through the uterus, showing the placenta and the fetus shortly before birth.

ninth months, fat is deposited under the skin, causing the wrinkles partially to smooth out; the limbs become rounded, the nails project at the finger tips, the original coat of hair is shed, and the fetus is "at full term," ready to be born (Fig. 250). The total **gestation period,** or time of development, for human beings is about 280 days from the beginning of the last menstrual period before conception, until the time of birth.

265. FORMATION OF THE HEART

In contrast to many organs which develop in the embryo without having to function at the same time, the heart and circulatory system must function while still developing. The heart forms first as a simple tube from the fusion of two thin-walled tubes beneath the developing head. In this early condition it is essentially like a fish heart, consisting of four chambers arranged in a series: the **sinus venosus,** which receives blood from the veins; the single **atrium;** the single **ventricle;** and the **arterial cone,** which leads to the aortic arches.

In the beginning the heart is a fairly straight tube, with the atrium lying posterior to the ventricle; but since the tube grows faster than the points to which its front and rear ends are attached, it is forced to bulge out to one side (Fig. 251). The ventricle then twists in an S-shaped curve down and in front of the atrium, coming to lie posterior and ventral to it as it does in the adult. The sinus venosus gradually becomes incorporated into the atrium as the latter grows around it, and most of the arterial cone is merged with the wall of the ventricle.

The embryonic heart, when it first appears, is a single structure with only one of each chamber, whereas the adult heart is a double pump, with separate right and left atria and ventricles. This separation prevents the mixing of aerated blood from the lungs with nonaerated blood from the rest of the body. The lungs are nonfunctional in the embryo, and not much blood passes through them. The heart begins separating into four chambers at an early stage. The two ventricles are completely separated by the end of the second month. The atria are partly separated, but complete separation does not occur until after birth, when the **oval window** between them finally closes. Before birth, this must be kept open to permit blood to get into the left side of the heart, for in the fetus only a small amount of blood passes through the lungs to the left atrium. Without this "window" the left side of the heart would be nearly empty, and most of the blood would pass through the right side only.

266. DEVELOPMENT OF THE DIGESTIVE TRACT

The digestive tract is first formed as a separate foregut and hindgut by the growth and folding of the body wall, which cut them off as two simple tubes from the original yolk sac. These tubes grow as the rest of the embryo grows, becoming greatly elongated. The lungs, liver and pancreas originate as hollow, tubular outgrowths from the original foregut, and hence are composed of entoderm. But these outgrowths always are associated with some mesodermal tissue, which forms the blood and lymph vessels, connective tissue and muscles of these organs. The entoderm forms only the internal epithelium of the digestive tract and lungs, and the actual secretory cells of the pancreas and liver. Both the enzyme-secreting cells and those of the islets of Langerhans in the pancreas are derived from tubular outgrowths from the foregut. The lung first develops as a single median outgrowth from the ventral side of the foregut. This single tube, which is the forerunner of the trachea, soon branches into two, the rudiments of the bronchi, which in turn divide repeatedly and eventually grow into the complex structure which is the adult lung.

The most anterior part of the foregut flattens out to become, in cross section, a flattened oval, rather than a circle, and develops into the pharynx. In the pharyngeal region a series of five paired pouches, the gill pouches, bud out laterally from the entoderm and meet a corresponding set of inpocketings from the overlying ectoderm. In lower vertebrates such as fish, the two sets of pockets fuse to make a continuous passage from the pharynx to the outside—the **gill slits,** which function as respiratory organs.

In the higher vertebrates this normally does not occur; the pouches exist, but are nonfunctional vestiges which give rise to other structures or disappear. For example, the first pouch becomes the cavity of the middle ear and its connection with the pharynx, the eustachian tube. The

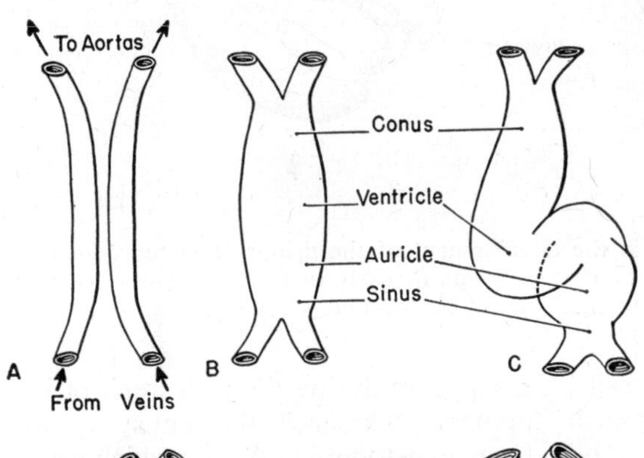

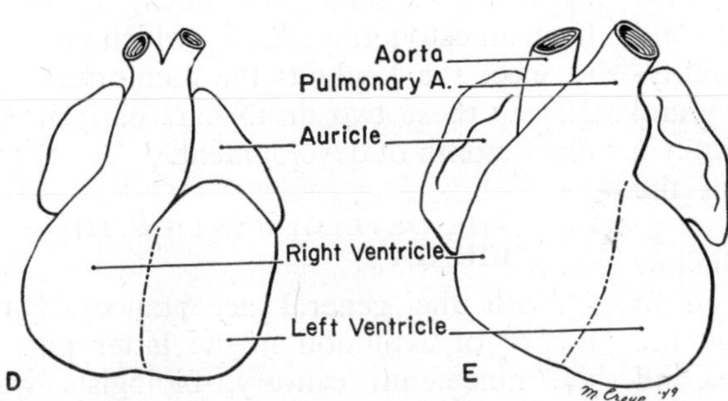

Figure 251. Ventral views of successive stages in the development of the heart. See text for discussion.

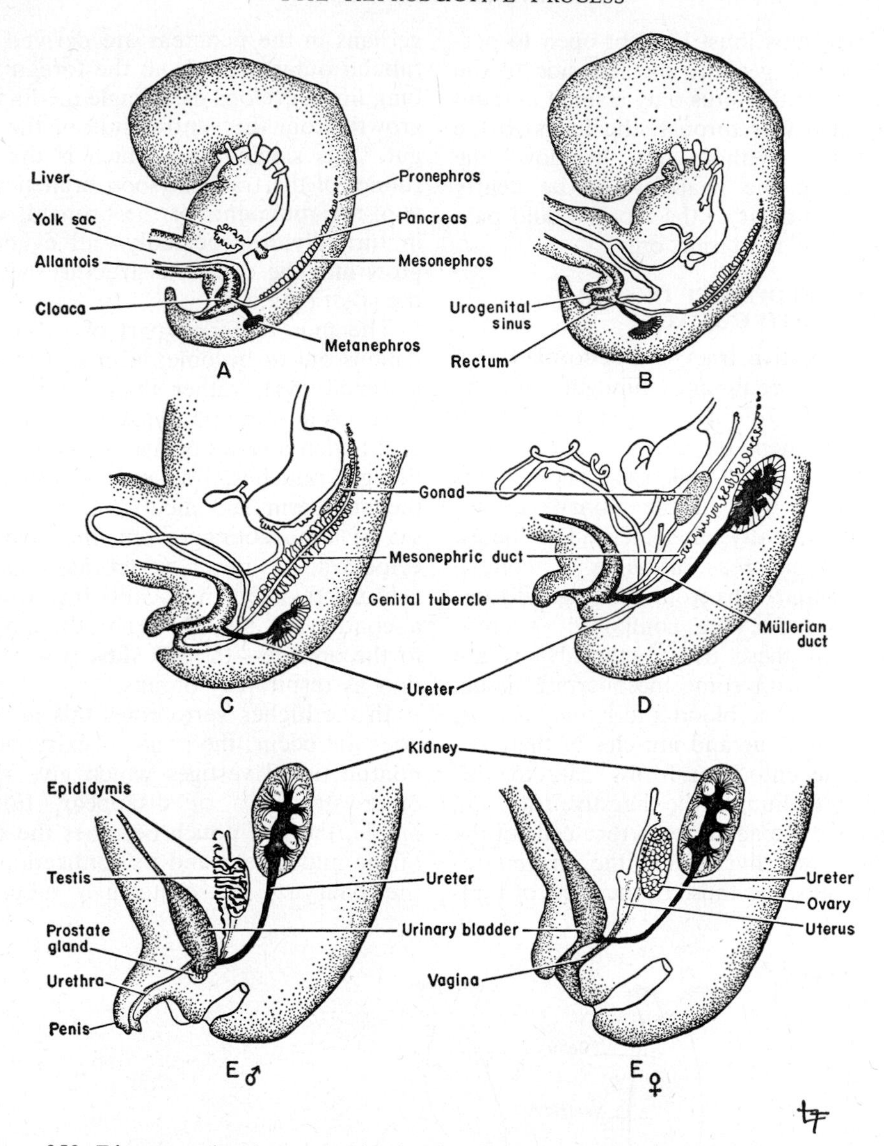

Figure 252. Diagrams showing stages in the development of the urinary and reproductive systems in man. *A,* Early in the fifth week of development; *B,* early in the sixth week; *C,* seventh week; *D,* eighth week. *E* ♂, Three months male; *E* ♀, three months female.

second pouch becomes a pair of tonsils, while parts of the third and fourth pouches become the thymus gland, and other parts of them become the parathyroids. The fifth pouch is rudimentary and disappears. The thyroid develops from a separate outgrowth on the floor of the pharynx.

The mouth cavity arises as a shallow pocket of ectoderm which grows in to meet the anterior end of the foregut; the membrane between the two ruptures and disappears during the fifth week of development. Similarly, the anus is formed from an ectodermal pocket which grows in to meet the hindgut; the membrane separating these two disappears early in the third month of development.

267. THE DEVELOPMENT OF THE KIDNEY

With the general acceptance of the theory of evolution in the latter part of the nineteenth century, biologists were stimulated to study the development of a wide variety of species to see how they resembled each other. These studies showed that the early development of all vertebrates is similar, and that, in general, those forms believed, from other evidence, to be most closely related have the most similar developmental processes. Moreover, it was found that the higher animals go through forms in development which resemble lower animals. From such observations arose the **theory of recapitulation.** This states that each individual in its embryonic growth recapitulates, or exhibits in structure, the steps in the evolution of its species. It has since been shown that the individual does not go through the *adult* stage of any of its ancestors, but repeats only the *embryonic* stages of those predecessors. Thus the human embryo develops gill pouches, like the fish embryo, but these never become gill slits, as they do in the adult fish.

The development of the kidney is one of the finest and most clear-cut examples of the principle of recapitulation. Within the subphylum of vertebrates are three different types of kidney. The earliest—the fish kidney—called the **pronephros,** is the adult kidney of certain primitive fishes. The second, or **mesonephros,** is the adult kidney of amphibians and the higher fishes. The third, the **metanephros,** is the adult kidney of reptiles, birds and mammals. But in development, each of the higher animals repeats the evolutionary sequence of this organ. Thus frog embryos first develop a fish kidney, which functions during early embryonic life, before the permanent frog kidney develops. And man develops first a nonfunctional fish kidney, then a frog kidney, which may be functional during fetal life, and finally the permanent, third kidney or metanephros (Fig. 252). The three kidneys develop one after another in both time and space, each new kidney lying posterior to the previous one.

The pronephros, which in the human embryo consists of about seven pairs of rudimentary kidney tubules, develops in the mesoderm and degenerates during the fourth week of embryonic life. From the tubules a pair of ducts grows back to the hindgut and connects with it.

The tubules of the mesonephros originate during the fourth week, reach their height at the end of the seventh week, and degenerate by the sixteenth week. These tubules connect with the ducts left by the degenerated pronephros, and empty into them. In the female the frog kidney and its ducts degenerate completely except for a few nonfunctional remnants, but in the male some of the tubules remain and are converted into the epididymides, while the ducts become the vas deferens.

The final kidney of reptiles, birds and mammals develops as a pair of buds from the ducts of the frog kidney. The ureter and collecting tubules of the kidney develop from these buds, while the Bowman's capsules and convoluted tubules develop from the same sort of mesoderm that formed the tubules of the pronephros and mesonephros at a more anterior point. Later, these two portions unite to form the kidney tubules of the adult. The metanephros begins forming during the fifth week and is practically complete by sixteen weeks.

268. THE CONTRIBUTIONS OF EACH GERM LAYER

The contributions of each of the three germ layers to the developing animal are summarized in the following table.

ECTODERM	ENTODERM	MESODERM
Epidermis of the skin	Lining of gut	Muscles—smooth, skeletal and cardiac
Hair and nails	Lining of trachea, bronchi and lungs	Dermis of skin
Sweat glands	Liver	Connective tissue, bone and cartilage
Entire nervous system: brain, spinal cord, ganglia, nerves	Pancreas	Dentin of teeth
Receptor cells of sense organs	Lining of gallbladder	Blood and blood vessels
Lens of eye	Thyroid, parathyroid and thymus glands	Mesenteries
Lining of mouth, nostrils and anus	Urinary bladder	Kidneys
Enamel of teeth	Urethra lining	Testes and ovaries

269. CONTROL OF DEVELOPMENT

One of the important, unsolved problems of modern biology is the nature of the mechanisms that cause development to proceed in a regular way, so that the organs appear at the proper time and in the proper relations to each other. How can a single cell give rise to many different kinds of cells that differ markedly in their structure, functions and chemical properties?

Early embryologists thought that development was simply a matter of increasing size, that the germ of the adult was fully formed and differentiated with head, arms and legs, before growth began. One theory stated that this tiny creature was to be found in the egg, another that it was in the sperm, which simply had to enter an egg to begin growth. These **preformation** theories really explained development by denying that there was such a thing, and led to some amusing calculations. If all the individuals of one generation are contained within the eggs of their mothers, and these in turn are contained within the eggs of their mothers, and so on, like a series of Chinese boxes, then, scientists seriously questioned, how many series of such boxes within boxes could have been accommodated in the ovaries of Eve, and how soon would those eggs be used up and the human race end?

With the development of better microscopes, it became apparent that there were no little people inside either sperm or eggs. It is now well established that each individual originates, as we have seen, from a single egg fertilized by a single sperm, and that development occurs by cell division and the progressive differentiation of these cells. This view is known as **epigenesis.**

With the development of experimental methods of study, investigators have found that development is not *simply* epigenetic. There are, it seems, certain *potentialities,* though not structures, localized in certain regions of the egg and early embryo. When experimentally separated at the two-cell stage or even the four-cell stage, the embryos of some species continue to develop normally. Each of the separated cells has the power to form an embryo complete in all details, although they are smaller than normal. Embryos of other species separated at the two-cell stage, however, demonstrate that certain potentialities are localized, for neither cell can develop into a whole embryo. Each half develops only those structures it normally would have—half an embryo, perhaps, or some part of one. This localization of potentialities *eventually* occurs in the development of *all* eggs; it simply occurs earlier in some than in others. From this we can deduce that some sort of chemical or physiologic differentiation is present before any structural differentiation is visible, but the basic problem of how the chemical or physiologic differentiation occurs remains unsolved. By a series of experiments it has been possible to map out the location of these potentialities in the primitive streak stage of the chick (Fig. 253, *B*). Presumably, the potentialities in the human embryo at the comparable stage are similarly placed.

270. MORPHOGENESIS

The mechanism of differentiation may depend upon (1) a segregation of potentialities during cell division, (2) the establishment of chemical gradients within the developing embryo, (3) somatic mutations, (4) the action of chemical "organizers" or (5) the induction of adaptive enzymes. The last theory, advanced recently by Jacques Monod of Paris, suggests that since the enzyme complement of a cell can be changed to some extent by extra- or intracellular influences, the gradients established by growth and cell multiplication could result in quantitative and even qualitative differences in enzymes. As a result of the stimulation or inhibition of one enzyme, a chemical substance could accumulate which would induce the formation of a new enzyme and thus confer a new functional ability on these cells. Morphogenesis is probably too complicated a phenomenon to be explained in terms of a single process such as enzyme induction. It can be shown experimentally that an enzyme can be induced in an embryo by the injection of a suitable substance.

Evidence of a different type of mechanism for differentiation has been obtained from experiments in which microsurgical instruments are used to cut out a bit of tissue from one embryo and transplant it to another one. When a piece of the dorsal lip of the blastopore of a frog gastrula is implanted beneath the ectoderm of a second gastrula, the tissue heals in place and causes the development of a second brain, spinal cord and other parts at the site, so that a double embryo or closely joined Siamese twins result. Many tissues show similar abilities to organize the development of an adjoining structure. The eye cup will initiate the formation of a lens from overlying ectoderm even if it is transplanted to the belly region, where the cells normally would form belly epidermis. Such experiments indicate that development is a coordinated series of chemical stimuli and responses, each step regularly determining the succeeding one. The term **"organizer"** is applied to the region of the embryo with this property and also to the chemical substance given off by that region which passes to the adjoining tissue and directs its development. There is evidence which suggests that organizers are nucleoproteins.

It had been widely accepted that organizers can transmit their inductive stimuli only when in direct physical contact with the reactive cells, but evidence obtained by Dr. Victor Twitty suggests that induction may be mediated by diffusible substances which can operate without direct physical contact of the two tissues. Twitty grew small clusters of frog ectoderm, mesoderm and entoderm cells in tissue culture and found that ectoderm alone would never differentiate into nerve tissue. Ectoderm cells placed in a medium in which mesoderm cells had been grown for the previous week did differentiate into chromatophores and nerve fibers. No comparable differentiation occurred when ectoderm cells were placed in cultures of entoderm cells. Twitty concluded that inductor tissues, such as notochord and mesoderm cells, contain and release diffusible substances which can operate at a distance and induce the differentiation of ectoderm. These substances also appear to be nucleoproteins.

Evidence that steroids, as well as nucleoproteins, may play a morphogenetic role in development has been obtained by Dr. Dorothy Price of the University of Chicago. When the reproductive tract of a fetal rat is dissected out and grown in tissue culture, development occurs normally if the testis or ovary is left in place. If both testes are removed, there is no development and differentiation of the accessory organs—the vas deferens, seminal vesicles and prostate gland. However, if both testes are removed and a pellet of testosterone, the male sex hormone, is implanted, development proceeds normally (Fig. 254). Thus the steroid testosterone can diffuse across a limited space and induce the development of male characters.

Some interesting data bearing on the problem of morphogenesis have been obtained recently by Drs. Briggs and King of the Lankenau Institute, Philadelphia. They have been able to transplant a nu-

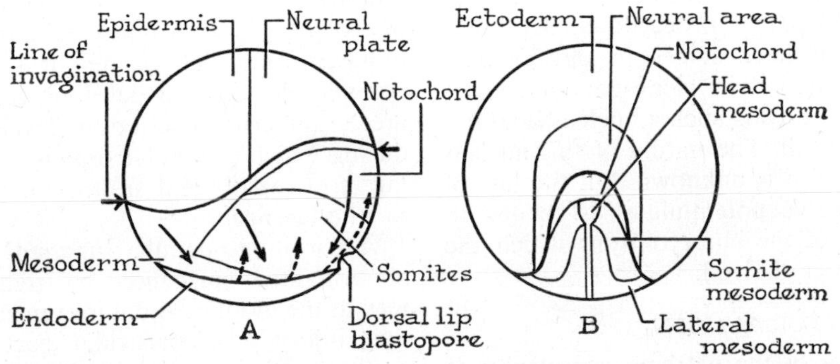

Figure 253. Embryo maps. *A*, Lateral view of a frog gastrula showing the presumptive fates of its several regions. *B*, Top view of a chick embryo showing location in the primitive streak stage of the cells which will form particular structures of the adult.

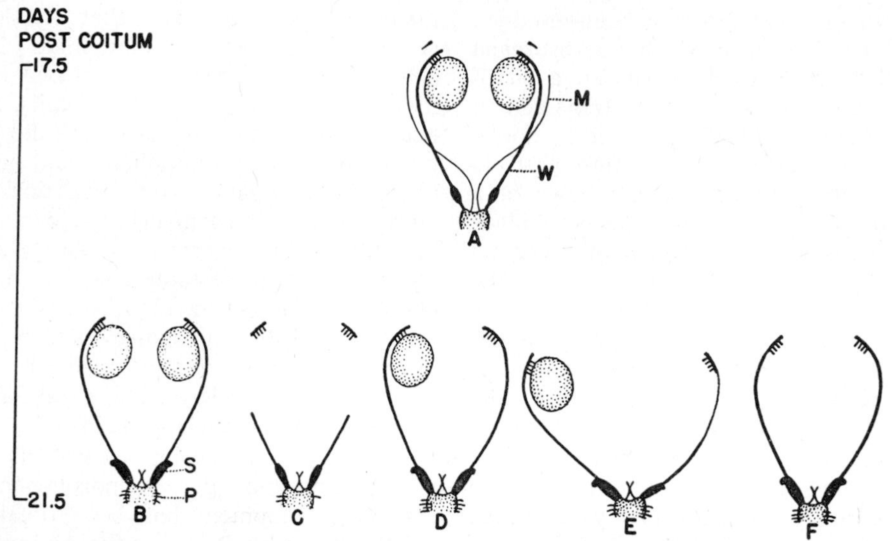

Figure 254. Schematic representation of fetal male reproductive tracts. *A,* At 17½ days, the age at explantation; *B, C, D, E,* cultured for 4 days on standard medium; *F,* cultured on medium containing testosterone micropellets. M: Müllerian duct; W: Wolffian duct; S: seminal vesicle; P: prostate gland. (Reprinted by permission, from Price and Pannabecker: Ciba Foundation Colloquia on Ageing: Ageing in Transient Tissues. G. E. W. Wolstenholme, Editor. The Blakiston Co., 1956.)

cleus from one of the cells of an early blastula of a frog into an enucleated egg. This egg will subsequently cleave, gastrulate and develop normally. However, if a nucleus is taken from a cell of the late gastrula, a mesoderm or midgut cell, and transplanted into an enucleated egg, abnormal development results. Development is arrested in the blastula or early gastrula stage. Transplanted mesoderm nuclei result in embryos with deficient or absent nervous systems, and transplanted midgut nuclei form embryos with thin or absent epidermis and no nervous system. Some change occurs in the intrinsic differentiative properties of the nucleus as cleavage and development proceed. Nuclei taken from even later stages in development cannot function in cleavage; an enucleated egg receiving such a nucleus does not develop at all. The nature of this nuclear specialization is unknown, but the loss of differentiative potentialities is related to the part of the embryo from which the nucleus was derived.

271. MALFORMATIONS

In view of the extreme complexity of the developmental process it is indeed remarkable that it occurs so regularly and that so few malformations occur. About one child in one hundred is born with some major defect such as a cleft palate, club foot or spina bifida. Some of these are inherited; others result from environmental factors. Experiments with fruit flies, frogs and mice have shown that x-rays, ultraviolet rays, temperature changes and a variety of chemical substances will induce alterations in development. The kind of defect produced depends on the time in development at which the environmental agent is applied, and does not depend to any great extent on the kind of agent used. For example, x-rays, the administration of cortisone and the lack of oxygen will all produce similar defects —harelip and cleft palate—if applied at comparable times in development. There are certain critical periods in development, during which particular organs are growing most rapidly and are most susceptible to interference.

According to popular superstition, such abnormalities are caused by fright or injury to the mother, and it is even supposed that injury to a particular part of the mother's body results in the malformation of that part of the fetus. Another unfounded belief is that reading or listening to music during pregnancy will develop literary or musical abilities in the offspring. Such superstitions have no basis in fact; there is no connection whatsoever between the nervous systems of mother and child. The injuries commonly believed to cause malformations usually occur later in pregnancy, long after the organs have been formed.

In man, apes and monkeys, as well as in many other species of mammals, a single offspring usually is produced at one time, although in other animals, more than one (up to twenty-five in the pig) are produced in a single litter. But about once in every eighty-eight human births, two individuals are delivered at the same time. The number of twins born appears to be gradually declining; in 1952 only one birth in ninety-seven was a twin. More rarely, three, four, five and even six children are born simultaneously. These multiple births all belong to one of two types, due to two basically different situations. About three fourths of the twins (and quadruplets, quintuplets, and so on) are the result of the simultaneous release of two eggs, one from each ovary, both of which are fertilized and develop. Such **fraternal twins** may be of the same or different sexes, and have only the same degree of family resemblance that brothers and sisters born at different times have, for they are entirely independent individuals with different hereditary characteristics. In contrast, **identical twins** (or triplets, and so on) are formed from a *single* fertilized egg which at some early stage of development divided into two (or more) independent parts, each of which develops into a separate fetus. Such twins, of course, are the same sex, have identical hereditary traits, and are so similar that it is difficult to tell them apart.

Occasionally, identical twins are not separated completely, but are born joined together and are known as **Siamese twins.** All grades of union have been known to occur, from almost complete separation, to fusion throughout most of the body, so that only the head or the legs are double. One twin may become sick independently and even die before the other, but the second will die within a few hours unless he is separated surgically from his twin. Sometimes the two twins are of different sizes and degrees of development, one being quite normal, while the second is only a partially formed parasite on the first. Such errors of development usually die during or shortly after birth.

272. THE CHANGES AT BIRTH

Great changes take place within a short time after a child is born. Hitherto it has received both food and oxygen from the placenta; now its own digestive and respiratory systems must function. Correlated with these changes are major changes in the circulatory system: the umbilical arteries and veins are cut off, and the blood flow through the heart and lungs is altered.

It is believed that the first breath of the newborn infant is initiated by the accumulation of carbon dioxide in the blood after the umbilical cord is cut, which stimulates the respiratory center in the medulla. The resulting expansion of the lungs enlarges the blood vessels therein, which previously were partially collapsed, and blood from the right ventricle thus flows in ever increasing amounts through the lung vessels, instead of through the arterial duct which connected the pulmonary artery and aorta during fetal life. The resulting increase in blood returning from the lungs to the left atrium results in the closing of the valvelike opening of the oval window (Fig. 255). These changes take place within a short period after birth, and eventually the flap of the oval window grows into place and the arterial duct degenerates, so that the adult type of circulation is established. Occasionally the oval window fails to close or the arterial duct fails to degenerate, and there is a mixing of oxygenated and nonoxygenated blood which results in a "blue baby," a child whose skin has a purplish hue because of the inadequate oxygenation of its blood. Delicate surgical procedures, developed after years of practice operations on dogs, now make it possible to operate on the heart itself and cure this condition.

Development does not cease at birth, of

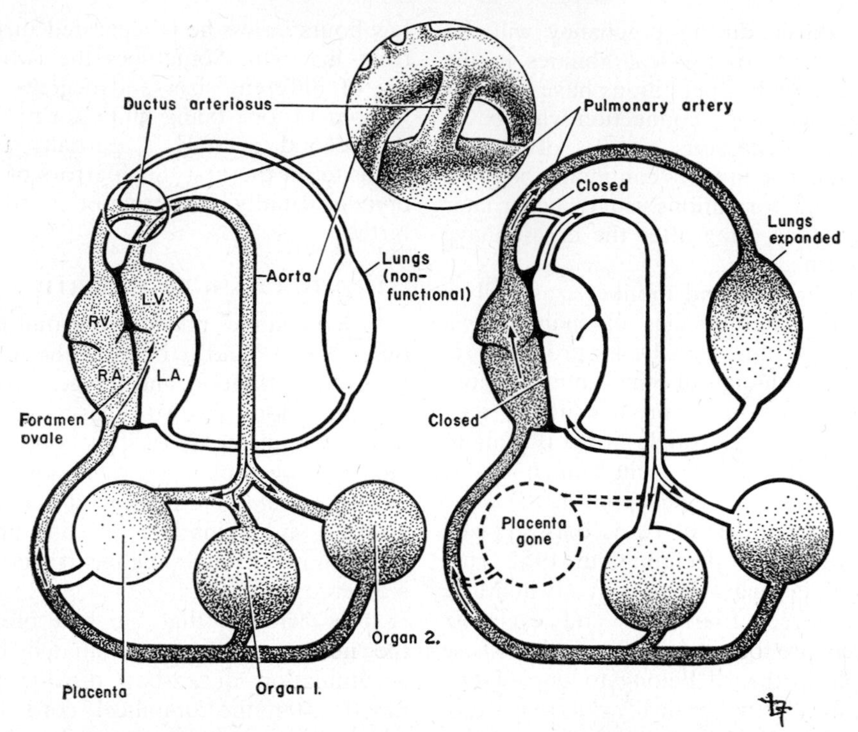

Figure 255. Changes in the human circulatory system at birth. *Left,* the circulatory system of the fetus; *right,* the circulatory system of the newborn child. Aerated blood is shown in white, nonaerated blood in dark stippling, and a mixture of the two in lighter stippling. Note that in the embryo little of the blood goes through the lungs; most of the blood entering the right auricle gets into the aorta either via the oval window (*foramen ovale*) between the right and left auricles or via the arterial duct (*ductus arteriosus*) between the pulmonary artery and aorta (inset). The changes at birth are: (1) the absence of the placenta, (2) the expansion of the lungs, (3) the closing of the foramen ovale, and (4) the closing and degeneration of the arterial duct.

course. At that time, such structures as the teeth and genital organs are only partly formed and the body proportions are quite different from those of the adult. The head, proportionally much larger than the rest of the body in early development, makes up about half the length of the two-month fetus, but its growth terminates in early childhood, so that the adult head is proportionately smaller than that of the newborn. The arms attain their proportionate size shortly after birth, but the legs attain their proportionate size only at the end of about ten years. The last organs to mature are the genitals, which do not begin to grow rapidly until between twelve and fourteen years after the infant is born.

Reproduction involves many complex and interdependent processes: the elaboration of hormones which regulate the development of the gonads, of secondary sex structures and the production of gametes in the parents; behavior patterns which bring the parents together to release their gametes at such a time and in such a place that their union is probable; the union of male and female pronuclei followed by cleavage, gastrulation and morphogenesis; and devices for the care and protection of the developing young.

QUESTIONS

1. What is the difference between isolecithal and telolecithal eggs?
2. What parts of the body develop from the ectoderm? The mesoderm? The entoderm?
3. What is the primitive streak? What happens to it during development?
4. What becomes of the notochord in the vertebrates?
5. How is the spinal cord formed?

6. What becomes of the gill pouches in the human embryo?
7. What are "organizers"? How do they control development?
8. Differentiate between fraternal and identical twins. What are Siamese twins?
9. What causes a "blue" baby?
10. Describe the sequence of events from fertilization to the completion of neurulation in the frog.
11. Trace the development of the kidney in the human embryo.

SUPPLEMENTARY READING

An interesting elementary presentation of human embryology is given by George W. Corner in his *Ourselves Unborn.* Detailed discussions of human embryology at an advanced level are L. B. Arey's *Developmental Anatomy* and B. M. Patten's *Human Embryology.* Research on the dynamics of development is summarized in Roberts Rugh's *Experimental Embryology.* Comparative vertebrate embryology is well presented in Emil Witschi's *Development of Vertebrates.*

Part Six

The Mechanism of Heredity

Chapter 30

The Physical Basis of Heredity

IT MUST have been thousands of years ago when man first made one of the fundamental observations of heredity, that "like begets like" and that new types of animals and plants may result when unlike forms are crossed. From prehistoric times, men have tried to breed better horses, dogs, cattle, vegetables and fruits, but as late as 1900 the scientific basis of inheritance was largely unknown.

New organisms closely resemble their parents: the offspring of a bean plant are always beans and nothing else; the mating of two cats always produces cats and no other animal. Furthermore, the mating of two Siamese cats always produces Siamese cats and not a different variety. Particular characters frequently appear in generation after generation in man as well as in other animals. This resemblance of individuals to their progenitors is called **heredity.**

273. HEREDITY AND VARIATION

The fact of heredity is so familiar that its significance is frequently overlooked. The sperm cells and the parts of eggs effective in inheritance, the nuclei, are both exceedingly small. All the sperm which gave rise to the present human population of the world could be contained in a single drop of water, and another drop could hold all the egg nuclei necessary. The sperm and egg nuclei transmit all the characters which the new individual inherits from his parents. The qualities themselves—color, size, shape, and so forth—are not present in the germ cells, but something representing them and capable of producing them in the new individual is present. The eggs and sperm contain factors which interact with each other and with the environment to produce the adult characteristics.

Although resemblances between parents and offspring are close, they are usually not exact. The offspring of a particular set of parents differ from each other and from their parents in many respects and to different degrees. This is clearly evident among human beings; in other animals and among plants it is not always immediately obvious, but detailed studies usually bring the dissimilarities to light. These differences, termed **variations,** are characteristic of living things. It is often said that variation is the most invariable thing in nature. Some variations are inherited, that is, caused by the segregation of hereditary factors among the offspring; others are not inherited, but are due to the effects of temperature, moisture, food, light and other factors in the environment on the development of the organism. Most characteristics are strongly influenced by the environment in which the individual develops.

The branch of biology concerned with the phenomena of heredity and variation and the study of the laws governing similarities and differences between individuals related by descent is called **genetics.** Since its inception at the beginning of this century, the science has advanced rapidly and is still developing.

Although rather hit-or-miss efforts at breeding domesticated animals and plants had been carried on for centuries, and breeding procedures, such as the artificial pollination of the date palm, were performed in Egypt and Mesopotamia centuries before the Christian era, scientific understanding of the processes of heredity and variation had to wait until knowledge of the details of sexual reproduction existed. Only after the invention of the microscope and the recognition of spermatozoa by early microscopists, the demonstration early in the eighteenth century that spermatozoa initiate development, and the demonstration in the latter half of the nineteenth century that eggs and sperm are single cells whose nuclei fuse in fertilization, could the science of genetics arise.

Throughout the eighteenth and nineteenth centuries many investigators tried to discover how characters are transmitted from generation to generation. From about 1760 to 1770, the German botanist Kölreuter experimented with crossing two species of tobacco by placing pollen from one species on the stigmas of the other. The resulting hybrids had characters intermediate between those of the two parents. This indicated that parental characters are transmitted through both pollen and ovule. Kölreuter and the plant hybridizers who followed him failed to discover the nature of the hereditary mechanism partly because the cytologic basis of heredity was unknown and partly because they tried to study the inheritance of all the characters of the plants simultaneously.

Gregor Mendel, an Austrian abbot who performed experiments in the garden of his monastery at Brunn, Austria, succeeded where previous hybridizers failed because he studied the inheritance of single contrasting characters (such as green versus yellow seed color, wrinkled versus smooth seed coat in peas), counted the number of each type, and kept accurate records of his crosses and counts. His experiments showed that inheritance is subject to certain laws, and that if the pedigrees of two individuals are known, the types of offspring produced by mating them can be predicted with a high degree of accuracy (Table 9). His results, published in an obscure Austrian journal in 1868, aroused little interest in the scientific world, and were neglected for more than thirty years.

In 1900, after the details of mitosis, meiosis and fertilization had been discovered, three investigators, de Vries in Holland, Correns in Germany, and von Tschermak in Austria, independently rediscovered the laws of inheritance described by Mendel. In searching through

Table 9. AN ABSTRACT OF THE DATA OBTAINED BY MENDEL FROM HIS BREEDING EXPERIMENTS WITH GARDEN PEAS

PARENTAL CHARACTERS	FIRST GENERATION	SECOND GENERATION	RATIOS
Yellow seeds × green seeds	all yellow	6022 yellow : 2001 green	3.01 : 1
Round seeds × wrinkled seeds	all round	5474 round : 1850 wrinkled	2.96 : 1
Green pods × yellow pods	all green	428 green : 152 yellow	2.82 : 1
Long stems × short stems	all long	787 long : 277 short	2.84 : 1
Axial flowers × terminal flowers	all axial	651 axial : 207 terminal	3.14 : 1
Inflated pods × constricted pods	all inflated	882 inflated : 299 constricted	2.95 : 1
Red flowers × white flowers	all red	705 red : 224 white	3.15 : 1

the literature, they found his paper, and so, sixteen years after Mendel's death, the two fundamental laws of inheritance were named after him. To understand these laws and their application to heredity, we must first consider the mechanism inside the cell which is responsible for inheritance.

From the chapter on development, you will remember that the growth of an individual plant or animal is due to cell division plus the growth of each cell comprising the organism. This division of cells is a regular process, exactly following a certain plan; it is known as **mitosis.** Under rare, usually pathologic, conditions, a cell may split irregularly, without following the exact plan; this is called **amitosis.**

274. CHROMOSOMES AND GENES

Chromosome Structure. When a dividing cell is stained and examined under the microscope, dark-staining bodies, called **chromosomes,** are visible within the nucleus. Each one consists of a central thread called a **chromonema,** along which lies a series of beadlike structures, the chromomeres. The chromosomes are distinctly visible as just described only during cell division; at other times they appear as fine strands called **chromatin,** which are also dark-staining. Research has shown that the chromosomes are present as distinct physiologic and structural entities between successive cell divisions even though in most forms they are not visible.

When the chromomeres were first seen, many biologists believed them to be the **genes,** the hereditary factors which previous breeding experiments had shown to lie within the chromosome in a linear order. More recent research has shown, however, that there is no one-to-one correspondence between chromomeres and genes; that is, there is not a single gene for each chromomere. Instead, some chromomeres contain several genes, and some genes have been located between chromomeres. The exact significance of these swellings along the chromonema is not clear.

Chromosome Number. Each cell of every organism of a given species contains a definite number of chromosomes. The number for man is forty-eight*—thus, every cell in the body of every human being has exactly forty-eight chromosomes (Fig. 256). Many other species of animals and plants also have forty-eight; so it is not the *number* of chromosomes alone that differentiates the various species of animals, but chiefly the nature of the hereditary factors in the chromosomes. A certain species of roundworm has only two chromosomes in each cell, and some crabs have as many as 200 in each cell. The highest chromosome number reported so far is about 1600, found in a radiolarian, a microscopic, single-celled marine animal. Most species have a chromosome number between ten and fifty; numbers above and below this are comparatively rare. *Chromosomes always exist in pairs;* there are invariably two of each kind. Thus, the forty-eight chromosomes of man consist of two of each of twenty-four different kinds. They differ in length, shape and the presence of knobs or constrictions along their length. In most species the chromosomes vary enough in these morphologic features that the cytologist can pick out the pairs quite easily.

* Careful studies published in 1956 indicate that man has 46 rather than 48 chromosomes per cell.

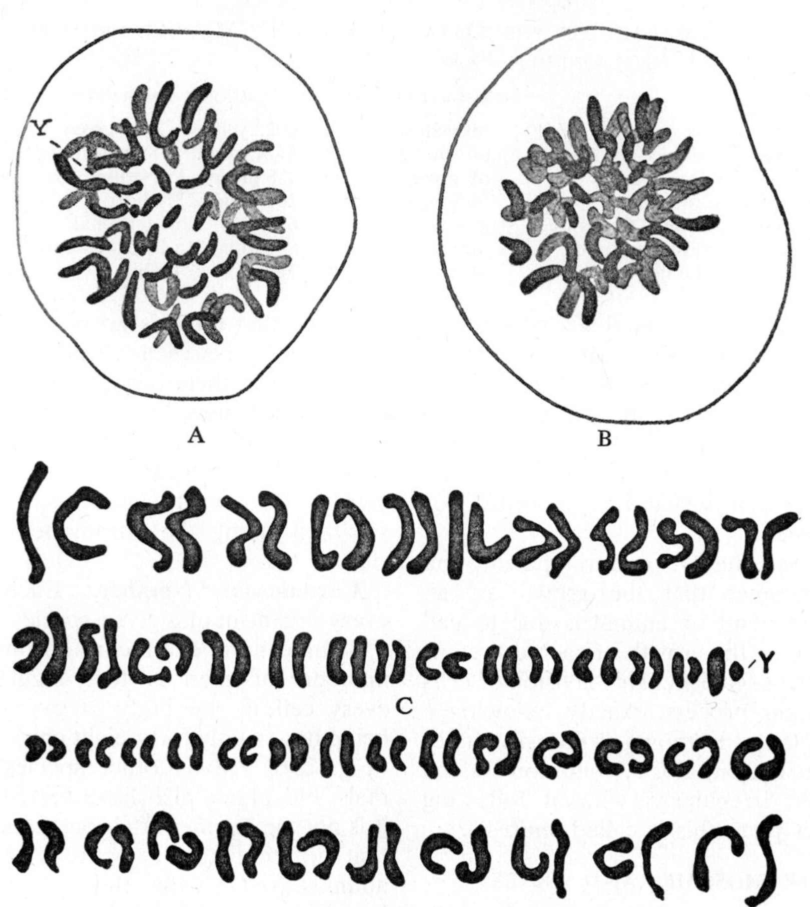

Figure 256. Human chromosomes. *A,* Equatorial plate from a mesenchyme cell (male) showing the forty-eight chromosomes. The Y chromosome is indicated. *B,* Equatorial plate from a mesenchyme cell (female), showing the forty-eight chromosomes. *C,* The chromosomes from the prophase nucleus of a male, lined up to show the homologous pairs. *D,* The chromosomes from the prophase nucleus (female), lined up to show the homologous pairs. (Evans and Swezy: The Chromosomes in Man, Memoirs of the Univ. of California.)

275. MITOSIS

Cell division must be a regular phenomenon so that each of the daughter cells formed by a cell division will receive exactly the right kind and number of chromosomes. If we tamper with the mechanism of cell division in an experimental animal or plant so that cells receive more or less than the proper number of chromosomes, great abnormalities and even death may result. *Mitosis, therefore, is the regular division of a cell in which each of the two daughter cells receives exactly the same number and the same kind of chromosomes that the parent had.* This involves what appears and was long believed to be a longitudinal splitting of each chromosome into two halves. There is abundant evidence that no such splitting occurs; instead each original chromosome brings about the synthesis of an exact replica immediately beside itself (a process known as **autocatalysis).** The new chromosome is manufactured from raw materials present in the nucleus some time before the visible mitotic process begins. The old and new chromosomes are identical in structure and function and at first lie so close to each other that they appear to be one. As cell division occurs, and the chromosomes contract, the line of cleavage between them becomes visible and the chromosomes appear to split. In man, then, each of the forty-eight chromosomes synthesizes an exact replica of itself, so that for a time there are ninety-six chromosomes in the cell. Then cell division is completed, and forty-eight go to one, and forty-eight to the other daughter cell. A complicated mechanism is necessary to insure the equal division of the chromosomes.

Each mitosis is a continuous process, one stage merging imperceptibly into the next. However, for descriptive purposes, biologists have arbitrarily divided it into four stages: **prophase, metaphase, anaphase** and **telophase.** Between mitoses, a cell is said to be in the **resting stage** (Fig. 257). Mitosis is an extremely active process. It is difficult to realize from a description, diagram or even from a prepared slide of cells undergoing mitosis just how active it is.

Prophase. In the prophase the chromatin threads begin to condense and form chromosomes, appearing as a tangled mass of threads within the nucleus. At first the chromosomes are stretched maximally and the individual chromomeres are clearly visible. In favorable conditions the chromomeres can be seen to differ one from another in size and shape, so that individual ones are recognizable. Later, the chromosomes contract and the chromomeres lie so close together that individual ones cannot be distinguished. Each chromosome has previously undergone autocatalysis and in certain species its double nature is apparent.

The cell cytoplasm contains a small, granular structure called a **centriole.** At the beginning of prophase the centriole divides, and the daughter centrioles migrate to opposite sides of the cell. Between the separating centrioles a spindle forms, composed of a number of protoplasmic threads called **spindle fibers.** These are arranged like two cones put together base to base, so that the spindle is narrow at the ends or poles near the centrioles, and broad at the center or equator. The spindle fibers stretch from equator to pole and are composed of a denser protoplasm than the surrounding nucleoplasm. The spindle is a definite structure; by using a micromanipulator it is possible to introduce a fine needle into a cell and push the spindle around. By special techniques it is possible to isolate spindle fibers from dividing cells for study (Fig. 258). While the centrioles have been separating and the spindle forming, the chromosomes in the nucleus have been contracting, getting shorter and thicker. Their double nature, which may not have been visible before, can now be clearly seen.

Metaphase. When the chromosomes have contracted to their fullest extent, forming short, dark-staining, rodlike bodies, the nuclear membrane disappears (dissolves) and the chromosomes line up across the equatorial plane of the spindle, which has been forming around them (Fig. 257). At this point the prophase is complete, and the short period during which the chromosomes are in the equatorial plane constitutes the metaphase. When the division of living human cells is observed under the microscope, the prophase lasts from thirty to sixty minutes and the metaphase from two to six minutes. The times vary for different tissues and different species.

Anaphase. The chromosomes immediately begin to separate, one of the separating members (daughter chromosomes) going to each pole. The events from the time when the chromosomes first begin to move apart until they reach the poles constitute the anaphase, a period of three to fifteen minutes. The exact mechanism causing the chromosomes to move poleward is unknown. The earliest suggestion was of a pull exerted by the spindle fibers. According to another theory, the spindle fibers merely act as tracks or guiderails along which the chromosomes glide in going to the poles. The motive power is supplied by protoplasm which gets between the daughter chromosomes, absorbs water, and swells, thereby pushing the chromosomes apart. If it were not for the spindle fibers, the chromosomes would be pushed at random in all directions. But with them, all of one set of daughter chromosomes are gathered at one pole, and all of the other set at the other pole.

Telophase. When the chromosomes

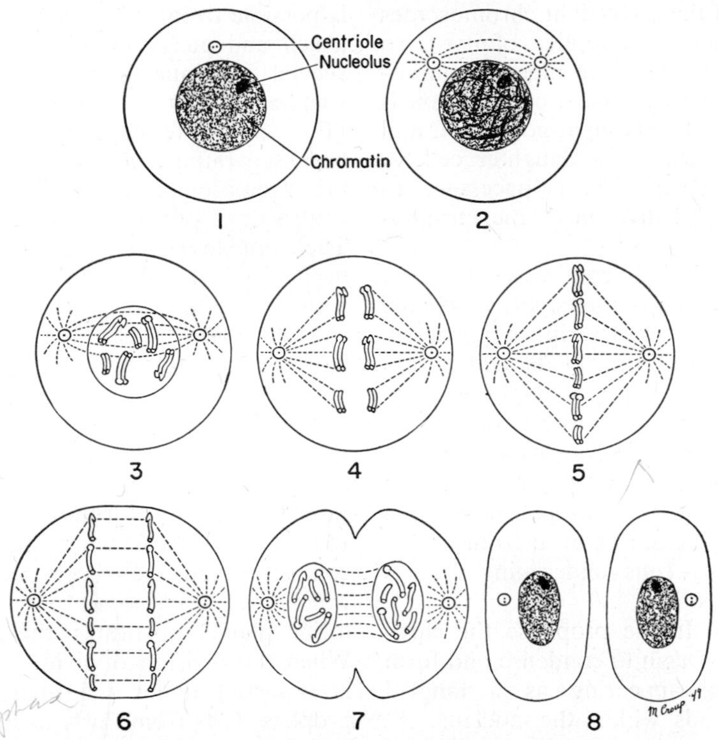

Figure 257. Mitosis in a cell of a hypothetical animal with a diploid number of six (haploid number = 3); one pair of chromosomes short, one pair long and hooked, and one pair long and knobbed. *1,* Resting stage. *2,* Early prophase: centriole divided and chromosomes appearing. *3,* Later prophase: centrioles at poles, chromosomes shortened and visibly doubled. *4,* Late prophase: nuclear membrane dissolved, spindle present. *5,* Metaphase: chromosomes arranged on equator of spindle. *6,* Anaphase: chromosomes migrating toward poles. *7,* Telophase: nuclear membranes formed; chromosomes elongating; cytoplasmic divisions beginning. *8,* Daughter cells: resting phase.

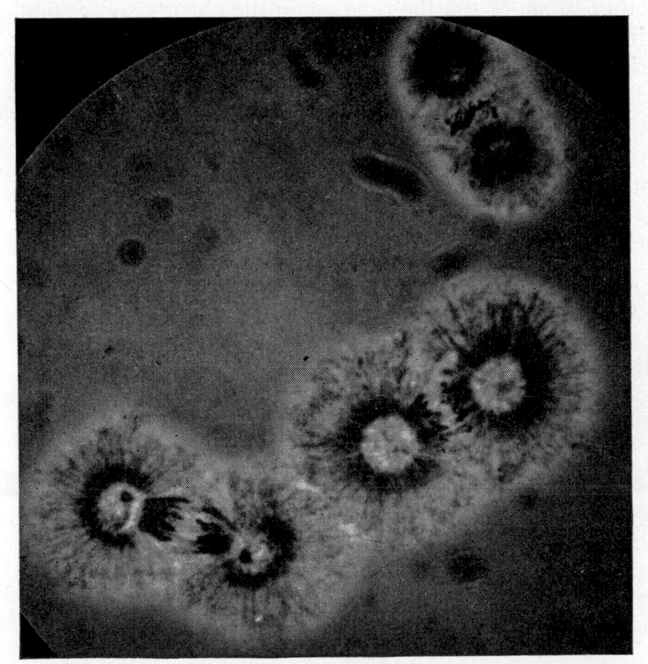

Figure 258. Photomicrograph of isolated spindle fibers of dividing cells from a sea urchin embryo. A metaphase figure appears in the upper right and two anaphase figures below. (Photograph courtesy of Dr. Daniel Mazia.)

reach the poles, the telophase begins. This is a period roughly equal to the prophase, lasting thirty to sixty minutes. The chromosomes elongate, lose their dark-staining capacity, and return to the resting condition in which only chromatin threads or granules are visible. While this is happening, a nuclear membrane forms around each daughter nucleus, and, simultaneously, the cytoplasm begins to divide. In animal cells the division is accomplished by a furrow encircling the surface of the cell in the plane of the equator. The furrow gradually deepens and separates the cytoplasm into two halves, the daughter cells, each of which has a nucleus. In most kinds of cells the entire process of mitosis consumes from one to two hours. In plants, division occurs by the formation of a **cell plate,** a partition of cytoplasm which forms first in the equatorial region of the spindle and then grows laterally to the cell wall. The net result, in any case, is the creation of two cells from a previous one. Each daughter cell has exactly the same number and kind of chromosomes that the parent cell had; hence, through mitosis, each cell in the body comes to have the hereditary material necessary for every characteristic of the organism. Mitosis occurs with widely different frequencies in various tissues, and in different species. In the red bone marrow, for example, where 10,000,000 red blood cells are produced per second, 10,000,000 mitoses must occur per second. In other tissues, such as those of the nervous system, mitoses occur very rarely. During the early development of an organism, cell divisions take place extremely rapidly and may occur every thirty minutes.

Control of Mitosis. The factors which initiate mitosis are not known exactly, but the ratio between the volume of nucleus and cytoplasm (the nucleoplasmic ratio) is believed to be involved. The increase in the size of the cell is effected by the older protoplasm manufacturing new protoplasm from simpler raw materials. This involves the transport of substances back and forth through both nuclear and cell membranes. Although the volume of a sphere increases as the cube of its radius, the surface increases only as the square of the radius, so that, as the cell grows, the volume increases more rapidly than the surface of the nuclear membrane. Beyond a certain point, then, the surface of the nucleus is insufficient to allow the necessary exchange of materials between nucleus and cytoplasm to provide for future growth. A division of the nucleus greatly increases the surface without increasing the volume, and it is believed that this limiting factor in the nucleoplasmic ratio somehow initiates mitosis.

Some biologists have suggested that a hormone is involved in starting mitosis. Indeed, the mitoses of various cells in a cleaving egg occur simultaneously for a long time; perhaps a hormone is released periodically to control them. Even in adult organisms, mitosis in tissue cells may occur in waves, suggesting that some inductive substance, or hormone, is present. Haberlandt found evidence that dying cells may produce a substance to stimulate cell division. In his experiments he cut a potato in half and examined the cut edges for mitosis. He found that if he cleaned off the cut edge, few mitoses took place; if he did not clean off the cut edge, cell divisions were more frequent; and if he put a mash of cut-up potato cells on the cut edge, many more occurred. He therefore concluded that the cut-up and dying potato cells produced a "wound hormone" which stimulated cell division and the formation of scar tissue. Marshak and Walker prepared an extract of the nuclei of rat liver cells and by fractionation obtained two fractions, one of which increased, and the other decreased, the rate of division of liver cells.

276. MEIOSIS

Constancy of the chromosome number in successive generations is brought about by the process of **meiosis,** which occurs during the formation of gametes, either eggs or sperm. Meiosis is essentially a pair of cell divisions during which the chromosome number is reduced to half, so that the gametes receive only half as many chromosomes as other cells in the body. When two gametes unite in fertilization, the normal chromosome number is reconstituted. The reduction of the chromosome number in meiosis does not occur at random, but in a regular way; the members of pairs of chromosomes separate and pass to different daughter cells. *As a result of meiosis, each gamete contains one and only one of each kind of chromosome—one complete set of chromosomes.* This is accomplished by the pairing or **synapsis** of the like chromosomes, and a separation of the members of the pair, one going to each pole. The like chromosomes which synapse during meiosis are called **homologous chromosomes.** They are identical in size and shape, have identical chromomeres along their lengths, and contain similar hereditary factors or genes. A set of one of each kind of chromosome is called the **haploid** number; a set of two of each kind is called the **diploid** number. For man, then, the haploid number is twenty-four, the diploid, forty-eight. Gametes have the haploid number; fertilized eggs and all the cells of the body developing from the zygote have the diploid number. A fertilized egg gets exactly half its chromosomes (and half its genes) from its mother, and half from its father. Only the last two cell divisions which result in mature, functional eggs or sperm are meiotic; all other cell divisions are mitotic.

The process of meiosis consists of two cell divisions which occur in quick succession, called, respectively, the first and second meiotic divisions (Fig. 259). Each of these has the same four stages, prophase, metaphase, anaphase and telophase, found in mitosis. There are, however, important differences between mitosis and meiosis within these stages, particularly in the prophase of the first meiotic division. In this, the chromosomes appear as long thin threads condensing from chromatin, as in mitosis. They begin to contract, getting shorter and thicker. Early in this prophase, while the chromosomes are still elongated and thin, the homologous chromosomes undergo synapsis; that is, they pair longitudinally, coming to lie close together, side by side along their entire length and twisted around each other. After synapsis, the chromosomes continue to shorten and thicken. Each one then becomes visibly double, consisting of two threads, as in mitosis. Again, the doubling has occurred by autocatalysis some time before meiosis begins. At the end of the prophase, then, the chromosomes have doubled and synapsed, yielding a bundle of four homologous chromosomes called a **tetrad.** Since each pair of chromosomes gives rise to a bundle of four, there are as many tetrads as the haploid number of chromosomes. In human cells there are twenty-four tetrads and a total of ninety-six chromosomes at this stage.

While these events are taking place, others progress as in the mitotic prophase: the centriole divides and the two centrioles go to opposite poles, the spindle forms between the centrioles, and finally the nuclear membrane dissolves. The tetrads then line up around the equator of the spindle, as the individual chromosomes do in mitosis, and the cell is said to be in metaphase. The homologous chromosomes now separate from each other and move to the poles. The doubled chromosomes do not separate; only the homologous ones which underwent synapsis do. Thus the chromosomes moving to the poles in the anaphase of the first meiotic division are double, and in the telophase of the first meiotic division in man there are twenty-four double chromosomes at each pole. The division of the cytoplasm follows, but in most animals and plants there is no resting stage between the two meiotic divisions. The chromosomes do not dissolve and form chromatin threads; instead, the centriole divides, a new spindle forms in each cell (at right angles to the spindle of the first division), and the haploid number of double chromosomes lines up on the equator of this spindle. Because of this, the telophase of the first meiotic division and the prophase of the second meiotic division are rather short. The lining up of the double chromosomes on the equator of the spindles constitutes the metaphase of the second meiotic division. It is possible to differentiate the metaphases of the first and second meiotic divisions when seen under the microscope, because in the first meiotic division the chromosomes are arranged on the equator in bundles of four (the tetrads), and in the second meiotic division in bundles of two. There is no further splitting or doubling of the chromosomes; they simply separate, so that in the anaphase of the second meiotic division twenty-four single chromosomes (one of each kind) arrive at each pole. In the telophase that follows, the cytoplasm divides, the chromosomes gradually change back into chromatin, and a nuclear membrane forms. From these two successive divisions, four cells emerge, each with the haploid number of chromosomes, each with one and only one of each kind of chromosome. These are now mature gametes, and do not undergo any further mitotic or meiotic divisions.

Fundamentally, the same process occurs in both the meiotic divisions in the testis resulting in sperm and the meiotic divisions in the ovary resulting in eggs,

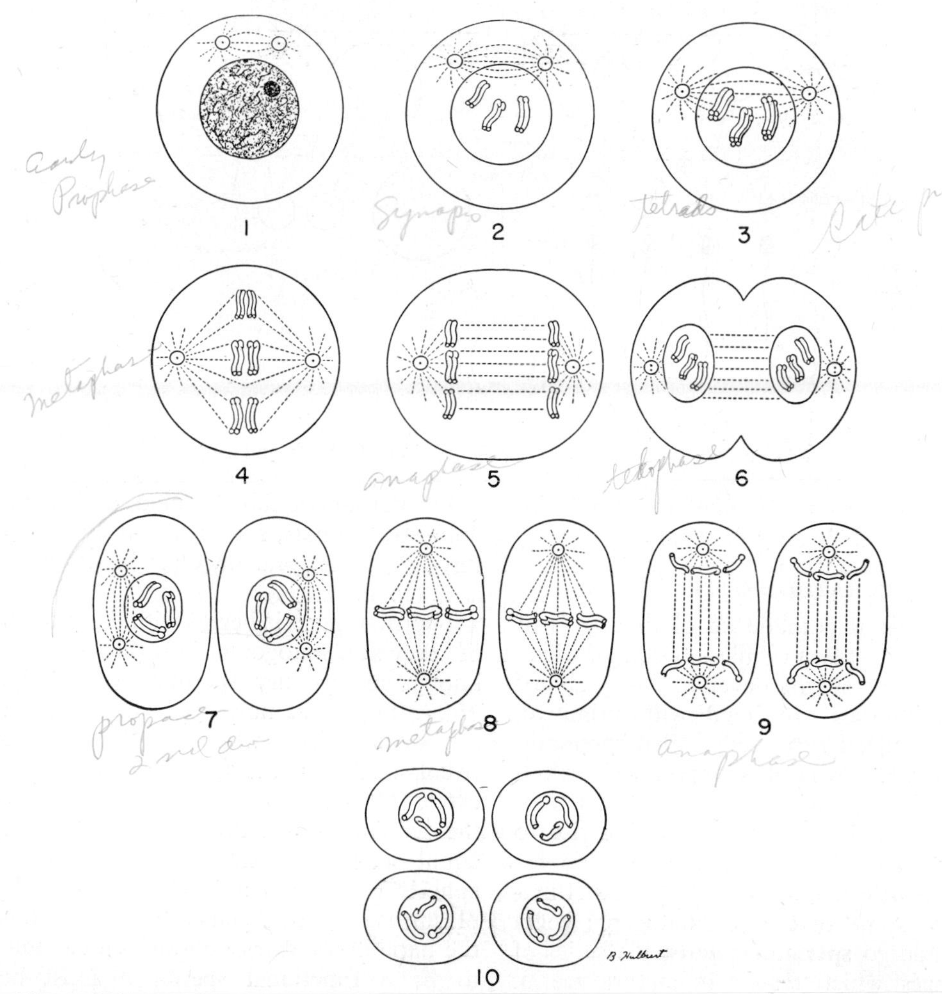

Figure 259. Meiosis in a hypothetical animal with a diploid chromosome number of six. It has three pairs of chromosomes, of which one is short, one is long with a hook at the end, and one is long and knobbed. *1,* Early prophase of the first meiotic division: chromosomes begin to appear. *2,* Synapsis: the pairing of the homologous chromosomes. *3,* Apparent doubling of the synapsed chromosomes to form groups of four identical chromosomes, tetrads. *4,* Metaphase of the first meiotic division, with the tetrads lined up at the equator of the spindle. *5,* Anaphase of the first meiotic division: the chromosomes migrating toward the poles. *6,* Telophase of the first meiotic division. *7,* Prophase of the second meiotic division. *8,* Metaphase of the second meiotic division. *9,* Anaphase of the second meiotic division. *10,* Mature gametes, each of which contains only one of each kind of chromosome.

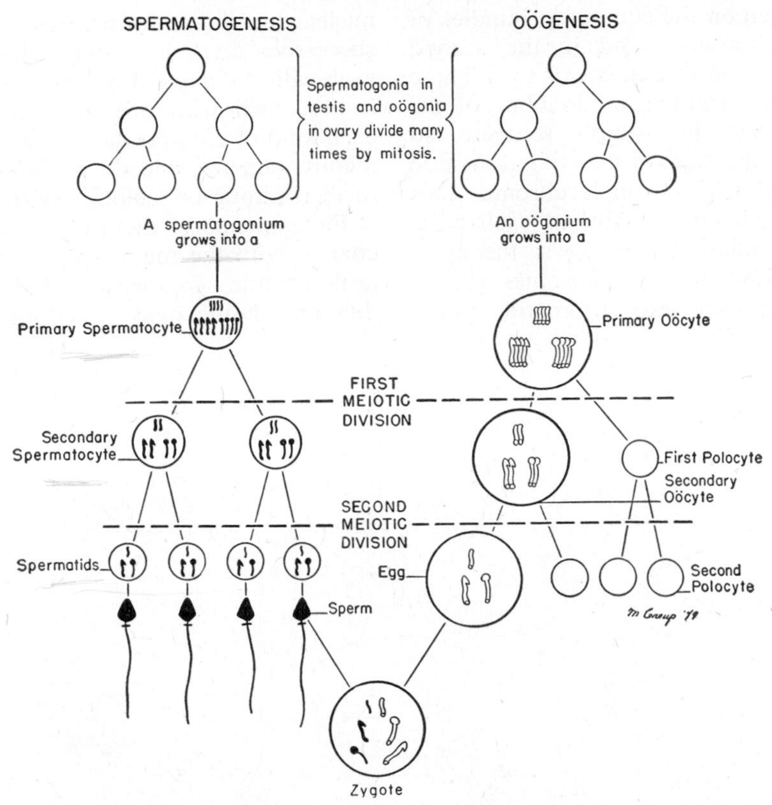

Figure 260. Comparison of the formation of sperm and eggs.

but there are some differences in detail which we shall discuss separately.

277. SPERMATOGENESIS

The testis is made up of thousands of cylindrical sperm tubules, in each of which millions of sperm develop. The walls of these tubules are lined with primitive, unspecialized germ cells called **spermatogonia.** Throughout embryonic development and during childhood the spermatogonia divide mitotically, giving rise to additional spermatogonia to provide for the growth of the testis. After sexual maturity, some of the spermatogonia begin to undergo **spermatogenesis,** the series of changes which results in mature sperm. Other spermatogonia continue to divide mitotically and produce more spermatogonia for later spermatogenesis. In most wild animals there is a definite breeding season, either in spring or fall, during which the testis increases in size and spermatogenesis occurs. Between breeding seasons the testis is small and contains only spermatogonia. In man and most domestic animals, spermatogenesis occurs throughout the year once sexual maturity is reached.

Spermatogenesis begins with the growth of the spermatogonia into larger cells known as **primary spermatocytes** (Fig. 260). These are now ready for the first meiotic division, which results in two equal-sized cells, the **secondary spermatocytes.** They in turn undergo division by the second meiotic division to form four equal-sized **spermatids.** The spermatid, a spherical cell with a good deal of cytoplasm, is a mature gamete because it has the haploid number of chromosomes, but, to be a functional sperm, it must be streamlined for swimming to meet the egg. This requires a complicated process of growth and change (though not cell division). The nucleus shrinks in size and becomes the head of the sperm (Fig. 261), while the sperm sheds most of its cytoplasm. Some of the Golgi bodies from the cytoplasm congregate at the front end of the sperm and form a point which may aid the sperm in puncturing the egg cell membrane. Part of the cytoplasm forms a long, flexible tail which can beat to drive the sperm ahead. The mitochondria move to the point where head and tail meet, and form a small middle piece; this is believed to enable the tail to beat.

The spermatozoa of various animal species may be quite different. Nearly all sperm have a tail, but there are great variations in its size and shape, as well as in the characteristics of the head and middle piece (Fig. 262). A few animals, such as the parasitic roundworm *Ascaris,* have sperm without a tail, which move by ameboid movement instead. Crabs and lobsters have a curious tailless sperm with three pointed projections on the head, which stick to the surface of the egg, holding the sperm securely in place. The middle piece uncoils like a spring, and pushes the nucleus of the sperm into the egg cytoplasm, thus accomplishing fertilization.

278. OÖGENESIS

The ova or eggs form in the ovary, developing from immature sex cells called **oögonia.** During an organism's early life the oögonia undergo many successive mitotic divisions, just as the spermatogonia do. All these immature cells have the diploid number of chromosomes. At the time of puberty, in each menstrual month one or more oögonia stop divid-

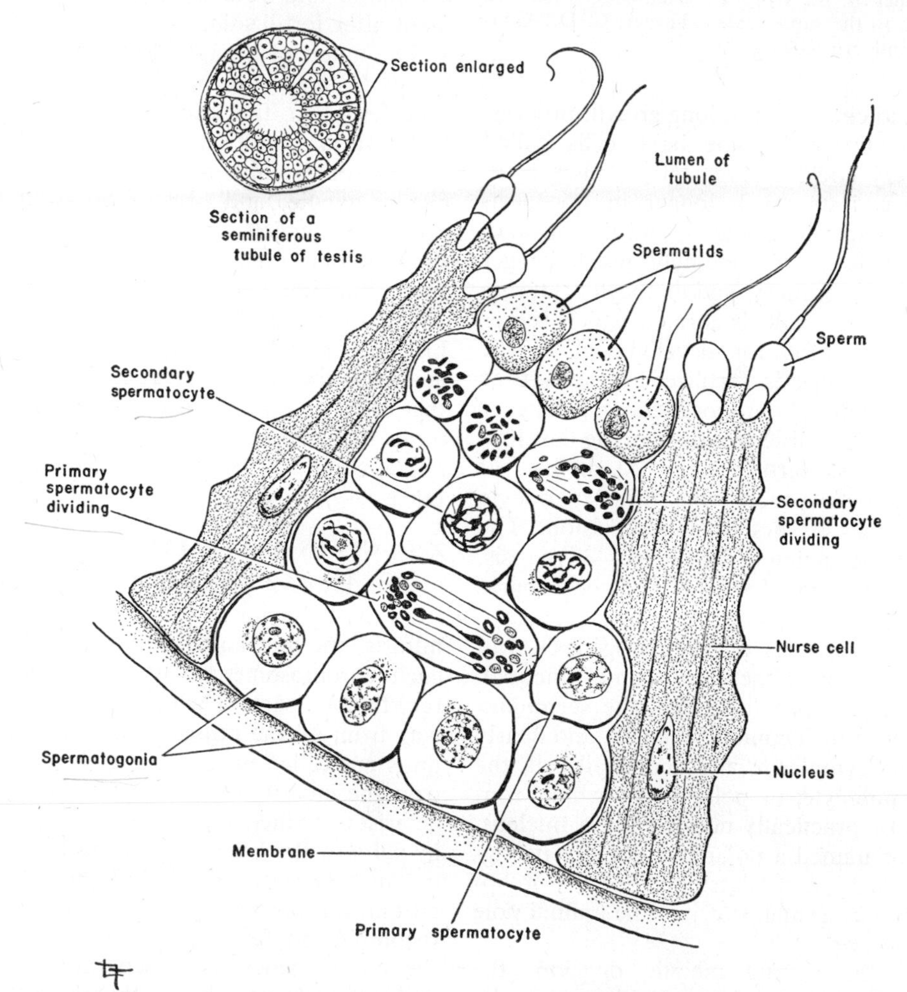

Figure 261. Diagram of part of a section of a human seminiferous tubule to show the stages in spermatogenesis and in the transformation of a spermatid into a mature sperm.

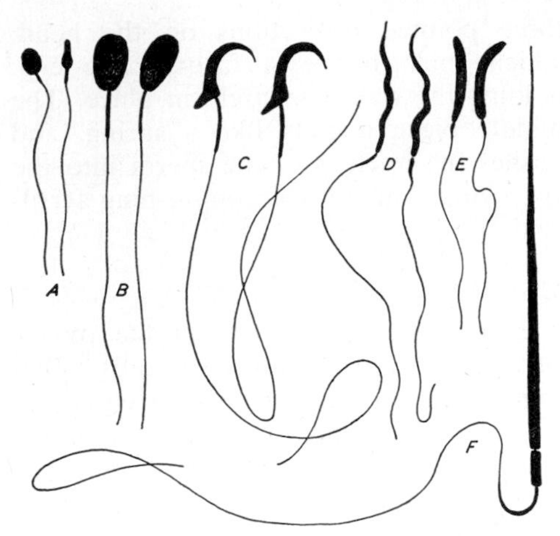

Figure 262. Spermatozoa from different species of vertebrates, illustrating the differences in size and shape. *A,* Human; *B,* sheep; *C,* rat; *D,* chicken; *E,* frog; *F,* salamander. All are drawn to the same scale. (Turner, C. D.: General Endocrinology.)

ing and enter upon a long growth process, eventually to become large cells called **primary oöcytes** (Fig. 260). This growth process lasts much longer in oögenesis than in spermatogenesis, and the primary oöcyte is correspondingly much bigger than the primary spermatocyte. During the process, **yolk** is formed which serves as stored food for development in the event the egg is fertilized. At this time many of the "morphogenetic substances" arise which subsequently regulate development in the fertilized egg. At the end of the growth process the primary oöcyte undergoes its first meiotic division. The events occurring in the nucleus—synapsis, the formation of tetrads and the separation of the homologous chromosomes—are the same as in spermatogenesis, but the division of the cytoplasm is unequal, resulting in one large cell, the **secondary oöcyte,** which contains the yolk and nearly all the cytoplasm, and one small cell, the first **polocyte,** or polar body, which consists of practically nothing but a nucleus. It was named a polar body before its significance was understood, because it appeared as a small speck at the animal pole of the egg.

In the second meiotic division, the secondary oöcyte again divides unequally into a large **oötid** and a small second polocyte, both of which have the haploid chromosome number. The first polocyte may divide into two additional second polocytes. Now the oötid can grow into a mature ovum. The three small polocytes soon disintegrate, so that each primary oöcyte gives rise to just one ovum, in contrast to the four sperm formed from each primary spermatocyte. The formation of polocytes enables the maturing egg to get rid of its excess chromosomes. The unequal cytoplasmic division insures that the mature egg will have enough cytoplasm and stored yolk to survive, if fertilized; the primary oöcyte puts all its yolk in one ovum. Thus, the egg has neatly solved the problem of reducing its chromosome number without losing the cytoplasm and yolk needed for development after fertilization.

The union of a haploid set of chromosomes from the sperm with another haploid set from the egg which occurs in fertilization reestablishes the diploid chromosome number. Thus the fertilized egg or zygote, and—by mitosis—all the body cells which develop from it, have the diploid number of chromosomes. Each individual gets exactly half of his chromosomes and half of his genes from his mother and half from his father. Because of the nature of gene interaction, the offspring may resemble one parent more than the other, but the two parents make equal contributions to its inheritance.

279. GENES AND ALLELES

The laws of heredity follow directly from the behavior of the chromosomes in mitosis, meiosis and fertilization. Within each chromosome are numerous hereditary factors, each of which differs in some way from all the others. These hereditary units are called **genes,** and each one controls the inheritance of one or more characteristics. Since the chromosomes exist in pairs in the body cells, *each body cell contains two of each kind of gene.* As the chromosomes separate in meiosis and recombine in fertilization, so, of course, the paired genes must separate and recombine. We now know that each chromosome is composed of a linear string of genes, and that the members of a homologous pair of chromosomes have genes arranged in similar order. Genes for particular traits occur at definite points, called **loci** (singular, *locus*), along the chromosome. When the chromosomes undergo synapsis in meiosis, the homologous chromosomes become attached point by point and, presumably, gene by gene.

The mode of inheritance of any given trait can be studied only when there are two contrasting conditions, such as Mendel's yellow and green peas, or brown eyes versus blue eyes in man. Such contrasting conditions are called **allelomorphs** or **alleles.** Alleles are defined as two contrasting traits, inherited in such a way that an individual may have either but not both of them. It follows that curly and straight hair are alleles, for a person has either one or the other. But curly hair and blonde hair are not alleles, because a person may have hair that is both blonde and curly. At the specific locus in the particular chromosome controlling coat color in guinea pigs there may be a gene for brown coat color or a gene for black coat color, but not both. Since there are two homologous chromosomes in body cells and therefore two of each kind of gene, a "pure" black guinea pig (one of a pedigreed race of black guinea pigs) has two genes for black coat—one in each chromosome—and a "pure" brown guinea pig has two genes for brown coat. It is essential to remember that the genes themselves have no color, are neither brown nor black; the brown gene controls certain chemical reactions so that a brown pigment is formed in the hair cells, whereas the black gene alters these chemical reactions, creating black pigment.

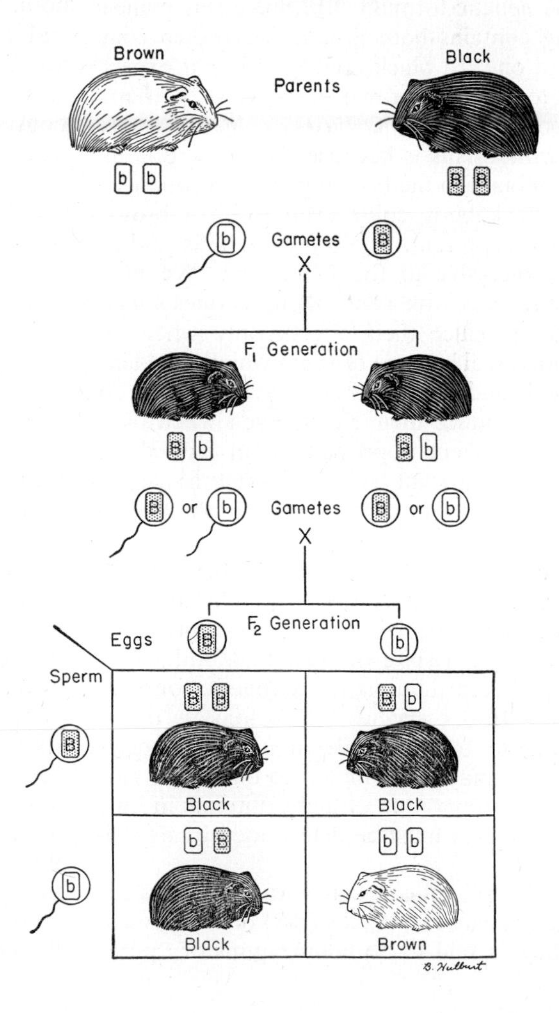

Figure 263. A monohybrid cross, the mating of a brown with a black guinea pig.

In working genetic problems, letters are used as symbols for the genes. A pair of genes for black pigment may be represented by **BB,** and a pair for brown pigment by **bb.** In choosing symbols, a capital letter is used for one gene, and the corresponding lower case letter is used to represent the gene for the contrasting trait or allele.

280. A MONOHYBRID CROSS

If we mate a "pure" brown, male guinea pig with a "pure" black, female guinea pig, the following events occur (Fig. 263): During meiosis in the male the two **bb** genes separate so that each sperm formed has ony one **b** gene, and in meiosis in the female the **BB** genes separate so that each ovum has only one **B** gene. Therefore, the fertilization of this egg by a **b** sperm results in an animal with the genetic formula, **Bb.** Since this guinea pig contains both a gene for brown coat and one for black coat, you might expect it to be dark brown or gray, or perhaps spotted; but *it is just as black as its mother.* This is because the black gene is **dominant** to the brown gene and produces a black body color even when a brown gene is present. The brown gene is said to be **recessive** to the black one. We may define recessive genes then, as ones which will produce their effects only when an individual has two of them which are identical, and dominant genes as those which will produce their effects even when only one of them is present in an individual. (The dominant gene is given a capital letter and the recessive gene the corresponding lower-case letter.) The phenomenon of dominance supplies part of the explanation as to why an offspring may resemble one of its parents more than the other despite the fact that both make equal contributions to its genetic constitution. In one species, black may be dominant to brown, while in another species brown may be dominant to black; the particular condition which obtains in any given form must be determined by experiment.

An animal or plant with two genes exactly alike, two blacks **(BB)** or two browns **(bb),** is said to be **homozygous** or "pure" for the character. An organism with one dominant and one recessive gene **(Bb),** is said to be **heterozygous** or **hybrid.** In the cross we have been considering, both the brown and black parents were homozygous for their respective conditions **(BB** or **bb),** and the offspring was heterozygous **(Bb).** Using these terms, we can now formulate better definitions for dominant and recessive: A recessive gene is one which will produce its effect only when homozygous; a dominant gene is one which will produce its effect whether it is homozygous or heterozygous.

If two of the heterozygous black guinea pigs are mated, the following events occur: In meiosis, the chromosome containing the **B** gene first synapses with and then separates from the chromosome containing the **b** gene, so that each sperm or egg has either a **B** gene or a **b** gene, but never both. Sperm and eggs containing **B** genes and those with **b** genes are formed in equal numbers. Since there are two types of eggs and two types of sperm, four combinations in fertilization are possible, and these occur in equal numbers. There is neither any special attraction nor repulsion between an egg and sperm containing the same genes; a **B** egg is just as likely to be fertilized by a **B** sperm as by a **b** sperm.

To see at a glance the possible zygote combinations, a "checkerboard" is made up, in which the possible types of eggs are represented along the top, the possible types of sperm are indicated along the left side, and the squares are filled in, showing the possible zygote combinations (Fig. 263). Three fourths of all the offspring will be **BB** or **Bb,** and therefore black, and one fourth will be **bb,** and appear brown. This 3 to 1 ratio is characteristic of the second generation offspring of a cross of two individuals which differ in a single trait governed by a single pair of genes. The genetic mechanism responsible for the 3:1 ratios obtained by Mendel in his pea breeding experiments is now evident. The generation with which a particular experiment is begun is called the $\mathbf{P_1}$, or **parental generation.** Offspring of this generation are called the $\mathbf{F_1}$, or **first filial generation.** Those resulting when two $\mathbf{F_1}$ individuals are bred constitute the $\mathbf{F_2}$, or **second filial generation** (grandchildren); those from mating of two F_2 individuals make up the F_3 generation, and so on.

281. PHENOTYPE AND GENOTYPE

The appearance of an individual with respect to a certain trait is known as its **phenotype,** while the organism's genetic constitution is called its **genotype.** The phenotypic ratio of the cross we have been discussing is 3 black: 1 brown, and the genotypic ratio is 1 **BB:** 2 **Bb:** 1**bb.** The phenotype may be defined as the external appearance of a given trait expressed in words such as black, brown, and so on. The genotype may be defined as the genetic constitution of an individual expressed in symbols, such as **BB, Bb** or **bb.** Guinea pigs which are **BB** and **Bb** are alike phenotypically, that is, both *appear* black, but they differ genotypically. The genotypes could be determined only by breeding tests. It is also possible for organisms to have like genotypes but different phenotypes.

It is important to realize that *all genetic ratios are probability ratios.* This means that if two heterozygous black guinea pigs are bred and produce exactly four offspring, although it is *probable* that three will be black and one brown, all *might* be black, or all might be brown, though the latter possibility would rarely occur. Or, the combinations of 3 blacks: 1 brown; 2 blacks: 2 brown; or 1 black: 3 brown also might occur. If 100 such matings were made, however, approximately 300 of the offspring would be black and 100 would be brown. The ratio of black to brown approaches more and more closely to an exact 3:1 as larger numbers of offspring are produced. This probability ratio can be stated another way: in mating two heterozygous individuals, **(Bb),** there are three chances out of four that any *particular* offspring will show the dominant character, and one chance out of four that it will show the recessive character.

In man, eye color is inherited by a pair of genes, brown eye color **(B)** being dominant to blue **(b).** If two heterozygous, brown-eyed people mate, there are three chances out of four that their child will have brown eyes, and one chance out of four that it will have blue eyes. Each mating is a separate, independent event, not affected by the results of previous matings. Thus, if these two heterozygous, brown-eyed parents have three brown-eyed children and are expecting a fourth child, the fourth one, too, has three chances out of four of being brown-eyed, and only one chance out of four of having blue eyes.

282. MENDEL'S FIRST LAW

The mating of black and brown guinea pigs previously described illustrates Mendel's First Law, sometimes called the "Law of the Purity of Gametes" or the Law of Segregation. This may be stated as follows: Genes exist in individuals in pairs, and in the formation of the gametes, each gene separates or segregates from the other member of the pair and passes into a different gamete, with the result that each gamete has one and only one of each kind of gene. The brown gene is not affected in any way by existing beside the black gene in the heterozygous black guinea pig. When gametes are formed it segregates from the black gene, and if it unites with a gamete containing another brown gene will produce a brown guinea pig.

283. TEST CROSSES

In the F_2 of the black-brown guinea pig mating, we found a ratio of three black: 1 brown guinea pig. Although the black guinea pigs are alike phenotypically, they differ genotypically. One third of them are homozygous black **(BB);** the other two thirds are heterozygous black **(Bb).** These can be differentiated by a breeding test, the "test cross" or "back cross." To perform the test, each black guinea pig is bred with a homozygous recessive (brown) one (Fig. 264). If all the offspring are black, the parent being tested is homozygous **(BB),** and if any of the offspring are brown, the parent is heterozygous **(Bb).**

This sort of testing is of the greatest importance in the commercial breeding of animals or plants, where the breeder is trying to establish a strain which will

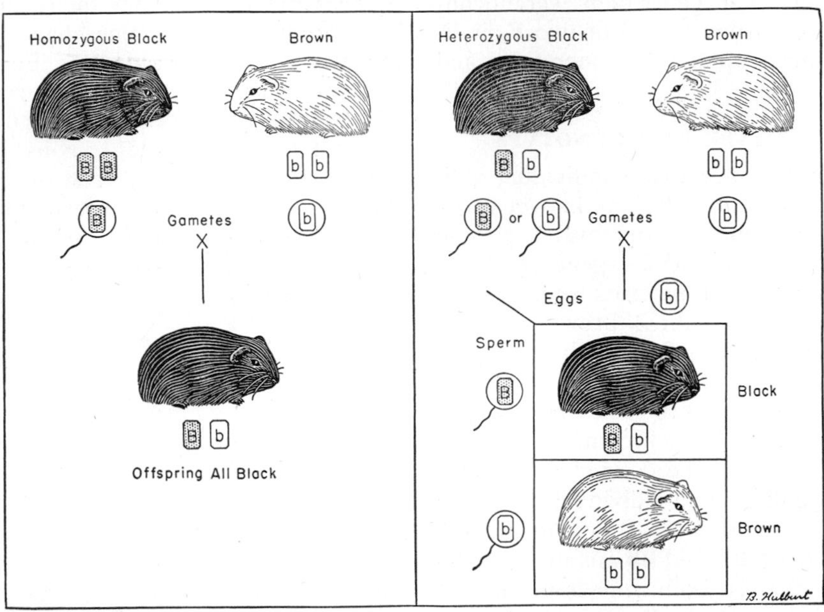

Figure 264. Diagram of test crosses, whereby a heterozygous black guinea pig may be distinguished from a homozygous black one. If a homozygous black is mated with a brown one (left), all of its offspring are black. If a heterozygous black is mated with a brown (right), half of the offspring are black, and half brown.

"breed pure" for a certain character. Formerly, farmers and stock breeders could pick only what appeared to be the best animals for breeding stock, or what seemed to be the best plants in a field, to bear the seed for the next year's crop. This selection by phenotype alone was not very successful because it was not based on knowledge of which animals and plants were homozygous and which heterozygous. As a result, the heterozygous individuals continued to bear some offspring with the less desirable qualities.

The more modern method of breeding, **progeny selection,** is one by which a breeder, in effect, tests the genotypes of his breeding stock by making matings and observing the offspring. If these are superior in respect to the desirable trait, the parents are thereafter used regularly for breeding. Two bulls may both look healthy and vigorous, yet one will have daughters with qualities of milk production that are distinctly superior to those of the daughters of the other bull. By observing the progeny, and using regularly for breeding only those animals that produce superior offspring, the desirable qualities of a strain of animals can be increased rapidly. The method of progeny selection enabled one geneticist to raise the average annual egg production of a flock of hens from 114 to 200 eggs in eight years.

284. INCOMPLETE DOMINANCE

In some species one gene is not completely dominant to the other, heterozygotes having a phenotype intermediate between the two parents. This condition, known as **incomplete dominance,** is in sharp contrast to the complete dominance of black over brown in the guinea pig cross. In sweet peas, for instance, the offspring of a cross of a red-flowered plant with a white-flowered plant all have pink flowers. When two pink-flowered plants are crossed, the offspring are one fourth red, one half pink, and one fourth white. In such crosses the phenotypic and genotypic ratios are identical. In cattle and horses, red coat color is incompletely dominant to white, the heterozygote being roan.

285. DEDUCING GENOTYPES

Another type of problem which a knowledge of genetics enables us to solve is that of deducing the genotypes of parents (and from that the probable character of additional offspring) from the results of a particular cross. In chickens, for example, the gene for rose comb **(R)** is dominant to the gene for single comb **(r).** Suppose that a cock is mated to three different hens (Fig. 265): The cock and hens A and C have rose combs; hen B has a single comb. Breeding the cock with hen A produces a rose-combed chick; with hen B, a single-combed chick; and with hen C, a single-combed chick. What offspring can be expected from further matings of the cock with these hens?

Since the single-combed condition is recessive, all the hens and chicks phenotypically single-combed must be homozygous. We can deduce, then, that hen B and the offspring of hens B and C are genotypically **rr.** All those phenotypically rose-combed have at least one **R** gene, and the cock, and hens A and C, are therefore **R?.** The fact that the offspring of the cock and hen B was single-combed proves that the cock is heterozygous. The single-combed chick received one **r** gene from its mother, but must have received the other from its father. The fact that the offspring of the cock and hen C had a single comb proves that hen C also was heterozygous. We now know the genotypes of all the individuals concerned except hen A, and it is impossible to decide from the data given whether she is homozygous or heterozygous. Further breeding would be necessary to determine this. Additional matings of the cock with hen B would result in one half rose-combed: one half single-combed individuals; additional matings with hen C would produce three fourths rose-combed: one fourth single-combed chicks.

The science of genetics resembles mathematics in that it consists of a few basic principles which, once grasped, enable the student to solve a wide variety of problems. Let us summarize the fundamental concepts. Inheritance is biparental: Both parents contribute to the genetic constitution of the offspring. Genes are

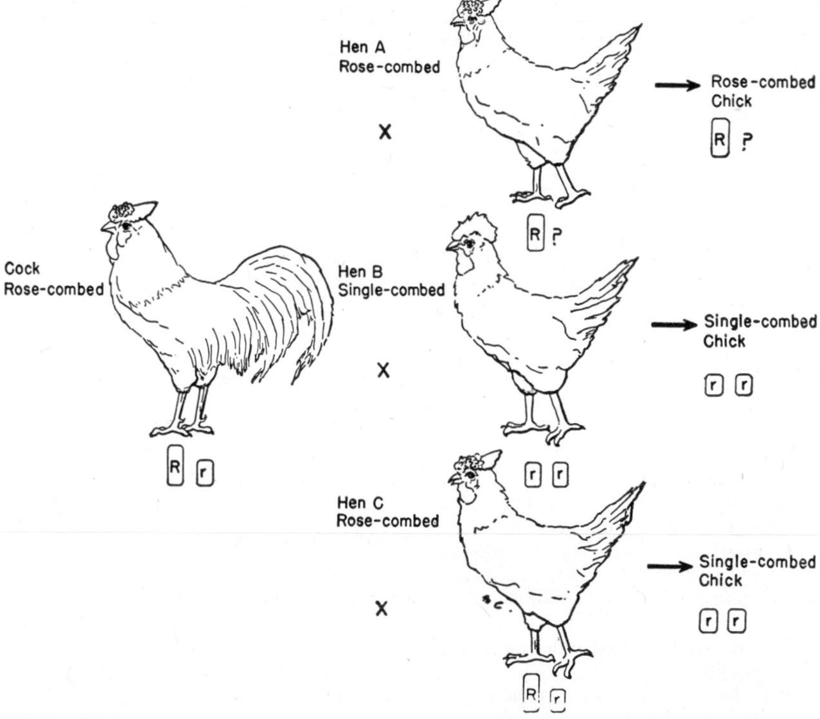

Figure 265. Deducing the parental genotypes from the phenotypes of the offspring. See text for discussion.

not altered by existing together in a heterozygote. Genes must be single in a gamete; either one or the other, but never both of a pair of genes is present. The gametes mate at random, irrespective of the genes they contain, and a sperm containing a **B** gene is equally likely to fertilize a **B** or a **b** egg; there is no specific attraction or repulsion between a **B**-containing egg and a **B**-containing sperm.

In working genetics problems, it is well to follow the procedure given here, in order to avoid errors:

1. Write down the symbols you are using for each gene.
2. Determine the genotypes of the parents, deducing them from the phenotypes of the offspring, if necessary.
3. Derive all possible kinds of gametes from these genotypes.
4. Set up a "checkerboard," putting all the possible types of sperm along the side and all the possible kinds of eggs along the top.
5. Fill in the checkerboard and read off the genotypic and phenotypic ratios among the offspring.

QUESTIONS

1. What is the difference between heredity and genetics?
2. What are the two chief causes of variations in living things?
3. What is accomplished by mitosis? By meiosis?
4. What is a chromosome? Describe its structure.
5. In a human cell undergoing mitosis, how many chromosomes are present in the metaphase? In the anaphase? In the resting daughter cell?
6. Outline briefly the events occurring in each stage of mitosis. Illustrate your discussion with a series of diagrams of mitosis in an organism with a haploid number of four.
7. What causes a cell to divide?
8. Compare mitosis with meiosis.
9. Compare spermatogenesis and oögenesis.
10. In an animal with a haploid number of 10, how many chromosomes are present in a spermatogonium? In the tetrad stage? In the anaphase of the first meiotic division? In the metaphase of the second meiotic division? In a first polocyte? In a second polocyte?
11. If a particular character in a certain species of animal were always transmitted from the mother to the offspring, but never from the father to the offspring, what would you conclude about its mode of inheritance?
12. How many human sperm will be produced from 100 primary spermatocytes? How many eggs will be produced from 100 primary oöcytes?
13. What do the following genetic symbols mean: **A, a, AA, aa, Aa?**
14. Define: gene, locus, allelomorph, dominant, recessive, homozygous, heterozygous, phenotype, genotype, back-cross.
15. In peas, yellow color is dominant to green. What will be the colors of the offspring of homozygous yellow × green? Heterozygous yellow × green? Heterozygous yellow × homozygous yellow? Heterozygous yellow × heterozygous yellow?
16. Could two blue-eyed parents have a brown-eyed child? Could two brown-eyed parents have a blue-eyed child?
17. If two animals heterozygous for a single pair of genes are mated and have 200 offspring, about how many will have the dominant phenotype?
18. Two long-winged flies were mated and the offspring included 77 with long wings and 24 with short wings. Is the short-winged condition dominant or recessive? What are the genotypes of the parents?
19. A blue-eyed man, both of whose parents were brown-eyed, marries a brown-eyed woman whose father was blue-eyed and whose mother was brown-eyed. This man and woman have a blue-eyed child. What are the genotypes of all the individuals mentioned?
20. Outline a breeding procedure whereby a true breeding strain of red cattle could be established from a roan bull and a white cow.
21. Suppose you learned that "shmoos" may have long, oval, or round bodies and that matings of shmoos gave the following:
    long × oval gave 52 long: 48 oval
    long × round gave 99 oval
    oval × round gave 51 oval: 50 round
    oval × oval gave 24 long: 53 oval: 27 round

    What hypothesis about the inheritance of shmoo shape would be consistent with these results?

SUPPLEMENTARY READING

There are several good elementary textbooks of genetics which provide further reading for those interested in the subject: L. H. Snyder, *The Principles of Heredity,* E. O. Dodson's *Genetics* and R. B. Goldschmidt's *Understanding Heredity.*

Chapter 31

Genetics

286. MENDEL'S SECOND LAW

The type of mating discussed in the previous chapter, where only one pair of genes is considered, is called a **monohybrid cross.** For the sake of simplicity it is best to begin the study of genetics with such a problem, but it is often necessary for the geneticist to consider the simultaneous inheritance of two, three, or even more pairs of genes. Such crosses are called **dihybrid, trihybrid,** and so on. The mating of individuals that differ in two or more traits follows the same principles as those of the simpler monohybrid cross, but since there is a greater number of types of gametes, the number of different types of zygotes is correspondingly larger.

When two pairs of genes lie in different (nonhomologous) chromosomes, *each pair of genes is inherited independently of the other pair;* that is, each pair separates during meiosis independently of the other. In writing the genetic formula of an organism in which we are considering two pairs of genes, we must write two of each kind of symbol. Thus, a black, short-haired guinea pig has the symbol **BBSS** (short hair being dominant to long hair) and a brown, long-haired one has the symbol **bbss.** The gametes of the **BBSS** animal are all **BS.** Those of the **bbss** guinea pig are all **bs.** Now let us suppose that the **B-b** genes are located in a rod-shaped chromosome, and the **S-s** genes in a hooked chromosome. During meiosis, the respective homologous chromosomes synapse and then separate so that each gamete gets one and only one of each kind. Since both parents are homozygous all the gametes produced by each are identical. All the offspring, therefore, are **BbSs,** heterozygous for hair color and for hair length, and all are phenotypically black and short-haired.

Each of these F_1 individuals produces *four* kinds of gametes in equal numbers—**BS, Bs, bS,** and **bs**—so that when two of them are mated, sixteen combinations are possible in the F_2 generation (see checkerboard, Fig. 266). Adding them together, there are nine black, short: three black, long: three brown, short: one brown, long. *This 9:3:3:1 ratio is characteristic of the second generation of a cross of two individuals differing in two characters whose genes are located in nonhomologous chromosomes.* Keep in mind that this is a *probability* ratio, which means that there are nine chances out of sixteen that any particular offspring will be black and short-haired, and only one chance out of sixteen that it will be brown and long-haired. A great many crosses probably would have to be made before the results would approach this theoretical ratio closely. This ratio illustrates Mendel's

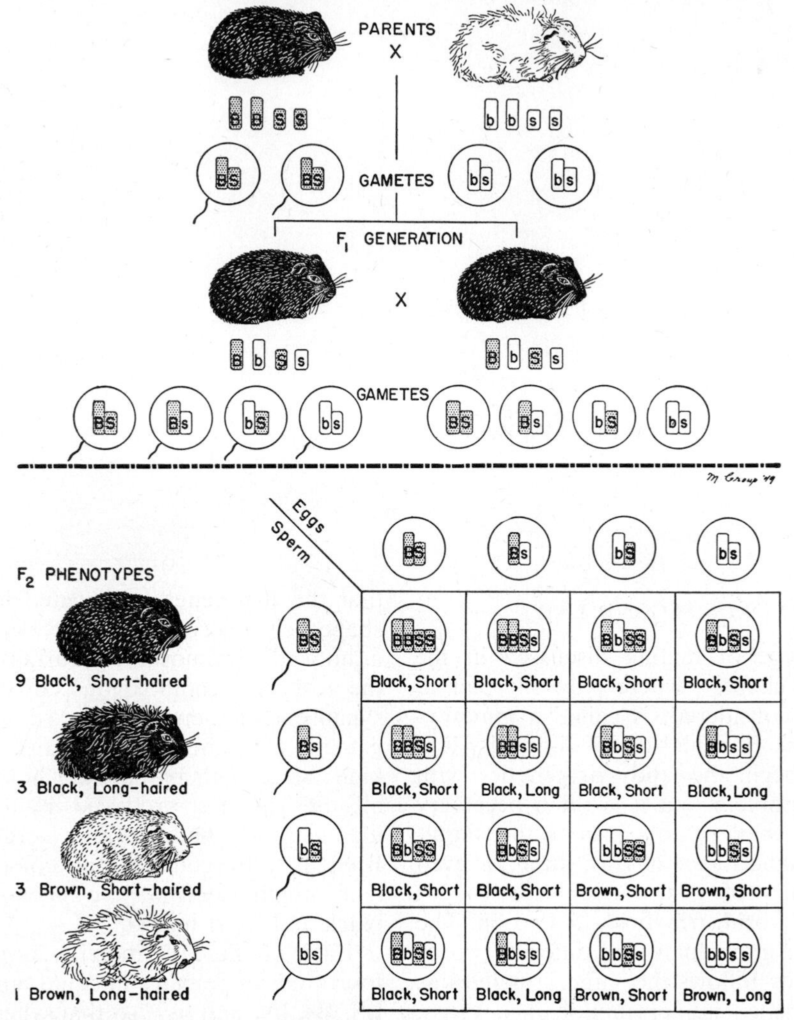

Figure 266. Diagram of a cross between a black, short-haired guinea pig and a brown, long-haired one, illustrating independent assortment.

Second Law, the Law of Independent Segregation, which states that the members of one pair of genes separate (segregate) from each other in meiosis independently of the members of other pairs of genes and come to be assorted at random in the resulting gametes.

In a similar fashion, problems involving three different pairs of genes may be solved. Of course, in a trihybrid cross the three heterozygous pairs of genes form *eight* types of gametes in equal numbers, and the union of these gives sixty-four possible zygotes in the F_2 generation. In peas, yellow **(Y)** seed color is dominant to green **(y)**, smooth **(S)** seeds are dominant to wrinkled **(s)**, and tall plants **(T)** are dominant to dwarf **(t)**. A cross of a homozygous yellow, smooth, tall **(YYSS TT)** with a homozygous green, wrinkled, dwarf **(yysstt)** produces offspring all of which are yellow, smooth and tall **(YySsTt)**. When two of these F_1 plants are mated, F_2 offspring are produced in the ratio of twenty-seven yellow, smooth, tall: nine yellow, smooth, dwarf: nine yellow, wrinkled, tall: nine green, smooth, tall: three yellow, wrinkled, dwarf: three green, wrinkled, tall: three green, smooth, dwarf: one green, wrinkled, dwarf. Work out the checkerboard to prove this.

287. INTERACTIONS OF GENES

The relationship between the genes and traits discussed so far is simple and clear: each gene produces a single trait. However, genetic research with many animals and plants has revealed that the relationship of gene to trait may be quite complex. Several pairs of genes may interact to affect the production of a single trait; one pair may inhibit or reverse the effect of another pair; or a given gene may produce different effects when the environment is changed in some way. The genes are inherited as units but may interact in some complex fashion to produce the trait. The study of the relation between gene and trait, known as **physiological genetics,** is being very actively pursued at present.

One of the simpler types of interaction is illustrated by the inheritance of combs in poultry. We mentioned previously that the gene for rose comb **(R)** is dominant to that for single comb **(r).** Another pair of genes governs the inheritance of pea comb **(P)** versus single comb **(p).** The genotype of a single-combed fowl, then, must be **pprr;** a pea-combed fowl is either **PPrr** or **Pprr;** and a rose-combed fowl is either **ppRR** or **ppRr** (Fig. 267). Investigators found that when a homozygous, pea-combed fowl is mated to a homozygous, rose-combed one, the offspring have neither pea nor rose combs, but a completely different type called walnut comb. Therefore, the walnut-combed condition is produced whenever a fowl has one or two **R** genes plus one or two **P** genes, and the genotypes **PPRR, PpRR, PPRr,** and **PpRr,** are all phenotypically walnut-combed. When two heterozygous, walnut-combed fowls are mated, offspring in the ratio of nine walnut: three rose: three

Figure 267. Heads of roosters, showing the different types of genetically determined combs. *1, 5,* and *6,* Single combs; *2* and *3,* pea combs; *4* and *8,* rose combs; *7,* V-shaped; *9,* strawberry. (Courtesy of the United States Department of Agriculture.)

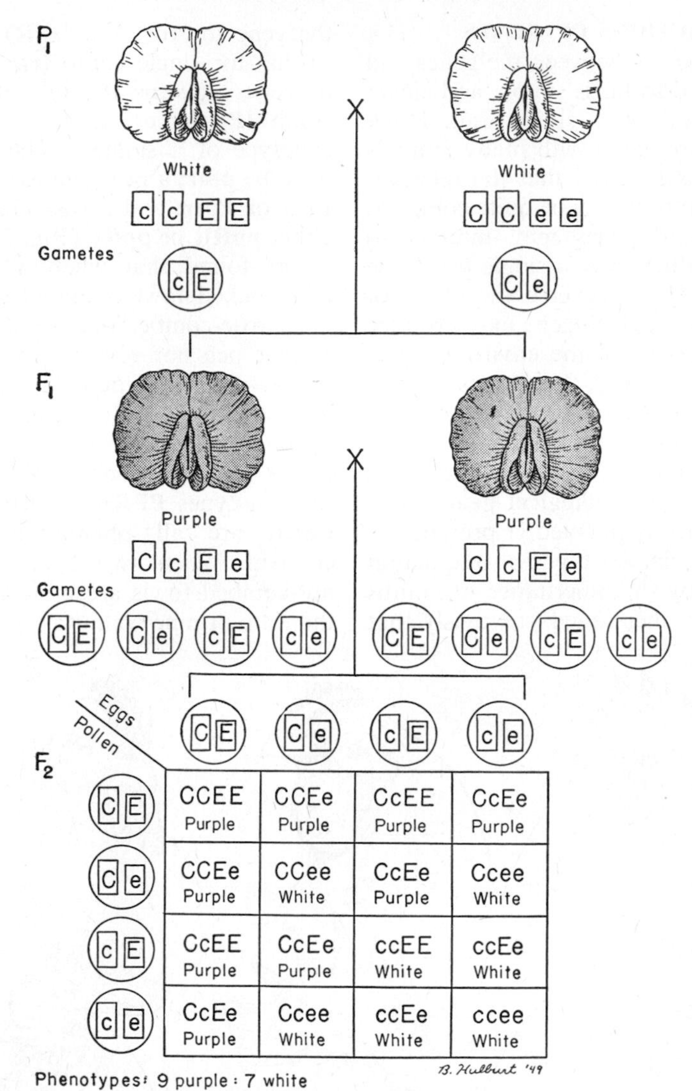

Figure 268. Diagram of a cross illustrating the action of complementary genes, the two pairs of genes which regulate flower color in sweet peas. Note that for a colored flower, the plant must have at least one C and one E.

pea: one single are produced. Make a checkerboard to demonstrate this.

Complementary Genes. Two pairs of independent genes may be related to each other in such a way that neither dominant can produce its effect unless the other is present. Such pairs of genes are called **complementary genes;** the action of each one complements the action of the other in the production of the phenotype. The presence of both dominants produces one character; the alternate character is produced by the absence of either one or both. In the course of breeding experiments with cultivated sweet peas, Bateson and Punnett were amazed to find that a cross of two white-flowered races gave offspring all of which had purple flowers! Crossing two of the F_1 purple plants produced an F_2 generation in the ratio of nine purple: seven white (Fig. 268). Two pairs of factors proved to be involved: one gene **(C)** regulates some essential step in the production of a white raw material from which a purple pigment can be made by an enzyme produced by the second gene **(E).** The homozygous recessive **cc** is unable to synthesize the raw material and the homozygous recessive **ee** lacks the enzyme to convert the raw material into purple pigment. In this particular cross, one of the races of white sweet peas lacked the raw material gene, being genotypically **ccEE,** the other lacked the enzyme gene and was genotypically **CCee.** Mating **ccEE** with **CCee** resulted in F_1 offspring all of which were **CcEe**—phenotypically purple—because both the raw material and the enzyme were present. When these F_1 purple plants were mated, purple and white plants appeared in the F_2 generation, in the ratio of 9:7. Nine out of sixteen had at least one **C** and one **E** and were purple; seven out of the sixteen lacked either a **C** or an **E** (or both) and were white. This 9:7 ratio is characteristic of the F_2 generation of a cross involving complementary genes. A race of sweet peas breeding true for purple could be set up by selecting two **CCEE** plants and mating them.

Supplementary Genes. A slightly different sort of interrelationship between genetic factors is demonstrated by **supplementary genes.** These are two independent pairs which interact in such a way that one dominant will produce its effect whether the other is present or not, but the second can produce its effect only in the presence of the first. The inheritance of coat color in guinea pigs supplies us with an example of supplementary genes. In addition to the **B** gene for black coat and the **b** gene for brown coat mentioned previously, there is a gene, **C,** which produces an enzyme which converts a colorless precursor into the pigment, **melanin,** and hence is required for the production of any pigment at all. The homozygous recessive, **cc,** lacks the enzyme, no melanin is produced, and the animal is a white-coated, pink-eyed **albino,** no matter what combination of **B** and **b** genes may be present. Because of this, mating an albino of the genotype **ccBB** with a brown guinea pig, **CCbb,** will produce an F_1 generation all of which are black! They will have the genotype **CcBb,** and when two such animals are crossed, their offspring will be in the ratio of nine black: three brown: four albino. Make a checkerboard to show this. This 9:3:4 ratio is characteristic of the F_2 generation in a cross involving supplementary factors. The color gene **(C)** will produce its effect no matter which of the **B-b** genes is present, but the **B** or **b** gene can produce its effect only when at least one **C** is present.

It is possible for both complementary and supplementary genes to be involved in one cross. In maize, for example, there are **C** and **R** genes necessary for red kernels, the absence of either causing white kernels. Hence they are complementary genes. In addition, there is a **P** gene which produces purple kernels, but only if both a **C** and an **R** are present. This, then, is supplementary to the other two pairs of genes.

It is beyond the scope of this book to discuss further complexities of gene interaction. The problems indicated here are the simplest with which the geneticist deals, and most traits are governed by a multitude of genes which interact with each other and with the environment to produce the final phenotype. The unscrambling of these interactions provides the geneticist with many difficult problems. W. E. Castle, of Harvard University, has found that more than twelve pairs of genes interact in various ways to produce just the coat color of rabbits, and the many geneticists working with fruit flies have found that over 100 genes are concerned with the color and shape of the eyes.

288. MULTIPLE FACTORS

Many human characteristics—height, body form, intelligence and skin color—and many commercially important characters such as milk production in cows, egg production in hens, the size of fruits, etc., are not separable into distinct alternate classes, and are not inherited by single pairs of genes. There are several, perhaps many, different pairs of genes which affect the same characteristic. The term **multiple factors** is applied to *two or more independent pairs of genes which affect the same character in the same way and in an additive fashion.* The investigation of the inheritance of skin color was pursued by Davenport in Jamaica, where there is no "color line" and matings be-

Table 10. MULTIPLE FACTOR INHERITANCE OF SKIN COLOR IN MAN

Parents	AaBb (Mulatto)	×	AaBb (Mulatto)
Gametes	AB Ab aB ab		AB Ab aB ab

Offspring:
- 1 with 4 dominants—AABB—phenotypically Negro
- 4 with 3 dominants—2 AaBB and 2 AABb—phenotypically "dark"
- 6 with 2 dominants—4 AaBb, 1 AAbb, 1 aaBB—phenotypically mulatto
- 4 with 1 dominant—2 Aabb, 2 aaBb—phenotypically "light"
- 1 with no dominants—aabb—phenotypically white

tween Negro and white occur freely. He found that two pairs of genes are involved, which he called **A-a** and **B-b**. The capital letters stand for genes causing dark skin—the more capital letters, the darker the skin. That is, the genes affect the character in an additive fashion. A full Negro has all four dominant genes, **AABB** (the genes are incompletely dominant, of course), and a white has all four recessive genes, **aabb**. The F_1 offspring of a mating of **aabb** with **AABB** are all **AaBb** and have intermediate (mulatto) skin color. A mating of two such mulattoes produces a wide variety of skin color in the offspring, ranging from as dark as the original Negro parent to as white as the original white parent (Table 10). Multiple-factor crosses are characterized by producing an F_1 generation intermediate between the two parents and showing little variation, and an F_2 generation showing a wide variation between the two parental types, but with a greater number of the intermediate forms, and only a few as extreme as either grandparent. Of the sixteen possible zygote combinations from a cross of **AaBb** with **AaBb** only one **(AABB)** will be as dark as the Negro grandparent, and only one **(aabb)** will be as light as the white grandparent. Since the genes **A** and **B** produce about the same amount of darkening of the skin, the genotypes **AaBb, AAbb** and **aaBB** all produce the same phenotype, mulatto. The genotype of a given mulatto can be determined only by observing his offspring.

Skin color in man is a rather simple example of multiple-factor inheritance, because only two pairs of genes are involved. Height in man is inherited by a much more complex set of multiple factors, possibly ten or more pairs of genes being involved. Since tallness is recessive to shortness, the more capital letters in the genotype, the shorter the individual. Because of the many pairs of genes involved and because height is modified by environmental conditions, we do not have the alternative conditions, short versus tall, or even short, medium and tall. Instead, there are people of every height from about 55 inches to about 84 inches. If we measured the heights of a thousand American adults taken at random, we would find the majority of average height—perhaps 68 inches—and only a few as tall as 80 inches, or as short as 55. In a diagram, when the number of people of each height is plotted against the height in inches, and the points connected, the result is a bell-shaped curve, called a "normal curve" or curve of normal distribution (Fig. 269).

All living things vary, and their variations usually follow this normal curve. If you measure the length of 1000 seashells of the same species, or count the number of kernels on 1000 ears of corn, or the number of pigs per litter in 1000 litters, or weigh 1000 hen's eggs, you will find the normal curve in each case. The variation resulting in this distribution may be caused by hereditary differences, by environmental differences or by a combination of the two. Commercial breeders who are trying to breed cows that will give more milk, or hens that will lay bigger eggs, or ears of corn with more kernels per ear, are working with multiple factors. By selecting organisms which approach the phenotype they are seeking, and using them in further matings, they gradually produce true breeding strains with the commercially desirable trait; that is, they select a strain homozygous for all the dominant (or recessive) factors involved. It is evident that there is a limit to the effectiveness of breeding by selection. When the strain becomes homozygous for all the factors involved, further selective breeding cannot increase the desirable quality.

The inheritance of certain traits is governed by a single pair of genes which determine the presence or absence of the trait plus a number of multiple factors which determine the extent of the trait. In most mammals which have been tested, the presence or absence of spots in the coat is determined by a single pair of factors. The gene for the presence of spots **(s)** is recessive to the gene for solid color **(S)**. The size and distribution of the spots varies widely among different animals. It is possible, by selective breeding, to produce animals with more or fewer spots, which shows that hereditary factors,

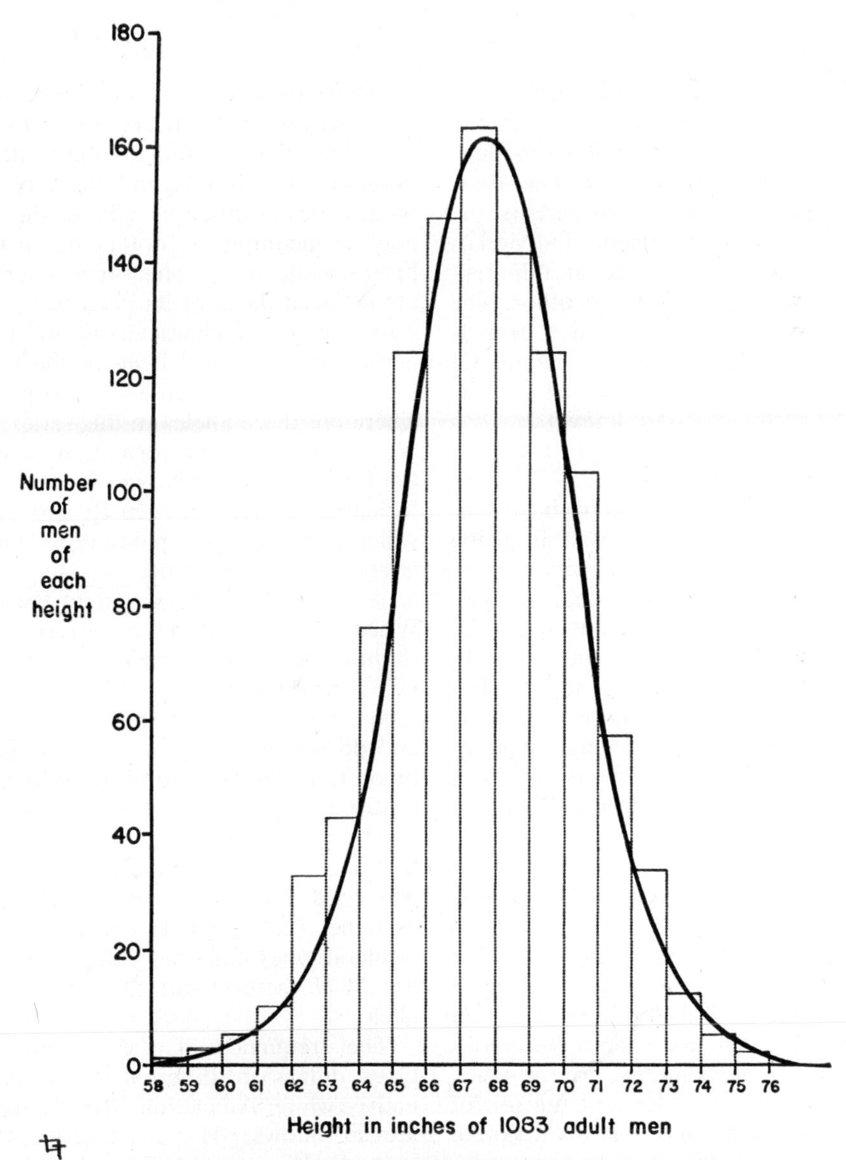

Figure 269. An example of a "normal curve," or curve of normal distribution, the heights of 1083 adult white males. The blocks indicate the actual number of men whose heights were within the unit range; for example, there were 163 men between 67 and 68 inches in height. The smooth curve is a normal curve based on the mean and standard deviation of the data.

Table 11. EXCLUSION OF PATERNITY BASED ON BLOOD TYPE TESTS

CHILD	MOTHER	FATHER MUST BE OF TYPE	FATHER CANNOT BE OF TYPE
O	O	O, A, B	AB
O	A	O, A, B	AB
O	B	O, A, B	AB
A	O	A, AB	O, B
A	A	A, B, AB, O	—
B	B	A, B, AB, O	—
A	B	A, AB	O, B
B	A	B, AB	O, A
B	O	B, AB	O, A
AB	A	B, AB	O, A
AB	B	A, AB	O, B
AB	AB	A, B, AB	O

and not prenatal environment, are mainly responsible for the amount of spotting. The type of inheritance involved is characteristic of multiple factors: The F_1 is intermediate between the two parents; the F_2 is variable, including some individuals with as many spots as one grandparent and some with as few as the other. The term **modifying factors** is applied to multiple factors which affect the *degree* of expression of another gene. Many examples of such factors are known.

289. MULTIPLE ALLELES

In our previous discussion we have considered only genes which exist in pairs, one dominant and one recessive. Many genes, however, have more than one allele; in addition to a dominant and a recessive gene, there may be one or more intermediate genes forming a series. Multiple alleles are three or more conditions of a single locus, producing different phenotypes. Among the members of a species, the alleles are inherited in such a way that a particular individual receives any two of the genes and never more than two. In rabbits, for example, there is a **C** gene which causes colored coat, while the homozygous recessive **(cc)** causes albino coat. Two other genes located at the same locus and inherited allelomorphically, are $\mathbf{c^h}$ and $\mathbf{c^{ch}}$. The gene $\mathbf{c^h}$, when homozygous causes the "Himalayan" pattern, in which the rabbit's body is white, but the tips of the ears, nose, tail and legs are colored. The $\mathbf{c^{ch}}$ gene, when homozygous, produces the "Chinchilla" pattern, a light gray color all over the body. These genes may be arranged in a series, **C**, $\mathbf{c^{ch}}$, $\mathbf{c^h}$, **c**, in which each gene is dominant to the genes following it, and recessive to those preceding it. From this it follows that $\mathbf{c^hc}$ produces a "Himalayan" phenotype. In other series of multiple alleles, the genes may be incompletely dominant, so that a heterozygote has a phenotype intermediate between those of its parents.

An important characteristic in man inherited by multiple alleles is the type of blood: O, A, B or AB (see also p. 238). There are three alleles in this series: gene **A,** which causes the formation of agglutinogen **A;** gene $\mathbf{A^B}$, which causes the formation of agglutinogen **B;** and gene **a,** which produces no agglutinogen. Gene **a** is recessive to the other two, but neither gene **A** nor $\mathbf{A^B}$ is dominant to the other. When both **A** and $\mathbf{A^B}$ are present, both agglutinogens are formed and the individual has blood group AB. Because the blood types are determined genetically and do not change in a person's lifetime, blood tests can be helpful in settling cases of disputed parentage. It must be emphasized that a blood test can never prove that a certain man *is* the father of a certain child, but only whether he *could be* its father (see Table 11). Two other sets of blood types inherited independently are the **M-N** factors and the series of **Rh** alleles.

The frequency of the various blood types differs in different races. Among native white Americans the frequencies are as follows: 41 per cent, O; 45 per cent, A; 10 per cent, B; and 4 per cent, AB. Among American Indians, the frequencies of A and B are almost reversed. Some interesting work combining anthropology and genetics has been done in tracing the relationships of different races as determined by the relative frequency of these blood types.

290. LETHAL GENES

Certain genes known as lethal genes cause such a tremendous deviation from the normal development of an organism that it is unable to survive. The existence of these genes is detected by upsets in genetic ratios. For example, in a certain strain of mice, although yellow coat color was exhibited, experimenters found it impossible to establish a true breeding strain for yellow coat. Instead, when two yellow mice were bred, offspring in the ratio of two yellow: one nonyellow occurred! A yellow mouse bred to a black mouse gave 50 per cent yellow and 50 per cent black offspring. The investigators noticed that the litters of yellow × yellow matings were only three quarters as large as other mice litters. Evidently, one quarter of the embryos (those homozygous for yellow) did not develop. Using $\mathbf{A}^{Y}$ for the yellow gene and **a** for its recessive allele (black), the yellow × yellow mating has the results shown in Figure 270.

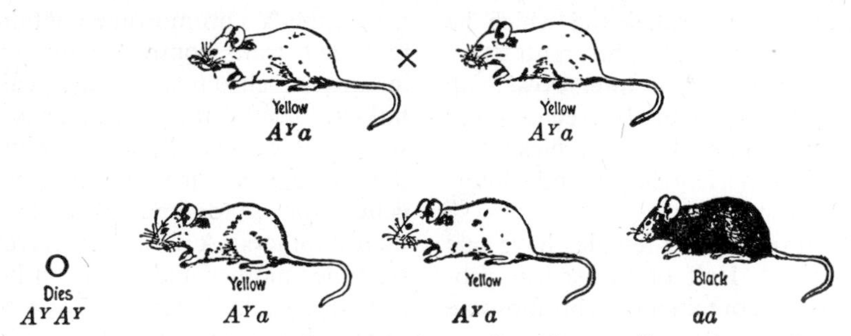

Figure 270. A cross involving lethal genes: The mating of two yellow mice produces offspring in the ratio of two yellow to one black. (Sinnott and Dunn: Principles of Genetics, McGraw-Hill Book Co., Inc.)

Later research showed that the homozygous, yellow mice began development, then stopped, died and were resorbed. If the uterus of the mother is opened in pregnancy, the abnormal embryos are visible.

An exactly similar situation occurs in chickens. "Creeper" fowl have short legs and shorter wings than normal. When two "creepers" are bred, the offspring are two "creepers": one normal. One quarter of the eggs—those homozygous for "creeper" have marked abnormalities of the vertebrae and spinal cord, and die without hatching.

These lethal genes, yellow and "creeper," produce a phenotypic effect when heterozygous and so are considered dominant. Many—perhaps most—lethal genes have no effect when heterozygous, but cause death when homozygous, and therefore are called recessive lethal genes. These can be detected only by special genetic techniques. Analyses of wild populations of fruit flies (*Drosophila*), made by these special techniques, have revealed many recessive lethals.

Several genes are suspected of being lethal in human embryos, one of which is a dominant gene for brachyphalangy (short fingers). Mohr and Wriedt reported a case in which two people with this defect married and produced a child with no fingers or toes, who died immediately.

291. THE GENETIC DETERMINATION OF SEX

The sex of an organism is another trait which is determined genetically. There is one exception to the general rule that all homologous pairs of chromosomes are identical in size and shape: the **sex chromosomes.** In females there are two identical sex chromosomes, called **X** chromosomes, but in males there is only one **X** chromosome, and a smaller one called **Y.** The latter contains few or no genes and is, in most species, of a different size and shape from the **X** chromosome, yet in meiosis in the male, **X** and **Y** undergo synapsis and then separate. Men, therefore, have twenty-three pairs of ordinary chromosomes, plus an **X** and a **Y** chromosome; women have twenty-three pairs of ordinary chromosomes, plus two **X** chromosomes. But it is not the **Y** chromosome which determines maleness. In a number of species the male has no **Y** chromosome, just a single **X** chromosome. Whether an organism is male or female is determined by whether there is one or two **X** chromosomes.

The work of C. B. Bridges showed that in fruit flies, sex is determined by the ratio of the number of **X** chromosomes to the number of haploid sets of other chromosomes. Thus, in males, there is one **X** to two haploid sets of other chromosomes, a ratio of 1:2 or 0.5. In females there are two **X** chromosomes to two sets of other chromosomes, a ratio of 1.0. Now, by certain genetic techniques possible in fruit flies, Bridges set up abnormal flies with a ratio of 0.33 which had all their male characteristics exaggerated, and which he named "supermales." He also set up abnormal individuals with the ratio of 1.5 sex chromosomes to sets of ordinary chromosomes, which had all the female characteristics exaggerated and which he named "superfemales." The ratio 0.67 produced animals called **intersexes,** with characters intermediate between males and females. These abnormal flies, "supermales," "superfemales," and intersexes, are all sterile.

All eggs produced by females have one **X** chromosome. Half of the sperm produced by males contain an **X** chromosome and half contain a **Y.** The fertilization of an **X**-bearing egg by an **X**-bearing sperm results in an **XX**—female—zygote. The fertilization of an **X**-bearing egg by a **Y**-bearing sperm results in an **XY**—male—zygote. Since there are equal numbers of **X**- and **Y**-bearing sperm, about equal numbers of each sex are born. Actually, about 107 boys are born to every 100 girls, and the ratio of males to females at conception is even higher. One possible explanation of this numerical difference is that the **Y** chromosome is smaller than the **X** chromosome, so that sperm containing a **Y** chromosome are lighter and can swim a little faster than **X**-bearing sperm; consequently, they are able to win the race to the egg slightly more than half the time. The ratio of males to females at conception is about 114:100. But during intrauterine development more boys than girls die, so that the ratio at birth is 107 to 100. After birth this differential death rate continues, and by the age of ten or twelve there are equal numbers of males and females, after which there is a majority of females.

This mechanism is believed to operate in all species with separate sexes. In some animals—birds and butterflies (Lepidoptera)—the mechanism is reversed: males are **XX** and females **XY.** Sex chromosomes have been detected in some plants—notably, strawberries—and probably exist in other plants with separate sexes. Remember that the members of many species have the organs of both sexes. In such organisms (called hermaphroditic if animals, and monoecious if plants) sex chromosomes have not been found.

292. SEX-LINKED CHARACTERS

The **X** chromosome contains many genes; the **Y** chromosome contains only a few. The traits controlled by genes lying in the **X** chromosome are called **sex-linked,** because their inheritance is linked with that of sex. Male offspring receive their single **X** chromosome (and all the genes for sex-linked characters) from their mothers. A female receives one **X** from her mother and one from her father. A male zygote, having but one **X** chromosome, has only one of each gene located in the **X** chromosome. To avoid confusion, its genotype is written with the **Y**. Thus if **A** is a sex-linked gene, the genotype for a female is **AA** (she has two **X** chromosomes and each contains an **A** gene). The genotype for a male is **AY** (he has one **X** chromosome containing the **A** and one **Y** chromosome which does not contain an **A** gene).

In fruit flies, one of the classic subjects of genetic research because the time between successive generations is only ten days, the normal eye color is dark red, but white-eyed forms do occur. The genes for red or white eye color are located in the **X** chromosome and hence are sex-linked. Red eye color **(R)** is dominant to white **(r).** In sex-linked inheritance the traits of the offspring depend upon the sex of the parent which had a particular trait. Crossing a homozygous, red-eyed female with a white-eyed male **(RR × rY)** produces all red-eyed offspring **(Rr** females and **RY** males). Crossing a white-eyed female with a red-eyed male **(rr × RY)** produces red-eyed female offspring **(Rr)** and white-eyed males **(rY).** Make the proper checkerboard to prove this.

Many sex-linked traits are known. In man, hemophilia (bleeder's disease) and color blindness are sex-linked. Because of this, a color-blind man with a color-blind father and a normal-visioned mother inherits his color blindness from his mother (who would have to be heterozygous) and not from his father!

293. SEX-INFLUENCED CHARACTERS

Not all the characters which differ in the two sexes are sex-linked; some, known as sex-influenced traits, are inherited by genes which, though not located in the sex chromosomes, are altered or influenced by the sex of the animal. Thus the same genotype may produce a different phenotype in a male from that produced in a female. In sheep, a single pair of genes determines the presence or absence of horns. The gene **H,** for the presence of horns, is dominant in males, but recessive in females, and its allelomorphic gene, **h,** for hornless is recessive in males, but dominant in females. Therefore, the genotype **HH** produces a horned animal, regardless of sex; **Hh** produces the horned phenotype if the animal is male, the hornless one if the animal is female; and **hh** produces a hornless animal whether it is a ram or a ewe.

In man the gene for pattern baldness is sex-influenced, its expression being altered by the amount of male sex hormone present. It is dominant in males and recessive in females. There are many more bald men than women because only one gene for baldness will cause a man to lose his hair, while two genes are needed to produce a bald woman. Supposing **B** to be the gene for baldness, and **b** to be the gene for nonbaldness, the genotype **BB** produces baldness in men and women; **Bb** produces baldness in men but not in women; and **bb** causes baldness in neither. Not all baldness is hereditary, of course; some types are caused by disease or other factors.

In working a problem involving a sex-influenced gene, the result must be stated separately for the two sexes. Therefore, the offspring of a heterozygous bald man **(Bb)** and a heterozygous nonbald woman **(Bb)** will be in the ratio of three bald to one nonbald, if male, but three nonbald to one bald, if female.

294. LINKAGE AND CROSSING OVER

In discussing the simultaneous inheritance of two pairs of genes, we stressed the point that to get a 9:3:3:1 ratio, the pairs of genes must lie in different (nonhomologous) chromosomes. Since there are only twenty-four pairs of chromosomes and many hundreds of different inherited traits, it is obvious that there must be many genes in each chromosome. All the genes in each chromosome tend to be inherited as a group and are said to be *linked.* This is due to the fact that in meiosis the homologous pairs of chromosomes separate *as units* and go to opposite poles. Therefore, all the genes lying in one chromosome go to one pole to become incorporated in one gamete, and all the genes contained in the other member of the homologous pair go to the other pole and become incorporated in another gamete.

Linkage between the genes in a given chromosome is usually not complete, however. During the process of synapsis the homologous chromosomes frequently exchange whole segments of chromosome material and the genes located in them. The exact mechanism of this exchange is still unknown, but it occurs at random along the length of the chromosomes. Therefore, the greater the distance between the loci of any two given genes in a chromosome, the greater is the chance that an exchange of segments will occur between them. The exchange of segments between homologous chromosomes, called

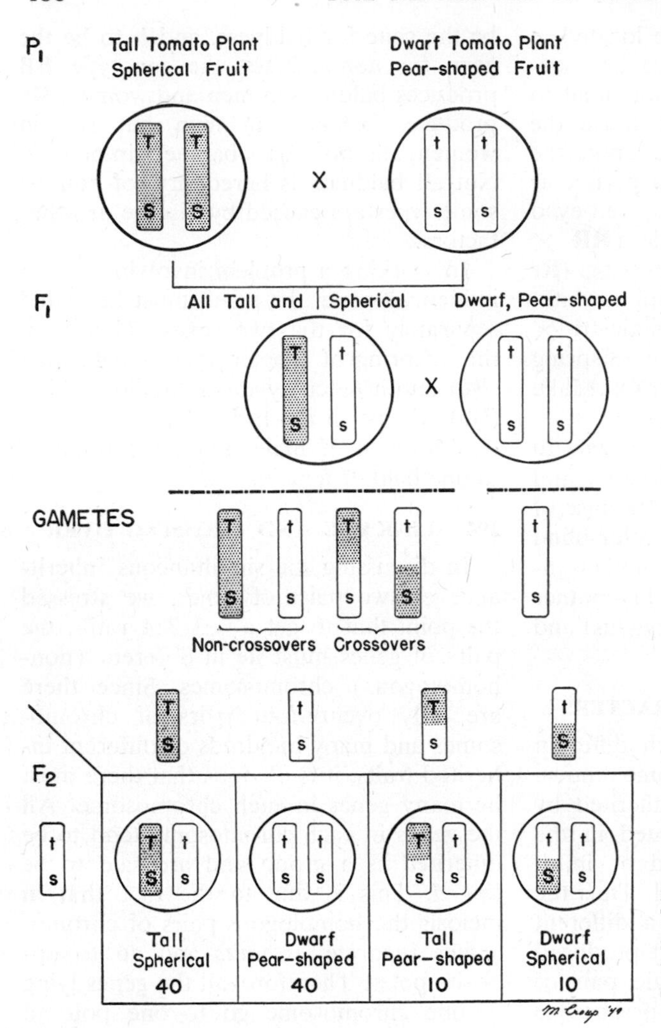

Figure 271. Diagram of a cross involving linkage and crossing over. The genes for tall vs. dwarf, and spherical vs. pear-shaped fruit in tomatoes, are linked; they are located in the same chromosome.

crossing over, makes possible new combinations of linked genes.

In tomatoes, the genes **T** (tall) and **t** (dwarf), and **S** (spherical fruits) and **s** (pear-shaped fruits) are both located in the same chromosome and are linked, **T** lying near one end, and **S** close to the other end of the chromosome. If a homozygous **TTSS** plant is crossed with a homozygous **ttss** plant (Fig. 271), the offspring are all tall plants with spherical fruits, but genetically heterozygous, **TtSs.** So far, there would appear to be no difference from the case in which the genes lie in different chromosomes. However, when one of the F_1 heterozygotes is crossed with a homozygous recessive **ttss**, the difference is apparent. If the genes were not linked, but were in different chromosomes, the offspring would be in the ratio of one tall, spherical: one dwarf, spherical: one tall, pear-shaped: one dwarf, pear-shaped. If the genes were completely linked and no crossing over occurred, only tall, spherical and dwarf, pear-shaped plants would appear, and these would be in equal numbers. Even with crossing over occurring, most of the offspring are either tall and spherical, or dwarf and pear-shaped (noncrossovers), and only those few in which an exchange took place in the chromosome between **T** and **S** are tall and pear-shaped, or dwarf and spherical (crossovers). This crossing over occurs 20 per cent of the time, and offspring are in the ratio of forty tall, spherical: forty dwarf, pear-shaped; ten tall, pear-shaped: ten dwarf, spherical.

The distance between two genes in a chromosome is measured in units of the percentage of crossing over that occurs between them; **T** and **S** are said to be 20 units apart.

The facts of crossing over provide proof that the genes lie in a linear order in the chromosomes. If three genes **A, B** and **C** lie in the same chromosome, and tests show that crossing over between **A** and **B** occurs 5 per cent of the time, and crossing over between **B** and **C** occurs 3 per cent of the time, then, when the percentage of crossing over between **A** and **C** is tested, it will be found to be either 8 per cent or 2 per cent. If it is 8 per cent, **C** lies to the right of **B** and the order is as follows:

8%
A B C. If it is 2 per cent, then
5% 3%

C lies between **A** and **B** and the order is

5%
like this: A C B. In all tests of
2% 3%

crossing over the percentage of crossing over between the first and third gene is either the sum of or the difference between the percentages of crossing of the first and second, and the second and third. These facts are only explained by assuming that the genes lie in a linear order in the chromosome.

All the genes in a particular chromosome constitute a **linkage group.** These groups, which, of course, are determined by genetic tests, always equal the haploid number of chromosomes. This is additional confirmation of the fact that the genes are located in the chromosomes. Linkage groups remain constant and are only changed by a **translocation** in which a piece of one chromosome breaks off and becomes attached to another, nonhomologous chromosome (see p. 485). The linkage between two particular genes, such as the linkage between tall and spherical, is called a **specific linkage.** This may be changed by crossing over so that tall becomes linked to pear-shaped. That is, through an exchange of segments between homologous chromosomes, the gene for tall plants and the gene for pear-shaped fruit come to lie in the same chromosome.

By putting together the data on the amount of crossing over between the various genes in a given chromosome, one can draw a **chromosome map.** The most detailed chromosome maps are those for fruit flies, which have only four pairs of chromosomes and have been studied most intensively. Other species with fairly well-mapped chromosomes are corn and mice; a beginning has been made toward mapping the **X** chromosome of man.

295. PENETRANCE AND EXPRESSIVITY

In all the genes discussed so far, each recessive gene produces its trait when it is homozygous, and each dominant gene produces its effect when it is homozygous or heterozygous. This is not true for all genes—geneticists have found some which do not always produce their phenotypes when they should. Genes such as the ones we have discussed so far, which always produce the expected phenotype, are said to have complete or 100 per cent **penetrance.** If only 70 per cent of the individuals of a stock homozygous for a certain recessive gene show the character phenotypically, however, then the gene is said to show 70 per cent penetrance. Penetrance is essentially a statistical concept of the regularity with which a gene produces its effect when present in the requisite homozygous (or heterozygous) state. The percentage of penetrance may be altered by changing the conditions of temperature, moisture, nutrition, and so forth, under which the organism develops.

Some stocks which are homozygous for a certain gene may exhibit wide variations in the appearance of the character. Fruit flies of a stock homozygous for a gene producing shortening and scalloping of the wings may exhibit wide variations in the degree of shortening and scalloping. Such differences are known as variations in the **expressivity,** or **expression,** of the gene. The expressivity of a gene may be altered also by changing environmental conditions during the organism's development.

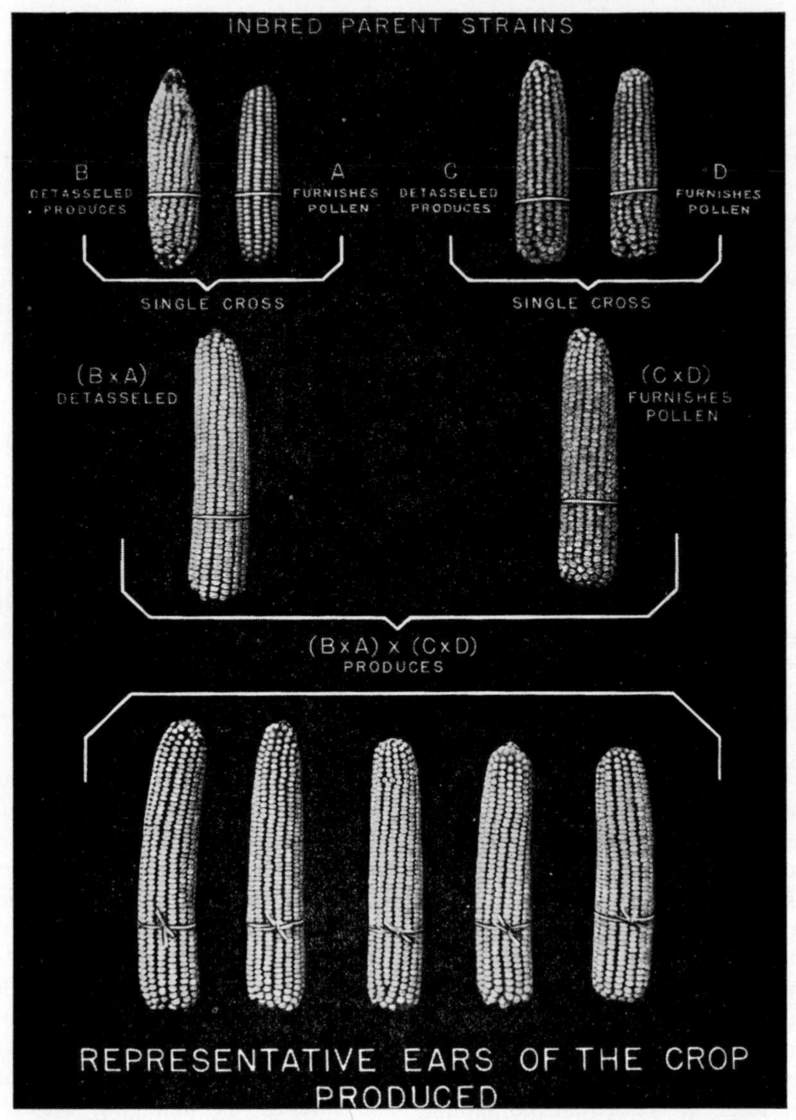

Figure 272. The mating of inbred strains of corn to produce the commercial variety with strong hybrid vigor. (Courtesy of the United States Department of Agriculture.)

296. INBREEDING AND OUTBREEDING

It is commonly believed that **inbreeding**—the mating of two closely related organisms such as brother and sister—is harmful and leads to the production of idiots and monstrosities. Even the marriage of first cousins is forbidden by law in some states. There is nothing harmful, however, in the act of inbreeding itself. Commercial breeders use inbreeding constantly to improve their strains of cattle, corn, cats or cantaloupes. It is not necessarily a bad practice in human beings; it simply provides a greater chance for recessive genes to become homozygous and express themselves phenotypically. All organisms are heterozygous for many characters; some of the recessive genes so hidden are for desirable characters, while others are for undesirable ones. If a stock is good, inbreeding will improve it; if a stock has many undesirable recessive traits, inbreeding will cause some of them to appear phenotypically.

The opposite of inbreeding is **outbreeding,** in which totally unrelated strains are bred. Frequently, as a result of this, the offspring are much better than either parent, a phenomenon known as **hybrid vigor.** A mule, the hybrid resulting from the mating of a horse and a donkey, is a strong and sturdy beast, more vigorous than either parent. A large part of the corn grown in the United States is a special hybrid developed by the United States Department of Agriculture from a mating of four different strains (Fig. 272). Each year, the seed to grow this uniformly fine hybrid corn must be obtained by mating the original strains, for the hybrid, being heterozygous, would give rise to a great variety of forms, probably none of which would be as good as the original hybrid.

The explanation of hybrid vigor is as follows: In the corn, each of the parental strains is homozygous for certain undesirable recessive genes, but any two strains are homozygous for *different* genes, and each has dominant ones to mask the undesirable recessive genes of the other. Let us consider just four pairs of characters, **A, B, C** and **D,** the capital letters standing for some dominant desirable trait, the lower-case letters for their recessive, undesirable alleles. One strain, then, might have the genotype **AAbbCCdd,** and the other, **aaBBccDD.** The hybrid offspring would be **AaBbCcDd,** combining all the desirable and none of the undesirable traits. The actual situation in corn is, of course, much more complicated than this.

297. POPULATION GENETICS

A question that sometimes puzzles beginning geneticists is why, if brown eye genes are dominant to blue eye genes, haven't all the latter disappeared? The answer lies partly in the fact that a recessive gene, such as the one for blue eyes, is not changed in any way by having existed for a generation next to a brown eye gene. The rest of the explanation lies in the fact that as long as there is no selection for either eye color, as long as people with blue eyes are just as likely to marry and to have as many children as people with brown eyes, successive generations will have the same proportions of blue- and brown-eyed people as the present one.

A brief excursion in mathematics will show why this is true. If we consider in a population (of men, animals or plants) the distribution of a single pair of genes, **A** and **a,** any member of the population will have the genotype **AA, Aa,** or **aa.** Let us suppose that these genotypes are present in the population in the ratio ¼ **AA : ½ Aa : ¼ aa.** (The point of the argument, that there is no change in the proportion in successive generations, will be the same no matter what initial ratio we assume.) If all the members of the population select their mates at random, without regard to whether they are **AA, Aa,** or **aa,** and if all the pairs produce comparable numbers of offspring, the succeeding generations will also have genotypes in the ratio ¼ **AA : ½ Aa : ¼ aa.** This can be demonstrated by setting down all the possible types of matings, the frequency of their random occurrence, and the kinds and proportions of offspring produced by each type of mating, then adding up the kinds of offspring (Table 12).

Hardy, a mathematician, and Weinberg, a physician, independently observed in 1908 that the frequencies of the members of a pair of allelic genes in a population are described by the expansion of a binomial equation. In the example just discussed, we assumed that the population contained initially ¼ **AA,** ½ **Aa** and ¼ **aa.** We can generalize this relationship if we let p be the proportion of **A** genes in the population and let q be the proportion of **a** genes in the population. Since the gene must be either **A** or **a,** then $p + q = 1$, and if we know either p or q we can calculate the other.

When we consider all the matings of any generation, a p number of **A**-containing eggs and a q number of **a**-containing eggs are fertilized by a p number of **A**-containing sperm and a q number of **a**-containing sperm: $(p\mathbf{A} + q\mathbf{a}) \times (p\mathbf{A} + q\mathbf{a})$. The proportion of the types of offspring of all these matings is described by the algebraic product: $p^2\,\mathbf{AA} + 2\,pq\,\mathbf{Aa} + q^2\,\mathbf{aa}$. This formula, and its implication of genetic stability in a population in the absence of selection, is known as the Hardy-Weinberg Law.

Table 12. THE OFFSPRING OF THE RANDOM MATING OF A POPULATION COMPOSED OF 1/4 **AA,** 1/2 **Aa** AND 1/4 **aa** INDIVIDUALS

MATING MALE	FEMALE	FREQUENCY	OFFSPRING
AA ×	AA	1/4 × 1/4	1/16 AA
AA ×	Aa	1/4 × 1/2	1/16 AA + 1/16 Aa
AA ×	aa	1/4 × 1/4	1/16 Aa
Aa ×	AA	1/2 × 1/4	1/16 AA + 1/16 Aa
Aa ×	Aa	1/2 × 1/2	1/16 AA + 1/8 Aa + 1/16 aa
Aa ×	aa	1/2 × 1/4	1/16 Aa + 1/16 aa
aa ×	AA	1/4 × 1/4	1/16 Aa
aa ×	Aa	1/4 × 1/2	1/16 Aa + 1/16 aa
aa ×	aa	1/4 × 1/4	1/16 aa
			Sum: 4/16 AA + 8/16 Aa + 4/16 aa

In studies of human genetics, where test matings are impossible, statistical methods based on this law have enabled investigators to determine the method of inheritance of many traits and to predict the proportion of types of offspring. For example, albinism, white skin color due to the absence of pigment, is a rare condition in man that is inherited by a single pair of genes. The gene **a** for albinism is recessive to the gene **A** for normal pigmentation. Surveys have shown that the frequency of albinos (genetically **aa**) in the population is about 1 in 20,000. Substituting in the Hardy-Weinberg equation, q^2, the frequency of **aa** individuals, is 1/20,000. The square root of 20,000 is about 141, so $q = 1/141$. Since $p + q$ equals 1, $p = 1 - q$ or $1 - 1/141$, or 140/141. Now we can calculate $2pq$, the frequency of occurrence of people genetically **Aa,** heterozygous for albinism: $2 \times 140/141 \times 1/141$, or about one person in 70 is a carrier for albinism. It may come as a surprise to you that there are so many carriers for such a rare trait. H. J. Muller has calculated that each of us is, on the average, heterozygous for 8 undesirable genes.

298. BIOCHEMICAL GENETICS

For many years, geneticists have not been content with the statement that a particular gene "produces" blue or brown eyes in man, or round or wrinkled peas, but have endeavored to discover the actual chemical and physical mechanisms involved. Today, there is enough information about this field—which involves a combination of genetics, embryology and biochemistry—to form the basis of a new science, that of **biochemical genetics.** This science concerns two principal, interrelated problems: (a) the chemical and physical nature of genes, and (b) the mechanisms whereby the genes control the development and maintenance of the organism.

The Chemical Nature of the Gene. Many attempts have been made to observe the genes within the chromosomes, but to date not even electron microscopy has revealed them. Chromosomes have been isolated from ground-up cells and found to be composed of proteins and nucleic acids. One of the two kinds of nucleic acid, **desoxyribonucleic acid** (abbreviated DNA), is found only within the chromosomes. These analyses have shown that the amount of DNA, like the number of genes, is the same in all the somatic cells of a given species, and that there is only half as much DNA in an egg or sperm as there is in a somatic cell of the same species. These observations have led to the conclusion that DNA is an integral part of the gene. There are other facts which lead to the conclusion that DNA is responsible for the transmission of genetic information from one generation to the next. Substances called "transforming agents" can be isolated from certain strains of bacteria, such as the one causing pneumonia, which will transform one strain of bacteria into another. These agents, with properties quite similar to those of genes, are composed solely of DNA. DNA appears to be the carrier of genetic information in bacteriophages, for when a bacteriophage enters a bacterium its protein coat remains outside; only the core of nucleic acid enters. This nucleic acid core produces, within the bacterium, both the nucleic acid cores and the protein coats of many additional bacteriophage particles which are released when the bacterial cell bursts.

The separation of the nucleic acid from the protein part of a plant virus has been achieved by W. M. Stanley. Neither part alone has viral activity but this reappears when the two parts are recombined. Stanley added the nucleic acid isolated from one virus to protein obtained from a different strain of virus and found that the new "hybrid" virus had the genetic properties only of the strain that supplied the nucleic acid; it did not resemble the strain that contributed the protein. From this Stanley concluded that the nucleic acid determines the biologic properties of the virus and the protein forms a protective coat which stabilizes the nucleic acid. Evidence from the experimental production of gene mutations also favors the concept that DNA is an essential component of the gene, for the physicochemical properties of the substance which mutates and those of DNA are very similar.

Estimates of the Number and Size of Genes. There are fairly reliable estimates of the number of genes per unit of length in the chromosomes of corn and fruit flies. If we assume that the number of genes per chromosome in man is comparable, man has about 25,000 genes in the nucleus of each cell. The error in this estimate is probably no more than fivefold, and the true number of genes lies between 5,000 and 125,000.

Early estimates of the size of the gene suggested that it was a very large particle, with a molecular weight in the range of 40,000,000 to 60,000,000. Hemoglobin, an average-sized protein, has a molecular weight of 68,000. More recent estimates place the gene size at about half of the original value. At one time it was believed that a gene was a true indivisible unit, and it was enthusiastically hailed as the "basic unit of life." It was believed that the unit of crossing over in the chromosome, the unit which undergoes mutation to form new types of genes, and the fundamental unit which regulates the phenotypic appearance of the character are all the same unit, the gene. Newer experiments make it clear that these units have quite different sizes, the mutation unit being much smaller and the functional unit perhaps larger than the unit of crossing over. Our concept of the intimate nature of the gene is undergoing constant revision as new experimental evidence appears.

Changes in Genes: Mutations. Although genes are remarkably stable and are transmitted to succeeding generations with great fidelity, they do, from time to time, undergo changes called **mutations.** After a gene has mutated to a new form, this new form is stable and usually has no greater tendency to change again than the original gene. **Chromosomal mutations** are accompanied by some visible change in the structure of the chromosome—the deletion or duplication of a small segment of the chromosome, the translocation of a segment of one chromosome to a new position on a different chromosome, or the inversion, turning end for end, of a segment of chromosome. **Point mutations** produce no visible change in chromosome structure and we assume that these involve such small alterations at the molecular level that they are not visible. From our current theory that genes are complex nucleic acid molecules we can guess that mutations involve some change in the order or arrangement of the nucleotide units which make up the DNA molecule.

Gene mutations can be induced by exposing the cell to certain chemicals or to radiation; x-rays, gamma rays, cosmic rays, ultraviolet rays and all the types of radiation that are byproducts of atomic power are effective mutation agents. Mutations do occur spontaneously at low but measurable rates which are characteristic of the species and of the gene; some genes are much more "mutable" than others. Natural radiations such as cosmic rays probably play some role in causing spontaneous mutations, but there are undoubtedly other important factors. The rates of spontaneous mutation of different human genes range from 1×10^{-5}

to 1 x 10^{-3} mutations per gene per generation. Since man has a total of some 2.5 x 10^{4} genes, this means that the total mutation rate is on the order of one mutation per person per generation. Each one of us, in other words, has some mutant gene that neither of our parents had.

Gene Action. According to current theory, genes act as catalysts for the production of enzymes. Enzymes are believed to owe their specificity to the specific configuration of the surface of the enzyme molecule. Only those substances whose molecules have the proper shape can fit on the surface of the enzyme, make contact at a number of points, and form an enzyme-substrate complex. The surface of the gene is believed to have a comparable specific conformation, and this specific conformation is transferred either directly or via an intermediate template to the enzyme. This theory requires that there be a separate gene for each type of enzyme, and there is quite a bit of experimental evidence which supports this view.

Our current concept of gene function may be summarized as follows: The materials transferred from one generation to the next in the nucleus of the egg and sperm, the genes, are templates composed of DNA and protein. By cell division these templates are duplicated and distributed to all the daughter cells that make up the animal or plant body. In each cell the DNA, either alone or in combination with protein, produces an intermediate template (called a **plasmagene** by some) made of ribonucleic acid and protein. The plasmagene in turn impresses this specific surface conformation onto a protein molecule as it is synthesized and converts it into the specific enzyme.

If we assume that a specific gene may indeed produce a specific enzyme by this or some other method, we must next inquire how the presence or absence of this specific enzyme may affect the development of the zygote. The expression of any trait is the result of a number of chemical reactions which occur in series, with the product of each reaction serving as the substrate for the next: A⟶ B⟶ C⟶ D. The dark color of most mammalian skin and fur is due to the pigment **melanin** (D), produced from dihydroxyphenylalanine (dopa) (C), produced in turn from tyrosine (B) and phenylalanine (A). Each of these reactions is controlled by a particular enzyme; the conversion of dopa to melanin, for example, is mediated by the enzyme dopa oxidase. **Albinism,** characterized by the absence of melanin, results from the absence of dopa oxidase. The gene for albinism, **a,** does not produce the enzyme dopa oxidase, but its normal allele, **A,** does.

In most animals and plants it is difficult to investigate the stepwise control of the expression of a character except those in which some colored product is formed. This difficulty was overcome when George Beadle and Edward Tatum conceived the idea of irradiating the bread mold, *Neurospora,* and looking for mutations which interfere in some way with the normal reactions by which the chemicals essential for its growth are produced. The normal bread mold requires as raw materials only sugar, salts, inorganic nitrogen and biotin, the so-called "minimal" medium (Fig. 273). By exposing the mold to x-rays or ultraviolet rays, a great many mutations are produced. After irradiation the mold is supplied with "complete" medium, an extract of yeast which contains all the known amino acids, vitamins, and so on. Any nutritional mutant produced by the irradiation is thus able to survive and reproduce to be tested subsequently.

A bit of the irradiated mold is then placed on minimal medium. If it is unable to grow we know that a mutant has been produced which interferes with the production of some compound essential for growth. Then, by trial and error, by adding substances to the minimal medium in groups or singly, the nature of this missing substance is determined. In each instance genetic tests show that the mutant strain produced by irradiation differs from the normal wild mold by a single gene, and chemical tests show that if a single chemical substance is added to the minimal medium the mutant strain is able to grow normally. The inference is that each gene produces a single enzyme which regulates one step in the biologic synthesis of this chemical. It has been possible in some instances to show that the particular enzyme can be extracted from the cells of normal *Neurospora* but not from the cells of the mutant strain. The synthesis of each substance includes a number of separate steps, each mediated by a separate gene-controlled enzyme. An estimate of the minimal number of steps involved can be obtained from the number of different mutants which interfere with its production.

Similar one-to-one relationships of gene, enzyme and biochemical reaction in man were first described by the English physician, A. E. Garrod, in 1909. **Alkaptonuria** is a trait, inherited by a recessive gene, in which the patient's urine turns black on exposure to air. The urine contains homogentisic acid; the tissues of normal people have an enzyme which oxidizes homogentisic acid so that it is excreted as carbon dioxide and water. Alkaptonurics lack this enzyme because they lack the gene which produces it. As a result, homogentisic acid accumulates in the tissues and blood and spills over into the urine. Garrod used the term "inborn errors of metabolism" to describe alkaptonuria and comparable conditions such as phenylketonuria and albinism.

It has recently been found that when a wing bud from a creeper chick is transplanted onto normal chick blastoderm it will develop into a normal wing, not a creeper wing. Evidently the creeper gene interferes with the production of some substance required for normal wing development, a substance which can be supplied by the enzyme systems of the normal tissue. If this substance could be identified and supplied in suitable amount to a fertilized creeper egg, the egg would presumably develop into a normal rather than a creeper chick.

The identification of the chemical and biologic mechanisms which underlie differentiation remains one of the major unsolved problems of biology. The regularity of the mitotic process appears to assure

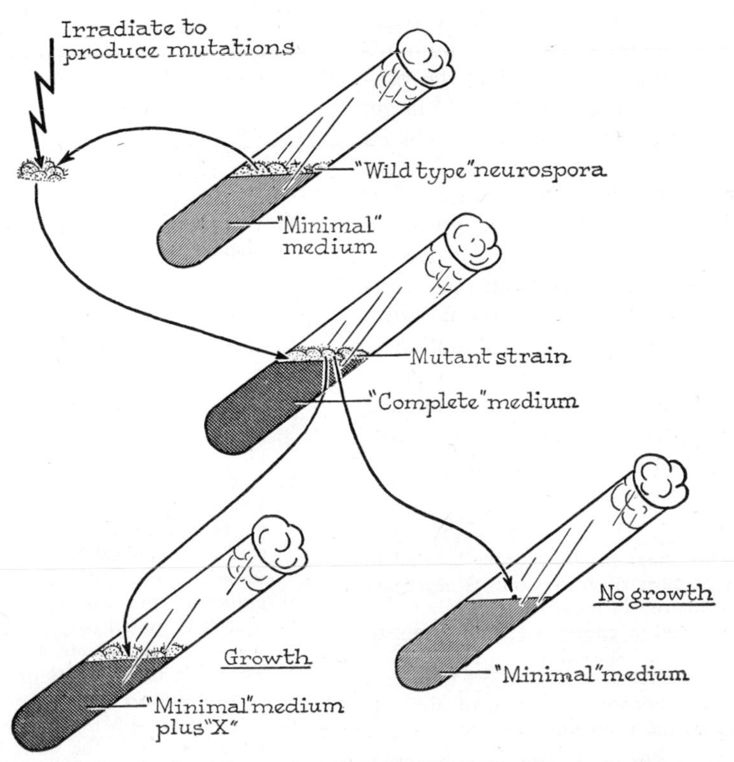

Figure 273. **The method of producing and testing biochemical mutants in *Neurospora*. See text for discussion.**

every cell of the body the same number and kinds of genes as every other cell, yet the tissues have marked differences in their physical, chemical and biologic properties. These differences apparently result from the different metabolic effects of similar genes working in different cytoplasmic environments.

One of the clearest demonstrations that the same genes working in dissimilar environments do have different effects was provided by experiments with three races of frogs found naturally in Florida, Pennsylvania and Vermont. Each of these races normally develops at a speed which is adapted to the normal length of the spring and summer seasons, Southern frogs developing slowly, and Northern frogs more rapidly. Northern frogs raised in Southern conditions were overaccelerated in development, while Southern frogs raised in Northern conditions were overretarded. By fertilizing an egg with sperm from a different race, and then removing the egg nucleus before the sperm nucleus united with it, it was possible to set up "Northern" genes operating in "Southern" cytoplasm. This resulted in poorly regulated development, and the animal's head, growing more rapidly than the posterior region, was disproportionately large. When "Southern" genes were introduced into "Northern" cytoplasm, again there was poorly regulated development, but with the head, rather than the posterior region, retarded, causing it to be disproportionately small. Genes from the Pennsylvania race acted as "Northern" with Florida cytoplasm, and as "Southern" with Vermont cytoplasm. Thus, exactly the same set of genes produced opposite morphologic effects when operating in different cytoplasmic environments.

QUESTIONS

1. Show by diagrams how genes located in different pairs of chromosomes segregate independently in meiosis.
2. In peas, yellow color (**Y**) is dominant to green (**y**), and smooth peas (**S**) are dominant to wrinkled ones (**s**). Give the genotypic and phenotypic ratios of the offspring of the following crosses:

    a. **YYss** × **yySS**
    b. **YySs** × **YySs**
    c. **YySs** × **yyss**
    d. **Yyss** × **yySs**

3. In rabbits, spotted coat (**S**) is dominant to solid color (**s**), and black (**B**) is dominant to brown (**b**). A brown spotted rabbit is mated to a solid black one and all the offspring are black spotted. What are the genotypes of the parents? What would be the appearance of the F_2 if two of these F_1 black spotted rabbits were mated?
4. The long hair of Persian cats is recessive to the short hair of Siamese cats, but the black coat color of Persians is dominant to the black-and-tan coat of Siamese. If a pure black, long-haired Persian is mated to a pure black-and-tan, short-haired Siamese, what will be the appearance of the F_1? If two of these F_1 cats are mated, what is the chance of obtaining in the F_2 a long-haired, black-and-tan cat?
5. In peas, tall plants (**T**) are dominant to dwarf (**t**). What would be the phenotypes of the offspring of the following matings?

    a. **TtYySs** × **ttyyss**
    b. **TtyySs** × **ttYySs**

6. Distinguish between: complementary genes and supplementary genes; multiple factors and multiple alleles; sex-linked character and sex-influenced character; penetrance and expressivity.
7. A walnut-combed rooster is mated to three hens. Hen *A,* which is walnut-combed, has offspring in the ratio of 3 walnut: 1 rose. Hen *B,* which is pea-combed, has offspring in the ratio of 3 walnut: 3 pea: 1 rose: 1 single. Hen *C,* which is walnut-combed, has only walnut-combed offspring. What are the genotypes of the rooster and the three hens?
8. What conditions result in the following phenotypic ratios:

    a. 3 : 1
    b. 1 : 2 : 1
    c. 9 : 3 : 3 : 1
    d. 9 : 7
    e. 1 : 4 : 6 : 4 : 1
    f. 2 : 1
    g. 47 : 47 : 3 : 3

9. The weight of the fruit in one variety of squash is determined by three pairs of genes, **AABBCC,** producing 6-pound squashes, and **aabbcc,** producing 3-pound squashes. Each dominant gene adds ½ pound to the weight. When a 6-pound squash is mated to a 3-pound squash, all the offspring weigh 4½ pounds. What would be the weights of the F_2 fruits if two of these F_1 plants were mated?
10. Mrs. Doe and Mrs. Roe had babies at the same hospital at the same time. Mrs. Doe took home a girl and named her Nancy. Mrs. Roe received a boy and named him

Richard. However, she was sure she had had a girl and brought suit against the hospital. Blood tests showed that Mr. Roe was type O, Mrs. Roe was type AB, Mr. and Mrs. Doe were both type B. Nancy was Type A and Richard type O. Had an exchange occurred?

11. A gene **l** is known in fruit flies which is sex-linked, recessive and lethal. If a normal male is mated to a female heterozygous for this gene, what will be the sex ratio of their offspring?
12. Explain the mechanism of the genetic determination of sex.
13. One pair of genes for coat color in cats is sex-linked. The gene **B** produces yellow coat, **b** produces black coat and the heterozygote **Bb** produces tortoise-shell color. What kind of offspring result from the mating of a black male and a tortoise-shell female?
14. The barred pattern of chicken feathers is inherited by a pair of sex-linked genes, **B** for barred and **b** for no bars. If a barred female is mated to a nonbarred male, what will be the appearance of the progeny? What commercial usefulness does this have?
15. A bald, normal-visioned man whose father was not bald, marries a nonbald, normal-visioned woman whose father was color-blind and whose mother was bald. What types of offspring may they have?
16. What is meant by linkage? By crossing over?
17. What are the advantages and disadvantages of inbreeding?
18. What is meant by hybrid vigor? What is the genetic explanation of this phenomenon?
19. In one large eastern city, 11 babies out of 100,000 were born with Cooley's anemia. This disease is inherited by a single pair of genes, the recessive gene producing the disease and the dominant one the normal condition. Calculate the number of people in this population that are heterozygous for this gene, i.e., are "carriers" of the trait.

SUPPLEMENTARY READING

Genetics in the Twentieth Century, edited by L. C. Dunn, is a collection of papers presented at the Golden Jubilee of Genetics on the 50th anniversary of the rediscovery of Mendel's work. Some articles of special interest to the beginning student are ones on the history of genetics by H. Iltis, C. Zirkle, W. E. Castle, and H. J. Muller, one by R. B. Goldschmidt on the relations of genetics to other sciences, one by G. W. Beadle on chemical genetics, and ones on practical applications of genetic knowledge by L. H. Snyder, J. W. Gowen, C. C. Little, A. Müntzing, J. L. Lush, J. C. Walker and P. C. Mangelsdorf. Another discussion of biochemical genetics by G. W. Beadle is given in Baitsell's *Science in Progress,* volume 6. Two more advanced texts of genetics are Sturtevant and Beadle, *An Introduction to Genetics,* and Sinnott, Dunn and Dobzhansky, *Principles of Genetics.*

Chapter 32

Inheritance in Man

THE past decade has seen great advances in our knowledge of human inheritance and problems in this field are being studied intensively at present with the newer methods of statistical analysis. Centers of research in human genetics are located in Ann Arbor, Chicago, Columbus, Minneapolis, New York City, Salt Lake City and Winston-Salem in this country, and in London, Stockholm and Copenhagen. The study of inheritance in man is more difficult than in plants or animals not only because such techniques as test crosses and progeny tests cannot be used, but also because (1) human beings have so few offspring per generation, (2) there is a long time between generations so that usually data are available from only two or a few generations, and (3) human beings are, in general, highly heterozygous.

299. HUMAN PEDIGREES

After the rediscovery of Mendel's Laws in 1900, it became apparent that these principles applied to the inheritance of human traits, and that the latter may be controlled by multiple factors, multiple alleles, sex-linked genes, and so on. The early studies usually concerned the inheritance in individual families of readily identified and fairly conspicuous traits. It was sometimes possible to collect data for several generations and to set up pedigrees such as the one in Figure 277. In this way the inheritance of several hundred human traits was shown to be determined by single dominant or recessive genes. Since 1925 the development of the statistical methods of population genetics has enabled geneticists to determine modes of inheritance with data from only two generations or even, in certain instances, from a single generation. In such studies, geneticists lump together the results of similar matings and calculate the relative frequencies of dominant and recessive alleles. Although in such studies classic mendelian ratios are not found, predictable ratios, based on the frequency of the alleles, do occur and a comparison of the observed and predicted ratios in a given population can be used to determine the number and kind of genes that determine a given trait. Such methods are not as efficient as classic mendelian genetics but are extremely valuable in studying human inheritance, where test crosses between genotypically defined individuals are impossible.

L. H. Snyder's study of the inheritance of the inability to taste phenylthiocarbamide is an example of the use of the principles of population genetics. Some people find that this substance has a bitter taste, others report it to be completely tasteless. Snyder examined 3643 people and found that 70.2 per cent were "tasters" and 29.8 per cent were "nontasters." If this trait is inherited by a single pair of genes, with "tasting" dominant to "nontasting," then the mathematics of population genetics predicts that in the population of marriages of tasters with tasters, 12.4 per cent of the children will be nontasters. It similarly predicts that in the population of marriages of tasters with nontasters, 35.4 per cent of the children will be nontasters. In Snyder's study, the percentages actually found were 12.3 per cent and 33.6 per cent. From this close agreement we may conclude that our original assumption is correct and that the tasting-nontasting trait is inherited by a single pair of alleles.

In pedigree diagrams, males are indicated by squares, and females by circles. Individuals showing the trait in question are indicated by black symbols, those not showing the character by white symbols. Relationship is indicated by connecting lines and all the members of the same generation are placed in the same row. Thus the pedigree in Figure 274 tells us that No. 11 is a blue-eyed girl, while her two sisters (12 and 13) and her brother (10), as well as her mother (8) and father (7) all have brown eyes. Her two aunts (5) and (6), an uncle (9), her paternal grandfather (1) and maternal grandmother (4) are also brown-eyed, but her paternal grandmother (2) and maternal grandfather (3) are blue-eyed. Her parents are, of course, heterozygous for blue eyes. Pedigrees for recessive traits are characterized by this "skipping of a generation."

300. THE INHERITANCE OF PHYSICAL TRAITS

The development of every organ of the body is regulated by genes, some of which are now known in spite of the difficulties involved in the study of human inheritance. The age at which a particular gene expresses itself phenotypically varies widely with different characteristics. Most characteristics develop long before birth although some, such as hair and eye color, may not appear until shortly after birth. Others such as amaurotic idiocy become evident in childhood and still others such as glaucoma and Huntington's chorea develop only after the individual has reached maturity. Some of the human traits known to be inherited are listed in Table 13. Some of these conditions may be due to two or more different genes, so that they are inherited by dominant genes in one pedigree and by recessive genes in another.

Human hair color is regulated by several genes. Blond hair is recessive to dark hair, but the production of different shades of dark or light hair is due to the interactions of several pairs of modifying factors. Red hair is caused by a single gene which is recessive to the gene for nonred hair. In a person homozygous for red hair who also has genes for dark hair, the red pigment is masked by the dark pigment and dark hair with a reddish tinge results. Curly hair is not completely dominant to straight hair—the heterozygous condition produces wavy hair.

Eye color depends on the amount and location of pigment in the iris. The pink eyes of albinos are really colorless; the apparent pink color is due to a reflection from blood vessels in the iris and in the retina of the eye. Nor do blue eyes contain blue pigment; there is a dark pigment in the back of the iris and when light is reflected from this an impression of blue is produced. If such an iris had in addition a bit of yellow pigment in the front of the iris the eye would appear to be green. A scattering of dark pigment in the front of the iris produces a gray eye and a brown or black eye is due to a heavy concentration of dark pigment in the front of the iris.

The slant-eyed appearance of Chinese and Japanese is due to the presence of a fold of skin extending from the upper lid

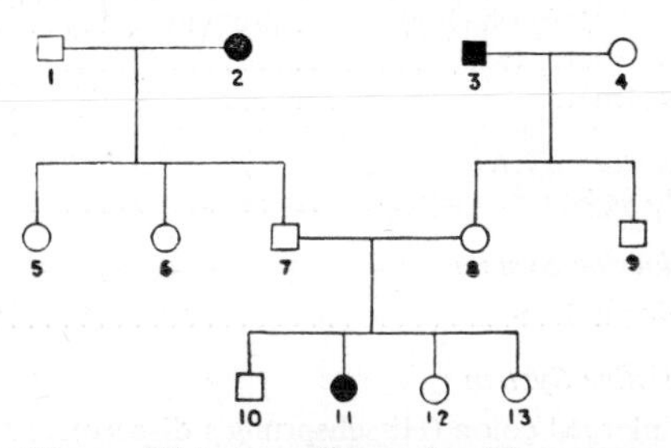

Figure 274. A pedigree for blue eyes.

Table 13. SOME TRAITS IN MAN KNOWN TO BE INHERITED

DOMINANT	RECESSIVE
Hair, Skin, Nails, Teeth	
Dark hair	Blond hair
Nonred hair	Red hair
Curly hair	Straight hair
Abundant body hair	Little body hair
Early baldness (dominant in male)	Normal
White forelock	Self-color
Piebald (skin and hair spotted with white)	Self-color
Pigmented skin, hair, eyes	Albinism
Black skin (2 pairs of genes, dominance incomplete)	White skin
Ichthyosis (scaly skin)	Normal
Epidermis bullosa (sensitiveness to slight abrasions)	Normal
Absence of enamel of teeth	Normal
Normal	Absence of sweat glands
Eyes	
Brown	Blue or gray
Hazel or green	Blue or gray
"Mongolian fold"	No fold
Congenital cataract	Normal
Nearsightedness	Normal vision
Farsightedness	Normal vision
Astigmatism	Normal vision
Glaucoma	Normal
Aniridia (absence of iris)	Normal
Congenital displacement of lens	Normal
Normal	Optic atrophy (sex-linked)
Normal	Microphthalmus
Features	
Free ear lobes	Attached ear lobes
Broad lips	Thin lips
Large eyes	Small eyes
Long eyelashes	Short eyelashes
Broad nostrils	Narrow nostrils
High, narrow bridge of nose	Low, broad bridge
"Roman" nose	Straight nose
Skeleton and Muscles	
Short stature (many genes)	Tall stature
Achondroplasia (dwarfism)	Normal
Ateliosis (midget)	Normal
Polydactyly (more than 5 digits on hands or feet)	Normal
Syndactyly (webbing of 2 or more fingers or toes)	Normal
Brachydactyly (short digits)	Normal
Cartilaginous growths on bones	Normal
Progressive muscular atrophy	Normal
Circulatory and Respiratory Systems	
Hereditary edema (Milroy's disease)	Normal
Blood groups A, B, and AB	Blood group O
Hypertension (high blood pressure)	Normal
Normal	Hemophilia (sex-linked)
Normal	Sickle cell anemia
Excretory System	
Polycystic kidney	Normal
Endocrine System	
Normal	Diabetes mellitus
Digestive System	
Enlarged colon (Hirschsprung's disease)	Normal

	DOMINANT	RECESSIVE
Nervous System		
	Tasters (of phenylthiocarbamide)	Nontasters
	Normal	Congenital deafness
	Normal	Spinal ataxia
	Huntington's chorea	Normal
	Normal	Amaurotic idiocy
	Migraine (sick headache)	Normal
	Normal	Dementia praecox (several pairs of genes involved)
	Normal	Phenylketonuria
	Paralysis agitans	Normal
Cancers		
	Normal	Xeroderma pigmentosum
	Recklinghausen's disease	Normal
	Normal	Retinal glioma

down over the inner corner of the eye. The gene for this "Mongolian fold" is dominant to the one for no fold.

Several skin conditions are known to be inherited. In *epidermis bullosa* the skin is very sensitive to slight abrasions and blisters will appear from the slight friction of shoes and clothing. Another inherited condition is the complete absence of sweat glands; the same gene also causes scanty hair and a deficient number of teeth. A person with this abnormality obviously suffers terribly in hot weather. A dominant gene causes hypertrichosis, in which the embryonic hair, which is normally shed before birth, remains and grows excessively throughout life. Another dominant gene produces ichthyosis, a dry skin covered with flakes or scales that are constantly being shed.

In discussing all these traits it is important to keep in mind the distinction between congenital and inherited traits. By a **congenital trait** we simply mean one that is present at birth. It may or may not be inherited. Some congenital abnormalities are inherited while others are caused by accidents in the developmental process or by such things as a woman contracting German measles some time during the first three months of her pregnancy. Conversely, not all inherited conditions are congenital, evident at birth; some appear later in life.

301. HEREDITY AND ENVIRONMENT

At one time a bitter argument raged as to whether heredity or environment was more important in determining human traits. Further research has made it abundantly clear that both physical and mental characters are the result of the interplay of genetic and environmental factors. Some genes, such as those which determine the blood groups, produce their effects regardless of the environment. The expression of other genes may be markedly affected by the environment. Recent studies in biochemical genetics suggest that the greater the number of biochemical steps that occur between a given gene and the final trait, the greater is the opportunity for environmental influences to affect the trait.

One widely held fallacy is that if a certain trait has a genetic basis, it cannot be affected by changing the environment; that is, it cannot be alleviated or cured by medical treatment. During World War II one group of experimenters reported that feeding large doses of vitamin A would cure red-green color blindness. Color blindness might indeed be curable, for the fact that it is inherited does not eliminate the possibility of cure. Since vitamin A is known to be a constituent of **visual violet,** the light sensitive substance in cones, it was not unreasonable that vitamin A might cure color blindness. The gene for color blindness might simply cause a higher requirement of vitamin A. The experiments were repeated by several other groups of investigators, none of whom could find any effect of vitamin A on color blindness. The original authors had stated that since color blindness is curable, it is not the simple mendelian trait popular theories assume it to be. The critics, who found negative results, argued that the disease is inherited and therefore incurable. Both of these arguments are false, because inherited diseases can be alleviated. It seems clear now that vitamin A in the doses given will not cure color blindness, but the fact that color blindness is inherited does not preclude the possibility of finding some way of enabling color blind people to see color. If the color blind gene blocks some step in the synthesis of visual violet, supplying the substance normally made by this step should "cure" color blindness.

With the realization that both genetic and environmental factors are involved in producing many traits, research has been undertaken to determine the relative importance of the two in determining particular traits. This cannot be done simply by collecting data on the similarity or variability of individual members of families, for these may be due to either genetic or nongenetic factors. The problem has been attacked in two ways, by studying twins and by studying children brought up in foster homes or in orphanages.

Identical twins, coming from a single fertilized egg, have identical genes; any differences between them are due to environmental factors. Fraternal twins are no more alike genetically than ordinary brothers and sisters born separately. Studies of feeble mindedness have shown that in fraternal twins, when one is feeble minded the other is also feeble minded in about 25 per cent of the cases. In identical twins, when one is feeble minded, the other has the trait in nearly every case. Identical twins are much more similar in intelligence than are fraternal twins; indeed, identical twins reared apart are more similar in intelligence than fraternal twins reared together.

Children reared together in an orphanage, where the environment is fairly constant, show just as wide variability in intelligence as children reared separately in their own homes. Even when children are adopted early in infancy there is a much greater correlation between the intelligence of the child and its true parents than between the child and its foster parents.

It may be concluded, therefore, that the upper limit of a person's mental ability is determined genetically, but how fully he develops these inherited abilities is determined by environmental influences—training and experience.

It is easy to understand why the offspring of intelligent parents are sometimes less intelligent than either parent. Since the coordinated action of many pairs of genes is involved in intelligence, the fortuitous combination of genes which produced the intelligent parents may be broken up by genic segregation. Conversely, the chance recombination of favorable genes may produce a brilliant child from average parents (but geniuses are never produced by feeble minded parents).

302. INHERITANCE OF MENTAL DISORDERS

Spinal ataxia is a condition in which the sensory nerve tracts degenerate, cutting off the sensory impulses on which equilibrium depends. A person suffering from this cannot stand without swaying, and in the advanced stages loses all power of independent movement. The condition is inherited through a recessive gene in some families, through a dominant one in others. A single dominant gene causes the degeneration of the large ganglion cells of the floor of the cerebrum. This results in **paralysis agitans,** characterized by involuntary movements, especially of the hands. **Huntington's chorea,** inherited by a dominant gene, is characterized by progressive degeneration of the nervous system, leading to involuntary twitchings of the head, arms and legs and finally to mental impairment and death. Another inherited degenerative disease of the nervous system is juvenile **amaurotic idiocy,** a defect which appears within the first two years of life. This degeneration of the retina and nervous system leads to blindness, paralysis, mental failure and death. **Phenylketonuria,** a type of imbecility inherited by a recessive gene, is accompanied by the excretion of phenylpyruvic acid in the urine. The mutant gene blocks a certain chemical reaction and phenylpyruvic acid, which normal people metabolize via a series of steps to carbon dioxide and water, accumulates in the blood and is excreted in the urine. The connection between this chemical block and the imbecility is not understood. In these mental disorders, which are fortunately quite rare, the mode of inheritance is clear. In some other more common types of insanity, the method of inheritance is more obscure.

The commonest type of insanity, affecting almost 1 per cent of the population, is **dementia praecox** (or schizophrenia) in which the patient lives in a world of his own. Losing interest in the normal pattern of life, he becomes withdrawn and incoherent. The disease definitely runs in families, its incidence among the children of an affected person being 20 times greater than in the general population. Clear evidence of its hereditary nature comes from twin studies. If one member of a pair of identical twins has dementia praecox, the other also has it in 68 per cent of cases. Environmental factors affecting the penetrance of the genes are presumably responsible for the absence of the disorder in the remainder. In fraternal twins, the second member of the pair develops dementia in only 11 per cent of twins with one member affected. Similarly, about 11 per cent of brothers and sisters of affected people are affected. A committee of the American Neurological Association has examined the evidence and concluded that heredity plays an important role in producing dementia praecox, but that hereditary factors are sometimes inadequate by themselves to produce the disease. It may even appear in the absence of such factors, so that they are not invariably essential. Statistical analyses of the data on the occurrence of dementia praecox have suggested that it is inherited by recessive genes.

Manic-depressive insanity is a condition in which the individual is abnormally sensitive to his surroundings and undergoes exaggerated changes of mood from the manic (excited) to the depressed state. In identical twins, if one member of the pair is affected, the other also develops the insanity in 70 to 96 per cent of cases. But with fraternal twins, only in 7 to 14 per cent of pairs do both members have the condition. Family histories of manic-depressive insanity suggest that several pairs of genes may be involved, some of which are dominant.

Epilepsy, in which a person suffers from periodic convulsions and loss of consciousness, may be caused either by genetic factors or by injury or infection of the brain. The cases caused by hereditary factors seem to depend on several pairs of recessive genes. Recent work with the electroencephalograph indicates that the relatives of epileptics sometimes show disordered brain waves but not the actual epileptic seizures. Such people may possibly have some, but not all, of the genes necessary for the manifestation of epilepsy. Two people with such disordered brain waves probably should avoid having children, because their offspring may be epileptics.

303. INHERITANCE OF GENERAL AND SPECIAL ABILITIES

The inheritance of mental ability or intelligence is one of the most important, and one of the most difficult, problems of human genetics. The interpretation of genetic facts can easily become colored by political, sociologic, psychologic or educational theories.

Because of the difficulties surrounding the measurement of intelligence, some investigators deny the validity of all intelligence tests. Within limits, however, they seem to have definite value. In recent years psychologists have improved these tests so as to recognize and measure such primary abilities as the ability to reason inductively, to memorize, and to visualize objects in three dimensions. It has been found that individuals with similar over-all intelligence may differ considerably in their primary abilities. The older tests, such as the Stanford-Binet, are prepared by setting tasks for children, and then determining what capacities can be expected normally of children of each age. Then in taking the test, a child is given progressively more difficult problems until finally he is unable to solve them. When such tests are given, a wide range of mental ability from complete incompe-

Table 14. CLASSIFICATION OF INTELLIGENCE

I.Q.	DESIGNATION	
140 or over	"Gifted"	
120-140	Very superior	
110-120	Superior	
90-110	Normal	
80- 90	Dull	
70- 80	Borderline	
50- 70	Moron	"Feeble minded"
25- 50	Imbecile	"Feeble minded"
0- 25	Idiot	"Feeble minded"

tence to excellent comprehension has been found for each age group. A child of six who can solve problems ordinarily solved by children eight years old obviously is superior to one of six who can do only those normally done by six year olds. This "mental age," as determined by the test, is divided by the actual chronologic age and the quotient multiplied by 100 to give the "Intelligence Quotient" or I.Q. When the I.Q.'s of a large number of people are measured, they form a normal curve of distribution from 0 to over 140, with the largest number of scores in the normal class, and progressively fewer scores in the classes farther from normal (Table 14).

Individuals with an I.Q. less than 70 are termed "feeble minded," and three grades of feeble mindedness are recognized. The highest class, called **morons,** with I.Q.'s of 50 to 70, are unable to govern their affairs satisfactorily, but can perform simple tasks and may be partially self-supporting. People with an I.Q. of 25 to 50 have a mental age of three to seven years and are termed **imbeciles.** They are unable to earn a living but can take care of themselves. Those with a mental age of one or two, or an I.Q. below 25, known as **idiots,** cannot perform even the simple tasks of feeding and taking care of themselves.

The fact that the mental capacities of different people form a continuous series from idiot to genius with a distribution of I.Q.'s conforming to a normal curve, suggests the operation of multiple factors, and other evidence substantiates this hypothesis. Some of the earliest studies on the inheritance of intelligence attempted to divide people into two classes, normal and feeble minded, and explained the inheritance of feeble mindedness as due to a single recessive gene. We now know that the inheritance of mental defect is much more complex, and that such reports as those of the "Jukes" and "Kallikaks" are not valid genetically. Feeble mindedness may be caused by diseases such as syphilis or meningitis, by injuries sustained during birth, or by other environmental agents, but the majority of cases are due to hereditary factors, as shown by studies on twins.

Even before the recent studies subdividing intelligence into a number of primary abilities, it was recognized that such special capacities as musical, artistic, mechanical and mathematical talents have a hereditary basis. The inheritance of musical ability has been investigated more thoroughly than the others, chiefly because it can be measured more accurately. Philiptschenko studied the inheritance of this trait and found evidence that four pairs of genes are involved. Musical ability is a complex of pitch discrimination, tone memory, and a sense of rhythm, melody and harmony, so that it is not surprising that its inheritance is complex. Special abilities are at least partially independent of general intelligence, and investigations of feeble minded persons have revealed that some have marked musical or mechanical talents.

304. EUGENICS

The term "eugenics" was coined by Francis Galton to refer to the study of the influences that may improve the hereditary qualities of future generations. Eugenic practices may be negative, concerned with preventing the reproduction of individuals with undesirable hereditary traits, or positive, concerned with encouraging the reproduction of individuals with superior hereditary characters.

A discussion of eugenics might be subdivided into three main questions: (1) Are there undesirable hereditary traits in man and, if so, are they increasing in successive generations? (2) Would it be possible by methods now available to effect a significant decrease in these traits? (3) If it is possible to affect the genetic composition of future generations, do we have the moral and legal right to do so?

It is now known that there are hundreds of human abnormalities and defects, of every organ system of the body, that are wholly or largely the result of the actions of genes. There can be no argument about the undesirability of traits such as feeble mindedness, insanity, or epilepsy, and the more appalling physical abnormalities such as **acheiropodia,** in which the arms and legs are short stumps devoid of hands and feet.

Probably the biggest single eugenic problem is feeble mindedness; the number of feeble minded in the population is estimated to be about 3,000,000, some 2 per cent of the population. The lowest grades, idiots and imbeciles, are usually confined to institutions and unable to reproduce, but the morons, in the words of Frederick Osborn, are "fertile, sexually receptive and apt to bear children prolifically in and out of wedlock." These unfortunates contribute little or nothing to society and many are burdens as paupers, criminals or inmates of institutions. Observation has clearly shown that almost without exception two feeble minded parents will produce only feeble minded children. Many different studies have shown that the less intelligent members of the population have more children than the more intelligent members of the population, whether intelligence is measured as I.Q., years of schooling, socio-economic level, or in some other way. Since intelligence is determined, at least in part, by genetic factors, and since the less intelligent are reproducing more rapidly than the more intelligent, it is clear that the average intelligence of the population is decreasing. Estimates of the amount of this decrease vary and they all indicate that it is very small in terms of the decrease *per generation,* but in terms of the long range future of the human race it is highly significant.

Several methods are possible for preventing the reproduction of the mentally unfit. They can be segregated in institutions, but this is unkind and unfair as well as very expensive; less than 10 per cent of all the feeble minded can be accommodated in our state institutions. These people could be sterilized and permitted to live as members of society. Sterilization is a simple operation, and involves only the cutting of the vas deferens in the male and tying off the oviducts in the female. In no way does it alter the person's sexual desires or experiences. Although twenty-seven states have enacted laws providing for the sterilization of the feeble minded, the insane, or both, only in California has the program been carried out extensively. The total number sterilized in the United States for eugenic reasons up to 1955 was 57,000. A survey of the sterilizations performed in California showed that the results are satisfactory both to the people sterilized and to society.

The methods of birth control, or **contraception,** which involve the use of some device, mechanical or chemical, to prevent the sperm from reaching the egg, are not as effective as sterilization in preventing the reproduction of the feeble minded, for many of them are too dull or too careless to use contraceptives successfully. There have been suggestions from some recent research that in time chemists may devise an effective contraceptive that can be taken orally. This might act by preventing ovulation, by rendering sperm unable to penetrate the membrane around the egg, or in some other way.

The effectiveness of selection in ridding a population of undesirable traits varies with the method of inheritance involved. A trait due to a single dominant gene is the easiest to eliminate; one inherited by a sex-linked recessive would be more difficult to eliminate, and one inherited by an autosomal recessive gene would be even more difficult. The most difficult are those due to the interaction of several pairs of genes. Osborn estimates that by preventing births among all definitely feeble minded persons, between one third and one tenth of the total feeble mindedness in the next generation could be prevented. Other authorities have estimated that such selection would reduce the number of feeble minded by only a few per cent. But even these few per cent would mean that some tens of thousands of unfortunates would not be born.

The right of society to use sterilization to prevent the reproduction of the mentally unfit was affirmed by Justice Oliver Wendell Holmes, Jr., in the case of Buck vs. Bell. In this Supreme Court decision on May 2, 1927, he pointed out that since the needs of the public welfare may call upon the best citizens for their lives, it would be strange if it could not call upon these others for this lesser sacrifice.

For the limited objective of reducing feeble mindedness a program of sterilization would be effective, though it could not eliminate it completely. The application of such negative eugenic measures to other problems has not been seriously suggested. Although there can be no doubt that feeble mindedness is undesirable, the problem of just which traits are desirable is more difficult. Obviously, before beginning any program of positive eugenics, some agreement must be reached as to what traits are desirable. A society composed solely of geniuses would undoubtedly have great difficulty in surviving. Adaptability, emotional stability, health and other qualities are as desirable as intelligence. Modern society is tremendously complex, with a wide range of jobs to be done, and so it requires people with a correspondingly wide range of temperaments, intelligence, skills and strength to fill all the positions.

QUESTIONS

1. What difficulties are inherent in the study of human inheritance?
2. Distinguish between congenital and hereditary.
3. Discuss the statement, "If a certain disease is shown to have a hereditary basis, medical research on that disease is useless."
4. Could two dark-haired parents have a red-haired child? What genotypes must the parents have to have a red-haired child?
5. How are studies of twins useful in supplying data about the relative importance of heredity and environment in determining a given trait?
6. Explain genetically how a child might be more intelligent than either parent. Under what circumstances might it be less intelligent than either parent?
7. What is the evidence that intelligence is inherited?
8. What can be done to decrease the number of mentally or physically unfit individuals in a population?
9. In each of the following pedigrees, determine the method of inheritance of the trait, and as far as possible, fill in the genotypes of each individual. Assume that any individual who marries into the family and does not exhibit the trait does not carry any recessive genes for it.

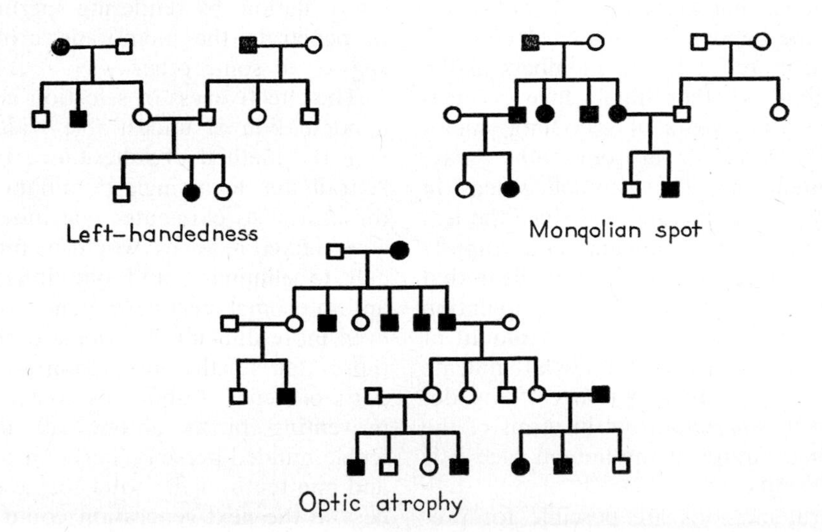

SUPPLEMENTARY READING

You and Heredity, by A. Scheinfeld, is a popular account of the inheritance of human characters. Curt Stern's *Principles of Human Genetics* is a clear, well-written text of general genetics with special emphasis on human inheritance. A discussion of twinning and the evidence from studies of fraternal and identical twins on the effects of heredity and environment in man is given in H. H. Newman, *Multiple Human Births.* A basic discussion of the principles of eugenics is S. J. Holmes, *Human Genetics and Its Social Import.* A thorough, rational and balanced discussion of the modern view of the eugenics problem is presented in Frederick Osborn's *Preface to Eugenics.* Descriptions of some of the inherited abnormalities are to be found in L. H. Snyder's *Medical Genetics* or in Muller, Little and Snyder, *Genetics, Medicine and Man.* Reed's *Counseling in Medical Genetics,* though written to tell physicians how to advise patients with genetic problems, is of great interest to any biologist.

Part Seven

Evolution

Chapter 33

Principles and Theories of Evolution

306. ORGANIC EVOLUTION

In the preceding chapters we have described some of the immense variety of living things that inhabit every conceivable place on land and in the water. These have a wide range of sizes, shapes, complexity of structure, and methods of getting food, evading predators, and reproducing. From the detailed comparison of the structures of living and fossil forms, from the sequence of the appearance and extinction of species in times past, from the physiologic and biochemical similarities and differences between species, and from the analysis of heredity and variation in many different plants and animals, has come one of the great unifying concepts of biology, that of **evolution.** Evolution is not a new topic at this point, however, for it has been fundamental, both implicitly and explicitly, to many of the subjects discussed previously.

The term evolution means an unfolding or unrolling—a gradual, orderly change from one condition to another. The planets and stars, the earth's topography, the chemical compounds of the universe, and even the chemical elements, made up of electrons, protons, neutrons and other subatomic particles, have undergone gradual, orderly changes sometimes called **inorganic evolution.** The principle of **organic evolution,** which is simply the application of this concept to living things, states that all the various plants and animals existing at the present time have descended from simpler organisms by gradual modifications which have accumulated in successive generations.

It is important to realize that evolution has not ceased, but is occurring more rapidly today than in many of the past ages. In the last few hundred thousand years, hundreds of species of animals and plants have become extinct and other hundreds have arisen. Although the process is usually too gradual to be observed, there are notable examples of evolutionary change within the time of recorded history. For example, early in the fifteenth century a litter of rabbits was released on Porto Santo, a small island near Madeira. Since there were no other rabbits and no carnivorous enemies on the island, the rabbits multiplied with amazing speed and by the nineteenth century were strikingly different from the ancestral European stock. They were only half as large as their European relatives, had a different color pattern, and were more nocturnal. More importantly, they could not produce offspring when bred with members of the European species. Within four hundred years, then, a new species of rabbit had developed.

The idea that the present forms of life arose from earlier, simpler ones was not new when Darwin's *The Origin of Species* was published in 1859. Elements of the theory of organic evolution are to be found in the writings of the early Greek philosophers, Thales (624–548 B.C.), Anaximander (588–524 B.C.), Empedocles (495–435 B.C.) and Epicurus (341–270 B.C.) The spirit of this age of Greek philosophy was somewhat like our own, for simple, natural explanations were sought for all phenomena. Little biology was known and their ideas about evolution were extremely vague; they can scarcely be said to foreshadow our present theory of organic evolution. Aristotle (384–322 B.C.), who was a great biologist as well as a philosopher, worked out an elaborate theory of gradually evolving life forms, according to the metaphysical belief that nature strives to change from the simple and imperfect to the more complex and perfect. The Roman poet Lucretius (99–55 B.C.) also gave an evolutionary explanation of the origin of plants and animals in his poem *De Rerum Natura*. With the Renaissance, interest in the natural sciences quickened, and from the fourteenth century on an increasingly large number of people found the concept of organic evolution reasonable. In *The Origin of Species* Darwin lists about twenty thinkers who had considered the theory seriously, among them his own grandfather, Erasmus Darwin (1731–1802), and the French scientist Lamarck (1744–1829).

Long before the Renaissance men had discovered odd fragments resembling bones, teeth and shells in the ground. Some of these corresponded to parts of familiar, living animals, but others were strangely unlike any known form. Many of the objects found in rocks high in the mountains resembled parts of marine animals. Leonardo da Vinci, in the fifteenth century, correctly interpreted these curious finds, and gradually people accepted his explanation that they were the remains of once-existing animals that had become extinct. Such evidence of former life suggested the theory of **catastrophism,** the idea that a succession of fires and floods have periodically destroyed all living things and necessitated the repopulation of the world by acts of special creation.

Three Englishmen in the eighteenth and early nineteenth centuries laid the foundations of modern geology, and by their careful, cogent arguments advanced the theory of **uniformitarianism.** James Hutton, in 1785, developed the concept that the geologic forces at work in the past were the same as those of the present. He arrived at this conclusion after a careful study of the erosion of valleys by rivers, and the formation of sedimentary deposits at the river mouths. He demonstrated that the processes of erosion, sedimentation, disruption and uplift, over long periods of time, could account for the formation of the fossil-bearing rock strata. In 1802 John Playfair's *Illustrations of the Huttonian Theory of the Earth* was published, in which he gave further explanation and examples of the idea of uniformitarianism in geologic processes. Sir Charles Lyell, one of the most influential geologists of his time, did much in his *Principles of Geology* (1832) to establish the principle of uniformity. By demonstrating the validity of geologic evolution he proved irrefutably that the earth is much older than a few thousand years, old enough for the processes of organic evolution to have occurred. He was a close personal friend of Darwin's, with great influence on his thinking, and through his work he paved the way for, and made possible the ideas presented in, *The Origin of Species*.

Jean Baptiste de Lamarck. The earliest theory of organic evolution to be logically developed was that of Jean Baptiste de Lamarck, the great French zoologist whose *Philosophie Zoologique* was published in 1809. Lamarck, like most biologists of his time, was a vitalist; he believed that organisms are guided through their lives by an innate and mysterious force which enables them to overcome handicaps in the environment. He believed too that these adaptations once made are transmitted from generation to generation—that is, that acquired characteristics are inherited. Developing the notion that new organs arise in response to demands of the environment, he went on to state that their size is proportional to their "use or disuse," and that changes in size also pass to successive generations. To illustrate this, Lamarck explained the evolution of the giraffe's long neck by suggesting that an ancestor took to browsing on the leaves of trees, instead of on grass, and in reaching up, stretched and elongated its neck; its offspring then supposedly inherited the longer neck.

The Lamarckian theory nicely explains the complete adaptation of many plants and animals to the environment, but it is unacceptable because overwhelming genetic evidence indicates that *acquired characteristics cannot be inherited*. Many experiments have been performed in an attempt to demonstrate the inheritance of acquired traits, but all have ended in failure. From what we now know about the mechanism of heredity, it is obvious that acquired traits cannot be inherited, for such characteristics are in the *body* cells only, whereas an inherited trait is transmitted by the *gametes*—the eggs and sperm.

Background for The Origin of Species. Darwin's contribution to the body of scientific knowledge was twofold: he presented a mass of detailed evidence and cogent argument to prove that organic evolution has occurred, and he devised a theory—that of **natural selection**—to explain how it operates.

Although his university training was in theology, Darwin was extremely interested in both biology and geology, and, while at Cambridge, became acquainted with Professor Henslow, the naturalist. Through him, shortly after leaving college, Darwin was appointed to the position of naturalist on the ship *Beagle*, which was to make a five-year cruise around the world so that oceanographic charts could be made for the British navy. Darwin studied the animals, plants and geologic formations of the East and West coasts of South America, making extensive collections and notes. The *Beagle* then went to the Galápagos Islands, west of Ecuador, where Darwin continued his observations of the flora and fauna, comparing them to the ones on the South American mainland. It was these observations which led him to reject the theory of special creation.

According to his journal, the idea of natural selection occurred to Darwin shortly after his return to England in 1836, but he spent the next twenty years or so accumulating the vast body of facts which eventually became *The Origin of Species*. In 1858 he received a manuscript from Alfred Russell Wallace, a young naturalist who was studying the distribution of plants and animals in the East Indies and the Malay Peninsula. Wallace set forth in this paper the idea of natural selection, which he had reached independently, stimulated, as Darwin had been, by Malthus' book on population growth and pressure, and the struggle for existence. By mutual agreement, Darwin and Wallace presented a joint paper on their theory at the meeting of the Linnean Society in London in 1858, and Darwin's monumental work was published the next year.

307. THE THEORY OF NATURAL SELECTION

Darwin's explanation for the way in which evolution takes place can be summarized as follows:

1. Variation is characteristic of every group of animals and plants, and there are many ways in which organisms differ. (Darwin did not know the cause of variation, and assumed it was one of the innate properties of living things. We now know that inherited variations are produced by mutations.)

2. More organisms of each kind are born than can possibly obtain food and survive. Yet, since the number of each species remains fairly constant under natural conditions, it must be assumed that most of the offspring in each generation perish. If all the offspring of any species remained alive and reproduced, they would soon crowd all other species from the earth.

3. Since more individuals are born than can survive, there is a struggle for survival, a competition for food and space. This may be an active kill-or-be-killed contest, or one less immediately apparent but no less real, such as the struggle of plants or animals to survive drought or cold.

4. Of the many variations exhibited by living things, some make it easier to survive in the struggle for existence, while others cause their owners to be eliminated. This idea of "the survival of the fittest" is the core of natural selection.

5. The surviving individuals will give rise to the next generation, and in this way the "successful" variations are transmitted to the next generation, and the next.

Successive generations in this way tend to become better adapted to their environment; as the environment changes further adaptations follow. As natural selection continues to operate over many years, later descendants may be quite different from their ancestors, different enough to be a separate species. Furthermore, certain members of a population with one group of variations may become adapted to environmental changes in one way, while other members, with a different set of variations, may become adapted in a different way, so that two or more species may arise from a single ancestral species.

Animals and plants exhibit many variations which are neither a help nor a hindrance to them; these will not be affected directly by natural selection, but, of course, will be transmitted to succeeding generations.

Darwin's theory of natural selection was so reasonable and well supported that most biologists soon accepted it. One of the early objections was that the theory did not explain the appearance of many apparently useless structures in an organism. It is now known that many of the visible differences between species are not important for survival, but are simply incidental effects of genes that have invisible, physiologic effects of great survival value. Other nonadaptive differences may be controlled by genes closely linked in the chromosome to other genes controlling characteristics important for survival.

Another objection was that new variations would be lost by "dilution" as the individuals showing them bred with others without them. We now know that although the phenotypic expression of a gene may be altered when it exists in combination with certain other genes, the fundamental nature of the gene itself is not altered or diluted, and the genes emerge unchanged in succeeding generations.

308. MODERN CHANGES IN THE THEORY OF NATURAL SELECTION

Rediscovery of Mendel's laws in 1900 made necessary two significant corrections of Darwin's theory of natural selection: (1) only *inherited* variations can provide the raw material for natural selection, and (2) there must be geographic or genetic *isolation* of incipient species to prevent interbreeding.

Modifications and Mutations. Darwin did not recognize that the variations of living things have their roots in one of two sources: some (physical or chemical) action of the environment on the developing embryo, or some alteration of the hereditary materials—the genes or chromosomes. Variations resulting from the first of these two situations—called **modifications**—are not inheritable and are not significant for evolution, but variations arising from changes in the genes or chromosomes—called **mutations**—are the raw materials for evolution by natural selection. Obviously, then, evolution cannot take place without mutations, and although natural selection does not create new characteristics, it plays an important part in determining which of them shall survive.

Isolation. Darwin realized that for a new species to come into existence a group of individuals must be differentiated from others of their kind by the accumulation of new characteristics, but just as he did not understand the significance of heredity in the process, so he overlooked another important factor—**isolation.** It is now well established that for a group to become differentiated, the organisms must first be prevented from breeding with their relatives and so passing on to them whatever new genes may have appeared. The only way to prevent interbreeding is by some form of isolation.

Perhaps the most common type of isolation is **geographic,** whereby groups of related organisms become separated by some physical barrier such as a sea, mountain, desert, glacier or river. In mountainous regions the individual ranges provide effective barriers between the valleys, and there are usually more different species for a given area than on the plains. In the mountains of western United States twenty-three species and subspecies of rabbits are known, while in the larger plains area of the Midwest and East there are only eight species. Valleys only a short distance apart, but separated by ridges perpetually covered with snow, have species of plants and animals peculiar to them. One of the most striking examples of geographic isolation is provided by the area divided by the Isthmus of Panama. On either side of the Isthmus, the phyla and classes of marine invertebrates are made up of different, but closely related species—a situation brought about by the fact that for some 16,000,000 years, during the Tertiary period, there was no connection between North and South America. This made it possible for animals to migrate freely between what is now the Gulf of Mexico and the Pacific Ocean. With the emergence of the Isthmus of Panama the closely related groups of animals were isolated, and the differences between the two fauna today represent the subsequent accumulation of hereditary differences.

Geographic isolation is usually not permanent; hence two previously isolated groups may come into contact again and interbreed unless **genetic isolation** or sterility has arisen in the meantime. The various races of man are the result of isolation and the accumulation of chance mutations, but since interracial sterility has not developed, the differences begin to disappear rapidly when geographic isolation breaks down. That they do not disappear even more quickly and completely is due largely to social taboos against intermarriage—itself a form of isolation.

Because genetic isolation is due to one or more mutations occurring by chance, independently of other mutations, it may arise only after long geographic isolation has produced marked differences between two groups of organisms, or it may arise within a single, otherwise homogeneous group. Such a mutation occurred in a species of fruit fly, *Drosophila pseudoobscura,* producing two groups of flies, externally indistinguishable, yet completely sterile when mated with each other. The two groups are isolated as effectively as if they lived on different continents, and as generations pass and different mutations accumulate by chance and selection, they will undoubtedly become visibly different. Biologists usually do not consider two closely related but different groups of organisms to be different species unless this genetic isolation has arisen.

Another type of isolation, called **ecologic,** depends upon the fact that two groups of animals living in the same geographic area may occupy different habitats. Marine animals living only in the intertidal zone are effectively isolated from others living only a few feet away, below the low-tide mark. Or ecologic isolation may be due to the fact that two groups breed at different times of the year.

Darwin originally assumed that in any group the variation of a particular character would continue, so that natural selection would operate indefinitely. But from the facts of heredity presented previously it should be clear that selection can operate only until the group becomes homozygous for the genes for that trait—for example, large body size. After that condition has been reached, neither artificial nor natural selection can do more until further mutations for large body size have occurred.

It must be emphasized that natural selection operates upon the organism as a whole, rather than on individual traits. One organism may survive despite obviously disadvantageous characters, while another may be eliminated despite traits extremely advantageous for getting along in life. The animals and plants that win the struggle for existence are usually not perfectly adapted to their environment, but have qualities the sum total of which renders them a little better able to survive and reproduce than their competitors. But since the environment itself changes from time to time, a characteristic of adaptive value at one period may be useless or even detrimental at another.

Chance, as well as natural selection, plays a part in determining whether a new trait will be passed from the single individual having it to the members of the group as a whole. During the process of meiosis the gene responsible for the character may or may not be included in the gametes. Even if it is included, a series of unlucky accidents may eliminate the few, early organisms having it, so that in spite of high survival value it disappears.

309. GENETIC DRIFT

One of the ways chance may play a role in evolution has been described by Sewall Wright as "genetic drift." This is the tendency, within small interbreeding populations, for heterozygous gene pairs to become homozygous for one allele or the other by chance rather than by selection. This may lead to the accumulation of certain disadvantageous characters and the subsequent elimination of the group. Investigations have shown that many animal populations in nature are subdivided into subgroups small enough to be affected by genetic drift.

Genetic drift is an exception to the Hardy-Weinberg Law (p. 483), the tendency for a population to maintain its proportion of homozygous and heterozygous individuals. The Hardy-Weinberg Law is based on statistical events and, like all statistical laws, does not hold true for small numbers. Genetic drift may explain the common observation that closely related species in different parts of the world frequently differ in curious ways which have no evident adaptive value.

310. PREADAPTATION

One of the recent modifications of the theory of natural selection is the theory of **preadaptation.** Because mutations occur at random, some result in characteristics either unimportant or disadvantageous to the organism in its usual environment. If, however, the environment changes or the organism migrates to a new location, these same traits may be of marked value for survival. In effect, an animal or plant may be adapted to an environment before being exposed to it. Suppose that a mutation in a fish occurs causing both eyes to be on one side of the skull. Obviously, if the fish continues its old habits, this will be a handicap. But if the animal takes to lying on its side on the bottom of the sea and grubbing in the mud for food, the new arrangement will be advantageous. This mutation actually has occurred in the flounder.

Preadaptation reasonably explains such occurrences as the evolution of land forms. For example, in a species of fish inhabiting a lake or river of the Devonian period, some 350,000,000 years ago, mutations may have occurred for the formation of primitive lungs and for changing the fan-shaped fins to sturdier, limblike fins with a fleshy lobe at the base. Such changes would have had no survival value for the fish and might have been deleterious—by interfering with its ability to swim—as long as it lived in a lake or stream. The Devonian age was one of violent climatic changes, in which seasons of drought alternated with seasons of rain. As the streams dried up, the water became stagnant and lacked enough dissolved oxygen for gills to function in respiration. An animal with lungs, however, could come to the surface, take a gulp of air, and obtain oxygen by diffusion across the membrane lining the lungs. Then, when the pond or stream dried up completely, he could use his sturdy, lobe-shaped fins to walk across the intervening land to some other stream. Some process such as this probably began the conquest of the land by vertebrates. Certainly the first vertebrates to venture out of the water onto land were not seeking air, for their ancestors already had lungs and could get air by coming to the surface; nor were they fleeing predators, for they were among the largest animals alive at the time. Since they ate other fish, and the only food on land was insects and plants, it can hardly be supposed that they were seeking food. Paradoxically, the first vertebrates to come out on land were probably looking for water, their own pond having just dried up.

311. MUTATIONS: THE RAW MATERIAL OF EVOLUTION

Mutation Theory of de Vries. The Dutch botanist Hugo de Vries, one of the three rediscoverers of Mendel's laws, was the first to emphasize the importance in evolution of sudden, large changes rather than the gradual accumulation of many small changes postulated by Darwin. De Vries experimented with a number of plants, especially the evening primrose, which grew wild in Holland. When he transplanted these plants into his garden and bred them, he found that many unusual forms, differing markedly from the original wild plant, appeared and bred true thereafter. For these sudden changes in the characters of an organism (which had been called "sports" by earlier breeders) he coined the name **"mutation."** Darwin had referred to a number of such changes, but believed that they occurred too rarely to be important in evolution, and that they might upset the harmonious relations between the various parts of the organism, and its adaptation to the environment. Countless breeding experiments with plants and animals since 1900 have shown that such mutations occur constantly and that their effects may be of adaptive value. With the development of the gene theory the term "mutation" has come to refer to sudden, discontinuous, random changes in the genes and chromosomes, although it is still used to some extent to refer to the new type of plant or animal.

In the plants and animals most widely used in breeding experiments—corn and fruit flies—some 400 and 600 mutations, respectively, have been observed in the past forty years. The fruit fly mutations are tremendously varied, including all shades of body color from yellow through brown and gray to black; red, white, brown or purple eyes; peculiarly shaped wings (crumpled, curled or shortened) and a complete absence of wings; oddly shaped legs and bristles; and such extraordinary arrangements as a pair of legs growing from the forehead in place of the antennae (Fig. 275). Among domestic animals, mutations are no less common; the six-toed cats of Cape Cod and the short-legged breed of Ancon sheep are two of many examples of the persistence of a single mutation. Short legs were useful to the breeder for preventing the sheep from leaping over fences; hence the trait was selected for in breeding.

Early in the present century a heated discussion arose as to whether evolution is the result of natural selection or of mutations. As more was learned about heredity, it became clear that natural selection can operate only when there is something to be selected—in other words, when mutations present alternate ways of coping with the environment. The evolu-

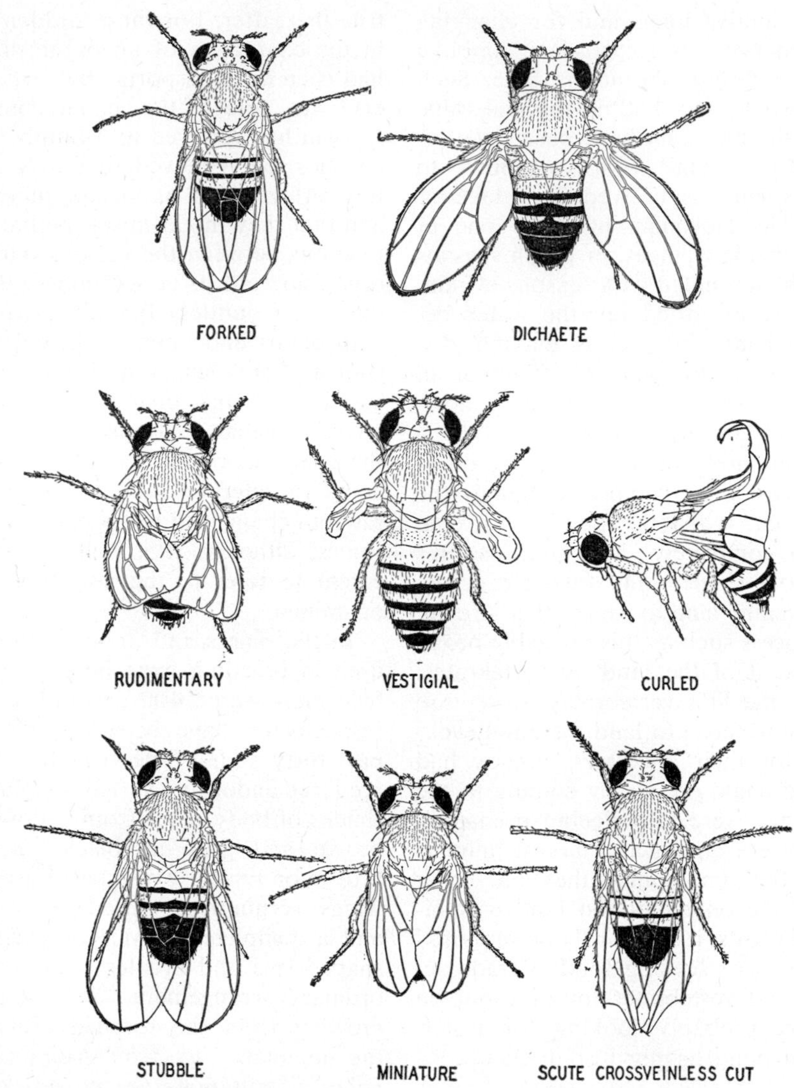

Figure 275. Some wing and bristle mutants in the fruit fly, *Drosophila melanogaster.* (Drawn by E. M. Wallace; from Sturtevant and Beadle: An Introduction to Genetics.)

tion of new species, then, involves both mutation and natural selection.

A similar argument has continued to the present day between the Neo-Darwinists, who believe that new species evolve by the gradual accumulation of small mutations, and another group, who believe that new species and genera arise in one step by a **macromutation** or major change in the genetic system. Such a macromutation, by producing a major change early in development, would result in an adult form vastly different from its parents—the new species or genus. Many major changes result only in monsters,* which die almost immediately, though some give rise to what Richard Goldschmidt of the University of California calls "hopeful monsters"—forms enabled by their mutation to occupy some new environment. Such a situation, he thinks, is responsible for the evolution of the extinct ancestral bird, *Archaeopteryx,* into the modern bird. The *Archaeopteryx* had a long, reptile-like (though feathered) tail. If, by a single mutation, that tail was

* The term "monster" refers to any form which departs markedly in structure from the usual type of the species, and does not necessarily imply ugliness.

greatly shortened, a "hopeful monster" with the fan-shaped arrangement of feathers might well have been the result. The new tail, then, which was better suited for flying than the old, long one, might have been selected, until today all birds have fan-shaped tails. There is, of course, no proof that this is how our present-day birds evolved, but there is much evidence that similar large skeletal changes do occur as the result of a single mutation. The Manx cat, for example, owes its stubby tail to a mutation in some ancestor, which caused the shortening and fusing of the tail vertebrae. Goldschmidt does not deny the accumulation of small mutations, but holds that they can lead only to varieties or geographic races and not to species, genera and the higher taxonomic divisions.

Types of Mutations. The distinction between chromosomal and point mutations is whether there is a visible change in the structure of the chromosome (p. 485). A chromosomal mutation may involve a change in the visible structure of one chromosome or in the total number of chromosomes per cell. There may be an addition or subtraction of a single chromosome from the usual diploid set, or there may be a doubling or tripling of the entire set, producing organisms called **polyploids,** which are usually larger and more robust than their parents. Changes in chromosome number are more frequent in plants than in animals, because, owing to the nature of the reproductive process in plants, there is a greater chance of their being passed from one generation to the next. Some of the cultivated varieties of tomatoes, corn and other plants owe their vigor and the large size of their fruit to the fact that they are polyploids.

Changes which may occur in the visible structure of a chromosome are the deletion of a small part, doubling of a small part, and the inverting (turning end for end) of a small or large section. Sometimes there is an interchange of segments, called a **translocation,** between two different chromosomes. Each of these changes may produce phenotypic changes and may be inherited in a mendelian fashion. In fact, many traits previously thought to be inherited by true genes have been found to arise from chromosomal mutations. This has led some geneticists to conclude that there are no genes in the sense of individual structures in the chromosome, but that the factors called genes are really chromosomal mutations.

Causes of Mutations. The causes of natural or spontaneous mutations are unknown. Both gene and chromosome mutations can be produced artificially by a variety of agents: x-rays; the alpha, beta and gamma rays emitted by radioactive elements; neutrons; heat and cold; ultraviolet rays; chemicals such as the war gas known as nitrogen mustard; and, in plants, the aging of seed. Cosmic and other rays bombarding the earth may cause some natural mutations, but since genes are complex, unstable molecules, it is possible that metabolic processes in the cell bring about some spontaneous mutations, without the intervention of external agents.

Whatever their cause, the role of mutations as the raw material for natural selection and evolution is now universally recognized. Both chromosomal and point mutations occur in wild populations as well as in laboratory stocks. They occur with a frequency great enough to account for the known rates of evolutionary change. It was formerly argued that the mutations observed in the laboratory have nothing to do with evolution, because (1) almost all are detrimental, and (2) the differences between species are usually slight variations, affecting many different parts of the organism and inherited by means of multiple factors, whereas the mutations observed in the laboratory are usually large variations, involving a single organ and inherited by single gene differences. However, in wild populations, as in the laboratory stocks, mutations at the present time are usually detrimental. The animals and plants living today are the result of a long process of natural selection which preserved the most beneficial mutations; they are, therefore, highly specialized and well adapted to their environment. Further mutations are much more likely to be disadvantageous than helpful.

A few laboratory mutations are beneficial and have survival value.

The second objection has been eliminated by further studies that showed that the slight variations differentiating species do appear in the laboratory, but, being difficult to detect, were missed in some of the earlier work. Indeed, we now know that the rate of such mutations is greater than that of the larger, more obvious ones.

312. STRAIGHT-LINE EVOLUTION

Many of the earlier paleontologists and other students of evolution were led to the conclusion that there are trends in evolution, that evolution tends to progress in a straight line. The term **orthogenesis** was coined to refer to straight-line evolution; some investigators had the somewhat mystical belief that organisms have an inherent tendency to evolve in a predetermined direction. Fuller examination of the accumulating fossil data has shown that many of the instances often quoted as examples of orthogenesis are not truly straight-line evolution at all. The horse is often said to have evolved in a straight line from the primitive *Hydracotherium* (a small animal, the size of a fox, with four toes on the front feet and three toes on the hind feet) to the modern *Equus,* but the complete fossil record shows that there were many side branches. The evolution of the horse was said to be characterized by an increase in size, a lengthening of the legs, an enlargement of the third digit and reduction of the others, an increase in the size of the molar teeth and in the complexity of the pattern of ridges on their crowns, and increases in the size of the lower jaw and skull. So many exceptions to each of these trends have been found that the concept of straight-line evolution of the horse has been abandoned.

The term orthogenesis is sometimes applied to the evolutionary overdevelopment of some characteristic. In successive generations of the now extinct Irish deer, for example, the antlers became larger and larger. Although this may have originally had adaptive value, the antlers eventually became so large, with a total spread of 11 feet, that the deer could not support them and the species became extinct.

The growth of our knowledge of how genes act in controlling development has enabled us to explain whatever straight-line trends in evolution may be real in terms of conventional evolution by mutation and selection. Many different types of developmental patterns may arise by random mutation, but most of them will result in unharmonious processes, which will not interdigitate properly and will lead to the death of the organism. Others, with no particular value for survival, will remain or be eliminated by chance. The ones most likely to survive are those which provide for further improvement in some peculiar adaptive structure already present. Thus orthogenetic series can be explained as the result of random mutation and selection occurring along one of the few possible lines of development. It now seems evident, too, why certain structures become too large and result in the extinction of the species. For we know that genes do not function independently, but must operate against the background of many other genes. Those controlling larger tusks, for instance, may function by causing the tusks to be *proportionately* larger than the rest of the body. If other genes cause the entire body size to increase, the tusks may become so large as to be unmanageable and, eventually, lethal to the animals possessing them.

313. THE ORIGIN OF SPECIES BY HYBRIDIZATION

A simpler and less spectacular way than mutation and natural selection by which species originate is through simple hybridization or the crossing of two different varieties or species.* Through such a mating the best characters of each of the original species may be combined into a single form, thereby creating a new type better able to survive than either of its

* In the chapter on plants we found that, generally speaking, species are not interfertile. But occasionally two different species can interbreed, a fact which makes a hard-and-fast definition of "species" difficult. A variety is a race within a species.

parents. Animal and plant breeders routinely use this method to establish new combinations of desirable characters.

When different species are crossed—ones with different chromosome numbers, for example—the offspring are usually sterile. The unlike chromosomes cannot pair properly in the process of meiosis, and the resulting eggs and sperm do not receive the proper assortment of chromosomes. But if, in such interspecific hybrids, a chromosome mutation occurs and doubles the chromosome number, meiosis can take place normally and normal, fertile eggs and sperm are produced. The hybrid will breed true thereafter and generally will not produce fertile offspring when bred with either of the parental species. Many related species of higher plants have chromosome numbers which are multiples of some basic number; there are species of wheat with fourteen, twenty-eight and forty-two chromosomes, species of roses with fourteen, twenty-eight, forty-two and fifty-six chromosomes, and species of violets with every multiple of six from twelve to fifty-four.

Evidence that such natural series probably arose by hybridization and chromosome doubling is supplied by laboratory experiments which have resulted in many similar series. One of the most famous of these experimental crosses is that made by Karpechenko, who crossed the radish with the cabbage, hoping perhaps to get a plant with a cabbage top and a radish root. Although radishes and cabbages belong to different genera, both have eighteen chromosomes, and the resulting hybrid also had eighteen chromosomes—nine from its radish parent and nine from its cabbage parent. The radish and cabbage chromosomes were unlike, and could not pair in meiosis, so the hybrid was almost completely sterile. By chance, however, a few of the eggs and sperm formed contained all eighteen chromosomes, and a mating between two of these resulted in a plant with thirty-six chromosomes. This new plant was fertile, for in meiosis the radish chromosomes paired and the cabbage ones paired; it exhibited some of the characteristics of each of its parents and bred true for them. Unfortunately, it had a radish-like top and a cabbage-like root. Since it could not be crossed readily with either of its parent species, it was in effect a new species, produced by hybridization and the doubling of the chromosome number.

An interesting example of a similar occurrence in nature is provided by a species of marsh grass, *Spartina townsendi.* This type of grass first appeared about a hundred years ago in the harbor at Southampton, England, in company with two other species, *S. stricta and S. alterniflora.* The new species, much more vigorous than either of the others, was soon widespread, and since it was especially valuable in collecting and holding soil, was transplanted to the Dutch dikes and other parts of the world. Because it was intermediate in many characteristics between the two species with which it was first found, it was believed to have originated as a hybrid. This was confirmed by an examination of the chromosome numbers which showed *S. townsendi* to have 126 chromosomes, *S. stricta,* fifty-six, and *S. alterniflora,* seventy. There is no doubt that the new species arose by hybridization and chromosome doubling.

314. THE ORIGIN OF LIFE

The modern theories of mutation, natural selection and population dynamics provide us with a satisfactory explanation of how the present-day animals and plants evolved from previous forms by descent with modification. The question of the ultimate origin of life on this planet has been given serious consideration by many different biologists. Some have postulated that some kind of spores or germs may have been carried through space from another planet to this one. This is unsatisfactory, not only because it begs the question of the ultimate source of these spores, but because it is extremely unlikely that any sort of living thing could survive the extreme cold and intense irradiation of interplanetary travel.

The concept that the first living things *did* evolve from nonliving things has been put forward by Pflüger, J. B. S. Haldane, R. Beutner, and especially by the Russian biochemist, A. I. Oparin, in his book, *The*

Origin of Life (1938). The earth originated some 2.5 billion to 4.5 billion years ago, either as a part broken off from the sun or by the gradual condensation of interstellar dust. Most authorities seem agreed that the earth at first was very hot and molten, and that conditions consistent with life arose only one billion or perhaps a billion and a half years ago. At that time the earth's atmosphere contained essentially no free oxygen; all the oxygen was combined as water and as oxides.

A number of reactions by which organic substances can be made from inorganic ones are known. It is believed that originally much of the earth's carbon was in the form of metallic carbides, which could react with water to form acetylene. This could subsequently polymerize to form compounds containing long chains of carbon atoms. It has been shown experimentally that high energy radiation, such as that of cosmic rays, can produce organic compounds. This has been demonstrated by M. Calvin, who irradiated solutions of carbon dioxide and water in a cyclotron and obtained formic, oxalic and succinic acids, which contain one, two and four carbons respectively. These are important intermediates in the metabolism of living organisms. Irradiation of solutions with ultraviolet light, or with electric charges to simulate lightning, also produces organic compounds. Harold Urey and Stanley Miller, at the University of Chicago, showed in 1953 that amino acids such as glycine and alanine, and even more complex organic substances, can be formed *in vitro* by exposing a mixture of water vapor, methane, ammonia and hydrogen gases to electric discharges for a mere week. All of these gases are believed to have been present in adequate amounts in the earth's atmosphere in prebiotic times.

The spontaneous origin of living things at the present time is believed to be extremely improbable, yet that this same event occurred in the past is quite probable. The difference lies in the conditions existing on the earth: the accumulation of organic molecules was possible before there were living things because there were no molds, no bacteria, no living things of any kind to bring about their decay. Furthermore, there was little or no oxygen in the atmosphere to bring about their spontaneous oxidation.

The details of the chemical reactions which would give rise, without the intervention of living things, to carbohydrates, fats and amino acids have been worked out by Oparin and extended by Calvin and others. Most of the reactions by which the more complex organic compounds were formed probably occurred in the sea, in which were dissolved and mixed the organic molecules formed. The sea, we may postulate, became a sort of dilute broth in which these molecules collided, reacted and aggregated to form new molecules of increasing size and complexity. The known forces of intermolecular attraction, and the tendency for certain molecules to form liquid crystals, provide us with means by which large, complex, specific molecules can form spontaneously. Oparin suggested that natural selection can operate at the level of these complex molecules, before anything recognizable as life is present. As the molecules came together to form colloidal aggregates, these aggregates began to compete with one another for raw materials. Some of the aggregates, which had some particularly favorable internal arrangement, acquired new molecules more rapidly than others and eventually became the dominant types.

Once some protein molecules had formed and had achieved the ability to catalyze reactions, the rate of formation of additional molecules would be greatly stepped up. Next, these complex protein molecules acquired the ability to catalyze the synthesis of molecules like themselves; they became autocatalytic. This hypothetical, autocatalytic particle would have some of the properties of a virus, or of a free-living gene. The next step in the development of a living thing is the addition of the ability of the autocatalytic particle to undergo inherited changes, to mutate. Then, if a number of these free "genes" had joined to form a single larger unit, the resulting organism would have been similar to certain present-day viruses. All the known viruses are parasites that can live only within the cells of higher animals and plants. However, a little reflection will suggest that free-living viruses, those which do not produce a disease, would be very difficult to detect; such organisms may indeed exist.

The first living organisms, having arisen in a sea of organic molecules and in contact with an atmosphere free of oxygen, presumably obtained energy by the fermentation of certain of these organic substances. These heterotrophs could survive only as long as the supply of organic molecules in the sea broth, accumulated from the past, lasted. Before the supply was exhausted, however, the heterotrophs evolved further and became autotrophs, able to make their own organic molecules by chemosynthesis or photosynthesis. One of the byproducts of photosynthesis is gaseous oxygen, and it is likely that all the oxygen in the atmosphere was produced and is still produced in this way. It is estimated that all the oxygen of our atmosphere is renewed by photosynthesis every 2000 years and all the carbon dioxide molecules pass through the photosynthetic process every 300 years. All the oxygen and carbon dioxide in the earth's atmosphere are the products of living organisms and have passed through living organisms over and over again in times past.

An explanation of how an autotroph may have evolved from these primitive heterotrophs was presented by N. H. Horowitz in 1945. This, simply stated, is that an organism might acquire, by successive mutations, the enzymes needed to synthesize complex from simple substances, in the reverse order to the sequence in which they are normally used. A model will make this clearer. Suppose that our first primitive heterotroph required organic substance A for its growth. Substance A, plus a variety of other organic compounds, B, C, D, etc., were present in the environment, having previously been made by the action of the nonliving factors of the environment. The heterotroph would survive nicely as long as the supply of substance A in the environment held out. If, by a mutation, a new enzyme appeared which enabled the heterotroph to make substance A out of the simpler substance B, the strain with this mutation would obviously be able to survive when the supply of substance A was exhausted. A second mutation, establishing an enzyme catalyzing the reaction by which substance B is made from substance C, even simpler and also abundant in the environment, again would have had great survival value when the supply of substance B in the environment was used up. Similar mutations, setting up enzymes enabling the organism to use successively simpler substances D, E, F, . . . down to some inorganic substance Z, would finally result in an organism able to make substance A, which it needs for growth, out of substance Z by way of all the intermediate compounds. When, by other series of mutations, the organism was finally able to synthesize all of its requirements from simple inorganic substances, as the green plants can, it would be an autotroph. And once the first simple autotrophs had evolved, the way was clear for the evolution of the vast variety of green plants, bacteria, molds and animals that inhabit the world today.

These considerations lead us to the conclusion that the origin of life, as an orderly natural event on this planet, was not only possible, it was almost inevitable. Furthermore, with the vast number of planets in all the known galaxies of the universe, many of them must have conditions which permit the origin of life. It is probable that there are many other planets on which life as we know it exists. Wherever life is possible, it should, if given enough time, appear and ramify into a wide variety of types. Some of these may be quite dissimilar from the ones on this planet, but others may be quite like those found here; some may, perhaps, be like ourselves.

It seems unlikely that we will ever know how life originated, whether it happened only once or many times, or whether it might happen again. This theory, that (1) organic substances were formed from inorganic substances by the action of physical factors in the environment; (2) that they interacted to form more and more complex substances, finally enzymes and then self-reproducing enzyme systems ("free genes"); (3) that these "free genes" diversified and united to form a primitive virus-like heterotroph; and (4) that autotrophs evolved from these primitive heterotrophs, at least has the virtue of being plausible, and possible. Certain parts of it are subject to experimental verification.

315. PRINCIPLES OF EVOLUTION

There is a great difference of opinion among investigators as to the nature of mutations, the kind of mutations involved in evolution, and the degree to which such factors as natural selection, isolation, genetic recombination, hybridization and the size of the breeding population affect the evolution of some particular organism, but there are several fundamental facts about which they are agreed. Let us review briefly what is actually known about how evolution operates. To begin with, we found that the basis of evolution is the changes within the genes and chromosomes, that some sort of isolation is necessary for the setting up of new species, and that natural selection is involved in the survival of some, but not all, of the mutations which occur. In addition, there are five principles of evolution to which nearly all scientists subscribe:

1. Evolution occurs more rapidly at some times than at others. At the present time it is occurring rapidly, with many new forms appearing and many old ones becoming extinct.

2. Evolution does not proceed at the same rate among different types of organisms. At one extreme are the lampshells or Brachiopods, some species of which have been exactly the same for the last 500,000,000 years at least, for fossil shells found in rocks deposited at that time are identical with those of animals living today. In contrast, several species of man have appeared and become extinct in the past few hundred thousand years. In general, evolution occurs rapidly when a new species first appears, and then gradually slows down as the group becomes established.

3. New species do not evolve from the most advanced and specialized forms already living, but from relatively simple, unspecialized forms. Thus the mammals did not evolve from the large, specialized dinosaurs, but from a group of small, unspecialized reptiles.

4. Evolution is not always from the simple to the complex. There are many examples of "regressive" evolution, in which a complex form has given rise to simpler ones. Most parasites have evolved from free-living ancestors which were more complex than the present forms; wingless birds, such as the cassowary, have descended from birds that could fly; many wingless insects have evolved from winged ones; the legless snakes came from reptiles with appendages; the whale, which has no hind legs, evolved from a mammal that had two pairs of legs. These are all reflections of the fact that mutations occur at random, and not necessarily from the simple to the complex or from the imperfect to the perfect. If there is some advantage to a species in having a simpler structure, or in doing without some structure altogether, any mutations which happen to occur for such conditions will tend to accumulate by natural selection.

5. Evolution occurs by populations, not by individuals, by the processes of mutation, natural selection and genetic drift.

QUESTIONS

1. What is the theory of uniformitarianism?
2. Explain the concept of organic evolution.
3. What is wrong with Lamarck's theory of adaptation?
4. If Darwin did not originate the theory of evolution, what contributions to it did he make?
5. Explain the theory of natural selection.
6. What is meant by the "survival of the fittest"? Do you think it applies to human populations?
7. Why are only inherited changes important in the evolutionary process?
8. State the theory of preadaptation.
9. After a mutation has occurred, what must occur for that trait to become established in the population?
10. What is meant by "genetic drift"? What role does it play in evolution?
11. From what you have learned of heredity and evolution, how do you think new species arise, by the accumulation of small mutations or by a few mutations with

large phenotypic effects? Give your reasons for thinking as you do.
12. Differentiate the several types of mutation. How may mutations be produced?
13. Why do nearly all the mutations occurring at the present time have a detrimental effect?
14. Contrast hybridization with other ways in which new species may be produced.
15. What do you think about the belief, held by many people, that the human race is becoming better and better all the time, and is gradually evolving toward an ultimate goal of physical and moral excellence?

SUPPLEMENTARY READING

Darwin's classic, *The Origin of Species,* is available in a number of modern editions and is well worth sampling for its clear logical arguments and wealth of examples. The impact of the theory of evolution on Victorian England and a vivid portrayal of Thomas Huxley's championing of Darwin's theory is presented in William Irvine's *Apes, Angels and Victorians.* Henry Fairfield Osborn's *From the Greeks to Darwin* is an interesting history of ideas on evolution. A. D. White's *History of the Warfare of Science with Theology* describes some of the reactions to the statement of the theory of evolution. A good, nontechnical presentation of our present ideas of evolution is George Gaylord Simpson's *The Meaning of Evolution.* Technical books on special phases of evolution are Carter's *Animal Evolution: A Study of Recent Views on Its Causes,* Stebbins' *Variation and Evolution in Plants,* Simpson's *The Major Features of Evolution,* which discusses the paleontologic and genetic aspects of evolution, Dobzhansky's *Genetics and the Origin of Species,* which presents the Neo-Darwinian viewpoint of the importance of natural selection, and Goldschmidt's *The Material Basis of Evolution,* which gives the detailed argument for the importance of large mutations in evolution. Theories of the origin of life are discussed in Oparin's *The Origin of Life* and Blum's *Time's Arrow and Evolution.* Two very readable, short discussions of the origin of life are given in Melvin Calvin's *Chemical Evolution and the Origin of Life,* and George Wald's *The Origin of Life.*

Chapter 34

The Fossil Evidence for Evolution

THE EVIDENCE that organic evolution has occurred is so overwhelming that no one who is acquainted with it has any doubt that new species are derived from previous ones by descent with modification. The fossil record provides direct evidence of organic evolution and gives the details of the evolutionary relationships of many lines of descent. There are, moreover, a great many facts from all of the subdivisions of biological science which acquire significance, and make sense, only when viewed against the background of evolution.

316. PALEONTOLOGY

The science of **paleontology** deals with the finding, cataloguing and interpretation of the abundant and diverse evidence of life in former times. The term **fossil** (Latin, *fossilium,* something dug up) refers not only to the bones, shells, teeth and other hard parts of a plant or animal body which have been preserved, but to any impression or trace left by some previous organism. Footprints or trails made in soft mud, which subsequently hardened, are a common type of fossil. From such remains one can tell something of the structure and body proportions of the animals which made them. In 1948, tracks were discovered near Pittsburgh of an amphibian from the Pennsylvanian period, some 250,000,000 years ago. That the animal moved by hopping, rather than by walking, was evident from the fact that the footprints lay opposite each other, in pairs.

Most of the vertebrate fossils are skeletal parts, from which it is possible to deduce the animal's posture and style of walking. From the bone scars, indicating muscle attachments, paleontologists can deduce the general position and size of the muscles and, from this, the contours of the body. On the basis of such considerations, reconstructions are made of how the animals looked in life. Such qualities as the texture and color of the fur or scales must be guessed at.

An interesting and striking type of fossil is that in which the original hard parts and, more rarely, soft tissues have been replaced by minerals—a process known as **petrifaction.** The replacing minerals may be iron pyrites, silica, calcium carbonate or other substances. The petrified muscle from a shark, more than 300,000,000 years old. was so well preserved by this process that not only individual muscle fibers, but their cross striations could be observed in thin sections under the microscope. The Petrified Forest in Arizona is a famous example of the process of petrifaction.

Molds and casts are superficially similar to petrified fossils but are produced differently. **Molds** were formed by the hardening of the material surrounding the buried organism, followed by the decay and removal of the organisms by seepage of the ground water. Sometimes the molds were subsequently filled with minerals which, in turn, hardened to form **casts**—replicas of the original structures.

Occasionally, paleontologists are fortunate enough to find organisms frozen in the soil or ice of the far North, usually in Siberia and Alaska. The remains of woolly mammoths more than 25,000 years old have been found so well preserved that their flesh was eaten by dogs. Other forms—plants, insects and spiders—have been preserved in amber, a fossil resin from pine trees. Originally the resin was a sap soft enough to engulf the fragile insect and penetrate every part; then it gradually hardened, preserving the animal intact (Fig. 276).

The formation and preservation of a fossil require that some structure be buried. This may take place at the bottom of a body of water, or on land by the accumulation of wind-blown sand, soil or volcanic ash. Many of the men and animals living in Pompeii were preserved almost perfectly by the volcanic ash from the eruption of Vesuvius. Sometimes animals were trapped and entombed in a bog, quicksand or asphalt pit, such as the famous La Brea tar pits in Los Angeles which have provided superb fossils of Pleistocene animals.

317. THE GEOLOGIC TIME TABLE

Geologists have found five major rock strata, each of which is subdivided into lesser strata, and on this basis the geologic time table of eras, periods and epochs was constructed (Table 15). The strata were formed by the accumulation of sediment—sand or mud—at the bottom of lakes, seas or oceans. The duration of each period or epoch can be estimated from the relative thickness of the sedimentary deposits, although, obviously, the rate of deposition varied at different times and in different places.

The layers of sedimentary rock should occur in the sequence of their deposition, with the newer, later strata on top of the older, earlier ones, but subsequent geo-

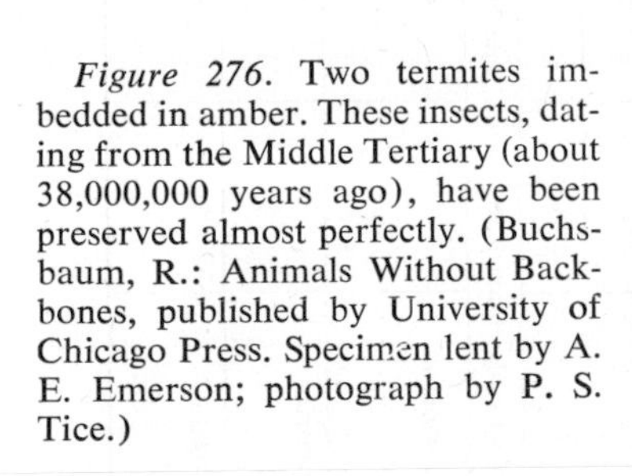

Figure 276. Two termites imbedded in amber. These insects, dating from the Middle Tertiary (about 38,000,000 years ago), have been preserved almost perfectly. (Buchsbaum, R.: Animals Without Backbones, published by University of Chicago Press. Specimen lent by A. E. Emerson; photograph by P. S. Tice.)

Table 15. GEOLOGIC TIME TABLE

ERA	PERIOD	EPOCH	DURATION IN MILLIONS OF YEARS	TIME FROM BEGINNING OF PERIOD TO PRESENT (MILLIONS OF YEARS)	GEOLOGIC CONDITIONS	PLANT LIFE	ANIMAL LIFE
Cenozoic (Age of Mammals)	Quaternary	Recent	0.025	0.025	End of last ice age; climate warmer	Decline of woody plants; rise of herbaceous ones	Age of man
		Pleistocene	1	1	Repeated glaciation; 4 ice ages	Great extinction of species	Extinction of great mammals; first human social life
	Tertiary	Pliocene	19	20	Continued rise of mountains of western North America; volcanic activity	Decline of forests; spread of grasslands; flowering plants, monocotyledons developed	Man evolved from manlike apes; elephants, horses, camels almost like modern species
		Miocene	15	35	Sierra and Cascade mountains formed; volcanic activity in northwest U. S.; climate cooler		Mammals at height of evolution; first manlike apes
		Oligocene	10	45	Lands lower; climate warmer	Maximum spread of forests; rise of monocotyledons, flowering plants	Archaic mammals extinct; rise of anthropoids; forerunners of most living genera of mammals
		Eocene	20	65	Mountains eroded; no continental seas; climate warmer		Placental mammals diversified and specialized; hoofed mammals and carnivores established
		Paleocene	10	75			Spread of archaic mammals
	Rocky Mountain Revolution (Little Destruction of Fossils)						
Mesozoic (Age of Reptiles)	Cretaceous		60	135	Andes, Alps, Himalayas, Rockies formed late; earlier, inland seas and swamps; chalk, shale deposited	First monocotyledons; first oak and maple forests; gymnosperms declined	Dinosaurs reached peak, became extinct; toothed birds became extinct; first modern birds; archaic mammals common
	Jurassic		30	165	Continents fairly high; shallow seas over some of Europe and western U. S.	Increase of dicotyledons; cycads and conifers common	First toothed birds; dinosaurs larger and specialized; insectivorous marsupials
	Triassic		60	225	Continents exposed; widespread desert conditions; many land deposits	Gymnosperms dominant, declining toward end; extinction of seed ferns	First dinosaurs, pterosaurs and egg-laying mammals; extinction of primitive amphibians

	Appalachian Revolution (Some Loss of Fossils)						
Paleozoic (Age of Ancient Life)	Permian		15	240	Continents rose; Appalachians formed; increasing glaciation and aridity	Decline of lycopods and horsetails	Many ancient animals died out; mammal-like reptiles, modern insects arose
	Pennsylvanian		35	275	Lands at first low; great coal swamps	Great forests of seed ferns and gymnosperms	First reptiles; insects common; spread of ancient amphibians
	Mississippian		50	325	Climate warm and humid at first, cooler later as land rose	Lycopods and horsetails dominant; gymnosperms increasingly widespread	Sea lilies at height; spread of ancient sharks
	Devonian		50	375	Smaller inland seas; land higher, more arid; glaciation	First forests; land plants well established; first gymnosperms	First amphibians; lungfishes, sharks abundant
	Silurian		50	425	Extensive continental seas; lowlands increasingly arid as land rose	First definite evidence of land plants; algae dominant	Marine arachnids dominant; first (wingless) insects; rise of fishes
	Ordovician		80	505	Great submergence of land; warm climates even in Arctic	Land plants probably first appeared; marine algae abundant	First fishes, probably fresh-water; corals, trilobites abundant; diversified molluscs
	Cambrian		80	585	Lands low, climate mild; earliest rocks with abundant fossils	Marine algae	Trilobites, brachiopods dominant; most modern phyla established
	Second Great Revolution (Considerable Loss of Fossils)						
Proterozoic			1000	1500	Great sedimentation; volcanic activity later; extensive erosion, repeated glaciations	Primitive aquatic plants—algae, fungi	Various marine protozoa; towards end, molluscs, worms, other marine invertebrates
	First Great Revolution (Considerable Loss of Fossils)						
Archeozoic			2000	3500	Great volcanic activity; some sedimentary deposition; extensive erosion	No recognizable fossils; indirect evidence of living things from deposits of organic material in rock	

logic events may have changed the relationship of the layers. Not all the strata occur in any one region, for some lands were exposed when others were submerged. In some regions the strata formed previously have subsequently emerged and eroded away so that relatively recent strata were then deposited directly upon very ancient ones. Moreover, certain sections of the earth's crust have undergone tremendous foldings and splittings, so that early layers may now rest on top of newer ones. Sometimes the age of a rock stratum can be determined by a study of its fossil content, for some kinds of fossils were deposited in only one era or period.

Rock deposits are now dated largely by taking advantage of the fact that certain radioactive elements are transformed into other elements at rates which are slow and essentially unaffected by the temperatures and pressures to which the rock has been subjected. Half of a given sample of uranium will be converted into lead in 4.5 billion years and, by measuring the proportion of uranium and lead in a given rock, an accurate estimate of the absolute age of the rock can be made. By this method the oldest rocks of the earliest geologic period are calculated to be about 3,500,000,000 years old, and the latest Cambrian rocks to be 500,000,000 years old. Events in more recent times can be dated by the decay of radioactive carbon-14, which has a half-life of 5568 years.

Relatively short periods of geologic time can be determined by measuring how rapidly waterfalls are receding (Niagara Falls is moving upstream at the rate of about 5 feet a year, as it wears away the rock over which it tumbles) or by counting the yearly deposits of clay on the bottom of lakes and ponds. Advances in isotopic techniques have made possible some astonishing conclusions in the field of geology. For example, it has been found that the proportion of the various oxygen isotopes in the calcium carbonate secreted by living organisms depends upon the temperature; consequently, by analyzing the calcium carbonate of fossil shells, it is possible to estimate the temperature of the sea in which those animals lived, hundreds of millions of years ago.

Between the major eras there were widespread geologic disturbances, called **revolutions,** which raised or lowered vast regions of the earth's surface and created or eliminated shallow inland seas. These revolutions changed the distribution of sea and land organisms and wiped out many of the previous life forms. The era known as the Paleozoic ended with the revolution that raised the Appalachian mountains and, it is thought, killed all but 3 per cent of the then existing forms of life. Similarly, the Rocky Mountain Revolution, which raised the Andes, Alps and Himalayas, as well as the Rockies, resulted in the annihilation of most of the reptiles of the Mesozoic era.

318. THE GEOLOGIC ERAS

Archeozoic Era. The rocks of the oldest geologic era are very deeply buried in most parts of the world, but are exposed at the bottom of the Grand Canyon and along the shores of Lake Superior. The oldest era does not begin with the origin of the earth but with the formation of the earth's crust, when rocks and mountains were already in existence and the processes of erosion and sedimentation had begun. This era, which lasted about two billion years, was as long as all the other eras combined. Apparently it was marked by catastrophic and widespread volcanic activity and deep-seated upheavals which climaxed in the raising of mountains. The heat, pressure and churning associated with these movements probably destroyed most of the fossils, but some evidence of life still remains. Scattered throughout the Archeozoic rocks are traces of graphite or pure carbon, which are probably the transformed remains of plant and animal bodies. If the amount of graphite in these rocks can be taken as a measure of the amount of living things (and this seems to be valid), then life must have been abundant in the Archeozoic, for there is more carbon in these rocks than in the coal beds of the Appalachians.

Proterozoic Era. The second era, some one billion years in length, was characterized by the deposition of large quantities of sediment, and by at least one great period of glaciation, during which ice sheets stretched to within 20 degrees of the equator. Only a few fossils have been found in Proterozoic rocks but they show not only that life was present but that evolution had proceeded quite far before the end of the era. Sponge spicules, jellyfish and the remains of fungi, algae, brachiopods, arthropods and worm tubes have been recovered from rocks of the era.

Paleozoic Era. Between the strata of the late Proterozoic and the earliest layers of the third major era, the Paleozoic, is a considerable gap, caused by a geologic revolution. During the 360,000,000 years of the Paleozoic, members of every phylum and class of animals appeared except the birds and mammals. Since some of these animals existed for a relatively short time, their fossils enable geologists to correlate rocks of the same era found in different localities.

The Cambrian Period. The earliest subdivision of the Paleozoic era, the Cambrian period, is represented by rocks rich in fossils, so that the reconstructions of what the world was like in those days are probably quite accurate. The forms living in this period were so varied and complex that they must have evolved from ancestors dating back to the Proterozoic period, at the latest, and possibly to the Archeozoic. All the present-day animal phyla, except the chordates, were represented, and all plants and animals lived in the sea. (The land must have been a weird, lifeless waste until the late Ordovician or Silurian when plants became established on land.) There were primitive, shrimplike crustaceans and arachnid-like forms, some of whose descendants exist almost unchanged today as the king crab. The sea floor was covered with simple sponges, corals, echinoderms growing on stalks, snails, pelecypods, primitive cephalopods, brachiopods and trilobites. **Brachiopods,** sessile, bivalved plankton feeders, flourished in the Cambrian and the rest of the Paleozoic. The **trilobites** were primitive arthropods with flattened, elongated bodies covered dorsally by a hard shell. The shell had two longitudinal grooves that divided the body into three lobes. There were a pair of legs on each somite but the last, and each leg had an outer gill branch and an inner walking or swimming branch. Most trilobites were two to three inches long but a few were as large as 24 inches. There were both unicellular and multicellular algae. One of the best-preserved collections of Cambrian fossils was found in the mountains of British Columbia; it includes annelids, crustacea, and a connecting link between annelids and arthropods, similar to the living peripatus.

Evolution since the Cambrian has not been marked by the establishment of entirely new body patterns, but by the ramification of the lines already present, and by the replacement of original, primitive forms with better-adapted ones. The fact that no new phyla have originated since the early Paleozoic does not necessarily mean that no other patterns of animal organization are possible, or that mutations for new patterns did not occur. It probably indicates only that by that time the already existing forms had reached a degree of adaptation to the environment which gave them a marked advantage over any new, unadapted types.

The Ordovician Period. During the Cambrian period the continents gradually had begun to be covered with water, and in the Ordovician period this submergence reached its maximum, so that much of what is now land was covered by shallow seas. Inhabiting the seas were giant cephalopods—squid or nautilus-like animals, with straight shells, 15 to 20 feet long, and a foot in diameter. The Ordovician seas were apparently quite warm, for corals, which grow only in warm waters, lived as far north as Ontario and Greenland. The first traces of the vertebrates are found in Ordovician rocks; these small animals, called **ostracoderms,** were jawless, armored, bottom-dwelling fishes, without fins (Fig. 277). Their armor consisted of a heavy, bony covering over the head and thick scales over the trunk and tail; otherwise they were similar to the jawless lamprey eels of today. Apparently, they lived in fresh water, and their armor plate was a defense against the carnivo-

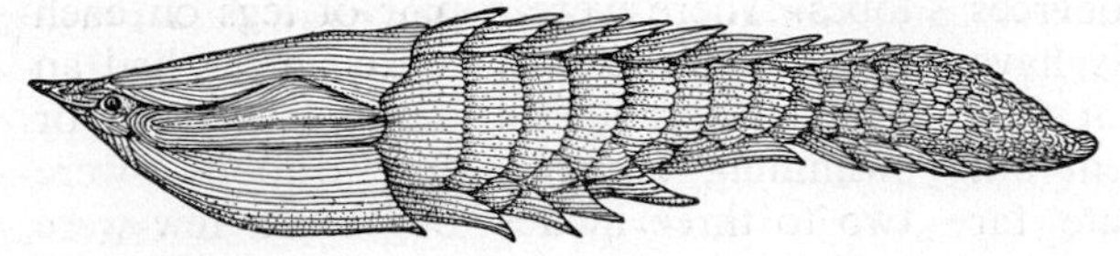

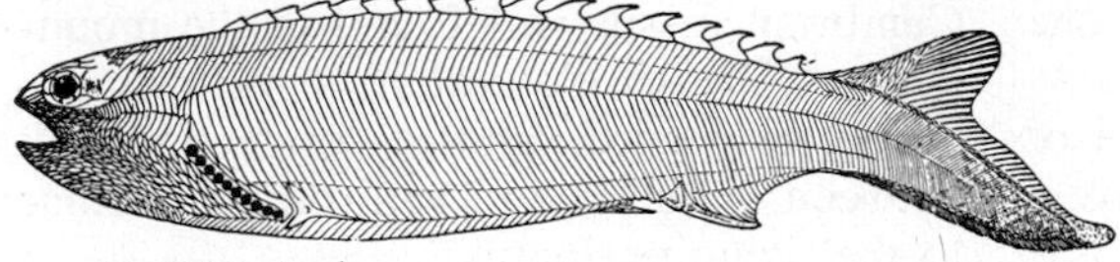

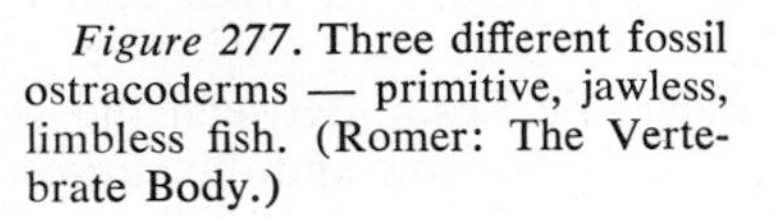
Figure 277. Three different fossil ostracoderms — primitive, jawless, limbless fish. (Romer: The Vertebrate Body.)

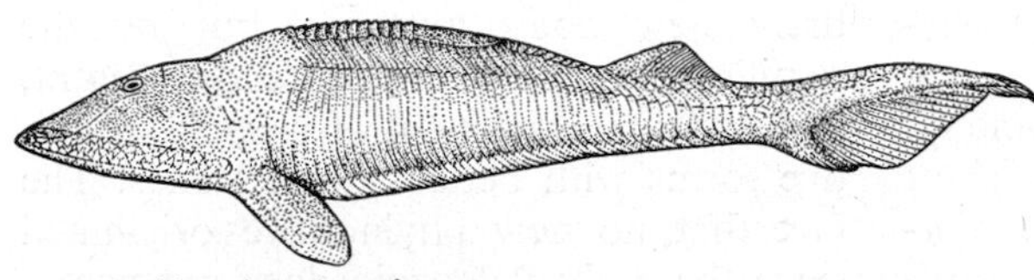

rous, giant water scorpions—sometimes 9 feet long—called **eurypterids,** which also inhabited fresh water.

The Silurian Period. Two events of great biologic importance occurred in the Silurian period: the land plants evolved, and air-breathing animals appeared. The first land plants seem to have resembled ferns rather than mosses, and ferns were the dominant plants of the following Devonian and Mississippian periods. The first air-breathing land animals were arachnids, resembling to some extent modern scorpions. The continental areas, which had been low during the Cambrian and Ordovician, rose—especially in Scotland and Northeastern North America—and the climate became much colder.

The Devonian Period. During the Devonian the original ostracoderms evolved into a great variety of fish, and the period is frequently referred to as the "age of fishes." The first to evolve jaws and paired fins were the **placoderms,** or spiny-skinned sharks (Fig. 278), which were small, armored, fresh-water forms. These animals had a variable number of paired fins; some had two, corresponding to the arms and legs of higher forms, and some had as many as five additional pairs between these two. True sharks appeared in fresh water during the Devonian, but tended to migrate to the ocean and to lose their cumbersome armor plate. The ancestors of the bony fishes also appeared in Devonian fresh-water streams, and had evolved by the middle of the Devonian into three main types: lungfish, lobe-finned fish and ray-finned fish. All had lungs and an armor of bony scales. A few lungfish have survived to the present, and the ray-finned fish, after undergoing a slow evolution in the remaining Paleozoic and early Mesozoic eras, ramified greatly in the latter part of the Mesozoic to give rise to the modern bony fish, or **teleosts.** The lobe-finned fish, which were the ancestors of the land vertebrates, were almost extinct by the end of the Paleozoic, and it was once believed that they had vanished with the end of the Mesozoic. But in 1939 and 1952 living coelacanths about 5 feet long were caught off the eastern coast of South Africa (Figs. 133, 279).

The latter part of the Devonian was marked by the appearance of the first land vertebrates—amphibians called **stegocephalians** (meaning roof-headed) (Fig. 280). These creatures, whose skulls were encased in bony armor, were similar in most respects to the lobe-fins, differing chiefly in having limbs instead of fins. The Devonian was the first period characterized by true forests; ferns, club mosses, horsetails and primitive gymnosperms—the "seed ferns"—all flourished. Insects and millipedes are believed to have originated in the late Devonian.

The Carboniferous Period. The Mississippian and Pennsylvanian periods are frequently grouped together as the Carboniferous, for during this time there flourished the great swamp forests whose remains gave rise to the major coal deposits of the world. The land was covered with low swamps, filled with horsetails, ferns, seed ferns and large-leaved evergreens (Fig. 281). The first reptiles, called **stem reptiles,** similar to their antecedent amphibians, appeared in the Pennsylvanian period, flourished in the final Paleozoic period—the Permian—and became extinct early in the Mesozoic era. It is a question whether the most primitive reptile known (called *Seymouria* for the town in Texas near which its fossil remains were found) should be considered an amphibian about to become a reptile, or a reptile just over the border from an amphibian. One of the main differences between reptiles and amphibians is the type of egg laid: Amphibians lay jelly-coated eggs in the water, and reptiles lay shell-covered eggs on land. Since no Seymourian eggs have survived, it may never be possible to decide to which class this animal belonged. *Seymouria* was a large, sluggish beast resembling a lizard. Its short, stubby legs extended laterally from the body as do a salamander's, instead of being closer together and extending directly down to form a pillar-like support for the body. Two important groups of winged insects evolved during the Carboniferous—the ancestors of the cockroaches, which reached a length of 4 inches, and ancestral dragonflies, some of which had a wingspread of 2½ feet.

The Permian Period. The final period of the Paleozoic was characterized by great changes in climate and topography. The level of the continents rose all over the world, so that the shallow seas which

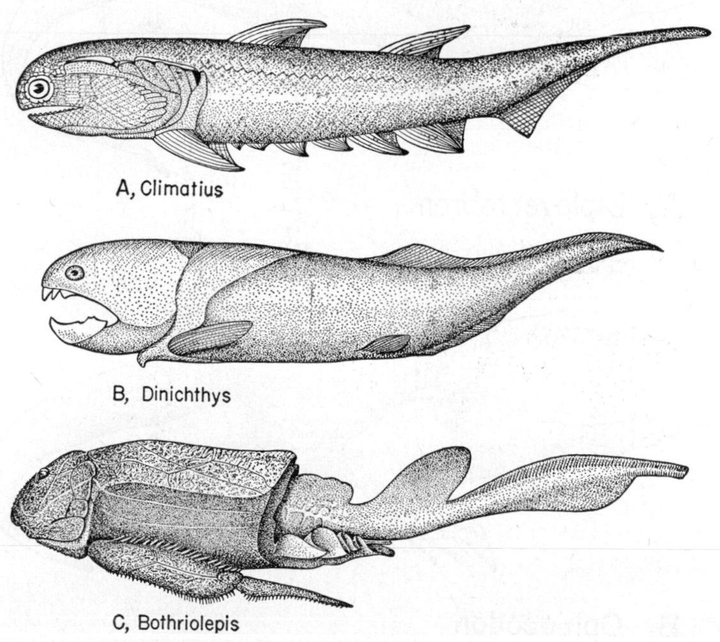

Figure 278. Three fossil placoderms from the Devonian Period. *A, Climatius,* a "spiny-skinned shark," with large fin spines and five pairs of accessory fins between the pectoral and pelvic pairs. *B, Dinichthys,* a giant arthrodire that grew to a length of 30 feet. Its head and thorax were covered by bony armor, but the rest of the body and tail were naked. *C,* Third type of placoderm that had a single pair of jointed flippers projecting from the body. (Romer: The Vertebrate Body.)

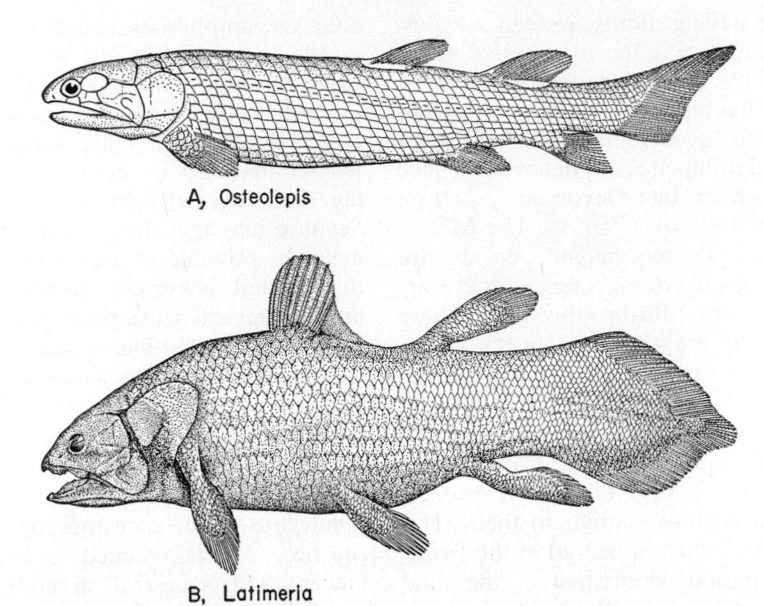

Figure 279. Crossopterygians. *A,* Typical fossil form from the Devonian. *B,* The living coelacanth found off Africa in 1939. (Romer: The Vertebrate Body.)

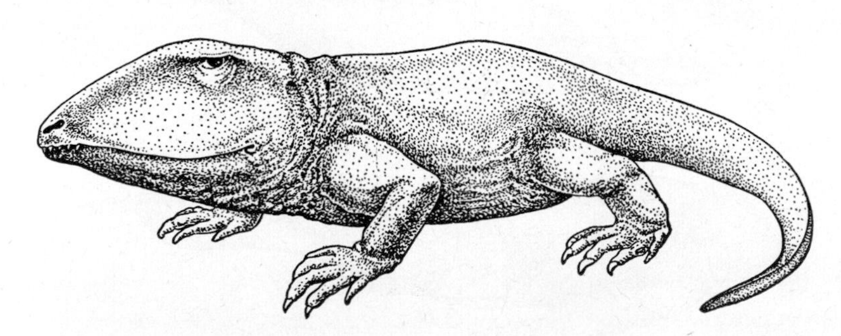

Figure 280. A, Diplovertebron, a primitive Paleozoic amphibian (stegocephalian), *B, Ophiacodon,* an early Permian pelycosaur. Although the pelycosaurs were primitive reptiles, they had certain characters indicating that they represent a first stage in the evolution of the mammals. (Romer: The Vertebrate Body.)

covered the region from Nebraska to Texas at the beginning of the period drained off, leaving the land a salt desert. At the end of the Permian a general folding of the earth's crust occurred, called the **Appalachian Revolution,** which raised the great mountain chain from Nova Scotia to Alabama. These mountains originally were higher than the present Rockies. Meanwhile, in Europe, other ranges were being brought into existence. A great glaciation, spreading from the Antarctic, covered most of the southern hemisphere, extending almost to the equator in Brazil and Africa; North America was one of the few parts of the world to escape glaciation at that time, but even its climate became much colder and drier than it had been during most of the Paleozoic. Many of the Paleozoic forms of life, apparently unable to adapt to the climatic changes, became extinct during the Appalachian Revolution. Even many of the marine forms became extinct, owing to the cooling of the water and the decrease in the amount of space available, caused by the diminishing of the shallow seas.

From the primitive stem reptiles there evolved during the late Carboniferous and early Permian the group of reptiles believed to be in the direct line which gave rise to the mammals. These were the **pelycosaurs** (Fig. 283), carnivorous reptiles that were more slender and lizard-like than the stem reptiles. In the latter part of the Permian there evolved, probably from the pelycosaurs, another group of reptiles with a few more mammalian characteristics—the **therapsids** (Fig. 281). One of these, *Cynognathus,* the "dog-jawed" reptile, was a slender, lightly built animal about 5 feet long, with a skull intermediate between that of a reptile and that of a mammal. The teeth, instead of being conical and all alike, as reptilian teeth are, were differentiated into incisors, canines and molars. In the absence of information about the animal's soft parts, whether it had scales or hair, whether or not it was warm blooded, and whether it suckled its young, it is called a reptile, but if more evidence were available, it might be classified as a very early mammal. The therapsids were widespread in the late Permian, but were crowded out early in the Mesozoic by the great variety of other reptiles.

The Mesozoic Era (The Age of Reptiles). The Mesozoic era, which began about 225,000,000 years ago and lasted some 150,000,000 years, is subdivided into three periods, the **Triassic, Jurassic** and **Cretaceous.** During the Triassic and Jurassic most of the continental areas were above water. The Triassic climate was dry, but warmer than the Permian, and the Jurassic was warmer and moister than the Triassic. The trees of the famous Petrified Forest in Arizona date from the Triassic.

During the Cretaceous Period the Gulf of Mexico expanded into Texas and New Mexico, and, in general, the sea gradually encroached upon the continental areas. There were great swamps, too, from Colorado to British Columbia. In the latter part of the Cretaceous the interior of the continent of North America was further submerged, and the bay from the Gulf of Mexico, meeting another from the Arctic, cut this continent into two parts. The Cretaceous ended in another great upheaval, called the **Rocky Mountain Revolution,** which raised the Rockies, Alps,

Figure 281. A mammal-like reptile (Lycaenops) from the late Permian of South Africa. (From Romer: The Vertebrate Body; after Colbert.)

Figure 282. Restoration of a swamp forest of the Carboniferous Period. The thick trunks at the left are giant club mosses, whose grasslike leaves and large cones can be seen in the upper left corner. To the left are seed ferns, which look rather like our present-day ferns but have seeds. The tree to the right is a Calamite, a giant horsetail, with leaves arranged in whorls. A large insect, an ancestor of the dragonflies, is visible just to the left of the Calamite. (Copyright, Chicago Natural History Museum.)

Figure 283. Texas in the Permian Period, about 230,000,000 years ago. Various pelycosaurs are shown. Some of them had large fins, others were essentially like lizards. In the right foreground are two salamander-like amphibians, with flat, triangular skulls. (Copyright, Chicago Natural History Museum, from the painting by Charles R. Knight.)

Figure 284. Giant dinosaurs from the Cretaceous Period of western North America. The largest flesh-eating dinosaur known was *Tyrannosaurus,* two of which are shown attacking the herbivorous, horned dinosaur, *Triceratops. Tyrannosaurus* reached a length of 47 feet and a height of 19 feet. Its head was as much as 6 feet long, and equipped with many sharp teeth. The front legs were small and completely useless. It walked on its powerful hind legs and balanced with its long tail. *Triceratops* was armed with a horn on the nose and a pair of horns over the eyes and was protected by a bony ruff covering the neck and shoulders. The rest of the body was covered with a leathery hide, so that it was vulnerable except when face-to-face with its opponent. (Copyright, Chicago Natural History Museum, from the painting by Charles R. Knight.)

Figure 285. During much of the Cretaceous Period, a shallow inland sea covered the western half of the Mississippi Valley. Three reptiles characteristic of this time and place are shown: in the center is a large mosasaur about 30 feet long; to the right is a giant, 8-foot long marine turtle; and flying in the left background are a number of reptiles of the genus *Pteranodon,* forms with short tails and a long crest extending back from the skull. (Copyright, Chicago Natural History Museum, from the painting by Charles R. Knight.)

Figure 286. Scene off the coast of North America in Jurassic times, about 155,000,000 years ago. Two types of marine reptiles are shown: plesiosaurs with long necks, broad, flat bodies, and sturdy, paddle-shaped limbs; and ichthyosaurs with fishlike fins and tails. Both types were fish-eaters. (Copyright, Chicago Natural History Museum, from the painting by Charles R. Knight.)

Figure 287. Western Canada in the Cretaceous Period, about 110,000,000 years ago. The land was low, well watered and covered with numerous swamps. Most of the dinosaurs were harmless, plant-eating forms of the Ornithischia group of reptiles, characterized by birdlike pelvic bones. Two types of duck-billed dinosaurs can be seen—three large, uncrested ones to the right, and two types of crested ones in the left background. In the middle foreground is a heavily armored, four-footed dinosaur covered with bony plates and spines. In the center background are two ostrich dinosaurs—tall, slender animals, with the general proportions of an ostrich, but with short forelegs and a long, slender tail. (Copyright, Chicago Natural History Museum, from the painting by Charles R. Knight.)

Himalayas and Andes and caused much volcanic activity in western North America.

Reptilian Evolution. The outstanding feature of the Mesozoic era was the origin, differentiation and final extinction of a great variety of reptiles, of which there were six main stocks.

The most primitive stock included, besides the ancient stem reptiles, the turtles, which originated in the Permian. The turtles have evolved the most complicated armor of any land animal, consisting of scales derived from the epidermis fused to the underlying ribs and breast bone. With this protection, both marine and land forms have survived with few structural changes since before the time of the dinosaurs. Their legs, which extend laterally, making locomotion difficult and slow, and their skulls, unpierced behind the eye sockets, have come down without alteration from the old stem reptiles.

A second group of reptiles to survive with relatively few changes from the ancestral stem reptiles are the lizards—the most abundant of living reptiles—and the snakes. For the most part the lizards have kept the primitive type of locomotion with the legs extended laterally, although many can run rapidly. Most of them are small, but the monitor lizard of the East Indies attains a length of 12 feet, and some fossil ones are 25 feet long. The **mosasaurs** (Fig. 285) of the Cretaceous were marine lizards which reached a length of 40 feet, and had a long tail useful in swimming. During the Cretaceous, snakes evolved from lizard ancestors. The important difference between snakes and lizards is not the loss of legs (some lizards are legless), but certain changes in the skull and jaws of the snake which enable it to open its mouth wide enough to swallow an animal larger than itself. A representative of an ancient line that has managed somehow to survive in New Zealand is the lizard-like **tuatera.** It shares several traits with the ancestral cotylosaurs, one of them being a third eye on the top of its head.

The main group of Mesozoic reptiles were the **archosaurs** or "ruling reptiles," of which the only living members are the alligators and crocodiles. At an early point in their evolution from stem reptiles the ruling reptiles, which were then about 3 feet long, became adapted to two-legged locomotion—their front legs became short, while the hind legs became long, stout and considerably modified. These animals rested or walked on all fours, but in emergencies they reared up and ran on the two hind legs, assisted by their fairly long tail, which served as a balance. From the early archosaurs developed many different, specialized forms, some of which continued to use two-legged locomotion, others of which reverted to walking on all fours. These descendents include the **phytosaurs**—aquatic, alligator-like reptiles, common during the Triassic; the **crocodiles,** which evolved during the Jurassic and replaced the phytosaurs as aquatic forms; and the **pterosaurs,** or flying reptiles, which included animals the size of a robin, as well as the largest animal ever to fly—*Pteranodon,* with a wingspread of 27 feet. There were two types of flying reptiles, one with a long tail that had a steering rudder at the end, the other with a short tail. Both these types, apparently, were fish-eaters, and they probably flew long distances over the water in search of food. Their legs were not adapted for standing, and it is believed that, like bats, they rested by clinging to some support and hanging suspended.

Of all the reptilian branches, the most famous are the **dinosaurs** (meaning terrible reptiles). These were divided into two main types: one with a birdlike pelvis, the other with a reptilian pelvis.

The **Saurischia** (meaning reptile pelvis) first evolved in the Triassic, and remained in existence until the Cretaceous. The early ones were fast, carnivorous, two-legged forms the size of a rooster, which probably preyed upon lizards and the archaic mammals then in existence. Throughout the Jurassic and Cretaceous there was a tendency in this group to grow larger, culminating in the gigantic carnivore of the Cretaceous, *Tyrannosaurus* (Fig. 284). Other Saurischia, beginning in the late Triassic, changed to a plant diet, reverted to a four-legged gait, and, during the Jurassic and Cretaceous, evolved into tremendous amphibious forms. Among these—the largest four-footed animals that ever lived—were *Brontosaurus,* with a length of 65 feet; *Diplodocus,* which reached a length of 87 feet, and *Brachiosaurus,* the biggest of them all, with an estimated weight of 50 tons.

The other group of dinosaurs, the **Ornithischia** (meaning birdlike pelvis), were vegetarians, probably from the beginning of their evolution. Although some of them walked upright, the majority had a four-legged gait. Having lost their front teeth, they developed a stout, horny, birdlike beak, which in some forms was broad and ducklike (hence the name "duck-billed" dinosaurs). Webbed feet were characteristic of this type; other species developed great armor plates as protection against the carnivorous saurischians. *Ankylosaurus* (Fig. 287), which has been dubbed "the reptilian tank," had a broad, flat body, covered with an armor plate and with large, laterally projecting spines. Still other ornithischians of the Cretaceous period developed bony plates around the head and neck. One of these, *Triceratops,* had two horns over its eyes and another over its nose—all 3 feet long.

Two other groups of Mesozoic reptiles, separate from each other and from the dinosaurs, were the marine **plesiosaurs** and **ichthyosaurs** (Fig. 286). The former were characterized by an extremely long neck, which took up over half of their total length. The trunk was broad, flat and rather turtle-like, the tail was small, and the animal paddled along by means of finlike arms and legs. Often it was 45 feet long. The ichthyosaurs (fish reptiles) had a body form superficially like that of a fish or a whale, with a short neck, a large dorsal fin, and a sharklike tail. They swam by wiggling their tails, using their feet only for steering. It is believed that the ichthyosaur young were born alive, after having hatched from eggs within the mother, for the adults were too specialized to come out on land, and a reptile egg will drown in water. The presence of skeletons of the young within the body cavity of adult fossils has strengthened this theory.

At the end of the Cretaceous a great many reptiles became extinct; they were apparently unable to adapt to the marked changes brought about by the Rocky Mountain Revolution. As the climate became colder and drier many of the plants which served as food for the herbivorous reptiles disappeared. Some of the herbivorous reptiles were too large to walk about on land when the swamps dried up. The smaller, warm-blooded mammals which had appeared were better able to compete for food and many of them ate reptilian eggs. The demise of the many kinds of reptiles was probably the result of a combination of a whole host of factors, rather than any single one.

Other Mesozoic Evolutionary Trends. Although the reptiles were the dominant animals of the Mesozoic, many other important organisms evolved during that time: snails and bivalves increased in number and kind; sea urchins reached their peak; mammals originated in the Triassic; and teleost fishes and birds arose during the Jurassic. Most of the modern orders of insects appeared early in the Mesozoic. During the early Triassic the most abundant plants were seed ferns, cycads and conifers; but by the Cretaceous, many others, resembling present-day species, had appeared—sycamores, magnolias, palms, maples and oaks.

From the Jurassic, excellent fossils have been preserved, showing even the outlines of feathers, of the earliest species of bird. This creature, called *Archeopteryx,* was about the size of a crow, had rather feeble wings, jawbones armed with teeth, and a long, reptilian tail, covered with feathers (Fig. 288). Fossils have been found in Cretaceous rocks of two other bird species—*Hesperornis,* an aquatic diving bird that had lost the ability to fly, and *Ichthyornis,* a powerful flying bird, about the size of a pigeon, with reptilian teeth. Modern toothless birds evolved early in the following era.

The Cenozoic Era (the Age of Mammals). With equal justice, the Cenozoic could be called the Age of Birds, the Age

Figure 288. Restoration of two of the earliest birds, of the genus *Archeopteryx.* (After Heilmann; Lull, R. S.: Organic Evolution, The Macmillan Co.)

of Insects or the Age of Flowering Plants, for it is marked by the evolution of all these forms as much as by the evolution of the mammals. It extends from the Rocky Mountain Revolution, some 75,000,000 years ago, to the present, and is subdivided into two periods, the **Tertiary,** which lasted some 74,000,000 years, and the **Quaternary,** which includes the last million, or million and a half, years.

The Tertiary Period. The Tertiary is subdivided into five epochs, named, from earliest to latest, **Paleocene, Eocene, Oligocene, Miocene** and **Pliocene.** The Rockies, formed at the beginning of the Tertiary, were considerably eroded by the time of the Oligocene, giving the North American continent a gently rolling topography. In the Miocene another series of uplifts raised the Sierra Nevadas and a new set of Rockies, and resulted in the formation of the western deserts. The climate was milder in the Oligocene than it is at present, and palms grew as far north as Wyoming. The uplift begun in the Miocene continued in the Pliocene and, coupled with the ice ages of the Pleistocene, killed many of the mammals and other forms that had evolved. The final elevation of the Colorado Plateau, which also caused the cutting of Grand Canyon, occurred almost entirely in the short Pleistocene and Recent epochs.

The earliest fossils of true mammals were deposited late in the Triassic, but by the Jurassic there were four orders of mammals, all about the size of a rat or small dog. These earliest mammals **(monotremes)** were egg-laying animals, and their only living survivors are the duck-billed platypus and the spiny anteater (Fig. 289) of Australia. Both have fur and suckle their young, but like turtles they lay eggs. The ancestral, egg-laying mammals certainly must have differed from the specialized platypus and anteater, but the fossil records of those early forms are incomplete. The present-day monotremes have been able to survive this long only because they lived in Australia, which, until recently, had no placental mammals to offer competition.

By the Jurassic and Cretaceous most mammals were advanced enough to bring forth their young alive, although the most primitive of them—the **marsupials**—gave birth to underdeveloped young which had to remain for several months in a pouch of the mother's abdomen, containing the nipples. The Australian marsupials, freed, like the monotremes, of competition from better-adapted placental mammals, which were responsible for the extinction of their cousins on other continents, evolved into a wide variety of types that superficially resemble some of the placentals;

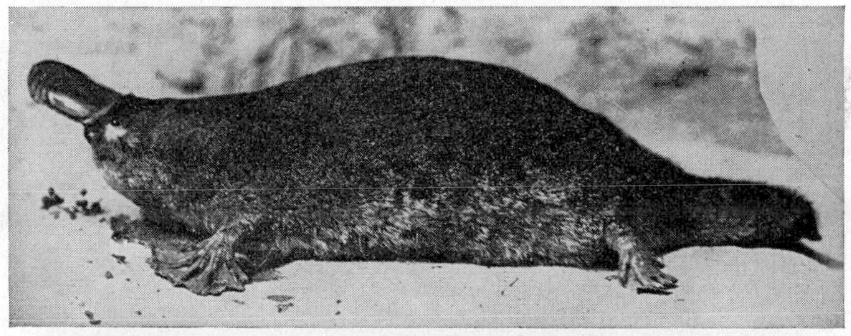

A

B

Figure 289. Two living examples of monotremes—mammals that lay eggs. *A,* The duck-billed platypus. Note the short fur, webbed feet, horny, duck-shaped beak and unusual tail. *B,* The spiny anteater, about 18 inches long, is covered with strong, pointed spines, yellow with black tips. Its narrow black snout is cylindrical, and it captures its food (mostly ants) with the long, protrusible tongue, which is covered with sticky saliva. (Photographs, courtesy of the Australian News and Information Bureau.)

Figure 290. Kangaroo mother with her young in her pouch. (Photograph, courtesy of the Australian News and Information Bureau.)

Figure 291. Restoration of an archaic meat-eating mammal, a creodont from the Eocene Period, eating a tiny ancestral horse, *Eohippus.* (Copyright, American Museum of Natural History, New York.)

Figure 292. Restoration of a scene at the Rancho La Brea tar pits (now a part of Los Angeles, California) in the Pleistocene. Many well-preserved specimens of animals now extinct have been found embedded in the asphalt. In the left foreground are two saber-toothed tigers; in the right foreground, three large ground sloths; the giant vultures, now extinct, had a wing-spread of 9 feet. In the background are mastodons and dire wolves. (Copyright, American Museum of Natural History, New York, from a painting by Charles R. Knight.)

there are marsupial mice, shrews, cats, moles, bears, and one species of wolf, as well as a number of forms with no placental counterparts, such as the kangaroo, wombat and wallaby (Fig. 290). During the Pleistocene there were giant kangaroos in Australia, and wombats the size of a rhinoceros. The opossum more closely resembles the primitive, ancestral marsupial type than do any of these more specialized forms; it is the only marsupial found outside Australia and South America.

The advanced, modern placental mammals, including man, which are distinctive in bringing forth their young alive and ready to live an independent existence, evolved from an insect-eating, tree-dwelling ancestor. Fossils of this animal in Cretaceous rocks show it to have been tiny and rather like the present-day shrew. Some of these ancestral mammals remained in the trees and gave rise, through a series of intermediate forms, to the primates — monkeys, apes and men. Others lived on or under the ground and, during the Paleocene, evolved into all the other mammals living today. The archaic mammals of the Paleocene had conical, reptilian teeth, five digits on each foot, and a small brain; they also walked on the soles of their feet instead of on their toes. During the Tertiary the evolution of grasses which served as food, and forests which afforded protection, were important factors in changing the mammalian body pattern. Concomitant with a tendency toward increased size, the mammals all displayed tendencies toward an increase in the relative size of the brain, and toward changes in the teeth and feet. As modern forms, better equipped for survival arose, the archaic mammals became extinct.

Although fossils of both marsupial and placental animals have been found in Cretaceous rocks, it is rather surprising to find the remains of highly developed mammals in strata of the early Tertiary. Whether they actually arose at that time or had existed before in the highlands, and had not been preserved, is unknown.

In the Paleocene and Eocene epochs the first carnivores, called **creodonts** (Fig. 291), arose from the primitive insect-eating placentals. They were replaced in the Eocene and Oligocene by more modern forms which eventually gave rise to the present-day carnivores, such as cats, dogs, bears and weasels, as well as to the web-footed, marine carnivores—the seals and walruses. One of the most famous fossil carnivores is the saber-toothed tiger (Fig. 292), which became extinct only recently in the Pleistocene. These animals had tremendously elongated, knifelike upper canine teeth, and a lower jaw that could be swung down and out of the way, allowing the teeth to be used as sabers for stabbing the prey.

The larger herbivorous mammals, most of which have hooves, are sometimes referred to as the **ungulates.** They do not form a single, natural group, but consist of several independent lines, so that although both horses and cows have hooves, they are no more closely related than either one is to a tiger. The molar teeth of ungulates are flattened and enlarged to facilitate the chewing of leaves and grass. Their legs have become elongated, and adapted for the rapid movement necessary to escape predators. The earliest ungulates, called **condylarths,** appeared in the Paleocene; they had long bodies and tails, flat, grinding molars, and short legs ending in five toes, each of which bore a hoof. Corresponding to the archaic carnivores, or creodonts, were the archaic ungulates called **uintatheres.** During the Paleocene and Eocene some of these were as large as elephants, and some had three large horns projecting from the top of the head.

The fossil records of several ungulate lines—the horse, the camel and the elephant—are complete, and it is possible to trace the evolution of these animals from small, primitive, five-toed creatures. The chief evolutionary tendencies in the ungulates have been toward an increase in the overall size of the body and a decrease in the number of toes. The ungulates were early divided into two groups, one characterized by an even number of toes, and including the cow, sheep, camel, deer, giraffe, pig and hippopotamus; the other characterized by an odd number of toes, and including the horse, zebra, tapir and rhinoceros. The elephants and their recently extinct relatives, the mammoths and mastodons, can be traced back to an Eocene ancestor the size of a hog which had no trunk. This primitive form, called *Moeritherium,* was close to the stem that also gave rise to such dissimilar creatures as the coney (a small, woodchuck-like animal found in Africa and Asia) and the sea cow.

The whales and porpoises descended from whalelike forms of the Eocene, called **zeuglodonts,** which in turn are believed to have evolved from the creodonts. The evolutionary history of the bats can be traced to ancestral, winged types, also of the Eocene, which descended from the primitive insectivores. The evolutionary history of some of the other mammals—the rodents, rabbits and edentates (anteaters, sloths and armadillos)—is less well known.

The Quaternary Period (the age of Man). The Quaternary Period, which includes the final million or million and a half years of the earth's history, is usually divided into two epochs, the **Pleistocene,** and the **Recent,** which began some 25,000 years ago with the recession of the last ice sheet. The Pleistocene was marked by four periods of glaciation, between which the sheets of ice retreated. At their greatest extent these ice sheets covered nearly 4,000,000 square miles of North America, extending south as far as the Ohio and Missouri rivers. The Great Lakes, which were carved out by the advancing glaciers, changed their outlines radically a number of times, and from time to time they emptied into the Mississippi. It is estimated that in the past, when the Mississippi drained lakes as far west as Duluth and as far east as Buffalo, its volume was more than sixty times as great as at present. During the Pleistocene glaciations enough water was removed from the sea and locked in the ice to lower the sea level 200 to 300 feet. This created land connections, highways for the dispersal of many land forms, between Siberia and Alaska at Bering Strait, and between England and the continent of Europe.

The plants and animals of the Pleistocene were similar to those alive today. It is sometimes difficult to distinguish between Pleistocene and Pliocene deposits, because the organisms were similar and nearly modern in form. A considerable number of mammals, including the saber-toothed tiger, mammoth and giant ground sloth, became extinct during the Pleistocene, after the appearance of primitive man. The Pleistocene was marked by the extinction of many species of plants, especially woody ones, and the appearance of numerous herbaceous forms.

The paleontologic record makes it impossible to doubt that the present species arose from previously existing, different ones. The fossil record is not equally clear for all lines of evolution. Most plant tissues are too soft to leave good fossil remains and the connecting links between the animal phyla were apparently soft-bodied forms that left no fossil traces. For many lines of evolution, especially the vertebrates, the successive steps are well known; other lines have some gaps which remain to be clarified by future paleontologists.

QUESTIONS

1. List the various kinds of paleontologic evidence. Why are fossils hard to find?
2. Name some of the forms of life of the Proterozoic Era.
3. What were the most common animals of the Cambrian period?
4. Explain how an estimate of the age of a rock is made on the basis of radioactive elements.
5. When did vertebrates first appear?
6. What were two great events of the Silurian period? What is the significance of the name Carboniferous?
7. What was the Appalachian Revolution? When did it occur and what were its effects?
8. What is meant by the term "stem reptile"? What relation do turtles bear to these animals?
9. What factors are believed to have caused the extinction of the dinosaurs?
10. When did the mammals originate? What are monotremes? What are marsupials?
11. Briefly trace the evolution of the various phyla from the beginning of life to the present.

SUPPLEMENTARY READING

The fossil evidence for evolution is clearly presented in Dodson's *Textbook of Evolution.*

A more advanced discussion of paleontology is R. S. Lull's *Organic Evolution.* A. S. Romer's *Man and the Vertebrates* is a readable description of the important fossil vertebrates, as is P. E. Raymond's *Prehistoric Life.* Colbert's *Evolution of the Vertebrates* is a well written elementary text for the general student. The interesting facts about fossil plants are discussed in H. N. Andrews' *Ancient Plants and the World They Lived In.* An excellent, recent book on the evolution of the invertebrates is Shrock and Twenhofel's *Principles of Invertebrate Paleontology.*

Chapter 35

The Living Evidence for Evolution

EVEN IF there were no fossil record, the evidence obtained from the study of the anatomy, physiology and biochemistry of modern plants and animals, their embryologic and genetic histories, and the manner in which they are distributed over the earth's surface would be enough to prove that organic evolution has occurred.

319. THE EVIDENCE FROM TAXONOMY

The science of naming, describing and classifying organisms, **taxonomy,** was outlined in Chapter 6. The founders of scientific taxonomy, Ray and Linnaeus, were firm believers in the unchanging nature of species, but present-day taxonomists are concerned with describing species primarily as a means of discovering evolutionary relationships. The fact that the characteristics of living things are such that they can be fitted into a hierarchical scheme of categories—species, genera, families, orders, classes and phyla—can best be interpreted as indicating evolutionary relationship. If the kinds of plants and animals were not related by evolutionary descent, their characters would be present in a confused, random pattern and no such hierarchy of forms could be established.

The basic taxonomic unit is the species, a population of closely similar individuals which are alike in their morphologic, embryologic and physiologic characters, which in nature breed only with each other, and which have a common ancestry. It is difficult to give a universally applicable definition of the term, for it must be modified to include those species whose life cycle includes two or more quite different forms (mosses, ferns, some algae, many coelenterates, worms, insects and amphibians, for example). A population that is spread over a wide territory may show local or regional differences which may be called subspecies. Many instances are known in which a species is subdivided into a chain of subspecies, each of which differs slightly from its neighbors but interbreeds with it. The groups at the two ends of the chain, however, may be so different that they cannot interbreed. Such a series of geographic subspecies is called a *Rassenkreis* (German, race-circle).

The classification of modern-day organisms into well defined groups is possible only because most of the intermediate forms have become extinct. If every animal and plant that ever lived were still living today, it would be a difficult matter to divide the living world into neat taxonomic categories, for there would be a continuous series of forms grading from lowest to highest. The species now living have been called "islands in a sea of death," and have been likened also to the terminal twigs of a tree of which the trunk and main branches have disappeared. The problem of the taxonomist is to reconstruct the missing branches and put each twig on the proper branch.

320. THE EVIDENCE FROM ANATOMY

Homologous Organs. Comparisons of the structure of groups of animals and plants show that each organ system has a certain basically similar pattern that is varied to some extent among the members of a given phylum. The skeletal, circulatory and excretory systems of vertebrates provide particularly clear illustrations of this. Only similarities based on homologous organs are valid in attributing evolutionary relationships. **Homologous organs,** you will recall, are basically similar in their structure, in their relationships to adjacent structures, in their embryonic development and in their nerve and blood supply. A seal's front flipper, a bat's wing, a cat's paw, a horse's front leg and the human hand and arm (Fig. 99), though superficially dissimilar and adapted for quite different functions, are homologous organs. Each consists of almost the same number of bones, muscles, nerves and blood vessels, arranged in the same pattern, and with very similar modes of development. The existence of such homologous organs is a strong argument for a common evolutionary origin.

Vestigial Organs. A second type of anatomic evidence for evolution is provided by the fact that in almost every species of plant and animal are organs or parts of organs which are useless and degenerate, often undersized or lacking some essential part, as compared to homologous structures in related organisms. In the human body there are more than 100 such **vestigial organs,** including the appendix, the coccyx (the fused tail vertebrae), the wisdom teeth, the nictitating membrane of the eye, the body hair, and the muscles that move the ears. Such organs are the remnants of ones which were functional in some earlier animal. Because of a change in the environment or mode of life of the species, the organ became unnecessary for survival and, gradually, nonfunctional. Mutations are constantly occurring which decrease the size and function of various organs, of course, and if the organs are necessary for survival, the organisms undergoing such mutations will be eliminated. The structures we recognize as vestigial organs are simply in the process of being eliminated.

The human appendix is the remnant of the blind pouch, the **cecum,** which is a large, functional structure in the digestive tract of herbivorous animals like the rabbit. Foods rich in cellulose require a long time for digestion, and the cecum provides a place where the food may be stored while the gradual process of digestion, mostly by intestinal bacteria, takes place. A long time ago in our evolutionary history, our ancestors changed to a diet containing more meat and less cellulose, and the cecum has gradually diminished to the present useless vestige, the appendix.

Man is not alone in having vestigial organs: both whales and pythons, for example, have bones embedded in the flesh of the abdomen which are vestigial hind legs; the wingless birds have vestigial wing bones; many blind, burrowing or cave-dwelling animals have vestigial eyes; and so on.

321. THE EVIDENCE FROM COMPARATIVE PHYSIOLOGY AND BIOCHEMISTRY

Perhaps no evidence of descent from a common ancestor could be more convincing than the similarity to our own bodies displayed by any of the mammals. It is startling to find, upon dissecting a rat for the first time, a replica of the human heart, lungs, stomach and most of the other organs, and to realize that these structures are functioning in a manner almost identical to our own. There are, of course, differences—in the vitamin requirements, in the arrangement of a few blood vessels, and in the workings of the reproductive system—but, in general, the important physiologic processes of respiration, digestion, circulation, excretion and nervous response are the same as ours. Yet rats and human beings belong to different orders of the class Mammalia, and frogs, whose internal workings also resemble ours remarkably, belong to a different class!

Blood chemistry has been a particularly fruitful field in yielding evidence of relationships. The degree of similarity between the proteins in the blood of various animals is tested by the antigen-antibody technique discussed previously (p. 237). An experimental animal, usually a rabbit, is given repeated injections of human serum, containing proteins foreign to the rabbit's blood. In response, the white cells of the rabbit produce antibodies specific for human, blood-protein antigens. The antibodies are then obtained by withdrawing blood from the rabbit and allowing it to clot (the antibodies are in the serum). Even a dilute sample of the serum, when mixed with human blood, results in a visible precipitation caused by the combination of antigens and antibodies. But nonhuman blood, in order to form such a precipitation, requires that the serum be much more concentrated, and even then the precipitation may be delayed or scanty. The blood of quite different animals may cause no precipitation at all. By using a series of rabbits, each injected with the blood of different species, it has been possible to obtain a series of antibodies, each specific for the blood proteins of a particular species of animal.

Thousands of tests involving different animals have revealed a basic similarity between the blood proteins of all the mammals, the degree of relationship being indicated by how much the antigen and antibody solutions can be diluted and still result in visible precipitation. Man's closest "blood relations," as determined in this way, are the great apes; then, in order, are the Old World monkeys, the New World prehensile-tailed monkeys and the tarsioids. Of all the types of primate blood, the lemur's results in the least precipitation when combined with antibodies specific for human serum.

The biochemical relationships of a variety of forms tested in this way correlate well with the relationships determined by other means. Cats, dogs and bears are more closely related to each other in this regard than to the other mammals, while cows, sheep, goats, deer and antelopes form another closely related group. Seals and sea lions are closer to the carnivores than to the other mammals. These tests reveal that there is a closer relationship among the modern birds than there is among the mammals, for all the several hundred species of birds tested so far give strong and immediate reactions with serum containing antibodies for chicken blood. The tests also indicate that birds are more closely related to the turtle-crocodile line of reptiles than to the snake-lizard line, which corroborates the paleontologic evidence. Similar differences in the proteins of the body cells can be detected by appropriate techniques. Recently the technique has been extended to plants by Mez and Königsberg; extracts of plant proteins are prepared and used as antigens in rabbits.

It might seem unlikely that an analysis of the urinary wastes of different species would provide evidence of evolutionary relationship, yet this is true. The kind of waste excreted depends upon the particular kinds of enzymes present and the enzymes are determined by genes which have been selected in the course of evolution. The waste products of the metabolism of purines (one of the constituents of nucleic acids) are excreted by man and other primates as uric acid, by other mammals as allantoin, by amphibians and most fishes as urea, and by most invertebrates as ammonia. Vertebrate evolution has been marked by the successive loss of enzymes required for the stepwise breakdown of uric acid. Joseph Needham made the interesting observation that the chick embryo in the early stages of development excretes ammonia, then urea, and finally uric acid. The enzyme uricase, which catalyzes the first step in the degradation of uric acid, is present in the early chick embryo but disappears later. The adult frog excretes urea but the tadpole excretes ammonia. These are biochemical examples of "recapitulation."

322. THE EVIDENCE FROM EMBRYOLOGY

The importance of the embryologic evidence for evolution was stressed by Darwin and brought into even greater prominence by Ernst Haeckel in 1866 when he developed his "Biogenetic Law," that embryos, in the course of development, repeat the evolutionary history of their ancestors in some abbreviated form. This idea, succinctly stated as "Ontogeny recapitulates phylogeny," stimulated research in embryology and focused attention on the general resemblance between embryonic development and the evolutionary process. It is now clear that the embryos of the higher animals resemble the *embryos* of lower forms, not the adults, as Haeckel had believed. The early stages of all vertebrate embryos are remarkably similar (Fig. 293) and it is not easy to differentiate a human embryo from the embryo of a pig, chick, frog or fish.

In recapitulating its evolutionary history in a few days, weeks or months the

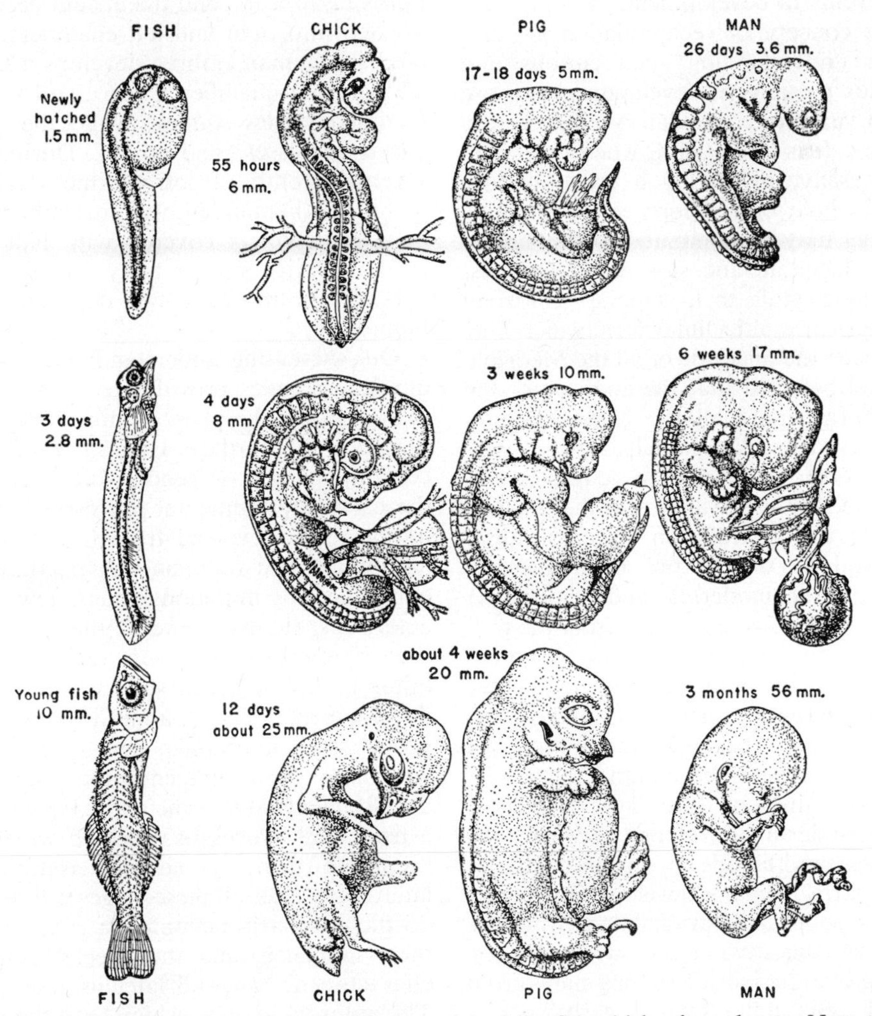

Figure 293. Stages in the embryonic development of the fish, chick, pig and man. Note that the earlier stages of development (top row) are remarkably similar and that differences become more marked as development proceeds. (Hunter and Hunter: College Zoology.)

embryo must eliminate some steps and alter and distort others. In addition, some new characters have evolved which are adaptive and enable the embryo to survive. Mammalian embryos have many early characteristics in common with those of fish, amphibia and reptiles but also have other structures which enable them to survive and develop within the mother's uterus rather than within an egg shell. These secondary traits may alter the original characters common to all vertebrates so that the basic resemblances are blurred. The concept of recapitulation must be used with due caution, rather than rigorously, but it does provide an explanation for many otherwise inexplicable events in development.

The concept of recapitulation is helpful in understanding such curious and complex patterns of development as those of the vertebrate circulatory or excretory systems. It is also useful, when not taken too literally, in getting a broad picture of the whole of development. The fertilized egg may be compared to the single-celled flagellate ancestor of all animals, and the blastula to a colonial protozoan or spherical multicellular form which may have been the ancestor of all the Metazoa. Haeckel believed that the ancestor of the coelenterates and all the higher animals was a gastrula-like animal, *Gastrea,* with two layers of cells and a central cavity connected by a blastopore to the outside.

After the gastrulation stage, development follows one of two main lines; in one (the echinoderms and chordates) the blastopore—the opening from the gastrular cavity—becomes the anus, or comes to lie near the anus; in the other (the annelids, molluscs, arthropods and others) the blastopore becomes the mouth, or comes to lie near the mouth.

In both lines, a third layer of cells—the mesoderm—develops between the ectoderm and entoderm. In the chordate-echinoderm line this develops, at least in part, as pouches or evaginations from the primitive digestive tract, while in the annelid line the mesoderm originates from special cells differentiated early in development.

Shortly after the appearance of the mesoderm all chordate embryos develop a dorsal, hollow nerve cord, as well as a notochord—the internal supporting rod for the body—and perforations in the pharynx (the gill slits). The early human embryo resembles a fish embryo, with gill slits, pairs of aortic arches, or blood vessels traversing the gill bars, a fishlike heart with a single atrium and ventricle, a primitive pronephros or fish kidney, and a tail, complete with muscles for wagging it. Later the human embryo resembles a reptilian embryo: Its gill slits close; the bones which make up each vertebra and which had been separate, as in fish embryos, fuse; a new kidney—the mesonephros—forms, and the pronephros disappears; and the atrium becomes divided into right and left chambers. Still later the human embryo develops a mammalian, four-chambered heart, and a third, completely new kidney, while the notochord regresses, and so on. During the seventh month of intrauterine development the human embryo resembles—in being completely covered with hair and in the relative size of body and limbs—a baby ape more than it does an adult human.

Our increasing understanding of physiological genetics provides us with an explanation for the phenomenon of recapitulation. All chordates have in common a certain number of genes which regulate the early developmental processes. But as our ancestors evolved from fish, through amphibian and reptilian phases, they accumulated, by mutations, many new characteristics, though some of the genes for the original fish form were retained. The latter group of genes still controls early development. Then, a little later, the genes which the human being still shares with amphibians influence the course of development, so that the animal resembles a frog. This process is repeated when the "reptilian" genes come into control still later. Only after all these stages are passed do the peculiarly mammalian genes assert their influence, and the special primate characteristics are still later in appearing. The anthropoid apes, which have the most immediate ancestors in common with us, naturally have the most genes in common with us too, and therefore their development is almost identical with ours except for the finer details. A pig or rat, whose ancestors are the same as ours only up to the primitive, placental mammal or insectivore stage, has fewer genes in common with us, and hence developmental processes that diverge at an earlier time. By and large during development, the general characters that distinguish phyla and classes appear before the special characters that distinguish genera and species.

323. THE GENETIC EVIDENCE

The selection and breeding of domesticated animals and cultivated plants for the past several thousand years provide us with models of how some of the evolutionary forces operate. All the varieties of present-day dogs are descended from one or a few related species of wild dog or wolf, and yet they vary tremendously in many characteristics. Compare, for example, the size of the chihuahua and the St. Bernard or Great Dane; the head shape of the bulldog and collie; the body proportions of the cocker, dachshund and Russian wolfhound. If these varieties were found in the wild, they would undoubtedly be assigned to different species, and perhaps even to different genera. But since all are known to come from common ancestors, and since all are interfertile, they are regarded as varieties or races of a single species.

The plant breeders who developed the present varieties of cultivated plants have similarly produced, by selection and interbreeding, a tremendous variety of plants from one or a few forms. The cliff cabbage, for example, which still grows wild in Europe, is the ancestor, not only of our cultivated cabbage, but of such dissimilar plants as cauliflower, kohlrabi, Brussels sprouts, broccoli and kale. Many varieties of wheat, too, have been produced by selection, each adapted for certain growing conditions. Thus there are winter wheats, spring wheats, wheats resistant to drought, to rusts, and to other pests. The cultivated species of tobacco has been traced back to a cross between two species of wild tobacco; corn has been traced to teosinte (a grass-like plant growing wild in the Andes and Mexico). Breeding experiments and observations indicate that species are not, as Linnaeus believed, unchangeable biologic entities, each of which was created separately, but groups of organisms which have arisen from other species and which can give rise to still others.

The number and the detailed structure of the chromosomes of related species can be compared by cytologic methods. Such studies have provided useful evidence concerning the evolutionary history of fruit flies, jimson weeds, primroses and many other plants and animals.

324. EVIDENCE FROM THE GEOGRAPHIC DISTRIBUTION OF ORGANISMS

Not all plants and animals are found in all parts of the world; they are not even found everywhere that they could survive, as one would expect if climate and topography were the only factors determining distribution. Central Africa, for example, has elephants, gorillas, chimpanzees, lions and antelopes, while Brazil, with a similar climate and other environmental conditions, has none of these, but does have prehensile-tailed monkeys, sloths and tapirs. The present distribution of organisms is understandable only on the basis of the evolutionary history of each species.

The *range* of a given species—that is, the portion of the earth in which it is found—may be only a few square miles or, as with man, almost the entire world. In general, closely related species do not have identical ranges, nor are their ranges far apart; they are usually adjacent, but separated by a barrier of some sort, such as a mountain or a desert. This generalization was formulated by David Starr Jordan and is known as **Jordan's Rule.** It follows from the role of isolation in species formation.

As one would expect, regions which have been separated from the rest of the world for a long time (such as Australia and New Zealand) have a flora and fauna peculiar to them. Australia has a mammalian population of monotremes and marsupials found nowhere else. This is because, during the Mesozoic, Australia was isolated from the rest of the world, so that its primitive mammals never had any competition from the better-adapted placental mammals which eliminated the monotremes and most of the marsupials everywhere else. The primitive mammals gave rise to a variety of forms which were able to take advantage of the different habitats available.

The situation of the oceanic islands is similar to that of Australia and New Zealand, except that these islands arose independently in the middle of the ocean and never have been attached to any continent. Because of their isolation, their floras and faunas are peculiar, made up to some extent of species found nowhere else. The plants and animals of the Galápagos Islands, studied by Darwin during the voyage of the *Beagle,* are similar to those on the nearest mainland—the coast of Ecuador, some 500 miles away, whence their ancestors had flown, swum or been transported. There are no frogs or toads on the Galápagos, even though there are woodland spots ideally suited for such creatures, because neither the animals nor their eggs can survive being exposed to sea water, an inevitable experience for any animal migrating to the islands. There are no terrestrial mammals, either, although there are many bats, as well as land and sea birds. A number of species of giant lizards and giant turtles also inhabit the island, some living on but a single island of the archipelago. The occurrence of these particular forms—closely related to, yet not identical with, those of the Ecuador coast—suggests strongly that after the first animals and plants arrived on the islands, mutations took place which changed the species slightly, and that these changes were retained because of isolation. These forms offer a good example of the evolutionary process.

There are many facts about the present distribution of animals and plants which can be explained only by their evolutionary history. Alligators are found only in the rivers of the southeastern United States, and in the Yangtse River of China, and sassafras, tulip trees and magnolias grow only in the eastern United States, Japan and eastern China. It is known that early in the Cenozoic the northern hemisphere was much flatter than it is now and the North American continent was connected with eastern Asia by a land bridge at Bering Strait, and possibly with Greenland. The climate of this region was much warmer than at present, and fossil evidence shows that alligators, magnolia trees and sassafras were distributed over the entire region. Later in the Cenozoic, as the Rockies increased in height, the western part of North America became colder and dry, causing the plants adapted to a warm, humid climate to become extinct. Then, with the Pleistocene glaciations, the ice sheets moving from the north met the desert and mountain regions in western North America, eliminating any surviving temperate plants; and in Europe the polar glaciations nearly met the glacier spreading from the Alps, so that many of the temperate plants there became extinct. In southeastern United States and eastern China there were regions untouched by the glaciation in which the magnolia trees and alligators survived. Because the alligators and magnolia trees of the two regions have been separated for several million years, they have followed separate evolutionary pathways, and so are slightly different, but they are still closely related species of the same genera.

The facts about the distribution of plants and animals constitute the science of **biogeography,** one of the basic tenets of which is that each species of animal and plant originated only once. The particular place where this occurred is known as its **center of origin.** The center of origin is not a single point, but the range of the population when the new species was formed. From its headquarters each species spreads out until halted by a barrier of some kind—physical, such as an ocean or mountain, environmental, such as an unfavorable climate, or biologic, such as the absence of food or the presence of enemy organisms which prey upon it or compete with it for food or shelter.

The Biogeographic Realms. Careful studies of the distribution of plants and animals over the earth have revealed the existence of six major biogeographic realms, each characterized by certain unique organisms. Although the divisions were originally based on mammalian distribution, they have since been found valid for many other classes of plants and animals. The areas of any one division are often widely separated, with great variations of climate and topography, but it has been possible, during most geologic ages, for organisms to pass more or less freely from one part of the realm to another. In contrast, the realms are separated from each other by major physical barriers.

The **Palearctic** realm comprises Europe, Africa north of the Sahara Desert, and Asia north of the Himalaya and Nan-Ling mountains, plus Japan, Iceland, the Azores and the Cape Verdes Islands. Some of the indigenous animals are moles, deer, oxen, sheep, goats, robins and magpies. A few species of some of these forms are also found in the Nearctic realm.

The **Nearctic** realm includes only Greenland and North America as far south as the northern plateau of Mexico. Besides many of the same forms characteristic of the Palearctic, it supports species of mountain goats, prairie dogs, opossums, skunks, raccoons, bluejays and turkey buzzards found nowhere else. The land bridge connecting North America and Asia at Bering Strait in former geologic times was used by many migrating animals and plants; hence the similarity of the fauna and flora of the Palearctic and Nearctic realms. Because the two regions have such similar forms, they are often referred to in combination as the **Holarctic** realm.

The **Neotropical** realm consists of South America, Central America, southern Mexico and the islands of the West Indies. It has a distinctive fauna, including alpacas, llamas, prehensile-tailed monkeys, bloodsucking bats, sloths, tapirs, anteaters and a host of bird species—toucans, puff birds, tinamous and others—found nowhere else in the world.

The part of Africa south of the Sahara, plus the island of Madagascar, makes up the **Ethiopian** realm. The gorilla, chimpanzee, zebra, rhinoceros, hippopotamus, giraffe, aardvark and many birds, reptiles and fishes live only in this region.

The **Oriental** realm includes India, Ceylon, Indo-China, southern China, the Malay peninsula and some of the islands of the East Indies. Outstanding of the animals peculiar to it are the orang-utan, black panther, Indian elephant, gibbon and tarsier.

The sixth and last realm, called the **Australian,** includes Australia, New Zealand, New Guinea and other islands of the East Indies. The imaginary dividing line between the Oriental and Australian realms, known as **Wallace's Line,** separates Bali and Lombok, goes through the straits of Macassar between Borneo and Celebes, and then passes east of the Philippines. Although the islands of Bali and Lombok are separated by a channel only 20 miles wide, their respective animals and plants are more unlike than are those of England and Japan—almost half a world apart. Native to the Australian realm are the duck-billed platypus, kangaroo, wombat, koala bear and other marsupials. And among its strange assortment of birds are the emu and cassowary (both flightless birds), the lyre-bird and cockatoo.

QUESTIONS

1. Give, briefly, the various types of evidence from living organisms for evolution. Which do you think is the strongest?
2. What is the theory of recapitulation? How has it been modified, and what is its significance in its present form?
3. Discuss the genetic basis of the theory of recapitulation.
4. Explain why the marsupials are so widespread in Australia and almost nonexistent elsewhere.
5. How do you explain the fact that the animals and plants of England and Japan are so similar, despite the fact that they lie nearly on opposite sides of the world?
6. Name some vestigial organs found in the human body. What functional organs are they the remnants of?
7. If, in tracing evolutionary relationships, anatomic evidence pointed one way and biochemical evidence the other, which do you think would be the more reliable? Why?

SUPPLEMENTARY READING

Ernest Baldwin's short *An Introduction to Comparative Biochemistry* is an interesting account of some of the biochemical similarities in different animals which point to evolutionary relationships. A rather detailed but readable discussion of the biochemical facts bearing on evolutionary theories is Marcel Florkin's *Biochemical Evolution.* A shorter account of the same general topic is George Wald's *Biochemical Evolution,* in E. S. G. Barron's *Trends in Physiology and Biochemistry.*

Chapter 36

The Evolution of Man

THE LINE of evolution leading from the ostracoderms, the primitive jawless fishes, to the mammals was traced in Chapter 34. The fossil records of horses, elephants, camels and many other mammals are quite complete, but those of the primates are regrettably fragmentary. Most of our primate ancestors lived in tropical forests, where animal remains are likely to undergo rapid decay before they can be fossilized. We can get some idea of what our ancestral primates looked like from the representatives of the primitive primates that have survived to the present. The earliest placental mammals were small, tree-dwelling, insect-eating animals; from these primitive, shrew-like insectivores have evolved all the kinds of placental mammals alive today.

325. MAN AND THE OTHER PRIMATES

The primate line appears to have begun with the **tree shrews,** which have characters intermediate between those of the insectivores and the primates. Fossil tree shrews have been found in oligocene deposits and a few tree shrews such as *Tupaia* (Fig. 294) still survive in the forests of Malaya and the Philippines. The tree shrews resemble squirrels with a long snout and tail, but have opposable first toes and their toes are tipped with flattened nails instead of claws. Primate evolution is in general characterized by adaptations for arboreal life; only in some of the larger apes and man has this been reversed.

The three suborders of the primates are the **lemuroids** (tree shrews, lemurs and lorises), **tarsioids** (tarsiers), and **anthropoids** (monkeys, apes and man). Primates are rather unspecialized mammals; they show some adaptations for arboreal life: prehensile hands and feet with opposable thumbs and great toes; digits tipped with flattened nails; long, flexible, mobile arms and legs; well developed brains; and binocular vision.

The **lemurs** (Fig. 295) live in the tropics of Africa and Asia and are especially abundant on the island of Madagascar. They are small, nocturnal, tree-living animals, superficially rather like squirrels, with a fairly long muzzle, eyes directed more to the side than forward, and a well-developed tail. Although some of the toes end in claws, the thumbs and big toes are covered by nails and are widely separated from the other digits. A complete skeleton of the Eocene lemur, *Notharctus,* has been found; it is very similar to the present-day lemurs.

The second suborder of the primates

Figure 294. Tupaia, a primitive tree shrew found in Malaya and the Philippines. (Photograph by F. W. Bond.)

is represented today by a single species of arboreal, nocturnal animal, called *Tarsius,* found in the East Indies. *Tarsius* is about the size of a rat, but has long hind legs and moves by hopping. It is characterized by enormous eyes, directed forward, and a relatively large brain. The muzzle is short, illustrating the beginning of the kind of face common to the higher primates. Its toes are long, slender and supplied with adhesive pads.

The third suborder, **Anthropoidea,** includes all the rest of the primates—monkeys, apes and man. These animals have a larger and more complicated brain than the other primates, eyes directed forward to provide stereoscopic vision, and a spot (the fovea) in the center of the retina where vision is especially acute. Most of the anthropoids normally walk on all fours, but all of them tend to sit upright, so that the hands are free to handle objects.

The suborder Anthropoidea is subdivided into two groups, the **Platyrrhines,** or broad-nosed monkeys, found in South and Central America, which have widely separated nostrils, directed forward and sideward, giving the nose a broad, flat appearance; and the **Catarrhines,** characterized by a much narrower nose with the nostrils closer together and directed downward.

The platyrrhines represent a group of primates, isolated in South America during the Tertiary, that underwent an evolution independent of the other primates. Living platyrrhines are the marmoset, or squirrel monkey; the capuchin monkey—the organ-grinder's companion; and the spider monkey (Fig. 296). Most of the platyrrhines have a well-developed prehensile tail used as a fifth hand for grasping objects and hanging from trees, and the group is sometimes called the prehensile-tailed monkeys.

The catarrhines include three groups—Old World monkeys, anthropoid apes and man—all of which have the same dental formula, a large brain, toes and fingers with flattened nails, and either a short, nonprehensile tail or none. The group of Old World monkeys is large, including the macaque, guenon (Fig. 297), mandrill, mangabey, baboon, langur, proboscis monkey and many others. All these monkeys tend to sit upright, and all have buttocks with bare, hardened sitting-pads, called **ischial callosities,** which are frequently a brilliant red or blue. The mandrills and baboons have taken to living on the ground instead of in trees, but they walk on all fours, have an elongated snout and large canine teeth. Baboons are intelligent animals which travel in troops and cooperate in obtaining food and protecting the females and young.

The second group of catarrhines—the anthropoid or manlike apes—includes four living types: the gibbon, orang-utan, chimpanzee and gorilla. These apes have rudimentary tails or none at all, arms longer than their legs, a semierect posture, opposable thumbs or great toes, and chests that are broad like man's rather than thin and deep like that of most mammals. Their brains are larger than those of the lower primates and more like man's in the pattern and relative size of the parts. They range in size from the 3-foot gibbon to the gorilla, which is 6 feet tall and weighs as much as 600 pounds.

Gibbons (Fig. 298), found in Malaya, have arms so long that they reach the ground when the animal walks erect. The usual mode of locomotion, called brachiating ("arming"), is by swinging through the trees from limb to limb, which allows the animal to clear as much as 40 feet at a time. The spectacular aerial acrobatics of the gibbon require great agility, coordination, good eyesight and the ability to make rapid judgments of distance and possible landing sites.

The orang-utan (Fig. 299), a native of Borneo and Sumatra, is a bulky, powerful animal covered with long, reddish brown hair. Although it is short-legged and scarcely 5 feet tall, it may weigh as much as 160 pounds. Orangs have enormously long arms, with a span of 7 or 8 feet, and long, slender hands and feet. They are successful arboreal animals but because of their weight they move more deliberately than the gibbons do. Orangs eat fruit and leaves and build nests in trees.

Chimpanzees and gorillas both live in Africa and have many characteristics in common. An adult chimpanzee (Fig. 300) weighs about 110 pounds, and is between 5 and 5½ feet tall. Like the orang-utans, chimpanzees build nests in trees at night and for noonday siestas. Although primarily tree-dwellers, they are quite at home on the ground for their legs are longer and their arms shorter than those of an orang. Psychological studies of chimpanzees and gorillas have shown that they are curious, perceptive, able to reason, and have strong emotions and social instincts. Gorillas are less docile, imitative and suggestible than chimpanzees.

Gorillas (Fig. 301) are not only the largest of the primates; they are, weight for weight, the strongest—several times stronger than man. The gorilla's legs are relatively short, the arms are more human in their proportions than those of the other apes, and the hands are relatively short and wide—quite like human hands. The massive head has large, bony crests on top of the skull for the attachment of the neck and jaw muscles. Gorillas are ground-dwellers, although they occasionally build sleeping nests in low trees, and have feet adapted for walking rather than for swinging through the trees. Like man, they walk on the soles of the feet with the toes extended, rather than curled under in the fashion of the other apes. Gorillas normally walk on all fours, but usually rear up on their hind legs when attacking. Although they will accept meat, they are ordinarily vegetarians, preferring such foods as bananas, carrots and nuts.

Man is more like the chimpanzee and gorilla than the other primates but differs in enough characters to be placed in a separate family, the **Hominidae.** The differences between the other great apes and ourselves are rather small differences in the proportion of parts, correlated with our adaptation for a terrestrial rather than an arboreal life. Almost every bone, muscle, internal organ and blood vessel of the ape is repeated in man. Some of the characters which distinguish man are: (1) Man's brain is larger, being two and a half to three times as large as the gorilla's. (2) The human nose has a prominent bridge and a peculiar, elongated tip. (3) Man's upper lip has a median furrow, and his lips are rolled outward, revealing the mucous membranes. (4) Man has a jutting chin, and

Figure 295. Ring-tailed lemur, one of the most primitive of living primates. (Courtesy of the San Diego Zoo

apes have none. (5) Man's great toe is not opposable, but is in line with the others. (6) The human foot is adapted for bearing weight by being arched both lengthwise and crosswise. (7) Man is relatively hairless. (8) Man's canine teeth project little, if at all, beyond the line of the other teeth. (9) Man has an erect posture. (10) Man's legs are longer than his arms.

No single ape resembles man in all traits more than do the rest. The gorilla, for instance, has hands, feet and a pelvis more like man's than are those of any other ape, but the chimpanzee's skull and skin color are the most like the human being's. The orang-utan is the only ape to have exactly the same number of ribs that we have, and it also has our high forehead, while the gibbon most closely resembles us in the relative length of its legs, in its posture and its gait on the ground. With respect to any structure or proportion of parts, the difference between man, the gorilla and the chimpanzee is less than the difference between any of these animals and the monkeys.

326. FOSSIL PRIMATES

The earliest known primate fossils are those of lemuroids and tarsioids, found in Paleocene rocks in both North America and Europe. These animals are believed to have descended from the tree-living insectivores of the late Cretaceous, which resembled present-day tree shrews. Since that time the lemurs have evolved separately and have not given rise to any other forms. The tarsioids, in contrast, gave rise to both New and Old World monkeys, each of which evolved separately from different tarsioids during the Eocene. The New World monkeys have had a separate evolutionary history since the Eocene and represent side branches rather than segments of the main evolutionary trunk leading to man.

The oldest Old World monkey, *Parapithecus,* fossils of which have been found in Lower Oligocene rocks in Egypt, was smaller than any modern monkey or ape, and was near the stem leading to man (it had the same dental formula). This animal probably represents the common ancestor of today's Old World monkeys, anthropoid apes and man. From this point, the evolution of the modern Old World monkeys diverged from that of the higher primates.

The rocks of the Lower Oligocene also contain fossils of the first anthropoid ape, named *Propliopithecus,* which probably descended from *Parapithecus* and was close to the common ancestor of man and all anthropoid apes. This little gibbon-like animal showed the first adaptation toward an upright sitting posture. In all apes the tail has disappeared, and the muscles which formerly wagged the tail have spread out to cover the floor of the pelvis and help support the abdominal organs. Another human-like characteristic of *Propliopithecus* was the balanced position of the skull on the end of the vertebral column.

During the Miocene, giant primates evolved from these primitive anthropoids, and by the middle of that epoch the evolutionary lines leading to the various types of modern anthropoids were distinct. The orang-utans probably differentiated from the common stock earlier than the others, for fossils of the ancestor of the orang-utan, *Paleosimia,* have been found in India in Miocene deposits. Fossils of *Limnopithecus,* believed to be ancestral to the gibbons, and *Proconsul,*

Figure 296. Spider monkey, a New-World monkey with a strong prehensile tail, used in swinging from tree to tree. (Courtesy of the San Diego Zoo.)

Figure 297. Old World monkeys (De Brazza guenons); an adult pair and a six month old infant. (Courtesy of the San Diego Zoo.)

Figure 298. An anthropoid, the white-banded gibbon. These apes use their long arms to swing from tree to tree with great agility. (Courtesy of the San Diego Zoo.)

Figure 299. An eight year old male orang-utan. Note the opposable big toe on the handlike foot. (Courtesy of the San Diego Zoo.)

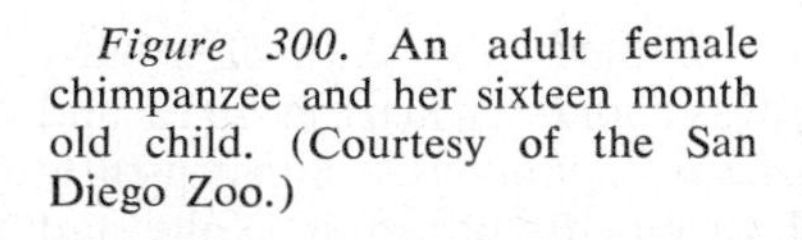

Figure 300. An adult female chimpanzee and her sixteen month old child. (Courtesy of the San Diego Zoo.)

Figure 301. Mbongo, an adult male mountain gorilla. When this picture was taken, he was thirteen years old and weighed 602 pounds. Gorillas normally assume the four-footed posture illustrated when walking on the ground. The massive crest on the top of the head, peculiar to the mountain gorilla, is absent from the lowland forms of western Africa. (Courtesy of the San Diego Zoo.)

on the line of evolution of the other apes, have been found in lower Miocene deposits in Africa. The animal which comes closest to being the common ancestor of the gorilla, chimpanzee and man is *Dryopithecus* (Fig. 302), an anthropoid ape of the Miocene with characteristics found today only in these three groups.

327. THE MAN-APES

Fossil anthropoids that almost bridge the gap from ape to man have been found in Pleistocene cave deposits in South Africa. These man-apes probably lived too recently to be man's ancestors but they illustrate the kind of changes by which the transition from ape to man was made. They are now regarded as "progressive apes," adapted for walking upright on the ground, that evolved independently of the human line from common dryopithecine ancestors in the Miocene.

The first of these, the skull of a baby man-ape, was found by Dart in 1925 and named *Australopithecus*. Subsequently Dart and Broom found adult skulls and parts of skeletons. Although these were given separate names, *Plesianthropus* and *Paranthropus,* they probably represent animals very closely related to, if not identical with, the original *Australopithecus*. These australopithecines have an interesting mélange of human and apelike characters. The skull and face were apelike, with a low-vaulted skull, protruding muzzle and heavy jaws. The brain capacity was 650 ml., larger than any known ape and almost as large as the earliest ape-men. The cheek bone, jaw hinge and teeth were very similar to man's, with small canine teeth and molars that resembled ours. These man-apes lived in caves, hunted animals, and may have learned how to use fire. The structure of the pelvis and leg bones and the location of the foramen magnum in the skull suggest that these creatures had a fairly erect posture. The largest australopithecine, the Swartkrans man-ape found in 1949, was a veritable giant, larger and heavier than the largest gorillas.

328. FOSSIL APE-MEN

The human stock appears to have diverged from the other apes some time after the Miocene. The remains of a number of creatures with characters intermediate between the fossil apes and living man have been found in Pliocene and Pleistocene deposits in widely scattered parts of Europe, Asia and Africa. These remains are mostly pieces of skull, jawbones and teeth, although some other skeletal parts have been found. It is understandable that the fossil record of man's immediate ancestors is incomplete, for those animals were too intelligent to be caught in quicksands or tarpits, and, being primarily forest-dwellers, their dead bodies must have been quickly and completely devoured by other animals. Furthermore, at about this stage, burial customs arose involving the cremation of the dead. The characteristics which distinguish man from the apes did not appear simultaneously in a single form, for although the ape-men are essentially human in some respects, they are apelike in others.

Whether these were apes or men is, perhaps, a matter of definition, but they were large-brained anthropoids who walked erect, had well formed hands and used tools. We have a fairly clear idea of what these ape-men looked like from their fossil remains, and we also know quite a bit about how they lived from the tools, weapons, ornaments and other cultural remains that have been found.

The Java Man (*Pithecanthropus erectus*). One of the most primitive ape-men was *Pithecanthropus,* the Java man whose remains were found in 1891 in Pleistocene deposits on the banks of the Solo River in eastern Java. He had a brain capacity of about 940 ml., intermediate between modern man's 1200 to 1500 ml. and the 600 ml. of the gorilla. By studying casts of the interiors of these skulls, the contours and relative proportions of the various parts of the brain can be determined, and apparently *Pithecanthropus* had the part of the brain which controls speech, though whether he actually did speak is unknown. Java man's brain was much larger and more convoluted than that of any of the primitive or living apes and was more human than simian. Reconstructions from the skeletal parts that have been found indicate that an adult was about 5 feet 8 inches tall, weighed 154 pounds and walked erect. His face was projecting and chinless, his jaws were massive and equipped with a huge set of teeth (although the canine teeth were not enlarged tusks as in the apes), he had a broad, low-bridged nose and a heavy, bony ridge over the eyes. It is probable that he had learned to use and make tools, for the rock strata bearing his remains also contain primitive stone implements. Java men probably traveled in small, family groups, living in caves and hunting in the forests. *Pithecanthropus* is believed to represent the remains of an archaic type which originated in the Pliocene and was once widespread throughout Asia.

Since 1891 when the first specimens of *Pithecanthropus* were found, excavations have turned up several more *Pithecanthropus* fossils, as well as the remains of a larger and apparently earlier ape-man, called *Meganthropus,* which is possibly a direct ancestor of the so-called Java man.

Peking Man (*Sinanthropus pekinensis*). Investigations during the early 1920's of certain limestone caves near Peking, China, revealed many animal fossils, among them two teeth which had belonged to a primitive ape-man of the middle Pleistocene, about 500,000 years

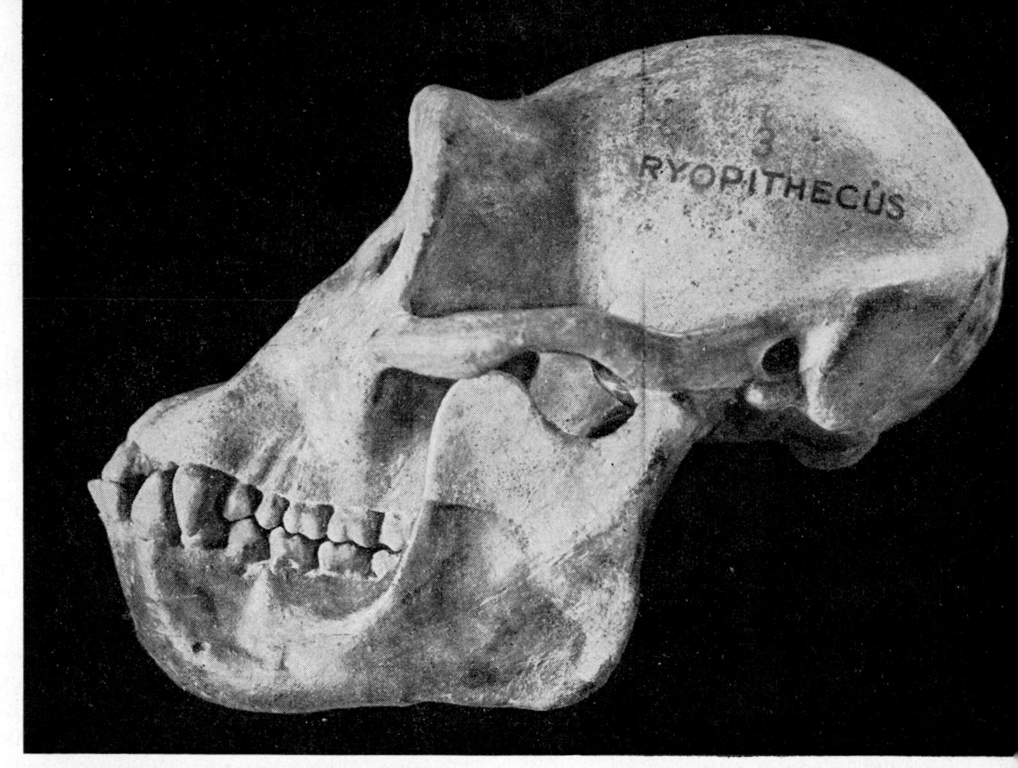

Figure 302. Restored skull of *Dryopithecus,* the fossil "oak ape," possibly the ancestor common to gorillas, chimpanzees and man. (Courtesy of the American Museum of Natural History, New York.)

ago. Their discoverer, Davidson Black, named them *Sinanthropus pekinensis.* Further excavations in these caves revealed parts of more than forty individuals—males and females, old and young—of the same species. It is now possible to reconstruct fairly accurately what this ancient ape-man looked like. *Sinanthropus* had a skull much like that of *Pithecanthropus,* with heavy, bony ridges over the eyes, and a low, slanting forehead (Fig. 303). His jaw was massive, chinless and rather apelike, and his nose was broad and flat. The fact that the remains fall into two distinct groups, one composed of much larger specimens than the other, suggests that the difference between the size of males and females was greater than in modern man. *Sinanthropus* had a distinctly larger brain than *Pithecanthropus,* the capacity ranging from 850 to 1300 ml. and averaging 1075 ml. The condition of the skulls with their bases broken open suggests that Peking man was a cannibal. As more specimens of Java and Peking men have turned up, it has become clear that the two were similar, probably representing two races of a single species. The anthropologist who has studied *Sinanthropus* most intensively, Franz Weidenreich, found that Java and Peking man are identical in fifty-seven out of seventy-four characters of the skull, and that there are clear differences in only four characters, one of which is the difference in size. He has suggested that they be renamed *Homo erectus erectus* (Java man) and *Homo erectus pekinensis.*

A curious anthropologic story is that of *Gigantopithecus,* the "Hong Kong Drugstore Giant." In the late 1930's von Koenigswald bought a number of fossil teeth in a Chinese apothecary shop, some of which he identified as those of an orang-utan. Three of them, however, were larger than those of any known primate. Their discoverer and Weidenreich believe them to be the remains of a giant ape-man which they named *Gigantopithecus,* related to *Meganthropus* and the Java and Peking men.

Piltdown Man (*Eoanthropus dawsoni*). Between 1908 and 1915 a British lawyer named Dawson, whose hobby was collecting fossils, found a number of skull pieces and a lower jaw in gravel pits on Piltdown common, Sussex (Fig. 304). The gravel deposits dated from the early

Pleistocene, more than 800,000 years ago, and the name *Eoanthropus* (dawn man) seemed appropriate for the remains. The skull bones were thick—twice as thick as those of modern man—but in other respects the skull was similar to modern man's: The forehead was high, there were no ridges over the eyes, and the brain capacity was about 1350 ml. In sharp contrast, the jaw was quite ape-like—chinless and with projecting canine teeth. At the time of the discovery many anthropologists thought that the skull and jaw actually belonged to two different animals and had merely been found in the same gravel pit by chance. The existence of these remains, of such antiquity, led many anthropologists to believe that modern man was descended from Piltdown man rather than from Java, Peking, or Neanderthal man.

Because of the disparity between the skull and jaw, however, other authorities doubted the authenticity of the remains. Bones that remain in the ground for a long time gradually accumulate fluorine. In 1953, Oxford Professors W. E. Le Gros Clark and J. E. Weiner analyzed bits of these remains and found that there was not enough fluorine present for the bones to be very ancient; Piltdown man is, in fact, one of the major scientific hoaxes of the century! The skull fragments proved to be perhaps 50,000 years old, genuine fossils but no older than many bones found throughout Europe. But the jawbone was that of a present-day ape, probably an orang-utan. The jaw had been artificially colored with an iron salt and bichromate to make it look yellow and old and the teeth had been filed to make them look more human. Since Dawson died in 1916 it is impossible to know whether he himself faked his finds to gain fame or whether someone else planted them to fool the experts. In contrast the remains of the other ape-men are too numerous and too well authenticated to be the creation of some scientific joker.

329. FOSSIL MEMBERS OF THE GENUS HOMO

The fossils of primitive man found in Europe, Asia and Africa are slightly different, but similar enough to be grouped together as the Neanderthaloids. This group, which includes Heidelberg man, Neanderthal man, Solo man and Rhodesian man, probably descended from the pithecanthropoids.

Heidelberg Man (*Homo heidelbergensis*). Heidelberg man is known only from a massive lower jaw found buried under 80 feet of sand in a pit near Heidelberg, Germany. The jaw is large and heavy and lacks a chin but the teeth are of moderate size and generally like modern man's teeth. The canines do not protrude beyond the other teeth and the pulp cavities are large and extend into the jaw. In many respects the jaw resembles that

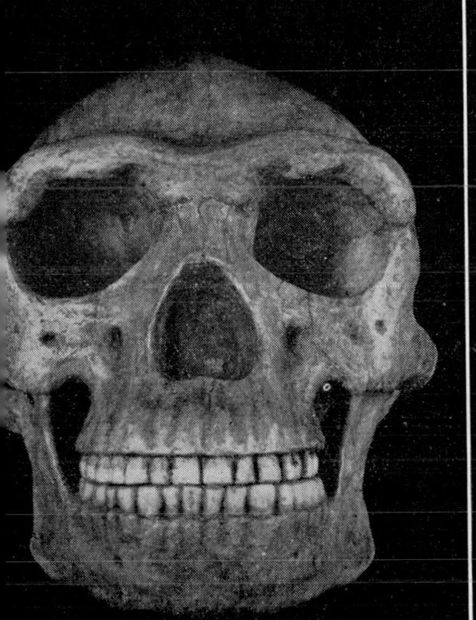

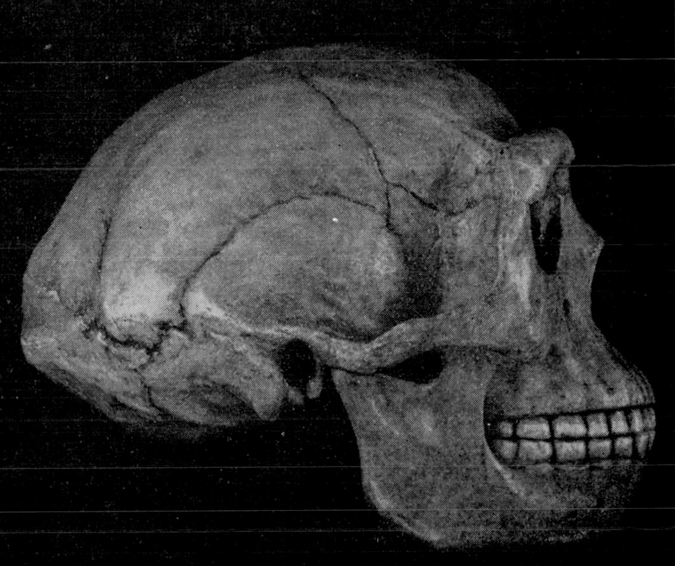

Figure 303. Front and side views of a reconstructed skull of Peking man, *Sinanthropus pekinensis*. Note the massive bony ridges over the eyes, the low, retreating forehead, the protruding jaws and the absence of a chin. (Courtesy of the American Museum of Natural History, New York.)

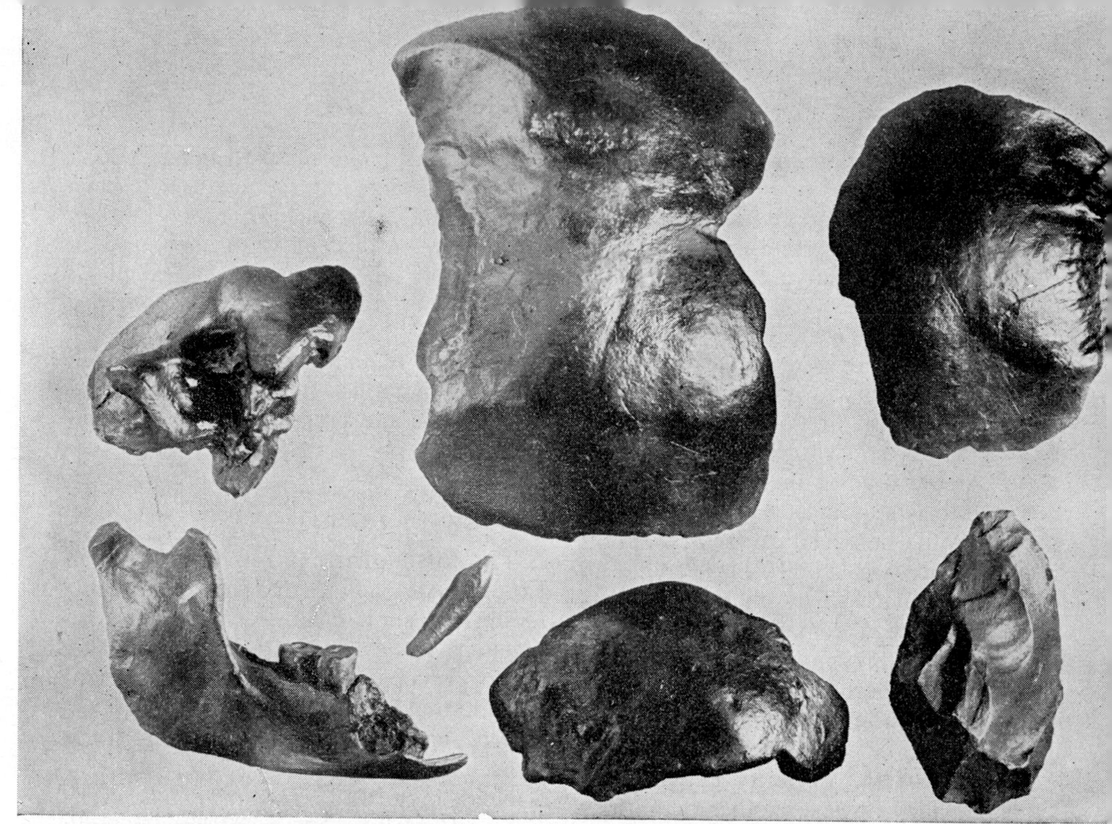

Figure 304. The Piltdown man remains, shown in 1953 to be a major scientific hoax. The skull fragments are actually from a skull, some 50,000 years old, of a man not too different from modern man. The jaw (lower left) was shown to be that of a present-day ape doctored to look like a fossil. The object in the lower right is a flint instrument. (Courtesy of the American Museum of Natural History, New York.)

of Neanderthal man, and Heidelberg man is generally regarded as an ancestor of the Neanderthals. Since the fossils associated with the jaw are those of warm-climate animals—rhinoceroses and lions—Heidelberg man must have lived during one of the interglacial periods, but it is not clear whether it was the first or second of these; in any case he lived over 500,000 years ago.

Neanderthal Man (*Homo neanderthalensis*). The first human fossil discovered was found in 1856 in a cave of the Neander valley near Düsseldorf, and it was given the name *Homo neanderthalensis,* or Neanderthal man. The skull excited a lively controversy at first, some scientists correctly guessing it to be the remains of a primitive man, others guessing it to be the skull of a congenital idiot, and one surmising that it was simply the skull of a Russian soldier killed in the Napoleonic wars. Since that time similar skulls have turned up in widely separated parts of Europe, Asia Minor, North Africa, Siberia and the islands of the Mediterranean. These remains are always associated with a particular Stone Age culture known as the *Mousterian* (named after le Moustier cave on the bank of the Vézère River in France). Neanderthals lived in Europe for thousands of years during and after the third and final interglacial period (about 150,000 years ago), and became extinct only about 25,000 years ago.

Neanderthal man was between 5 and 5½ feet tall, and powerfully built; he walked upright, with bent knees. His skull was large and massive with a thick, bony ridge over the eyes, and a receding forehead (Fig. 305). His nose was broad and short, and he had almost no chin at all. Despite these rather apelike features, Neanderthal man had a brain as large as or larger than that of modern man, its capacity being about 1550 ml. The proportions of the parts of the brain, estimated from casts, indicate that, in general intelligence, he was probably quite similar to modern man, although his frontal lobes were smaller. Neanderthals lived primarily in caves, used fire, made flint weapons, and buried their dead reverently with food and ornaments.

Solo Man. On the banks of the Solo River in Java, just a few miles from the site where the *Pithecanthropus* remains were buried, have been found fossils of a second human type, which used both stone and bone implements. These men, considerably more advanced than *Pithecanthropus,* probably were descendants of that species. Eleven skulls of this new type (*Homo soloensis*) have been discovered since 1936, all with their bases smashed in, suggesting that Solo man, like his Peking cousin, was a cannibal who considered brains a delicacy. Like the Neanderthals, Solo men had a heavy ridge over their eyes and a receding forehead; the brain capacity was about 1300 ml., below the average for Neanderthals, but above that of the earlier Java man. Weidenreich has made a detailed study of these skulls and concludes that they are more primitive than Neanderthal skulls.

It is believed that the Australian bushmen are the present-day descendants of Solo man. In 1940 two skulls definitely of Pleistocene age were found at Keilor, near Melbourne, Australia, and although they have a large cranial capacity (nearly 1600 ml.), they are intermediate in other respects between Solo man and the modern aboriginal Australian.

Rhodesian Man (*Homo rhodesiensis*). Still another primitive skull, to which the name Rhodesian man has been given, was found in 1921 at Broken Hill, Rhodesia, in a large limestone cave. The skull, which was in an almost perfect state, has an extremely large eyebrow ridge and a receding forehead like a gorilla's, but a cranial capacity of about 1300 ml. The teeth are large, but definitely human rather than apelike, and are badly decayed—a condition never found in apes and only rarely in primitive man. The relationship of *Homo rhodesiensis* to the other extinct and living species of man is obscure; he lived either in the late Pliocene or the early Pleistocene, and in some respects is more primitive than Java man.)

Modern Man (*Homo sapiens*). The species *Homo sapiens* includes not only all the living races of man, but also some extinct ones, such as the Cro-Magnon. The idea that *Homo sapiens* appeared relatively recently in the late Pleistocene is no longer valid, for the antiquity of the Swanscombe skull is now well authenticated by the fluorine test. *Homo sapiens* was a contemporary of Neanderthal man and perhaps antedates him.

The **Galley Hill** fossil, found in 1888 in the Thames valley below London, is a nearly complete human skeleton, and the gravel pit containing it is of the Middle Pleistocene (about 500,000 years ago). Galley Hill man was short (5 feet 3 inch-

Figure 305. Skull cap of Neanderthal man. Note the heavy eyebrow ridges to the right and the extreme thickness of the bone. (Courtesy of the American Museum of Natural History, New York.)

es) and stocky, with no feature more apelike than some exhibited by living human races. His brain capacity was about 1400 ml. and the cast of his brain shows an essentially modern development of the various brain regions. The skull bones are quite thick, but the eyebrow ridges are not excessively large. The actual age of these remains is in dispute because the skeleton was removed before a qualified geologist could attest to the antiquity of the deposit. The finding of the **Swanscombe** skull, which is essentially modern in shape and size, though differing in the greater thickness of the bones, in nearby Middle Pleistocene deposits renders likely the opinion of Sir Arthur Keith and Earnest A. Hooton that Galley Hill man also dates from the Middle Pleistocene. These investigators regard the Swanscombe and Galley Hill fossils as definitely of the species *Homo sapiens*.

Since 1868 about 100 unmistakably *Homo sapiens* fossils, dating back 15,000 to 60,000 years, have been found in western and central Europe, especially in France. In the valley of the Vézère River in south central France the first well-preserved fossils of this type were discovered in the so-called **Cro-Magnon** rock shelters. At first these remains were thought to be representative of a single race, and were referred to collectively as the Cro-Magnon remains. Indeed, they do share certain characteristics, such as a long, massive skull without eyebrow ridges and with a prominent chin and a high forehead, and a rather large brain capacity (as great as 1800 ml.). But further studies indicated that the fossils were not homogeneous enough to belong to a single race. Whatever their origins, they were probably contemporaries of the Neanderthal species, and may have hastened its extinction.

Excavations during the early 1930's, near Mt. Carmel in Palestine, disclosed a number of skeletons of Upper Pleistocene age, associated with Mousterian stone implements. The skeletons show an odd mixture of Neanderthal characteristics with others like those of the Cro-Magnon remains. The eye ridges are smaller than those of Neanderthal skulls, the forehead is less receding, the face is neither so large nor so long, the nose is narrower, and there is a fairly well-developed chin. The brain capacity varies from 1300 ml. to 1500 ml. and, in the development of the frontal lobes, resembles that of modern man more than that of the Neanderthal species. It appears that these remains are the result of hybridization between *Homo neanderthalensis* and early members of *Homo sapiens,* and many anthropologists believe that the Neanderthals did not become extinct in the strict sense of the term, but were absorbed by interbreeding with the various progressive and genetically dominant races of *Homo sapiens*.

Recent excavations in South Africa have revealed several fossils of *Homo sapiens* who died 20,000 or so years ago. In the upper strata of the same caves near Peking in which the ape-man, *Sinanthropus,* was discovered, the remains of *Homo sapiens* of 75,000 years ago were uncovered. Among the latter, seven skeletons, belonging perhaps to members of the same family, were found together, apparently the victims of a human raid, for the skulls are all fractured. These are especially interesting because of the great variation among them, one of the men being like the European Cro-Magnon, one of the women resembling a modern Eskimo, and another a Melanesian. This heterogeneity suggests extensive hybridization among the various races of the time.

The center of origin of modern man appears to have been in Asia, in the general region of the Caspian Sea. The white races spread westward around both shores of the Mediterranean to Europe, Southwestern Asia and North Africa, displacing the Cro-Magnons who had in turn displaced the earlier Neanderthalers. Some of the inhabitants of Ireland and Scandinavia, and the Basques of Southern France and Northern Spain, show marked similarities to Cro-Magnons and may represent their descendants who were pushed westward by the migrating Neolithic men.

The Negroid races spread south on both sides of the Indian Ocean to Africa and Melanesia. It appears that they displaced more primitive races and pushed the Bushmen to the tip of South Africa and the Australoids into Australia and Tasmania.

The Mongoloids spread east and north, occupying Siberia and China. About 40,000 years ago they crossed the Bering Straits and occupied North and South America.

In the course of his evolution from apemen man has not increased greatly in height, and his frame has become less massive. He now stands completely erect, and his head is balanced on a relatively slender neck, instead of being held in place by a massive set of neck muscles, and jutting forward from the shoulders. His cranial capacity has increased, the forehead has become more vertical, the bony ridges over the eyes have diminished, and the face—particularly the jaws—has become smaller in relation to the rest of the skull (Fig. 307). Correlated with the reduction in the jaw size has been a reduction in the size and complexity of the teeth, and now there is a strong tendency for the third molars (the wisdom teeth) to be vestigial. The evolutionary trend toward greater intelligence made man less dependent upon sheer physical strength

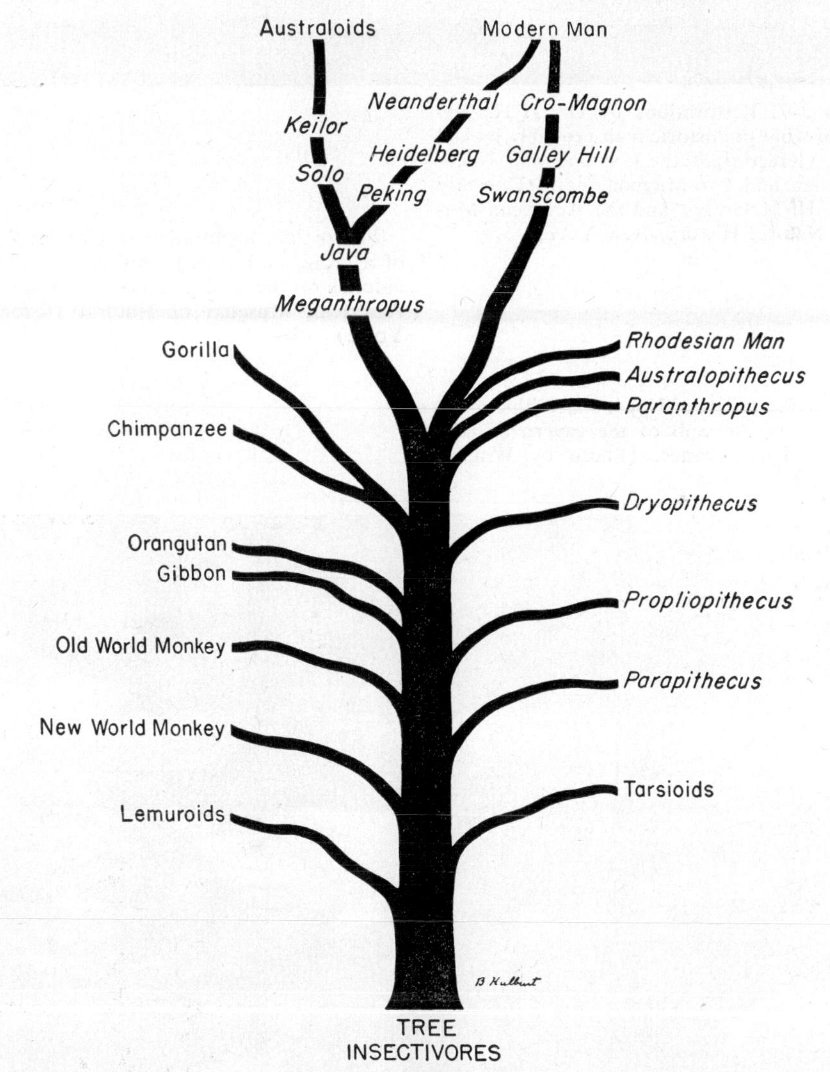

Figure 306. An evolutionary tree of the primates, beginning with the primitive tree insectivores. The forms known only as fossils are indicated in italics.

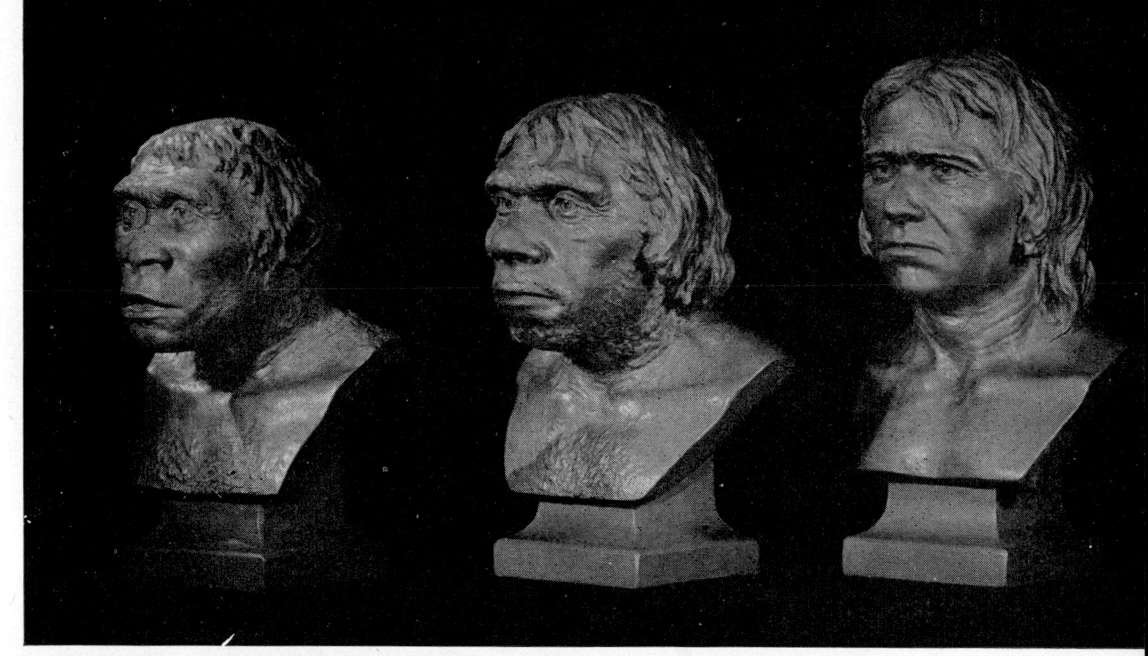

Figure 307. Restorations by Dr. J. H. McGregor of what prehistoric men probably looked like. From left to right, the Java ape-man, Neanderthal man and Cro-Magnon man. (Courtesy of Dr. J. H. McGregor and the American Museum of Natural History, New York.)

Figure 308. Restoration by Charles R. Knight of a group of Upper Paleolithic artists drawing animals on the wall of a cave. (Courtesy of the American Museum of Natural History, New York.)

Figure 309. The art of Upper Paleolithic man: paintings from the wall of the cavern of Lascaux, Dordogne, France. (Photo by Windels Montignac.)

for getting food and fighting enemies. Speech, tools and weapons were developed and man began to live in clans and tribes, completing the transition from his ancestral solitary arboreal primates to a ground-dwelling, civilized animal (Fig. 306).

330. CULTURAL EVOLUTION

Although most of the evidence of the path of human evolution comes from the actual fossils just discussed, some corroborative evidence comes from cultural implements, such as tools, weapons, cook-

ing utensils and ornaments. The science of **archeology,** concerned with the cultural significance of such objects, is complex and fascinating, and we can do no more here than indicate its importance.

Like fossils, the objects made and used by man, called **artifacts,** were deposited at widely separated times, and so are found in different layers of the soil, the later ones usually lying above the earlier ones. Thus, when fossils and artifacts are found together, and the date of the culture associated with the artifacts is known, the anthropologist is able to determine the age of the fossils at once.

Although early man must have learned to pick up and use stones of a convenient size and shape, it was not until the middle Pleistocene, apparently, that he learned how to chip pieces of flint to make hand axes. The culture characterized by these chipped stone tools is called the **Lower Paleolithic,** and was the culture of Java and Peking man. These men made their homes in caves and were hunters and food gatherers with no knowledge of agriculture or animal domestication. They did, however, understand the value of fire and knew how to make it. The association of certain kinds of axes and scraping tools with the Java and Peking fossils enables investigators to study the distribution of this type of ape-man, for similar artifacts, unaccompanied by skeletal remains, have been found in India and Burma.

In deposits dating from the third interglacial and the last glacial periods, different, more advanced tools have been found, and the culture they represent has been named the **Middle Paleolithic.** The European branch of this culture, called the **Mousterian,** was associated with Neanderthal man. The implements of the Mousterian culture were made by chipping flakes from a piece of flint and then sharpening the edges by removing more flakes with a bone tool. The most common weapon seems to have been a triangular piece of stone, the forerunner of the spear and the arrowhead.

Later, in the **Upper Paleolithic** culture, an improved method of tool making was discovered. Flakes were removed from the pieces of flint by steadily and carefully applied pressure rather than by blows. This produced long, slender, knifelike blades, many of which were elaborately and skillfully carved, and were true works of art. These Upper Paleolithic men, Cro-Magnons and others, were painters as well as skilled craftsmen, for their cave paintings, found in France and Spain, show a remarkable grasp of the principles of design (Figs. 308, 309). New caves found in the valley of the Dordogne in 1948 and 1956 have a wealth of beautifully preserved paintings of contemporary animals.

The **Mesolithic,** or **Middle Stone Age,** culture shows no important advance over the Paleolithic cultures. Mesolithic man was still a hunter and food gatherer, although he did domesticate the dog. His tools were smaller and poorer in quality than those of his Upper Paleolithic predecessors. Perhaps, however, he attempted to write, for pebbles of this age have been found marked with red ochre dots, bars and crosses. He lived in small, isolated breeding groups which would favor the occurrence of genetic drift and lead to the formation of divergent groups.

The **Neolithic** or **New Stone Age** culture originated in the Near East, somewhere between Egypt and India. It is characterized by implements which bear the marks of careful grinding and polishing and by the beginning of agriculture and animal husbandry. The earliest animals to be domesticated, after the dog, were the pig, sheep, goat and cow; the horse was not domesticated until much later. Man gradually changed from a wandering hunter and food gatherer to a settled food producer, raising grain, making pottery and cloth, and living in villages. The increase in the food supply led to an increase in the size of the population, breeding groups became larger and interbred with neighboring ones, and the tendency toward genetic drift was greatly decreased. The remains of bowls, pitchers and other utensils have been extremely helpful to archeologists, for each cultural group since that time has used certain peculiar methods in making and decorating such things. Other inventions of the ingenious Neolithic people are the dugout canoe and the wheel.

With the Neolithic age we come to historical times, for the oldest Egyptian and Mesopotamian cultures were Neolithic. The use of metals—first copper, then bronze (a copper-tin alloy)—for making vessels, tools and weapons began sometime between 4000 and 3000 B.C., and about 1400 B.C. men of the Near East initiated the Iron Age by mastering the technique of deriving iron from its ores.

331. THE PRESENT RACES OF MAN

The differences between the various races of man living today are just as great as those between a number of other related animals generally regarded as different species, and a few anthropologists and geneticists—for example, R. R. Gates (*Human Ancestry,* 1948)—argue that they are in fact different species. There can be no final answer to a question which is largely a matter of definition, but since the word "species" is usually used to imply interfertility, and since all the modern races of man are interfertile—as far as anyone knows—most biologists agree to consider them all members of *Homo sapiens.*

The qualities which distinguish races are physical and physiologic ones, similar to but smaller than those distinguishing the members of the larger taxonomic categories, such as phyla or classes. A race may be defined genetically as a population which differs significantly from other populations with respect to the frequency of one or more genes. It may also be defined phenotypically as a population whose members, though varying individually, are distinguished as a group by a certain combination of morphologic and physiologic traits which they share because of their common descent.

It should be clear by now that, ultimately, all the present races had a common origin, and that probably there is no such thing as a "pure" race. Anthropologists themselves are not agreed as to where the lines should be drawn, and, indeed, any individual may be phenotypically so unlike his parents as to be in a separate group. Furthermore, any particular quality, such as skin color, varies tremendously within each race, so that a member of the white race may have skin as dark as a typical Negro's and a Chinese may have skin as white as a Caucasian's.

Skin color, the color of the hair and eyes, the waviness and texture of the hair, the shape of the head and its features, the arrangement of the whorls and loops on the skin of the fingertips, and the proportions of the various parts of the body, are some of the important characters differentiating human races. Anthropologists pay particular attention to the ratio between the breadth and length of the head. When measurements of a living person's head are made, this ratio is called the **cephalic index;** when measurements are made on a skull, the ratio is called the **cranial index.** By convention, a skull with a breadth less than 75 per cent of its length is called longheaded or **dolichocephalic,** one with a cranial index of 80 or more is said to be broad, roundheaded or **brachycephalic,** and a skull with a cranial index between 75 and 80 is said to be **mesocephalic.** Similarly, noses are classified according to the ratio of breadth to length—the **nasal index**—and faces are classified on the basis of the **facial index,** the ratio of length to breadth.

Some of the physiologic characteristics in which the people of the world differ significantly are the chemistry of the blood, the rate of the basal metabolism and susceptibility to certain diseases.

The White Race. The white or Caucasian race probably includes a greater diversity of types than does either of the other two main races; in fact, so great is the diversity that only a few characteristics are shared by all members of the race. Generally speaking, however, the following traits are typical of this division of the human species: a fairly light skin, ranging in color from light olive-brown through white to ruddy; medium to fine wavy or straight hair; fairly abundant facial and body hair; medium to thin lips; a moderately pronounced chin; and a broad pelvis. Women of the white race have hemispherical breasts and rather prominent buttocks.

There is no single method for classifying the members of the white race into subgroups, and no universally accepted opinion as to exactly how many subgroups there are. But most anthropologists agree that three main divisions do exist, called **Mediterranean, Nordic** and **Alpine.***

The Mediterranean group includes many Egyptians, Arabians, Portuguese, Spaniards, southern Italians and others, and is centralized between the Mediterranean and East Africa, the Indian Ocean and the Caspian Sea and between the Egyptian Delta and India. Typically, the people of this group have long heads, slender, slight bodies, olive or light-brown skin, black or brown hair that is straight or slightly wavy, and dark brown eyes. The Mediterranean race is probably the oldest and least specialized of any of the modern races.

The Nordic race, which is a partially depigmentized branch of the Mediterranean group, includes many Swedes, Norwegians, Englishmen, Scotsmen and north Germans. Its members are characterized by long heads, red, blonde or light-brown hair, blue or gray eyes, long, narrow faces with high, narrow noses, thin lips, ruddy skin and tall, slender bodies.

In contrast to the groups just described, our third main group, the Alpine, consists of broadheaded people whose ancestry includes Upper Paleolithic strains, and may well include some Neanderthal intermixture. Since the skulls of fossil men are longheaded almost without exception, brachycephaly apparently arose recently in human evolution; certainly it is not characteristic of the old Mediterranean stock. Unlike some Upper Paleolithic survivors, the Alpine people show a reduction in head and body size, being of medium stature, with round faces as well as heads, and of moderately stocky build. In hair and eye color they are usually intermediate between the dark Mediterraneans and the blonde Nordics. Mem-

* There are different names for them; these terms and the descriptions that follow are based on C. S. Coon's *The Races of Europe* and E. A. Hooton's *Up from the Ape*.

bers of the Alpine group are numerous in France, southern Germany, Switzerland and Greece and are found in Asia as far east as the Pamir Mountains.

Two similar, composite white races are so important in the racial composition of modern Europe that they must be described. These races, called the **Armenoid** and the **Dinaric,** were produced by the interbreeding of the various main subraces and show a combination of their characteristics. The Armenoid is produced by a blending of Mediterranean and Alpine and is typically broadheaded, with a high, sloping forehead, high, pointed skull and hooked, convex nose. A great many of the residents of Turkey, Syria and Palestine are Armenoid. The Dinaric is a mixture of Mediterranean, Alpine and Nordic and, like the Armenoid, is roundheaded with a long face and prominent nose. Dinarics are common in Yugoslavia, the Austrian Tyrol, western Asia and Asia Minor, though they are found almost everywhere in the world.

The Negroid Race. It is generally believed that the original skin color of the human race was white, but just when the mutation for black or dark brown skin occurred in man's evolution no one knows. Nor is it known how the specialization of the Negro race came about or what relationships the Negroid subraces bear to each other.

Modern Negroids occur natively in two widely separated parts of the world: Africa and the islands of the Pacific from the Fijis to New Guinea. To explain this, anthropologists have postulated that the race originated in Asia and from there migrated east and west to those locations. Evidence that Negroids were once present along the southern and southeastern coasts of Asia is supplied by the many Negroid traits of several native tribes in India, Burma, Persia and Arabia. But the relationship between the Negros and the Negritos—the black pygmy race—is still undetermined.

Most Negroids have black, woolly or frizzy hair, dark brown or black skin, brown or black eyes, a low, broad nose, thick, puffy lips, a long head, sparse body and facial hair, relatively long shanks and forearms, a narrow pelvis, large feet and type O blood. The women usually have conical breasts and less protrusive buttocks than do white women.

The **Negritos** or **pygmies** are extremely short people, averaging less than 4 feet 9 inches in height, with round heads, narrow shoulders, pot bellies and short legs. They live in the Congo basin and in the interior of the Philippines, New Guinea, the Malay Peninsula and the Andaman Islands of the Indian Ocean. The Negritos are probably the most primitive culturally of living men, for they do not carry on agriculture, have no domesticated animals, and speak no language of their own. Many full-sized Negroes keep them as pets. It was thought at one time that they represent an earlier stage in the evolution of man, but there is no fossil evidence that man went through a pygmy stage; indeed, most of the prehistoric men were larger than the ones living today.

The **Oceanic Negroids** or **Melanesians** strongly resemble the Africans, but have thinner lips, less protrusive jaws, slightly lighter hair and skin, and their hair is less kinky.

The black-skinned race which originally occupied the island of Tasmania was destroyed by the white men who arrived there a little more than a century ago. The **Tasmanians** were quite short with woolly hair, short, flat noses, and abundant beards and body hair. They may have been an early offshoot of the Negritos, or they may have represented a type of Negro close to the primitive human stock. At any rate, they were living in a primitive Stone Age culture.

The **Bushmen** of South Africa are a rapidly vanishing race, which probably originated in that spot. They, too, are in a primitive Stone Age culture. Like the Negroes farther north, they have black, kinky hair, and broad, flat noses, but their skin is much lighter—yellow to brown—and they average under 5 feet in height. The condition of steatopygia, which is an excessive accumulation of fat on the buttocks and thighs, is characteristic of these people, particularly of the women. Their extremely kinky hair is of a peculiar type, known as "pepper corn," because of the way it grows in tiny clumps of spiral coils, separated by bare spaces.

The **Hottentots** of South Africa resemble the Bushmen in many respects and are believed to have arisen by hybridization between Bushmen and whites.

The Mongoloid Race. The Mongolians are characterized by straight, coarse, black hair, yellow to yellow-brown skin, brown eyes, sparse beards and body hair, a broad face with projecting cheek bones, and a rather flat nose of medium breadth. Their most striking feature, of course, is the epicanthic fold of the upper eyelid, which is chiefly, but not wholly, responsible for the slanting appearance of their eyes. Actually, the eye slit itself is narrower and more slanting than that of the Caucasian eye. A high percentage of Mongolians have type B blood.

The classic Mongoloids, found in Mongolia, Eastern Siberia, and, slightly mixed with other types, in China, Korea and Japan, are roundheaded and vary in height from short to moderately tall. In contrast, the Arctic Mongoloids — the **Eskimos**—are short, long-headed people with a less pronounced epicanthic fold and more prominent cheek bones. The Arctic Mongoloids are found from Greenland to Northeastern Asia.

Composite races that are predominantly Mongoloid are the **Indonesians** and **American Indians.** The Indonesians, believed to have arisen by hybridization of the Mongoloids with Mediterraneans, Ainus and Negritos, are found in South China, Indo-China, Siam, Burma, Malaya and parts of the East Indies. Some of the Japanese belong to this group. The American Indians are the descendants of Mongoloids who entered America via Bering Strait in a number of waves of migration beginning some 20,000 years ago. Like the classic Mongolian, the Indians have round heads, prominent cheek bones and black, straight, coarse hair, but their skin has a reddish rather than a yellowish cast, the Mongolian eye fold is absent or small, and the nose is sometimes long, convex and high-bridged. The many subraces of the American Indian have not yet been thoroughly studied and classified, but even now it is apparent that they exhibit more variations in skin color, facial features and head shape than any other group of Mongolians. This is probably because by the time of the later invasions, Mediterraneans, Negritos and Australoids had interbred with the original Mongolian stock in Asia.

The Ainus, the Veddoids and the Australoids are groups which do not seem to belong the any of the three main races nor do they appear to be the result of hybridization.

The Ainus are an ancient and remarkably primitive group which has lived for as far back as there is any record, on the islands of Northern Japan. They appear not to be of Mongolian origin, for they have white skin and wavy, black hair, which grows profusely on the face and body—hence they are usually called the "Hairy Ainu." There is no trace of the Mongolian epicanthic fold. There are geographic and historical reasons for doubting that they are Caucasian, and some anthropologists believe them to be Mongolians who underwent mutations which became established because of the former isolation of the Japanese islands.

The Australoids, or Australian aborigines, have chocolate-colored skin, dark eyes, black hair, protruding jaws and receding chins. Because of their heavy eyebrow ridges, they are generally considered the most primitive of living races. But in spite of their many Negroid characteristics, they are not, apparently, of pure Negroid ancestry, for they have wavy rather than kinky hair which grows abundantly on both face and body. Their status, like that of the Ainu, has not been determined.

Throughout India and the East Indies, as well as in southeastern Arabia, occurs the third of our groups of unknown origin—the Veddoids. These are dark, slender and long-headed; they have broad noses and wavy hair, and the beard is moderately developed. In some ways, these people are similar to the Caucasian race, in others they resemble the Australoids, but their true origin is unknown. It is the intermixture with the Mediterranean race of the Veddoid element, rather than the Negroid, which is responsible for the dark skin color of so much of the population of India.

It should be clear why it is difficult to make generalizations about the superiority of any modern race. Firstly, no race is "pure"; man's evolutionary history is one of continuous intermixture of races as peoples migrated, invaded and conquered their neighbors or were conquered by them. Secondly, the testing of intelligence and psychological traits is vastly complicated by the difficulty of differentiating between inherited and environmental influences. Psychologists have found, for instance, that city children do better on intelligence tests than country children. The advantages of an adequate diet, the prompt correction of physical disabilities, and a feeling of security—in short, the benefits of a good environment—are incalculable.

There is no reason for supposing that a pure race, if one did exist, would indeed be superior. On the contrary, the principle of hybrid vigor works just as well for human populations as it does for animals and plants. Thus, crosses between Indians and whites, Hottentots and whites, Chinese and Negroes often produce offspring which are taller, more vigorous and longer-lived then either of the parents.

At present, the only conclusion that anthropologists can offer concerning the capabilities of the various human races is that all of them have great potentialities and all have made important contributions to civilization. In time to come, investigators may find solutions for the problems of testing, and may be able to show correlations between certain racial strains and certain abilities. Whether that time comes or not, however, the person who can look upon the tangled web of humanity with any feeling of superiority at whatever his own little place in it may be, is indeed a victim of pride and prejudice.

QUESTIONS

1. What does it mean to say that an animal is unspecialized?
2. What are the ways in which man differs from the great apes?
3. Why is it incorrect to say that man came from monkeys? What did he come from?
4. Discuss how man's physical features have altered during his evolution from the ancestral lemuroids.
5. How does the science of archaeology help that of anthropology?
6. Do you think any of the known fossil primates should be called a "missing link"? If so, which one?
7. What is the relationship of Neanderthal to modern man?
8. Do you believe the various races of men living today constitute one species or several? Why?
9. Discuss the main subdivisions of the white race and list the characters which differentiate them.
10. What is meant by dolichocephalic? How is the cephalic index measured?

SUPPLEMENTARY READING

Romer's *Man and the Vertebrates,* H. F. Osborn's *Men of the Old Stone Age,* Howell's *Mankind So Far,* and W. E. L. Clark's *History of the Primates* give fine descriptions of prehistoric men. E. A. Hooton's *Up from the Ape* is an amusing and informative discussion of the primates, of human evolution and of the present races of man. Read Weidenreich's *Apes, Giants and Man* for a fascinating account of the ape-men by one of the major researchers in the field. R. R. Gates' *Human Ancestry* is an advanced discussion of human evolution from the viewpoint that the present races of man are in fact different species. C. S. Coon's *The Races of Europe* is a definitive but understandable treatise of the many subdivisions of the white race.

Part Eight

Ecology

Chapter 37

Principles of Ecology

WHENEVER A close study is made of some particular plant or animal, the investigator is struck by the remarkable fitness of the organism for the place in which it lives. This fitness of structure, of function, even of behavior pattern, has arisen, as we have seen, in the course of evolution by natural selection. The outcome of evolution is a population of organisms, a species, adapted to survive in a certain type of environment. The species shows adaptations both to the physical environment—wind, sun, moisture, temperature, and so on—and to the biotic environment, which includes all the plants and animals living in the same region. Some of the fundamentals of **ecology,** the study of the interrelations between living things and their physical and biotic environment, were discussed in Chapter 6. Now that we have considered some of the details of plant and animal structure and function, and have gained some idea of how these forms arose in evolution, we are ready to return to the problems of ecology and consider them in more detail.

332. FACTORS REGULATING THE DISTRIBUTION OF PLANTS AND ANIMALS

Probably no species of plant or animal is found everywhere in the world; some parts of the earth are too hot, too cold, too wet, too dry, or too something else for the organism to survive there. The environment may not kill the adult directly, but effectively keeps the species from becoming established if it prevents its reproducing or kills off the egg, embryos, or some other stage in the life cycle. Most species of organisms are not even found in all the regions of the world where they could survive. The existence of barriers prevents their further dispersal and enables us to distinguish the major biogeographic realms (p. 547) characterized by certain assemblages of plants and animals.

Biologists were aware more than 100 years ago that each species requires certain materials for growth and reproduction, and can be restricted if the environment does not provide a certain minimum amount of each one of these materials. V. E. Shelford pointed out in 1913 that *too much* of a certain factor would act as a limiting factor just as well as too little of it, and that the distribution of each species is determined by its **range of tolerance** to variations in each of the environmental factors. Much work has been done to define the limits of tolerance, the limits within which species can exist, and the results have been very helpful in understanding the distribution of organisms. It has usually been found that certain stages in reproduction are critical in limiting organisms—seedlings and larvae are usually more sensitive than adult plants and animals. Thus, although adult blue crabs can survive in water with a low salt content and can migrate for some distance up river from the sea, their larvae cannot, and the species cannot become permanently established there.

Some organisms have very narrow ranges of tolerance to environmental factors, others can survive within much broader limits. Any given organism may have narrow limits of tolerance for one factor and wide limits for another. Ecologists use the prefixes **steno-** and **eury-** to refer to organisms with narrow and wide, respectively, ranges of tolerance to a given factor. A stenothermic organism is one which will tolerate only narrow variations in temperature. The housefly is a eurythermic organism, for it can tolerate a range of temperatures from 43 to 113° F.

Temperature is an important limiting factor, as the relative sparseness of life in the desert and arctic demonstrates. Most of the animals that do live in the desert have adapted to the rigors of the environment by living in burrows during the day and coming out to forage only at night. Many animals escape the bitter cold of the northern winter not by migrating south but by burrowing beneath the snow. Measurements made in Alaska show that when the surface temperature is –68° F., the temperature 2 feet under the snow, at the surface of the soil, is + 20° F. Although the ring-necked pheasant has been introduced into the southern states a number of times and the adults survive well, the developing eggs are apparently killed by the high daily temperatures and are unable to complete development.

The role of light in controlling plants and animals has been described previously; the role of the photoperiod in determining the time of flowering of plants (p. 97) and the migration of birds (p. 223) is mediated in both by some hormonal mechanism. Despite the fact that plants must have light for photosynthesis—much of plant evolution has been guided by a competition for light—an excess of light is lethal and both plants and animals have had to evolve mechanisms for protection against too much or too little light.

Water is a physiologic necessity for all protoplasm, but is a limiting factor primarily for land organisms. The amount of rainfall, its seasonal distribution, the humidity, and the ground supply of water are some of the limiting factors in distribution. Although desert animals can escape the high temperature and low humidity by retreating to underground burrows (a kangaroo rat's burrow 2 feet underground can have a temperature of 60° F. when the surface temperature is over 100° F.), the desert plants must stay on the surface and have had to evolve structures to prevent water loss and to resist high temperatures.

Knowledge of the limits of water tolerance can be used by man to regulate insect pests. For example, wire worms, pests attacking West Coast crops, were found to have rather narrow limits of tolerance to water and are most sensitive as larvae and pupae. They can be destroyed by exceeding the maximum limit of tolerance—by flooding irrigated fields—or by planting alfalfa or wheat, which dry out the soil below the limit of tolerance of the larvae.

Atmospheric gases are usually not limiting for land organisms except for forms living deep in the soil or on mountain heights, but in aquatic environments the amount of dissolved oxygen present may vary considerably and be the limiting factor for certain forms. The trace elements necessary for plant and animal life may be present in too small amounts and be limiting; deficiencies of cobalt and copper produce severe deficiency diseases in plants and grazing animals—certain regions of Australia are unsuitable for raising cattle or sheep because of this. Water currents are limiting factors of certain aquatic plants and animals—there are marked differences in the flora and fauna of a still pond and a swiftly-flowing creek. The type of soil present (p. 105), the amount of topsoil, its *p*H, porosity, slope, and so on are limiting factors for many plants. Even fire may be a factor of ecologic importance. The fine forests of longleaf pines in the southeast are due to their superior resistance to fire. In the absence of occasional small ground fires, these pines are gradually replaced by small hardwoods, much less valuable as timber.

333. FOOD CHAINS

The need of living things for energy and the ultimate source of energy—sunlight—were discussed in Chapter 6. Only a small fraction, about 3 per cent, of the light energy striking a green plant is transformed by photosynthesis into the potential energy of food substances; the rest escapes as heat. When an animal eats a plant, much of the energy again is dissipated as heat and only a fraction is used to synthesize the animal's protoplasm. When a second animal eats the first, there is a further loss of energy as heat. The transfer of food energy from its ultimate source in plants, through a series of organisms each of which eats the preceding and is eaten by the following, is known as a **food chain.** The number of steps in the series is limited to perhaps four or five because of the great decrease in available energy at each step. The percentage of food energy consumed that is converted to new protoplasm (and thus is available as food energy for the next animal in the food chain) is known as the per cent efficiency of energy transfer.

The first step in any chain, the capture of light energy by photosynthesis and the formation of energy containing foods by plants, is relatively inefficient; only about 0.2 per cent of the incident light energy is stored as food. The efficiency of energy transfer when one animal eats a plant or another animal is higher, ranging from 5 to 20 per cent. Man is the end of a number of food chains; for example, man eats big fish, which ate little fish which ate small invertebrates which ate algae. The ultimate size of the human population is limited by the length of our food chain, the per cent efficiency of energy transfer at each step in the chain, and by the amount of light energy falling on the earth. Since man can do nothing about increasing the amount of incident light energy, and very little about the per cent efficiency of energy transfer, he can increase his supply of food energy only by shortening his food chain, i.e., by eating the primary producers, plants, rather than animals. In overcrowded countries such as India and China men are largely vegetarians because this food chain is shortest and a given area of land can in this way support the greatest number of people. Steak is a luxury ecologically as well as economically!

Parasites may also exist as members of food chains; for example, mammals and birds are parasitized by fleas and in the fleas live protozoa which are in turn hosts of bacteria. Since the bacteria might be parasitized by viruses, there could be a five-step parasite food chain.

Since, in any food chain, there is a loss of energy at each step, it follows that there will be a smaller amount of protoplasm in each successive step. For example, H. T. Odum has calculated that 17,850 pounds of alfalfa plants are required to provide the food for 2,250 pounds of calves, which provide enough food to keep one twelve year old, 105-pound boy alive. Although boys eat things other than veal and calves other things besides alfalfa, these numbers illustrate the principle of a food chain. A food chain can be visualized as a pyramid; each step in the pyramid is much smaller than the one on which it feeds. Since the predators are usually larger than the forms on which they prey, the pyramid of *numbers* of individuals in each step of the chain is even more striking than the pyramid of the protoplasmic mass of the individuals in successive steps—one boy requires 4.5 calves, which require 20,000,000 alfalfa plants.

The physical limiting factors described above are important in determining whether or not a given species can become established in a given region. Each region is inhabited by a host of animals and plants and there are many interrelationships—competition, commensalism, predation and other factors—between them that are also involved in determining whether or not some single species can survive there. The ecologist refers to the organisms living in any given area as a **biotic community;** this is composed of smaller groups, the members of which are more intimately associated, known as **populations.** There is no sharp distinction between a population and a community.

334. POPULATIONS AND THEIR CHARACTERISTICS

The population may be defined as a group of organisms of the same or similar species which occupy a given area. It has characteristics which are a function of the whole group and not of the individuals; these are **population density, birth rate, death rate, age distribution, biotic potential, rate of dispersion** and **growth form.** Although individuals are born and die, individuals do not have birth rates and death rates; these are characteristics of the population as a whole.

One important attribute of a population is its density—the number of individuals per unit area or volume, e.g., human inhabitants per square mile, trees per acre in a forest, millions of diatoms per cubic meter of sea water. This is a measure of the population's success in a given region. Frequently in ecologic studies it is important to know not only the population density, but whether it is changing and if so, what the rate of change is. Population density is often difficult to measure in terms of individuals, but measures such as the number of insects caught per hour in a standard trap, or the number of birds seen or heard per hour, are usable substitutes. A method that will give good results when used with the proper precautions is that of capturing let us say 100 animals, tagging them in some way, and then releasing them. On some subsequent day, another 100 animals are trapped and the proportion of tagged animals is determined. This assumes that animals caught once are neither more likely nor less likely to be caught again, and that both sets of trapped animals are random samples of the population. If the 100 animals caught on the second day include 20 tagged ones, the total population of tagged and untagged animals in the area of the traps is 500.*

A graph in which the number of organisms is plotted against time is a **population growth curve** (Fig. 310). Since such curves are characteristic of populations, rather than of a single species, they are amazingly similar for populations of almost all organisms from bacteria to man. From a study of the human population growth curve to date, and by comparing this curve to a general one (Fig. 311), Raymond Pearl estimated that the human population, now about 2.2 billion, would reach 2.65 billion in the year 2100 and would remain stable after that unless there was some change in the ability of the earth to support human life.

The birth rate, or natality, of a population is simply the number of new individuals produced per unit time. The **maximum birth rate** is the largest number of organisms that could be produced per unit time under ideal conditions, when there are no limiting factors. It is a constant for a species, determined by physiologic factors such as the number of eggs produced, the proportion of females in the species, and so on. The actual birth rate is usually considerably less than this, varying with the size and composition of the population and with environmental conditions. It is difficult to determine the maxi-

* $x/100 = 100/20$; $x = 500$.

mum natality, for it is difficult to be sure that all limiting factors have been removed. However, under experimental conditions, one can get an estimate of this value which is useful in predicting the rate of increase of the population and in providing a yardstick for comparison with the actual birth rate.

The mortality rate of a population refers to the number of individuals dying per unit time. There is a theoretical **minimum mortality,** somewhat analogous to the maximum birth rate, which is the number of deaths which would occur under ideal conditions, deaths due simply to the physiologic changes of old age. This, too, is a constant for a given population. The actual mortality rate will vary depending upon physical factors and on the size and composition of the population. By plotting the number of survivors in a population against time, one gets a

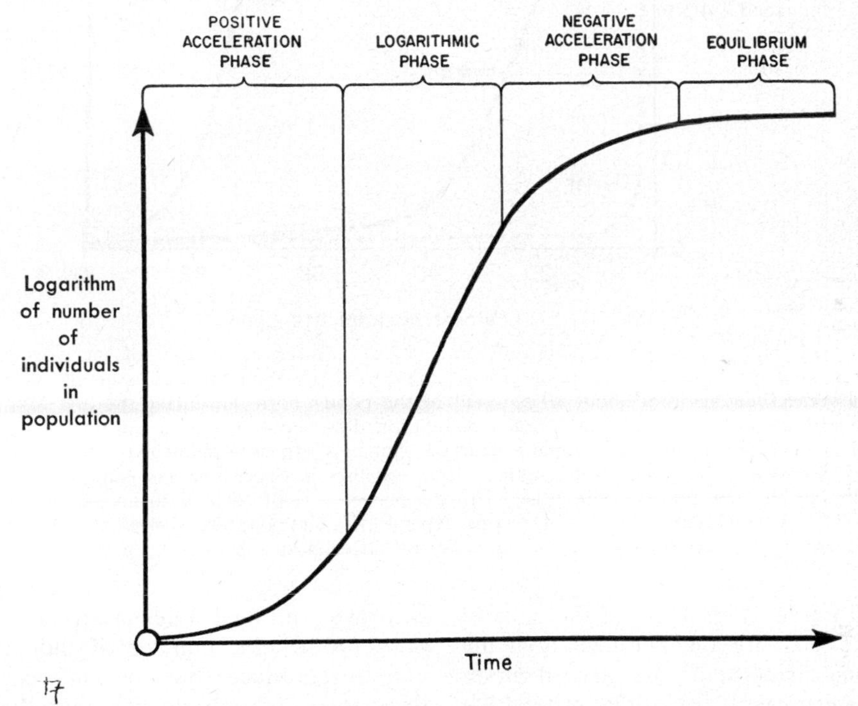

Figure 310. A typical growth curve of a population, one in which the logarithm of the total number of individuals is plotted against the time. The absolute units of time and the total number in the population would vary from one species to another, but the shape of the growth curve would be similar for all populations.

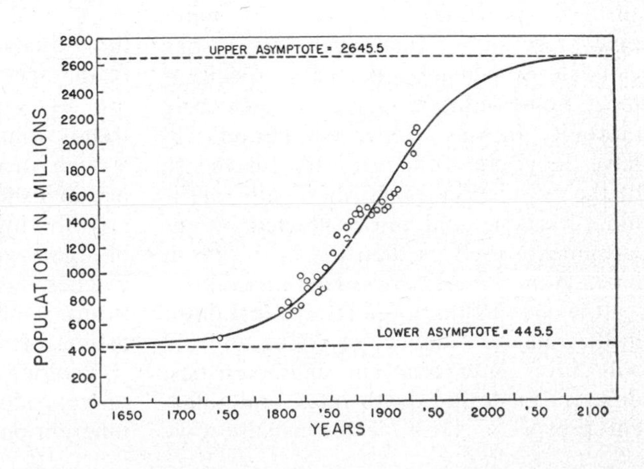

Figure 311. The growth curve of the human population of the world. The circles indicate census totals, the solid line the mathematically fitted, smooth, S-shaped curve which predicts an equilibrium world population of 2,645,500,000, to be reached in the year 2100 A.D. (After Pearl and Gould, from Allee et al.: Principles of Animal Ecology.)

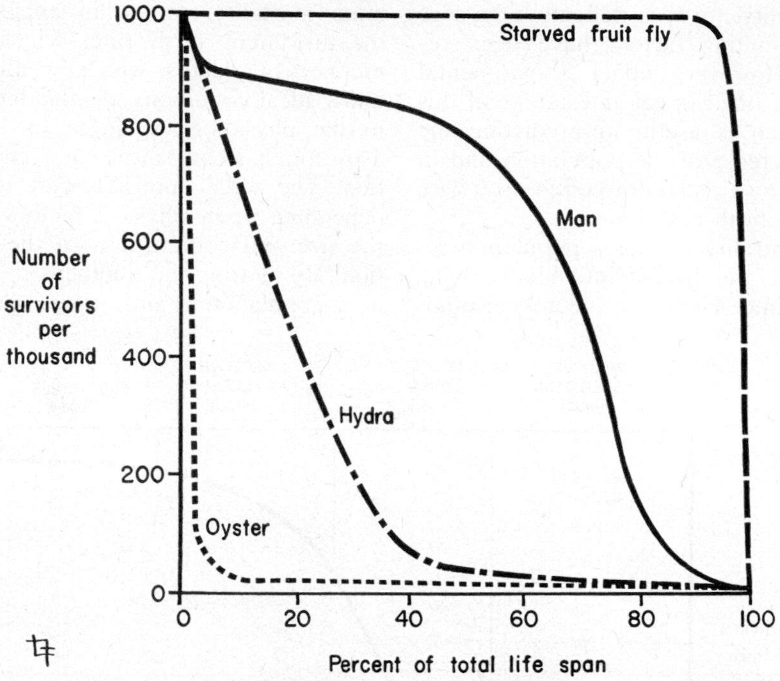

Figure 312. Survival curves of four different animals, plotted as number of survivors left at each fraction of the total life span of the species. The total life span for man is about 100 years; the solid curve indicates that about 10 per cent of the babies born die during the first few years of life. Only a small fraction of the human population dies between ages 5 and 45 but after 45 the number of survivors decreases rapidly. Starved fruit flies live only about five days, but almost the entire population lives the same length of time and dies at once. The vast majority of oyster larvae die but the few that become attached to the proper sort of rock or to an old oyster shell survive. The survival curve of hydras is one typical of most animals and plants, in which a relatively constant fraction of the population dies off in each successive time period.

survival curve (Fig. 312). If the units of the time axis are the per cent total life span, one can compare the survival curves for organisms with very different total life spans. Civilized man has improved his average life expectancy greatly by modern medical practice, and the curve for human survival approaches the curve for minimum mortality. From such curves one can determine when a particular species is most vulnerable; reducing or increasing mortality in this vulnerable period will have the greatest effects on the future size of the population. Since the death rate is more variable and more affected by environmental factors than the birth rate, it has a primary role in population control.

It is obvious that populations that differ in the relative numbers of young and old will have quite different characteristics, different birth and death rates, and different prospects. Death rates usually vary with age, and birth rates are usually proportional to the number of individuals able to reproduce; thus one can recognize three ages—prereproductive, reproductive and post-reproductive. A. J. Lotka has shown from theoretical considerations that a population will tend to become stable and have a constant proportion of individuals of these three ages. Censuses of the ages of plant or animal populations are of value in predicting population trends; rapidly growing populations have a high proportion of young forms. The age of fishes can be determined from the growth rings on their scales, and studies of the age ratios of commercial fish catches are of great use in predicting future catches and in preventing overfishing of a region.

Ecologists use the term **biotic potential** or reproductive potential to express the inherent power of a population to increase in numbers when the age ratio is stable and all environmental conditions are optimal.* Under usual conditions, when the environment is less than optimal, the rate of population growth is less and the difference between the potential ability of a population to increase and the actual observed performance is a measure of environmental resistance. Even when a population is growing rapidly in number, each *individual* organism of the reproductive age carries on reproduction at the same rate as before; the increase in numbers is due to increased survival. At a conservative estimate, one man and one woman, with the cooperation of their children and grandchildren, could give rise to 200,000 progeny within a century, and a pair of fruit flies could multiply to give 3368×10^{52} offspring in a year. Since optimum conditions are not maintained, such biologic catastrophes do not occur, but there are actual situations, such as in India and China, which indicate the tragedy implicit in the tendency toward overpopulation.

The sum of the physical and biologic factors which prevent a species from reproducing at its maximum rate is termed the **environmental resistance.** Environmental resistance is often low when a species is first introduced into a new territory, so that the species increases in number at a fantastic rate, as when the rabbit was introduced into Australia and the English sparrow and Japanese beetle were brought into the United States. But as a species increases in number the environmental resistance to it also increases, in the form of organisms that prey upon it or parasitize it, and the competition between the members of the species itself for food and living space.

Population growth curves have a characteristic shape (Fig. 310). When a few individuals enter a previously unoccupied area, growth is slow at first (called the positive acceleration phase), then becomes rapid and increases exponentially (the logarithmic phase) and eventually slows down as environmental resistance gradually increases (the negative acceleration phase) and finally reaches an equilibrium or saturation level.

* For the mathematically inclined: The biotic potential is the slope of the population growth curve during the logarithmic phase of growth.

335. POPULATION CYCLES

Once a population becomes established in a certain region and has reached the equilibrium level, the numbers will vary up and down from year to year, depending on variations in environmental resistance or on factors intrinsic to the population. Some of these population variations are completely irregular, but others are regular and cyclic. One of the best known of these is the regular 9 to 10 year cycle of abundance and scarcity of the snowshoe hare and the lynx in Canada which is based on the records of the number of pelts received by the Hudson's Bay Company. The peak of the hare popula-

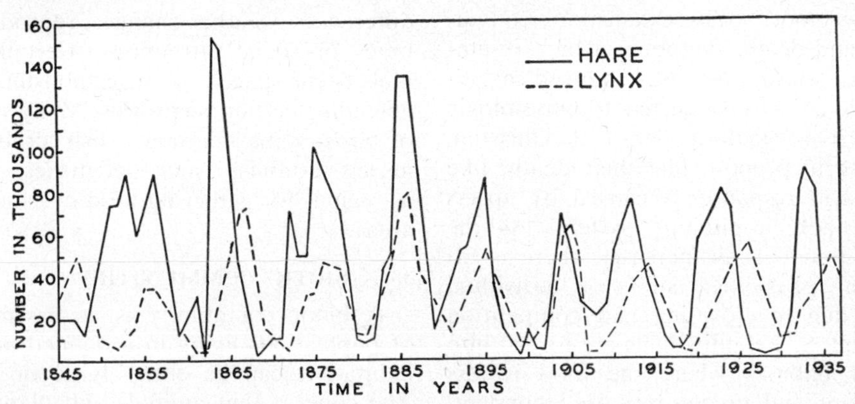

Figure 313. Changes in the abundance of the lynx and the snowshoe hare, as indicated by the number of pelts received by the Hudson's Bay Company. This is a classic case of cyclic oscillation in population density. (Redrawn from MacLulich, 1937.)

tion comes about a year before the peak of the lynx population (Fig. 313). Since the lynx feeds on the hare, it is obvious that the lynx cycle is related to the hare cycle. Another cycle, recurring every 3 to 4 years, is shown by lemmings and voles, small micelike animals living in the northern tundra region. Every three or four years there is a great increase in the number of lemmings; they eat all the available food in the tundra and then migrate in vast numbers looking for food. They may invade villages in hordes and finally reach the sea and drown. The numbers of foxes and snowy owls, which feed on lemmings, increase similarly and when the lemming population decreases, the foxes starve and the owls migrate south—thus there is an invasion of snowy owls in the United States every three or four years.

Attempts to explain these vast oscillations in numbers on the basis of climatic changes have been unsuccessful. At one time it was believed that these were caused by sunspots, and the sunspot and lynx cycles do appear to correspond during the early part of the 1800's. However, the cycles are of different lengths and by 1920 were completely out of phase, sunspot maxima corresponding to lynx minima. Attempts to correlate these cycles with other periodic weather changes and with cycles of disease organisms have been unsuccessful.

The hares die off cyclically even in the absence of predators and in the absence of known disease organisms. The animals apparently die of "shock," characterized by low blood sugar, exhaustion, convulsions and death, symptoms which resemble the "alarm response" induced in laboratory animals subjected to physiologic stress. This similarity led J. J. Christian (1950) to propose that their death, like the alarm response, is caused by upsets in the adrenal-pituitary system. As the population density increases, there is increasing physiologic stress on individual hares due to crowding and competition for food. Some individuals are forced into poorer habitats where the food is less abundant and predators more abundant. The physiologic stresses stimulate the adrenal medulla to secrete epinephrine, which stimulates the pituitary to secrete more ACTH (adrenocorticotropic hormone). In the latter part of the winter of a peak year, with the stress of cold weather, lack of food and the onset of the new reproductive season putting additional demands on the pituitary to secrete gonadotropins, the adrenal-pituitary system breaks down, carbohydrate metabolism (normally under its control) is upset, and low blood sugar, convulsions and death ensue.

Populations have a tendency to disperse, or spread out in all directions until some barrier is reached. Within the area, the members of the population may occur at random (rarely), they may be distributed uniformly (when there is competition or antagonism to keep them apart), or they may occur in small groups or clumps (most common). Aggregation in clumps may increase the competition between the members of the group for food or space, but this is more than counterbalanced by the greater survival power of the group during unfavorable periods. It can be shown experimentally that a group of animals has much greater resistance than a single individual to adverse conditions such as desiccation, heat, cold or poisons. The combined effect of the protective mechanisms of the group is effective in countering the adverse environment whereas that of a single individual is not. Allee has called this tendency to form aggregations for group survival "unconscious cooperation." Aggregation may be caused by local habitat differences, weather changes, reproductive urges or social attractions. Certain animals occur spaced apart, establishing and defending certain **territories.** Many species of birds, some mammals, fish, crabs and insects establish such territories, either as regions for gathering food or as nesting areas.

336. BIOTIC COMMUNITIES

A biotic community is an assemblage of populations living in a defined area or habitat; it can be either large or small. The concept that animals and plants live together in an orderly manner, not strewn haphazardly over the surface of the earth, is one of the important principles of ecology. Sometimes adjacent communities are sharply defined and separated from each other, more frequently they blend imperceptibly together. The unraveling of why certain plants and animals comprise a given community, how they affect each other, and how man can control them to his advantage are some of the major problems of ecologic research. In trying to control some particular species, it has frequently been found more effective to modify the community rather than to attempt direct control of the species itself. For example, the most effective way to increase the quail population is not to raise and release birds, nor even to kill off predators, but to maintain the particular biotic community in which quail are most successful. Detailed studies of simpler biotic communities such as those of the arctic or desert, where there are fewer organisms and their interrelations are more evident, have provided a basis for studying and understanding the much more varied and complex forest communities.

Although each community may contain hundreds or thousands of species of plants and animals, most of these are relatively unimportant and only a few, by their size, numbers or activities, exert a major control. In land communities, these major species are usually plants, for they both produce food and provide shelter for many other species, and many land communities are named for their dominant plants — sagebrush, oak-hickory, pine, and so on.

337. COMMUNITY SUCCESSION

Any given area tends to have an orderly sequence of communities, which change together with the physical conditions and lead finally to a stable mature community or **climax community.** These series are so regular that an ecologist, recognizing the particular community present in a given area, can predict the sequence of future changes. The ultimate causes of these successions are not clear. Climate and other physical factors play some role, but the succession is directed in part by the nature of the community itself, for the action of each community is to make the area less favorable for itself and more favorable for other species. One of the classic studies of ecologic succession was made on the shores of Lake Michigan (Fig. 314). As the lake has become smaller it has left successively younger sand dunes, and one can study the stages in ecologic succession as one goes away from the lake. The youngest dunes, nearest the lake, have only grasses and insects; the next older ones have shrubs, then evergreens and finally there is a beech-maple climax community, with deep rich soil full of earthworms and snails. As the lake retreated it also left a series of ponds. The youngest of these contains little rooted vegetation and lots of bass and bluegills. Later the ponds become choked with vegetation and smaller in size as the basins fill. Finally the ponds become marshes and then dry ground, invaded by shrubs and ending in the beech-maple climax forest. Man-made ponds, such as those behind dams, similarly tend to become filled up.

Ecologic succession can be demonstrated in the laboratory. If a few pieces of dry grass are placed in a beaker of water, a population of bacteria will appear in a few days. Next, flagellates appear and eat the bacteria, then ciliated protozoa such as paramecia appear and eat the flagellates. Finally predator protozoa such as *Didinium* will appear and eat the paramecia. All the protozoa were present as spores or cysts attached to the grass.

Biotic communities show marked **vertical stratification.** In a forest there will be successive strata of plants: mosses and herbs, shrubs, low trees and high trees. Each of these strata has distinctive animal populations; even such highly motile animals as birds have been found to be restricted to certain layers—some are found only in shrubs, others only in the tops of tall trees.

338. APPLICATIONS OF ECOLOGIC PRINCIPLES

The most important application of ecology is the rational conservation of our natural resources. **Conservation** does not

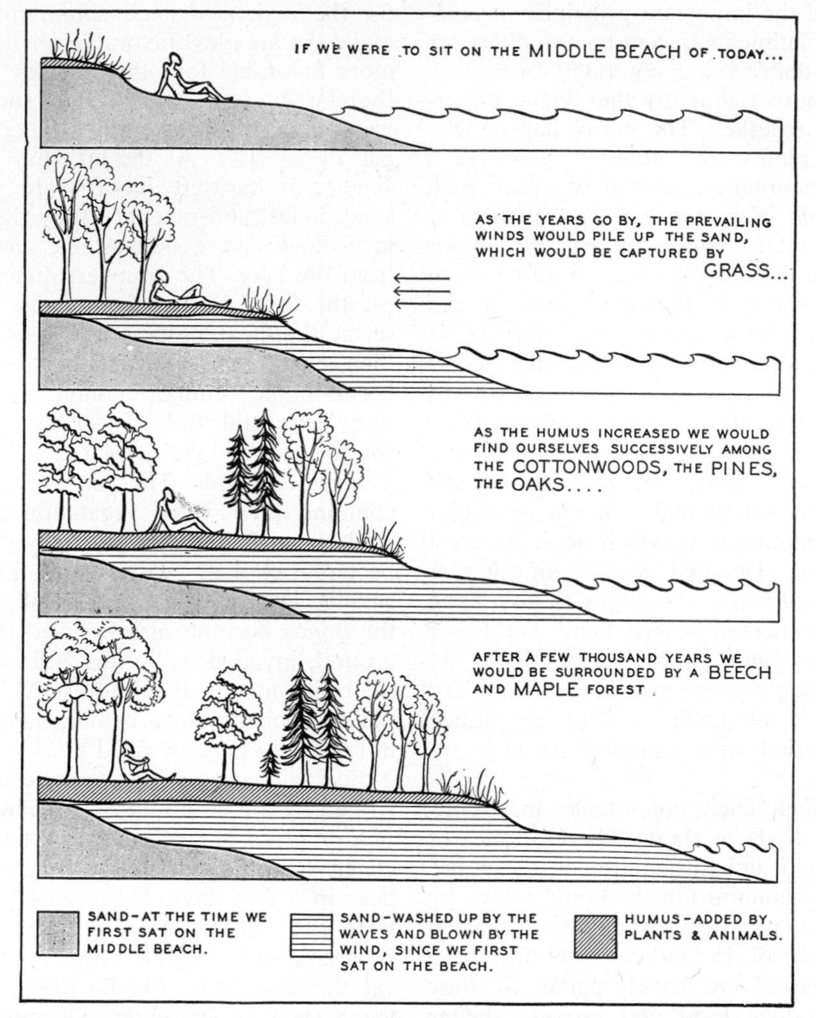

Figure 314. Diagram of the succession of communities with time along the shores of Lake Michigan in northern Indiana. (Allee et al.: Principles of Animal Ecology.)

mean simply hoarding, but the establishing of a balance of harvest and renewal so that there will be a continuous yield of useful plants, animals and materials. Man apparently has not yet learned that he is part of a complex environment which must be studied and treated as a whole, and not in terms of isolated "projects," for in attempting to carry out one project he may nullify or completely overcome the results of another one.

The control of insects by chemicals such as DDT must be carried out cautiously with possible ecologic upsets in mind. Spraying orchards, forests and marshes may destroy not only pests but useful insects such as honeybees, which pollinate many kinds of fruit trees and crops, and useful insect parasites. In some cases the insect pests have actually increased after the use of DDT because the chemical killed off greater numbers of insect enemies of the pest than of pests!

The soil conservation program, carried on jointly by Federal and local agencies, is effective because it is based on sound ecologic principles; crop rotation, contour farming, the establishing of wind breaks, and the use of proper fertilizers to renew the soil are all measures which maintain a balanced ecosystem. Successful farming must follow the principles of good land use. If the grasslands of regions of slight rainfall are plowed and planted with wheat, a "dust bowl" will inevitably follow, but if it is kept in grass and grazed in moderation, no dust bowl will develop and the land can be used economically year after year.

The management of our forests is a field of applied ecology. Since in some regions the desirable timber trees are members of the climax community, the problem is simply the best way to speed the return of the climax community after the trees have been cut. In other regions the desirable trees are earlier members of the ecologic succession and the problem is how to prevent the succession from proceeding. With proper forest management, new trees are grown as rapidly as trees are removed by cutting, fires and diseases and a sustained yield is obtained.

Other fields of applied ecology are the management of our fish and wildlife resources. "Wildlife" usually means game and fur-bearing animals. Since the various types of wildlife are adapted to different stages of ecologic succession, their management requires a knowledge of and the proper use of these stages. Of the three general methods used to increase the population of game animals—laws restricting the number killed, artificial stocking, and the improvement of the habitat—the latter is the most important. If the game habitats are destroyed or drastically altered, protective laws and artificial stocking are useless. Protective laws must operate to prevent a population getting too large as well as too small. Deer populations, in the absence of natural predators but subject to a constant, moderate amount of hunting, may increase to a point where they actually ruin the vegetation of the forest. Hunting should be restricted when populations are small and increased when they are larger; this necessitates an accurate estimate of the population density of the game. Stocking a region artificially with game animals is effective only if they are being introduced into a new region or into one from which they had been killed off. The principles of population growth make it clear that if game animals of a certain species are already present, artificially stocking a region with additional members of the species will be futile. Stocking a region with a completely new species must be done cautiously, for the species may succeed so well as to become a pest and upset the biotic community, as rabbits have in Australia.

The management of the fish in a pond may be directed toward providing sport, or toward raising a "crop" of food fish and draining the pond at regular intervals to harvest the crop. It has been found that sport fishing with hook and line is not likely to overfish a lake; the lake is more likely to be underfished and the resulting crowding tends to stunt the growth of the members of the fish population.

The building of dams raises intricate ecologic problems, for dams may be intended for power, for flood control, for the prevention of soil erosion, or for the creation of recreational areas. Since no one dam can satisfactorily accomplish all of these objectives, the primary objective must be clearly delineated and the secondary results must be understood. A contrast of two proposals for dealing with the same watershed (Table 16) shows that the multiple dam plan costs less, de-

Table 16. A COMPARISON OF A SINGLE MAIN RIVER RESERVOIR PLAN WITH A PLAN FOR MULTIPLE SMALLER HEADWATERS RESERVOIRS

	MAIN STREAM RESERVOIR	MULTIPLE HEADWATERS RESERVOIRS
Number of reservoirs	1	34
Drainage area, square miles	195	190
Flood storage, acre feet	52,000	59,100
Surface water area for recreation, acres	1,950	2,100
Flood pool, acres	3,650	5,100
Bottom farm land inundated, acres	1,850	1,600
Bottom farm land protected, acres	3,371	8,080
Total cost	$6,000,000	$1,983,000

From E. P. Odum: Fundamentals of Ecology.

stroys less productive farm land, impounds more water and is more effective in controlling floods and soil erosion.

The application of ecologic principles is also essential in dealing with such problems as the management of rangelands (the prevention of overgrazing) and the management of marine fish and shellfish populations. For example, oysters were killed off in a number of Long Island bays by the raising of ducks on the shore! Oysters feed primarily on diatoms, and these were replaced by other algae when the ecosystem of the bay was changed by the large amount of waste materials washed in from the duck farms. Once an oyster bed has been seriously depleted, it may fail to recover, for oyster larvae require the shells of old oysters for attachment. Providing the larvae with artificial attachment sites is the practice in commercial oyster "farms."

Since many kinds of animals serve as vectors in transmitting disease germs, the field of public health is, in part, applied ecology. Careful studies of the rat population in Baltimore revealed that traps, poisons, cats and professional rat exterminators were much less effective in permanently reducing the rat population than a city-wide sanitation program by which the food and hiding places of the rats were greatly decreased.

A little reflection will lead the student to the realization that the ecologic principles discussed in this chapter apply to human populations as well as to plant and animal ones, and in recent years the ecologic approach to a study of human societies has led to the development of the field of **human ecology.** This can contribute to the social sciences by pointing out man's myriad relations to other living things and to the physical environment. Although man can control his environment to a considerable extent, this control is far from complete and man must adapt to those situations he cannot change. An understanding of, and cooperation with the various cycles of nature are better for man's future survival than a blind attempt to change and control them.

Experts in the field differ as to whether the human population is in danger of multiplying beyond the ability of the earth to support it. The disagreement is probably due to the lack of sufficient data to provide the basis for a sound prediction. Much more study of man and nature as an ecologic unit is needed.

QUESTIONS

1. How would you define ecology?
2. Discuss the factors that may prevent a given species of animal or plant from becoming established in a particular region.
3. Why is the number of steps in a food chain limited?
4. What is meant by a biotic community? Give examples from your experience.
5. What characteristics are peculiar to a population as a whole and not to its individual members?
6. What is meant by a survival curve? Discuss the importance of such curves to a life insurance company.
7. Which vary more, birth rates or death rates? Why?
8. Define the terms biotic potential and environmental resistance. Draw a population growth curve and indicate graphically the relation of these terms to the curve.
9. Explain why there is a tendency for there to be an orderly sequence of communities leading to a climax community. What is the climax community in your region?
10. Discuss the measures that could be taken to increase the number of beavers in Pennsylvania. The number of quail in Virginia.

SUPPLEMENTARY READING

The principles of ecology are clearly and interestingly presented in E. P. Odum's *Fundamentals of Ecology*. For a detailed discussion of plant ecology consult Weaver and Clement's *Plant Ecology*. The definitive, thorough treatise on animal ecology is *Principles of Animal Ecology* by Allee, Emerson, Park, Park, and Schmidt. The problem of the conservation of our natural resources is considered in Fairfield Osborn's *Our Plundered Planet,* Paul Sears' *Deserts on the March,* and William Vogt's *Road to Survival.* Harrison Brown's *The Challenge of Man's Future* is an able and fascinating discussion of some important aspects of human ecology.

Chapter 38

The Outcome of Evolution: Adaptation

SOME OF the details of the physical and biotic factors affecting populations of organisms were described in the previous chapter. A complete discussion of the many ways in which living things have adapted to neutralize or even take advantage of these factors would fill a large library, but we can describe and give examples of some of the general types of adaptations shown by plants and animals. In the course of time organisms have had to become readapted many times as their environment changed or as they migrated to a new environment. As a result, many organisms today have structures or physiologic mechanisms that are useless or even somewhat deleterious, but which were useful for survival in earlier times when the organism was adapted to a rather different environment.

339. ADAPTIVE RADIATION

Because of the competition for food and living space, there is a tendency for each group of organisms to spread out and occupy as many different habitats as they can reach and which will support them. This evolution, from a single ancestral species, of a variety of forms which occupy different habitats is called **adaptive radiation.** It is obviously advantageous in enabling organisms to tap new sources of food and to escape from some of their enemies. The placental mammals provide a classic illustration of the process, for from a primitive, insect-eating, five-toed, short-legged creature that walked with the soles of its feet flat on the ground, have evolved all the present-day types of placental mammals (Fig. 315). There are dogs and deer, adapted for terrestrial life in which running rapidly is important for survival; squirrels and primates, adapted for life in the trees; bats, equipped for flying; beavers and seals, which maintain an amphibious existence; the completely aquatic whales, porpoises and sea cows; and the burrowing animals, moles, gophers and shrews. The number and shape of the teeth, the length and number of leg bones, the number and attachment sites of muscles, the thickness and color of the fur, and so on, are some of the structures which are involved in adaptation.

340. CONVERGENT EVOLUTION

Conversely, many of the animals inhabiting the same type of habitat have developed similar structures which make

Figure 315. Adaptive radiation. All the various mammals shown have evolved from the common ancestor shown in the center, but have become adapted to a wide variety of environments.

them superficially alike, even though they may be but distantly related. This evolution of similar structures by animals adapting to similar environments is known as **convergent evolution.** The dolphins and porpoises (which are mammals), the extinct ichthyosaurs (which were reptiles) and both bony and cartilaginous fishes have all evolved streamlined shapes, dorsal fins, tail fins, and flipper-like fore and hind limbs which make them look much alike (Fig. 316). The moles and gophers, in adapting to a burrowing life, have evolved similar fore and hind leg structures adapted for digging, but the mole is an insectivore and the gopher is a rodent.

341. STRUCTURAL ADAPTATIONS

In many animals, the specialized adaptation to a certain way of life is simply the latest stage in a series of adaptations. For example, both man and the baboon, whose immediate ancestors were tree-dwellers, have returned to the ground and have become readapted to walking rather than to climbing trees. Readaptation may be a very complicated process. The present-day Australian tree-climbing kangaroos are the descendants of an original ground-dwelling marsupial. From the ground-dwellers evolved forms which, in adaptive radiation, took to the trees and developed limbs adapted to tree climbing (or perhaps the sequence of events was the reverse—first the evolution of specialized limbs, then the adoption of a tree-dwelling mode of life). Some of these tree-dwellers eventually left the trees and became readapted to ground life, and the hind legs became lengthened, strengthened, and adapted for leaping. But finally some of these kangaroos went back to the trees, although now their legs were so highly specialized for leaping that they could not be used for grasping a tree trunk. Consequently the present-day tree kangaroos must climb like bears, by bracing their feet against the tree trunk. A comparison of the feet of existing Australian marsupials reveals all the stages in this complicated, shifting process of adaptation.

342. PHYSIOLOGIC ADAPTATIONS

Since one of the major struggles among organisms stems from the competition for food, a mutation enabling an animal to use a new type of food is extremely advantageous. This may be accomplished in a number of ways—by the evolution of a new digestive or energy-liberating enzyme system, for example. It was a mutation resulting in a new energy-liberating enzyme which enables the sulfur bacteria to obtain energy from hydrogen sulfide, a substance which is poisonous to almost all other organisms.

Another type of favorable mutation is one which decreases the growing season of a plant or the total length of time required for an insect to develop. Such mutations enable the organism to survive farther from the equator and open up new areas of living space and new sources of food.

Other organisms have solved the problem of living in arctic regions by becoming dormant during the cold season or by migrating. Many birds, but only a few mammals, migrate south to avoid the cold northern winter. A number of kinds of mammals—monotremes, shrews, rodents and bats—hibernate over the winter. It is doubtful whether carnivores such as bears and skunks actually hibernate; they simply sleep for long periods at a time. In true **hibernation** the body temperature falls to just a degree or two higher than the surrounding air temperature, metabolism decreases greatly, and the heart rate and rate of respiration become very slow. No food is eaten and the animal uses up his stores of body fat, awakening in the spring in an emaciated condition (and probably "as hungry as a bear"). The factors which induce hibernation and which wake the animal in the spring are not completely clear; changes in the environmental temperature are, of course, important, but changes within the animal body are also involved.

Any mutation that increases the limits of temperature tolerance of a species (makes it more eurythermic) may enable it to inhabit a new part of the earth. Birds and mammals are unique in possessing mechanisms for controlling body temperature, keeping it constant despite wide fluctuations in the environmental temperature. These thermostated animals are said to be **homoiothermic** ("warm-blooded" is not quite correct; they are really "constant-temperature-blooded"). Fish, amphibians, reptiles and all the invertebrates are **poikilothermic**—their body temperature fluctuates with that of the environment (again, "cold-blooded"

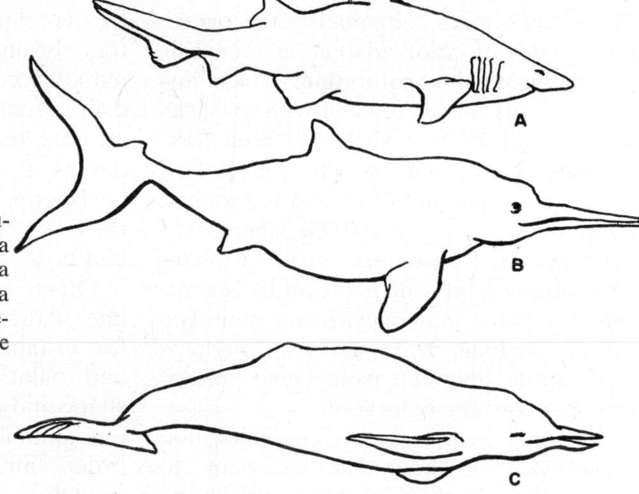

Figure 316. Convergent evolution. *A,* Shark; *B,* ichthyosaur (a fossil reptile); and *C,* dolphin (a mammal), all of which have a marked superficial similarity because of their adaptation to the same environment.

is not exactly descriptive; their body temperature is simply determined by the environmental temperature).

Marine fish are usually adapted to survive within a certain range of pressures and hence at a particular depth. Surface animals are crushed by the terrific pressures of the deep and deep-sea animals usually burst when brought to the surface. The whale, however, is able to withstand great changes in pressure, and can dive to depths of 2500 feet without injury. Presumably its lung alveoli collapse when the pressure on the body reaches a certain point and gases are no longer absorbed into the blood. A man can survive pressures as high as six atmospheres if the pressure is increased and then decreased slowly. The increase in pressure causes an increase in the amount of gases dissolved in the blood. If the pressure is decreased suddenly, these gases come out of solution and form bubbles in the blood which impede circulation and bring about the symptoms of diver's disease, or "the bends." The pilot of a jet plane may climb so rapidly that the atmospheric pressure is reduced fast enough to bring bubbles of gas out of solution in his blood and cause a type of the bends.

343. COLOR ADAPTATIONS

Adaptations for survival are evident in the color and pattern of plants and animals as well as in their structure and physiologic processes. Ecologists recognize three types of color adaptation: concealing or **protective coloration,** which enables the organism to blend with its background and be less visible to predators; **warning coloration,** which consists of bright, conspicuous colors and is assumed by poisonous or unpalatable animals to warn potential predators not to eat them; and **mimicry,** in which the organism resembles some other living or nonliving object—a twig, leaf, stone, or perhaps some other animal which, being poisonous, has warning coloration.

In some cases, concealing coloration serves to hide an animal that wants to escape the notice of potential predators; in other cases it is assumed by a predator in order to be unnoticed by his potential prey. Examples of such coloration are legion—the white coats of arctic mammals, and the stripes and spots of tigers, leopards, zebras and giraffes which, though conspicuous in a zoo, blend imperceptibly with the moving pattern of light and dark typical of their native savanna. Some animals—frogs, flounders, chameleons, crabs and others—can change their color and pattern as they move from a dark to a light background or from one that is uniform to one that is mottled (Fig. 317).

To demonstrate experimentally that concealing coloration does have survival value, investigators fastened grasshoppers with different body colors to plots of different colored soils—light, dark, grassy, sandy, and so on. After these plots had been exposed to the predatory activities of wild birds or chickens for a given length of time, the survivors were tabulated. It was found that there was a significantly higher percentage of survivors among those grasshoppers which matched their background.

When an animal is equipped with poison fangs, a stinging mechanism, or some chemical which gives it a noxious taste, it is to its advantage to have the fact widely advertised, and in fact, many animals of this type do have warning colors. An interesting example is a species of European toad with a bright scarlet belly. The toad has certain chemicals secreted by its skin glands which make it extremely unpalatable, and whenever a potential predator, such as a stork, swoops over a congregation of toads, they flop on their backs, exposing their scarlet bellies as a warning. The storks and other birds apparently become conditioned by the association of the red color and the bad taste, and avoid the toads assiduously.

Other animals survive by mimicking one of these protectively colored animals; for instance, some harmless, defenseless and palatable animals are identical in shape and color with a poisonous or noxious animal of quite a different family or order, and, being mistaken for it by predators, are left alone. Many tropical insects have evolved this type of protection. The arrangement is only possible where there are many more genuinely disagreeable or dangerous organisms than forms which mimic them, for obviously, if half the time or more, a predator finds that animals with a particular shape and color *are* palatable, he will not become conditioned to avoid them.

The reality of the selective advantage of color adaptations has been debated. It has been argued that animal vision may be so different from human vision (certain animals may be colorblind, or may be able to see ultraviolet or infrared light) that an animal that appears to be protectively colored to a man may be quite evident to its natural predators. However, many experimental studies, such as the grasshopper experiment mentioned previously, have shown that protective coloration does have survival value.

Color and pattern are also used to attract other organisms when that is necessary for survival. The red and blue ischial callosities of monkeys, and the gay, extravagant plumage of various birds apparently have an attraction for members of the opposite sex. And vividly colored flowers seem to attract the birds and insects whose activities are needed to insure the pollination of the plants.

344. ADAPTATIONS OF SPECIES TO SPECIES

The evolution and adaptation of each species have not occurred in a biologic vacuum, independent of other forms; instead, many species have had a marked influence on the adaptation of other species. As a result of this, many types of cross-dependency between species have arisen. Some of the clearest and best understood of these involve insects. Insects are necessary for the pollination of a great many plants; the plants are so dependent on certain insects that they are unable to survive in a given region unless those particular insects are present. For example, the Smyrna fig could not be grown in California, even though all climatic conditions were favorable, until the fig insect, which pollinates the plant, was introduced. Birds, bats, and even snails serve as pollen transporters for some plants, but insects are the prime pollinators. Flowering plants have developed bright colors and fragrances, presumably to attract insects and birds and ensure pollination. There has been some doubt as to whether they can detect them at all. However, the experiments of Karl von Frisch show that honeybees, at least, are able to differentiate colors and scents and that they are guided in their visits to flowers by these stimuli.

Some of the species-to-species adaptations are so exact that neither form can exist in a region without the other. The yucca plant and the yucca moth, like the fig and fig insect, have evolved to a point of complete interdependence (Fig. 318). The yucca moth, by a series of instinctive acts, goes to a yucca flower, collects some pollen, and takes it to a second flower. There it pushes its ovipositor (egg-laying organ) through the wall of the ovary of the flower and lays an egg. It then carefully places some pollen on the stigma. The yucca plant in this way is sure to be fertilized and produce seeds; the larva of the yucca moth feeds on these yucca seeds. The yucca produces a large number of seeds and is not injured by the loss of the few seeds eaten by the moth larva.

Other examples of species-to-species

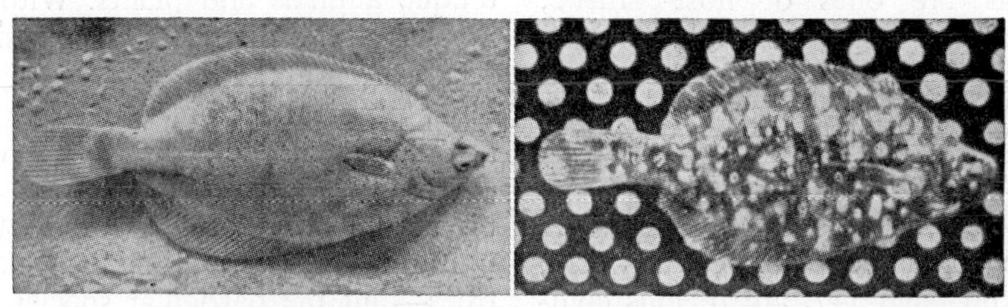

Figure 317. An experiment to show the remarkable ability of the flounder to change its color and pattern to conform with its backgraund. *Left,* A flounder on a uniform, light background; *right,* the same fish after being placed on a spotted, darker background.

Figure 318. The yucca plant (left) is pollinated only by the yucca moth, one of which is shown in the open flower at the right. (Weatherwax: Botany.)

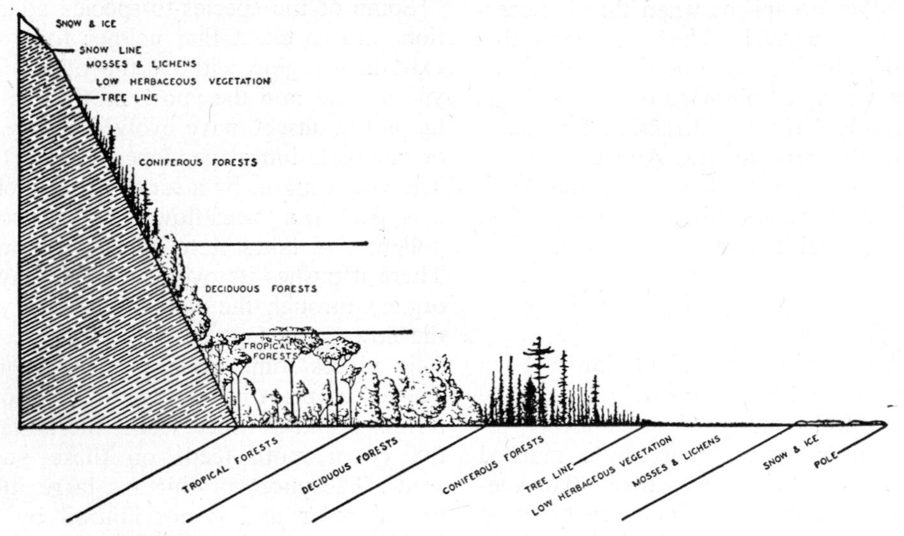

Figure 319. Correspondence of latitudinal and altitudinal life zones in North America. (Allee et al.: Principles of Animal Ecology.)

adaptations are ones of host-parasite, prey-predator, commensals and mutualistic interdependence. These were discussed in Chapter 6.

345. TERRESTRIAL LIFE ZONES: BIOMES

The biogeographic regions of the world, discussed on page 547, are regions composed of a whole continent or a large part of it, and characterized by certain unique animals and plants. Within these biogeographic divisions, and arising as a result of complex interactions of climate, other physical factors and biotic factors, are large, distinct, easily differentiated community units called **biomes.** In each biome the *kind* of climax vegetation is uniform—grasses, conifers, deciduous trees—but the particular species of plant may vary in different parts of the biome. The kind of climax vegetation depends upon the physical environment and the two together determine the kind of animals present. The definition of biome includes not only the actual climax community of a region, but also the several intermediate communities that precede the climax community.

Some of the biomes recognized by ecologists are **tundra, coniferous forest, deciduous forest, broad-leaved evergreen subtropical forest, grassland, desert, chaparral,** and **tropical rain forest.** These biomes are distributed, though somewhat irregularly, as belts around the world, and as one travels from the equator to the pole he may traverse tropical rain forest, grassland, desert, deciduous forest, coniferous forest and finally reach the tundra in Northern Canada and Alaska. Since climatic conditions at higher altitudes are in many ways similar to those at higher latitudes, there is a similar succession of biomes on the slopes of high mountains (Fig. 319). For example, as one goes from the San Joaquin Valley of California into the Sierras, one passes

Figure 320. The tundra biome. *Above:* View of the low tundra near Churchill, Manitoba, in July. Note the numerous ponds. *Below:* View of tundra vegetation showing "lumpy" nature of low tundra and a characteristic tundra bird, the willow ptarmigan. (Lower photo by C. Lynn Haywood.)

Figure 321. The grassland biome; characteristic animals of the African grasslands, zebra and wildebeest, Kruger National Park, Transvaal. (Photograph by Herbert Lang.)

from desert through deciduous forest and coniferous forest to, above timberline, a region resembling the tundra of the Arctic.

Tundra. The tundra biome (Fig. 320), found in northern North America, northern Europe, and Siberia, is characterized by low temperatures and a short growing season. The plants are lichens, mosses, grasses, and a few low shrubs. The animals present are caribou or reindeer, arctic hare, arctic fox, lemmings, snowy owls, and during the summer, swarms of flies and mosquitoes and a host of migratory birds.

Northern Coniferous Forest. This biome, stretching across both North America and Eurasia, just south of the tundra, is characterized by spruce, fir and pine trees, and animals such as the snowshoe hare, lynx and wolf.

Temperate Deciduous Forest. The areas with abundant, evenly distributed rainfall and moderate temperatures with distinct summers and winters originally were covered with forests of beech, maple, oak, hickory or chestnut trees. These covered most of Eastern North America and much of Europe, but most of them have been replaced by cultivated fields. The animals present originally were deer, bear, squirrels, wild turkeys and woodpeckers.

Broadleaved Evergreen Subtropical Forest. In regions of fairly high rainfall but where temperature differences between winter and summer are less marked, as in Florida, the vegetation includes live oaks, magnolias, tamarinds and palm trees, with many vines and epiphytes such as orchids and Spanish moss.

Grasslands. This biome (Fig. 321) occurs where rainfall is about 10 to 30 inches per year, insufficient to support a forest, yet greater than that of a true desert. Grasslands typically occur in the interiors of continents—the prairies of western United States, and those of Argentina, Australia, southern Russia and Siberia. The animals are either grazing or burrowing mammals—bison, antelope, zebras, rabbits, ground squirrels, prairie dogs and gophers—and birds such as prairie chickens, meadow larks and rodent hawks. There is a broad belt of tropical grassland or **savanna** in Africa lying between the Sahara desert and the tropical rain forest of the Congo basin. Although the annual rainfall is high, up to 50 inches, there is a distinct dry season from June to August which prevents the development of forests. In this region are great numbers and many kinds of grazing animals.

Deserts. In regions with less than 10

inches of rain per year, vegetation is sparse and consists of greasewood, sagebrush or cactus (Fig. 322). In the brief rainy season, the California desert becomes carpeted with an amazing variety of wild flowers and grasses, most of which complete their life cycle from seed to seed in a few weeks. The animals present are reptiles, insects and burrowing rodents such as the kangaroo rat and pocket mouse, both of which are able to live without drinking water, extracting water from their food—seeds and succulent cactus.

Tropical Rain Forest. Low-lying regions near the equator, with annual rainfalls of 90 inches or more, are characterized by thick rain forests, with a tremendous variety of plants and animals (Fig. 323). No single species is present in large enough numbers to be dominant. The valleys of the Amazon, Orinoco, Congo and Zambesi rivers and parts of Malaya and New Guinea are covered with

Figure 322. Two types of desert in western North America, a "cool" desert in Idaho dominated by sagebrush (above) and (below) a rather luxuriant "hot" desert in Arizona, with giant cactus (Saguaro) and palo verde trees, in addition to creosote bushes and other desert shrubs. In extensive areas of desert country the desert shrubs alone dot the landscape. (Upper photograph by U. S. Forest Service, lower by U. S. Soil Conservation Service.)

Figure 323. The rain forest biome: border of a clearing in the Ituri Forest of Nala, Belgian Congo. (Photograph by Herbert Lang; courtesy of The American Museum of Natural History.)

tropical rain forests. The vegetation is very thick and vertically stratified—tall trees, shrubs, vines and epiphytes crowd together—and many animals live in the upper layers of the vegetation. Here are found monkeys, sloths, termites, anteaters, many reptiles, and many brilliantly colored birds.

346. MARINE LIFE ZONES

Like the land, the ocean has clearly demarcated regions characterized by different physical conditions, and consequently inhabited by different kinds of plants and animals. These are: (1) the **tidal zone,** the beach between the high and low tide marks; (2) the **shallow sea,** the region lying over the continental shelf and extending out to a depth of about 500 feet; (3) the **pelagic zone,** the open ocean down as far as sunlight can penetrate (some 500 to 1000 feet); and (4) the **abyssal zone,** the ocean beyond the continental shelf and beneath the pelagic zone.

The tidal zone is one of the most favorable of all the habitats in the world and many biologists believe that life originated here. The abundance of water, light, oxygen, carbon dioxide and minerals makes it extremely salutary for plants. The dense growth of plants, providing food and shelter, makes it an excellent habitat for animals. The plants of the region are primarily a wide variety of algae plus a few grasses. Members of every phylum of the animal kingdom are present, a situation which prevails only here and in the nearby shallow sea. There is keen competition among the plants for space, and among the animals for space and food, so the forms living here have had to evolve special adaptations to survive.

Since the intertidal zone is exposed to air twice daily, its inhabitants have had to develop some sort of protection against drying up. Some animals avoid this by burrowing into the damp sand or rocks until the tide returns; others have developed shells which can be closed, retaining a supply of water inside. Many plants contain jelly-like substances such as agar, which absorb and retain large quantities of water while the tide is out.

One of the outstanding characteristics of this region is the ever-present action of the waves, and the organisms living on a sandy or rocky beach have had to evolve ways of resisting wave action. The many seaweeds have tough pliable bodies, able to bend with the waves without breaking, while the animals are either encased in hard calcareous shells, such as those of molluscs, bryozoa, starfish, barnacles and crabs, or are covered by a strong leathery skin that can bend without breaking, like that of the sea anemone and octopus.

The shallow sea region is also thickly populated, for it has plenty of light and other things required by plants. The absence of the periodic exposure to air and the diminished wave action permit many plants and animals to live here that could not survive in the tidal zone. Here live many species of fish and many single-celled algae; the larger seaweeds, which require a substrate for attachment, are found only in the shallower parts of the region.

The pelagic region, distinguished by the presence of sunlight and the absence of a substrate, is populated by "swimmers" and "floaters." There are no large seaweeds here, except occasional pieces torn from their anchorage in the shallow sea, and fewer microscopic algae than in the zones nearer shore. There are many varieties of microscopic animals: protozoa such as foraminifera and radiolaria; small crustacea; and the larvae of many forms. A few large animals live here—jellyfish, the Portuguese man-of-war, squid, and a few fishes and whales. Some whales feed upon the microscopic forms of life and are equipped with strainers to remove them from the water; others, which prey upon fish, squid and other whales, are equipped with teeth.

Below the pelagic is the abyssal region, characterized by the absence of light and the consequent absence of plants. The waters are quiet and cold, and the pressure is stupendous. The animals here must feed upon each other or on the bodies of dead plants and animals that are constantly settling down from above. Most of the fish of the abyssal region are rather small and peculiarly shaped; many are equipped with luminescent organs, which may serve as lures for the forms preyed upon. The majority of the deep-sea creatures are related to shallow-sea forms, and they must have migrated to their present habitat recently (geologically speaking) for none is older than the Mesozoic.

Since the number of members of any one species in these vast depths is small, reproduction is more of a problem than in any other region, and some fish have evolved a curious adaptation to ensure that reproduction will occur. At an early age the male becomes attached to and fuses with the head of the female, where he continues to live as a small (inch-long) parasite (Fig. 324). In due course he becomes mature and when the female lays

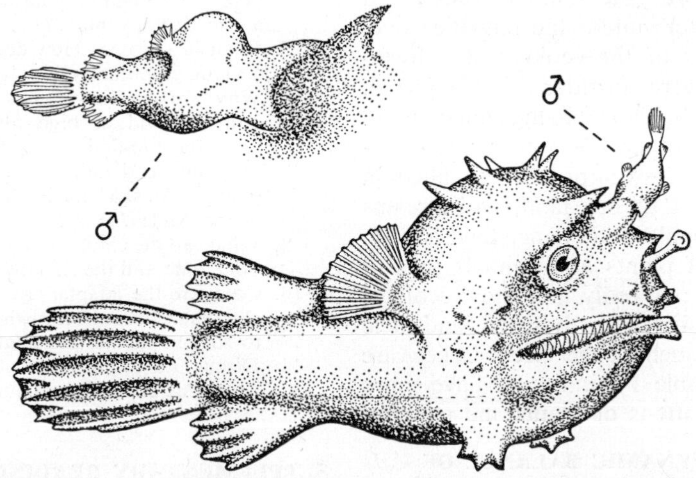

Figure 324. Sexual parasitism in the deep-sea angler fish, *Photocorynus spiniceps,* in which the difficulty of one sex finding the other is met by permanent attachment of the much smaller male to the female. The union is so complete that the male has no independent existence at all, being nourished by the blood of the female to which he is attached. (After Norman, from Allee et al.: Principles of Animal Ecology.)

her eggs, he releases his sperm into the water to fertilize them.

The bottom of the sea is a soft ooze, made of the organic remains and shells of foraminifera, radiolaria, and other animals and plants. Many invertebrates live on the ocean floor at great depths, and are usually characterized by thin, almost transparent shells, whereas the related shallow-water forms, exposed to wave action, have thick, hard shells. Apparently even the greatest "deeps" are inhabited, for tube-dwelling worms have been dredged from depths of 24,000 feet, and sea urchins, starfish, bryozoa and brachiopods have been found at depths of 18,000 feet.

347. FRESH-WATER LIFE ZONES

Fresh-water habitats can be divided into **standing water**—lakes, ponds and swamps—and **running water**—rivers, creeks and springs—each of which can be further subdivided. The biologic communities of fresh-water habitats are in general more familiar than the salt-water ones, and many of the animals used as specimens in biology classes are from fresh water—amebas, hydras, planarias, crayfish and frogs. Standing water, such as a lake, can be divided (much as the zones of the ocean were distinguished) into the shallow water near the shore (the **littoral zone**), the surface waters away from the shore (the **limnetic zone**), and the deep waters under the limnetic zone. Some aspects of the ecology of a fresh-water pond were considered in Chapter 6. Fresh-water habitats change much more rapidly than other life zones; ponds become swamps, swamps become filled in and converted to dry land, and streams erode their banks and change their course. The kinds of plants and animals present may change markedly and show ecologic successions similar to those on land. The large lakes, such as the Great Lakes, are relatively stable habitats and have more stable populations of plants and animals.

348. THE DYNAMIC BALANCE OF NATURE

The concept of the dynamic state of the body constituents was discussed in Chapter 5 and we learned that the protein, fat, carbohydrate and other constituents of the body are constantly being broken down and resynthesized. From the discussion of ecology, it is evident that communities are constantly undergoing an analogous reshuffling and that the concept of the **dynamic state of communities** is a valid one. Plant and animal populations are constantly subject to changes in their physical and biotic environment and must adapt or die. A population may vary in size but if it outruns its food supply, like the Kaibab deer or the lemmings, equilibrium is quickly restored. Communities of organisms are comparable in many ways to a many-celled organism, and exhibit growth, specialization and interdependence of parts, characteristic form, and even development from immaturity to maturity, old age and death.

QUESTIONS

1. Contrast adaptive radiation and convergent evolution.
2. What is meant by hibernation? How could you determine whether an animal was hibernating or simply asleep?
3. Why is poikilothermic a better term for a lizard than cold-blooded?
4. What is the cause of "the bends"? What measures can be taken to prevent it?
5. Give examples from your experience of concealing coloration and warning coloration.
6. What experiments would you devise to determine whether color adaptations have a selective advantage?
7. What is a biome? How does it differ from a biotic community?
8. Why are similar biomes found at high latitudes and at high altitudes? Would you expect to find *exactly* the same species of plants and animals in the tundra region of Alaska and in the tundra region of the Andes? Why?
9. What are the chief differences between the tidal zone and the shallow sea?
10. Compare the adaptations made by barnacles, snails and starfish which enable them to survive in the intertidal zone.
11. Discuss the implications of the phrase, the "dynamic state of communities."

SUPPLEMENTARY READING

A wonderfully illustrated account of animal camouflage is to be found in H. B. Cott's *Adaptive Coloration in Animals.*

Appendix

A Survey of the Plant and Animal Kingdoms

THE SYSTEM of classifying animals and plants—cataloguing them by phylum, class, order, family, genus and species—was discussed in Chapter 6. In the synoptic survey given here the phyla within the kingdoms and the classes within the phyla are arranged in the order of increasing complexity as far as possible, but since evolution has proceeded in a branching fashion, it is impossible to arrange the animals and plants rigidly in order, from simple to complex. The numbers given are estimates of known species in the phylum; for many of the groups there are many additional species as yet undescribed.

I. THE PLANT KINGDOM

Organisms classified as plants usually have stiff cell walls and chlorophyll.

SUBKINGDOM THALLOPHYTA:

Plants not forming embryos. The simplest plants, without true roots, stems or leaves; the body is either a single cell or an aggregation of cells with little differentiation into tissues. (107,000)

Phylum Cyanophyta. The blue-green algae, with no distinct nuclei or chloroplasts; probably the most primitive of existing plants. (2,500)

Phylum Euglenophyta. Euglenoids.

Phylum Chlorophyta. The green algae, with definite nuclei and chloroplasts. (5,000)

Phylum Chrysophyta. The yellow-green algae, the golden-brown algae, and the diatoms.

Phylum Pyrrophyta. The cryptomonads and dinoflagellates.

Phylum Phaeophyta. The brown algae, with multicellular, often large bodies—the large seaweeds. (1,000)

Phylum Rhodophyta. The red algae. Multicellular, usually marine plants, sometimes impregnated with calcium carbonate. (3,000)

Phylum Schizomycophyta. The bacteria. (3,000)

Phylum Myxomycophyta. The slime molds. The body consists of a mass of protoplasm containing many nuclei, but not sharply divided into cells. Movement by ameboid motion.

Phylum Eumycophyta. The true fungi. (70,000)

Class Phycomycetes. The algal fungi—bread molds and leaf molds.

Class Ascomycetes. The sac fungi—yeasts, mildews and cheese molds.

Class Basidiomycetes. Mushrooms, toadstools, rusts and smuts.

Class Fungi Imperfecti. A heterogeneous collection of fungi in which sexual reproduction is unknown, and which are not easily assigned to one of the other classes.

SUBKINGDOM EMBRYOPHYTA:

Plants forming embryos.

Phylum Bryophyta. Embryophyte plants without conducting tissues. Multicellular plants, usually terrestrial, with a marked alternation of sexual and asexual generations. The prominent plant is the gametophyte (sexual generation), on which the sporophyte is dependent. (23,000)

Class Musci. Mosses. The gametophyte plant has an erect stem, and leaves arranged in a spiral.

Class Hepaticae. The liverworts. Usually simple, flat plants, living in moist, shady places.

Class Anthocerotae. Hornworts.

Phylum Tracheophyta. Vascular plants.

Subphylum Psilopsida. Leafless and rootless vascular plants.

Class Psilophytinea.

Order Psilophytales.*

Order Psilotales.

Subphylum Lycopsida. The club-mosses with simple conducting systems and small green leaves. (900)

Class Lycopodineae. The club-mosses and quillworts.

Order Lycopodiales. Club-mosses.

Order Selaginellales. Small club-mosses.

Order Lepidodendrales.* Giant club-mosses.

Order Pleuromeiales.*

Order Isoetales. Quillworts.

Subphylum Sphenopsida. Horsetails with simple conducting systems, jointed stems, and reduced, scalelike leaves.

Class Equisetineae.

Order Hyeniales.*

Order Sphenophyllales.

Order Equisetales. Horsetails.

Subphylum Pteropsida. Complex conducting systems, large conspicuous leaves.

Class Filicinae. Ferns. (9,000)

Order Coenopteridales.*

Order Ophioglossales.

Order Marattiales.

Order Filicales.

Class Gymnospermae. Conifers, cycads, and most other evergreen trees and shrubs. No true flowers or ovules are present; the seeds are born naked on the surface of the cone scales. (640)

Subclass Cycadophytae.

Order Cycadofilicales.* The seed ferns—the most primitive of the seed plants—known only from fossils (Late Paleozoic Era).

Order Bennettitales.*

Order Cycadales. The cycads, the most primitive living seed plants, found in tropical and subtropical regions.

Subclass Coniferophytae.

Order Cordaitales.* Extinct, large-leaved evergreen trees. Fossil remains of these have been found in deposits from the Devonian to the Permian.

Order Ginkgoales. The ginkgo, or maidenhair tree, is the only living member of this group.

Order Coniferales. The conifers, the common evergreen trees and shrubs, with needle-shaped leaves.

Order Gnetales. Climbing shrubs or small trees found in tropical and semitropical regions, with many characteristics in common with the angiosperms.

Class Angiospermae. Flowering plants, with seeds enclosed in an ovary. (200,000)

Subclass Dicotyledoneae. Most flowering plants. Embryos with two cotyledons or seed-leaves; vascular bundles in a ring in the stem; leaves with netlike venation; flower parts (sepals, petals, stamens and carpels) in fives, fours or twos. (125,000)

Subclass Monocotyledoneae. The grasses, lilies and orchids. Leaves with parallel veins, stems in which the vascular bundles are scattered, and flower

* Known only as fossils; no living representatives.

parts in threes or sixes. The embryo has only one seed-leaf. (75,000)

II. THE ANIMAL KINGDOM

Organisms classified as animals usually lack stiff cell walls and never have chlorophyll; nutrition is either holozoic or parasitic.

Phylum Protozoa. Microscopic, unicellular animals, which sometimes aggregate in colonies. Some are free-living; others are parasitic. (15,000)

Class Flagellata. Protozoa which move by whiplike protrusions of protoplasm called flagella. Primitive animals, probably the group most closely related to the one-celled plants.

Class Sarcodina. Protozoa which move by pseudopodia.

Class Sporozoa. Parasitic protozoa which reproduce by spores and have no method of locomotion.

Class Ciliata. Protozoa which move by means of cilia.

Class Suctoria. Protozoa with cilia only in young stages, adults attached by stalk to substrate. Some parasitic.

Phylum Porifera. The sponges, both fresh-water and marine. The lowest of the many-celled animals, resembling in many respects a protozoan colony. The body is perforated with many pores to admit water, from which food is strained. Three classes: Calcarea (calcareous spicules), Hexactinellida (siliceous spicules) and Demospongiae (protein spicules—bath sponges). (3,000)

Phylum Coelenterata. Radially symmetrical animals with a central gastrovascular cavity. The body wall consists of two layers of cells, in the outer of which are stinging cells, nematocysts. (4500)

Class Hydrozoa. Hydra-like animals, either single or colonial. There is usually an alternation of a hydra-like (asexual) generation with a jellyfish (sexual) generation.

Class Scyphozoa. True jellyfishes.

Class Anthozoa. The corals and sea anemones, which have no alternation of generations. The digestive cavities of these animals are divided by mesenteries to increase the effective surface.

Phylum Ctenophora. The comb jellies or sea walnuts. These animals lack the stinging capsules of coelenterates and move by means of eight comb-like bands of cilia. (100)

Phylum Platyhelminthes. The flatworms, with flat, and either oval or elongated, bilaterally symmetrical bodies, and three cell layers. The excretory organs are flame cells. There is a true central nervous system. (6000)

Class Turbellaria. Nonparasitic flatworms with a ciliated ectoderm.

Class Trematoda. The flukes, parasitic flatworms with nonciliated ectoderm, and one or more suckers. Many are internal parasites with complicated life cycles.

Class Cestoda. The tapeworms, parasitic flatworms with no digestive tract; the body consists of a head and a chain of "segments" or individuals which bud from the head.

Phylum Nemertea. The proboscis worms. Nonparasitic, usually marine animals with a complete digestive tract and a protrusible proboscis armed with a hook for capturing prey. The lowest animals with a blood vascular system. (400)

Phylum Nematoda. The roundworms. An extremely large phylum. Characterized by elongated, cylindrical, bilaterally symmetrical bodies; they live as parasites in plants and animals, or are free-living in the soil or water. (80,000)

Phylum Acanthocephala. The hook-headed worms. Parasitic worms with no digestive tract and a head armed with many recurved hooks. (100)

Phylum Chaetognatha. Free-swimming marine worms, with a body cavity (coelom) which develops from pouches of the digestive tract (as in the echinoderms and lower chordates). (30)

Phylum Nematomorpha. The horse-hair worms. Extremely thin, brown or black worms about 6 inches long, resembling a horsehair. The adults are free-living, but the larvae are parasitic in insects. (200)

Phylum Rotifera. Small, wormlike animals, commonly called "wheel animalcules," with a complete digestive tract, flame cells, and a circle of cilia on the head, the beating of which suggests a wheel. (1200)

Phylum Gastrotricha. Microscopic, wormlike animals resembling the rotifers, but lacking the crownlike circle of cilia. (100)

Phylum Bryozoa. "Moss" animals. Microscopic organisms, usually marine, which form branching colonies. Characterized by a U-shaped row of ciliated tentacles, the lophophore, by means of which they capture food. (1200)

Phylum Brachiopoda. The lamp shells. Marine animals with two hard shells (one dorsal and one ventral), superficially like a clam. They obtain food by means of a lophophore. (200 at present; 3000 extinct)

Phylum Phoronida. Wormlike, marine forms which secrete and live in a leathery tube; they have a U-shaped digestive tract and a lophophore. (10)

Phylum Annelida. The segmented worms. There is a distinct head, digestive tract, coelom, and—in some—non-jointed appendages. The digestive system is divided into specialized regions. (8000)

Class Polychaeta. Mostly marine worms. Each segment of their bodies has a pair of paddle-like structures (parapodia) for swimming. Some burrow in sand and mudflats; some live in calcareous tubes which they secrete; others swim freely in the ocean.

Class Oligochaeta. Fresh-water or terrestrial worms, with no parapodia and few bristles per segment.

Class Archiannelida. Primitive annelids without bristles or external segmentation.

Class Hirudinea. The leeches—flattened annelids lacking bristles and parapodia, but with suckers at anterior and posterior ends.

Phylum Onychophora. Rare, tropical animals, structurally intermediate between annelids and arthropods, with an annelid-like excretory system and an insect-like respiratory system. Only a few species known.

Phylum Arthropoda. Segmented animals with jointed appendages and a hard, chitinous skin, with a body divided into head, thorax and abdomen. (650,000)

Class Trilobita. Trilobites, primitive marine arthropods that originated in the Cambrian and became extinct in the Permian. Segmented body divided by two longitudinal furrows into three lobes. All segments except last had a pair of biramous appendages.

Class Crustacea. Lobsters, crabs, barnacles, water fleas, and sowbugs. Animals that are usually aquatic, have two pairs of antennae, and respire by means of gills.

Class Chilopoda. The centipedes. Each body segment, except the head and tail, has a pair of legs.

Class Diplopoda. The millipedes. Each external segment (really two segments fused) bears two pairs of legs.

Class Arachnoidea. Spiders, scorpions, ticks, mites and king crabs. Adults have no antennae; the first pair of appendages ends in pincers, the second pair is used as jaws, and the last four pairs are used for walking.

Class Insecta. The largest group of animals, mostly terrestrial. The body is divided into a distinct head, with four pairs of appendages; the thorax has three pairs of legs and usually two pairs of wings; the abdomen has no appendages. Respiration by means of tracheae. There are about twenty-four different orders of insects, of which the following are common:

Order Orthoptera. Grasshoppers and cockroaches.

Order Isoptera. Termites.

Order Odonata. Dragonflies and damsel flies.

Order Anopleura. Lice.

Order Hemiptera. Water boatmen, bedbugs and back-swimmers.

Order Homoptera. Cicadas, aphids and scale insects.

Order Coleoptera. Beetles, weevils and fireflies.

Order Lepidoptera. Butterflies and moths.

Order Diptera. Flies, mosquitoes and gnats.

Order Hymenoptera. Ants, wasps, bees and gallflies.

Phylum Mollusca. Unsegmented, soft-bodied animals, usually covered by a shell, and with a ventral, muscular foot. Respiration is by means of gills, protected by a fold of the body wall—the mantle. (80,000)

Class Amphineura. Chitons, marine forms with a shell composed of eight plates.

Class Scaphopoda. Tooth shells, marine forms living in sand or mud; tubular shells open at both ends.

Class Gastropoda. Snails, slugs, whelks, abalones; asymmetrical animals with a single spiral shell, or no shell.

Class Pelecypoda. Clams, mussels, oysters, scallops. These lack a head and have a hatchet-shaped foot for burrowing. The shell consists of two plates or valves (the animals are called "bivalves"), one on each side of the body.

Class Cephalopoda. Squids, cuttlefish, octopuses. Marine animals having a well-developed "head-foot," with eight or ten tentacles, and well-developed eyes and nervous system.

Phylum Echinodermata. Marine animals which are radially symmetrical as adults, bilaterally symmetrical as larvae. The skin contains calcareous, spine-bearing plates. The animals have a unique water vascular system of canals, and tube feet for locomotion. Respiration is by skin gills or by out-pocketings of the digestive tract. (6000)

Class Asteroidea. The starfishes. The body is a central disc with broad arms (usually five) not sharply marked off from the disc.

Class Ophiuroidea. The brittle stars and serpent stars. The body is a central disc with narrow arms sharply marked off.

Class Echinoidea. The sea urchins and sand dollars. Spherical or flattened oval animals with many long spines.

Class Holothuroidea. Sea cucumbers. Long, ovoid, soft-bodied echinoderms, usually with a ring of tentacles around the mouth.

Class Crinoidea. Sea lilies and feather stars. The body is cup-shaped and attached by a stalk to the subtrate. Most of these are known only as fossils; only a few survive.

Phylum Chordata. Bilaterally symmetrical animals with a notochord, gill clefts in pharynx, and a dorsal, hollow neural tube. (70,000)

Subphylum Hemichorda. The acorn worms. Marine, wormlike forms resembling larval echinoderms during development.

Subphylum Urochorda. The tunicates or sea squirts. The adults are saclike, attached animals, which often form colonies, but the larval forms are free-swimming and have a notochord in the tail region.

Subphylum Cephalochorda. Amphioxus. Marine animals with a segmented, elongated, fishlike body. They burrow in the sand and take in food by the beating of cilia on the anterior end. They have a notochord extending from the tip of the head to the tip of the tail.

Subphylum Vertebrata. Animals having a definite head, a backbone of vertebrae, a well-developed brain, and, usually, two pairs of limbs. They have a ventrally located heart, and a pair of well-developed eyes.

Class Agnatha. Lampreys, hagfishes and fossil ostracoderms. Vertebrates without jaws or paired fins.

Class Placodermi. The spiny-skinned sharks. The earliest fishes with jaws, known only from fossils.

Class Chondrichthyes. Sharks, rays, skates and chimaeras. Fishes with a cartilaginous skeleton and scales of dentin and enamel imbedded in the skin.

Class Osteichthyes. The bony fishes. The sturgeon, bowfin, salmon and lungfishes.

Class Amphibia. Frogs, toads, salamanders and the extinct forms, stegocephalians. As larvae these forms breathe by gills; as adults they breathe by lungs. There are two pairs of five-toed limbs; the skin is usually scaleless.

Class Reptilia. Lizards, snakes, turtles, crocodiles, the extinct dinosaurs and other forms. The body is covered with scales derived from the epidermis of the skin. The animals breathe by means of lungs and have a three-chambered heart.

Class Aves. The birds. Warm-blooded animals whose skin is covered with feathers. Present-day birds are toothless, but the primitive ones had reptilian teeth. The forelimbs are modified as wings.

Class Mammalia. Warm-blooded animals whose skin is covered with hair. The females have mammary glands which secrete milk for the nourishment of the young.

Subclass Prototheria. The monotremes, primitive forms that lay eggs. Most of them are extinct; only two species survive, the duckbilled platypus and the spiny anteater.

Subclass Metatheria. The pouched mammals. The young are born alive, but in a very undeveloped state. They complete development in a pouch on the mother's abdomen. The subclass includes opossums, and a variety of forms found only in Australia: kangaroos, wallabies, koala bears, wombats, and so on.

Subclass Eutheria. The placental mammals. The young develop within the uterus of the mother, obtaining nourishment via the placenta.

Order Insectivora. Primitive, insect-eating mammals; moles and shrews.

Order Chiroptera. Bats.

Order Carnivora. Dogs, cats, bears, sea lions and seals.

Order Rodentia. Rats, squirrels, beavers and porcupines.

Order Lagomorpha. Rabbits and hares.

Order Primates. Monkeys, apes and man.

Order Artiodactyla. Even-toed ungulates—cattle, deer, camels and hippopotamuses.

Order Perissodactyla. Odd-toed ungulates—horses, zebras and rhinoceroses.

Order Edentata. Armadillos, sloths and anteaters.

Order Proboscidea. Elephants.

Order Cetacea. Whales.

Order Sirenia. Sea cows—large, plant-eating aquatic mammals.

Bibliography

Adrian, Edgar: *The Physical Background of Perception*. London, Oxford University Press, 1947.

Alexopoulos, Constantine: *Introductory Mycology*. New York, John Wiley, 1952.

Allee, W. C., Emerson, A. E., Park, O., Park, T., and Schmidt, K. P.: *Principles of Animal Ecology*. Philadelphia, W. B. Saunders Co., 1949.

Anderson, E. A.: *Plants, Life and Man*. Boston, Little, Brown & Co., 1952.

Andrews, Henry N.: *Ancient Plants and the World They Lived in*. Ithaca, Comstock, 1947.

Arey, L. B.: *Developmental Anatomy*. 6th Ed. Philadelphia, W. B. Saunders Co., 1954.

Armstrong, E. A.: *Bird Display and Behavior*. New York, Oxford University Press, 1947.

Baerg, W. J.: *Introduction to Applied Entomology*. Ann Arbor, Edwards, 1948.

Bainbridge, F. A., Bock, A. V., and Dill, D. B.: *The Physiology of Muscular Exercise*. 3rd Ed. London, Longmans, Green and Co., 1931.

Baitsell, George (ed.): *Science in Progress*. Vol. 6. New Haven, Yale University Press, 1949.

Baldwin, E. B.: *An Introduction to Comparative Biochemistry*. 3rd Ed. Cambridge, Cambridge University Press, 1949.

Baldwin, E. B.: *Dynamic Aspects of Biochemistry*. 2nd Ed. Cambridge, Cambridge University Press, 1952.

Barbour, Thomas: *Reptiles and Amphibians*. Boston, Houghton Mifflin, 1926.

Barron, E. S. G. (ed.): *Trends in Physiology and Biochemistry*. New York, Academic Press, 1952.

Bartley, S. H.: *Vision*. New York, D. Van Nostrand Company, 1941.

Beach, Frank: *Hormones and Behavior*. New York, Paul B. Hoeber, 1947.

Beaumont, William: *Experiments and Observations on the Gastric Juice and the Physiology of Digestion*. Cambridge, Harvard University Press, 1929.

Best, C. H., and Taylor, N. B.: *The Living Body*. New York, Henry Holt and Company, 1952.

Blum, Harold F.: *Time's Arrow and Evolution*. Princeton University Press, 1951.

Bonner, J., and Galston, A. W.: *Principles of Plant Physiology*. San Francisco, W. H. Freeman, 1952.

Boynton, Holmes: *The Beginnings of Modern Science*. New York, Classics Club, 1948.

Brown, H.: *The Challenge of Man's Future*. New York, Viking Press, 1954.

Buchsbaum, Ralph: *Animals Without Backbones*. Rev. Ed. Chicago, University of Chicago Press, 1948.

Bulloch, William: *The History of Bacteriology*. London, Oxford University Press, 1938.

Calvin, M.: *Chemical Evolution and the Origin of Life*. American Scientist, July 1956, Vol. 44, pp. 248–263.

Cannon, W. B.: *The Way of an Investigator*. New York, Norton, 1945.

Cannon, W. B.: *The Wisdom of the Body*. New York, Norton, 1932.

Carson, Rachel L.: *The Sea Around Us*. New York, Oxford University Press, 1951.

Carter, G. S.: *Animal Evolution: A Study of Recent Views of Its Causes*. London, Sidgwick and Jackson, 1951.

Clark, W. E. L.: *The History of the Primates*. London, British Museum, 1949.

Cobb, Stanley: *Foundations of Neuropsychiatry*. 5th Ed. Baltimore, Williams and Wilkins, 1952.

Cohen, I. B.: *Science, Servant of Man*. Boston, Little, Brown & Co., 1948.

Colbert, E. H.: *Evolution of the Vertebrates*. New York, John Wiley & Sons, 1955.

Comstock, J. H., and Gertsch, W. J.: *The Spider Book*. 2nd Ed. New York, Doubleday Doran and Co., 1940.

Conant, J. B.: *On Understanding Science*. New Haven, Yale University Press, 1947.

Conant, J. B.: *Science and Common Sense*. New Haven, Yale University Press, 1951.

Coon, C. S.: *The Races of Europe*. New York, The Macmillan Company, 1939.

Corner, George W.: *Ourselves Unborn*. New Haven, Yale University Press, 1944.

Corner, George W.: *The Hormones in Human Reproduction.* Princeton, Princeton University Press, 1942.

Cott, H. B.: *Adaptive Coloration in Animals.* Oxford, Oxford University Press, 1940.

Curtis, O. F., and Clark, D. G.: *An Introduction to Plant Physiology.* New York, McGraw-Hill Book Company, 1950.

Darwin, Charles: *The Origin of Species.* 1859. Available in a number of recent reprint editions.

Dennis, W.: *Readings in the History of Psychology.* New York, Appleton-Century-Crofts, 1948.

DeRobertis, E. D. P., Nowinski, W. W., and Saez, F. A.: *General Cytology.* 2nd Ed. Philadelphia, W. B. Saunders Co., 1954.

Dobell, Clifford: *Antony van Leeuwenhoek and His "Little Animals."* London, John Bale Medical Pub., 1932.

Dobzhansky, T.: *Genetics and the Origin of Species.* 3rd Ed. New York, Columbia University Press, 1951.

Dodson, Edward O.: *A Textbook of Evolution.* Philadelphia, W. B. Saunders Co., 1952.

Dodson, Edward O.: *Genetics.* Philadelphia, W. B. Saunders Co., 1956.

Dubos, René J.: *Louis Pasteur, Free Lance of Science.* Boston, Little, Brown & Co., 1950.

Dunn, L. C. (ed.): *Genetics in the Twentieth Century.* New York, The Macmillan Co., 1951.

Florkin, Marcel: *Biochemical Evolution.* New York, Academic Press, 1949.

Frobisher, Martin: *Fundamentals of Microbiology.* 6th Ed. Philadelphia, W. B. Saunders Co., 1957.

Fuller, H. J., and Tippo, O.: *College Botany.* New York, Henry Holt and Co., 1954.

Fulton, John: *Selected Reading in the History of Physiology.* Springfield, Ill., Charles C Thomas, 1930.

Fulton, John: *Textbook of Physiology.* 17th Ed. Philadelphia, W. B. Saunders Co., 1955.

Gabriel, M. L., and Vogel, S.: *Great Experiments in Biology.* New York, Prentice-Hall, 1955.

Gaebler, O. H.: *Enzymes: Units of Biological Structure and Function.* New York, Academic Press, 1956.

Garrett, H. E.: *Great Experiments in Psychology.* 3rd Ed. New York, Appleton-Century-Crofts, 1951.

Gates, R. R.: *Human Ancestry.* Cambridge, Harvard University Press, 1948.

Gerard, Ralph: *Unresting Cells.* New York, Harper & Bros., 1940.

Goldschmidt, R. B.: *The Material Basis of Evolution.* New Haven, Yale University Press, 1940.

Goldschmidt, R. B.: *Understanding Heredity.* New York, John Wiley, 1952.

Grant, Madeleine: *Microbiology and Human Progress.* New York, Rinehart, 1953.

Guthrie, Douglas: *A History of Medicine.* Philadelphia, J. B. Lippincott Co., 1946.

Hall, Thomas S.: *A Source Book in Animal Biology.* New York, McGraw-Hill Book Co., 1951.

Harvey, E. N.: *Bioluminescence.* New York, Academic Press, 1952.

Harvey, E. N.: *Living Light.* Princeton, Princeton University Press, 1940.

Harvey, William: *Anatomical Studies on the Motion of the Heart and Blood.* (Translated by C. D. Leake.) Springfield, Ill., Charles C Thomas, 1931.

Hegner, R. W.: *Parade of the Animal Kingdom.* New York, The Macmillan Co., 1937.

Hegner, R. W., and Stiles, K. A.: *College Zoology.* 6th Ed. New York, The Macmillan Co., 1951.

Henderson, L. J.: *The Fitness of the Environment.* New York, The Macmillan Co., 1913.

Henrici, A. T., and Ordal, E. J.: *The Biology of the Bacteria.* Boston, D. C. Heath and Co., 1948.

Hill, A. V.: *Muscular Movement in Man.* New York, McGraw-Hill Book Co., 1927.

Holmes, S. J.: *Human Genetics and Its Social Import.* New York, McGraw-Hill Book Company, 1936.

Hooton, Ernest: *Up from the Ape.* Rev. Ed. New York, The Macmillan Co., 1945.

Howells, W. W.: *Mankind So Far.* New York, Doubleday, Doran & Co., 1944.

Hylander, C. J., and Stanley, O. B.: *The Plant World.* New York, The Macmillan Co., 1949.

Hyman, L. H.: *The Invertebrates.* Vols. 1, 2 and 3. New York, McGraw-Hill Book Co., 1940, 1950 and 1951.

Imms, A. D.: *Social Behavior in Insects.* London, Methuen, 1947.

Irvine, William: *Apes, Angels, and Victorians.* New York, McGraw-Hill Book Co., 1955.

Krueger, W. W.: *Principles of Microbiology.* Philadelphia, W. B. Saunders Company, 1953.

Leopold, A. C.: *Auxins and Plant Growth.* Berkeley, University of California Press, 1955.

Lorenz, Konrad: *King Solomon's Ring.* New York, Thomas Y. Crowell, 1952.

Lull, R. S.: *Organic Evolution.* Rev. Ed. New York, The Macmillan Co., 1947.

MacGinitie, G. E.: *Natural History of Marine Animals.* New York, McGraw-Hill, 1949.

Maximow, A. A., and Bloom, W.: *A Textbook of Histology.* 7th Ed. Philadelphia, W. B. Saunders Co., 1957.

Moulton, F. R. (ed.): *The Cell and Protoplasm.* Lancaster, Pa., The Science Press, 1940.

Muller, H. J., Little, C. C. and Snyder, L. H.: *Genetics, Medicine, and Man.* Ithaca, Cornell University Press, 1947.

Naylor, A. W.: *Physiology of Reproduction in Plants.* In *Survey of Biological Progress,* Vol. 2. New York, Academic Press, 1952.

Needham, Joseph, and Baldwin, E. B.: *Hopkins and Biochemistry.* Cambridge, W. Heffer and Sons, 1950.

Newman, H. H.: *Multiple Human Births.* New York, Doubleday Doran and Co., 1940.

Nickell, Louis G.: *The Control of Plant Growth by the Use of Special Chemicals.* In *Survey of Biological Progress,* Vol. 2. New York, Academic Press, 1952.

Nordenskiöld, Erik: *The History of Biology.* New York, Alfred Knopf, 1932.

Odum, E. P.: *Fundamentals of Ecology.* Philadelphia, W. B. Saunders Co., 1953.

Oparin, A. I.: *The Origin of Life.* New York, The Macmillan Co., 1938.

Osborn, Fairfield: *Our Plundered Planet.* Boston, Little, Brown & Co., 1948.

Osborn, Frederick: *Preface to Eugenics.* 2nd Ed. New York, Harper & Bros., 1951.

Osborn, Henry Fairfield: *From the Greeks to Darwin.* New York, The Macmillan Co., 1913.

Osborn, Henry Fairfield: *Men of the Old Stone Age.* 3rd Ed. New York, Charles Scribners Sons, 1918.

Patten, B. M.: *Human Embryology.* 2nd Ed. New York, The Blakiston Co., 1953.

Pavlov, I. P.: *Conditioned Reflexes.* New York, International Publishers, 1941.

Prosser, C. L., Brown, F. A., Bishop, D. W., Jahn, T. L., and Wulff, V. J.: *Comparative Animal Physiology.* Philadelphia, W. B. Saunders Co., 1950.

Pycraft, W. P.: *The Courtship of Animals.* 2nd Ed. London, Hutchinson and Co., 1933.

Rahn, Otto: *Microbes of Merit.* Lancaster, Pa., Jaques Cattell Press, 1945.

Ranson, S. W., and Clark, S. L.: *The Anatomy of the Nervous System.* 9th Ed. Philadelphia, W. B. Saunders Co., 1953.

Raymond, P. E.: *Prehistoric Life.* Cambridge, Harvard University Press, 1939.

Reed, S. C.: *Counseling in Medical Genetics.* Philadelphia, W. B. Saunders Co., 1955.

Ricketts, E. F., and Calvin, J.: *Between Pacific Tides.* Stanford, Stanford University Press, 1939.

Rolfe, R. T.: *Romance of the Fungus World.* Philadelphia, J. B. Lippincott Co., 1926.

Romer, A. S.: *Man and the Vertebrates.* 3rd Ed. Chicago, University of Chicago Press, 1941.

Romer, A. S.: *The Vertebrate Body.* 2nd Ed. Philadelphia, W. B. Saunders Co., 1956.

Rosebury, Theodor: *Peace or Pestilence.* New York, Whittlesey House, 1949.

Rugh, Roberts: *Experimental Embryology.* Ann Arbor, Edwards, 1948.

Scheer, Bradley T.: *Comparative Physiology.* Rev. Ed. New York, John Wiley and Sons, 1953.

Scheinfeld, A.: *You and Heredity.* 2nd Ed. New York, F. A. Stokes and Co., 1950.

Schoenheimer, Rudolf: *The Dynamic State of the Body Constituents.* Cambridge, Harvard University Press, 1949.

Schroedinger, E.: *What Is Life?* Cambridge, Cambridge University Press, 1944.

Schultz, L. P.: *The Ways of Fishes.* New York, Van Nostrand, 1948.

Sears, Paul B.: *Deserts on the March.* Norman, Okla., University of Oklahoma Press, 1935.

Sedgwick, W. T., Tyler, H. V., and Bigelow, R. P.: *A Short History of Science.* New York, The Macmillan Co., 1939.

Selye, Hans: *Textbook of Endocrinology.* 2nd Ed. Montreal, University of Montreal Press, 1949.

Sherrington, C. S.: *Integrative Action of the Nervous System.* New Haven, Yale University Press, 1947.

Shrock, R. R., and Twenhofel, W. H.: *Principles of Invertebrate Paleontology.* New York, McGraw-Hill Book Co., 1953.

Simmons, J. S.: *Global Epidemiology.* Philadelphia, J. B. Lippincott Co., 1944 and 1953.

Simpson, George Gaylord: *The Major Features of Evolution.* New York, Columbia University Press, 1953.

Simpson, George Gaylord: *The Meaning of Evolution.* New Haven, Yale University Press, 1950.

Singer, Charles: *A History of Biology.* Rev. Ed. New York, Harper & Bros., 1950.

Sinnott, E. W., Dunn, L. C., and Dobzhansky, T.: *Principles of Genetics.* 4th Ed. New York, McGraw-Hill Book Co., 1952.

Sinnott, E. W., and Wilson, K. S.: *Botany, Principles and Problems.* New York, McGraw-Hill Book Co., 1955.

Smith, Geeds: *Plague on Us.* New York, Commonwealth Fund, 1941.

Smith, G. M., Gilbert, E. M., Bryan, G. S., Evans, R. I., and Stauffer, J. F.: *A Textbook of General Botany.* 5th Ed. New York, The Macmillan Co., 1953.

Snyder, L. H.: *Medical Genetics.* Durham, N. C., Duke University Press, 1941.

Snyder, L. H.: *The Principles of Heredity.* 4th Ed. Boston, D. C. Heath and Co., 1951.

Solomon, A. K.: *Why Smash Atoms?* Cambridge, Harvard University Press, 1943.

Starling, E. H.: *Principles of Human Physiology.* 6th Ed. London, Churchill, 1947.

Stebbins, G. Ledyard, Jr.: *Variation and Evolution in Plants.* New York, Columbia University Press, 1950.

Stern, Curt: *Principles of Human Genetics.* San Francisco, W. H. Freeman, 1949.

Stevens, S. S., and Davis, H.: *Hearing: Its Psychology and Physiology.* New York, John Wiley and Sons, 1938.

Sturtevant, A. H., and Beadle, G. W.: *An Introduction to Genetics.* Philadelphia, W. B. Saunders Co., 1939.

Szent-Györgyi, Albert: *Chemical Physiology of Contraction in Body and Heart Muscle.* New York, Academic Press, 1953.

Thompson, D'Arcy W.: *On Growth and Form.* Rev. Ed. New York, The Macmillan Co., 1942.

Tiffany, L. H.: *Algae: The Grass of Many Waters.* Springfield, Ill. Charles C Thomas, 1939.

Tinbergen, N.: *The Study of Instinct.* Oxford, Oxford University Press, 1952.

Tullis, James (ed.): *Blood Cells and Plasma Proteins.* New York, Academic Press, 1953.

Turner, C. D.: *General Endocrinology*. 2nd Ed. Philadelphia, W. B. Saunders Co., 1955.

Vallery-Radot, R.: *Life of Pasteur*. New York, Doubleday, Doran, 1928.

Vogt, William: *Road to Survival*. New York, Sloane, 1948.

Von Frisch, Karl: *Bees, Their Vision, Chemical Senses and Language*. Ithaca, Cornell University Press, 1950.

Wald, G.: *The Origin of Life*, in *The Physics and Chemistry of Life*. New York, Simon and Schuster, 1955.

Walter, H. E. and Sayles, L. P.: *Biology of the Vertebrates*. 3rd Ed. New York, The Macmillan Co., 1950.

Weatherwax, Paul: *Botany*. 3rd Ed. Philadelphia, W. B. Saunders Co., 1956.

Weaver, J. E. and Clements, F. E.: *Plant Ecology*. 2nd Ed. New York, McGraw-Hill Book Co., 1938.

Weidenreich, F.: *Apes, Giants and Man*. Chicago, University of Chicago Press, 1947.

Weiner, A. S.: *Blood Groups and Transfusions*. 3rd Ed. Springfield, Ill., Charles C Thomas, 1943.

Wheeler, W. M.: *The Social Insects, Their Origin and Evolution*. New York, Harcourt, Brace and Co., 1928.

White, A. D.: *History of the Warfare of Science with Theology*. New York, Appleton, 1896.

White, P. R.: *Growth Hormones and Tissue Growth in Plants*. In *Survey of Biological Progress*, Vol. 1. New York, Academic Press, 1949.

Wiener, Norbert: *Cybernetics*. New York, John Wiley and Sons, 1948.

Wightman, W. P. D.: *The Growth of Scientific Ideas*. New Haven, Yale University Press, 1951.

Williams, Carroll: *Morphogenesis and the Metamorphosis of Insects*. Harvey Lectures, Vol. 47, New York, Academic Press, 1953.

Williams, R. H.: *Textbook of Endocrinology*, 2nd Ed. Philadelphia, W. B. Saunders Co., 1955.

Wilson, C. M.: *Trees and Test Tubes: The Story of Rubber*. New York, Henry Holt & Co., 1943.

Wilson, Douglas P.: *They Live in the Sea*. London, Collins, 1947.

Wilson, E. Bright: *An Introduction to Scientific Research*. New York, McGraw-Hill, 1952.

Witschi, E.: *Development of Vertebrates*. Philadelphia, W. B. Saunders Co., 1956.

Yonge, C. M.: *A Year on the Great Barrier Reef*. New York, Putnams, 1930.

Young, J. Z.: *Doubt and Certainty in Science*. Oxford, Oxford University Press, 1951.

Zilboorg, G., *A History of Medical Psychology*. New York, W. W. Norton and Co., 1941.

Zinsser, Hans: *Rats, Lice and History*. Boston, Little, Brown and Co., 1935.

Index

This index is intended to serve as a glossary as well. The page on which the term is defined is indicated in **boldface** type.

Abilities, inheritance of, 495
Abortion, 426
Abscission layer, **113**
Absorption, 279
Abstracts, 3
Abyssal zone, 592
Acetabularia, 33
Acetone bodies, **302**
Acetylcholine, 349
Acheiropodia, **497**
Acid, **23**
Acidosis, **273**
Acorn worms, 216
Acquired characteristics, **503**
Acromegaly, **390**
ACTH, 393, 578
Actin, **339**
Action currents, **248**
Adam's apple, 284
Adaptation, 19, 571, 583
Adaptive radiation, **583**
Addison's disease, 389
Adenosine triphosphate, 339
Adrenal cortex, 389
Adrenal glands, 388
Adrenal medulla, 388
Adrenal-pituitary axis, **399**
Adrenal-pituitary system, 578
Adrenocorticotropic hormone, 393, 578
African sleeping sickness, 187
Afterbirth, **425**
Agar, 122, 138
Age of fishes, 522
Agglutination, **238**
Aggregation, **578**
Ainus, 569
Air, amount respired, 270
Air sacs, 267, 277
Alarm response, **578**
Albinism, **484,** 486
Albino, 473
Albumin, 228
Alcohol, 68
Aldosterone, 389
Algae, **131**
  blue-green, 133
  green, 134
Alkaptonuria, **487**
All or none law, **337,** 347
Allantois, 422
Alleles, **463**
Allelomorphs, **463**
Allergy, 405
Alligators, 546
Alpha-tocopherol, 307
Alpha waves, **355**
Alpine race, 567
Alternation of generations, 133, 168
Alveolar air, composition of, 270
Alveoli, 267
Amaurotic idiocy, 494
Amber, 517
Ameba, 184, 261, 295
Amebic dysentery, 409
Amebocytes, **188**
Ameboid motion, **17,** 184
Amensalism, **76**
American Indians, **569**
Amino acids, **27,** 288
  essential, **28,** 300
Aminopterin, 237, 312
Amitosis, **453**
Ammocoetes, 220
Amnion, **422**
Amniotic fluid, 423
Amphibia, 222
Amphioxus, 218, 434
Ampullas, **214**
Anabolism, **17**
Analogous structures, **181**
Anaphase, **455**
  meiotic, 459
Anatomy, 541
Ancon sheep, 507
Androsterone, 389
Anemia, **236**
Angiosperm, **161**
  life cycle of, 173
Animals, classification of, 183
  invertebrate, 181
  warm-blooded, 223
Anions, **24**
Annelida, 200
Anterior, **51**
Anther, 174
Antheridia, **168,** 169, 171
Anthropoids, **550**
Antibiotics, 405
Antibodies, **233,** 403
Antidiuretic hormone, 322, 390
Antigen, **233,** 403
Antigen-antibody reaction, 237
Antimetabolite, **311**
Antitoxin, **403**
Aorta, 241
Ape-men, fossil, 557
Aphasia, **355**
Apical meristem, **48**
Appalachian revolution, 525
Appendicitis, 291
Appendix, 289, 541
Aqueous humor, **371**
Arachnids, 203
Archegonium, **169,** 171
Archenteron, 431
Archeology, 565
Archeopteryx, 508, 533
Archeozoic era, 520
Archiannelida, 202
Archosaurs, 532
Aristotle, 502
Armenoid race, 568
Arterial duct, **252**
Arteries, 241

Arterioles, 254
Arteriosclerosis, 259
Arthropoda, 202
Artifacts, **565**
Ascomycetes, 142
Ascorbic acid, 306
Ascospores, **143**
Asphyxia, **274**
Association areas, 355
Astaxanthin, 134
Asthma, 405
Astigmatism, **374**
Athlete's foot, 409
Atmospheric gases, 573
Atoms, **20**
Atrioventricular node, 245
Atrium, **244,** 440
Auditory canal, 375
Augmentation, **321**
Aureomycin, 406
Australian realm, 547
Australoids, 569
Australopithecus, 556
Autocatalysis, **454**
Autocatalytic particle, 512
Autoclave, 122
Autotrophic, **68**
Autotrophs, evolution of, 513
Auxins, **92**
Avicularia, **197**
Avidin, 311
Axon, **47,** 344
Axons, giant, 201
Aysheaia, 209

Baboons, 551
Bacilli, **118**
Bacteria, 117 ff., 290
  economic uses of, 123
  evolutionary relationships of, 127
  identification of, 122
  methods for studying, 122
  nucleus of, 120
  parasitic, 124
  reproduction of, 120
Bacteria-free animals, 70
Bacteriophages, **126,** 484
Basal metabolic rate, **298**
Base, **23**
Basidiomycetes, 144
Basidiospores, **144**
Basilar membrane, 377
Basophils, 232
Bast fibers, 50
Bath sponge, 187
Beriberi, **307**
Berries, 176
Bestiaries, **1**
Bicuspid valve, 244
Bilateral symmetry, **51**
Bile, 287
Bile pigments, 231, 287
Bile salts, 287
Binomial system, **66**
Bioassay, **96**
Biochemical genetics, 484
Biogenesis, 9
Biogenetic Law, **11,** 543
Biogeographic realms, **547**
Biogeography, 546
Biologic oxidation, 265
Biological journals, 3
Biology, **2**
  applications of, 6
  early history of, 1
Bioluminescence, **63**
Biome, **586**
Biotic community, **574,** 578
Biotic potential, **576**
Biotin, 69, 311
Bipinnaria larva, 219
Birds, 223
Birth, changes at, 447
Birth control, 497
Birth process, 424
Blade, 111
Blastocoele, **431**
Blastocyst, **432**
Blastopore, **431**
Blastula, **430**
Blind spot, 373
Blood, 227
  calcium content of, 386
  clotting of, 234
  concentration of glucose in, 301
  diseases of, 236
  functions of, 227
  tests for, 237
  volume of, 227
Blood bank, 239
Blood flow, rate of, 252
Blood platelets, 234
Blood pressure, 255
Blood proteins, 542
Blood transfusions, 238
Blood types, 238
Blood vessels, 241
Blue baby, 252, 447
Blue crabs, 572
Body fluids, osmotic pressure of, 321
Body form, development of, 437
Body plan, 51
Bolus, **280**
Bone, 44
Bowman's capsule, **319,** 443
Brachiating, **551**
Brachiopoda, 198, 521
Brachycephalic, **567**
Brachyphalangy, **477**
Bracket fungi, 148
Brain, 351
  ventricles of, 351, 436
Brain waves, **355,** 495
Bread mold, 142
Breathing, control of, 274
  mechanics of, 268
Broadleaved evergreen subtropical forest, 590
Bronchi, 267
Brownian movement, **39**
Bryophytes, 151
Bryozoa, 197
Bud, 110
Budding, **414**
Buffer, **28**
Burning foot syndrome, 311
Bushmen, 568

Calciferol, 306
Calcium, 303
Calcium ions, 235
Calorie, **54**
Cambium, 48, 104, **108**
Cambrian period, 521
Cancer, 261
Capillaries, **241**
  exchange of materials in, 256
Carbohydrate metabolism, 301
Carbohydrates, **25,** 299
Carbon cycle, 69
Carbon dioxide, 271
  transport of, 273
Carbon dioxide fixation, **70,** 84
Carbon monoxide, 230
Carbon monoxide poisoning, 274
Carbonic acid, 265
Carbonic anhydrase, 273
Carboniferous period, 523
Cardiac muscle, 243, 334
Caries, **313**
Carotene, **80,** 91, 305
Carotid arteries, 250
Carotid sinus, **275**
Carrageenin, **138**
Cartilage, 44, 220, 331
Casts, 517
Catabolism, **17**
Catalase, 55
Catalysis, **54**
Catalyst, **54**
Catarrhines, **550**
Catastrophism, **502**
Cathartics, 290
Cathepsins, **292**
Cations, **24**
Cats, six toed, 507
Caul, **423**
Cave paintings, 566
Cecum, 289
Cell body, 47, 344
Cell function, effects of salts on, 24
Cell furrow, 457
Cell plate, 457
Cell sap, 86
Cell theory, **9,** 32
Cell wall, 33
Cells, **15,** 32 ff.
  factors limiting size of, 37
  methods of studying, 38
  sizes of, 37
Cellulose, 26
Cenozoic era, 533
Center of origin, 546

Centipedes, 203
Centriole, 36, 455
Cephalic index, **567**
Cephalochordata, 218
Cephalopoda, 212
Cerebellum, 352
Cerebral cortex, localization of function in, 354
Cerebral hemispheres, 353
Cerebrospinal fluid, 351
Cerebrum, 353
Cervix, 420
Cesarean operation, **426**
Cestoda, 193
Chaetae, **201**
Chemoreceptor cells, 368
Chemosynthetic bacteria, 68
Chemotropism, **90**
Chimpanzees, 551
Chiton, 211
Chlamydomonas, 164
Chlorophyll, **80**, 186
Chlorophyta, 134
Chloroplast, **36**, 80
Cholecystokinin, 398
Cholesterol, 287
Choline, 311
Cholinesterase, 349
Chondrichthyes, 220
Chordata, 216
Chordates, origin of, 219
Chorion, **422**
Chromatin, **453**
Chromomeres, **453**
Chromonema, **453**
Chromoplast, **36**
Chromosome map, 481
Chromosome number, 453
Chromosome structure, 453
Chromosomes, **36**, 453
Chrysophyta, 135
Chyme, **285**
Cilia, **44**, 185
Ciliary body, 371
Ciliata, 185
Circulation in lower animals, 261
  in plants, 88
Circulatory system, 227, 241 ff.
  fetal, 250
Clams, 212
Cleavage, 429
Cliff cabbage, 545
Climax community, **579**
Clitoris, **420**
Clot, blood, 228
Cobalamin, 237
Cocci, 118
Cochlea, 218, **377**
Coelacanth, 221
Coelenterata, 188
Coelom, 183, **198**, 201
Coenzyme, 12, **56**, 304
Coenzyme A, 311
Cohesion theory, 114
Coleoptile, **94**
Collenchyma, **50**
Colloid, **29**
Colon, 289
Color adaptations, 586
Color blindness, 479
Comb jellies, 192
Comb types, inheritance of, 467
Commensalism, **76**
Communities, dynamic state of, 594
Community succession, **579**
Comparative physiology, 541
Competition, interspecific, 76
Compound, organic, 23
  pure, **22**
Concealing coloration, 586
Conceptual scheme, **4**
Condiments, **304**
Conditioned reflex, **360**
Condylarths, 537
Cones, 158, 160
  of eye, 371
  ovulate, **171**
  staminate, **171**
Congenital abnormalities, 493
Conifers, 159
Connective tissues, types of, 44
Conservation, **579**
Conservation of energy, Law of, 39, 72
Conservation of matter, Law of, 54, 69
Constipation, 290
Contour cultivation, 106, 580
Contraception, **497**
Contractile vacuole, 43, **185**, 323
Contraction period, 338
Control group, **6**
Conus, 262
Convergent evolution, **583**
Coordination in plants, 89
Copper, 303
Copulation, 415, 420
Corallines, 138
Corals, 192
Cork, 49
Cork cambium, **109**
Cornea, 371
Corona radiata, **420**
Corpus luteum, 396, 397
Cortisone, 389
Cotyledons, **177**
Coupled reactions, **55**
Cowper's glands, 418
Cowpox, 404
Cramps, 341
Cranial index, **567**
Cranial nerves, 357
Cranium, 329
Creatinine, 322
Creeper fowl, 477
Creodonts, 537
Cretaceous period, 525
Cretinism, **384**
Crocodiles, 532
Cro-Magnon man, 562
Crop, 201
Crop rotation, 580
Cross, dihibrid, **469**
  monohybrid, **464**
Crossing over, **479**
Crossopterygii, 221
Crustacea, 203
Ctenophores, 192
Curare, 334
Curve of normal distribution, 474
Cutin, **48**, 111
Cuttings, 164
Cyanide poisoning, 58, 274
Cyanophyta, 133
Cycads, 159
Cyclosis, **17**
Cytochromes, **59**, 232, 265
Cytoplasm, **36**

Dams, 581
Dandruff, 327
Dark reaction, **83**
Darwin, Charles, 94, 503
Dead space, **270**
Deafness, 378
Deamination, 28, **302**
Decarboxylation, **61**, 265
Defecation, 290, **317**
Dementia praecox, 495
Dendrite, **47**, 344
Dentin, 281
Dermis, **327**
De Saussure, 80
Deserts, 590
Desmids, 135
Desoxycorticosterone, 389
Desoxyribose nucleic acid (DNA), 28, 484
Development, control of, 444
Devonian period, 507, 522
de Vries, Hugo, 507
Dextran, 240
Diabetes, 302, 387
Diabetes insipidus, **322**, 390
Diabetes mellitus, **322**
Dialysis, **41**
Diapause, **207**
Diaphragm, 269
Diarrhea, **290**
Diastase, 293
Diastole, **246**
Diatoms, 135
Dicotyledoneae (dicots), 108, 161
Diet, 312
Diffusion, **39**
  speed of, 40
Digestion, **279**
  chemistry of, 291
  intracellular, 295
  plant, 88
Digestive glands, stimulation of, 294
Digestive system, 279 ff.
Digestive tract, development of, 441

Digestive tract, disorders of, 291
Digger wasp, 210
Digitigrade, **331**
Dinaric race, 568
Dinoflagellates, 136
Dinosaurs, 532
Diploid, **132,** 458
Disease, body defenses against, 402
 germ theory of, **401**
 humoral theory of, **401**
Diseases, virus, 124
DNA, 28, 484
Dolichocephalic, **567**
Dominance, incomplete, **466**
Dominant genes, **464**
Dorsal, **51**
Drosophila, 505
Drupes, **176**
Dryopithecus, 556
Duck-billed platypus, 224, 428, 535
Dugesia, 193
Dulse, 138
Duodenum, 286
Dust bowl, 581
Dynamic balance of nature, 594
Dynamic equilibrium, **231**

Ear, 375
Ear ossicles, 377
Earthworm, 200, 262, 295, 363, 370
Ecdysone, 207
Echinodermata, 213
Ecologic niche, **74**
Ecology, 12, **65**
 principles of, 571
Ecosystems, **73**
Ectoderm, 183, 188, 193, 431
Edema, **260,** 322
Effectors, **344**
Eggs, 47, 131, 451, 462
 centrolecithal, **428**
 isolecithal, **428**
 telolecithal, **428**
 types of, 428
Electric organ, 343
Electrocardiograph, **248**
Electroencephalograph, **355**
Electrolytes, **24**
Electron microscopy, 38
Electrons, **20**
Element, **20**
Elephantiasis, 260
Embolus, **235**
Embryo, 429
 nutrition of, 421
Embryonic development, 428 ff.
Embryonic membranes, 422
Embryophyta, 151
Emotions, 360
Emulsion, **29**
Enamel, 281
Endamoeba histolytica, 409
Endocrine glands, interrelationships between, 399
Endocrine system, 381 ff.
Endolymph, **377**
Endoskeleton, **199,** 328
Endosperm, **158**
Endosperm nucleus, 174
Endothelium, **241,** 243
Endotoxin, **402**
Energy, **20,** 38
Energy cycle, 72
Energy requirements, 299
Energy-rich phosphate bond, **340**
Energy transfer, efficiency of, **573**
Entelechies, **8**
Enterokinase, 293
Entoderm, 183, 188, 193, 431
Environment, effects on plants and animals, 571
Environmental resistance, **577**
Enzymes, 11, **54,** 291
 adaptive, 444
 factors affecting activity of, 57
 properties of, 55
 turnover number of, 55
Enzyme poisons, 58
Enzyme-substrate complex, **56**
Eocene epoch, 534
Eosinophils, 232
Epiboly, **432**
Epicanthic fold, **569**
Epicotyl, **177**
Epidermis, **327**
Epidermis bullosa, **493**
Epididymis, 418
Epigenesis, **444**
Epiglottis, 266
Epilepsy, 495
Epinephrine, 249, 254, 388
Epithelia, types of, 44
Erectile tissue, **418**
Ergot, 146
Erythroblastosis fetalis, **239**
Erythrocytes, 229
Eskimos, 569
Esophagus, 284
Estradiol, 396
Ethiopian realm, 547
Eugenics, **496**
Euglenophyta, 134
Eumycophyta, 140
Eunuch, **394**
Eurypterids, 522
Eurythermic, **572**
Eustachian tubes, 283, 376, 441
Evergreens, 159
Evolution, convergent, **213**
 cultural, 565
 genetic evidence for, 545
 fossil evidence for, 516
 living evidence for, 540
 of primates, 549
 principles and theories, 501, 514
Evolution, regressive, 514
 straight-line, **510**
Excretion, **317**
Excretory system, 317 ff.
Exercise, effects of, 245
Exophthalmic goiter, **385**
Exoskeleton, 199, **328**
Exotoxin, **402**
Experiment, controlled, 33
Experimental method, 2
Expressivity, **481**
Extensor, **335**
Eye, human, 369
 structure of, 371
Eye color, inheritance of, 491
Eye spots, 370

Facial index, **567**
Facilitation, **360**
Fallopian tubes, 419
Fat deposits, 299
Fat metabolism, 302
Fats, **26,** 288, 300
Fatigue, **341**
Fatty acids, essential, **300**
Feces, 290
Feeble mindedness, 496
Fermentation, **58,** 121
Fern, life cycle of, 170
Ferns, 156
Fertile period, **398**
Fertilization, **420**
 types of, 416
Fertilizers, 580
Fetus, **423**
Fibrin, 234
Fibrinogen, 228, 234
Filicinae, 156
Fire blight, 124
Fishes, age of, 522
Fishing, 581
Fission, **414**
Flagella, 120, 186
Flagellata, 186
Flame cells, **193,** 195, 323
Flatworms, 192
Flexor, **335**
Flower, 173
 pistillate, **174**
 staminate, **174**
Flower-producing hormone, 98
Fluids, body, osmotic pressure of, 321
Folic acid, 311
Follicle-stimulating hormone, 392
Food, absorption of, 288
Food chains, **573**
Food vacuole, 279, 295
Foods, analyses of, 314–316
Foraminifera, 185
Foregut, 438
Fossil, **516**
Fovea, **373**
Fracastorius, 408
Fragmentation, **414**

Free genes, 512
Free-living viruses, 513
Freshwater life zones, 594
Frog, 434
Fronds, **156**
Fructose, 25
Fruit, **175**
  accessory, **176**
  parthenocarpic, **97**
  seedless, 97
  true, **176**
  types of, 176
Fruit flies, 507
Fruit scars, 111
Fucoxanthin, 136
Functional diseases, **381**
Fungi, **131,** 140
  economic importance of, 146
Fungi Imperfecti, **148**
Fungicide, 148

Galápagos Islands, 546
Galen, 1
Gallbladder, 287
Galley Hill man, 561
Gallstone, 231, 287
Gametes, 131, 164, **413**
Gametophyte, **133,** 168
Gamma globulin, **228,** 407
Gamma rays, 234
Ganglia, 344
  sympathetic, 437
Gastrea, 544
Gastric juice, 284, 292
Gastrin, 294
Gastrocolic reflex, **290**
Gastropoda, 212
Gastrotricha, 197
Gastrovascular cavity, 188
Gastrula, **431**
Gastrulation, **431**
Geiger counter, 22
Gel, **29**
Gemma cups, 152
Gene, 36, **453,** 462
  chemical nature of, 484
Gene action, 486
Gene theory, **9**
Genes, complementary, **472**
  interactions of, 471
  number and size of, 485
  supplementary, **473**
Genetic drift, **506,** 566
Genetic isolation, 505
Genetic stability, 483
Genetics, **452,** 469
Genotype, **465**
Genus, **66**
Geographic distribution, 545
Geographic isolation, 505
Geologic time table, 517
Geotropism, **90**
Germ layers, derivatives of, 443
Germinal epithelium, **396**
Germination, 177
Germplasm, continuity of, 9
Gestation, **440**
Gibbons, 551
Gigantopithecus, 558
Gills, 220, 266, 276, 277
Gill slits, 441
Ginkgo, 160
Gizzard, 201
Glaciation, 520, 525, 538
Glass sponge, 187
Glomerular filtrate, 319
Glomerulus, **319**
Glottis, 283
Glucagon, **387**
Glucocorticoid, **389**
Glucose, 25, 59, 85, 228, 288, 301
Glycogen, 26, 228, 301, 339
Glycolytic cycle, **59,** 84
Gnetales, 160
Goiter, 383
Golgi bodies, **37**
Gonadotropic hormone, 392
Gorillas, 551
Grana, **80**
Graphite, 520
Grasslands, 590
Gray matter, **350**
Greek philosophy, 502
Growth, **19**
Growth and differentiation hormone, 207
Growth hormone, 390
Guano, **224**
Guard cells, 111
Gymnosperm, 158
  life cycle of, 171

Habitat, **73**
Hagfishes, 219
Hair, 328
Hair color, inheritance of, 491
Haploid, **132,** 458
Hardy-Weinberg Law, **10,** 483
Hearing, mechanism of, 377
Heart, 243
  disorders of, 258
  evolution of, 262
  formation of, 440
  oval window of, 251, 440, 447
Heart attack, 258
Heart beat, 245
  changes with body activity, 248
Heart cycle, 246
Heart murmur, 248
Heart sounds, 248
Heart strings, 244
Heartwood, **108**
Heidelberg man, 559
Hemacytometer, 229
Hemocoel, **219**
Hemocyanin, 232
Hemoglobin, **27,** 195, 229, 272
Hemophilia, **235,** 479
Heparin, 235
Hepatic portal vein, 250
Herbals, **1**
Heredity, **451**
  physical basis of, 451
Heredity and environment, 493
Hermaphroditism, **415**
Heterogamy, **165**
Heterotrophic, **68**
Heterozygous, **464**
Hibernation, **585**
High energy phosphate bonds, 55
Hindgut, 437
Hirudinea, 202
Holarctic realm, 547
Holdfast, **138**
Holozoic, **68**
Homeostasis, **12**
Homo sapiens, 561
Homogentisic acid, 487
Homoiothermic, **585**
Homologous chromosomes, **458**
Homologous organs, **541**
Homologous structures, **181**
Homozygous, **464**
Honeybees, 209, 210, 416
Hopeful monsters, 508
Hormone, **12,** 381
Horsetails, 155
Hottentots, 569
Human ecology, 582
Human pedigrees, 490
Humus, **105**
Hunger, 368
Huntington's chorea, 494
Hutton, James, 502
Hyaluronidase, 420
Hybrid corn, 483
Hybrid vigor, **483**
Hybrid virus, 485
Hybridization, 510
Hydra, 188, 261, 295
Hydrocortisone, 389
Hydrophytes, **114**
Hydroponics, **106**
Hymen, 420
Hypersensitivity, **405**
Hypertonic solution, **42**
Hypertrichosis, **493**
Hyphae, **140**
Hypocotyl, **177**
Hypothalamus, 393
Hypothesis, **3**
Hypotonic solution, **42**

Ichthyosaurs, 533
Ichthyosis, **493**
Idiot, **496**
Imbecile, **496**
Immune carrier, **406**
Immunity, acquired, 403
  natural, 404
Implantation, **421**
Impulses, transmission of, 92
Inborn errors of metabolism, **487**
Inbreeding, **482**
Indians, American, **569**

Indigestion, 291
Indoleacetic acid, 93
Indonesians, 569
Infant, nutrition of, 426
Infectious diseases, 401 ff.
Inferior vena cava, 244
Inflammation, **233**, 403
Ingenhousz, Jan, 79
Ingestion, **279**
Inguinal canal, 417
Inheritance of physical traits, 491
Inhibition, **350**
Inoculation, 118, 125, **401**, 403
Inositol, 311
Insanity, 356
Insect behavior, 210
Insect metamorphosis, hormone control of, 207
Insects, 203
  colonial, 209
Insight, **360**
Instincts, **210**
Insulin, 27, **288**, 387
Integration, **344**
Integument, **326**
Intelligence, 353
Intelligence Quotient, **496**
Intermedin, **222**
Interrelationships, biologic, 65
Intersexes, **478**
Intestinal juice, 293
Intestinal movements, 286
Intracellular energy wheel, 59
Invagination, **432**
Involution, **432**
Iodine, 43, 303, 384
Ion, **22**
Iris, 371
Iron, 303
Irritability, 17
Ischial callosities, **550**
Islets of Langerhans, **288**, 387
Isogamy, **164**
Isolation, **505**
  ecologic, 505
Isotonic solution, **42**
Isotopes, **21**

Jaundice, 288
Java man, 557
Jellyfish, 191
Joint, 330
Jordan's Rule, **545**
Jugular veins, 250
Jurassic period, 525
Juvenile hormone, 208

Kangaroos, 537
Kelps, 136
Ketone bodies, 387
Kidney, 318
  development of, 442
  diseases of, 322
  regulatory function of, 442
  structure of, 319
Kidney stones, 323
Kidney threshold, **321**
Kinesthesis, **367**
Kinetic energy, **38**
Koch's postulates, **118**
Krebs citric acid cycle, **59**

Labia, 420
Labor pains, 424
La Brea tar pits, 517
Lactic acid, 339
Lactogenic hormone, 392
Lactose, 25
Lamarck, Jean Baptiste de, 503
Lampreys, 219
Lampshells, 198
Larva, **205**
Larynx, 266
Latent period, 338
Lateral, **51**
Latex, **88**
Leaf, 111
Leaf scars, 111
Learning, 360
Leather, 44
Leeches, 202
Leeuwenhoek, 117
Lemmings, 578
Lemurs, 549
Lens, 371
Lenticels, **110**
Leonardo da Vinci, 502
Lethal genes, **477**
Leukemia, **237**
Leukocytes, 232
Leukoplast, **36**
Lichens, **145**
Life cycle, **168**
Life on other planets, 513
Life, origin of, 511
Ligaments, 44
Light reaction, **83**
Limnetic zone, **594**
Limnopithecus, 553
Linkage, **479**
Linkage group, **481**
Lipase, 287, 293
Litmus, **145**
Littoral zone, **594**
Liver, 287, 441
  metabolic functions of, 301
Liver flukes, 194
Liverworts, 151
Living things, characteristics of, 16
Lizards, 532
Loam, **106**
Locomotion, types of, 331
Locus, **463**
Lophophore, **197**
Lower Paleolithic culture, 565
Luciferase, 63
Luciferin, 63
Lucretius, 502
Lungfish, 222, 276, 522
Lungs, 441
  evolution of, 276
  exchange of gases in, 271
Luteinizing hormone, 392
Lycopodium, 153
Lycopsida, 153
Lyell, Sir Charles, 502
Lymph, flow of, 260
  functions of, 260
Lymph nodes, 259
Lymph system, 259
Lymphocytes, 232
Lynx, 577
Lysergic acid, 146

Macromutation, **508**
Magnesium, 303
Magnolias, 546
Malaria, 230, 415
Malformations, 446
Malpighian tubules, 324
Maltase, 282
Maltose, 25
Mammals, 224
Mammary glands, 327
Man, distinguishing characters, 551
  races of, 566
Man-apes, 556
Manic-depressive insanity, **357**, 495
Mantle, 211
Marine life zones, 592
Marrow cavity, 44
Marsupials, 224, 535
Mass spectrometer, 22
Matrix, **44**
Matter, cyclic use of, 69
  structure of, 20
Maximum birth rate, **574**
Mechanistic theory of life, **16**
Medial, **51**
Mediterranean race, 567
Medulla, 274, 351
Medusa, **191**
Megagametophyte, **171**, 174
Megaspores, **155**, 171
Meiosis, 132, **457**
Melanesians, 568
Melanin, 473, 486
Membrane, nuclear, 35
  permeability of, 41
Membrane theory of nerve transmission, 348
Mendel, Gregor, 452
Mendel's Law of Independent Segregation, **10**, 469
Mendel's Law of Segregation, **10**, 465
Meninges, **351**
Menopause, **398**
Menstrual cycle, 396
Menstruation, 235, 397
Mental ability, inheritance of, 495

Mental disorders, inheritance of, 494
Meristem, apical, 48
Mesocephalic, **567**
Mesoderm, 183, 193
  formation of, 433
Mesoglea, **188**
Mesolithic age, 566
Mesonephros, **443**
Mesophytes, **115**
Mesozoic era, 525
Metabolism, 11, **16,** 298 ff.
  bacterial, 121
  genic control of, 12
Metamorphosis, 222, 385, 417
  insect, 205
Metanephros, **443**
Metaphase, **455**
  meiotic, 458
Method of agreement, 5
Method of concomitant variation, 5
Method of difference, 5
Microgametophyte, **171,** 174
Micronucleus, 186
Microorganisms, spread of, 406
Microphonic, **377**
Microsomes, **37,** 56
Microspores, **155,** 171
Microtome, 38
Midbrain, 352
Middle Paleolithic culture, 565
Middle Stone Age, 566
Midget, 390
Migration of birds, 223
Milk, 426
  pasteurized, 406
Millipedes, 203
Mimicry, 587
Mineralocorticoids, **389**
Minerals, 72, 105, 302
  daily requirements of, 303
Minimum mortality, **575**
Miocene epoch, 534
Miscarriage, 426
Misreference, **366**
Mistletoe, 69
Mitochondria, **37,** 56, 84, 461
Mitosis, 132, **454**
Mitotic spindle, 455
Mixture, **23**
Modern man, origin of, 562
Modification, **505**
Modifying factors, **476**
Mold, 142
  fossil, **517**
Molecular motion, 39
Molecule, **22**
Mollusca, 211
Molting, 204
Mongolian fold, 493
Mongoloid race, 569
Monkeys, 550
Monocotyledoneae (monocots), 108, 161
Monocytes, 232
Monotremes, 224, 535
Mons veneris, 420
Moods, 360
Moron, **496**
Morphogenesis, **444**
Morphogenetic substances, 462
Mortality, minimum, **575**
Mosaic eyes, 370
Mosasaurs, 532
Moss, 151
  life cycle of, 169
Mother-of-pearl, 212
Mousterian culture, 565
Mouth cavity, structures in, 279
Movement, 17
Mucus, 200
Multiple alleles, **476**
Multiple factors, **473**
Muscle, types of, 45
Muscle contraction, chemistry of, 339
  types of, 337
Mushroom, 144
Musical ability, inheritance of, 496
Mutations, **485,** 505, 507
  causes of, 509
  chromosomal, **485**
  point, **485**
  types of, 509
Mutualism, **76,** 145
Mycelium, **140**
Myelin, 344
Myofibrils, **45**
Myosin, **339**
Myxedema, **383**
Myxomycophyta, 140

Nails, 328
Nares, 266
Nasal chamber, 266
Nasal index, **567**
Natural selection, theory of, 10, 503, 504
    modern changes in, 504
Nausea, **368**
Nautilus, 213
Navel, 425
Neanderthal man, 560
Nearctic realm, 547
Negritos, 568
Negroid race, 568
Nematocysts, **188**
Nematoda, 195
Nemertea, 194
Neotropical realm, 547
Nephridium, **201,** 324
Nephritis, 322
Nerve centers, 352
Nerve impulse, 345
Nerves, 344
  roots of, 358
Nervous system, 344 ff.
  autonomic, 361
  central, 350
Nervous system, development of, 436
  peripheral, 357
Neural crest, **436**
Neural folds, 436
Neurillemma, 344
Neurohumor, **349,** 393
Neuromotor fibers, 363
Neurons, **47**
  types of, 344
Neurosecretion, **204**
Neuroses, **356**
Neurospora, 69, 143, 304, 486
Neutralism, **75**
Neutrons, **20**
Neutrophils, 232
New Stone Age, 566
Niacin, 309
Night blindness, 305
Nitrate bacteria, 71
Nitrogen balance, **300**
Nitrogen cycle, 70
Nodal tissue, **245**
Node, 110
Nodule bacteria, 71
Nonelectrolytes, **24**
Nordic race, 567
Northern coniferous forest, 590
Notochord, **218,** 435
Nuclear membrane, 455
Nuclei, transplantation of, 446
Nucleic acids, 28
Nucleolus, 36
Nucleoplasmic ratio, **457**
Nucleus, **16,** 33
Nurse cells, 418
Nut, 176
Nutrition, 68, 298 ff.

Obesity, 312
Occam's razor, **4**
Oceanic islands, 546
Octopus, 213
Oedogonium, 167
Oligocene epoch, 534
Oligochaeta, 202
Oöcytes, primary, 462
  secondary, 462
Oögenesis, **461**
Oögonia, 461
Oögonium, **168**
Oötid, 462
Operculum, **221**
Optic lobes, 352
Orang-utan, 551
Orchil, **145**
Ordovician period, 521
Organ systems, types of, 51
Organic evolution, theory of, 10, 501
Organic substances, synthesis of, 85
Organizer, **445**
Oriental realm, 547
Origin of Species, 503
Ornithischia, 533

Orthogenesis, **510**
Osmosis, **41**
Osmotic pressure, **41**
Osteichthyes, 220
Ostracoderms, 219, 521
Outbreeding, **482**
Ovarian follicle, 397
Ovary, 174, 395, 419
Overgrazing, 582
Oviducts, 419
Ovulation, **396**
Ovule, **171,** 174
Ovum, **131,** 462
Oxygen, 271
  transport of, 229, 271
Oxygen consumption, 299
Oxygen debt, **340**
Oxyhemoglobin, 230, 272
Oxytocin, 390

Pacemaker, 245
Pair-feeding, 5
Palate, 279
Palearctic realm, 547
Paleocene epoch, 534
Paleolithic cultures, 565
Paleontology, 516
Paleosimia, 553
Paleozoic era, 521
Palisade cells, **113**
Pancreas, 288, 441
Pancreatic juice, 288, 293
Pangenes, **9**
Pantothenic acid, 311
Para-aminobenzoic acid, 311
Paralysis agitans, **494**
Paramecium, 185, 261
Paramylum, **134**
Paranoia, **357**
Parapithecus, 552
Parapodia, **202**
Parasites, 573
Parasitism, **69**
Parasympathetic system, 363
Parathormone, 386
Parathyroid glands, 385, 442
Parenchyma, **49**
Parotid glands, 282
Parthenogenesis, **415**
Parturition, 424
Pasteurization, 117
Paternity tests, 476
Pavlov, 294
Peat, 70, 151
Pectoral girdle, 330
Peking man, 557
Pelagic zone, **592**
Pelecypoda, 212
Pellagra, 309
Pelvic girdle, 330
Pelycosaurs, 525
Penetrance, **481**
Penicillin, 76, 142, 405
Penis, 223, 418
Pepper corn hair, 569
Pepsin, 292
Peptidases, 293
Perception of pitch, 377
Pericardium, **243**
Pericycle, **104,** 108
Peripatus, 202, 209
Peristalsis, **284**
Peritonitis, 291
Permian period, 523
Pernicious anemia, 236
Perspiration, 317
Petals, **174**
Petiole, 111
Petrifaction, 516
Petrified forest, 516
pH, **23**
Phaeophyta, 136
Phagocytosis, **233**
Pharynx, 266, 283, 441
Phenotype, **465**
Phenylketonuria, 494
Phenylthiocarbamide, 369, 490
Phloem, **50,** 88, 104
Phosphocreatine, 339
Phosphoglyceric acid, 84
Phosphorus, 303
Phosphorus cycle, 72
Photoperiod, 572
Photoperiodism, **97**
Photosynthesis, 79, 513
Phototropism, **90**
Phycocyanin, 133
Phycoerythrin, **80,** 138
Phycomycetes, 142
Phylum, **67**
Physiologic adaptations, 585
Physiological genetics, 471
Physiological saline, 42
Phytoplankton, **73**
Phytosaurs, 532
Piltdown man, 558
Pineal gland, 398
Pistil, 174
Pitch, perception of, 377
Pith, 108
Pithecanthropus, 557
Pituitary, 389
  anterior lobe, 390
  posterior lobe, 390
Placebo, **6**
Placenta, **224,** 398, 421, 423
Placental mammals, orders of, 224
Placoderms, 220, 522
Plagues, 401
Planaria, 261, 295, 370
Plant diseases, 148
Plant hormones, 93
Plant reproduction, 163
Plant sap, 89
Plantigrade, **331**
Plants, aquatic, 150
  economic importance of, 115
  embryonic development of, 176
  etiolated, **91**
  evolutionary trends in, 178
Plants, herbaceous, **106**
  long-day, **97**
  short-day, **97**
  skeletal system of, 86
Plasma, **228**
Plasma membrane, **16,** 33, 41
Plasmagene, **486**
Plasmolysis, **88**
Platyhelminthes, 192
Platyrrhines, **550**
Playfair, John, 502
Pleistocene epoch, 538
Plesiosaurs, 533
Pleura, 268
Pleurisy, 268
Pliny, 1
Pliocene epoch, 534
Pneumotaxic center, 274
Poikilothermic, **585**
Poliomyelitis, 407
Pollen grain, 171, 174
Polocyte, **462**
Polychaeta, 202
Polycythemia, **237**
Polyploid, **509**
Polyps, **191**
Pons, 352
Population, **574**
Population cycles, 577
Population density, **574**
Population genetics, **483**
Population growth curve, **574**
Porifera, 187
Portuguese man-of-war, 191
Posterior, **51**
Posture, 335
Potassium, 303
Potato blight, 146
Potential energy, **38**
Preadaptation, **506**
Predator-prey relationship, 77
Preformation theory, **444**
Pregnancy, 398
Prehensile tail, 550
Pressure, diastolic, **255**
  pulse, 256
  systolic, **255**
Priestley, Joseph, 79
Primates, 549
  fossil, 552
Primitive streak, 432, 435
Probability, 465
Proboscis, **216**
Proboscis worms, 194
Proconsul, 553
Progeny selection, **466**
Progesterone, 396
Proglottids, **194**
Pronephros, **443**
Prophase, **455**
  meiotic, 458
Propliopithecus, 553
Proprioceptors, **367**
Prostate glands, 418
Proteins, **27,** 300
  biologically adequate, **300**

Protein metabolism, 302
Proterozoic era, 520
Prothrombin, **235,** 307
Protocooperation, **76**
Protonema, **151,** 170
Protons, **20**
Protoplasm, **15** ff., 32
  dynamic state of, 61
  organic compounds in, 24
  physical characteristics of, 29
  water content of, 22
Protozoa, 184
  mating groups of, 415
Pseudopods, **184**
Psilophytales, 153
Psilopsida, 153
Psychoses, **357**
Pteranodon, 532
Pteropsida, 156
Pterosaurs, 532
Ptyalin, **282,** 291
Public health, 582
Pulse, **253**
Pupa, 205
Pus, 233
Putrefaction, **121**
Pygmies, 568
Pyocyanin, 405
Pyridoxine, 310
Pyrrophyta, 136
Pyruvic acid, 59, 308

Quarantines, 406
Quaternary period, 538
Quillworts, 155
Quinine, 451

Rabbits, Porto Santo, 502
Race, **566**
Radial symmetry, **51**
Radicle, **177**
Radioactive carbon, 84
Radioactive clock, 520
Radiolaria, 185
Radula, **211**
Range, **545**
Range of Tolerance, **572**
Rangelands, 582
Rassenkreis, **540**
Rays, medullary, **108**
Recapitulation, theory of, 11, 443, **543**
Recent epoch, 538
Receptacle, 173
Receptors, **344**
Recessive genes, **464**
Recovery period, **338**
Rectum, 289
Red blood cells, 229
  life history of, 230
Red bone marrow, 230
Red cell count, 229
Redi, Francesco, 19
Reflex, **358**
Reflex arc, **358**
Refractory period, **342,** 347
Regeneration, 192
Reindeer moss 145
Reinforcement, **350**
Relaxation period, **338**
Relaxin, **398**
Renin, 323
Rennin, 292
Reproduction, 19
  asexual, 163, 413
  human, 417
  sexual, 415
Reptilia, 223
Respiration, **58,** 265
  artificial, 274
  direct, **265**
  external, **266**
  in plants, 85
  indirect, **266**
  internal, **266**
Respiratory center, 274
Respiratory system, 265
  structure of, 266
Resting stage, **455**
Resurrection plant, 155
Retina, 371
Reverberating circuit, **274**
Revolutions, 520
Rh factor, 239
Rhizoids, **151**
Rhizomes, **106**
Rhodesian man, 561
Rhodophyta, 138
Riboflavin, 308
Ribose nucleic acid (RNA), 28
Ribs, 329
Rickets, 306
Rickettsias, **127**
Rigor mortis, **30**
Rings, annual, 108
RNA, 28
Rock deposits, dating of, 520
Rocky mountain revolution, 525
Rods, 371
Root, adventitious, **105**
  cortex of, 102
  diffuse, **105**
  endodermis of, 102
  epidermis of, 102
  functions of, 100
  tap, **105**
Root hairs, **101**
Root pressure, **105,** 114
Rootstocks, 164
Rotifera, 196
Runners, 164
Running water, 594

Saber-toothed tiger, 537
Saccharin, 26
Saccule, **379**
Sago palm, 160
Saliva, 282, 291
Salivary glands, 282
Salk vaccine, 407
Salt, 24
Salvarsan, 408
Sand dunes, 579
Sap, ascent of, 114
Saprophytic, **68**
Sapwood, **108**
Sarcodina, 185
Saurischia, 532
Savanna, 590
Schizophrenia, **357,** 495
Scientific method, 3
Sclerenchyma, **50**
Scouring rushes, 155
Scrotal sac, 417
Scurvy, 306
Sea anemones, 192
Sea cucumbers, 214
Sea lettuce, 135
Sea lilies, 214
Sea sickness, 380
Sea urchins, 214
Sea within us, 242
Seaweeds, 43
Second wind, 340
Secretin, 294, 381, 398
Secretion, **317**
Sedimentary rock, 517
Seed, **158,** 172
  economic importance of, 178
  germination of, 176
Seed ferns, **158,** 523
Seed plants, **158**
Segmentation, 183, 200
Selaginella, 155
Selection, limits of, 506
Semicircular canals, 379
Semilunar valves, 244
Seminal fluid, 418
Seminal vesicles, 418
Seminiferous tubules, 417
Semipermeable membrane, **41**
Sensations, perception of, 366
Sense organs, 365
Senses, tactile, **367**
Sensitive plant, 93
Sepals, **174**
Serial dilution of bacteria, 122
Serpentstars, 214
Serum, **229,** 234
Sewage, 124
Sex, evolution of, 164
  genetic determination of, 477
Sex chromosomes, 477
Sex-influenced characters, **479**
Sex-linked characters, **478**
Sexual parasitism, 593
Shallow sea zone, 592
Shipworm, 212
Shock, **258**
Sickle cell anemia, 236
Sieve tubes, 51, 104
Silurian period, 522
Sinanthropus, 557
Single twitch, **338**
Sino-atrial node, **245**
Sinus gland, 204
Sinus venosus, 245, 262, 440
Skeletal muscle, 334

Skeleton, 328
  parts of, 329
Skin, functions of, 326
Sleep, 355
Sleep movements, 99
Sleeping sickness, African, 187
Slime molds, **140**
Small intestine, 285
Smallpox, 404
Smell, 369
Smooth muscle, 334
Snowshoe hare, 577
Snowy owls, 578
Sodium chloride, 303
Soil, 105
Sol, **29**
Solo man, 561
Solution, **29**
Somatoplasm, **9**
Somites, 438
Species, **66,** 540
  interactions between, 75
Species specificity, theory of, 27
Specific dynamic action, **301**
Specific linkage, **481**
Sperm, 47, 132, 451
Spermatids, 460
Spermatocytes, primary, 460
  secondary, 460
Spermatogenesis, **460**
Spermatogonia, 460
Spermatophore, **416**
Spermatozoa, 461
Sphagnum, 151
Sphenopsida, 155
Sphincter, **284**
Sphygmomanometer, 255
Spinal ataxia, 494
Spinal cord, 350
Spinal nerves, 357
Spindle fibers, 455
Spiny anteater, 535
Spirochetes, 118
Spirogyra, 165
Spleen, 230
Sponges, 187
Spontaneous generation, **9,** 117
Sporangium, **142,** 170
Spores, bacterial, 121
Sporophyte, **133,** 168
Sporozoa, 186
Stamens, 174
Standing water, 594
Staphylococci, 118
Starch, 26
Starfish, 214
Statistical analysis, 5
Steatopygia, **568**
Stegocephalians, 522
Stele, **104**
Stem reptiles, 523
Stems, 106
Stenothermic, **572**
Steroids, **27**
  morphogenetic role of, 445
Stigma, 174
Stimuli, localization of, 366
Stipe, **138**
Stomach, structure of, 284
Stomata, **111**
Stone cells, 50
Streptococci, 118
Streptomycin, 406
Stress, 399
Structural adaptations, 584
Strychnine, 350
Style, 174
Suberin, 49
Subgerminal space, **432**
Sublingual glands, 282
Submaxillary glands, 282
Subspecies, **540**
Substrate, **11**
Sucrose, 25, 85
Suctorians, 186
Sulfanilamide, 311
Superfemales, **478**
Superior vena cava, 244
Supermales, **478**
Survival curve, **576**
Survival of the fittest, 11, 504
Suspension, **29**
Suspensor, **177**
Swallowing, 283
Swamp forests, 523
Swanscombe skull, 562
Sweat, 317
Sweat glands, 327
Swim bladder, **221,** 276
Symbols, genetic, 464
Symmetry, types of, 51
Sympathetic system, 361
Sympathin, 349
Synapse, **47,** 347
  transmission across, 349
Synapsis, **458**
Syphilis, 407
Systole, **246**

Tadpole, 222, 417
Tapeworm, 194, 409
Tarsius, 550
Tasmanians, 568
Taste bud, 280, 368
Taxis, **90**
Taxonomy, 540
Teeth, 281
  canine, 281
  incisor, 281
  molar, 281
  premolar, 281
  wisdom, 282
Teleosts, 522
Telophase, **455**
  meiotic, 459
Temperate deciduous forest, 590
Temperature, 572
Template, **486**
Tendons, 44
Teosinte, 545
Termites, 76, 209
Terrestrial life, adaptations for, 199
Territories, 578
Tertiary period, 534
Test crosses, **465**
Testis, interstitial cells of, 394
Testosterone, 12, 394, 445
Tetanus, **338**
Tetany, 386
Tetrad, **458**
Thalamus, 353
Thallophytes, **131**
Thallus, **131**
Thecodont, 223
Theory, **4**
Therapsids, 224, 525
Thermodynamics, laws of, 72
Thiamine, 307
Thigmotropism, **90**
Thiouracil, 385
Thirst, 368
Thrombin, 235
Thromboplastin, **235**
Thrombus, **235**
Thymus, 398, 442
Thyroglobulin, 383
Thyroid, 383, 442
Thyrotropic hormone, 393
Thyroxine, 249, 383
Tidal zone, 592
Tissue, connective, 44
  meristematic, 48
  muscular, 45
  nervous, 47
  plant, 48
  reproductive, 47
Tissue culture, 33, 38, 236
Tissue fluid, **242,** 259
Tissue transplantation, 445
Tissues, 43
  animal, 43
  conductive, 50
  epithelial, 43
  fundamental, 49
  protective, 48
Tobacco mosaic virus, **125**
Tongue, 279
Tonus, **335,** 339
Topsoil, 106, 573
Tornaria larva, 219
Toxoid, **404**
Trace elements, 56, 573
Trachea, 267
Tracheal tubes, 278
Tracheids, 50, 104
Tracheophyta, 152
Training, effects of, 275
Trait, congenital, **493**
  inherited, 463
Transforming agents, **484**
Translocation, **481,** 509
Transpiration, **113**
Tree kangaroos, 584
Tree shrews, 549
Trematoda, 193
Triassic period, 525

Trichinosis, 409
Trichocysts, **186**
Tricuspid valve, 244
Trilobites, 203, 521
Trochophore larva, 202, **211**
Tropical rain forest, 591
Tropism, **90**
Truffle, 142
Trypsin, 293
Tsetse fly, 187
Tuatera, 532
Tube feet, 214
Tubers, **106,** 164
Tubules, kidney, 319, 320
Tundra, 590
Tunicates, 43, 216
Turgor pressure, 43, 86, **87**
Turtles, 532
Twins, fraternal, **447**
  identical 413, **447**
  Siamese, **447**
Tympanic canal, 377
Typhus, 407

Uintatheres, 537
Ulcer, 291
Ulothrix, 165
Umbilical arteries and veins, 251
Umbilical cord, 423, 438
Unconscious cooperation, **578**
Ungulates, 537
Unguligrade, **331**
Uniformitarianism, **502**
Universal donors, 238
Universal recipients, 238
Upper Paleolithic culture, 566
Urea, 28, 302, 321, 322
Ureter, 318, 443
Urethra, 418
Uric acid, 322
Urinalysis, 542
Urinary bladder, 318
Urine, composition of, 322
  formation of, 319
Urochrome, 322
Urogenital system, 324
Use or disuse, 503
Useless structure, 504
Uterine milk, 421
Uterus, 419
Utricle, 379

Vaccine, **403**
Vacuoles, **37**
Vagina, 420
Vagus, 357
Vanadium, 43
van Helmont, 79
Variation, **452**
Varicose veins, 259
Vascular bundles, 108
Vas deferens, 418
Vas efferens, 418
Vasopressin, 390
Veddoids, 569
Veins, 241
Ventral, **51**
Ventricle, 244, 440
Venus flytrap, 17
Vertebrae, **218,** 329
Vertebrates, 218
Vertical stratification, 579
Vestibular canal, 377
Vestigial organs, **541**
Villus, **288**
Viruses, **124**
  components of, 125
Visceral mass, 211
Vision, 369
  defects in, 374
Visual purple, **305,** 373
Visual violet, 373, 493
Vital capacity, **270**
Vitalism, **16**
Vitamin A, 305, 373
Vitamin B_1, 307
Vitamin B_2, 308
Vitamin B_6, 310
Vitamin B_{12}, 311
Vitamin B complex, 307
Vitamin C, 306
Vitamin D, 306
Vitamin E, 307
Vitamin K, 235, 307
Vitamins, 12, **304**
Vitreous humor, 371
Volvox, 167
Vomiting, 285
Vulva, **420**

Wallace, Alfred Russell, 503
Wallace's line, 547
Wassermann test, 408
Water, 303, 572
  properties of, 22
Water cycle, 72
Water vascular system, 214
Weed killers, 96
Wheat rust, 148
Wheel animals, 196
White blood cells, 232
  functions of, 233
  life history of, 234
White matter, **350**
White race, 567
Wildlife, 581
Wind breaks, 106, 580
Wire worms, 572
Womb, 419
Wombats, 537
Woody perennials, 107
Wooly mammoths, 517
Wound hormone, 457

X organ, 204
Xanthophyll, **80,** 91
Xerophthalmia, **305**
Xerophytes, **115**
Xylem, **50,** 88, 104

Yeasts, 68, 142
Yellow mice, 477
Yolk, **48,** 462
Yolk sac, 422

Zeuglodonts, 538
Zoospores, 164
Zygote, 131, 164, **413**

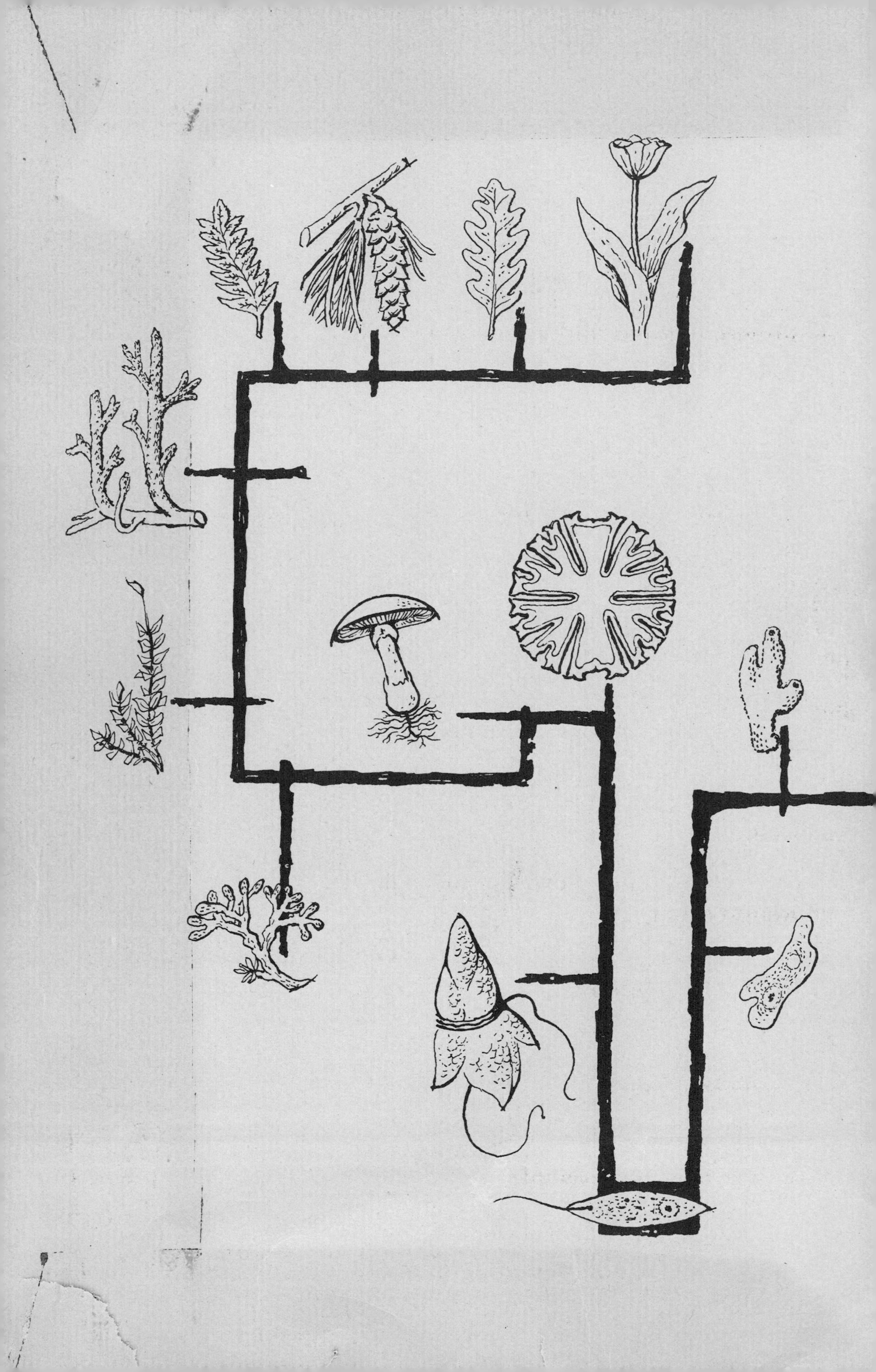